Bacterial Pathogenesis

Molecular and Cellular Mechanisms

Edited by

Camille Locht and Michel Simonet

Center for Infection and Immunity of Lille
University of Lille Nord de France
Institut Pasteur de Lille
France

Caister Academic Press

Caister Academic Press
Norfolk, UK

www.caister.com

British Library Cataloguing-in-Publication Data
A catalogue record for this book is available from the British Library

ISBN: 978-1-904455-91-2

Cover design adapted from Figure 9.1

Printed and bound in Great Britain

Contents

Contributors

Joseph E. Alouf
Domaine de la Ronce
Ville d'Avray
France

joseph.alouf@wanadoo.fr

Edward N. Baker
School of Biological Sciences and Maurice Wilkins Centre for Molecular Biodiscovery
University of Auckland
Auckland
New Zealand

en.baker@auckland.ac.nz

Sophie de Bentzmann
UPR CNRS 9027-LISM
Marseille
France

bentzman@ifr88.cnrs-mrs.fr

Christophe S. Bernard
UPR CNRS 9027-LISM
Marseille
France

cbernard@ifr88.cnrs-mrs.fr

Marta Biedzka-Sarek
Department of Chronic Disease Prevention
National Institute for Health and Welfare
Helsinki
Finland

biedzka@gmail.com

Olivia L. Champion
School of Biosciences
University of Exeter
Exeter, Devon
UK

O.L.Champion@exeter.ac.uk

Pascale Cossart
Inserm U604 & INRA USC2020
Institut Pasteur
Paris
France

pcossart@pasteur.fr

Frank R. DeLeo
Laboratory of Human Bacterial Pathogenesis
Rocky Mountain Laboratories
National Institute of Allergy and Infectious Diseases
National Institutes of Health
Hamilton, MT
USA

fdeleo@niaid.nih.gov

Teresa Frisan
Department of Cell and Molecular Biology
Karolinska Institutet
Stockholm
Sweden

Teresa.Frisan@ki.se

Ana Rita Furtado
CNRS URA2582
Institut Pasteur
Paris
France

anaritafurtado@yahoo.com

Caroline Giraud
UPR CNRS 9027-LISM
Marseille
France

cgiraud@ifr88.cnrs-mrs.fr

Lina Guerra
Department of Cell and Molecular Biology
Karolinska Institutet
Stockholm
Sweden

Lina.Guerra@ki.se

Riccardo Guidi
Department of Cell and Molecular Biology
Karolinska Institutet
Stockholm
Sweden

Riccardo.Guidi@ki.se

John D. F. Hale
Facility for Anti-infective Drug Development and Innovation
Monash Institute of Pharmaceutical Sciences
Faculty of Pharmacy and Pharmaceutical Sciences
Monash University
Parkville
Melbourne, VIC
Australia

john.hale@monash.edu

Klaus Hantke
IMIT
Interfakultäres Institut für Mikrobiologie und Infektionsmedizin Tübingen
Universität Tübingen
Tübingen
Germany

klaus.hantke@mikrobio.uni-tuebingen.de

Hae Joo Kang
School of Biological Sciences and Maurice Wilkins Centre for Molecular Biodiscovery
University of Auckland
Auckland
New Zealand

h.kang@imperial.ac.uk

Scott D. Kobayashi
Laboratory of Human Bacterial Pathogenesis
Rocky Mountain Laboratories
National Institute of Allergy and Infectious Diseases
National Institutes of Health
Hamilton, MT
USA

kobayashis@niaid.nih.gov

Frank Lafont
CNRS UMR8204
Inserm U1019
Institut Pasteur de Lille
Lille
France

frank.lafont@pasteur-lille.fr

Duncan J. Maskell
Department of Veterinary Medicine
University of Cambridge
Cambridge
UK

djm47@cam.ac.uk

Kevin Moreau
CNRS UMR8204
Inserm U1019
Institut Pasteur de Lille
France

km510@cam.ac.uk

Serge Mostowy
Inserm U604 & INRA USC2020
Institut Pasteur
Paris
France

smostowy@pasteur.fr

Gavin K. Paterson
Department of Veterinary Medicine
University of Cambridge
Cambridge
UK

gavin.paterson@hotmail.co.uk

Thomas Proft
School of Medical Sciences and Maurice Wilkins Centre for Molecular Biodiscovery
University of Auckland
Auckland
New Zealand

t.proft@auckland.ac.nz

Kevin M. Rigby
Laboratory of Human Bacterial Pathogenesis
Rocky Mountain Laboratories
National Institute of Allergy and Infectious Diseases
National Institutes of Health
Hamilton, MT
USA

rigbyk@niaid.nih.gov

Joseph W. St. Geme, III
Department of Pediatrics
Duke University Medical Center
Durham, NC
USA
j.stgeme@duke.edu

Amanda J. Sheets
Department of Pediatrics
Duke University Medical Center
Durham, NC
USA

amanda_sheets@med.unc.edu

Mikael Skurnik
Department of Bacteriology and Immunology
Haartman Institute
University of Helsinki
Helsinki University Central Hospital Laboratory Diagnostics
Helsinki
Finland

mikael.skurnik@helsinki.fi

Jennifer Spagnolo
UPR CNRS 9027-LISM
Marseille
France

jspagnolo@ifr88.cnrs-mrs.fr

Agathe Subtil
CNRS URA 2582
Institut Pasteur
Paris
France

agathe.subtil@pasteur.fr

Richard W. Titball
School of Biosciences
University of Exeter
Exeter, Devon
UK
R.W.Titball@exeter.ac.uk

Raphael H. Valdivia
Department of Molecular Genetics and Microbiology and Center for Microbial Pathogenesis
Duke University Medical Center
Durham, NC
USA

valdi001@mc.duke.edu

Current Books of Interest

Small DNA Tumour Viruses	2012
Extremophiles: Microbiology and Biotechnology	2012
Bacillus: Cellular and Molecular Biology	2012
Microbial Biofilms: Current Research and Applications	2012
Bacterial Glycomics: Current Research, Technology and Applications	2012
Non-coding RNAs and Epigenetic Regulation of Gene Expression	2012
Brucella: Molecular Microbiology and Genomics	2012
Molecular Virology and Control of Flaviviruses	2012
Bacterial Pathogenesis: Molecular and Cellular Mechanisms	2012
Bunyaviridae: Molecular and Cellular Biology	2011
Emerging Trends in Antibacterial Discovery: Answering the Call to Arms	2011
Epigenetics: A Reference Manual	2011
Metagenomics: Current Innovations and Future Trends	2011
Nitrogen Cycling in Bacteria: Molecular Analysis	2011
The Biology of Paramyxoviruses	2011
Helicobacter pylori	2011
Microbial Bioremediation of Non-metals: Current Research	2011
Lactic Acid Bacteria and Bifidobacteria: Current Progress in Advanced Research	2011
Viruses and Interferon: Current Research	2011
Essentials of Veterinary Parasitology	2011
Hepatitis C: Antiviral Drug Discovery and Development	2011
Streptomyces: Molecular Biology and Biotechnology	2011
Alphaherpesviruses: Molecular Virology	2011
Recent Advances in Plant Virology	2011
Vaccine Design: Innovative Approaches and Novel Strategies	2011
Salmonella: From Genome to Function	2011
PCR Troubleshooting and Optimization: The Essential Guide	2011
Insect Virology	2010
Environmental Microbiology: Current Technology and Water Applications	2011
Sensory Mechanisms in Bacteria: Molecular Aspects of Signal Recognition	2010
Bifidobacteria: Genomics and Molecular Aspects	2010
Molecular Phylogeny of Microorganisms	2010
Nanotechnology in Water Treatment Applications	2010
Iron Uptake and Homeostasis in Microorganisms	2010
Caliciviruses: Molecular and Cellular Virology	2010
Epstein-Barr Virus: Latency and Transformation	2010
Anaerobic Parasitic Protozoa: Genomics and Molecular Biology	2010
Lentiviruses and Macrophages: Molecular and Cellular Interactions	2010
Microbial Population Genetics	2010
Borrelia: Molecular and Cellular Biology	2010
Influenza: Molecular Virology	2010

Preface

One of the greatest public health achievements during the twentieth century was the dramatic reduction in the incidence of infectious diseases owing to the development of improved hygiene, vaccines and antimicrobial agents. Nevertheless, infectious diseases today still represent the second cause of global mortality, accounting for 25% of all annual deaths world-wide, i.e. 14,000,000 deaths per year, or 1 every 2 s. As examples of bacteria-induced illnesses, tuberculosis kills every year about 2 million individuals, whooping cough and typhoid fever cause around 300,000 and 200,000 annual deaths. A particular feature of the past four decades is the emergence of new infectious diseases and the discovery that pathogenic bacteria may be the cause of diseases for which an infectious aetiology had previously not been suspected. Undoubtedly, one of the most spectacular example is the gastro-duodenal ulcer, which was not suspected to be caused by a bacterium (*Helicobacter pylori*) until the pioneering work of the Australian physicians Robin Warren and Barry Marshall during the early eighties, to whom the Nobel Prize of Physiology and Medicine was awarded in 2005. Ascribing to a newly discovered microorganism, a causal role in a particular disease requires the potential infectious agent to fulfil the criteria formulated by Robert Koch in order to establish the aetiologies of anthrax and tuberculosis, and known as the (Henle)-Koch's postulates:

1. The organism must be regularly associated with the disease and its characteristic lesions.
2. The organism must be isolated from the diseased host and grown in culture.
3. The disease must be reproduced when a pure culture of the organism is introduced into a healthy, susceptible host.
4. The same organism must be re-isolated from the experimentally infected host.

Koch should probably have added the supplementary criterion, that the disease must be cured when the organism is eradicated from the host, if antimicrobial agents had been available at his time. However, today we know that several genuine pathogens fail to satisfy all of Koch's postulates.

Since the beginning of the 1970s, the development of molecular biology and genetic tools led to extensive studies on the molecular and cellular elements that support the virulence properties of pathogenic bacteria. Almost 100 years after the Koch's postulates had been formulated, Stanley Falkow, often referred to as the father of molecular microbial pathogenesis, proposed by analogy a set of experimental criteria that must be satisfied to show that a gene from a pathogenic microorganism contributes to the disease triggered by the pathogen:

1. The virulence/pathogenicity gene must be present in pathogenic strains of the genus or species.
2. Inactivation of the gene in question must be associated with loss of bacterial virulence/pathogenicity.

3 Reversion of the mutated gene or its replacement by an active allele must restore bacterial pathogenicity.
4 The gene must be expressed during infection.

Beyond cognitive science, why is it important to define bacterial pathogenicity/virulence factors? The fight against pathogens by means other than antibiotics to which bacteria today resist with increasing frequently, may strongly benefit from the understanding of their strategies to infect the host: inhibition of essential pathogenicity determinants may therefore prevent the occurrence of an infectious process or abrogate its course. As an example, the demonstration that filamentous haemagglutinin, pertussis toxin and pertactin from *Bordetella pertussis* are critical for whooping cough development and are protective antigens led to the replacement of the heat-killed whole-cell vaccine by acellular vaccines for the prevention of this childhood disease, thereby substantially decreasing the frequency of unwanted adverse events associated with the former vaccines. Another example is illustrated by the type III secretion system, used by many pathogenic bacteria to inject virulence effectors into the host cell cytosol. Its blockade is now possible by chemical compounds. As a consequence, these pathogens become non-pathogenic and are subsequently cleared by the host immune system. Determinants of bacterial pathogenicity may also be helpful for the diagnosis of infectious diseases, especially when they are caused by slow-growing, fastidious or non-cultivable bacteria: their rapid detection in clinical specimens by molecular analytical methods, such as the polymerase chain reaction, allows medical staff to early initiate specific therapy and, if needed, to quickly isolate patients in order to avoid disease transmission. Similarly, the fast release of pathogen-free perishable foodstuffs in large-scale food retailing is a major challenge for manufacturers: it can be met by the molecular diagnosis of pathogens instead of conventional detection by culture, followed by identification of the pathogen, which is time-consuming.

When Horizon Scientific Press asked us to edit a new compendium on bacterial pathogenesis, we felt that it would be interesting to not simply produce yet another book with a series of pathogens, one after the other, but, instead, to overview the subject with respect to the different steps of an infectious process. With this aim in mind, we organized *Bacterial Pathogenesis* in five sections, with each two to four chapters written by specialists in the fields in question:

1 approaches to study bacterial pathogenesis;
2 bacterial adhesion to the cell surface and extracellular matrix of host tissues;
3 poisoning the host by toxins;
4 cellular invasion by bacterial pathogens; and
5 bacterial evasion of host defences.

This has only been possible with the enthusiastic involvement of distinguished scientists from eight different countries and three continents. All of them have made an important effort to present the most relevant and up-to-date pathogenic features across the bacterial world in the most didactic way to address the entire scientific community from students to senior scientists and physicians. We therefore hope that this book will reach a broad range of people interested in the mechanisms of bacterial infectious diseases and express our deep gratefulness to all contributors.

Dr Camille Locht
Professor Michel Simonet

Part I

Approaches to the Study of Bacterial Pathogenesis

Models for Studying Bacterial Pathogenesis

1

Richard W. Titball and Olivia L. Champion

Abstract

Research into mechanisms of virulence underpins work to devise improved control measures for infectious diseases. However, the ability of a pathogen to cause disease is implicitly dependent not only on the bacterial species but also on the host. Without the availability of a suitable experimental host, studies on mechanisms of virulence either cannot proceed or cannot be interpreted in an appropriate contextual manner. This chapter considers the advantages and disadvantages of different hosts, including mammals, zebra fish, plants, *Caenorhabditis elegans*, insects such as *Drosophila melanogaster* and *Galleria mellonella*, and cell culture systems for investigating mechanisms of virulence of pathogens which cause disease in humans. The potential for gaining additional information and using fewer animals in experiments involving mammals is reviewed, with a particular emphasis on the potential for new imaging techniques to provide additional information. Much of this information will of course be of relevance to pathogens which are associated with diseases in non-human hosts.

Introduction

Research into mechanisms of virulence of bacterial pathogens underpins work to devise improved control measures for infectious diseases. However, the ability of a pathogen to cause disease is implicitly dependent not only on the bacterial species but also on the host. Pathogens able to cause disease in one host may be avirulent in other host species. Moreover, within outbred populations there will be large differences in the ways in which pathogens behave. Without the availability of a suitable experimental host, studies on mechanisms of virulence either cannot proceed or cannot be interpreted in an appropriate contextual manner. This chapter considers the advantages and disadvantages of different hosts for investigating mechanisms of virulence of pathogens which, cause disease in humans. Much of this information will of course be of relevance to pathogens associated with diseases in non-human hosts.

Mammalian models of infection

Mammalian models of infection continue to be the most widely used and characterized models of human disease. These models are of particular value because, unlike the plant and invertebrate models outlined below, mammals have innate and adaptive immune systems. Therefore, these models allow an understanding of the host–pathogen relationship at a systems level.

Although mammalian models of infection do have a clear value to the study of bacterial pathogenicity, it is also important to understand the limitations of these models from ethical, social and scientific viewpoints. The use of mammalian models of infection is tightly regulated in most countries and the use of mammals for experimental purposes is dependent on the appropriate training of staff, the availability of suitable experimental facilities and the rigorous justification of the work to be undertaken. Particular ethical and social concerns are associated with the use of non-human primates and domesticated mammals such as cats and dogs. Finally, although mammals can often serve as valuable models of human infectious disease, the mammalian species in question may lack the molecular features of humans, such as cell surface receptors, which means that the pathogenesis of disease in these species is different to that in humans. Collectively, these concerns have driven work to reduce, refine and replace mammalian models of infection.

Assays for measuring the interaction of mammals with bacteria and the effects of pathogens on mammals

A range of host species have been used to model disease caused by bacterial pathogens (Table 1.1). In many cases, several different species have been shown to be susceptible to infection. The choice of which model to use (or to develop) is dictated by a number of factors. The animal species must be both available, and ideally, bred specifically for this purpose to ensure freedom from pathogens which might be found in wild animals. The presence of such pathogens in experimental animals might complicate the interpretation of results and could present a significant risk to the researcher. The availability of reagents to monitor the immune response in animals is also an important consideration. For mice, a vast array of immunological reagents is available; for other species, the range is limited. An additional concern is the genetic diversity of the animals. Genetic diversity of the animal population might arguably provide for robust models since it will reflect the diversity of the human population. However, for studies on the immune response to infection, genetically defined populations are preferred. Finally, ethical and public concerns mean that some animals (such as non-human primates) should be used only for experiments where there is no other alternative available. Whatever the species of mammal chosen, it is typical to use animals of a similar age and sex to minimize animal-to-animal variation, and therefore to avoid the need for large experimental groups.

The route of infection would ideally mimic the natural route of infection, since this would ensure that the pathogen met appropriate barriers to infection. For some infections, models have been developed which closely reflect the natural routes of dissemination and infection. For example, infection models have been established where *Citrobacter rodentium* is transmitted from infected to uninfected animal via shedding in faeces (Wiles *et al.*, 2005), and there are good models of infections in humans which are associated with synthetic implants (Zimmerli, 1993).

Other infection models go some way towards mimicking the natural infection. For example, pathogenic micro-organisms which infect by the oral or inhalational route or via the genital tract can often be delivered by these routes. However, considerable skills and, in some cases, specialist equipment are required to successfully infect by these routes. For oral infection, pathogens are typically delivered by gavage where the inoculum is given using a blunt-ended needle into the stomach (Lockman and Curtis, 1990; Chatfield *et al.*, 1992). In this artificial infection model, it might be necessary to give the bacterium in a suspension, which includes components such as bicarbonate able to protect against the effects of stomach acidity (Lockman and Curtis, 1990).

Inhalational infection models require specialist equipment, such as the Henderson apparatus (Thomas *et al.*, 2009), which allows pathogens to be aerosolised and delivered into the respiratory tract in a safe and controllable manner. The particle size has a profound impact on the site of deposition – typically particles in the 1–3 μm range are deposited in the lower respiratory tract, whereas larger particles are deposited in the upper respiratory tract (Thomas *et al.*, 2009), and smaller particles may not be deposited at all (Bakker-Woudenberg, 2003). Animals are typically exposed via the nose only, so that airborne pathogens are not deposited on the body surfaces. The calculation of dose received by individual animals requires measurement of the breathing rate and depth (Guyton, 1947; Bakker-Woudenberg, 2003). Not surprisingly, in the light of these technical challenges, many researchers use intra-tracheal installation in place of inhalation exposure. However, it is difficult to ensure delivery of bacteria to the deep lung using this method, and it is likely that at least some of the inoculum enters the gastrointestinal tract (Bakker-Woudenberg, 2003).

Table 1.1 Examples of the range of species of mammals used in bacterial infection models

Host species	Pathogen	Reference
Armadillos	*Mycobacterium leprae*	Job *et al.* (1985)
Cattle	*Escherichia coli* O157	Dean-Nystrom (2003)
Cats	*Helicobacter pylori*	Lee (1998)
Cotton rat	*Francisella tularensis*	Ellis *et al.* (2002)
Dogs	*H. pylori*	Lee (1998)
Ferrets	*H. pylori*	Lee (1998)
Gerbils	*H. pylori*	Hirayama *et al.* (1996)
Goats	*Brucella melitensis*, *Clostridium perfringens*	Uzal and Kelly (1998), Fernandez Miyakawa and Uzal (2003), Kahl-McDonagh *et al.* (2006)
Guinea-pigs	*Yersinia pestis*; *F. tularensis*	Oyston *et al.* (1996), Ellis *et al.* (2002)
Hamsters	*F. tularensis*	Ellis *et al.* (2002)
Horses	*Clostridium difficile*, *Burkholderia mallei*, *Streptococcus equi*	Ensink *et al.* (2003), Lopez *et al.* (2003), Arroyo *et al.* (2004)
Macaques	*Burkholderia pseudomallei*, *Bartonella quintana*, *Mycobacterium tuberculosis*, *Mycobacterium leprae*, *Bacillus anthracis*, *Y. pestis*, *F. tularensis*, *Brucella* spp., *Coxiella burnettii*, *E. coli*, *Campylobacter jejuni*, *Listeria monocytogenes*, *Legionella pneumophila*, *Shigella* spp., *Streptococcus* spp., *Rickettsia prowazekii*	Gardner and Luciw (2008)
Marmosets	*B. anthracis*, *F. tularensis*, *Chlamydia trachomatis*	Johnson *et al.* (1980), Lever *et al.* (2008), Nelson *et al.* (2009)
Mice	*Salmonella enterica* serovar Typhimurium; *Y. pestis*, *Yersinia pseudotuberculosis*, *Yersinia enterocolitica*, *F. tularensis*; *Staphylococcus aureus*, *Vibrio cholerae*, *Brucella abortus*, *Brucella melitensis*, *H. pylori*	Eubanks *et al.* (1976), Baselski *et al.* (1977), Lockman and Curtis, 3rd (1990), Brubaker (1991), Chatfield *et al.* (1992), Oyston *et al.* (1996), Ellis *et al.* (2002), Brouillette and Malouin (2005), Rabelo-Gonçalves *et al.* (2005), Smither *et al.* (2009), Thomas *et al.* (2009)
Rabbits	*F. tularensis*	Ellis *et al.* (2002)
Rats	*E. coli*, *Neisseria menigitidis*, *Pseudomonas aeruginosa*	Ryan (1970), Cash *et al.* (1979), Saukkonen *et al.* (1987)
Pigs	*Chlamydia trachomatis*, *H. pylori*	Lee (1998), Vanrompay *et al.* (2005)
Sheep	*Clostridium perfringens*	Fernandez Miyakawa *et al.* (2003)

Finally, we should not overlook that some models mimic transmission by an insect vector. For example, *Yersinia pestis* or *Francisella tularensis* can be delivered subcutaneously into rodents mimicking the bite from an infected insect.

Whilst studies in the older literature often used death of the animal as the end point of the experiment, more recently there has been a move towards the use of alternative and more humane end points (Bast *et al.*, 2004; Stokes *et al.*, 2007; Thomas *et al.*, 2009). Humane end points involve euthanizing animals when they display signs which are indicative of impending death and before animals display signs of significant suffering. The use of humane end points does, of course, first require that a study is undertaken to map the progression signs of disease and to determine at what point the signs of disease indicate a fatal outcome. These signs might include body temperature changes, piloerection, posture (hunching), and mobility problems (Bast *et al.*, 2004; Stokes *et al.*, 2007). A more recent development has been the use of implanted chips, which are able to record changes in the temperature and behaviour of animals and therefore allows for repeated monitoring of infected animals (Nelson *et al.*, 2009). These experimental approaches allow additional information to be obtained from each study, and therefore may contribute to a longer term reduction in use of mammals for experimental purposes.

An important development, which has the potential to both reduce the use of mammals and increase the amount of experimental information obtained, is the imaging of infected animals. Bacteria expressing bioluminescent markers can be visualized within animals using highly sensitive charge coupled device (CCD) imaging systems (Contag *et al.*, 1995; Doyle *et al.*, 2004). Bioluminescence might be generated by expressing, in the bacterium, the *luxCDABE* operon or a luciferase (Jawhara and Mordon, 2004). Expression of the *lux* operon has the advantage that it is not necessary to provide additional substrate to generate light, whereas for bioluminescence to be generated by luciferase, it is necessary to add luciferin (Jawhara and Mordon, 2004). This might limit the use of luciferase reporters to the imaging of infections at a body surface (Jawhara and Mordon, 2004). Light is emitted from the bacteria only when they are metabolically active. However, the bacterial density estimated from bioluminescence intensity and the number of viable cells determined by the direct plating of tissues, closely parallel each other (Kadurugamuwa *et al.*, 2003; Doyle *et al.*, 2004). Because of the requirement for oxygen bioluminescence might not be expected to work in sites with low oxygen tensions, such as within the gut, but in practice, this does not appear to be a problem (Wiles *et al.*, 2005). Bacteria growing within tissues as well as on artificial implanted devices have been imaged using this approach (Kadurugamuwa *et al.*, 2003).

In vivo bioluminescent imaging offers several advantages over conventional approaches to monitor infection. The progress of an infection within individual animals can be tracked, reducing the need for large cohorts because of animal-to-animal variation (Fig. 1.1). The ability to follow the temporal pattern of infection is limited only by the need to anaesthetise animals before each imaging event – in practice up to once or twice daily. However, there are some potential limitations. For some pathogens, it is not possible to express bioluminescent reporters at a sufficiently high level to allow imaging and even for those pathogens which can be successfully imaged, there appears to be a lower limit of detection of around 10^3–10^4 cfu at any given site. Although there are reports that bacteria such as *Staphylococcus aureus* and *Pseudomonas*

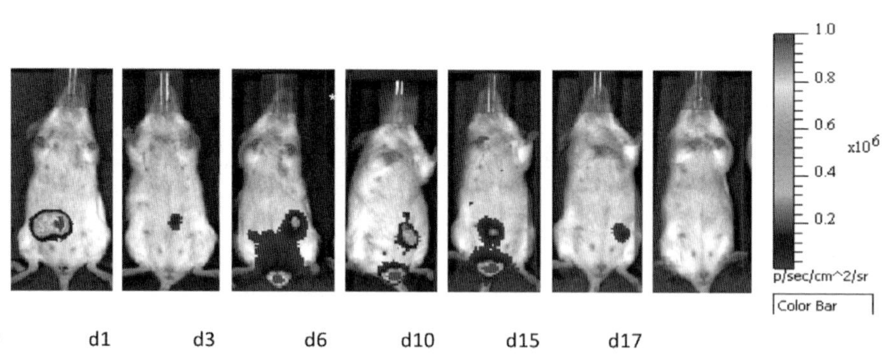

Figure 1.1 *In vivo* colonization and clearance dynamics of bioluminescent *Citrobacter rodentium*. Mice were inoculated with *C. rodentium* ICC180 (Wiles *et al.*, 2004) and biophotonic imaging performed immediately after oral gavage (first panel) and at regular intervals over a 17-day period using an IVIS Spectrum (Caliper Life Sciences, Hopkinton, MA, USA) after gaseous anaesthesia with isoflurane. Representative images from the same animal show colonization of the caecal patch at day 1, followed by the colon and rectum by 3 days post infection. Images displayed as peak bioluminescence. The bar indicates relative signal intensity [as photons cm/s/steridian (sr)]. Mice were imaged with an integration time of 1 min (except *10s) at medium binning. Image kindly provided with permission by Siouxsie Wiles.

aeruginosa retain the plasmid-borne *lux* operon for at least 20 days *in vivo* (Kadurugamuwa *et al.*, 2003), suggesting that the metabolic burden is relatively low, it is possible that the expression of reporters influences the pathogenesis of disease. Finally, it is important to bear in mind that the intensity of the light detected outside of the animal is dependent on the depth of the infection below the skin surface. Thus, bacteria located in deep tissues and/or in large animals might not be detectable. There are also some exciting prospects on the horizon for improved imaging techniques, which could overcome these limitations, and for multiplexed imaging, which could allow a range of molecular events to be interrogated in real time (Eliceiri and Contag, 2009). It seems likely that the bioimaging of infection is to become a standard methodology in many infectious disease laboratories.

Identification of virulence-associated genes

Mammalian models of infection have been used extensively to identify virulence genes. Within this chapter, it is not possible to cover the range of results that have been obtained using mammalian models of infection, and many of these results are detailed in subsequent chapters. However, it is worth addressing two important points here. First, genes involved in virulence in one species may not play roles in another. Therefore, virulence-associated genes can only be identified within the context of the model. Genes associated with host specificity have been elegantly revealed from signature-tagged mutagenesis studies (Autret and Charbit, 2005). One study with *Salmonella enterica* serovar Typhimurium identified three genes, which played a role in infection in mice but not in calves, and one gene contributing to infection in calves but not in mice (Tsolis *et al.*, 1999).

Secondly, the route and site of infection can have a profound influence on genes necessary for virulence. Examples of genes associated with tissue tropism have again been revealed from whole-genome mutagenesis studies with pathogens as diverse as *Streptococcus pneumoniae*, *S. aureus*, *Yersinia pseudotuberculosis* and *Burkholderia pseudomallei* (Autret and Charbit, 2005; Cuccui *et al.*, 2007).

Evaluation of vaccines and antibacterials

Mammalian infection models continue to be central to programmes of research aimed at developing and evaluating vaccines and antibacterials. Clearly, in the case of vaccines, it is essential to use a host, which has an immune system capable of responding to the immunogen in a similar way to the response in humans. The immunizing dose and route are both important considerations and should mimic the likely concept of use in humans. There is a general tendency to use mice for vaccination studies, and this is often a good starting point. However, immune responses to an immunogen in mice are often much more pronounced than responses in humans. In some cases, this can lead to the identification of candidates which are effective in mice but which fail to elicit protective responses or show much reduced protective responses in other animal species. Consequently, there is often a requirement to evaluate vaccines in more than one animal model of disease, especially when human clinical trials are difficult (e.g. biodefence vaccines) (Barrett and Beasley, 2009). Separately, the safety of vaccines is also determined by giving human doses of the vaccine to animals. Typically, one rodent species and one non-rodent (e.g. non-human primate) are used for these studies (Barrett and Beasley, 2009). These studies are animal intensive, and there has been some progress in developing alternative methods, such as computational models, which predict the effects of vaccines (Querec *et al.*, 2009).

Mammalian models of infection have been also been used extensively to screen antimicrobials. The ability to follow the course of infection in individual animals significantly reduces animal-to-animal variation (Rocchetta *et al.*, 2001), and could provide new insight into the efficacy of antibiotics in different tissue sites (Jawhara and Mordon, 2004) and at different stages in the infection process. Clearly, to obtain these data without *in vivo* bioluminescent imaging would require large numbers of experimental animals.

Identification of host genes playing roles in resistance to infection

Mice are amenable to forward and reverse genetic analyses to understand the role of host genes in the infection process. Forward genetic studies seek to understand the genetic basis of a phenotype or trait, such as resistance to infection, and have been applied to mice since long before the development of reverse genetic tools that target individual genes for characterization (Skamene *et al*, 1982; Shultz, 1991). One host gene, natural resistance-associated macrophage protein-1 (Nramp1), also known as solute carrier family 11a member 1 (Slc11a1), was identified in mouse strains with natural resistance to pathogens with an intracellular stage in their lifecycle, namely *Mycobacterium bovis* BCG, *S. enterica* serovar Typhimurium and *Leishmania donovanii*. Mutations that conferred resistance to these intracellular pathogens were mapped back to the same locus that was consequently named *Bcg/lty/lsh* (Bradley *et al.*, 1979; Plant and Glynn, 1979; Gros *et al.*, 1983). The gene *Nramp1/Slc11a1* encodes a ≈100-kDa phosphoglycoprotein with 12 transmembrane domains expressed exclusively on the phagolysosome of macrophages, controlling intracellular replication of pathogens (Searle *et al.*, 1998). Nramp1 acts as a divalent cation transporter, but has multiple pleiotropic effects on macrophage activation. Strain to strain variation in Nramp1/Slc11a1 expression has been demonstrated; for example, C57BL6/J and BALB/c mouse strains possess an *Nramp1* gene, but do not express a mature functional Nramp1 protein (Vidal *et al.*, 1996). However, an increasing list of diverse pathogens has been shown to be controlled by Nramp1 (Champion *et al.*, 2008; Forbes and Gros, 2001). The significant effect of a single host gene in the control of such important bacterial pathogens in mice has led to multiple studies analysing the role of Nramp1/Slc11a1 in control of infectious and autoimmune diseases in humans (Blackwell *et al.*, 2001). Polymorphisms in Nramp1 were mapped in a case control study of tuberculosis in the Gambia, revealing that genetic variations in Nramp1 affect susceptibility to tuberculosis in West Africans (Bellamy *et al.*, 1998). Nramp1 has also been shown to play a role in regulating the acquired immune response to human infection with *Mycobacterium leprae* (Alcaïs *et al.*, 2000).

Other mouse genes identified by forward genetic analysis include *Naip5/Birc1e*. *Naip5* was identified through the study of mouse strains with increased resistance to Legionnaire's disease, a severe pneumonia. The aetiological agent of Legionnaire's disease is *Legionella pneumophila*, an intracellular pathogen that survives in host macrophages and modulates host cellular processes. Most inbred mouse strains, including C57BL6/J, are resistant to *L. pneumophil*a, but the A/J mouse strain is susceptible. This difference was mapped back to a single locus on chromosome 13 called *Lgn1*. This locus contains multiple repeats of the gene baculoviral inhibitor of apoptosis protein repeat-containing 1 (*Birc1/Naip*). Complementation with *Birc1e/Naip5* conferred resistance to susceptible A/J mice to *L. pneumophila* through an unknown mechanism (Diez *et al.*, 2003; Wright *et al*, 2003).

Mammalian Toll-like receptors recognize microbial cell components such as lipopolysaccharide (LPS) and peptidoglycan, inducing the innate immune response. Inbred C3H/HeJ mice are non-responsive to LPS challenge, impairing macrophage activation, thereby resulting in increased susceptibility to infection with Gram-negative bacteria, such as *S. enterica* serovar Typhimurium (Watson and Riblet, 1974). C3H/HeJ mice possess a point mutation in the cytoplasmic Toll/interleukin-1 receptor (TIR) of the *Tlr4* gene. Analysis of murine *Tlr4* has revealed the mechanism of TLR4 signalling in mammalian host defence (Poltorak *et al.*, 1998; Qureshi *et al.*, 1999).

Reverse genetic screens have also revealed genes involved in the mouse response to infection. Knock-out or knock-in technologies have facilitated the transgenesis of hundreds of different genes in mice, through which the molecular basis of host resistance to bacterial infection continues to be dissected. Transgenic mice may be divided into two main categories, humanized immune system and receptor transgenics. The publication of the first transgenic mouse and the molecular cloning of human major histocompatibility complex (MHC) antigens published in the 1980s paved the way for the development of humanized transgenic mice (Brinster *et al.*, 1981; Ploegh *et al.*, 1981). A mouse gene may be replaced by its human orthologue to produce an immunologically humanized strain. Transgenic mice with humanized immune systems may be analysed to characterize T cell responses, such as cytotoxic T cell immunity that develops in response to foreign antigens (Pascolo, 2005; Tammiruusu *et al.*, 2005). Severe combined immunodeficient (SCID) mice lack functional T and B cells but possess an intact innate immune response. SCID mice therefore accept transplantation with heterologous cells, tissue and organs without rejection. The prototype human-mouse chimera used transplanted human peripheral blood leucocytes into SCID mice forming hu-PBL-SCID mice (Mosier *et al.*, 1988). Human adaptive immune cells transplanted into SCID mice retain their antigen specificity and this system has been exploited to analyse the role of components of the human adaptive immune response (Davis and Stanley, 2003).

However, the majority of reverse genetic screens in mice have focussed on the characterization of cytokines and their receptors. Many cytokine pathway knock-out mice demonstrate altered resistance to bacterial infection (Flynn *et al.*, 1993, 1995; Kopf *et al.*, 1994; Zheng *et al.*, 1995; Hess *et al.*, 1996; Pasparakis *et al.*, 1996; Dai *et al.*, 1997; Ladel *et al.*, 1997; McKenzie *et al.*, 1998; Brombacher *et al.*, 1999; Reznikov *et al*, 1999; Wei *et al.*, 1999; Kamradt *et al.*, 2000; Sugawara *et al.*, 2000; Yamada *et al.*, 2000; Hultgren *et al.*, 2002; Jacobs *et al.*, 2008). Receptor transgenics have been reviewed by Buer and Balling (2003).

Non-mammalian vertebrate models of infection – zebra fish

Zebra fish (*Danio rerio*) is a well-established model of vertebrate cellular and genetic development and has gained attention as an infection model. Zebra fish embryos develop innate immunity, the first line of defence against infection mediated through recognition of microbes via a limited number of pathogen recognition receptors, a system that is highly conserved in eukaryotes (Krimbrell and Beutler, 2001; Akira *et al.*, 2006). Adaptive immunity comprises pathogen-specific antigen recognition, immunological memory of pathogens and pathogen-specific antibodies. Adaptive immunity represents a novel evolutionary feature of vertebrates, including zebra fish, which is absent in invertebrates. The immunological response of zebra fish to challenge with bacterial pathogens may be used to model the mammalian response. For example, *D. rerio* and mammals both possess phagocytic cells, Toll-like receptors, complement, pro-inflammatory cytokines, interferon-dependent defence functions, acute-phase response proteins, leucocytes, antigen-processing cells, T cells, B cells and immunoglobulins (Trede *et al.*, 2001; Zarkadis *et al.*, 2001; Yoder *et al.*, 2002; Meijer *et al.*, 2004; van der Sar *et al.*, 2006; Lin *et al.*, 2007; Sieger *et al.*, 2009). The complex cross talk between innate and adaptive immune responses is difficult to study (Germain, 2004; Beutler *et al.*, 2006). In zebra fish, the innate immune system develops in the embryo and the adaptive immune response at later developmental stages. Owing to the temporal separation of development of the innate and adaptive immune responses in zebra fish embryos, cross talk between these systems may be studied.

A growing number of genetic tools are available for the zebra fish system. Characterization of the role of individual components of the zebra fish immune system is possible through the creation of selected homozygous mutations (Wienholds *et al.*, 2002, 2003). Host genes may also be inactivated using a knock-down approach in which translation of the target gene is inhibited through an injection of morpholino-modified antisense oligonucleotides (morpholinos) designed against the leader sequence of the target mRNA (Nasevicius and Ekker, 2000). Using these techniques, large scale, economical forward and reverse genetic screens are possible. Zebra fish embryos also have many practical advantages over mammalian infection models; they are small, economical, prolific and easily confined and maintained. Zebra fish embryos also lend themselves to the study of host pathogen interactions, as they are transparent for the first three weeks of development, thus constituting an ideal system to monitor real-time infections and fluorescent transgenic immune cells (Davis *et al.*, 2002). Disadvantages of the zebra fish embryo as an infection model for understanding human bacterial pathogens include the inability of zebra fish embryos to survive at 37°C; thus, temperature-regulated bacterial virulence genes may not be induced during infection. Zebra fish may not possess the mammalian target organ or receptor and finally, zebra fish may not exhibit the complete mammalian immune response. However, no infection model is perfect but must be selected based on the research question being asked. The validity of zebra fish as an infection model has been verified by elucidating complex host–pathogen interactions that could not be answered using traditional animal infection models (Hamm *et al.*, 2006; Volkman *et al.*, 2010).

Assays for measuring the interaction of zebra fish with bacteria

The use of both zebra fish embryos and adult fish has been described as infection models for bacterial pathogens. Zebra fish embryos, prior to becoming able to feed independently, are not protected under current UK legislation.

Fertilized eggs are collected from aquaria, stored at 29°C and hours post fertilization (hpf) staged visually using established developmental criteria (Kimmel *et al.* 1995). Two common developmental stages at which embryos are challenged is (i) 28 hpf, when macrophage-like cells appear in the circulation (Herbomel *et al.*, 1999) and (ii) 50 hpf, when granulocytes may be observed accumulating at sites of infection (Lieschke *et al.*, 2001). Embryos must be dechorionated, either manually or enzymatically, and may be anaesthetised with tricaine prior to infection. The site of infection must be visualized using a stereomicroscope, and embryo cell division occurs approximately once every 30 minutes at 30°C (rate of development increases with temperature). Pathogen challenge by microinjection into the blood, either blood island or axial vein, or into the yolk circulation valley results in a systemic infection. Challenge into an individual embryonic cell when the embryo is multi-cellular often results in a localized infection. Manual microinjection of zebra fish embryos is time-consuming, and to ensure embryos in a single experiment are all infected at the same developmental stage, the number of larvae that can be infected in one experiment is limited. Time constraints imposed by the developmental stage of the zebra fish embryo used in an experiment and human fatigue owing to the labour-intensive nature of zebra fish embryo microinjections make this a bottleneck in the scale of experiments that may be performed. Robotic microinjectors overcome this bottleneck (Wang *et al.*, 2007). Automation of bacterial challenge increases the number of embryos that may be infected to 15 zebra fish embryos per minute (chorion unremoved) with a survival rate of 98%, but also reduces experimental reproducibility as the site of infection may vary between embryos. Reported challenge doses range between 50 and 2400 colony-forming units (cfu) per embryo in a 0.5–2 nl volume, and a dose–response is usually observed. The challenge dose may be retrospectively enumerated by plating out embryos immediately after infection or by microinjecting the dose directly onto agar plates. Embryos are returned to the embryo medium post infection, often in individual wells in 24-well plates, and observed at regular time points. Host–pathogen interactions may be observed in real time owing to the optically accessible nature of the zebra fish embryo. The most common end point in zebra fish embryo infections is embryo death, defined as no visible heartbeat or circulation. Alternatively, embryos may be homogenised mechanically by pipetting and treatment with detergent and, at defined time points post infection, the bacterial loads enumerated (Vergunst *et al.*, 2010). Replicate embryos per time point range between 5 and 50, depending on the type of experiment.

Alternative published assays use zebra fish embryos to model neutrophil migration. Embryos are collected, staged and anaesthetised as described above, then wounded at a specific site on the embryo anatomy. The injured tissue is exposed to a treatment, e.g. inflammatory modulators (Mathias *et al.*, 2006). Zebra fish embryos are then fixed and stained for whole-cell mount immunofluorescence, and neutrophil migration to the wounded site as well as other sites in the embryo are monitored (Szabady *et al.*, 2009).

Assays using adult zebra fish require compliance with local animal welfare legislation and require diurnal light. Intraperitoneal challenge with the test pathogen is a common route of infection, with infection doses depending on the pathogen. Adult zebra fish have been used in research to understand bacterial virulence and pathology (Neely *et al.*, 2002; Swaim *et al.*, 2006), as well as the host response to bacterial infection through transcriptome analysis (van der Sar *et al.*, 2009).

Effects of pathogens on zebra fish

Zebra fish are susceptible to challenge with pathogens including *P. aeruginosa, S. enterica* serovar Typhimurium, *Mycobacterium* spp., *Burkholderia cenocepacia, Edwardsiella tarda, S. aureus, Streptococcus iniae, Leptospira interrogans* and *F. tularensis* (Neely *et al.*, 2002; van der Sar *et al.*, 2003, 2004; Pressley *et al.*, 2005; Wareham *et al.*, 2005; Prajsnar *et al.*, 2008; Sullivan and Kim, 2008; Brannon *et al.*, 2009; Davis *et al.*, 2009; Phelps *et al.*, 2009; Vojtech *et al.*, 2009; Vergunst *et al.*, 2010). However, this list is growing rapidly as more pathogens are tested in this model. Zebra fish respond to bacterial challenge in four ways: acute lethal infection, chronic infection, asymptomatic colonization and control, and clearance of bacteria (van der Sar *et al.*, 2004; Davis *et al.*, 2009; Vergunst *et al.*, 2010). Susceptibility to pathogens is often infection-dose dependent, and the developmental stage at which zebra fish embryos are infected may influence the outcome of infection. For example, at 28 hpf, macrophages only are present in the circulation system, and zebra fish embryos have a lower LD_{50} to challenge with *P. aeruginosa* than when challenged 50 hpf, when both macrophages and neutrophils have developed (Clatworthy *et al.*, 2009). Acute lethal infection or chronic infection may be established within the zebra fish model depending on the bacterial pathogen and strain used (van der Sar *et al.*, 2004). Differences in the outcome of infection into zebra fish may be exploited to identify strain-to-strain variations in virulence of a pathogen (Menudier *et al.*, 1996; Levraud *et al.*, 2009; Vergunst *et al.*, 2010). For example, the epidemic ET12 lineage of *B. cenocepacia*, a pathogen associated with cystic fibrosis, is highly virulent in zebra fish with IV injections of >10 bacteria, resulting in 100% embryo mortality in less than 3 days. Significant differences in zebra fish survival were observed between *B. cenocepacia* ET12 and K56-2, J2315 and BC7 clonal lineages. Other pathogens associated with cystic fibrosis, such as *Burkholderia vietnamiensis* FC441 and *Burkholderia stabilis* LMG14294, were avirulent in the zebra fish infection model (Vergunst *et al.*, 2010). Bacterial pathogenicity in zebra fish has also been shown to correlate with pathogenicity in mammalian infection models. For example, *Listeria monocytogenes* is virulent in zebra fish with an infection process identical to that seen in mammals, whilst *Listeria innocua* is avirulent in zebra fish and mammalian infection models (Levraud *et al.*, 2009). Pathogen strains that are avirulent in zebra fish can be determined by lack of mortality and enumeration of bacteria indicating whether a successful immune response has been mounted to clear the infection. Importantly, correlation between the infection process of pathogens in zebra fish and mammals has been demonstrated. For example, *L. monocytogenes* escapes from the phagosome through production of listeriolysin, invades the cytosol and forms ActA-dependent actin-based comet tails (Levraud *et al.*, 2009).

Identification of virulence-associated genes

Virulence genes of bacterial pathogens may be identified following challenge with the wild-type and deficient mutant strains into zebra fish, with reduced mortality in the mutant, demonstrating a role in virulence for the gene of interest (van der Sar *et al.*, 2004; Brannon *et al.*, 2009; Vergunst *et al.*, 2010). However, one of the key strengths of the zebra fish model lies in the embryo transparency, allowing virulence mechanisms to be imaged and elucidated in real time, in an *in vivo* system. For example, *Clostridium difficile* toxin B (TcdB) had been extensively studied in tissue culture and traditional animal models, but the *in vivo* target remained elusive until the transparency of the zebra fish embryo infection model was exploited to image TcdB localization and organ-specific tissue damage, revealing cardiotoxic activity (Hamm *et al.*, 2006). Another elegant study used the zebra fish embryo infection model to image mycobacteria–host interactions, revealing that granuloma formation is induced by bacteria rather than as a host response to infection. The bacterial secreted protein ESAT-6 (early secreted antigenic target) induces matrix metalloproteinase-9 (MMP9) in epithelial cells neighbouring infected macrophages. MMP9 enhances recruitment of macrophages, contributing to nascent granuloma maturation and bacterial growth. Disruption of MMP9 function resulted in attenuated granuloma formation and bacterial growth (Davis and Ramakrishnan, 2009; Volkman *et al.*, 2010). Another secreted metalloprotease, *Escherichia coli* O157:H7 StcE, has been shown to modulate neutrophil-induced inflammation during infection in zebra fish (Szabady *et al.*, 2009). The optical accessibility of zebra fish embryos has also been exploited to understand the role of other bacterial virulence genes, including a type three secretion system in *P. aeruginosa* (Brannon *et al.*, 2009) and the role of CepR quorum-sensing regulator in *B. cenocepacia* replication and dissemination during infection (Vergunst *et al.*, 2010). Importantly, zebra fish infection model data have been shown to correlate with mouse infection data for virulence-associated genes tested in both models (Neeley *et al.*, 2002).

Identification of novel antibacterials

Zebra fish embryos are the only vertebrate infection model for which high-throughput *in vivo* screening of small molecules has been possible (Murphey and Zon, 2006). The ease and economy with which zebra fish embryos can be produced, along with their optical accessibility, have led to combined animal testing and high-throughput screening of embryos in which complex whole system phenotypes can be assayed in a 48- or 96-well plate format. In one study, antibiotics added to the water surrounding zebra fish embryos were able to rescue embryos from lethal challenge with *P. aeruginosa* (Clatworthy *et al.*, 2009). The identification of antibiotics that perturb phenotypes in an *in vivo* system has several advantages, including the selection of targets that attenuate infection, are cell permeable, have no gross toxic side effects and have acceptable pharmacokinetic profiles (MacRae and Peterson, 2003). Embryos are also treatable by waterborne exposure using only tiny amounts of the test compound. Moreover, drug activity in zebra fish and mammals is conserved, and the immunological reaction to drug toxicity and environmental pollutants are well correlated between humans and zebra fish (Mandrekar and Thakur, 2009).

Identification of host genes which play roles in resistance to infection

Similarities between zebra fish and mammalian immune systems, the availability of genetic tools to disrupt zebra fish gene function and the optical transparency of the embryo make this an excellent system to characterize the role of host genes involved in resistance to infection. Zebra fish embryos have been reported as a real time model for understanding host–pathogen interactions during mycobacterial infections, with correlation to mammalian infection model data (Clay *et al.*, 2008). *Mycobacterium marinum* infection in zebra fish mimics some of the key features of clinical tuberculosis (Swaim *et al.*, 2006; Clay *et al.*, 2007). For example, mycobacteria replicate within zebra fish macrophages and recruit uninfected macrophages to form granulomas, as in mammalian infection. The role of phagocytic cells in the establishment of infection has been characterized in zebra fish using specimens with defined inactivated genes. For example, zebra fish lacking functional macrophages (pu.1 mutants) challenged with *Mycobacterium* revealed a dichotomous role for macrophages, in control of infection and dissemination of mycobacteria throughout the body (Clay *et al.*, 2007). Zebra fish with inactivated TNF receptor 1 revealed that TNF restricts *Mycobacterium* growth and necrosis within macrophages, maintaining granuloma integrity (Clay *et al.*, 2008). The essential role of the adaptive immune response in the control of mycobacterial infection has also been explored using Rag1 fish that have no functional T or B cells (Swaim *et al.*, 2006). Transcriptome analysis of *Mycobacterium*-infected adult zebra fish to understand the host response to infection has been reported (Hegedus *et al.*, 2009; van der Sar *et al.*, 2009).

Invertebrate models of infection, *Caenorhabditis elegans* and other nematodes

Caenorhabditis elegans, a free-living nematode worm, was originally proposed as a simple model for neurobiology and developmental biology research (Sulston and Brenner, 1974). It is easily maintained in the laboratory on agar plates previously seeded with *E. coli* or *Bacillus subtilis* as a food source (Darby, 2005), and the small genome size, along with ease of culture, has permitted large-scale genetic screens to identify *C. elegans* gene functions (Jorgensen and Mango, 2002). Subsequently, *C. elegans* has been used extensively as an infection model for a range of bacterial pathogens. However, it is worth noting that other species of nematodes such as *Panagrellus redivivus* might also be useful as infection models (Laws *et al.*, 2005).

There are several different developmental stages of *C. elegans*. The young larvae emerge from eggs and pass through four distinct larval stages (LI–L4). When stressed, the L3 stage can enter an alternative stress-resistant stage termed dauer larvae. Differences in the susceptibility of the different developmental stages to infection have been reported. For example, L1 stage larvae are resistant to infection with *Serratia marcescens* but L4 adults are highly susceptible (Kurz *et al.*, 2003; Darby, 2005).

Assays for measuring the interaction of nematodes with bacteria

In its natural environment, the soil, *C. elegans* feeds on micro-organisms, including bacteria (Hodgkin, 2001), and some of the bacteria encountered are potentially hazardous to *C. elegans* (Couillault and Ewbank, 2002). In principle, the assays for assessing the effects of pathogens on *C. elegans* are simple. The pathogen of interest is cultured on agar and nematodes are placed onto the bacterial lawn and allowed to ingest the bacteria. Nematode survival is then scored at intervals. Nematode death might be evidenced in a number of ways. Dead nematodes fail to show any movement even when gently prodded with a blunt instrument and after exposure to some pathogens, the remnants of the dead nematode appear as a ghost-like impression in the agar (O'Quinn *et al.*, 2001). Other workers have proposed alternative assays, such as a feeding inhibition assay, which involves measuring a decrease in optical density of a suspension of bacteria as the nematodes ingest and digest the bacteria (Smith *et al.*, 2002). Recently, other elegant assays have been described, using bacteria tagged with fluorescent reporters (Fuhrman *et al.*, 2008): this allows the high-throughput screening of nematodes.

Effects of pathogens on *C. elegans*

C. elegans was first used as a model to investigate host–pathogen interactions by Tan *et al.*, who showed susceptiblity to killing by *P. aeruginosa* (Tan *et al.*, 1999a). Subsequently, the survival of *C. elegans* has been shown to be detrimentally affected by many pathogenic bacteria (Table 1.2). Some pathogens, like *Enterococcus faecium* and *Streptococcus pyogenes* are reported to be pathogenic for *C. elegans* by some groups (Jansen *et al.*, 2002; Moy *et al.*, 2004) but not by others (Garsin *et al.*, 2001). This might reflect differences in the strains tested–strain-dependent virulence in *C. elegans* is frequently reported (Cardona *et al.*, 2005; Sotto *et al.*, 2008). Some human pathogens do not affect viability of *C. elegans*. *Burkholderia mallei*, a clonal derivative of *B. pseudomallei*, did not affect viability of *C. elegans* (O'Quinn *et al.*, 2001). It is suggested that this might reflect the different lifestyles of these pathogens. Both are pathogens of humans, but *B. mallei* is unable to survive outside of a mammalian host, whereas *B. pseudomallei* can also survive and replicate in soil and water courses.

C. elegans might also be of value for the investigation of *in vivo* interactions between different bacterial species and might therefore provide insight into the ways in which virulence might be influenced in complex communities of bacteria. Co-infection with *Enterococcus faecalis* and *E. coli* appeared to enhance virulence. In contrast, virulence of *Candida albicans* was reduced in the presence of *Acinetobacter baumannii* or *S. enterica*, probably as a consequence of reduced fungal filamentation (Peleg *et al.*, 2008; Tampakadis *et al.*, 2009).

The mechanisms by which bacterial pathogens kill *C. elegans* appear to be quite diverse and can depend on the strain and growth conditions tested. However, in general three mechanisms have been recognized. Rapid killing of *C. elegans* is often attributable to the production of toxins, such as pyocyanins (Mahajan-Miklos *et al.*, 1999) or hydrogen cyanide (Gallagher and Manoil, 2001) of *P. aeruginosa*, hydrogen peroxide by pathogenic streptococci and *E. faecium* (Jansen *et al.*, 2002;

Table 1.2 Bacterial pathogens shown to affect *Caenorhabditis elegans*. Adapted from Darby (2005)

Species	Effect on *C. elegans*	Reference
Aeromonas hydrophila	Reduced lifespan	Couillault and Ewbank (2002)
Agrobacterium tumefaciens	Reduced lifespan	Couillault and Ewbank (2002)
Burkholderia cenocepacia	Gut infection; toxin	Coenye *et al.* (2001), Köthe *et al.* (2003), Huber *et al.* (2004)
Burkholderia pseudomallei	Diminished lifespan toxin	O'Quinn *et al.* (2001), Gan *et al.* (2002)
Enterococcus faecaelis	Gut infection	Garsin *et al.* (2001), Kurz and Ewbank (2007), Maadani *et al.* (2007), Moy *et al.* (2009)
Erwinia carotovora	Diminished lifespan	Couillault and Ewbank (2002)
Erwinia chrysanthemi	Diminished lifespan	Couillault and Ewbank (2002)
EPEC	Gut infection	Couillault and Ewbank (2002), Anyanful *et al.* (2005, 2009)
Photorhabdus luminescens	Diminished lifespan	Couillault and Ewbank (2002)
Pseudomonas aeruginosa	Diminished lifespan, fast killing mediated by toxins	Mahajan-Miklos *et al.* (1999), Tan *et al.* (1999a,b), Gallagher and Manoil (2001), Hendrickson *et al.* (2001), Yorgey *et al.* (2001), Choi *et al.* (2002), Alegado *et al.* (2003), Wareham *et al.* (2005), Laws *et al.* (2006), Lee *et al.* (2006), Zlosnik *et al.* (2006), Garvis *et al.* (2009)
Salmonella enterica serovar Typhimurium	Gut infection. exceptionally germline cell death	Aballay and Ausubel (2001), Aballay *et al.* (2000), Darby (2005), Tenor *et al.* (2004)
Serratia marcescens	Gut infection, tissue degradation	Kurz *et al.* (2003), Darby (2005)
Staphylococcus aureus	Gut infection	Garsin *et al.* (2001), Sifri *et al.* (2003), Sotto *et al.* (2008)
Streptococcus agalactiae (Group B)	Toxin (hydrogen peroxide)	Bolm *et al.* (2004)
Streptococcus dysgalactiae	Toxin (hydrogen peroxide)	Bolm *et al.* (2004)
Streptococcus mitis	Toxin (hydrogen peroxide)	Bolm *et al.* (2004)
Streptococcus oralis	Toxin (hydrogen peroxide)	Bolm *et al.* (2004)
Streptococcus pneumoniae	Toxin (hydrogen peroxide)	Garsin *et al.* (2001), Bolm *et al.* (2004)
Streptococcus pyogenes	Toxin (hydrogen peroxide)	Jansen *et al.* (2002), Bolm *et al.* (2004)
Yersinia pestis	Biofilm on cuticle, feeding blocked or gut infection	Darby *et al.* (2002), Sun *et al.* (2008), Styer *et al.* (2005)
Yersinia pseudotuberculosis	Biofilm on cuticle, feeding blocked	Darby *et al.* (2002), Erickson *et al.* (2006)

Bolm *et al.*, 2004; Moy *et al.*, 2004), a tryptophan metabolite by enteropathogenic *E. coli* (EPEC) (Anyanful *et al.*, 2005) and an unidentified toxin of *B. pseudomallei* (O'Quinn *et al.*, 2001; Gan *et al.*, 2002). Slower killing often involves a true infection process. Killing by pathogens such as *S. aureus*, *E. faecalis*, *S. enterica* or *P. aeruginosa* occurs over several days and is correlated with an increase in bacterial numbers in the gut (Tan *et al.*, 1999a; Aballay *et al.*, 2000; Garsin *et al.*, 2001; Sifri *et al.*, 2003). This killing mechanism of *P. aeruginosa* has been termed slow killing (Tan *et al.*, 1999a; Kurz and Ewbank, 2000) to differentiate it from the toxin-mediated or fast killing outlined above. Elegant experiments where *C. elegans* is exposed to pathogens for a short period of time and then moved onto a lawn of non-pathogenic *E. coli* have shown that *E. faecalis* and *S. enterica* proliferate in the gut, rather than simply accumulate as a consequence of ingestion, and the number of bacteria in the gut reaches 10^4 cfu (Aballay *et al.*, 2000; Garsin *et al.*, 2001). The gut often becomes distended as a consequence of the increasing burden of bacteria (Garsin *et al.*, 2001). However, it is not simply the bacterial burden that causes death, since *E. faecium* is readily able to colonize the gut at similar levels to that found with pathogens, yet without detrimental effects on the nematode (Garsin *et al.*, 2001). Crossing of the gut wall and invasion of the body cavity appears to be an infrequent event. Finally, some pathogens appear to colonize the surface of *C. elegans*, for example *Y. pestis* and some strains of *Y. pseudotuberculosis* form biofilms on the head of *C. elegans* (Darby *et al.*, 2002; Erickson *et al.*, 2006). These biofilms are reminiscent of the biofilms that form in the gut of fleas infected with *Y. pestis*, but there is no reported correlation between the formation of biofilms and colonization of fleas (Lee *et al.*, 2006).

Identification of virulence-associated genes

The extent to which *C. elegans* can be used as a model for the identification of virulence determinants playing a role in disease in humans (or other mammals) has been the subject of much debate. It is clear that *C. elegans* has a number of limitations as a model. One is that *C. elegans* cannot be maintained at 37°C – even at 30°C the lifespan is markedly reduced. Therefore, virulence determinants, which are expressed only at 37°C, cannot be identified in *C. elegans*. Nematode species, which are able to survive and replicate at 37°C, might potentially serve as infection models, but many of these species are themselves pathogenic towards humans. In an attempt to resolve this limitation, Laws *et al.* (2005) showed that the *P. redivivus* nematode was able to survive for several days but replicate at 37°C. This nematode species was susceptible to killing by a range of pathogens, and interestingly some strains of these pathogens killed only at 37°C.

Another limitation is the inability to determine the dose received by individual worms, or to quantify the degree of virulence or attenuation of mutants. This limitation stems from the difficulties associated with measuring the intake of bacteria by individual nematodes, and means that realistically it is not possible to obtain LD_{50} values.

Notwithstanding these limitations, a number of experimental studies confirm that some virulence-associated genes are required for disease in both *C. elegans* and in mammalian hosts. For example, the quorum-sensing system and cytolysin of *E. faecalis* (Garsin *et al.*, 2001) are important for virulence in both hosts. However, one study also indicates that, at least in *P. aeruginosa*, the role of individual genes in virulence may be dependent on their genetic context, so that genes that play a role in virulence of some strains may not contribute to virulence in others (Lee *et al.*, 2006). In addition, it is clear that many genes or gene clusters, which play key roles in virulence in mammals, do not play significant roles in virulence in *C. elegans*. In some cases, this may reflect a low level of expression in the nematode. However, this is not always the case. The type III secretion systems of *P. aeruginosa* and EPEC are key virulence mechanisms in mammalian models of disease but appear to play no role in disease in *C. elegans* (Darby, 2005; Davis *et al.*, 2009).

The finding that genes can play a role in virulence of mammals and nematodes has allowed *C. elegans* to be used as a screen for novel virulence genes. Pyocyanins were originally identified as virulence factors in *C. elegans* (Mahajan-Miklos *et al.*, 1999) and subsequently shown to promote human neutrophil apoptosis (Usher *et al.*, 2002).

Similarly, a sucrose-6-phosphate hydrolase (*scrB*) mutant of *E. faecalis* was initially identified as attenuated in *C. elegans* and later, shown to also be attenuated in mice (Garsin *et al.*, 2001). This approach to the identification of virulence-associated genes has been developed by several groups for the high-throughput screening of panels of mutants. The pioneering high-throughput study by Tan *et al.* (1999b) involved the screening of 2400 transposon mutants of *P. aeruginosa* and identified eight which were attenuated for slow killing in *C. elegans*. Of these, five were also attenuated in mice, including several potentially novel virulence-associated genes (Tan *et al.*, 1999b). A screen of 984 *Y. pestis* transposon mutants identified six which showed reduced killing of *C. elegans*, and two of these, including one encoding a completely novel virulence factor, were shown to play a role in disease in mice (Styer *et al.*,2005). A similar hit rate has been reported for *E. faecalis* – of 540 mutants screened, nine were attenuated in *C. elegans*, and five of these were subsequently shown to be attenuated in mice (Maadani *et al.*, 2007).

Most of these studies have relied on the exposure of nematodes to mutants individually grown on agar and the recorded subsequent mortality. However, a number of alternative formats suitable for high-throughput screening have since been described. For example, liquid-based assays for the high-throughput screening of mutants have been suggested (Smith *et al.*, 2002). Mutants with reduced virulence might be expected to be ingested in larger numbers by *C. elegans*, resulting in a greater reduction in optical density of the suspension (Smith *et al.*, 2002).

Identification of novel antibacterials

High-throughput screens with *C. elegans* have also been used successfully to identify novel antibacterial compounds. In one recently reported study, a robotic screening system was used to evaluate 37,200 compounds for activity towards *E. faecalis*. Twenty-eight novel antimicrobials were identified, and importantly, some were active against *E. faecalis in vivo* but not *in vitro* (Moy *et al.*, 2009). This finding highlights the potential for *C. elegans*-based screens to provide paradigm shifts in this field.

Identification of nematode genes which play roles in resistance to infection

Whilst the potential use of *C. elegans* to identify virulence factors and novel antimicrobials is now well established, the use of *C. elegans* to study the role of host genes in resistance (or susceptibility) to infection is still largely unexplored. However, reported work indicates the potential for these studies.

The mechanisms by which *C. elegans* actively resists infection fall broadly into two areas. Firstly, there is a significant body of evidence indicating that it is able to sense the presence of potential pathogens and physically avoid encounters with these bacteria. The ability of *C. elegans* to respond in this way is not surprising considering that the soil would contain both bacteria that are food sources and potential pathogens. *C. elegans* uses a variety of signals to locate food, and it is becoming clear that the sensing of bacteria involves the oxygen sensing system. Mutants defective in oxygen-sensing guanylate cyclase gcy-35 spend greater time on a lawn of bacteria, resulting in increased exposure (Reddy *et al.*, 2009). A homologue of the mammalian neuropeptide Y receptor gene (*npr-1*), which encodes a G protein–coupled receptor, is also implicated in oxygen-dependent avoidance of bacteria (Reddy *et al.*, 2009). These findings highlight how *C. elegans* might regulate exposure to bacteria, but they do not explain how the nematodes distinguish different species and strains of bacteria with different pathogenic potentials (Laws *et al.*, 2006).

Secondly, *C. elegans* possesses an innate immune system, which shows many similarities with the innate immune systems of mammals. Its innate immune system involves signalling cascades, which include a Toll-like receptor (TOL-1), mitogen-activated protein kinases (MAPK), transforming growth factor-beta (TGF-β), insulin-like receptor (ILR), and the programmed cell death (PCD) pathway. The range of innate effectors include lysozymes and antimicrobial peptides (Evans *et al.*, 2008). There is evidence that the *C. elegans* TOL-1 receptor responds to Gram-negative pathogens; that it is required for the production of defensins and resistance to *S. enterica* infection (Tenor and Aballay, 2008), and for avoidance of *S. marcescens* (Pujol *et al.*, 2001). There is no good evidence that TOL-1 responds to Gram-positive pathogens (Tenor and Aballay, 2008). Innate signalling occurs via p38 MAPK, the Sma/TGF-β, and the DAF-2/DAF-16 insulin-like signalling pathways rather than the NF-κB pathway, which would be involved in Toll signalling in mammals (Ewbank, 2006; Gravato-Nobre and Hodgkin, 2005). Not surprisingly there is evidence that at least some pathogens are able to modulate these signalling pathways to down-regulate the *C. elegans* innate response to infection (Evans *et al.*, 2008).

Two other mechanisms that appear to regulate resistance to pathogens are the induction of apoptosis and resistance as a consequence of the modulation of lifespan. Programmed cell death is a normal component of development of *C. elegans* and might also play a role in resistance to infection. *C. elegans ced-3* and *ced-4* mutants in which developmentally regulated cell death is blocked are hypersensitive to *S. enterica*-mediated killing. These results suggest that programmed cell death may be involved in the *C. elegans* defence response to pathogen attack (Aballay and Ausubel, 2001). The association of ageing with altered resistance to pathogens is thought to be linked to co-modulation of the innate immune system because the DAF-2 pathway appears to regulate both of these processes (Kurz and Tan, 2004). There may also be an association between ageing and hypoxic responses (Chen *et al.*, 2009). These findings indicate the intimate linkage of many of the responses that confer protection against pathogens. The ease with which *C. elegans* can be manipulated, coupled with the availability of high throughput screens, suggest that other genes associated with resistance to infection will be discovered in the future.

Galleria mellonella (and other insect larvae)

Innate immunity is the ability to recognize and eliminate pathogens via pattern recognition receptors (PRR), which bind microbial components, and is a highly conserved system in eukaryotes (Kimbrell and Beutler, 2001). Insects have an innate immune system comprising cellular and humoral responses (Hoffmann, 1995; Lavine and Strand, 2002; Kavanagh and Reeves, 2004). The cellular response of insects to challenge involves rapid production and mobilization of haemocytes, which engulf bacteria. The humoral response comprises activation of the Toll and Imd pathways leading to activation of transcription factors belonging to the NF-κB family, resulting in expression of antimicrobial peptide genes (Wang and Ligoxygakis, 2006; Marmaras and Lampropoulou, 2009). The potential to explore aspects of mammalian innate immunity in insect larvae combined with their economy, ease of maintenance and ethical acceptability has led to the development of wax moth larvae, *Galleria mellonella,* as an established infection model. *G. mellonella* have been used as an infection model to understand human bacterial pathogens since the 1980s (Morton *et al.*, 1987), and as more bacteria have been tested, the list of pathogens to which *Galleria* is susceptible has rapidly grown to include *Campylobacter jejuni, B. cenocepacia, F. tularensis* and *Listeria* spp. among others (Morton *et al.*, 1987; Jander *et al.*, 2000; Miyata *et al.*, 2003; Aperis *et al.*, 2007; Schell *et al.*, 2008; Champion *et al.*, 2009, 2010; Peleg *et al.*, 2009a; Mukherjee *et al.*, 2010). *G. mellonella* possess haemocytes that share a high degree of structural and functional similarity with mammalian phagocytic cells. For example, haemocytes ingest bacteria and generate bactericidal products, such as superoxide via a respiratory burst (Bergin *et al.*, 2005). The similarities between insect and mammalian phagocytic cells have opened a new arena for interrogation of bacteria with an intracellular lifecycle stage (Champion *et al.*, 2009; Mukherjee *et al.*, 2010). Insect larvae have the advantage that they can be infected at 37°C, ensuring expression of temperature-regulated virulence genes. Moreover, a defined infection site and the ability to challenge larvae with exact doses of bacteria make insect larvae an appealing infection model. Drawbacks of this model are the lack of an adaptive immune response, as well as mammalian target organs and receptors and the absence of genetic tools to characterize host genes involved in resistance to infection, since, to date, the *G. mellonella* genome has not been sequenced.

Assays for measuring the interaction of *G. mellonella* with bacteria

The most common *G. mellonella* assay involves challenge of larvae by subcutaneous micro-injection into a defined infection site (often a proleg) with a defined inoculum. The inoculum is estimated by optical density of the bacterial culture and may be retrospectively enumerated through colony counts. It is injected into larvae in a 5–10 µl volume. Infected larvae may be incubated at temperatures ranging from 15°C to 37°C, as required. The ability to incubate larvae at 37°C facilitates analysis of temperature-regulated virulence genes. Larvae are observed at defined time points post infection, and dose-dependent killing is often observed (Fig. 1.2). End points include survival with death measured as larvae unable to move, even when manipulated with a pipette. Survival is often scored 24 h, 48 h, 72 h or more post infection. More detailed pathological scoring systems could be established to tease out degrees of gross pathology, such as melanisation, reduced movement etc. Phosphate-buffered saline administration and uninfected controls are included in studies to ensure that larval death is not a result of trauma due to the injection. Groups of ten larvae are generally used as internal replicates in an experiment, with two or three experimental replicates providing large data sets for statistical analyses (Seed and Dennis, 2008; Champion *et al.*, 2009, 2010; Mukherjee *et al.*, 2010).

In addition to live/dead scoring of larvae, enumeration of bacterial load at given time points may also be calculated by homogenising larvae and plating out dilutions. The bacterial load in specific larval compartments, cadaver, haemolymph and haemocoel, may be established by draining haemolymph from infected larvae and using centrifugation to separate haemocytes from the haemolymph. The larval cadaver may then be homogenised and the homogenized cadaver serially diluted and plated out with the haemolymph and haemocytes.

Galleria may receive an oral challenge when this infection route is physiologically relevant. The oral route of infection has been used for *in vivo* expression technology using *Galleria* as a host. A site-specific recombinase was used to identify transient activation, and a clone library of DNA fragments upstream from the recombinase was screened in *Galleria*, of which 100 clones were selected from dead larvae. The inserts were sequenced to identify *in vivo*-activated genes (Fedhila *et al.*, 2006).

To investigate the fate of bacteria within the larvae, assays have been designed to observe the presence of green fluorescent protein-tagged bacteria within haemocytes. Observation of intracellular survival, using confocal fluorescence microscopy, combined with enumeration of bacterial loads can indicate whether bacteria are able to replicate within phagocytic cells in an *in vivo* system. These assays may be used to underpin the genetic basis for intracellular survival (Champion *et al.*, 2009). Finally, cell-free haemolymph may be collected from the larval abdomen using a needle to monitor innate immune responses to infection,

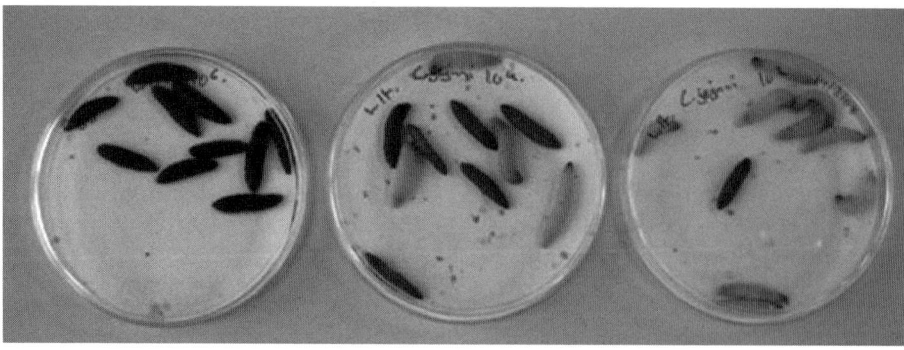

Figure 1.2 Appearance of groups of 10 *G. mellonella* larvae 24 h after injection with 1×10^6 CFU, 1×10^4 CFU or 1×10^2 CFU of broth grown *C. jejuni* 11168H (shown left to right). Larvae were incubated at 37°C following challenge (Champion *et al.*, 2010).

such as variations in lysozyme and apolipophorin-III (Andrejko *et al.* 2008a,b).

Effects of pathogens on *G. mellonella*

G. mellonella possess a cream coloured cuticle, which acts as a physical barrier to infection. Once breached, bacteria are faced with the cellular and humoral innate immune response of the insect. *Galleria* produce antimicrobial peptides, such as lysozyme and undergo a complex enzymatic cascade resulting in melanisation of haemolymph. The cellular response of *Galleria* to bacterial challenge involves subsets of haemocytes that phagocytose bacteria and form aggregates, or nodules (Li *et al.*, 2002; Theopold *et al.*, 2002; Andrejko *et al.*, 2008a,b). Cytotoxic reactions are then initiated (Bergin *et al.*, 2005). Haemocyte nodulation results in melanisation through the action of phenoloxidase (Lavine and Strand, 2002; Jiravanichpaisal *et al.*, 2006). The degree of melanisation in infected larvae ranges from mottled grey to uniform black, depending on the infecting bacteria, and usually indicates a lethal challenge. Bacteria that are not pathogenic to *Galleria* do not usually induce melanisation. Infected larvae become less motile, and faecal staining may be observed on filter paper discs on which larvae are stored. Survival of *Galleria* to bacterial challenge is dose-responsive, and inter- and intra-species variation in LD50 has been reported (Jander *et al.*, 2000; Seed and Dennis, 2008).

Identification of virulence-associated genes

Galleria is a simple *in vivo* system in which to characterize bacterial virulence genes with correlation to infection data in other animal models. For example, analysis of a panel of mutants in *P. aeruginosa* strain PA14 in murine and insect infection models (Jander *et al.*, 2000). *E. faecalis* is a major cause of hospital-acquired infection; a collagen-binding protein encoded by *ace* was demonstrated for the first time to be a virulence factor using *Galleria* as an infection model, correlating with murine infection data (Lebreton *et al.*, 2009). Virulence of *S. aureus* in *Galleria* is dependent on the functional status of the global virulence regulator, *agr* (Peleg *et al.*, 2009b). *Galleria* has been used as a host for *in vivo* expression technology (IVET) to identify 20 *in vivo*-induced (*ivi*) genes in *Bacillus cereus*, shown to be upregulated during infection. The most strongly induced *ivi* gene was characterized as an internalin-encoding gene (Fedhila *et al.*, 2006). Exploiting the homology between the respiratory burst and subsequent release of cytotoxic superoxide in both insect and mammalian phagocytic cells, *sod*C was identified as a virulence determinant in *Y. pseudotuberculosis* (Champion *et al.*, 2009). Lastly, *Galleria* has also been reported as an effective infection model for identification of *C. jejuni* virulence genes including *cj1416*, a critical gene in the biosynthetic pathway of O-methyl phosphoramidate (MeOPN). MeOPN is a capsule modification associated with more severe clinical outcomes of *C. jejuni* infection. Characterization of the virulence mechanism of MeOPN is required (Champion *et al.*, 2010).

Identification of novel antibacterials

G. mellonella is the first and, to date, only animal from which a specific peptidic inhibitor of microbial metalloproteinase has been reported. Inducible metalloprotease inhibitor (IMPI) inhibits thermolysin-like microbial metallopeptidases (Wedde *et al.*, 1998; Clermont *et al.*, 2004). IMPI shows promise as a template for second-generation antibiotics, as it has specific activity against bacterial metalloproteinases such as those of *Vibrio vulnificus* (Wedde *et al.*, 2007).

Galleria is amenable to antibiotic treatment. Exploiting this feature of the infection model, antimicrobial efficacy of agents against *A. baumannii* and *F. tularensis* have been tested and shown to rescue *Galleria* from lethal infection (Aperis *et al.*, 2007; Peleg *et al.*, 2009a). Furthermore, bacteriophage therapy has been tested as an alternative treatment for *B. cepacia* complex (BCC) infections in *G. mellonella* (Seed and Dennis, 2009). These studies indicate that *Galleria* is a simple system in which to analyse antimicrobial efficacy.

Identification of host genes which play roles in resistance to infection

The *Galleria* IMPI is currently the only specific inhibitor of microbial metalloproteinases isolated from animals to be reported. The *impi* gene encodes two recombinant structures: (i)

RIMP1, encoded by the N terminal section of the *impi* sequence, is glycosylated and shows variable fucosylation, (ii) RIMP2, encoded by the C terminal domain, is glycosylated and fully fucosylated. RIMP1 regulates thermolysin-like metalloproteinases produced by invading pathogens such as *V. vulnificus*. RIMPI has been proposed as a template for the design of second-generation antimicrobials. RIMP2 does not inhibit bacterial metalloproteinases but is active against human matrix metalloproteinases (Wedde *et al.*, 2007).

Drosophila

Recently, there has been interest in *Drosophila melanogaster* as an infection model for a diverse range of pathogens including *P. aeruginosa* (Chugani *et al.*, 2001; D'Argenio *et al.*, 2001; Kurz and Ewbank, 2007), *S. enterica* serovar Typhimurium (Frandsen *et al.*, 2008), *S. aureus* (Fleming *et al.*, 2006), *L. monocytogenes* (Ayres *et al.*, 2008), *F. tularensis* (Vonkavaara *et al.*, 2008), *S. pneumoniae* (Schneider *et al.*, 2007) and *Vibrio cholerae* (Park *et al.*, 2005). Infection assays with *Drosophila* are carried out at 20–30°C, and the model suffers from the same limitations as many other invertebrate models in that the effects of genes up-regulated at 37°C will be masked.

Assays for measuring the interaction of *Drosophila* with bacteria

Broadly, three different routes of inoculation of *D. melanogaster* with pathogens have been reported (Apidianakis and Rahme, 2009) (Table 1.3). The strain of flies used, their sex and age may all have a profound impact on the nature of the infection (Apidianakis and Rahme, 2009). Female flies are larger and easier to handle, but suffer from the disadvantage that fertilization can impact on immunity to infection (Apidianakis and Rahme, 2009).

One frequently used assay involves the localized delivery of pathogens into mature *D. melanogaster* flies with death of flies recorded as an end point. This route of delivery is sometime referred to as a fly nicking or needle pricking model. Flies are inoculated with a needle that has previously been dipped into a bacterial culture resulting in the introduction of bacteria at the wound site (D'Argenio *et al.*, 2001; Apidianakis and Rahme, 2009). The inoculum size can be varied by diluting the bacterial culture. Female flies are often used in the fly nicking model because the smaller males appear to be susceptible to death simply as a consequence of the inoculation procedure (D'Argenio *et al.*, 2001). A more precise method of delivery involves the direct injection of bacteria into the haemolymph of adult flies or 'injector pumping' (Apidianakis and Rahme, 2009). More recently, a *Drosophila* embryo infection model has been described (Vlisidou *et al.*, 2009). This approach allows the microinjection of small and defined volumes of bacterial culture (e.g. 50–4000 nl) into the fly (Ayres *et al.*, 2008; Vonkavaara *et al.*, 2008; Apidianakis and Rahme, 2009). The direct inoculation of bacteria into *Drosophila*, either by fly nicking or injector pumping, generally results in an acute infection. Flies inoculated with *P. aeruginosa* die 16–28 h after infection (D'Argenio *et al.*, 2001) and the direct inoculation of 50–100 cfu of *S. marcescens* results in deaths by day 1 (Nehme *et al.*, 2007). However, one exception involves *F. tularensis* where fly deaths occur up to 20 days post challenge (the latest time sampled) (Vonkavaara *et al.*, 2008).

Oral infection models with a range of pathogens have also been described, where bacteria in L-broth or in a sucrose solution are fed to *Drosophila*. It is difficult to estimate the dose received in these experiments, but in some studies a blue

Table 1.3 Routes of infection and disease of *D. melanogaster* caused by bacterial pathogens

Delivery method	Site of delivery	Dose delivered	Disease
Fly nicking/needle pricking	Local wound	Broadly defined	Typically acute
Injector pumping/microinjection	Haemolymph	Precisely defined	Typically acute
Feeding/oral dosing	Gut	Not controllable	Typically chronic

dye added to the foodstuff allows feeding to be monitored. Death of flies is the usual end point in these studies, but weight loss might also be used as an indicator of disease (Park *et al.*, 2005). Compared with disease that occurs following the direct inoculation in the body cavity, oral infections develop much more slowly. For example, flies die within 1 day of the direct inoculation of *S. marcescens,* whereas after feeding, the flies do not die until day 6 (Nehme *et al.*, 2007). The delayed time to death after feeding on *P. aeruginosa* has led to the suggestion that this model may represent chronic disease caused by this pathogen (Chugani *et al.*, 2001). The pathogenesis of disease caused by feeding on pathogens is complex and almost certainly varies from pathogen to pathogen. Infection of *Drosophila* with *V. cholerae* results in an infection that is localized to the gut, and cholera toxin appears to play a major role in fly death (Park *et al.*, 2005). *S. marcescens* is able to cross the gut wall and replicate in the body cavity (Nehme *et al.*, 2007). In this respect, oral infection models with *Drosophila* appear to contrast markedly with *C. elegans* infection models, where there is little evidence of the pathogens crossing the gut wall.

Effects of pathogens on *Drosophila*

Death is the usual end point of assays with *Drosophila,* but a range of different mechanisms might be involved in killing. In some cases, toxins play important roles in disease. For example, killing by *V. cholerae* involves cholera toxin. Other pathogens such as *F. tularensis* and *P. aeruginosa* appear to be able to cause a bacteraemia in the fly, and the pathogen is found throughout body tissues (Fauvarque *et al.*, 2002; Vonkavaara *et al.*, 2008). In general, the mechanisms by which these pathogens kill *Drosophila* are not well understood. More subtle indicators of pathogenicity might also be used. For example, the bacterial load at different times post challenge can be measured, and the expression profiles of bacterial genes in *Drosophila* determined (Fauvarque *et al.*, 2002; Schneider *et al.*, 2007).

The effects of pathogens on *Drosophila* embryos have received proportionally greater attention. Embryonic haemocytes are phagocytically competent and are able to phagocytose and kill bacteria such as *E. coli* (Vlisidou *et al.*, 2009). The killing mechanisms are broadly similar to those of *G. mellonella* haemocytes (see above). In contrast pathogens, such as *Photorhabdus asymbiotica,* cause a 'freezing' phenotype of the haemocytes, which is associated with significant rearrangement of the actin cytoskeleton (Vlisidou *et al.*, 2009). The behaviour of other pathogens in this system is not yet characterized.

Identification of virulence-associated genes

A key question relating to any invertebrate infection model is to what extent this model reflects disease in mammals, and to what extent can the model be used to identify virulence factors. Almost all of the studies that shed light on the utility of the *Drosophila* model for the identification of virulence genes have been carried out with *P. aeruginosa* infection. An overall conclusion is that, like many invertebrate models, there is some commonality in mechanisms of virulence, but also many differences. Specific virulence mechanisms, including the type III secretion system (Fauvarque *et al.*, 2002), peroxide-resistance proteins (Kim *et al.*, 2009) and virulence regulators such as OxyR (Lau *et al.*, 2005), play roles in virulence both in mammals and in *Drosophila.* There is less consensus in the conclusions reached from broader screens of mutants. In some studies, there is good agreement between genes required for virulence in *Drosophila* and in mammals. In others, limited overlap in these gene sets is reported. For example, Potvin *et al.* (2003) screened over 7000 *P. aeruginosa* mutants in a rat infection model and identified 148 attenuated mutants. However, of these only eight were reported to be attenuated in *Drosophila,* suggesting little overlap of virulence genes in these two species. This conclusion is supported by D'Argenio *et al.* (2001), who found that when several thousand mutants were screened, attenuation of bacterial virulence in *Drosophila* was associated with loss of twitching motility and with mutations of *pil* or *chp* genes. The reasons for these differences are not clear. It is suggested that the subtle differences in the assay procedure might have affected the results, and lower challenge doses might allow better discrimination of attenuated from virulent mutants (Kim *et al.*, 2008).

Identification of novel antibacterials

The *Drosophila* model provides some unique opportunities to investigate novel approaches to antibacterials. This might involve testing compounds in *Drosophila* models of infection or identifying targets for new antimicrobials (Potvin *et al.*, 2003). An alternative approach involves expressing genes encoding antibacterials in *Drosophila* and evaluating the susceptibility of transgenic *Drosophila* to infection. For example, *Drosophila* expressing paraoxonase are protected from the lethal effects of *P. aeruginosa*, because this enzyme disrupts quorum sensing signalling molecules (Stoltz *et al.*, 2008). The use of *Drosophila* for the screening of antimicrobials could have some value. However, the differences in metabolism of these compounds in humans and in *Drosophila* are likely to render anything other than crude analyses of efficacy of any value.

Identification of host genes which play roles in resistance to infection

The genetics of *D. melanogaster* is well understood, and therefore not surprisingly, much of the work with this infection model has been focussed on the identification of host genes modulating resistance to disease. Three types of response occur after infection (Ayres *et al.*, 2008). A cellular immune response depends on phagocytic cells, such as haemocytes, that can ingest and kill bacteria. Secondly, the melanisation response generates melanin and toxic reactive oxygen species at the infection site. Finally, humoral immunity involves the release of antimicrobial peptides. PRR function in much the same way as in mammals, recognizing bacterial lipopolysaccharides and peptidoglycans. These PRR trigger the production of antimicrobial peptides via the Toll and Imd signalling pathways (Hallem *et al.*, 2007; El Chamy *et al.*, 2008). The Imd pathway principally mediates the response against Gram-negative bacteria, whilst the Toll pathway regulates genes that mainly target Gram-positive bacteria (Nehme *et al.*, 2007). A range of antimicrobial peptides belonging to the attacin, cecropin, diptericin, defensin, drosocin, drosomycin, and metchnikowin classes may be produced and act by disrupting bacterial cell membranes. Differences in the susceptibility of different wild type strains of *Drosophila* to infection are at least partially due to polymorphisms in innate immunity genes associated with pathogen recognition and intracellular signalling (Lazzaro *et al.*, 2004).

In addition to the characterized mechanisms of resistance highlighted above, high-throughput screens in *Drosophila* have identified other genes, which play a role in resistance to infection. Some mutants were more susceptible to death, even though the wild type were able to control the proliferation of bacteria. These flies may have altered physiology, which makes them more susceptible to infection (Ayres *et al.*, 2008). These findings might indicate additional mechanisms of resistance to infection. Other factors also affect the resistance of flies to infection. The survival rates of *Drosophila* exposed to *P. aeruginosa* vary according to the time during the day that they are infected, peaking in the middle of the night. The relative resistance of flies infected during the night time appears to be related to a transient burst in the expression of a limited number of innate immunity genes (Lee and Edery, 2008). Depending on the exact mutation, flies with impaired circadian clocks show enhanced or reduced survival after challenge (Lee and Edery, 2008).

Collectively, these studies in *Drosophila* demonstrate the value of these models in identifying mechanisms of resistance to infection and might provide profound new insights into new strategies for controlling infection in humans.

Plant models of infection

Plant models of infection are clearly appropriate to study bacterial plant pathogens. These pathogens are not considered further in this chapter. However, it is only relatively recently that plants have been used as infection models for human disease. Probably, the pioneering study in this field was that reported by Rahme *et al.* (1995), who showed that virulence factors, which played roles in disease of mammals may also play a role in disease of plants. The value of these models is considered in more detail below. In many cases, these plant models are a consequence of the ability of some pathogens to survive in the environment in association with plants. Therefore, it appears that some bacterial pathogens have evolved to have the potential to cause disease in both plant and mammalian hosts.

Assays for measuring the interaction of plants with bacteria and effects of pathogens on plants

A number of assays have been used to measure pathogenicity towards plants. In general, the prior wounding of plant tissues is required to allow entry of bacteria into the host. The most frequently reported assay involves delivering a suspension of the pathogen into the leaf. In studies using *Arabidopsis thaliana* or alfalfa plants, *P. aeruginosa* or *E. faecalis* were injected into surface-sterilised leaves at a dose of approximately 10^3 cfu/cm^3 (Rahme *et al.*, 1997; Silo-Suh *et al.*, 2002; Jha *et al.*, 2005). However, *E. faecalis* also appears to be able to enter plant leaves via stomata (Jha *et al.*, 2005), and *B. cenocepacia* is reported to be able to cause disease in alfalfa plants after surface inoculation onto unwounded seedlings (Uehlinger *et al.*, 2009). The ability to induce a soft rot of plants is a common feature of plant infection models, and overt signs of rot are often preceded by water-soaking and chlorosis of tissues (Silo-Suh *et al.*, 2002).

Many of the pioneering studies on plant infection models have used clinical isolates of *P. aeruginosa*, but it is important to bear in mind that there are significant differences in the pathogenic potential of *P. aeruginosa* isolates towards plants. Some isolates are highly pathogenic (Rahme *et al.*, 1995, 1997; Silo-Suh *et al.*, 2002); others are reported to have growth promoting effects in the rhizosphere, for example, by modulating fungal growth (Steindler *et al.*, 2009). In *A. thaliana* or lettuce leaves, some strains of *P. aeruginosa* cause a soft rot infection, which is accompanied by a 4–5 log multiplication of the bacteria in the leaves. The *A. thaliana* and lettuce models appear to be substitutable because mutants attenuated in one model are attenuated in the other. A later method for the assessment of *P. aeruginosa* used lettuce plants instead of *A. thaliana*, where 10 μl volumes of a bacterial suspension were inoculated into the stems using a Pipetman (Rahme *et al.*, 1997). The use of lettuces in place of *A. thaliana* reportedly has several advantages, because commercially available lettuces can be used in these studies, and several mutants can be tested in each lettuce leaf. In the case of either the *A. thaliana* or lettuce models, signs of infection occur after several days of incubation at 28–30°C in a humid atmosphere.

More recently, root infection models have been developed for *P. aeruginosa* (Walker *et al.*, 2004) and *E. faecalis* (Jha *et al.*, 2005) using *A. thaliana* or *Ocimum basilicum* (sweet basil). These models might allow different mechanisms of virulence to be investigated, because the bacteria form biofilms on the root surface and are then able to invade plant tissues, typically causing plant death 7 days post inoculation. For *P. aeruginosa,* wounding of the root is required for infection – for example by severing the root tip (Walker *et al.*, 2004). However, *E. faecalis* appears to be able to invade roots without prior wounding (Walker *et al.*, 2004).

A final model, which is considered here, is the onion rot assay. The inoculation of onion slices or scales with some strains of the *B. cepacia* complex results in the maceration of onion tissues. The finding that virulence in onions is regulated by quorum sensing genes and *rpoS* suggest that it may be a useful screen for virulence regulators (Aguilar *et al.*, 2003). Maceration is dependent on a plasmid-encoded polygalacturonase (PehA) (Gonzalez *et al.*, 1997). However, strains lacking *pehA* may also cause a less pronounced effect on onion slices, termed water-soaking, which is characterized by partial tissue collapse and translucence of the tissue (Engledow *et al.*, 2004). The ability to cause water-soaking is dependent on a plasmid-borne gene cluster encoding a type IV secretion (Engledow *et al.*, 2004), and it is possible that this type IV secretion delivers effectors necessary for this phenotype. Strains of the *B. cepacia* complex capable of causing onion rot are often agricultural isolates. Therefore, the use of this assay for assessing virulence of human pathogenic strains is not clear (Jacobs *et al.*, 2008; Springman *et al.*, 2009).

Identification of virulence-associated genes

The pioneering studies by Rahme *et al.* (1995) showed that three virulence-associated genes of *P. aeruginosa* (*plcS*, encoding a phospholipase; *toxA*, encoding exotoxin A; *gacA*, encoding a regulator) played roles in disease in *A. thaliana* and disease in the burned mouse infection. Subsequently, this work was extended and modelled in both plant and mammalian hosts, and the lettuce-infection model was used to screen 2500 transposon mutants for reduced virulence (Rahme *et al.*, 1997). The finding that seven of nine mutants were also attenuated in the burned mouse model (Rahme *et al.*, 1997) demonstrates the remarkable potential for this screening method. The utility of the alfalfa infection model is less clear, because, with the exception of the RhlR quorum sensing system, many virulence regulatory mutants in mammals (*vfr*, *rpoS*, *pvdS*, or *lasR* mutants) were not attenuated in this model (Silo-Suh *et al.*, 2002). Fewer mutants have been tested in the root biofilm model, but, as might be expected quorum sensing systems, which regulate biofilm formation, play key roles in virulence (Walker *et al.*, 2004). Therefore, at least in the case of *P. aeruginosa*, virulence genes required for the infection of plants and mammals appear to significantly overlap.

The substantial overlap of *P. aeruginosa* genes required for infection of plants and mammals contrasts markedly with findings on *B. cepacia* and *E. faecalis*. It is clear that there is some overlap in the spectrum of *B. cepacia* virulence factors required for disease in plants and in other hosts such as *C. elegans*, *G. mellonella* and mammals. However, it is also apparent that many of the virulence factors are host specific (Uehlinger *et al.*, 2009). In the case of *E. faecalis*, five mutants known to be attenuated in mammalian models of infection were tested, and two (putative quorum-sensing system, *fsrB*; and serine protease, *sprE*) were also attenuated in *A. thaliana*.

Identification of host genes which play roles in resistance to infection

The mechanisms by which plants resist bacterial infections have received significant attention over the past few years. Innate immune responses, which are analogous to the innate responses occurring in animals, are one of the key responses of plants to pathogens. In some cases, there are remarkable similarities between mammalian and plant innate systems. In plants, the equivalent of TLR5, which senses bacterial flagellin, is FLS2, a leucine-rich repeat receptor kinase (Boller and He, 2009). The EFR receptor in brassicas recognizes bacterial elongation factor Tu (Boller and He, 2009). Other PRRs are less well characterized, but it is believed that a vast array of different bacterial components might be recognized by different plant PRRs. The responses triggered by the interactions between pathogen and PRR have been termed pathogen-associated molecular pattern (PAMP)-triggered immunity. The potential effectiveness of PAMP-triggered immunity in plants is highlighted by the findings that many pathogens are able to suppress these responses, for example by producing effectors that target mitogen-activated protein kinases in the PRR signalling cascades (Boller and He, 2009). Unlike mammals, the antimicrobial responses to bacteria in plants is largely dependent on the production of a range of antimicrobial chemicals, which are plant metabolites (Bednarek and Osbourn, 2009). Plant hormones such as jasmonate, auxin, abscisic acid, and gibberellin also play key roles in regulating the overall level of resistance or susceptibility to pathogens (Grant and Jones, 2009).

To date, plant models for disease caused by human pathogens have focussed almost exclusively on using these models to identify virulence-associated genes of the pathogens. Whilst the use of models in this way is valuable, it may be the case that plant infection models have a wider utility to examine the interaction of pathogens with the innate immune system of the plant.

Cultured cell infection models

Representative and robust models of human cells and tissues are essential for the interrogation of host–pathogen interactions. Whole system models, as described previously in this chapter, and organ explants and human trials provide the most information; but such models are not adapted for high throughput and are unsuitable for the analysis of bacterial virulence mechanisms and drug discovery screening. Cell culture provides researchers with a cost effective, high-throughput experimental system that facilitates the dissection of pathogen interactions with a single cell type. Cultured mammalian cells used for interrogating host–pathogen interactions include primary cells derived from a host, for example bone marrow-derived macrophages. Primary cell lines usually have a short life span but may be more representative of an *in vivo* response. Immortalized cell lines proliferate indefinitely through random mutation or genetic modification. Different cell lines have advantages and disadvantages in terms of their relevance and use as an infection model. For example, immortalized cell lines represent mammalian professional phagocytic cells, such as the murine macrophage-like cell lines J774A.1 and RAW264.1 or human monocyte cell line THP-1. Monkey- and canine-derived COS-1 and MDCK cell lines are readily transfected with genetic material but may be less physiologically relevant to clinical infection than HeLa and Hep-2 cells that are derived from human tissue. CaCo2 and T84 cells are difficult to transfect but are derived from human intestinal epithelium tissue, and may be more physiologically relevant for the analysis of gastrointestinal pathogens that must breach the gut wall (Hurley and McCormick, 2003). Stem cells are amenable to genetic manipulation and can differentiate into any cell type making them an attractive model in which to dissect host–pathogen interactions. Mutant embryonic stem (ES) cells can be immortalized and differentiated into desired cell types or used to produce mutant mice for *in vivo* studies (Yu *et al.*, 2009).

Cultured cell infections also include single cell organisms, such as amoebae, including waterborne *Acanthamoeba castellanii* and soil-dwelling *Dictyostelium discodium* (Kurz and Ewbank, 2007; Salah *et al.*, 2009). Amoebae predate upon bacteria in the environment, and phagocytose bacteria into a phagosome, which fuses with lysosomes, eradicating bacteria that do not possess intracellular defence mechanisms. Amoebae have therefore been proposed as a model for mammalian professional phagocytes. Hypotheses about host–pathogen interactions developed using a reductionist cell culture approach should be verified in a whole system to validate the data in the context of an infection.

Assays for measuring the interaction of cultured cells with bacteria

There are three main approaches to monitor the interaction of bacteria with cultured cells. The first involves imaging the interactions using a range of microscopy approaches, including electron and fluorescence microscopy. Cell monolayers are grown on an optically accessible surface, such as glass-bottomed tissue culture wells and infected with bacteria that may be fluorescently labelled to accommodate fluorescence microscopy. The specific interaction of the bacteria with a single cell type may be analysed. For example, non-typhoidal *Salmonella* induce remarkable cellular rearrangements in enterocytes lining the gastrointestinal tract during invasion (Takeuchi and Sprinz, 1967). Imaging these initial host–pathogen interactions and following the fate of the intracellular bacteria *in vitro* has been central to understanding the mechanism of pathogenesis of *Salmonella* (Galán and Curtis 3rd, 1989, 1991; Lee *et al.*, 1992; Wood *et al.*, 1996; Darwin and Miller, 1999; Lostroh and Lee, 2001; Zhou and Galán, 2001). The second way in which host–pathogen interactions may be measured is through enumeration of bacterial loads following challenge with a bacterial pathogen in an experiment designed to answer a specific question. For example, quantification of bacteria that have adhered to or invaded cell monolayers, such as intestinal epithelial cells, or quantification of bacteria that have survived within macrophage monolayers, are often performed using gentamicin protection assays. The basic principle of a gentamicin protection assay is that cell monolayers are challenged with bacteria; after a specified time, extracellular bacteria are washed away and cells are overlaid with tissue culture medium containing gentamicin, an antimicrobial that cannot penetrate into host cells. At specific time points, host cells are thoroughly washed to remove all traces of gentamicin and lysed, releasing bacteria that have survived/replicated within the host cells. Bacteria may then be enumerated by plating out serial dilutions of the host cell lysate containing bacteria. These assays are particularly useful for dissecting the genetic basis of a pathogenesis mechanism, such as adhesion, invasion and intracellular survival (Saini *et al.*, 1999; Friss *et al.*, 2005; Zhang *et al.*, 2007; Steele-Mortimer, 2008). The third way in which bacterial interactions with host cells may be measured is through analysis of the inflammatory response of host cells to bacterial challenge via cytokine and chemokine secretion profiles (Eckmann *et al.*, 1993; McCormick *et al.*, 1993; Hobbie *et al.*, 1997; Yang *et al.*, 1997; Eckmann and Kagnoff, 2001).

Standard cell culture uses two-dimensional cell cultures grown on an impermeable surface, such as glass or plastic. Cell culture models with a three-dimensional architecture, allowing cells to differentiate into the structure and function of the parent tissue have been developed. Three-dimensional cell cultures mimic the organization and differentiated function of an organ whilst remaining high throughput as a model (Abbott, 2003; Carterson *et al.*, 2005; Höner zu Bentrup *et al.*, 2006). Cell culture infection models are also amenable for bacteria with specific growth requirements, such as microaerobic *C. jejuni* and anaerobic *C. difficile*, as these pathogens may be co-cultured with aerobic cell monolayers using an Ussing chamber (Fernandez Miyakawa *et al.*, 2003; Beltinger *et al.*, 2008).

Effects of pathogens on cultured cells

Cultured cells challenged with bacterial pathogens respond in ways that are specific to the complex host–pathogen interactions that have evolved between a specific pathogen and its host. The host cell architecture may be rearranged to facilitate bacterial invasion or establishment of an intracellular niche, often in response to bacterial effectors translocated into cells through type III or IV secretion systems. For example, following adherence, enteropathogenic *E. coli* and *Helicobacter pylori* induce pedestal-like formation in the intestinal epithelium, and *B. pseudomallei* cause giant cell formation in phagocytic and non-phagocytic cells (Segal *et al.*, 1996; Kespichayawattana *et al.*, 2000; Vallance and Finlay, 2000). These specific host–pathogen effects may be imaged using appropriate imaging platforms. However, there are certain end points that can be standardized and compared in cell culture experiments, such as apoptosis and inflammatory responses measured by cytokine and chemokine profiles. These host cell effects are often induced in response to bacterial toxins (Oldani *et al.*, 2009; Lemichez *et al.*, 2010).

Bacteria may also induce autophagy in host cells. Autophagy is an innate immune response to a number of intracellular pathogens including *Mycobacterium tuberculosis* and group A *Streptococcus*, that has been studied using cell culture techniques (Gutierrez *et al.*, 2004; Nakagawa *et al.*, 2004). This is burgeoning area of research about which several comprehensive reviews have been written recently (Schmid and Münz, 2007; Huang and Brumell, 2009).

Identification of virulence-associated genes

Cell culture models have been instrumental in the identification of virulence-associated genes in many pathogenic bacteria. Physiologically relevant cell lines may be used as an infection model to answer questions about the pathogenesis mechanism of specific bacterial pathogens depending on their lifecycle. For example, cell culture assays using a number of different cell lines were used to characterize *Salmonella* pathogenicity islands I and II, which encode two different type three secretion systems (T3SS) having a central role in bacterial pathogenicity (Galán and Curtis 3rd, 1989, 1991; Lee *et al.*, 1992; Valdez *et al.*, 2009). Cell culture has been used to dissect the role of many T3SS effectors such as SipA and SipC that directly bind actin, enhancing uptake (Hayward and Koronakis, 1999; Zhou *et al.*, 1999). Expression profiling of pathogenic bacteria interacting with host cells has also been a useful tool to identify genes with a possible role in the infection process. In one such study, transcriptome profiling of *S.* Typhimurium surviving within murine macrophages revealed that 919 of 4451 genes were differentially regulated during macrophage intracellular survival, of which almost half were of unknown function (Eriksson *et al.*, 2003). Cell culture is therefore an excellent tool for hypothesis generation and virulence screening before testing hypotheses *in vivo*.

Identification of novel antibacterials

In vitro screens have been used heavily in the development of antimicrobial drugs. Cell culture is economical and well adapted for high-throughput screening of drug activity and toxicity. As antimicrobial resistance increases so too does the need for new drugs, such as the development of drugs that target bacterial virulence genes that are essential for virulence *in vivo* (Clatworthy *et al.*, 2007). *In vivo* screens for novel antimicrobials have identified active compounds that were not identified *in vitro*, suggesting that these compounds boosted the immune system or inhibited bacterial virulence factors. Thus, small high-throughput whole-system infection models, such as zebra fish embryos, *G. mellonella* and *C. elegans*, are now being applied to the field of drug discovery (Moy *et al.*, 2006; Clatworthy *et al.*, 2009; Peleg *et al.*, 2009a). Cell culture assays have been used to identify a novel antimicrobial compounds against *F. tularensis* and *S.* Typhimurium that act by boosting autophagy. One such compound, AR-12, increases autophagosome formation, and subsequent increased colocalization of *S.* Typhimurium with autophagosomes in macrophages was observed resulting in clearance of intracellular bacteria (Chiu *et al.*, 2009a,b; Welsh *et al.*, 2009).

Identification of host genes which play roles in resistance to infection

In the post genomics era, systems biology approaches have been adopted to map host responses to bacterial infection to identify genes with a role in resistance. For example, using a transcriptome profiling approach, uncharacterised urothelial cell host genes associated with the early response to infection with *E. faecalis* have been identified as targets for enhancing the host response to infection (Liu *et al.*, 2006; Dozmorov *et al.*, 2007; Waddell *et al.*, 2007). Cell culture is a useful model system for expression analyses, as the interaction between a bacterium and single cell can be explored at multiple time points. Clearly simplistic, cell culture models do not represent complex *in vivo* host pathogen interactions, but mechanistic insight into the interactions of bacteria with key cell types during the infection process may be gained. Eight main approaches have been used to monitor gene expression profiles associated with host response to bacterial infection; host cell challenge with wild type pathogens *in vitro* provides a global overview of the host transcriptome. Host cell challenge with a panel of different pathogens may illuminate pathogen-specific host responses, underlying specific clinical symptoms opposed to global pathogen responses. Host cell challenge with mutant bacteria may reveal pathogen components that elicit a specific host transcriptional response. The mutant bacteria approach may be verified through challenge of host cells with pathogen components, either applied extracellularly or expressed in the host cell. Mechanistic insight into host transcriptome response may be gleaned through challenge of mutant host cells. Similarly, signalling pathways of the host response may be inactivated through the use of small molecule inhibition. Host genes with a role in pathogen latency may be investigated through transcriptome analysis of cells infected *in vivo* and subsequently cultured *in vitro*. Finally, *in vivo* expression profiling provides a systems approach to identify host gene networks associated with the infection process. Jenner and Young (2005) have written a review and meta analysis of transcriptome profiles associated with host response to bacterial infection.

References

Aballay, A., and Ausubel, F.M. (2001). Programmed cell death mediated by *ced-3* and *ced-4* protects *Caenorhabditis elegans* from *Salmonella typhimurium*-mediated killing. Proc. Natl. Acad. Sci. U.S.A. *98*, 2735–2739.

Aballay, A., Yorgey, P., and Ausubel, F.M. (2000). *Salmonella typhimurium* proliferates and establishes a persistent infection in the intestine of *Caenorhabditis elegans*. Curr. Biol. *10*, 1539–1542.

Abbott, A. (2003). Cell culture: biology's new dimension. Nature *424*, 870–872.

Aguilar, C., Bertani, I., and Venturi, V. (2003). Quorum-sensing system and stationary-phase sigma factor (*rpoS*) of the onion pathogen *Burkholderia cepacia* genomovar I type strain, ATCC 25416. Appl. Environ. Microbiol. *69*, 1739–1747.

Akira, S., Uematsu, S., and Takeuchi, O. (2006). Pathogen recognition and innate immunity. Cell *124*, 783–801.

Alcaïs, A., Sanchez, F.O., Thuc, N.V., Lap, V.D., Oberti, J., Lagrange, P.H., Schurr, E., and Abel, L. (2000). Granulomatous reaction to intradermal injection of lepromin (Mitsuda reaction) is linked to the human *NRAMP1* gene in Vietnamese leprosy sibships. J. Infect. Dis. *181*, 302–308.

Alegado, R.A., Campbell, M.C., Chen, W.C., Slutz, S.S., and Tan, M.W. (2003). Characterization of mediators of microbial virulence and innate immunity using the *Caenorhabditis elegans* host–pathogen model. Cell. Microbiol. *5*, 435–444.

Andrejko, M., Mizerska-Dudka, M., and Jakubowicz, T. (2008a). Changes in *Galleria mellonella* lysozyme level and activity during *Pseudomonas aeruginosa* infection. Folia Microbiol. (Praha) *53*, 147–151.

Andrejko, M., Mizerska-Dudka, M., and Jakubowicz, T. (2008b). Changes in *Galleria mellonella* apolipophorin III level during *Pseudomonas aeruginosa* infection. J. Invertebr. Pathol. *97*, 14–19.

Anyanful, A., Dolan-Livengood, J.M., Lewis, T., Sheth, S., Dezalia, M.N., Sherman, M.A., Kalman, L.V., Benian, G.M., and Kalman, D. (2005). Paralysis and killing of *Caenorhabditis elegans* by enteropathogenic *Escherichia coli* requires the bacterial tryptophanase gene. Mol. Microbiol. *57*, 988–1007.

Anyanful, A., Easley, K.A., Benian, G.M., and Kalman, D. (2009). Conditioning protects *C. elegans* from lethal effects of enteropathogenic *E. coli* by activating genes that regulate lifespan and innate immunity. Cell Host Microbe *5*, 450–462.

Aperis, G., Fuchs, B.B., Anderson, C.A., Warner, J.E., Calderwood, S.B., and Mylonakis, E. (2007). *Galleria mellonella* as a model host to study infection by the *Francisella tularensis* live vaccine strain. Microbes Infect. *9*, 729–734.

Apidianakis, Y., and Rahme, L.G. (2009). *Drosophila melanogaster* as a model host for studying *Pseudomonas aeruginosa* infection. Nat. Protoc. *4*, 1285–1294.

Arroyo, L.G., Weese, J.S., and Staempfli, H.R. (2004). Experimental *Clostridium difficile* enterocolitis in foals. J. Vet. Intern. Med. *18*, 734–738.

Autret, N., and Charbit, A. (2005). Lessons from signature-tagged mutagenesis on the infectious mechanisms of pathogenic bacteria. FEMS Microbiol. Rev. *29*, 703–717.

Ayres, J.S., Freitag, N., and Schneider, D.S. (2008). Identification of *Drosophila* mutants altering defense of and endurance to *Listeria monocytogenes* infection. Genetics *178*, 1807–1815.

Bakker-Woudenberg, I.A. (2003). Experimental models of pulmonary infection. J. Microbiol. Methods. *54*, 295–313.

Barrett, A.D., and Beasley, D.W. (2009). Development pathway for biodefense vaccines. Vaccine *27* (Suppl. 4), D2–7.

Baselski, V., Briggs, R., and Parker, C. (1977). Intestinal fluid accumulation induced by oral challenge with *Vibrio cholerae* or cholera toxin in infant mice. Infect. Immun. *15*, 704–712.

Bast, D.J., Yue, M., Chen, X., Bell, D., Dresser, L., Saskin, R., Mandell, L.A., Low, D.E., and de Azavedo, J.C. (2004). Novel murine model of pneumococcal pneumonia: use of temperature as a measure of disease severity to compare the efficacies of moxifloxacin and levofloxacin. Antimicrob. Agents Chemother. *48*, 3343–3348.

Bednarek, P., and Osbourn, A. (2009). Plant–microbe interactions: chemical diversity in plant defense. Science *324*, 746–748.

Bellamy, R., Ruwende, C., Corrah, T., McAdam, K.P., Whittle, H.C., and Hill, A.V. (1998). Variations in the *NRAMP1* gene and susceptibility to tuberculosis in West Africans. N. Engl. J. Med. *338*, 640–644.

Beltinger, J., del Buono, J., Skelly, M.M., Thornley, J., Spiller, R.C., Stack, W.A., and Hawkey, C.J. (2008). Disruption of colonic barrier function and induction of mediator release by strains of *Campylobacter jejuni* that invade epithelial cells. World J. Gastroenterol. *14*, 7345–7352.

Bergin, D., Reeves, E.P., Renwick, J., Wientjes, F.B., and Kavanagh, K. (2005). Superoxide production in *Galleria mellonella* haemocytes: identification of proteins homologous to the NADPH oxidase complex of human neutrophils. Infect. Immun. *73*, 4161–4170.

Beutler, B., Jiang, Z., Georgel, P., Crozat, K., Croker, B., Rutschmann, S., Du, X., and Hoebe, K. (2006). Genetic analysis of host resistance: Toll-like receptor signaling and immunity at large. Annu. Rev. Immunol. *24*, 353–389.

Blackwell, J.M., Goswami, T., Evans, C.A., Sibthorpe, D., Papo, N., White, J.K., Searle, S., Miller, E.N., Peacock CS, Mohammed, H., *et al.* (2001). SLC11A1 (formerly NRAMP1) and disease resistance. Cell. Microbiol. *3*, 773–784.

Boller, T., and He, S.Y. (2009). Innate immunity in plants: an arms race between pattern recognition receptors in plants and effectors in microbial pathogens. Science *324*, 742–744.

Bolm, M., Jansen, W.T., Schnabel, R., and Chhatwal, G.S. (2004). Hydrogen peroxide-mediated killing of *Caenorhabditis elegans*: a common feature of different streptococcal species. Infect. Immun. *72*, 1192–1194.

Bradley, D.J., Taylor, B.A., Blackwell, J., Evans, E.P., and Freeman, J (1979). Regulation of *Leishmania* populations within the host. III. Mapping of the locus controlling susceptibility to visceral leishmaniasis in the mouse. Clin. Exp. Immunol. *37*, 7–14.

Brannon, M.K., Davis, J.M., Mathias, J.R., Hall, C.J., Emerson, J.C., Crosier, P.S., Huttenlocher, A., Ramakrishnan, L., and Moskowitz, S.M. (2009). *Pseudomonas aeruginosa* type III secretion system interacts with phagocytes to modulate systemic infection of zebrafish embryos. Cell. Microbiol. *11*, 755–768.

Brinster, R.L., Chen, H.Y., Trumbauer, M., Senear, A.W., Warren, R., and Palmiter, R.D. (1981). Somatic expression of herpes thymidine kinase in mice following injection of a fusion gene into eggs. Cell *27*, 223–231.

Brombacher, F., Dorfmüller, A., Magram, J., Dai, W.J., Köhler, G., Wunderlin, A., Palmer-Lehmann, K., Gately, M.K., and Alber, G. (1999). IL-12 is dispensable for innate and adaptive immunity against low doses of *Listeria monocytogenes*. Int. Immunol. *11*, 325–332.

Brouillette, E., and Malouin, F. (2005). The pathogenesis and control of *Staphylococcus aureus*-induced mastitis: study models in the mouse. Microbes Infect. *7*, 560–568.

Brubaker, R.R. (1991). Factors promoting acute and chronic diseases caused by Yersiniae. Clin. Microbiol. Rev. *4*, 309–324.

Buer, J., and Balling, R. (2003). Mice, microbes and models of infection. Nat. Rev. Genet. *4*, 195–205.

Cardona, S.T., Wopperer, J., Eberl, L., and Valvano, M.A. (2005). Diverse pathogenicity of *Burkholderia cepacia* complex strains in the *Caenorhabditis elegans* host model. FEMS Microbiol. Lett. *250*, 97–104.

Carterson, A.J., Höner zu Bentrup, K., Ott, C.M., Clarke, M.S., Pierson, D.L., Vanderburg, C.R., Buchanan, K.L., Nickerson, C.A., and Schurr, M.J. (2005). A549 lung epithelial cells grown as three-dimensional aggregates: alternative tissue culture model for *Pseudomonas aeruginosa* pathogenesis. Infect. Immun. *73*, 1129–1140.

Cash, H.A., Woods, D.E., McCullough, B., Johanson, W.G., Jr., and Bass, J.A. (1979). A rat model of chronic respiratory infection with *Pseudomonas aeruginosa*. Am. Rev. Respir. Dis. *119*, 453–459.

Champion, O.L., Valdez, Y., Thorson, L., Guttman, J.A., Menendez, A., Gaynor, E.C., and Finlay, B.B. (2008). A murine intraperitoneal infection model reveals that host resistance to *Campylobacter jejuni* is Nramp1 dependent. Microbes Infect. *10*, 922–927.

Champion, O.L., Cooper, I.A., James, S.L., Ford, D., Karlyshev, A., Wren, B.W., Duffield, M., Oyston, P.C., and Titball, R.W. (2009). *Galleria mellonella* as an alternative infection model for *Yersinia pseudotuberculosis*. Microbiology *155*, 1516–1522.

Champion, O.L., Karlyshev, A.V., Senior, N.J., Woodward, M., La Ragione, R., Howard, S.L., Wren, B.W., and Titball, R.W. (2010). Insect infection model for *Campylobacter jejuni* reveals that O-methyl phosphoramidate has insecticidal activity. J. Infect. Dis. *201*,776–782.

Chatfield, S.N., Strahan, K., Pickard, D., Charles, I.G., Hormaeche, C.E., and Dougan, G. (1992). Evaluation of *Salmonella typhimurium* strains harbouring defined mutations in *htrA* and *aroA* in the murine salmonellosis model. Microb. Pathog. *12*, 145–151.

Chen, D., Thomas, E.L. and Kapahi, P. (2009). HIF-1 modulates dietary restriction-mediated lifespan extension via IRE-1 in *Caenorhabditis elegans*. PloS. Genet. *5*, e1000486.

Chiu, H.C., Kulp, S.K., Soni, S., Wang, D., Gunn, J.S., Schlesinger, L.S., and Chen, C.S. (2009a). Eradication of intracellular *Salmonella enterica* serovar Typhimurium with a small-molecule, host cell-directed agent. Antimicrob. Agents Chemother. *53*, 5236–5244.

Chiu, H.C., Soni, S., Kulp, S.K., Curry, H., Wang, D., Gunn, J.S., Schlesinger, L.S., and Chen, C.S. (2009b). Eradication of intracellular *Francisella tularensis* in THP-1 human macrophages with a novel autophagy inducing agent. J. Biomed. Sci. *16*, 110.

Choi, J.Y., Sifri, C.D., Goumnerov, B.C., Rahme, L.G., Ausubel, F.M., and Calderwood, S.B. (2002). Identification of virulence genes in a pathogenic strain of *Pseudomonas aeruginosa* by representational difference analysis. J. Bacteriol. *184*, 952–961.

Chugani, S.A., Whiteley, M., Lee, K.M., D'Argenio, D., Manoil, C., and Greenberg, E.P. (2001). QscR, a modulator of quorum-sensing signal synthesis and virulence in *Pseudomonas aeruginosa*. Proc. Natl. Acad. Sci. U.S.A. *98*, 2752–2757.

Clatworthy, A.E., Pierson, E., and Hung, D.T. (2007). Targeting virulence: a new paradigm for antimicrobial therapy. Nat. Chem. Biol. *3*, 541–548.

Clatworthy, A.E., Lee, J.S., Leibman, M., Kostun, Z., Davidson, A.J., and Hung, D.T. (2009). *Pseudomonas aeruginosa* infection of zebrafish involves both host and pathogen determinants. Infect. Immun. *77*, 1293–1303.

Clay, H., Davis, J.M., Beery, D., Huttenlocher, A., Lyons, S.E., and Ramakrishnan, L. (2007). Dichotomous role of the macrophage in early *Mycobacterium marinum* infection of the zebrafish. Cell Host Microbe *2*, 29–39.

Clay, H., Volkman, H.E., and Ramakrishnan, L. (2008). Tumor necrosis factor signaling mediates resistance to mycobacteria by inhibiting bacterial growth and macrophage death. Immunity *29*, 283–294.

Clermont, A., Wedde, M., Seitz, V., Podsiadlowski, L., Lenze, D., Hummel, M., and Vilcinskas, A. (2004). Cloning and expression of an inhibitor of microbial metalloproteinases from insects contributing to innate immunity. Biochem. J. *382*, 315–322.

Coenye, T., Vandamme, P., Govan, J.R., and LiPuma, J.J. (2001). Taxonomy and identification of the *Burkholderia cepacia* complex. J. Clin. Microbiol. *39*, 3427–3436.

Contag, C.H., Contag, P.R., Mullins, J.I., Spilman, S.D., Stevenson, D.K., and Benaron, D.A. (1995). Photonic detection of bacterial pathogens in living hosts. Mol. Microbiol. *18*, 593–603.

Couillault, C., and Ewbank, J.J. (2002). Diverse bacteria are pathogens of *Caenorhabditis elegans*. Infect. Immun. *70*, 4705–4707.

Cuccui, J., Easton, A., Chu, K.K., Bancroft, G.J., Oyston, P.C., Titball, R.W., and Wren, B.W. (2007). Development of signature-tagged mutagenesis in *Burkholderia pseudomallei* to identify genes important in survival and pathogenesis. Infect. Immun. *75*, 1186–1195.

Dai, W.J., Köhler, G., and Brombacher, F. (1997). Both innate and acquired immunity to *Listeria monocytogenes* infection are increased in IL-10-deficient mice. J. Immunol. *158*, 2259–2267.

Darby, C. (2005). Interactions with microbial pathogens. WormBook *6*, 1–15.

Darby, C., Hsu, J.W., Ghori, N., and Falkow S. (2002). *Caenorhabditis elegans*: plague bacteria biofilm blocks food intake. Nature *417*, 243–244.

D'Argenio, D.A., Gallagher, L.A., Berg, C.A., and Manoil, C. (2001). *Drosophila* as a model host for *Pseudomonas aeruginosa* infection. J. Bacteriol. *183*, 1466–1471.

Darwin, K.H., and Miller, V.L. (1999). Molecular basis of the interaction of *Salmonella* with the intestinal mucosa. Clin. Microbiol. Rev. *12*, 405–428.

Davis, J.M., and Ramakrishnan, L. (2009). The role of the granuloma in expansion and dissemination of early tuberculous infection. Cell *136*, 37–49.

Davis, J.M., Clay, H., Lewis, J.L., Ghori, N., Herbomel, P., and Ramakrishnan, L. (2002). Real-time visualization of *Mycobacterium*–macrophage interactions leading to initiation of granuloma formation in zebrafish embryos. Immunity *17*, 693–702.

Davis, J.M., Haake, D.A. and Ramakrishnan, L. (2009). *Leptospira interrogans* stably infects zebrafish embryos, altering phagocyte behavior and homing to specific tissues. PLoS Negl. Trop. Dis. *3*, e463.

Davis, P.H., and Stanley, S.L., Jr. (2003). Breaking the species barrier: use of SCID mouse-human chimeras for the study of human infectious diseases. Cell. Microbiol. *5*, 849–860.

Dean-Nystrom, E.A. (2003). Bovine *Escherichia coli* O157:H7 infection model. Methods Mol. Med. *73*, 329–338.

Diez, E., Lee, S.H., Gauthier, S., Yaraghi, Z., Tremblay, M., Vidal, S., and Gros, P. (2003). *Birc1e* is the gene within the *Lgn1* locus associated with resistance to *Legionella pneumophila*. Nat. Genet. 33, 55–60.

Doyle, T.C., Burns, S.M., and Contag, C.H. (2004). *In vivo* bioluminescence imaging for integrated studies of infection. Cell. Microbiol. *6*, 303–317.

Dozmorov, M.G., Kyker, K.D., Saban, R., Shankar, N., Baghdayan, A.S., Centola, M.B., and Hurst, R.E. (2007). Systems biology approach for mapping the response of human urothelial cells to infection by *Enterococcus faecalis*. BMC Bioinformatics *8* (Suppl. 7), S2.

Eckmann, L., and Kagnoff, M.F. (2001). Cytokines in host defense against *Salmonella*. Microbes Infect. *3*, 1191–1200.

Eckmann, L., Kagnoff, M.F., and Fierer, J. (1993). Epithelial cells secrete the chemokine interleukin-8 in response to bacterial entry. Infect. Immun. *61*, 4569–4574.

El Chamy, L., Leclerc, V., Caldelari, I., and Reichhart, J.M. (2008). Sensing of 'danger signals' and pathogen-associated molecular patterns defines binary signaling pathways 'upstream' of Toll. Nat. Immunol. *9*, 1165–1170.

Eliceiri, K.W., and Contag, C.H. (2009). Integrated studies of biology: multiplexed imaging assays from molecules to man and back. Curr. Opin. Biotechnol. *20*, 1–3.

Ellis, J., Oyston, P.C., Green, M., and Titball, R.W. (2002). Tularemia. Clin. Microbiol. Rev. *15*, 631–646.

Engledow, A.S., Medrano, E.G., Mahenthiralingam, E., LiPuma, J.J., and Gonzalez, C.F. (2004). Involvement of a plasmid-encoded type IV secretion system in the plant tissue watersoaking phenotype of *Burkholderia cenocepacia*. J. Bacteriol. *186*, 6015–6024.

Ensink, J.M., Smit, J.A., and van Duijkeren, E. (2003). Clinical efficacy of trimethoprim/sulfadiazine and procaine penicillin G in a *Streptococcus equi* subsp. *zooepidemicus* infection model in ponies. J. Vet. Pharmacol. Ther. *26*, 247–252.

Erickson, D.L., Jarrett, C.O., Wren, B.W., and Hinnebusch, B.J. (2006). Serotype differences and lack of biofilm formation characterize *Yersinia pseudotuberculosis* infection of the *Xenopsylla cheopis* flea vector of *Yersinia pestis*. J. Bacteriol. *188*, 1113–1119.

Eriksson, S., Lucchini, S., Thompson, A., Rhen, M., and Hinton, J.C. (2003). Unravelling the biology of macrophage infection by gene expression profiling of intracellular *Salmonella enterica*. Mol. Microbiol. *47*, 103–118.

Eubanks, E.R., Guentzel, M.N., and Berry, L.J. (1976). Virulence factors involved in the intraperitoneal infection of adult mice with *Vibrio cholerae*. Infect. Immun. *13*, 457–463.

Evans, E.A., Kawli, T., and Tan, M.W. (2008). *Pseudomonas aeruginosa* suppresses host immunity by activating the DAF-2 insulin-like signaling pathway in *Caenorhabditis elegans*. PLoS Pathog. *4*, e1000175.

Ewbank, J.J. (2006). Signaling in the immune response. WormBook *23*, 1–12.

Fauvarque, M.O., Bergeret, E., Chabert, J., Dacheux, D., Satre, M., and Attree, I. (2002). Role and activation of type III secretion system genes in *Pseudomonas aeruginosa*-induced *Drosophila* killing. Microb. Pathog. *32*, 287–295.

Fedhila, S., Daou, N., Lereclus, D., and Nielsen-LeRoux, C. (2006). Identification of *Bacillus cereus* internalin and other candidate virulence genes specifically induced during oral infection in insects. Mol. Microbiol. *62*, 339–355.

Fernandez Miyakawa, M.E., and Uzal, F.A. (2003). The early effects of *Clostridium perfringens* type D epsilon toxin in ligated intestinal loops of goats and sheep. Vet. Res. Commun. *27*, 231–241.

Fernandez Miyakawa, M.E., Ibarra, C.A., and Uzal, F.A. (2003). *In vitro* effects of *Clostridium perfringens* type D epsilon toxin on water and ion transport in ovine and caprine intestine. Anaerobe *9*, 145–149.

Fleming, V., Feil, E., Sewell, A.K., Day, N., Buckling, A., and Massey, R.C. (2006). Agr interference between clinical *Staphylococcus aureus* strains in an insect model of virulence. J. Bacteriol. *188*, 7686–7688.

Flynn, J.L., Chan, J., Triebold, K.J., Dalton, D.K., Stewart, T.A., and Bloom, B.R. (1993). An essential role for interferon γ in resistance to *Mycobacterium tuberculosis* infection. J. Exp. Med. *178*, 2249–2254.

Flynn, J.L., Goldstein, M.M., Chan, J., Triebold, K.J., Pfeffer, K., Lowenstein, C.J., Schreiber, R., Mak, T.W., and Bloom, B.R. (1995). Tumor necrosis factor-α is required in the protective immune response against *Mycobacterium tuberculosis* in mice. Immunity *2*, 561–572.

Forbes, J.R., and Gros, P. (2001). Divalent-metal transport by NRAMP proteins at the interface of host–pathogen interactions. Trends Microbiol. *9*, 397–403.

Frandsen, J.L., Gunn, B., Muratoglu, S., Fossett, N., and Newfeld, S.J.(2008) *Salmonella* pathogenesis reveals that BMP signaling regulates blood cell homeostasis and immune responses in *Drosophila*. Proc. Natl. Acad. Sci. U.S.A. *105*, 14952–14957.

Friis, L.M., Pin, C., Pearson, B.M., and Wells, J.M. (2005). *In vitro* cell culture methods for investigating *Campylobacter* invasion mechanisms. J. Microbiol. Methods *61*, 145–160.

Fuhrman, L.E., Shianna, K.V., and Aballay, A. (2008). High-throughput isolation and mapping of *C. elegans* mutants susceptible to pathogen infection. PLoS One *3*, e2882.

Galán, J.E., and Curtiss, R., 3rd. (1989). Cloning and molecular characterization of genes whose products allow *Salmonella typhimurium* to penetrate tissue culture cells. Proc. Natl. Acad. Sci. U.S.A. *86*, 6383–6387.

Galán, J.E., and Curtiss, R., 3rd. (1991). Distribution of the *invA*, *-B*, *-C*, and *-D* genes of *Salmonella typhimurium* among other *Salmonella* serovars: *invA* mutants of *Salmonella typhi* are deficient for entry into mammalian cells. Infect. Immun. *59*, 2901–2908.

Gallagher, L.A., and Manoil, C. (2001). *Pseudomonas aeruginosa* PAO1 kills *Caenorhabditis elegans* by cyanide poisoning. J. Bacteriol. *183*, 6207–6214.

Gan, Y.H., Chua, K.L., Chua, H.H., Liu, B., Hii, C.S., Chong, H.L., and Tan, P. (2002). Characterization of *Burkholderia pseudomallei* infection and identification of novel virulence factors using a *Caenorhabditis elegans* host system. Mol. Microbiol. *44*, 1185–1197.

Gardner, M.B., and Luciw, P.A. (2008). Macaque models of human infectious disease. ILAR.J. *49*, 220–255.

Garsin, D.A., Sifri, C.D., Mylonakis, E., Qin, X., Singh, K.V., Murray, B.E., Calderwood, S.B., and Ausubel, F.M. (2001). A simple model host for identifying Gram-positive virulence factors. Proc. Natl. Acad. Sci. U.S.A. *98*, 10892–10897.

Garvis, S., Munder, A., Ball, G., de Bentzmann, S., Wiehlmann, L., Ewbank, J.J., Tümmler, B., and Filloux, A. (2009). *Caenorhabditis elegans* semi-automated liquid screen reveals a specialized role for the chemotaxis gene *cheB2* in *Pseudomonas aeruginosa* virulence. PLoS Pathog. *5*, e1000540.

Germain, R.N. (2004). An innately interesting decade of research in immunology. Nat. Med. *10*, 1307–1320.

Gonzalez, C.F., Pettit, E.A., Valadez, V.A., and Provin, E.M. (1997). Mobilization, cloning, and sequence determination of a plasmid-encoded polygalacturonase from a phytopathogenic *Burkholderia (Pseudomonas) cepacia*. Mol. Plant. Microbe Interact. *10*, 840–851.

Grant, M.R., and Jones, J.D. (2009). Hormone (dis) harmony moulds plant health and disease. Science *324*, 750 752.

Gravato-Nobre, M.J., and Hodgkin, J. (2005). *Caenorhabditis elegans* as a model for innate immunity to pathogens. Cell. Microbiol. *7*, 741–751.

Gros, P., Skamene, E., and Forget, A. (1983). Cellular mechanisms of genetically controlled host resistance to *Mycobacterium bovis* (BCG). J. Immunol. *131*, 1966–1972.

Gutierrez, M.G., Master, S.S., Singh, S.B., Taylor, G.A., Colombo, M.I., and Deretic, V. (2004). Autophagy is a defense mechanism inhibiting BCG and *Mycobacterium tuberculosis* survival in infected macrophages. Cell *119*, 753–766.

Guyton, A. (1947). Measurement of the respiratory volumes of laboratory animals. Am. J. Physiol. *150*, 70–77.

Hallem, E.A., Rengarajan, M., Ciche, T.A., and Sternberg, P.W. (2007). Nematodes, bacteria, and flies: a tripartite model for nematode parasitism. Curr. Biol. *17*, 898–904.

Hamm, E.E., Voth, D.E., and Ballard, J.D. (2006). Identification of *Clostridium difficile* toxin B cardiotoxicity using a zebrafish embryo model of intoxication. Proc. Natl. Acad. Sci. U.S.A. *103*, 14176–14181.

Hayward, R.D., and Koronakis, V. (1999). Direct nucleation and bundling of actin by the SipC protein of invasive *Salmonella*. EMBO J. *18*, 4926–4934.

Hegedus, Z., Zakrzewska, A., Agoston, V.C., Ordas, A., Rácz, P., Mink, M., Spaink, H.P., and Meijer, A.H. (2009). Deep sequencing of the zebrafish transcriptome response to *Mycobacterium* infection. Mol. Immunol. *46*, 2918–2930.

Hendrickson, E.L., Plotnikova, J., Mahajan-Miklos, S., Rahme, L.G., and Ausubel, F.M. (2001). Differential roles of the *Pseudomonas aeruginosa* PA14 *rpoN* gene in pathogenicity in plants, nematodes, insects, and mice. J. Bacteriol. *183*, 7126–7134.

Herbomel, P., Thisse, B., and Thisse, C. (1999). Ontogeny and behaviour of early macrophages in the zebrafish embryo. Development *126*, 3735–3745.

Hess, J., Ladel, C., Miko, D., and Kaufmann, S.H. (1996). *Salmonella typhimurium* aroA- infection in gene-targeted immunodeficient mice: major role of CD4 + TCR-αβ cells and IFN-γ in bacterial clearance independent of intracellular location. J. Immunol. *156*, 3321–3326.

Hirayama, F., Takagi, S., Yokoyama, Y., Iwao, E., and Ikeda, Y. (1996). Establishment of gastric *Helicobacter pylori* infection in Mongolian gerbils. J. Gastroenterol. *31* Suppl. 9, 24–28.

Hobbie, S., Chen, L.M., Davis, R.J., Galán, J.E. (1997). Involvement of mitogen-activated protein kinase pathways in the nuclear responses and cytokine production induced by *Salmonella typhimurium* in cultured intestinal epithelial cells. J. Immunol. *159*, 5550–5559.

Hodgkin, J. (2001). What does a worm want with 20,000 genes? Genome Biol. *2*, COMMENT2008, PMID 11737938.

Hoffmann, J.A. (1995). Innate immunity of insects. Curr. Opin. Immunol. *7*, 4–10.

Höner zu Bentrup, K., Ramamurthy, R., Ott, C.M., Emami, K., Nelman-Gonzalez, M., Wilson, J.W., Richter, E.G., Goodwin, T.J., Alexander, J.S., Pierson, D.L., *et al.* (2006). Three-dimensional organotypic models of human colonic epithelium to study the early stages of enteric salmonellosis. Microbes Infect. *8*, 1813–1825.

Huang, J., and Brumell, J.H. (2009). Autophagy in immunity against intracellular bacteria. Curr. Top. Microbiol. Immunol. *335*, 189–215.

Huber, B., Feldmann, F., Köthe, M., Vandamme, P., Wopperer, J., Riedel, K., and Eberl, L. (2004). Identification of a novel virulence factor in *Burkholderia cenocepacia* H111 required for efficient slow killing of *Caenorhabditis elegans*. Infect. Immun. *72*, 7220–7230.

Hultgren, O.H., Svensson, L., and Tarkowski, A. (2002). Critical role of signaling through IL-1 receptor for development of arthritis and sepsis during

Staphylococcus aureus infection. J. Immunol. *168*, 5207–5212.

Hurley, B.P., and McCormick, B.A. (2003). Translating tissue culture results into animal models: the case of *Salmonella typhimurium*. Trends Microbiol. *11*, 562–569.

Jacobs, J.L., Fasi, A.C., Ramette, A., Smith, J.J., Hammerschmidt, R., and Sundin, G.W. (2008). Identification and onion pathogenicity of *Burkholderia cepacia* complex isolates from the onion rhizosphere and onion field soil. Appl. Environ. Microbiol. *74*, 3121–3129.

Jacobs, M., Brown, N., Allie, N., Gulert, R., and Ryffel, B. (2000). Increased resistance to mycobacterial infection in the absence of interleukin-10. Immunology *100*, 494–501.

Jander, G., Rahme, L.G., and Ausubel, F.M. (2000). Positive correlation between virulence of *Pseudomonas aeruginosa* mutants in mice and insects. J. Bacteriol. *182*, 3843–3845.

Jansen, W.T., Bolm, M., Balling, R., Chhatwal, G.S., and Schnabel, R. (2002). Hydrogen peroxide-mediated killing of *Caenorhabditis elegans* by *Streptococcus pyogenes*. Infect. Immun. *70*, 5202–5207.

Jawhara, S., and Mordon, S. (2004). *In vivo* imaging of bioluminescent *Escherichia coli* in a cutaneous wound infection model for evaluation of an antibiotic therapy. Antimicrob. Agents Chemother. *48*, 3436–3441.

Jenner, R.G., and Young, R.A. (2005). Insights into host responses against pathogens from transcriptional profiling. Nat. Rev. Microbiol. *3*, 281–294.

Jha, A.K., Bais, H.P., and Vivanco, J.M. (2005). *Enterococcus faecalis* mammalian virulence-related factors exhibit potent pathogenicity in the *Arabidopsis thaliana* plant model. Infect. Immun. *73*, 464–475.

Jiravanichpaisal, P., Lee, B.L., and Söderhäll, K. (2006). Cell-mediated immunity in arthropods: hematopoiesis, coagulation, melanization and opsonization. Immunobiology *211*, 213–236.

Job, C.K., Sanchez, R.M., and Hastings, R.C. (1985). Manifestations of experimental leprosy in the armadillo. Am. J. Trop. Med. Hyg. *34*, 151–161.

Johnson, A.P., Hetherington, C.M., Osborn, M.F., Thomas, B.J., and Taylor-Robinson, D. (1980). Experimental infection of the marmoset genital tract with *Chlamydia trachomatis*. Br. J. Exp. Pathol. *61*, 291–295.

Jorgensen, E.M., and Mango, S.E. (2002). The art and design of genetic screens: *Caenorhabditis elegans*. Nat. Rev. Genet. *3*, 356–369.

Kadurugamuwa, J.L., Sin, L., Albert, E., Yu, J., Francis, K., DeBoer, M., Rubin, M., Bellinger-Kawahara, C., Parr, T.R., Jr., and Contag, P.R (2003). Direct continuous method for monitoring biofilm infection in a mouse model. Infect. Immun. *71*, 882–890.

Kahl-McDonagh, M.M., Elzer, P.H., Hagius, S.D., Walker, J.V., Perry, Q.L., Seabury, C.M., den Hartigh, A.B., Tsolis, R.M., Adams, L.G., Davis, D.S., *et al.* (2006). Evaluation of novel *Brucella melitensis* unmarked deletion mutants for safety and efficacy in the goat model of brucellosis. Vaccine *24*, 5169–5177.

Kamradt, A.E., Greiner, M., Ghiara, P., and Kaufmann, S.H. (2000). *Helicobacter pylori* infection in wild-type and cytokine-deficient C57BL/6 and BALB/c mouse mutants. Microbes Infect. *2*, 593–597.

Kavanagh, K., and Reeves, E.P. (2004). Exploiting the potential of insects for *in vivo* pathogenicity testing of microbial pathogens. FEMS Microbiol. Rev. *28*, 101–112.

Kespichayawattana, W., Rattanachetkul, S., Wanun, T., Utaisincharoen, P., and Sirisinha, S. (2000). *Burkholderia pseudomallei* induces cell fusion and actin-associated membrane protrusion: a possible mechanism for cell-to-cell spreading. Infect. Immun. *68*, 5377–5384.

Kim, S.H., Park, S.Y., Heo, Y.J., and Cho, Y.H. (2008). *Drosophila melanogaster*-based screening for multihost virulence factors of *Pseudomonas aeruginosa* PA14 and identification of a virulence-attenuating factor, HudA. Infect. Immun. *76*, 4152–4162.

Kim, S.H., Lee, B.Y., Lau, G.W., and Cho, Y.H. (2009). IscR modulates catalase A (KatA) activity, peroxide resistance and full virulence of *Pseudomonas aeruginosa* PA14. J. Microbiol. Biotechnol. *19*, 1520–1526.

Kimbrell, D.A., and Beutler, B. (2001). The evolution and genetics of innate immunity. Nat. Rev. Genet. *2*, 256–267.

Kimmel, C.B., Ballard, W.W., Kimmel, S.R., Ullmann, B., and Schilling, T.F. (1995). Stages of embryonic development of the zebrafish. Dev. Dyn. *203*, 253–310.

Kopf, M., Baumann, H., Freer, G., Freudenberg, M., Lamers, M., Kishimoto, T., Zinkernagel, R., Bluethmann, H., and Köhler G. (1994). Impaired immune and acute-phase responses in interleukin-6-deficient mice. Nature *368*, 339–342.

Köthe, M., Antl, M., Huber, B., Stoecker, K., Ebrecht, D., Steinmetz, I., and Eberl, L. (2003). Killing of *Caenorhabditis elegans* by *Burkholderia cepacia* is controlled by the *cep* quorum-sensing system. Cell. Microbiol. *5*, 343–351.

Kurz, C.L., and Ewbank, J.J. (2000). *Caenorhabditis elegans* for the study of host–pathogen interactions. Trends Microbiol. *8*, 142–144.

Kurz, C.L., and Ewbank, J.J. (2007). Infection in a dish: high-throughput analyses of bacterial pathogenesis. Curr. Opin. Microbiol. *10*, 10–16.

Kurz, C.L., and Tan, M.W. (2004). Regulation of aging and innate immunity in *C. elegans*. Aging Cell *3*, 185–193.

Kurz, C.L., Chauvet, S., Andrès, E., Aurouze, M., Vallet, I., Michel, G.P., Uh, M., Celli, J., Filloux, A., De Bentzmann, S., *et al.* (2003). Virulence factors of the human opportunistic pathogen *Serratia marcescens* identified by *in vivo* screening. EMBO J. *22*, 1451–1460.

Ladel, C.H., Blum, C., Dreher, A., Reifenberg, K., Kopf, M., and Kaufmann, S.H. (1997). Lethal tuberculosis in interleukin-6-deficient mutant mice. Infect. Immun. *65*, 4843–4849.

Lau, G.W., Britigan, B.E., and Hassett, D.J. (2005). *Pseudomonas aeruginosa* OxyR is required for full virulence in rodent and insect models of infection and for resistance to human neutrophils. Infect. Immun. *73*, 2550–2553.

Lavine, M.D., and Strand, M.R. (2002). Insect haemocytes and their role in immunity. Insect. Biochem. Mol. Biol. *32*, 1295–1309.

Laws, T.R., Smith, S.A., Smith, M.P., Harding, S.V, Atkins, T.P., and Titball, R.W. (2005). The nematode *Panagrellus redivivus* is susceptible to killing by human pathogens at 37°C. FEMS Microbiol. Lett. *250*, 77–83.

Laws, T.R., Atkins, H.S., Atkins, T.P., and Titball, R.W. (2006). The pathogen *Pseudomonas aeruginosa* negatively affects the attraction response of the nematode *Caenorhabditis elegans* to bacteria. Microb. Pathog. *40*, 293–297.

Lazzaro, B.P., Sceurman, B.K., and Clark, A.G. (2004). Genetic basis of natural variation in *D. melanogaster* antibacterial immunity. Science *303*, 1873–1876.

Lebreton, F., Riboulet-Bisson, E., Serror, P., Sanguinetti, M., Posteraro, B., Torelli, R., Hartke, A., Auffray, Y., and Giard, J.C. (2009). *ace*, which encodes an adhesin in *Enterococcus faecalis*, is regulated by Ers and is involved in virulence. Infect. Immun. *77*, 2832–2839.

Lee, A. (1998). Animal models for host–pathogen interaction studies. Br. Med. Bull. *54*, 163–173.

Lee, C.A., Jones, B.D., and Falkow, S. (1992). Identification of a *Salmonella typhimurium* invasion locus by selection for hyperinvasive mutants. Proc. Natl. Acad. Sci. USA *89*, 1847–1851.

Lee, D.G., Urbach, J.M., Wu, G., Liberati, N.T., Feinbaum, R.L., Miyata, S., Diggins, L.T., He, J., Saucier, M., Déziel, E., *et al.* (2006). Genomic analysis reveals that *Pseudomonas aeruginosa* virulence is combinatorial. Genome Biol. *7*, R90.

Lee, J.E., and Edery, I. (2008). Circadian regulation in the ability of *Drosophila* to combat pathogenic infections. Curr. Biol. *18*, 195–199.

Lemichez, E., Lecuit, M., Nassif, X., and Bourdoulous, S. (2010). Breaking the wall: targeting of the endothelium by pathogenic bacteria. Nat. Rev. Microbiol. *8*, 93–104.

Lever, M.S., Stagg, A.J., Nelson, M., Pearce, P., Stevens, D.J., Scott, E.A., Simpson, A.J., and Fulop, M.J. (2008). Experimental respiratory anthrax infection in the common marmoset (*Callithrix jacchus*). Int. J. Exp. Pathol. *89*, 171–179.

Levraud, J.P., Disson, O., Kissa, K., Bonne, I., Cossart, P., Herbomel, P., and Lecuit, M. (2009). Real-time observation of *Listeria monocytogenes*–phagocyte interactions in living zebrafish larvae. Infect. Immun. *77*, 3651–3660.

Li, D., Scherfer, C., Korayem, A.M., Zhao, Z., Schmidt, O., and Theopold, U. (2002). Insect haemolymph clotting: evidence for interaction between the coagulation system and the prophenoloxidase activating cascade. Insect. Biochem. Mol. Biol. *32*, 919–928.

Lieschke, G.J., Oates, A.C., Crowhurst, M.O., Ward, A.C., Layton, J.E. (2001). Morphologic and functional characterization of granulocytes and macrophages in embryonic and adult zebrafish. Blood *98*, 3087–3096.

Lin, B., Chen, S., Cao, Z., Lin, Y., Mo, D., Zhang, H., Gu, J., Dong, M., Liu, Z., and Xu, A. (2007). Acute phase response in zebrafish upon *Aeromonas salmonicida* and *Staphylococcus aureus* infection: striking similarities and obvious differences with mammals. Mol. Immunol. *44*, 295–301.

Liu, M., Popper, S.J., Rubins, K.H., and Relman, D.A. (2006). Early days: genomics and human responses to infection. Curr. Opin. Microbiol. *9*, 312–319.

Lockman, H.A., and Curtiss, R., 3rd. (1990). *Salmonella typhimurium* mutants lacking flagella or motility remain virulent in BALB/c mice. Infect. Immun. *58*, 137–143.

Lopez, J., Copps, J., Wilhelmsen, C., Moore, R., Kubay, J., St-Jacques, M., Halayko, S., Kranendonk, C., Toback, S., DeShazer, D., *et al.* (2003). Characterization of experimental equine glanders. Microbes Infect. *5*, 1125–1131.

Lostroh, C.P., and Lee, C.A. (2001). The *Salmonella* pathogenicity island-1 type III secretion system. Microbes Infect. *3*, 1281–1291.

Maadani, A., Fox, K.A., Mylonakis, E., and Garsin, D.A. (2007). *Enterococcus faecalis* mutations affecting virulence in the *Caenorhabditis elegans* model host. Infect. Immun. *75*, 2634–2637.

Macchiarini, F., Manz, M.G., Palucka, A.K., and Shultz, L.D. (2005). Humanized mice: are we there yet? J. Exp. Med. *202*, 1307–1311.

McCormick, B.A., Colgan, S.P., Delp-Archer, C., Miller, S.I., and Madara, J.L. (1993). *Salmonella typhimurium* attachment to human intestinal epithelial monolayers: transcellular signalling to subepithelial neutrophils. J. Cell Biol. *123*, 895–907.

McKenzie, G.J., Emson, C.L., Bell, S.E., Anderson, S., Fallon, P., Zurawski, G., Murray, R., Grencis, R., and McKenzie, A.N. (1998). Impaired development of Th2 cells in IL-13-deficient mice. Immunity *9*, 423–432.

MacRae, C.A., and Peterson, R.T. (2003). Zebrafish-based small molecule discovery. Chem. Biol. *10*, 901–908.

Mahajan-Miklos, S., Tan, M.W., Rahme, L.G., and Ausubel, F.M. (1999). Molecular mechanisms of bacterial virulence elucidated using a *Pseudomonas aeruginosa-Caenorhabditis elegans* pathogenesis model. Cell *96*, 47–56.

Mandrekar, N., and Thakur, N.L. (2009). Significance of the zebrafish model in the discovery of bioactive molecules from nature. Biotechnol. Lett. *31*, 171–179.

Marmaras, V.J., and M. Lampropoulou, M. (2009). Regulators and signalling in insect haemocyte immunity. Cell. Signal. *21*, 186–195.

Mathias, J.R., Perrin, B.J., Liu, T.X., Kanki, J., Look, A.T., and Huttenlocher, A. (2006). Resolution of inflammation by retrograde chemotaxis of neutrophils in transgenic zebrafish. J. Leukoc. Biol. *80*, 1281–1288.

Meijer, A.H., Gabby Krens, S.F., Medina Rodriguez, I.A., He, S., Bitter, W., Ewa Snaar-Jagalska, B., and Spaink, H.P. (2004). Expression analysis of the Toll-like receptor and TIR domain adaptor families of zebrafish. Mol. Immunol. *40*, 773–783.

Menudier, A., Rougier, F.P., and Bosgiraud, C. (1996). Comparative virulence between different strains of *Listeria* in zebrafish (*Brachydanio rerio*) and mice. Pathol. Biol. (Paris) *44*, 783–789.

Miyata, S., Casey, M., Frank, D.W., Ausubel, F.M., and Drenkard, E. (2003). Use of the *Galleria mellonella* caterpillar as a model host to study the role of the type III secretion system in *Pseudomonas aeruginosa* pathogenesis. Infect. Immun. *71*, 2404–2413.

Morton, D.B., Dunphy, G.B., and Chadwick, J.S. (1987). Reactions of haemocytes of immune and non-immune *Galleria mellonella* larvae to *Proteus mirabilis*. Dev. Comp. Immunol. *11*, 47–55.

Mosier, D.E., Gulizia, R.J., Baird, S.M., and Wilson, D.B. (1988). Transfer of a functional human immune system to mice with severe combined immunodeficiency. Nature *335*, 256–259.

Moy, T.I., Mylonakis, E., Calderwood, S.B., and Ausubel, F.M. (2004). Cytotoxicity of hydrogen peroxide produced by *Enterococcus faecium*. Infect. Immun. *72*, 4512–4520.

Moy, T.I., Ball, A.R., Anklesaria, Z., Casadei, G., Lewis, K., and Ausubel, F.M. (2006). Identification of novel antimicrobials using a live-animal infection model. Proc. Natl. Acad. Sci. USA *103*, 10414–10419.

Moy, T.I., Conery, A.L., Larkins-Ford, J., Wu, G., Mazitschek, R., Casadei, G., Lewis, K., Carpenter, A.E., and Ausubel, F.M. (2009). High-throughput screen for novel antimicrobials using a whole animal infection model. ACS Chem. Biol. *4*, 527–533.

Mukherjee, K., Altincicek, B., Hain, T., Domann, E., Vilcinskas, A., and Chakraborty, T. (2010). *Galleria mellonella* as a model system for studying *Listeria* pathogenesis. Appl. Environ. Microbiol. *76*, 310–317.

Murphey, R.D., and Zon, L.I. (2006). Small molecule screening in the zebrafish. Methods *39*, 255–261.

Nakagawa, I., Amano, A., Mizushima, N., Yamamoto, A., Yamaguchi, H., Kamimoto, T., Nara, A., Funao, J., Nakata, M., Tsuda, K., *et al.* (2004). Autophagy defends cells against invading group A *Streptococcus*. Science *306*, 1037–1040.

Nasevicius, A., and Ekker, S.C. (2000). Effective targeted gene 'knockdown' in zebrafish. Nat. Genet. *26*, 216–220.

Neely, M.N., Pfeifer, J.D., and Caparon, M. (2002). *Streptococcus*-zebrafish model of bacterial pathogenesis. Infect. Immun. *70*, 3904–3914.

Nehme, N.T., Liégeois, S., Kele, B., Giammarinaro, P., Pradel, E., Hoffmann, J.A., Ewbank, J.J., and Ferrandon, D. (2007). A model of bacterial intestinal infections in *Drosophila melanogaster*. PLoS Pathog. *3*, e173.

Nelson, M., Lever, M.S., Savage, V.L., Salguero, F.J., Pearce, P.C., Stevens, D.J., and Simpson, A.J. (2009). Establishment of lethal inhalational infection with *Francisella tularensis* (tularaemia) in the common marmoset (*Callithrix jacchus*). Int. J. Exp. Pathol. *90*, 109–118.

Oldani, A., Cormont, M., Hofman, V., Chiozzi, V., Oregioni, O., Canonici, A., Sciullo, A., Sommi, P., Fabbri, A., Ricci, V., *et al.* (2009). *Helicobacter pylori* counteracts the apoptotic action of its VacA toxin by injecting the CagA protein into gastric epithelial cells. PLoS Pathog. *5*, e1000603.

O'Quinn, A.L., Wiegand, E.M., and Jeddeloh, J.A. (2001). *Burkholderia pseudomallei* kills the nematode *Caenorhabditis elegans* using an endotoxin-mediated paralysis. Cell. Microbiol. *3*, 381–393.

Oyston, P.C., Russell, P., Williamson, E.D., and Titball, R.W. (1996). An *aroA* mutant of *Yersinia pestis* is attenuated in guinea-pigs, but virulent in mice. Microbiology *142*, 1847–1853.

Park, S.Y., Heo, Y.J., Kim, K.S., and Cho, Y.H. (2005). *Drosophila melanogaster* is susceptible to *Vibrio cholerae* infection. Mol. Cells *20*, 409–415.

Pascolo, S. (2005). HLA class I transgenic mice: development, utilisation and improvement. Expert. Opin. Biol. Ther. *5*, 919–938.

Pasparakis, M., Alexopoulou, L., Episkopou, V., and Kollias, G. (1996). Immune and inflammatory responses in TNFα-deficient mice: a critical requirement for TNFα in the formation of primary B cell follicles, follicular dendritic cell networks and germinal centers, and in the maturation of the humoral immune response. J. Exp. Med. *184*, 1397–1411.

Peleg, A.Y., Tampakakis, E., Fuchs, B.B., Eliopoulos, G.M., Moellering, R.C., Jr, and Mylonakis, E. (2008). Prokaryote–eukaryote interactions identified by using *Caenorhabditis elegans*. Proc. Natl. Acad. Sci. U.S.A. *105*, 14585–14590.

Peleg, A.Y., Jara, S., Monga, D., Eliopoulos, G.M., Moellering, R.C., Jr, and Mylonakis, E. (2009a). *Galleria mellonella* as a model system to study *Acinetobacter baumannii* pathogenesis and therapeutics. Antimicrob. Agents Chemother. *53*, 2605–2609.

Peleg, A.Y., Monga, D., Pillai, S., Mylonakis, E., Moellering, R.C., Jr, and Eliopoulos, G.M. (2009b). Reduced susceptibility to vancomycin influences pathogenicity in *Staphylococcus aureus* infection. J. Infect. Dis. *199*, 532–536.

Phelps, H.A., Runft, D.L., and Neely, M.N. (2009). Adult zebrafish model of streptococcal infection. Curr. Protoc. Microbiol. *13*, 9D.1.1–9D.1.27.

Plant, J., and Glynn, A.A. (1979). Locating *Salmonella* resistance gene on mouse chromosome 1. Clin. Exp. Immunol. *37*, 1–6.

Ploegh, H.L., Orr, H.T., and Strominger, J.L. (1981). Major histocompatibility antigens: the human (HLA-A, -B, -C) and murine (H-2K, H-2D) class I molecules. Cell *24*, 287–299.

Poltorak, A., He, X., Smirnova, I., Liu, M.Y., Van Huffel, C., Du, X., Birdwell, D., Alejos, E., Silva, M., Galanos, C., *et al.* (1998). Defective LPS signaling in C3H/HeJ and C57BL/10ScCr mice: mutations in *Tlr4* gene. Science *282*, 2085–2088.

Potvin, E., Lehoux, D.E., Kukavica-Ibrulj, I., Richard, K.L., Sanschagrin, F., Lau, G.W., and Levesque, R.C. (2003). *In vivo* functional genomics of *Pseudomonas aeruginosa* for high-throughput screening of new virulence factors and antibacterial targets. Environ. Microbiol. *5*, 1294–1308.

Prajsnar, T.K., Cunliffe, V.T., Foster, S.J., and Renshaw, S.A. (2008). A novel vertebrate model of *Staphylococcus aureus* infection reveals phagocyte-dependent resistance of zebrafish to non-host specialized pathogens. Cell. Microbiol. *10*, 2312–2325.

Pressley, M.E., Phelan, P.E., 3rd, Witten, P.E., Mellon, M.T., and Kim, C.H. (2005). Pathogenesis and inflammatory response to *Edwardsiella tarda* infection in the zebrafish. Dev. Comp. Immunol. *29*, 501–513.

Pujol, N., Link, E.M., Liu, L.X., Kurz, C.L., Alloing, G., Tan, M.W., Ray, K.P., Solari, R., Johnson, C.D., and Ewbank, J.J. (2001). A reverse genetic analysis of components of the Toll signaling pathway in *Caenorhabditis elegans*. Curr. Biol. *11*, 809–821.

Querec, T.D., Akondy, R.S., Lee, E.K., Cao, W., Nakaya, H.I., Teuwen, D., Pirani, A., Gernert, K., Deng, J., Marzolf, B., *et al.* (2009). Systems biology approach predicts immunogenicity of the yellow fever vaccine in humans. Nat. Immunol. *10*, 116–125.

Qureshi, S.T., Larivière, L., Leveque, G., Clermont, S., Moore, K.J., Gros, P., Malo, D. (1999). Endotoxin-tolerant mice have mutations in Toll-like receptor 4 (Tlr4). J. Exp. Med. *189*, 615–625.

Rabelo-Gonçalves, E.M., Nishimura, N.F., and Zeitune, J.M. (2005). Development of a BALB/c mouse model of *Helicobacter pylori* infection with fresh and frozen bacteria. Biol. Res. *38*, 101–109.

Rahme, L.G., Stevens, E.J., Wolfort, S.F., Shao, J., Tompkins, R.G., and Ausubel, F.M. (1995). Common virulence factors for bacterial pathogenicity in plants and animals. Science *268*, 1899–1902.

Rahme, L.G., Tan, M.W., Le, L., Wong, S.M., Tompkins, R.G., Calderwood, S.B., and Ausubel, F.M. (1997). Use of model plant hosts to identify *Pseudomonas aeruginosa* virulence factors. Proc. Natl. Acad. Sci. U.S.A. *94*, 13245–13250.

Reddy, K.C., Andersen, E.C., Kruglyak, L., and Kim, D.H. (2009). A polymorphism in *npr-1* is a behavioral determinant of pathogen susceptibility in *C. elegans*. Science 323, 382–384.

Reznikov, L.L., Fantuzzi, G., Selzman, C.H., Shames, B.D., Barton, H.A., Bell, H., McGregor, J.A., and Dinarello, C.A (1999). Utilization of endoscopic inoculation in a mouse model of intrauterine infection-induced preterm birth: role of interleukin 1β. Biol. Reprod. *60*, 1231 1238.

Rocchetta, H.L., Boylan, C.J., Foley, J.W., Iversen, P.W., LeTourneau, D.L., McMillian, C.L., Contag, P.R., Jenkins, D.E., and Parr, T.R., Jr. (2001). Validation of a noninvasive, real-time imaging technology using bioluminescent *Escherichia coli* in the neutropenic mouse thigh model of infection. Antimicrob. Agents Chemother. 45, 129–137.

Ryan, D.M. (1970). A method for inducing experimental urinary tract infections in the rat and its use in experimental chemotherapy. Postgrad. Med. J. *Suppl*, 19–24.

Saini, L.S., Galsworthy, S.B., John, M.A., and Valvano, M.A. (1999). Intracellular survival of *Burkholderia cepacia* complex isolates in the presence of macrophage cell activation. Microbiology *145*, 3465–3475.

Salah, I.B., Ghigo, E., and Drancourt, M. (2009). Free-living amoebae, a training field for macrophage resistance of mycobacteria. Clin. Microbiol. Infect. *15*, 894–905.

van der Sar, A.M., Musters, R.J., van Eeden, F.J., Appelmelk, B.J., Vandenbroucke-Grauls, C.M., and Bitter, W. (2003). Zebrafish embryos as a model host for the real time analysis of *Salmonella typhimurium* infections. Cell. Microbiol. *5*, 601–611.

van der Sar, A.M., Abdallah, A.M., Sparrius, M., Reinders, E., Vandenbroucke-Grauls, C.M., and Bitter, W. (2004). *Mycobacterium marinum* strains can be divided into two distinct types based on genetic diversity and virulence. Infect. Immun. *72*, 6306–6312.

van der Sar, A.M., Stockhammer, O.W., van der Laan, C., Spaink, H.P., Bitter, W., and Meijer, A.H. (2006). MyD88 innate immune function in a zebrafish embryo infection model. Infect. Immun. *74*, 2436–2441.

van der Sar, A.M., Spaink, H.P., Zakrzewska, A., Bitter, W., and Meijer, A.H. (2009). Specificity of the zebrafish host transcriptome response to acute and chronic mycobacterial infection and the role of innate and adaptive immune components. Mol. Immunol. *46*, 2317–2332.

Saukkonen, K., Abdillahi, H., Poolman, J.T., and Leinonen, M. (1987). Protective efficacy of monoclonal antibodies to class 1 and class 3 outer membrane proteins of *Neisseria meningitidis* B:15:P1.16 in infant rat infection model: new prospects for vaccine development. Microb. Pathog. *3*, 261–267.

Schell, M.A., Lipscomb, L., and DeShazer, D. (2008). Comparative genomics and an insect model rapidly identify novel virulence genes of *Burkholderia mallei*. J. Bacteriol. *190*, 2306–2313.

Schmid, D., and Münz, C. (2007). Innate and adaptive immunity through autophagy. Immunity *27*, 11–21.

Schneider, D.S., Ayres, J.S., Brandt, S.M., Costa, A., Dionne, M.S., Gordon, M.D., Mabery, E.M., Moule, M.G., Pham, L.N., and Shirasu-Hiza, M.M. (2007). *Drosophila eiger* mutants are sensitive to extracellular pathogens. PLoS Pathog. *3*, e41.

Searle, S., Bright, N.A., Roach, T.I., Atkinson, P.G., Barton, C.H., Meloen, R.H., and Blackwell, J.M. (1998). Localisation of Nramp1 in macrophages: modulation with activation and infection. J. Cell Sci. *111*, 2855–2866.

Seed, K.D., and Dennis, J.J. (2008). Development of *Galleria mellonella* as an alternative infection model for the *Burkholderia cepacia* complex. Infect. Immun. *76*, 1267–1275.

Seed, K.D., and Dennis, J.J. (2009). Experimental bacteriophage therapy increases survival of *Galleria mellonella* larvae infected with clinically relevant strains of the *Burkholderia cepacia* complex. Antimicrob. Agents Chemother. *53*, 2205–2208.

Segal, E.D., Falkow, S., and Tompkins, L.S. (1996). *Helicobacter pylori* attachment to gastric cells induces cytoskeletal rearrangements and tyrosine phosphorylation of host cell proteins. Proc. Natl. Acad. Sci. U.S.A. *93*, 1259–1264.

Shultz, L.D. (1991). Immunological mutants of the mouse. Am. J. Anat. *191*, 303–311.

Sieger, D., Stein, C., Neifer, D., van der Sar, A.M., and Leptin, M. (2009). The role of gamma interferon in innate immunity in the zebrafish embryo. Dis. Model Mech. *2*, 571–581.

Sifri, C.D., Begun, J., Ausubel, F.M., and Calderwood, S.B. (2003). *Caenorhabditis elegans* as a model host for *Staphylococcus aureus* pathogenesis. Infect. Immun. *71*, 2208–2217.

Silo-Suh, L., Suh, S.J., Sokol, P.A., and Ohman, D.E. (2002). A simple alfalfa seedling infection model for *Pseudomonas aeruginosa* strains associated with cystic fibrosis shows AlgT (sigma-22) and RhlR contribute to pathogenesis. Proc. Natl. Acad. Sci. U.S.A. *99*, 15699–15704.

Skamene, E., Gros, P., Forget, A., Kongshavn, P.A., St Charles, C., and Taylor, B.A. (1982). Genetic regulation of resistance to intracellular pathogens. Nature *297*, 506–509.

Smith, M.P., Laws, T.R., Atkins, T.P., Oyston, P.C., de Pomerai, D.I., and Titball, R.W. (2002). A liquid-based method for the assessment of bacterial pathogenicity using the nematode *Caenorhabditis elegans*. FEMS Microbiol. Lett. *210*, 181–185.

Smither, S.J., Perkins, S.D., Davies, C., Stagg, A.J., Nelson, M., and Atkins, H.S. (2009). Development and characterization of mouse models of infection with aerosolized *Brucella melitensis* and *Brucella suis*. Clin. Vaccine Immunol. *16*, 779–783.

Sotto, A., Lina, G., Richard, J.L., Combescure, C., Bourg, G., Vidal, L., Jourdan, N., Etienne, J., and Lavigne, J.P. (2008). Virulence potential of *Staphylococcus aureus* strains isolated from diabetic foot ulcers: a new paradigm. Diabetes Care *31*, 2318–2324.

Springman, A.C., Jacobs, J.L., Somvanshi, V.S., Sundin, G.W., Mulks, M.H., Whittam, T.S., Viswanathan, P., Gray, R.L., Lipuma, J.J., and Ciche, T.A. (2009). Genetic diversity and multihost pathogenicity of clinical and environmental strains of *Burkholderia cenocepacia*. Appl. Environ. Microbiol. *75*, 5250–5260.

Steele-Mortimer, O. (2008). Infection of epithelial cells with *Salmonella enterica*. Methods Mol. Biol. *431*, 201–211.

Steindler, L., Bertani, I., De Sordi, L., Schwager, S., Eberl, L., and Venturi, V. (2009). LasI/R and RhlI/R quorum sensing in a strain of *Pseudomonas aeruginosa* beneficial to plants. Appl. Environ. Microbiol. *75*, 5131–5140.

Stokes, M.G., Titball, R.W., Neeson, B.N., Galen, J.E., Walker, N.J., Stagg, A.J., Jenner, D.C., Thwaite, J.E., Nataro, J.P., Baillie, L.W., *et al.* (2007). Oral administration of a *Salmonella enterica*-based vaccine expressing *Bacillus anthracis* protective antigen confers protection against aerosolized *B. anthracis*. Infect. Immun. *75*, 1827–1834.

Stoltz, D.A., Ozer, E.A., Taft, P.J., Barry, M., Liu, L., Kiss, P.J., Moninger, T.O., Parsek, M.R., and Zabner, J. (2008). *Drosophila* are protected from *Pseudomonas aeruginosa* lethality by transgenic expression of paraoxonase-1. J. Clin. Invest. *118*, 3123–3131.

Styer, K.L., Hopkins, G.W., Bartra, S.S., Plano, G.V., Frothingham, R., and Aballay, A. (2005). *Yersinia pestis* kills *Caenorhabditis elegans* by a biofilm-independent process that involves novel virulence factors. EMBO Rep. *6*, 992–997.

Sugawara, I., Yamada, H., Mizuno, S., and Iwakura Y. (2000). IL-4 is required for defense against mycobacterial infection. Microbiol. Immunol. *44*, 971–979.

Sullivan, C., and Kim, C.H. (2008). Zebrafish as a model for infectious disease and immune function. Fish Shellfish Immunol. *25*, 341–350.

Sulston, J.E., and Brenner, S. (1974). The DNA of *Caenorhabditis elegans*. Genetics *77*, 95–104.

Sun, Y.C., Hinnebusch, B.J., and Darby, C. (2008). Experimental evidence for negative selection in the evolution of a *Yersinia pestis* pseudogene. Proc. Natl. Acad. Sci. USA *105*, 8097–8101.

Swaim, L.E., Connolly, L.E., Volkman, H.E., Humbert, O., Born, D.E., and Ramakrishnan, L. (2006). *Mycobacterium marinum* infection of adult zebrafish causes caseating granulomatous tuberculosis and is moderated by adaptive immunity. Infect. Immun. *74*, 6108–6117.

Szabady, R.L., Lokuta, M.A., Walters, K.B., Huttenlocher, A., and Welch, R.A. (2009). Modulation of neutrophil function by a secreted mucinase of *Escherichia coli* O157:H7. PLoS Pathog. *5*, e1000320.

Takeuchi, A., and Sprinz, H. (1967). Electron-microscope studies of experimental *Salmonella* infection in the preconditioned guinea pig: II. Response of the intestinal mucosa to the invasion by *Salmonella typhimurium*. Am. J. Pathol. *51*, 137–161.

Tammiruusu, A., Haveri, A., Pascolo, S., Lahesmaa, R., Stevanovic, S., Rammensee, H.G., Sarvas, M., Puolakkainen, M., and Vuola, J.M. (2005). Clearance of *Chlamydia pneumoniae* infection in H-2 class I human leucocyte antigen-A2.1 monochain transgenic mice. Scand. J. Immunol. *62*, 131–139.

Tampakakis, E., Peleg, A.Y., and Mylonakis, E. (2009). Interaction of *Candida albicans* with an intestinal pathogen, *Salmonella enterica* serovar Typhimurium. Eukaryot. Cell *8*, 732–737.

Tan, M.W., Mahajan-Miklos, S., and Ausubel, F.M. (1999a). Killing of *Caenorhabditis elegans* by *Pseudomonas aeruginosa* used to model mammalian bacterial pathogenesis. Proc. Natl. Acad. Sci. USA *96*, 715–720.

Tan, M.W., Rahme, L.G., Sternberg, J.A., Tompkins, R.G., and Ausubel, F.M. (1999b). *Pseudomonas aeruginosa* killing of *Caenorhabditis elegans* used to identify *P. aeruginosa* virulence factors. Proc. Natl. Acad. Sci. U.S.A. *96*, 2408–2413.

Tenor, J.L., and. Aballay, A. (2008). A conserved Toll-like receptor is required for *Caenorhabditis elegans* innate immunity. EMBO Rep. *9*, 103–109.

Tenor,.JL., McCormick, B.A., Ausubel, F.M., and Aballay A. (2004). *Caenorhabditis elegans*-based screen identifies *Salmonella* virulence factors required for conserved host–pathogen interactions. Curr. Biol. *14*, 1018–1024.

Theopold, U., Li, D., Fabbri, M., Scherfer, C., and Schmidt, O. (2002). The coagulation of insect haemolymph. Cell. Mol. Life Sci. *59*, 363–372.

Thomas, R.J., Webber, D., Collinge, A., Stagg, A.J., Bailey, S.C., Nunez, A., Gates, A., Jayasekera, P.N., Taylor,

R.R., Eley, S., *et al.* (2009). Different pathologies but equal levels of responsiveness to the recombinant F1 and V antigen vaccine and ciprofloxacin in a murine model of plague caused by small- and large-particle aerosols. Infect. Immun. *77*, 1315–1323.

Trede, N.S., Zapata, A., and Zon, L.I. (2001). Fishing for lymphoid genes. Trends Immunol. *22*, 302–307.

Tsolis, R.M., Townsend, S.M., Miao, E.A., Miller, S.I., Ficht, T.A., Adams, L.G., and Bäumler, A.J. (1999). Identification of a putative *Salmonella enterica* serotype Typhimurium host range factor with homology to IpaH and YopM by signature-tagged mutagenesis. Infect. Immun. *67*, 6385–6393.

Uehlinger, S., Schwager, S., Bernier, S.P., Riedel, K., Nguyen, D.T., Sokol, P.A., and Eberl, L. (2009). Identification of specific and universal virulence factors in *Burkholderia cenocepacia* strains by using multiple infection hosts. Infect. Immun. *77*, 4102–4110.

Usher, L.R., Lawson, R.A., Geary, I., Taylor, C.J., Bingle, C.D., Taylor, G.W., and Whyte, M.K. (2002). Induction of neutrophil apoptosis by the *Pseudomonas aeruginosa* exotoxin pyocyanin: a potential mechanism of persistent infection. J. Immunol. *168*, 1861–1868.

Uzal, F.A., and Kelly, W.R. (1998). Experimental *Clostridium perfringens* type D enterotoxemia in goats. Vet. Pathol. *35*, 132–140.

Valdez, Y., Ferreira, R.B., and Finlay, B.B. (2009). Molecular mechanisms of *Salmonella* virulence and host resistance. Curr. Top. Microbiol. Immunol. *337*, 93–127.

Vallance, B.A. and Finlay, B.B. (2000). Exploitation of host cells by enteropathogenic *Escherichia coli*. Proc. Natl. Acad. Sci. U.S.A. *97*, 8799–8806.

Vanrompay, D., Hoang, T.Q., De Vos, L., Verminnen, K., Harkinezhad, T., Chiers, K., Morré, S.A., and Cox, E. (2005). Specific-pathogen-free pigs as an animal model for studying *Chlamydia trachomatis* genital infection. Infect. Immun. *73*, 8317–8321.

Vergunst, A.C., Meijer, A.H., Renshaw, S.A., and O'Callaghan, D. (2010). *Burkholderia cenocepacia* creates an intra-macrophage replication niche in zebrafish embryos, followed by bacterial dissemination and establishment of systemic infection. Infect. Immun. *78*, 1495–1508.

Vidal, S.M., Pinner, E., Lepage, P., Gauthier, S., and Gros, P. (1996). Natural resistance to intracellular infections: Nramp1 encodes a membrane phosphoglycoprotein absent in macrophages from susceptible (Nramp1 D169) mouse strains. J. Immunol. *157*, 3559–3568.

Vlisidou, I., Dowling, A.J., Evans, I.R., Waterfield, N., Ffrench-Constant, R.H., and Wood, W. (2009). *Drosophila* embryos as model systems for monitoring bacterial infection in real time. PLoS Pathog. *5*, e1000518.

Vojtech, L.N., Sanders, G.E., Conway, C., Ostland, V., and Hansen, J.D. (2009). Host immune response and acute disease in a zebrafish model of *Francisella* pathogenesis. Infect. Immun. *77*, 914–925.

Volkman, H.E., Pozos, T.C., Zheng, J., Davis, J.M., Rawls, J.F, and Ramakrishnan L. (2010). Tuberculous granuloma induction via interaction of a bacterial secreted protein with host epithelium. Science *327*, 466–469.

Vonkavaara, M., Telepnev, M.V., Rydén, P., Sjöstedt, A., and Stöven, S. (2008). *Drosophila melanogaster* as a model for elucidating the pathogenicity of *Francisella tularensis*. Cell. Microbiol. *10*, 1327–1338.

Waddell, S.J., Butcher, P.D., and Stoker, N.G. (2007). RNA profiling in host–pathogen interactions. Curr. Opin. Microbiol. *10*, 297–302.

Walker, T.S., Bais, H.P., Déziel, E., Schweizer, H.P., Rahme, L.G., Fall, R., and Vivanco, J.M. (2004). *Pseudomonas aeruginosa*-plant root interactions. Pathogenicity, biofilm formation, and root exudation. Plant. Physiol. *134*, 320–331.

Wang, L., and Ligoxygakis, P. (2006). Pathogen recognition and signalling in the *Drosophila* innate immune response. Immunobiology *211*, 251–261.

Wang, W., Liu, X., Gelinas, D., Ciruna, B., and Sun, Y. (2007). A fully automated robotic system for microinjection of zebrafish embryos. PLoS One *2*, e862.

Wareham, D.W., Papakonstantinopoulou, A., and Curtis, M.A. (2005). The *Pseudomonas aeruginosa* PA14 type III secretion system is expressed but not essential to virulence in the *Caenorhabditis elegans-P. aeruginosa* pathogenicity model. FEMS Microbiol. Lett. *242*, 209–216.

Watson, J., and Riblet, R. (1974). Genetic control of responses to bacterial lipopolysaccharides in mice. I. Evidence for a single gene that influences mitogenic and immunogenic respones to lipopolysaccharides. J. Exp. Med. *140*, 1147–1161.

Wedde, M., Weise, C., Kopacek, P., Franke, P., and Vilcinskas, A. (1998). Purification and characterization of an inducible metalloprotease inhibitor from the haemolymph of greater wax moth larvae, *Galleria mellonella*. Eur. J. Biochem. *255*, 535–543.

Wedde, M., Weise, C., Nuck, R., Altincicek, B., and Vilcinskas, A. (2007). The insect metalloproteinase inhibitor gene of the lepidopteran *Galleria mellonella* encodes two distinct inhibitors. Biol. Chem. *388*, 119–127.

Wei, X.Q., Leung, B.P., Niedbala, W., Piedrahita, D., Feng, G.J., Sweet, M., Dobbie, L., Smith, A.J., and Liew, F.Y. (1999). Altered immune responses and susceptibility to *Leishmania major* and *Staphylococcus aureus* infection in IL-18-deficient mice. J. Immunol. *163*, 2821–2828.

Welsh, M., Mangravite, L., Medina, M.W., Tantisira, K., Zhang, W., Huang, R.S., McLeod, H., and Dolan, M.E. (2009). Pharmacogenomic discovery using cell-based models. Pharmacol. Rev. *61*, 413–429.

Wienholds, E., Schulte-Merker, S., Walderich, B., and Plasterk, R.H. (2002). Target-selected inactivation of the zebrafish *rag1* gene. Science *297*, 99–102.

Wienholds, E., van Eeden, F., Kosters, M., Mudde, J., Plasterk, R.H., and Cuppen, E. (2003). Efficient target-selected mutagenesis in zebrafish. Genome Res. *13*, 2700–2707.

Wiles, S., Clare, S., Harker, J., Huett, A., Young, D., Dougan, G., and Frankel, G. (2004). Organ specificity, colonization and clearance dynamics *in vivo* following oral challenges with the murine pathogen *Citrobacter rodentium*. Cell. Microbiol. *6*, 963–972.

Wiles, S., Dougan, G., and Frankel, G. (2005). Emergence of a 'hyperinfectious' bacterial state after passage of *Citrobacter rodentium* through the host gastrointestinal tract. Cell. Microbiol. *7*, 1163–1172.

Wood, M.W., Rosqvist, R., Mullan, P.B., Edwards, M.H., and Galyov, E.E. (1996). SopE, a secreted protein of *Salmonella dublin*, is translocated into the target eukaryotic cell via a *sip*-dependent mechanism and promotes bacterial entry. Mol. Microbiol. *22*, 327–338.

Wright, E.K., Goodart, S.A., Growney, J.D., Hadinoto, V., Endrizzi, M.G., Long, E.M., Sadigh, K., Abney, A.L., Bernstein-Hanley, I., and Dietrich, W.F. (2003). Naip5 affects host susceptibility to the intracellular pathogen *Legionella pneumophila*. Curr. Biol. *13*, 27–36.

Yamada, H., Mizumo, S., Horai, R., Iwakura, Y., and Sugawara, I. (2000). Protective role of interleukin-1 in mycobacterial infection in IL-1αβ double-knockout mice. Lab. Invest. *80*, 759–767.

Yang, S.K., Eckmann, L., Panja, A., and Kagnoff, M.F. (1997). Differential and regulated expression of C-X-C, C-C, and C-chemokines by human colon epithelial cells. Gastroenterology *113*, 1214–1223.

Yoder, J.A., Nielsen, M.E., Amemiya, C.T., and Litman, G.W. (2002). Zebrafish as an immunological model system. Microbes Infect. *4*, 1469–1478.

Yorgey, P., Rahme, L.G., Tan, M.W., and Ausubel, F.M. (2001). The roles of *mucD* and alginate in the virulence of *Pseudomonas aeruginosa* in plants, nematodes and mice. Mol. Microbiol. *41*, 1063–1076.

Yu, J., Rossi, R., Hale, C., Goulding, D., and Dougan, G. (2009). Interaction of enteric bacterial pathogens with murine embryonic stem cells. Infect. Immun. *77*, 585–597.

Zarkadis, I.K., Mastellos, D., and Lambris, J.D. (2001). Phylogenetic aspects of the complement system. Dev. Comp. Immunol. *25*, 745–762.

Zhang, M.J., Meng, F.L., Ji, X.Y., He, L.H., and Zhang, J.Z. (2007). Adherence and invasion of mouse-adapted *H. pylori* in different epithelial cell lines. World J. Gastroenterol. *13*, 845–850.

Zheng, H., Fletcher, D., Kozak, W., Jiang, M., Hofmann, K.J., Conn, C.A., Soszynski, D., Grabiec, C., Trumbauer, M.E., Shaw, A., *et al.* (1995). Resistance to fever induction and impaired acute-phase response in interleukin-1β-deficient mice. Immunity *3*, 9–19.

Zhou, D., and Galán, J. (2001). *Salmonella* entry into host cells: the work in concert of type III secreted effector proteins. Microbes Infect. *3*, 1293–1298.

Zhou, D., Mooseker, M.S., and Galán, J.E. (1999). Role of the *S. typhimurium* actin-binding protein SipA in bacterial internalization. Science *283*, 2092–2095.

Zimmerli, W. (1993). Experimental models in the investigation of device-related infections. J. Antimicrob. Chemother. *31 Suppl D*, 97–102.

Zlosnik, J.E., Tavankar, G.R., Bundy, J.G., Mossialos, D., O'Toole, R., and Williams, H.D. (2006). Investigation of the physiological relationship between the cyanide-insensitive oxidase and cyanide production in *Pseudomonas aeruginosa*. Microbiology *152*, 1407–1415.

Strategies for Identifying Bacterial Pathogenicity Genes

2

Raphael H. Valdivia

Abstract

The identification of factors that enhance the virulence of bacteria is a critical step in understanding the molecular basis for their pathogenesis. Over the last several years, molecular and genetic strategies have been developed to identify bacterial virulence determinants. These approaches, combined with rapidly advancing genome technologies, have allowed investigators to map the pathogen's transcriptional response to the host environment and to define the relative contribution of individual genes to the infectious process. This chapter highlights some of the many approaches currently under use.

Introduction

Genes that enhance the ability of bacteria to infect, to evade host immune responses, and to disseminate are commonly referred to as virulence genes. These genes may encode factors ranging from toxins and adhesins, with readily recognized roles in disease causation, to enzymes that enhance the pathogen's metabolic properties within the host's nutrient-limiting environment. Identifying virulence genes and determining the function of their encoded products is central to our understanding of microbial pathogenesis.

Before starting a discussion on how to identify these factors, it is useful to appreciate that what constitutes a *bona fide* 'virulence gene' is the subject of much debate. Disease, the medically relevant outcome of a host–microbe interaction, is the result of a complex interplay between the pathogen and the host's immune defences. For instance, factors that allow bacteria to be successful and innocuous on skin surfaces can turn deadly if the bacteria penetrate deeper tissues or if the host's immune system becomes compromised. In the laboratory, virulence factors are often identified by testing bacterial mutants in various infection models. Therefore, variables such as what animal model, route of infection or dose is administered can significantly alter our perception of how any individual gene contributes to pathogenesis. Similarly, the host's immune status or its genetic makeup can influence the outcome of a host–pathogen interaction and further obscure our ability to determine the relative contribution of individual virulence factors to disease. For the purposes of this chapter we will assign the term virulence genes to any genetic element that either directly or indirectly promotes (i) attachment and/or invasion, (ii) replication, (iii) tissue damage, (iv) immune evasion and (v) dissemination within a susceptible animal host.

The hunt for virulence genes dates back to the early days of molecular genetics. Many molecular genetic tools such as transposons, shuttle plasmids and generalized transducing phages were developed to generate and map mutations in pathogenic bacteria. These 'classical' tools have become increasingly more sophisticated and now enable the investigator to sample complex environments such as live animal hosts and to simultaneously monitor the fate of multiple mutants within one animal. In the following sections, I highlight some of the established methods that are currently used to identify virulence genes and some of the recent advances that have extended the power of these approaches.

Mutational approaches to identify virulence genes

The most efficient molecular tools to introduce random, loss-of-function mutations in bacteria are transposable elements. Transposases recognize specific inverted DNA repeats, excise the DNA flanked by these repeats and catalyse the insertion of the excised DNA into random sites within a target DNA molecule (Bennett, 2004). The insertion of these sequences often leads to the insertional inactivation of genes. The sequences flanked by the inverted repeats can be engineered to contain antibiotic resistance markers and/or reporter genes to generate a 'transposon'. Delivery of these genetic elements into bacterial cells followed by selection for the transposon-encoded antibiotic resistance marker allows for the selection and amplification of bacteria containing transposon insertions. Because the integration occurs at random sites on the chromosome, mutant 'libraries' of bacteria containing unique transposon insertions in all non-essential genes can be readily generated (Maloy, 2007).

Individual transposon insertion mutants can be screened for defects in their ability to infect and cause disease in laboratory animals. Assuming that the transposon has no DNA sequence bias for insertion, the number of mutants that need to be generated to ensure that all possible genes have been disrupted is generally 5–10-fold in excess to the number of genes present in the pathogen being studied. Individually testing thousands of such mutants (>25,000 for a pathogen with ~5000 genes) for virulence defects in animals is unrealistic given the expense and ethical issues behind the use of large number of laboratory animals. The development of surrogate infection models such as cultured mammalian cells has permitted more rapid and cost effective screening of complex mutant libraries for reduced pathogen adherence, invasion, or survival within host cells.

The bottleneck in the screening of individual mutants for virulence defects is the complexity of the virulence assay used and the number of mutants that need to be screened. The lower the number of mutants the more complex the virulence assay can become. Therefore, new techniques are constantly being developed to generate defined collections of bacteria in which all non-essential genes have been systematically inactivated. In this manner, an investigator can screen the phenotype of all possible mutants for defects in any particular virulence trait. Thousands of mutants can be screened in these whole cell-based assays either manually or semi-automatically with robotics dispensers and analysed with powerful software platforms that allows for the simultaneous monitoring of many parameters within a single assay.

High throughput mutant screens have been also extended to small non-vertebrate host systems. For example, the nematode *Caenorhabditis elegans* and the fruit fly *Drosophila melanogaster* are colonized and killed by a variety of human pathogens (Mylonakis and Aballay, 2005). Furthermore, many of the same virulence factors that are required for mammalian infections are also required to infect nematodes and flies. These host systems lack adaptive immune systems (e.g. lymphocytes, antibodies) and highly specialized cellular immune effectors (e.g. granulocytes, cytotoxic lymphocytes) so not all facets of the pathogenic process can be faithfully reproduced. Nonetheless, nematodes and flies have been very useful for identifying bacterial virulence factors required to cope with innate immune responses and for addressing basic questions in host–pathogen interactions, such as the molecular basis for the recognition of microbes by multicellular organisms, antimicrobial responses and induction of tolerance to microbial microflora (Mylonakis *et al.*, 2007; Irazoqui *et al.*, 2010). A further advantage of these model systems is that they have well-developed genetic and genomic tools permitting a detailed analysis of the host contribution to the pathogenic process (Dorer and Isberg, 2006). Experiments in nematodes and fruit flies have facilitated the identification of resistance and susceptibility loci to various pathogens.

Parallel analysis of mutants with virulence defects

To bypass the labour intensive and costly process of screening individual mutants in virulence assays, methodologies have been developed to monitor the fate of individual bacterial mutants within a pool (Mazurkiewicz *et al.*, 2006). This is achieved by tagging each mutant with unique molecular identifiers. In this manner, mutants with a virulence defect that are outcompeted by wild type counterparts from a pool of mutants can be identified by the loss of their unique molecular identifiers.

Signature-tagged mutagenesis

The first system for parallel analysis of mutants for virulence defects was termed signature-tagged mutagenesis (STM) (Hensel *et al.*, 1995). In STM, unique DNA 'tag' sequences are incorporated within the sequences of a transposon, which are then used to generate a collection of barcoded mutants and maintained as an array of individual clones. These mutants are then pooled and used to infect an animal. The bacteria are recovered from infected tissues, and the barcodes associated with transposon insertion in the surviving bacteria are amplified by polymerase chain reaction (PCR) and used as hybridization probes against an ordered array of DNA tags derived from the input bacteria (Fig. 2.1A). Tags associated with mutants that failed to replicate or showed decrease fitness in an animal are depleted in the output pool as compared with the input pool. This approach has been termed 'negative' selection, to highlight the fact that mutants that were selected against in the host environment can be readily identified. In its original application, STM led to the discovery of several new *Salmonella* virulence genes, including genes encoding a novel Type III secretion apparatus that is required for bacterial survival in host tissues (Hensel *et al.*, 1995, 1997). The basic concept of using DNA barcodes to identify missing members from among a pool of mutagenized microbes has been applied to a variety of Gram-negative, Gram-positive, protozoan and fungal pathogens to identify virulence factors (Mazurkiewicz *et al.*, 2006).

DNA-microarray assisted mapping of transposon mutants

With the increased abundance of microbial genomic sequences, it is now feasible to use the DNA sequences adjacent to transposon insertion sites as substitutes for the artificial barcodes used to track transposon insertions (Baldwin and Salama, 2007). A collection of approaches [e.g. transposon site hybridization (TraSH) (Sassetti and Rubin, 2003), microarray tracking of transposons (MATT) (Salama *et al.*, 2004)] were developed that use sequences unique to the transposons and random oligonucleotide primers to selectively amplify by PCR the sequences adjacent to the transposon insertion. In *Mycobacterium tuberculosis*, where such approaches were first pioneered, pools of mutagenized bacteria were used to infect mice, the sequences adjacent to insertion sites were amplified by PCR from surviving bacteria and the resulting DNA probes labelled with a fluorophore (Sassetti and Rubin, 2003; Rengarajan *et al.*, 2005). As a control, the insertion sites adjacent to the mutants in the input pool were also amplified and labelled with a different fluorophore. Both probes were mixed and hybridized to a DNA microarray representing the entire pathogen's genome. As with STM, probes adjacent to transposon insertions in genes important for virulence were underrepresented after passage through an animal. Since the location of all genes spotted in a microarray is known, all genes important for survival in an animal can be identified within a single experiment (Fig. 2.1B).

One major advantage of STM-like approaches is the ability to query many mutants simultaneously within one experimental animal. However, in some instances, aspects unique to the biology of the pathogen under study can limit the applicability of STM. For instance, some routes of infections have natural bottlenecks, wherein relatively few individual 'founder' bacteria are able to colonize or penetrate deeper tissues even when large numbers of bacteria are used (Mecsas, 2002). Therefore, the relative abundance of individual tags is subject to stochastic variation among the founding members. This problem is particular prominent in infection models where the pathogen needs to breach an epithelial barrier (e.g. gastrointestinal infections) and only a minority of the initial pool can access deeper tissues or become systemic. This low seeding rate necessitates using less complex mutant pools for each infection and larger number of animals per screen. An additional potential disadvantage of parallel screens within one animal is that some mutants in essential virulence genes may survive selection in an animal if their functions are complemented *in trans*. For instance, a mutant deficient in toxin

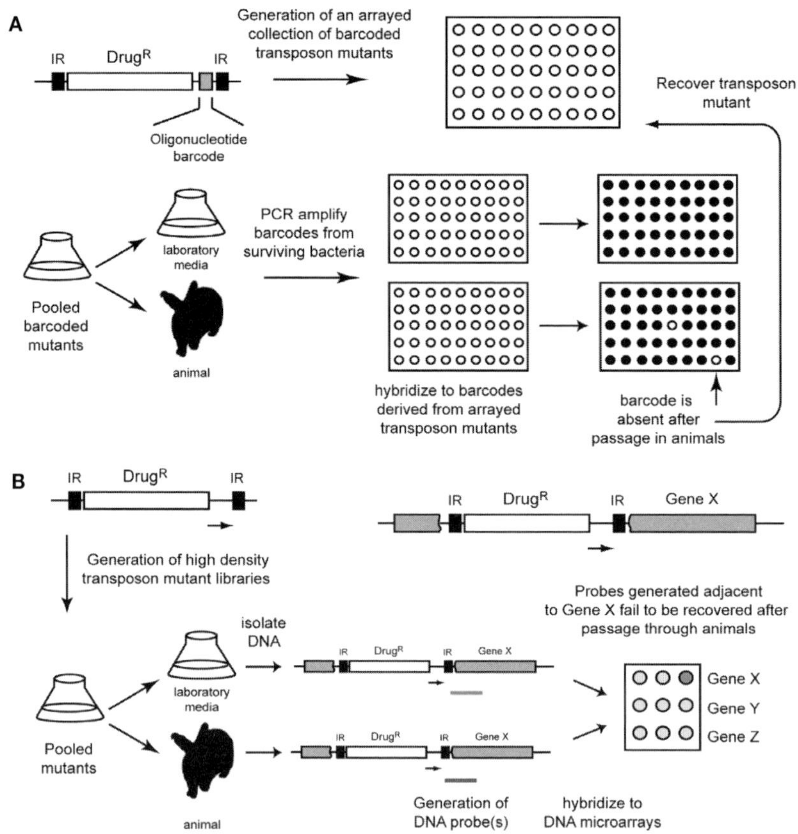

Figure 2.1 Parallel analysis of mutants with virulence defects. (A) Signature tagged mutagenesis (STM). Transposons are engineered to contain unique oligonucleotide barcodes that allow the tracking of individual mutants in a pool. After passage through an animal, surviving mutant are recovered, the transposon-associated DNA barcodes amplified and used as hybridization probes against synthetic barcodes that have been immobilized on filters. Loss of a particular mutant can be assessed by the loss of the corresponding barcode. (B) Tracking of transposon insertions with DNA microarrays. Large numbers of transposon mutants are generated and are either passaged in laboratory media or through an experimental animal. DNA immediately adjacent to the transposon insertion site is amplified by polymerase chain reaction or by *in vitro* RNA synthesis using promoters (small arrows) engineered near the transposons inverted repeats (IR), followed by treatment with reverse transcriptase. The DNA probes generated under the two experimental conditions (e.g. laboratory media vs. animal) are labelled with two different fluorophores and hybridized to a DNA microarray derived from the entire pathogens genome sequence. Genes that fail to hybridize to probes generated after passage through an animal, indicate that insertion in these genes are likely detrimental to bacterial survival in the host.

synthesis, which is expected to be cleared by the host, might survive if the secreted toxin is provided by neighbouring bacteria.

Identifying virulence factors by their selective expression in the host environment

Bacterial genes required to colonize, survive and replicate within an infected animal are preferentially expressed at the appropriate time in the host. Some simple stimuli such as increased temperature, iron deprivation or the presence of antimicrobial peptides can alert a bacterium that it has entered a host. Based on these observations, screens designed to identify genes expressed under such conditions has uncovered some key virulence factors. However, because gene expression is the result of a complex integration of many signalling events activated in response to multiple cues, it is often difficult to distil the host environment into such simple stimuli. Furthermore, the host environment is neither homogenous nor static, and different niches within an animal and changing immune status (e.g. inflammation) provide a range of stimuli that cannot be easily replicated in the laboratory.

To identify bacterial genes expressed in this complex environment, gene selection strategies have been devised wherein promoter elements are identified by fusing them to reporter genes that provide a selective advantage to the bacterium within the host. These bacteria are then screened outside the host (e.g. on a solid agar) for the absence of reporter gene activity. Genes regulated by promoter elements that are activated inside the host but not outside are termed '*in vivo*'-induced genes. In the past decade, different combinations of selectable markers, fluorescent proteins and other genetic reporters have expanded the application of promoter trap strategies to identify pathogen genes expressed within mammalian cells and organs of whole animals (Lee and Camilli, 2000).

In vivo expression technologies

To define the set of genes expressed by a pathogen upon entering an animal, Mahan *et al.* developed a promoter trap strategy termed '*in vivo* expression technology' (IVET) (Mahan *et al.*, 1993). A library of *S.* Typhimurium strains was created wherein a tandem two-reporter gene system was inserted randomly in the bacterial chromosome. The first reporter gene, *purA*, encodes an enzyme required for purine biosynthesis. The second reporter, *lacZ*, encodes an enzyme that catalyses the cleavage of lactose. Because the bacterial strain in which the promoter fusion library was constructed had a mutation in the chromosomal copy of *purA* and purine biosynthesis is necessary for survival in mice, only bacteria containing host-induced promoter fused to *purA* survived. These bacteria were harvested from the animal and tested for their ability to use lactose as a carbon source when grown on laboratory media. Those bacteria with no LacZ activity contained gene fusion that were active in the mouse but inactive on culture plates and thus were termed *in vivo*-induced (*ivi*) genes (Fig. 2.2A). This gene selection strategy led to the identification of numerous *ivi* genes in *Salmonella*, including many genes regulated by the transcriptional regulator PhoP – a central virulence factor (Heithoff *et al.*, 1999).

Recombination-based reporters: RIVET

The original IVET approach is not without pitfalls. An inherent bias in this selection strategy is that the *purA* reporter gene must be continuously active for survival in the mouse. As a result, the original IVET screen missed many of the promoters driving genes needed only during restricted times or at unique anatomical sites within an infected animal. IVET strategies involving reporter genes that are not required for growth in animals have extended the utility of this approach. In recombination-based IVET (RIVET), the promoter trap system uses a DNA resolvase as a reporter of gene activation (Camilli *et al.*, 1994). The bacterial strains bearing random promoter fusions to a resolvase gene harbour at a defined site in the chromosome an antibiotic resistance gene [e.g. tetracycline resistance pump (TetR)] flanked by DNA sequences that are recognized by the resolvase. Outside the host, if the promoter fusion is not expressed, the tetracycline resistance cassette is not excised. Therefore, growth of the library of promoter fusions in laboratory media supplemented with tetracycline selects for promoter fusions that are transcriptionally inactive. After passage through an animal, if the promoter element is activated, even transiently, the resolvase is expressed leading to the deletion of the TetR cassette (Fig. 2.2B). Strains recovered from the infected animal that have lost the antibiotic resistance marker harbour a promoter fusion that was activated within an animal.

Fluorescence-based reporters: differential fluorescence induction (DFI)

The green fluorescent protein (GFP) is a versatile reporter of gene expression that can also be used in IVET-like applications (Valdivia and Ramakrishnan, 2000). GFP is stable, requires no enzymatic substrate, and the fluorescent signal from individual GFP-expressing bacteria can be detected by various fluorescence-based detection

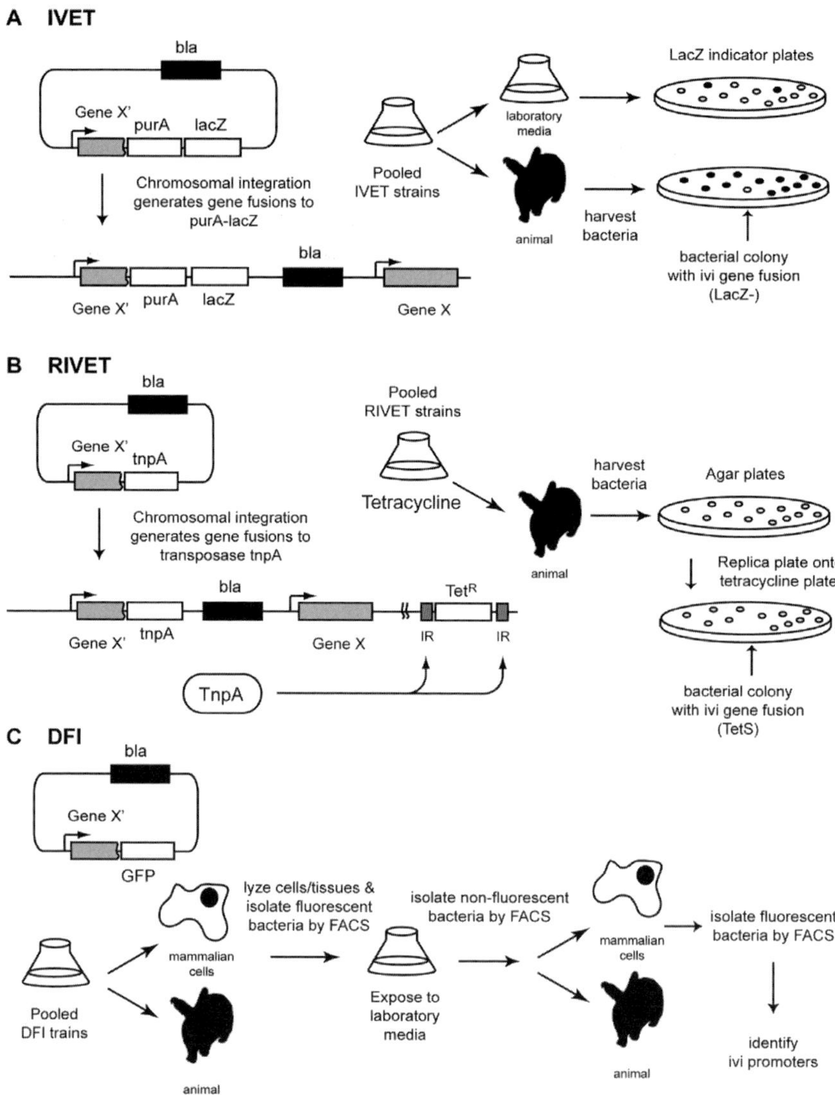

Figure 2.2 Identification of bacterial gene expressed in the host environment. (A) *In vivo* expression technologies (IVET). Strains are constructed bearing random promoter fusions to *purA-lacZ* reporter genes. Because *purA* expression is required for bacterial survival in the animal, passages of these IVET strains through an animal will select for bacteria bearing active promoter fusions. The recovered bacteria are then screened on indicator plates for LacZ activity. Most strains recovered from the animal will be LacZ$^+$, representing promoter fusions to constitutively expressed promoters. The rare LacZ$^-$ colonies represent bacteria with a promoter fusion that was expressed in the animal but not on agar plates. These are referred to as *in vivo*-induced (ivi) promoters. (B) Recombination-base IVET (RIVET). In RIVET, a transposase (TnpA) is used as a reporter of gene activity. Elsewhere in the chromosome, a tetracycline resistance cassette (TetR) is flanked by inverted repeats (IR) that are recognized of TnpA. Passage of RIVET strains in tetracycline – containing media selects for promoters that are silent outside the host. After passage through animals, any promoter activity – even transient – will lead to the expression of TnpA and the excision of the TetR cassette. Bacteria recovered from animals are screened for loss of tetracycline resistance. (C) Differential fluorescence induction (DFI). DFI uses a fluorescence-activated cell sorter (FACS) to isolate bacteria expressing different levels of the green fluorescent protein. Bacteria bearing random promoter fusion to *gfp* are passaged through cells or animals and the most fluorescent bacteria are isolated by FACS. These bacteria are then exposed to laboratory media and the least fluorescent bacteria isolated by FACS. A final passage of these bacteria through cells or animals followed by re-isolation of most fluorescent bacteria by FACS enriches for promoter elements that are preferentially activated *in vivo*.

systems including flow cytometers. Flow cytometers generate a thin stream of liquid containing single cells. After crossing a laser beam, the light scattering and fluorescence of each bacterium are recorded and individual cells can be separated and collected into pure populations (Bumann and Valdivia, 2007). In this manner, highly fluorescent bacteria can be separated from those with intermediate fluorescence or no fluorescence. Because the signal intensity does not affect the efficiency of cell sorting, bacteria bearing weak and strong promoters fused to *gfp* can be collected with similar efficiencies. This feature bypasses a problem with other IVET reporter genes where the strength of *ivi* promoters may influence the efficiency of *in vivo* selection.

The promoter trap methodology using GFP is known as differential fluorescence induction (DFI) (Valdivia and Falkow, 1997). In DFI, bacteria bearing promoters of various strengths fused to *gfp* are isolated from a library by flow sorting. In the original DFI methodology, macrophages were infected with a library of *Salmonella* bearing *gfp* promoter fusions and all cells that became fluorescent were collected. The bacteria within these cells were harvested, grown in tissue culture media, and the least fluorescent bacteria were collected again by flow cytometry. These low to non-fluorescent bacteria were used to re-infect macrophages and the infected cells with the highest fluorescence were sorted. The bacteria recovered after this last sort had promoter fusions that were preferentially activated within macrophages (Fig. 2.2C). This general strategy to isolate differentially expressed promoters by flow cytometry has been extended to sorting bacteria directly from infected tissues. For example, flow sorting of mycobacteria from infected animal tissues allowed the identification of gene expressed specifically in granulomas, a site of active mycobacterial replication (Ramakrishnan *et al.*, 2000).

Microarray-based transcriptional profiling of pathogen gene expression

The expression of pathogen genes can be assessed directly by monitoring mRNA abundance. This has been traditionally assessed by hybridization of labelled gene-specific DNA probes to total RNA (Northern hybridization) or by reverse transcriptase treatment of total RNA followed by PCR amplification with gene specific primers. Nowadays, it is increasingly common to monitor virtually all bacterial mRNA transcripts by hybridization to DNA microarrays (Harrington *et al.*, 2000). In a typical scenario, the investigator would harvest total RNA from bacteria grown under experimental condition 'A' or a control condition 'B' and synthesize cDNA with reverse transcriptase. The cDNA derived from the two experimental conditions are labelled with different fluorescent dyes, mixed and hybridized to DNA microarrays representing the entire pathogen's genome. The spots that hybridize more intensely to the labelled cDNAs under experimental condition A but not the control condition B, indicate that the genes represented by those spots are preferentially expressed under condition A. As with the genetic reporters described in the sections above, microarray-based transcriptional profiling can be used to identify genes transcribed under conditions mimicking the host environment (e.g. in the presence of blood serum or acidic pH of the stomach). Ideally, bacterial RNAs isolated directly from infected animals would be analysed. However, this is difficult given the overabundance of host-derived RNA and the technical challenges in isolating sufficient quantities of high quality bacterial mRNA from infected animal tissues. As a result, successful microarray-based identification of host-induced genes has been limited to simpler infection systems such as tissue culture cells (Waddell *et al.*, 2007).

With the advent of low-cost deep DNA sequencing technologies, identification of differentially expressed RNA transcripts will be increasingly feasible by sequencing total bacterial cDNAs isolated under different experimental conditions (Coppée, 2008). This approach has the advantage of identifying low abundance transcripts, alternative transcriptional start sites, non-coding RNAs – all of which can provide unique insights into the regulation of gene expression under host-like conditions or within the host environment itself. Furthermore, it also bypasses the need to synthesize and assemble pathogen-specific DNA microarrays. Similarly, microarray-based tracking of transposon mutants will probably be substituted by direct sequencing of sites adjacent to the transposon insertions (van Opijnen *et al.*, 2009).

All the approaches described above have significantly advanced our understanding of how pathogens sense and respond to the host environment. However, there are some important issues that need to be taken into consideration when interpreting findings from IVET-like screens and transcriptional profiling. Both genetic and biochemical experimental approaches to identify host-induced genes are largely framed by what we define as a 'non-host' environment. For instance, if one identifies a promoter that is active in the host, the next step is to check that the promoter is not expressed in laboratory media so as to exclude constitutively active 'house keeping' genes. But, if the counter-screen for *in vitro*-activated promoters is performed at 37°C, temperature-regulated promoters will no longer be considered to be '*in vivo*'-induced promoters. In the case of the original IVET screen in *Salmonella*, if the counter screens had been performed in Mg^{++} poor media, the entire PhoP regulon – which is induced under low Mg^{++} concentrations – would not have been classified as *ivi* genes.

Nonetheless, the concepts introduced by IVET fundamentally changed how investigators began to study the dynamics of pathogen gene expression within live animals and there has been a growing interest in what types of genes are activated inside the host, regardless of their behaviour in laboratory growth media. The genes identified by IVET and IVET-like strategies have provided unique insights into the ecology of the host environment, the demands it places upon bacterial metabolism and how pathogenic bacteria adapt (Mahan *et al.*, 2000).

Function-based approaches to identify effector proteins

A common theme among bacterial pathogens is the presence of specialized secretion systems (Types III and IV) that deliver bacterial proteins directly to host membranes and cytoplasm (Coburn *et al.*, 2007; Llosa *et al.*, 2009). These secreted 'effector' proteins are responsible for the manipulation of many host cellular functions. Genes encoding effectors are often linked to genes encoding secretion apparatus components or present in pathogenicity islands. In addition, effectors can sometimes be identified based on their homology to eukaryotic proteins known to regulate cellular functions. However, a large number of effectors remain to be identified. Current efforts at identifying effectors have focused on the functional attributes of these proteins.

Identification of effector proteins by their secretion signals

Targets of Type III (T3) and IV (T4) secretion contain minimal domains at their amino and carboxy termini, respectively, that can direct the translocation of these proteins into mammalian cells (Cambronne and Roy, 2006). Therefore, reporter proteins fused to the amino or carboxyl termini of pathogen proteins of unknown function can be used to monitor whether these proteins are potential transport substrates of these specialized secretion systems. In this manner, a 'signal sequence' trap approach can be used to identify proteins containing T3 or T4 secretion signals (Fig. 2.3A). The system is conceptually analogous to promoter traps. A library of bacteria expressing random translational gene fusions to a selectable marker or a reporter protein (e.g. β-lactamase), and the activity of secreted proteins can be monitored in bacterial mutants defective for T3 or T4 secretion systems to confirm that they are potential targets of these systems. Fusions to β-lactamase and fluorigenic lactamase substrates have been particularly useful to monitor the temporal hierarchy of translocation of effector proteins during infection (Mills *et al.*, 2008). Another widely used reporter is the calmodulin-dependent *Bordetella pertussis* adenylate cyclase. Because this enzyme is only active in the host cytoplasm, infection of mammalian cells with bacteria expressing putative T3 or T4 secreted proteins fused to adenylate cyclase will generate cAMP (the end product of the enzymatic reaction) if the fusion protein was secreted and translocated across the eukaryotic membranes (Nagai *et al.*, 2005).

Since the recognition of T3 and T4 secretion signals is relatively evolutionarily conserved, secretion substrates from one bacterial species can sometimes be recognized by the secretion system of a different species. In this manner, pathogens

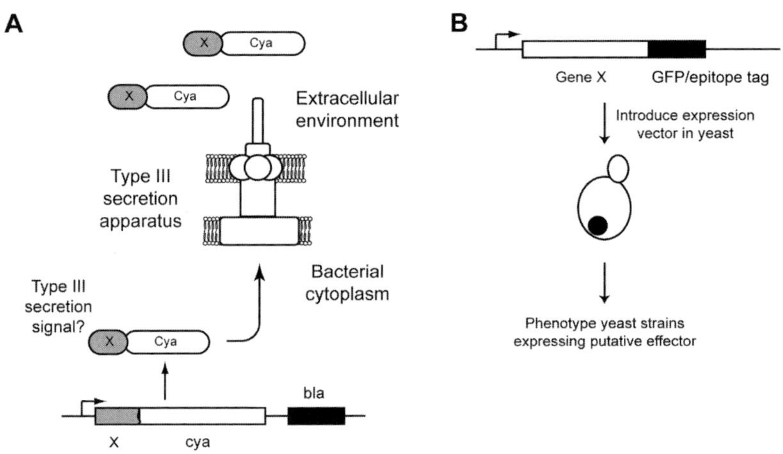

Figure 2.3 Functional approaches to identify bacterial modulators of host cellular functions. (A) Identification of secreted virulence proteins (effectors). Fusion to reporter proteins (e.g. adenylate cyclase – Cya) with activities that are dependent on export from the bacterial cell can be used to assess whether bacterial proteins contain virulence-associated secretion signals (e.g. Type III or Type IV secretion systems. (B) Phenotypic characterization of putative effector proteins expressed in yeast. Inducible expression of putative effector proteins in yeast or other model eukaryotes can be used to determine whether these proteins interfere with basic eukaryotic cellular processes. Addition of GFP or other epitope tags allows for the monitoring effector protein expression by immunoblots and also provides potentially important clues as to the effector protein's tropism within a eukaryotic cell.

with well characterized T3 and T4 secretion systems can be used to identify potential secretion substrates in other bacterial species. For example, putative targets of T3 and T4 secretion from genetically intractable pathogens such as *Chlamydia* and *Coxiella* have been identified by their ability to secrete reporter proteins in *Shigella* and *Legionella*, respectively (Subtil *et al.*, 2005; Pan *et al.*, 2008). As more T3 or T4 secretion substrates are being identified, increasingly sophisticated algorithms have been developed to predict bioinformatically whether a protein is a likely target of one of these secretion systems (Arnold *et al.*, 2009). However, experimental validation of many of these new targets of secretion is still lacking.

Identification of effector proteins by expression in eukaryotic cells

Effector proteins often have defined activities that allow them to modify or interrupt host cellular pathways. Therefore, ectopic expression of these proteins in eukaryotic cells, followed by a phenotypic characterization of the resulting cells, can provide important clues as to their function. For example, tropism of an effector protein expressed in a mammalian cell to distinct subcellular structures (e.g. plasma membrane, endosomes, nucleus) or gross morphological alterations and cytotoxicity associated with effector protein expression can provide insights as to their targets in eukaryotic cells. These activities can also be used as the basis for screens designed specifically to identify effectors (Fig. 2.3B). Since the mechanisms governing membrane transport, cytoskeletal dynamics and signalling – common targets of bacterial pathogens – are relatively conserved among eukaryotic cells, model eukaryotes like the yeast *Saccharomyces cerevisiae*, have been useful to screen for bacterial proteins with activities that interfere with eukaryotic cellular processes (Siggers and Lesser, 2008). In addition, whole cell-based phenotypic assays to monitor virtually all aspects of yeast signalling and cell biology permits the design of screens to identify factors involved in one particular facet of the host pathogen interaction. For example, phenotypic assays that monitor cytoskeletal function or the disruption of protein transport to endosomal compartments in yeast has been adapted to identify *Legionella* effectors that specifically disrupt this arm of vesicle transport (Heidtman *et al.*, 2009).

Identification of virulence factors by comparative genomics

In comparative genomics, biological function is inferred by comparing DNA sequences at the whole genome level. For example, a comparison of the genome sequences of pathogens to those of closely related species that are non-pathogenic or of pathogens that cause disease by different mechanisms, can narrow the compendium of genes that may be responsible for a pathogenic trait (Raskin *et al.*, 2006). Similarly, changes in promoter sequences can help identify instances where gene circuitry has bee re-wired to adapt to new environmental challenges (e.g. the host).

Comparative genomics between commensal and pathogenic enteric bacteria have revealed an abundance of clusters of genes in virulent strains ('pathogenicity islands') that are acquired in bulk as part of horizontal gene transfer events and that enhance the pathogens' virulence properties. These genome-scale comparisons are often performed by microarray-assisted comparative genome hybridization (CGH) (Sturdevant *et al.*, 2010) (Fig. 2.4). In CGH, total DNA from the query bacterial strains is labelled with a fluorescent dye and hybridized to an ordered DNA microarray made from the sequenced strain (see gene expression profiling by microarrays). Positions in the array that provide a positive fluorescence signal indicate a gene that is conserved between strains, while those that fail to give a signal indicate a gene that is either absent from the strain or so divergent that it cannot hybridize. This approach to studying pathogens is widespread, as it can be used to broadly survey genomic variation in bacterial species or groups of closely related species even when only one representative has been sequenced. Unfortunately, these technologies do not have the resolving capacity to determine differences at the nucleotide levels or identify new genes present in the query strain but not the reference one. With the advent of new deep DNA sequencing technologies, the generation of 'draught' genomes of new bacterial species and isolates has become routine and cost effective. As a result, comparative genome analysis in the near future will increasingly rely on *de novo* sequencing of the strains to be compared (Sturdevant *et al.*, 2010).

Complex traits are the result of the interplay of multiple genetic loci and the function of a particular allele of a gene can be heavily influenced by its genetic context. Therefore, the mere absence or presence of a gene may be insufficient to define the virulence properties of a pathogen. Rather, it is likely that many genes that are common to both pathogenic and non-pathogenic species are present as a combination of distinct alleles that can significantly modulate bacterial virulence. Similarly, the rewiring of genetic circuitry by promoter

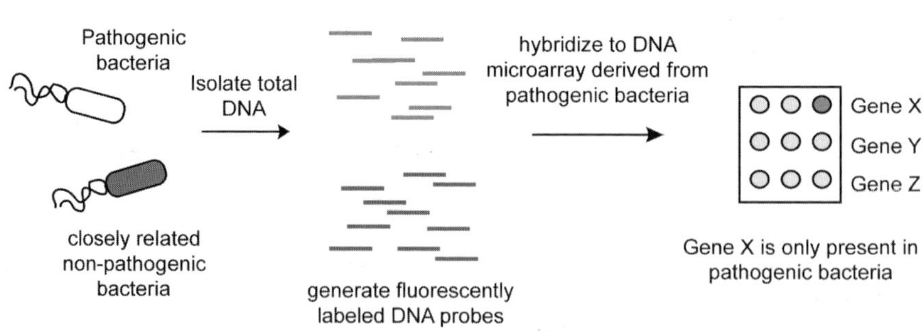

Figure 2.4 Comparative genome hybridization. The genomes of closely related bacterial species can be compared by the use of DNA microarrays. DNA probes derived from total DNA isolated from pathogenic and a closely related non-pathogenic species are labelled with different fluorophores and hybridized to DNA microarrays derived from the pathogenic strain. Absence of specific genes in the non-pathogenic strains may point to potential virulence determinants.

alterations can alter the transcriptional outcome upon exposure to the host environment and turn a commensal into a potential pathogen.

Conclusion

The preceding sections detailed a variety of molecular, genetics and genomic approaches to identify bacterial virulence genes. The success of any one approach is limited by the particular challenges inherent to each system. For example, genetic approaches have been very successful in pathogens amenable to routine genetic analysis. In contrast, molecular and comparative genomic approaches have been more fruitful in less genetically tractable pathogens. In general, a combination of approaches is recommended, as different techniques can yield unique sets of virulence factors. Ultimately, the technique used to identify virulence factors is only as good as the assay used to measure virulence. Appropriate cellular and animal models that faithfully reproduce the constraints placed upon an invading pathogen by a human host are crucial for the proper identification of pathogenicity determinants and their relative contribution to the aetiology of microbial disease.

References

Arnold, R., Brandmaier, S., Kleine, F., Tischler, P., Heinz, E., Behrens, S., Niinikoski, A., Mewes, H.W., Horn, M., and Rattei, T. (2009). Sequence-based prediction of type III secreted proteins. PLoS Pathog. *5*, e1000376.

Baldwin, D.N., and Salama, N.R. (2007). Using genomic microarrays to study insertional/transposon mutant libraries. Methods Enzymol. *421*, 90–110.

Bennett, P.M. (2004). Genome plasticity: insertion sequence elements, transposons and integrons, and DNA rearrangement. Methods Mol. Biol. *266*, 71–113.

Bumann, D., and Valdivia, R.H. (2007). Identification of host-induced pathogen genes by differential fluorescence induction reporter systems. Nat. Protoc. *2*, 770–777.

Cambronne, E.D., and Roy, C.R. (2006). Recognition and delivery of effector proteins into eukaryotic cells by bacterial secretion systems. Traffic *7*, 929–939.

Camilli, A., Beattie, D.T., and Mekalanos, J.J. (1994). Use of genetic recombination as a reporter of gene expression. Proc. Natl. Acad. Sci. U.S.A. *91*, 2634–2638.

Coburn, B., Sekirov, I., and Finlay, B.B. (2007). Type III secretion systems and disease. Clin. Microbiol. Rev. *20*, 535–549.

Coppée, J.Y. (2008). Do DNA microarrays have their future behind them? Microbes Infect. *10*, 1067–1071.

Dorer, M.S., and Isberg, R.R. (2006). Non-vertebrate hosts in the analysis of host–pathogen interactions. Microbes Infect. *8*, 1637–1646.

Harrington, C.A., Rosenow, C., and Retief, J. (2000). Monitoring gene expression using DNA microarrays. Curr. Opin. Microbiol. *3*, 285–291.

Heidtman, M., Chen, E.J., Moy, M.Y., and Isberg, R.R. (2009). Large-scale identification of *Legionella pneumophila* Dot/Icm substrates that modulate host cell vesicle trafficking pathways. Cell. Microbiol. *11*, 230–248.

Heithoff, D.M., Conner, C.P., Hentschel, U., Govantes, F., Hanna, P.C., and Mahan, M.J. (1999). Coordinate intracellular expression of *Salmonella* genes induced during infection. J. Bacteriol. *181*, 799–807.

Hensel, M., Shea, J.E., Gleeson, C., Jones, M.D., Dalton, E., and Holden, D.W. (1995). Simultaneous identification of bacterial virulence genes by negative selection. Science *269*, 400–403.

Hensel, M., Shea, J.E., Raupach, B., Monack, D., Falkow, S., Gleeson, C., Kubo, T., and Holden, D.W. (1997). Functional analysis of *ssaJ* and the *ssaK/U* operon, 13 genes encoding components of the type III secretion apparatus of *Salmonella* Pathogenicity Island 2. Mol. Microbiol. *24*, 155–167.

Irazoqui, J.E., Urbach, J.M., and Ausubel, F.M. (2010). Evolution of host innate defence: insights from *Caenorhabditis elegans* and primitive invertebrates. Nat. Rev. Immunol. *10*, 47–58.

Lee, S.H., and Camilli, A. (2000). Novel approaches to monitor bacterial gene expression in infected tissue and host. Curr. Opin. Microbiol. *3*, 97–101.

Llosa, M., Roy, C., and Dehio, C. (2009). Bacterial type IV secretion systems in human disease. Mol. Microbiol. *73*, 141–151.

Mahan, M.J., Slauch, J.M., and Mekalanos, J.J. (1993). Selection of bacterial virulence genes that are specifically induced in host tissues. Science *259*, 686–688.

Mahan, M.J., Heithoff, D.M., Sinsheimer, R.L., and Low, D.A. (2000). Assessment of bacterial pathogenesis by analysis of gene expression in the host. Annu. Rev. Genet. *34*, 139–164.

Maloy, S.R. (2007). Use of antibiotic-resistant transposons for mutagenesis. Methods Enzymol. *421*, 11–17.

Mazurkiewicz, P., Tang, C.M., Boone, C., and Holden, D.W. (2006). Signature-tagged mutagenesis: barcoding mutants for genome-wide screens. Nat. Rev. Genet. *7*, 929–939.

Mecsas, J. (2002). Use of signature-tagged mutagenesis in pathogenesis studies. Curr. Opin. Microbiol. *5*, 33–37.

Mills, E., Baruch, K., Charpentier, X., Kobi, S., and Rosenshine, I. (2008). Real-time analysis of effector translocation by the type III secretion system of enteropathogenic *Escherichia coli*. Cell Host Microbe *3*, 104–113.

Mylonakis, E., and Aballay, A. (2005). Worms and flies as genetically tractable animal models to study host–pathogen interactions. Infect. Immun. *73*, 3833–3841.

Mylonakis, E., Casadevall, A., and Ausubel, F.M. (2007). Exploiting amoeboid and non-vertebrate animal model systems to study the virulence of human pathogenic fungi. PLoS Pathog. *3*, e101.

Nagai, H., Cambronne, E.D., Kagan, J.C., Amor, J.C., Kahn, R.A., and Roy, C.R. (2005). A C-terminal translocation signal required for Dot/Icm-dependent delivery of the *Legionella* RalF protein to host cells. Proc. Natl. Acad. Sci. U.S.A. *102*, 826–831.

van Opijnen, T., Bodi, K.L., and Camilli, A. (2009). Tn-seq: high-throughput parallel sequencing for fitness and genetic interaction studies in microorganisms. Nat. Methods *6*, 767–772.

Pan, X., Lührmann, A., Satoh, A., Laskowski-Arce, M.A., and Roy, C.R. (2008). Ankyrin repeat proteins comprise a diverse family of bacterial type IV effectors. Science *320*, 1651–1654.

Ramakrishnan, L., Federspiel, N.A., and Falkow, S. (2000). Granuloma-specific expression of *Mycobacterium* virulence proteins from the glycine-rich PE-PGRS family. Science *288*, 1436–1439.

Raskin, D.M., Seshadri, R., Pukatzki, S.U., and Mekalanos, J.J. (2006). Bacterial genomics and pathogen evolution. Cell *124*, 703–714.

Rengarajan, J., Bloom, B.R., and Rubin, E.J. (2005). Genome-wide requirements for *Mycobacterium tuberculosis* adaptation and survival in macrophages. Proc. Natl. Acad. Sci. U.S.A. *102*, 8327–8332.

Salama, N.R., Shepherd, B., and Falkow, S. (2004). Global transposon mutagenesis and essential gene analysis of *Helicobacter pylori*. J. Bacteriol. *186*, 7926–7935.

Sassetti, C.M., and Rubin, E.J. (2003). Genetic requirements for mycobacterial survival during infection. Proc. Natl. Acad. Sci. U.S.A. *100*, 12989–12994.

Siggers, K.A., and Lesser, C.F. (2008). The Yeast *Saccharomyces cerevisiae*: a versatile model system for the identification and characterization of bacterial virulence proteins. Cell Host Microbe *4*, 8–15.

Sturdevant, D.E., Virtaneva, K., Martens, C., Bozinov, D., Ogundare, O., Castro, N., Kanakabandi, K., Beare, P.A., Omsland, A., Carlson, J.H., *et al.* (2010). Host–microbe interaction systems biology: lifecycle transcriptomics and comparative genomics. Future Microbiol. *5*, 205–219.

Subtil, A., Delevoye, C., Balaña, M.E., Tastevin, L., Perrinet, S., and Dautry-Varsat, A. (2005). A directed screen for chlamydial proteins secreted by a type III mechanism identifies a translocated protein and numerous other new candidates. Mol. Microbiol. *56*, 1636–1647.

Valdivia, R.H., and Falkow, S. (1997). Fluorescence-based isolation of bacterial genes expressed within host cells. Science *277*, 2007–2011.

Valdivia, R.H., and Ramakrishnan, L. (2000). Applications of gene fusions to green fluorescent protein and flow cytometry to the study of bacterial gene expression in host cells. Methods Enzymol. *326*, 47–73.

Waddell, S.J., Butcher, P.D., and Stoker, N.G. (2007). RNA profiling in host–pathogen interactions. Curr. Opin. Microbiol. *10*, 297–302.

Genetic Determinants of Bacterial Pathogenicity

3

Gavin K. Paterson and Duncan J. Maskell

Abstract

Determinants of bacterial pathogenicity are encoded by different types of genetic element. Of particular note are large loci including pathogenicity islands, bacteriophages, integrative and conjugative elements, plasmids and integrons. Gene loss can also play a significant role in determining bacterial pathogenicity and should not be disregarded. Here we discuss these main genetic determinants of bacterial pathogenicity and how they influence the behaviour of important human and veterinary pathogens. The chapter illustrates the vast diversity and adaptability of bacterial pathogens, features that will ensure their continued medical and veterinary importance for quite some time to come.

Pathogenicity and bacterial–host interactions

Bacterial pathogenicity is frequently defined as the ability of a bacterium to cause damage to a host. The medical and veterinary consequences of such an outcome understandably attract much attention. However, the vast majority of bacterial–host interactions do not result in overt disease and host damage. For many pathogenic bacteria, asymptomatic colonization and/or clearance without overt illness are the most common outcomes of interaction with a host. Typically, disease occurs relatively rarely, usually as a result of specific interactions between both host and bacterial factors. These factors are often variably expressed, and may vary in primary sequence between different strains of the same bacterium, or individuals of the same host species. This makes each encounter potentially unique, even between the same host and bacterium, producing a spectrum of possible outcomes, from immediate clearance of the bacterium, to long-term asymptomatic carriage, to host death, with potential for every nuance in between (Fig. 3.1).

This scenario includes even some of the most prevalent and medically important bacterial pathogens, such as *Streptococcus pneumoniae, Neisseria meningitidis, Staphylococcus aureus* and *Enterococcus faecalis* to list but a few. Even though these bacteria are responsible for significant burdens of death and disease, such outcomes are rare compared with exposure and colonization rates. Very often, asymptomatic carriage is a pre-requisite to disease, with the genetic elements that promote bacterial fitness and/or transmission during carriage in the absence of disease often being the same elements that mediate pathogenicity in a disease setting. It is likely that, in most cases, it is during colonization, rather than disease, that horizontal acquisition of genetic determinants of pathogenicity occurs and where selection pressures act. For these reasons, we do not restrict our discussion in this chapter to disease settings and pathogenicity per se and prefer to consider the genetic determinants of pathogenicity in the context of the entire life/transmission cycle of bacterial pathogens. Although we use the term determinants of pathogenicity, in the context of carriage this might be better considered to be determinants of bacterial fitness.

Genetic determinants of pathogenicity may be found as single genes or arranged in large loci,

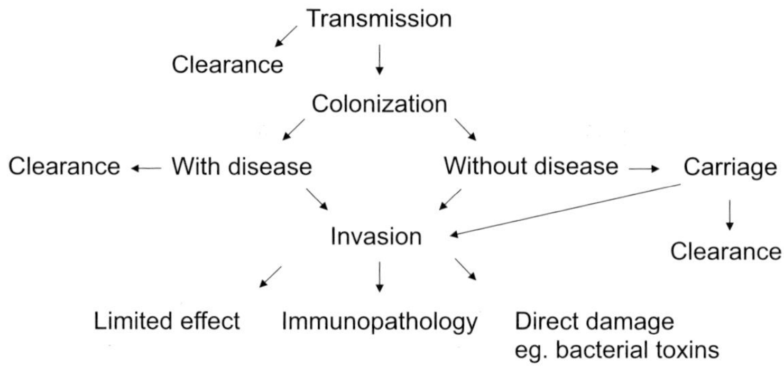

Figure 3.1 Outcomes of bacterial–host interactions.

often as operons. Alternatively they may be organized as regulons: unlinked genes regulated under the same conditions to allow a global response to stimuli. Genetic determinants of pathogenicity may be arranged in large and complex loci such as those encoding the genes for production of capsular polysaccharides or lipopolysaccharides, or large multi-protein assemblies such as type three secretion systems. These large loci may show features suggestive of horizontal gene transfer leading them to be classed as pathogenicity islands (or islets in the case of smaller regions). Other horizontally acquired features such as plasmids, bacteriophages, integrons and integrative and conjugative elements (ICE) can all carry important determinants of pathogenicity that influence the behaviour of their host bacterium. Various examples of genetic determinants of pathogenicity (or fitness) encoded by these different genetic features will be discussed here. It will become clear that the genetic determinants of bacterial pathogenicity are encoded by many different types of genetic element and that these elements contribute to numerous and diverse aspects of bacterial–host interactions. Often, the same function may be encoded by different types of element or, vice versa, different functions may be derived from the same type of element. Genetic diversity, in terms of individual gene sequences, total gene content per isolate and distribution of genes between strains, is a common feature of these elements, with a major role for horizontal gene transfer. This genetic diversity is becoming more apparent as whole genome sequencing expands but may only represent the 'tip of the iceberg' at present. Very often, the origins of horizontally acquired elements are obscure, but this may also become clearer as genome sequencing expands, especially with metagenomic approaches. Likewise these elements are often dynamic, able to move between species or strains, and to mutate or recombine to generate novel sequences. Finally, genome decay should not be ignored and can also have a significant effect on bacterial pathogenicity and fitness, especially in the context of host specificity. Below we discuss various classes of genetic element using specific examples from different bacterial pathogens.

Large loci encoding genetic determinants of pathogenicity

Pathogenicity genes are often organized into large loci. Genes within these loci usually contribute different elements to the biosynthesis of a structure that is required for successful host–pathogen interaction. Keeping the genes together in a single locus is probably the evolutionary consequence of needing to maintain linkage of the genes required to make a structure, given that if one of them is inappropriately expressed or lost by chromosomal rearrangement, then the whole structure is not made. Within large loci, there are often regions that are responsible for separate tasks in the biosynthesis of the structure. So for example, in loci required for the biosynthesis of a polysaccharide capsule there are often regions individually responsible for biosynthesis of the sugar polymer itself, for export of the polysaccharide and for assembly of the polymer in the correct form on the bacterial cell surface. Other large loci encoding pathogenicity genes include loci for the biosynthesis of lipopolysaccharide (LPS), and loci encoding the machinery for assembly of complex multi-protein structures on the cell surface, such as fimbriae, flagella and Type III secretion systems.

Pneumococcal capsule (*cps*) loci

The polysaccharide capsule of *S. pneumoniae* (the pneumococcus) is critical for the full expression of pathogenicity by the bacterium. Its main function within the host is to protect the pneumococcus from opsonophagocytosis. Almost all disease isolates express capsule, and capsule mutants are avirulent in animal models (Watson and Musher, 1990; Kerr *et al.*, 2005). It was this essential role for the capsule in virulence that allowed the classic experiments of Griffith to demonstrate the phenomenon of bacterial transformation (Griffith, 1928) and Avery *et al.* to identify the transforming principle as DNA (Avery *et al.*, 1944). Although the pneumococcus is an important pathogen, it is far more commonly encountered being carried asymptomatically in the human nasopharynx. The capsule contributes to the maintenance of carriage by preventing entrapment and clearance of bacteria in the luminal mucus (Nelson *et al.*, 2007b). The capsule is antigenically diverse and this forms the basis for pneumococcal serotyping with ninety-one serotypes currently recognized. This antigenic variability probably arose as a result of selective pressure from the human immune system, with antibodies against a given serotype offering protection against subsequent colonization with that same serotype, but little protection against other serotypes (Granat *et al.*, 2009). The chemical structures of more than half of these polysaccharides have been determined, most being complex, composed of various sugars, linkages and side-chains, while a few are relatively simple, consisting of only one or two sugars (Vandam *et al.*, 1990). With the exception of serotypes 3 and 37, which are synthesized via the synthase pathway, pneumococcal capsules are synthesized by the Wzx/Wzy-dependent pathway. The entire capsule (*cps*) locus has been sequenced for the all serotypes recognized at the time, allowing a complete picture of the genetics, evolution, biochemistry and immunology of pneumococcal capsule to be drawn (Bentley *et al.*, 2006; Aanensen *et al.*, 2007; Mavroidi *et al.*, 2007). This and earlier work on the sequence of a few selected capsule loci, shows that despite the genetic diversity in this region, the *cps* loci of different serotypes typically exhibit a similar overall organization. Always found on the same site in the genome, between *dexB* and *aliA* and in the same orientation as these flanking genes, the *cps* loci vary greatly in size from 10,337 base pairs (bp) in the case of serotype 3 up to 30,298 for serotype 38. The 5′-end of each *cps* locus is characterized by the regulatory and processing genes common to and conserved between serotypes: *cpsABCD* (also known as *wzg*, *wzh*, *wzd* and *wze*). Next to these are the type-specific genes encoding the various glycosyltransferases, polymerases, transporters and other enzymes required to synthesize a given capsule.

Sequencing of the *cps* loci has offered insight into the mechanisms responsible for the antigenic diversity displayed by pneumococcal capsular polysaccharides. These include divergence of an ancestral serotype into two related serotypes through accumulation of point mutations or the insertion/deletion of a single gene. The most common mechanism however appears to have been multiple episodes of horizontal transfer of genes encoding capsule biosynthesis enzymes into the pneumococcus from different and unknown sources. This supposition is supported by the high diversity of apparently non-homologous biosynthesis enzymes together with the low-percentage G + C content of their genes compared with the rest of the pneumococcal genome. Interestingly, the reliance of serological typing on immunodominant epitopes, to the exclusion of other capsule structural features, means that antigenically related serotypes can show little genetic similarity while conversely, apparently unrelated serotypes can share more similarity at the level of *cps* locus sequences than would be expected based on serology. The variety of biosynthesis enzymes in the *cps* loci offers the potential opportunity for transformation and recombination to continually generate new serotypes. However, sequencing of the *cps* loci offered no evidence for chimera serotypes arising in this manner. Limited homology between serotype-specific genes may limit the opportunity for such events to occur. Furthermore novel precursor molecules generated by such events may not be recognized by the resident enzymes and transporters causing abortive capsule production, which would be likely to lead to clearance and extinction of the strain.

Electron microscopy has revealed that during intimate contact and invasion into host cells, originally encapsulated pneumococci become devoid of capsule (Hammerschmidt *et al.*, 2005). The capsule is known to impede such interactions and presumably this down-regulation of capsule production acts to promote adherence and invasion by exposing bacterial adhesins. However, the signals and mechanisms underlying this reversible alteration in capsule expression are yet to be elucidated. The pneumococcus also undergoes reversible phase variation manifesting in opaque and transparent colonies, which involves differences in capsule level. These opacity variants behave very differently, with transparent variants showing increased adherence to host cells and enhanced colonization in infant rats (Weiser *et al.*, 1994; Cundell *et al.* 1995). In contrast, transparent variants are more resistant to opsonophagocytosis and more virulent in systemic infections (Cundell *et al.*, 1995; Kim and Weiser, 1998; Kim *et al.*, 1999). In patients with invasive disease, paired isolates had a predominately transparent colony phenotype when cultured from the nasopharynx, and an opaque colony phenotype when cultured from the blood. The capsule appears important in these differences, with transparent variants expressing increased levels compared with non-transparent variants (Kim and Weiser, 1998; Weiser *et al.*, 2001). This difference was associated with decreased tyrosine phosphorylation of CpsD (Weiser *et al.*, 2001), a tyrosine kinase encoded within the *cps* locus and a known regulator of capsule synthesis (Morona *et al.*, 2000). However, the full molecular basis of opacity variation is not clear (Saluja and Weiser, 1995) and other differences between the variants are also likely to contribute to their phenotypes (Overweg *et al.*, 2000; King *et al.*, 2004).

Pathogenicity islands

Pathogenicity islands (PAIs) are horizontally acquired elements found among many different Gram-negative and Gram-positive human, animal or plant pathogens. They encode virulence or fitness factors, which are often absent among closely related non-pathogenic strains and species. Although diverse in structure and function, PAIs frequently share a number of common features (Table 3.1). As noted above, they encode one or more virulence-related genes of diverse function and often they cover a large region of the bacterial chromosome, from 10 to 100 kilobases (kb). Smaller regions with the features of PAIs are often referred to as pathogenicity islets. Their acquisition by horizontal gene transfer is suggested by the fact that they often have a codon usage and G + C content that is different from the core genome, with the original source of the PAI DNA usually unknown. Likewise, the mobility of these regions is suggested by the presence of insertion sequences, direct repeats, and integrase and resolvase genes, which may be fully functional, or may be non-functional, in which case the PAI appears to be stably inherited as part of the genome. Nonetheless, even when inactivated, such features may mediate homologous recombination within the PAI allowing for intra-strain variation in the gene content of the same PAI. In other cases, for example in the PAIs encoding P-fimbriae and certain haemolysins in *Escherichia coli*, integration may be unstable and the PAI can

Table 3.1 Features of an ideal bacterial Pathogenicity Island (PAI)

Encompasses a large region of chromosome
Encodes known or putative virulence or fitness factors
Absent in related non-pathogenic strains and species
Evident of acquisition by horizontal gene transfer
Associated with mobile genetic elements (which may or may not be functional)
Integrated at site of tRNA gene

be excised and lost (Hacker *et al.*, 1990). Another feature shared among PAIs is that they are frequently integrated into the host chromosome at the site of tRNA genes. The 3′ ends of these genes often match well to the preferred integration sites of many bacteriophages and plasmids and this has led to the proposal that they serve as general points of integration of foreign DNA. The conserved sequences of these genes between bacteria may have been a driving force for such a role. However, these PAI characteristics are not stringent and actually very few PAIs fulfil all the features discussed.

There are numerous examples of PAIs playing a key role in bacterial pathogenesis and only two of many are discussed here.

Salmonella pathogenicity islands 1 and 2 (SPI-1 and SPI-2)

Salmonella enterica is a diverse Gram-negative bacterial species consisting of six subspecies which are further divided into >2400 serovars. Serovars have different host ranges and cause different clinical outcomes. Of particular note among the various subspecies and serovars are *S. enterica* subspecies *enterica* serovars Typhi, Typhimurium and Enteritidis (*S.* Typhi, *S.* Typhimurium and *S.* Enteritidis, respectively). *S.* Typhi is the causative agent of typhoid fever in humans, a disease with an estimated annual incidence of >20 million cases with 200,000 deaths (Crump *et al.*, 2004) while *S.* Typhimurium and *S.* Enteritidis are prevalent causes of gastroenteritis from food poisoning in humans. In addition *S.* Typhimurium, *S.* Enteritidis and other non-typhoidal *Salmonella* serovars can cause serious invasive disease in immunocompromised human hosts. Non-typhoidal salmonellae cause a wide range of other diseases in non-human hosts, and can also be carried apparently asymptomatically. *S.* Typhimurium produces a systemic infection in mice with similarities to human typhoid, which, owing to the narrow host specificity of *S.* Typhi makes this the main tractable model to study typhoid. *S. enterica* contains several pathogenicity islands of which *Salmonella* pathogenicity islands (SPI)-1 and -2 are the best characterized, with both playing critical roles in the biology of this bacterium.

SPI-1 was first identified in a transposon mutagenesis screen for mutants deficient for invasion of host cells *in vitro* (Galán and Curtiss, 1989). It covers a region of ~40 kb inserted between two genes, *mutS* and *fhlA*, that are consecutive in *E. coli* K12 (Mills *et al.*, 1995). It is not associated with tRNA genes nor does it carry sequences linked with DNA instability or mobility. However, the percentage of G + C content of SPI-1 (47%) is lower than that of the core genome (52%) suggesting acquisition by horizontal gene transfer. The absence of mobile elements suggests that following its acquisition, SPI-1 stabilized into the *S. enterica* genome and its presence in all subspecies and serovars tested so far would support this. However, SPI-1 is not invariably present or functional. Environmental isolates with deletions within SPI-1 have been described among *S. enterica* serovars Senftenberg and Litchfield (Ginocchio *et al.*, 1997) and an isolate of *S.* Senftenberg from a human food poisoning outbreak completely lacks SPI-1 with a 1-kb insertion sequence inserted at the SPI-1 deletion (Hu *et al.*, 2008).

SPI-1 encodes at least two distinct functions: a Type III secretion system (TTSS) and an iron uptake system. Like TTSS of other Gram-negative bacteria, the SPI-1-encoded system forms needle-like surface appendages (Kubori *et al.*, 1998) which mediate the delivery of bacterial proteins from the cytoplasm of extracellular *Salmonella* into the host cell cytosol. Over a dozen TTSS- secreted effectors have been identified, but importantly, not all of these are encoded within the SPI-1 locus. A key function of some of these effectors is the subversion of the host cytoskeletal structure to promote bacterial uptake by macropinocytosis into non-phagocytic cells such as intestinal epithelial cells. This process results in dramatic changes in the host cell surface that manifest as characteristic membrane ruffles. In contrast to the effects of SPI-1 on non-phagocytic cells, within macrophages and dendritic cells SPI-1 induces caspase-1-dependent pro-inflammatory cell death, a process termed pyroptosis. The SPI-1 effector SipB is central to this, activating host caspase-1 leading to pyroptosis and the cleavage of pro-IL-1B and pro-IL-18 into their active forms (Hersh *et al.*, 1999; van der Velden *et al.*, 2003). These events are critical to colonization of the mucosa. In mice lacking caspase-1, the LD_{50} of *Salmonella* infection is 1000-fold greater than is seen in the wild-type mice, and there is reduced colonization of organs following oral infection in the caspase-1 deficient animals (Monack *et al.*, 2000). In a mouse colitis model, inflammation induced by *S. enterica* allows it to compete successfully with the established intestinal microbiota, with SPI-1-induced pryoptosis likely to play an important role in the induction of this inflammation (Stecher *et al.*, 2007).

The regulation of SPI-1 is complex, influenced by a variety of environmental signals and involving five SPI-1-encoded regulators and global regulatory systems (reviewed by Altier, 2005). HilA, a member of the ToxR/OmpR family, is a key SPI-1-encoded regulator. HilA activates the *sip* operon (also known as the *ssp* operon), which encodes secreted proteins, and the *inv/spa* and *prg* operons, coding for the structural components of the TTSS apparatus. HilA is also an activator of a second SPI-2-encoded transcriptional regulator, InvF, which itself induces the expression of the secreted proteins of the *sip* operon. Thus, HilA both directly and indirectly controls expression of the secreted proteins. InvF, however, has functions independent of HilA, activating the expression of the effector SopB which resides in the chromosome outside of SPI-1 (Eichelberg and Galán, 1999). Therefore, the two regulators HilA and InvF have overlapping, but not identical, sets of target genes. Such complexity presumably allows for the proper timing and level of expression of SPI-1 genes in response to various environmental signals. Conditions likely to reflect those found at the preferred site of *Salmonella* invasion at the distal ileum such as low oxygen conditions, high osmolarity and near neutral pH all enhance SPI-1 gene expression. Likewise, acetate, produced by the anaerobes of the large intestine, induces SPI-1 expression at the concentrations (15–30 mM) at which it is found in the distal ileum.

Just as it appears imperative to ensure the appropriate induction of SPI-1 expression during invasion, repression is similarly important to avoid expression that would be unproductive and inefficient or even potentially detrimental. Several transcriptional repressors of SPI-1 have been identified including the two-component regulator PhoP/PhoQ and the small nucleoid-binding protein H-NS. Environmental signals that repress SPI-1 expression include the intracellular environment of macrophages where an invasive phenotype is not required. Bile, which is secreted into the proximal small intestine, and the fatty acids propionate and butyrate, which are increased in the caecum and colon, all sites where *Salmonella* does not invade, repress SPI-1 expression.

Salmonella pathogenicity island-2

SPI-2 was first identified by a signature-tagged mutagenesis (STM) screen as being required for systemic infection of mice by *S.* Typhimurium (Hensel *et al.*, 1995). The 40-kb locus is inserted adjacent to the tRNA gene *valV* and, as with SPI-1, SPI-2 lacks putative mobile elements, suggesting that it is now stably inserted into the *S. enterica* genome. In support of this, naturally occurring deletions of the SPI-2 pathogenicity island have not been reported. A key function of SPI-2 is the expression of a TTSS which promotes intracellular survival of salmonellae, with SPI-2 mutants having reduced survival and proliferation inside host cells (Ochman *et al.*, 1996; Cirillo *et al.*, 1998). The 25-kb portion of SPI-2 encoding the TTSS has a G + C content of 43% compared with the value of 52% for the core genome. The function of SPI-2 in infection and its absence from the related and phylogenetically older *Salmonella bongori*, which is rarely pathogenic, suggest that the acquisition of SPI-2 has been an important step towards systemic colonization of warm-blooded hosts by *S. enterica* (Baumler, 1997; Groisman and Ochman, 1997).

The SPI-2 TTSS is only distantly related to the SPI-1-encoded TTSS, suggesting that their presence in *S. enterica* resulted from two independent horizontal gene transfer events rather than duplication of an original TTSS. Mutants lacking this TTSS are highly attenuated in mouse models of infection (Hensel *et al.*, 1995; Shea *et al.*, 1999; Chaudhuri *et al.*, 2009), so much so in fact that a SPI-2 TTSS mutation (*ssaV*) has been incorporated into a live attenuated vaccine strain of *S.* Typhi currently being tested for use in humans against typhoid fever and as a vector for heterologous antigen delivery (Medina *et al.*, 1999; Hindle *et al.*, 2002). The expression of SPI-2 is directly controlled by SsrAB, a SPI-2-encoded two-component system, expression of which is controlled by the non-SPI-2-encoded OmpR/EnvZ two-component regulatory system. Expression of SPI-2 is activated by the intracellular environment and, although the exact signals involved are not known, defined media with limited concentrations of Mg^{2+} and inorganic phosphate, components that appear to be limited within host cells, activate SPI-2 expression (Deiwick *et al.*, 1999). Low pH, a feature of the *Salmonella*-containing vacuole, is also important for inducing SPI-2 function during bacterial growth in defined media. The role of low pH, however, appears to be a signal for assembly of the TTSS rather than for gene expression (Rappl *et al.*, 2003). Interestingly, SPI-1 regulators can regulate SPI-2 genes, with HilA binding and repressing the promoter of *ssaH* (Thijs *et al.*, 2007) and HilD binding and activating the promoter of the *ssrAB* operon (Bustamante *et al.*, 2008). Reciprocal regulation of SPI-1 genes by SsrAB has not been described. The cytoplasmic oxidoreductase Trx-1 can also influence the induction of SPI-2 gene expression with a deletion mutant in *trxA*, the gene encoding Trx-1, leading to a delay in expression of a *ssaG*::*gfp*$^+$ reporter. Trx-1 is also required for effector secretion via the SPI-2 TTSS, although the basis of these interactions between Trx-1 and SPI-2 is not yet clear (Negrea *et al.*, 2009).

The *rlrA* islet, a pilus-encoding pathogenicity island in *S. pneumoniae*

The ~14-kb *rlrA* islet in *S. pneumoniae* encodes a pilus made of three subunits: the main pilus subunit RrgB and two accessory subunits RrgA and RrgC (Barocchi *et al.*, 2006; Falker *et al.*, 2008; Hilleringmann *et al.*, 2009). The *rlrA* islet is also referred to as pilus islet 1 (PI-1) following the discovery of a second pilus-encoding islet in the pneumococcus, which was designated PI-2 (Bagnoli *et al.*, 2008). The Rrg proteins are assembled into a pilus by three islet-encoded sortase enzymes SrtC-1, SrtC-2 and SrtC-3 (formally designated as SrtBCD) (Falker *et al.*, 2008; LeMieux *et al.*, 2008). Also encoded in the islet is the transcriptional regulator RlrA which positively regulates its own expression and that of the islet sortase and pilus subunit genes (Hava *et al.*, 2003). The islet contributes to virulence and carriage in the mouse and was first identified when transposon insertions in *rlrA* and *srtD* where found to be attenuated in the mouse lung during an STM screen (Hava and Camilli, 2002; Barocchi *et al.*, 2006; Nelson *et al.*, 2007a; Rosch *et al.*, 2008). However, assessment of the importance and role of the *rlrA* islet has been complicated by some conflicting results with examples of hypervirulence being displayed by *rlrA* islet mutants in certain circumstances (Hava and Camilli, 2002; Rosch *et al.*, 2008). The pilus mediates adhesion to and invasion of host cells although the host cell receptor(s) and uptake pathway(s) have not yet been identified. However, the Rrg proteins share weak sequence similarities to so-called microbial surface components recognizing adhesive matrix molecules, and both purified pilus and recombinant RrgA bind to the extracellular matrix components collagen I, fibronectin and laminin *in vitro* (Hilleringmann *et al.*, 2008). Interestingly, deletion of the *rlrA* islet reduces the host inflammatory response to pneumococcal infection in a mouse model, an effect that was not simply due to lower numbers of the islet mutant surviving *in vivo* (Barocchi *et al.*, 2006). This inflammatory activity may aid the pneumococcus to compete with other microbes in the nasopharynx during carriage. Indeed, the presence of RlrA-piliated pneumococci is significantly associated with the absence of *Staphylococcus aureus* co-colonization (Regev-Yochay *et al.*, 2009). In the context of infection, pilus-induced inflammatory responses may be pathological and this has been proposed as a factor explaining differences in inflammatory potential between pneumococcal strains (Barocchi *et al.*, 2006).

The islet is flanked on both sides by insertion sequences (IS*1167*) although, within one of these, a frame shift mutation is predicted to inactivate the element. Even when inactivated, such an IS could serve as a site for homologous recombination. Indeed, sequence analysis of the islet among thirty strains revealed the absence of the 8-bp direct repeats typically generated by the duplication of the IS*1167* target sequence during transposition. This suggests the islet has been acquired by homologous recombination and not by active insertion by the IS (Moschioni *et al.*,

2008). The islet is present in only a minority of clinical isolates, ~17–30% (Paterson and Mitchell, 2006; Basset *et al.*, 2007; Aguiar *et al.*, 2008; Muzzi *et al.*, 2008; Regev-Yochay *et al.*, 2009), and is always found located between two conserved genes, which encode a formate acetyltransferase and a hypothetical protein. In the majority of *rlrA*-negative strains, the gene for a hypothetical protein lies between these two conserved genes in the absence of the two ISs. As mentioned above, *rlrA* islet transcription is positively regulated by RlrA itself. However, the *rlrA* islet is not autonomous and crosstalk occurs with conserved regulatory pathways within the cell.

Two-component systems (TCSs) are key regulators of pneumococcal virulence, particularly important given that streptococci lack the typical alternative sigma factors for regulating gene expression (Paterson *et al.*, 2006). The pneumococcal TCSs 03, 06 and 13 all negatively regulate *rlrA* islet expression (Rosch *et al.*, 2008; Song *et al.*, 2009) and in the case of TCS 03 and 06, their cognate response regulators bind directly and specifically to the *rlrA* promoter (Rosch *et al.*, 2008). TCS 09 also regulates *rlrA* islet expression, and it is interesting that the influence of TCS 09 depends on the bacterial growth phase (Hendriksen *et al.*, 2007). Deletion of *rlr09* resulted in a down-regulation of *rlrA* islet transcription during early growth stages but caused an up-regulation at later times (Hendriksen *et al.*, 2007). *rlrA* transcription is down-regulated in response to high manganese concentrations, an effect mediated largely via the a metal-dependent regulator PsaR (Johnston *et al.*, 2006). The regulators MerR – a metalloregulator responsive to zinc and other metals – and MgrA – a pneumococcal orthologue of the *Streptococcus pyogenes* Mga (Hemsley *et al.*, 2003), both also repress expression of the *rlrA* islet (Hemsley *et al.*, 2003; Rosch *et al.*, 2008). It thus appears that RlrA pilus regulation is important and complex but, despite these data, it remains unclear as to exactly when and how these different pathways interact to modulate pilus expression during carriage and infection.

Bacteriophage

Temperate or lysogenic bacteriophages can alter bacterial fitness and virulence in a number of ways (Brussow *et al.*, 2004). These include lysogenic conversion, generalized transduction, as sites for genomic rearrangements and by gene disruption. In the case of lysogenic conversion, some lysogenic phages carry 'cargo' genes that integrate with the prophage into the host bacterial chromosome. These cargo genes, also referred to as lysogenic conversion genes or, somewhat unfortunately given the dictionary definition of the word in English, morons, are apparently not required for the bacteriophage life cycle but encode fitness factors presumed to be of benefit to the host bacterium (the lysogen). The selective advantage conferred by these genes allows for the lysogen to be maintained or even to dominate in the population despite the potential fitness costs of replicating phage DNA and host cell lysis following phage induction. Such lysogenic conversion is in turn of direct benefit to the lysogenic phage by ensuring the success of the host bacterium upon which its survival depends. Examples of lysogenic conversion are widespread among bacterial pathogens, often with the prophage responsible for converting avirulent strains to virulent ones, with the cargo gene products having a key role in disease pathogenesis. Classic examples of the stark effect of lysogenic conversion include botulinum, cholera and diphtheria toxins, all of which are encoded by prophages and play a central role in the pathogenesis of their associated diseases. Although it was exotoxin production that was the first widely recognized as being linked to bacteriophage infection, prophages are now known to be associated with numerous effects on bacterial virulence, including bacterial adhesion, invasion, resistance to immune defences and transmissibility. In some cases the benefit of lysogenic conversion to the lysogen, if indeed there is one, is unclear. For example, prophage-encoded Shiga-like toxin (Stx) in *E. coli* strains (STEC) is only expressed when the prophage is induced into the lytic phase of its life cycle, and only secreted when the host bacterium is lysed as part of this induction. The individual lysogen therefore does not appear to benefit directly from lysogenic conversion and this may offer an intriguing illustration of the selfish gene concept, with the released Stx phage able to infect bystander intestinal *E. coli* allowing further amplification of the phage. Experiments with mice have shown that the resistance or susceptibility pattern of the intestinal flora towards the released Stx phage exerts either a protective or an enhancing effect on the severity of STEC infections (Gamage *et al.*, 2003). Clinically, this observation could contribute to the variability of the disease symptoms in infected patients. Furthermore, although a theoretical benefit to the lysogen may be obvious, in many cases it is difficult to establish firmly an altered phenotype, possibly owing to inadequate or inappropriate model systems (for example, using a mouse model for a human-specific bacterium), or looking at an infection model rather than a model of carriage, not to mention the general dearth of models available with which to investigate transmission. Even when strong epidemiological associations are seen between lysogens and clinical features such as those described for emergent *Streptococcus pyogenes* clones [Group A *Streptococcus* (GAS), discussed below], it can be difficult to show unequivocally a role for prophages in the bacterial phenotype in the presence of other intra-strain genetic differences.

Nonetheless the prominent role of prophages in GAS biology is worthy of discussion, given that prophage sequences make up the majority of the diversity that is observed between strains (Banks *et al.*, 2002). GAS, a human-restricted commensal of the skin and nasopharynx, is also the cause of various infections, ranging from those that are relatively minor to those that are life-threatening, and from those that are acute to those that are chronic, including those that result in severe post-infection sequelae. Analysis of the first three GAS genome sequences, from strains belonging to serotypes M1, M3 and M18 (Ferretti *et al.*, 2001; Beres *et al.*, 2002; Smoot *et al.*, 2002), revealed that all strains were poly-lysogenic and that, though prophages constituted < 13% of each genome, they were responsible for on average over 50% of the differences in gene content between these three strains (Banks *et al.*, 2002). Further microarray-based examination of the gene content of 36 M18 strains found that virtually all the variation in gene content was due to differences in prophage content, either to the complete absence of prophages to small differences in the gene content of the prophages (Smoot *et al.*, 2002). A similarly prominent role for prophages in intra-strain variation was also seen when assessing gene content by microarray in 30 M1 strains (Sumby *et al.*, 2005b) and 33 M3 strains (Beres *et al.*, 2004). Importantly, many putative or known virulence factors are encoded in GAS prophage elements, suggesting a key role for prophages in GAS fitness and virulence. Since the mid-1980s, the frequency and severity of infections caused by GAS strains have sharply increased, which appears to be driven in part by the acquisition of phage-encoded fitness or virulence factors. This includes the emergence and persistence of a so-called hyper-virulent M1T1 clone, an emergence that is linked with the acquisition of prophages encoding the virulence factors SpeA and SdaD2 (Aziz *et al.*, 2005; Sumby *et al.*, 2005b; Aziz and Kotb, 2008). It is suggested that the superantigen activity of SpeA participates in disease pathogenesis while *sdaD2* encodes an extracellular DNase that acts to degrade the extracellular net produced by neutrophils to entrap bacteria (Sumby *et al.*, 2005a). However, other non-phage-related genetic changes are seen in this lineage relative to ancestor M1T1 strains, making it difficult to evaluate the relative importance of phages in the emergence, virulence and persistence of these clones (Aziz *et al.*, 2005; Sumby *et al.*, 2005b; Aziz and Kotb, 2008). Likewise, a phage-encoded, secreted phospholipase A2, SlaA, and streptococcal pyrogenic exotoxin, SpeK, are implicated in the emergence of a virulent M3 clone (Beres *et al.*, 2002; Sitkiewicz *et al.*, 2006). The appearance of these features coincided temporally with the emergence of virulent M3 clones and, in the case of SlaA, there are convincing data for a role for this protein in GAS carriage and virulence (Sitkiewicz *et al.*, 2006). An *slaA* mutant is less able to attach to and kill human host cells and was attenuated for virulence in mouse septicaemia and skin lesion models, as well as in a *Cynomolgus* macaque pharyngitis model (Sitkiewicz *et al.*, 2006). Also, among M3 clones, there is an association with the presence of a prophage encoding streptococcal pyrogenic exotoxin A and the ability of a strain to persist between epidemics, with strains lacking this prophage failing to do so (Beres *et al.*, 2004, 2006). The presence of this prophage is also linked to recurrent episodes of pharyngitis (Musser *et al.*, 1992).

In addition to lysogeny, bacteriophages play other roles in determining virulence and in moving DNA between bacteria. An important mechanism for the latter is generalized transduction. This occurs when random bacterial DNA is incorporated into the phage during packaging of the phage head. This does not alter the functioning of the final phage, which can attach and deliver the DNA to a new host bacterium. The DNA, if not degraded, has the potential to recombine into the bacterial genome or, in the case of plasmids, to re-circularise; this appears to be an important process in the spread of drug resistance plasmids.

In addition to the gain of genes, gene loss also has an important impact in bacterial pathogenicity and fitness, a process in which lysogenic phages may play a role, an example being the insertional inactivation of β-toxin in *S. aureus* owing to chromosomal integration of prophage phi13 (Coleman *et al.*, 1991).

Integrative and conjugative elements

Integrative and conjugative elements (ICE) are diverse and widely distributed mobile elements that become integrated into their host bacterium's genome, and they are increasingly being recognized as important agents of horizontal gene transfer. They are able to excise themselves, form a circular intermediate, and pass to new host cells by single-stranded transfer via a self-encoded conjugative apparatus. Re-formation of a double-stranded circular intermediate occurs before re-integration into the genome of the original as well as the recipient host cells via site-specific recombination (Fig. 3.2). In addition to encoding the genes required for this cycle of excision, conjugative transfer and integration, ICE carry various 'cargo' genes that can have a significant impact on the host bacterium. One example is the ~76-kb ICEKp1 found in certain strains of *Klebsiella pneumoniae* (Lin *et al.*, 2008). The cargo genes carried by this element include a region with similarity to the high-pathogenicity island (HPI) of *Yersinia* spp., which carries the yersiniabactin siderophore system essential for virulence in these bacteria. ICEKp1 also carries genes homologous to the siderophore system encoded by the large plasmid, pLVPK, of *K. pneumoniae* and a regulator of the mucoid phenotype (*rmpA*) which is associated with virulence in *K. pneumoniae* (Nassif *et al.*, 1989). It therefore seems likely that the presence of this region may enhance the pathogenicity of host *K. pneumoniae* strains and, indeed, the prevalence of ICEKp1 was higher in sepsis strains where primary liver abscesses (PLA) were seen (38 of 42 strains) than in sepsis isolates where PLA were absent (5 of 32 strains), suggesting a role for ICEKp1 in tissue invasion (Lin *et al.*, 2008).

Antibiotic resistance genes are frequent among ICE cargoes in pathogens and the SXT-R391 family of ICEs found in *Vibrio* spp. offers a good example of this. SXT is a ~100-kb ICE originally

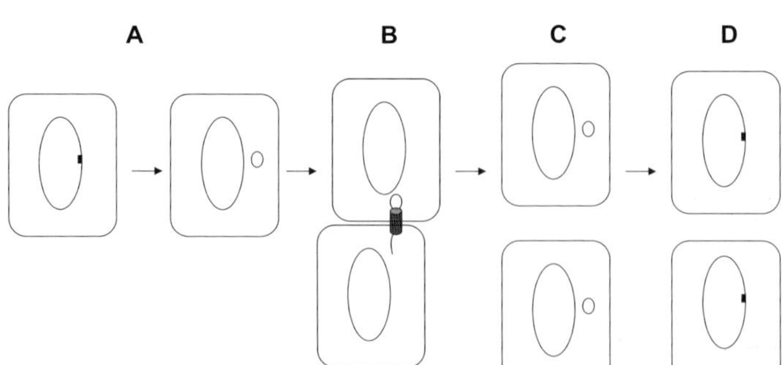

Figure 3.2 Schematic lifecycle of an ICE. The ICE is stably integrated into the genome of a host bacterium until excision is triggered (A). The circularized ICE intermediate replicates and is transferred to a recipient cell via an ICE-encoded conjugation apparatus (B). Circular ICE intermediates are now present in both cells (C) and integrate into the bacterial chromosome (D).

discovered in a clinical isolate of *Vibrio cholerae*. It and closely related ICEs encode resistance to antibiotics such as sulfamethoxazole and trimethoprim that have been used for the treatment of cholera (Burrus *et al.*, 2006). The maintenance of SXT, in the absence of selection for its resistance cargo genes, occurs through a self-encoded toxin-antitoxin system (Wozniak and Waldor, 2009). If bacterial cell division occurs when STX (or any ICE) is in its excised form, daughter cells could be generated that lack STX. However, STX-negative daughter cells are exceptionally rare, occurring at a frequency of ~1×10^{-7} even though cells that have excised STX occur at a frequency of ~2% of the bacterial population. The STX-encoded genes, *mosA* and *mosT* (maintenance of SXT antitoxin and toxin) which are transcriptionally derepressed when STX excises, play a major role in this maintenance. Deletion of these genes greatly increased the frequency (~1000-fold) of STX loss, compared with wild type (Wozniak and Waldor, 2009). Although expression of *mosT* kills host cells, and this is known to be neutralized by the presence of *mosA*, the mechanistic details of how this toxin-antitoxin system functions remain to be elucidated. Interestingly, *mosAT* are not conserved among the STX-R391 family and other mechanisms must operate to facilitate maintenance in this family of ICEs. Similarly acting toxin-antitoxin systems are encoded by other genetic elements, such as bacteriophages, plasmids and integrons, to promote their persistence in the host population.

Another example of an ICE in the context of antibiotic resistance is ICESpn23FST81, described after genome sequencing a *S. pneumoniae* strain belonging to the multidrug-resistant pandemic Spain 23F ST 81 lineage (Croucher *et al.*, 2009). A Tn*916*-like element within ICESpn23FST81 carries the tetracycline resistance gene *tetM* while a Tn*5252*-like component carries a gene for chloramphenicol acetyltransferase, which appears to have been acquired through wholesale integration of the pC194 plasmid originally identified in chloramphenicol-resistant *S. aureus*. Despite the fact that the Spain 23F ST 81 lineage is a frequent cause of pneumococcal disease, it actually has a low propensity for causing invasive disease relative to carriage rates (odds ratio of 0.4) (Brueggemann *et al.*, 2003). Its prevalence as a pathogenic strain comes from its success during carriage and ICESpn23FST81 may contribute in this context, as it encodes an apparently intact lantibiotic synthesis gene cluster. The structural gene appears to encode a novel Group II lantibiotic, which, based on its predicted ring structure, diglycine cleavage motif, and net neutral charge, is most closely related to mersacidin and lichenicidin. Adjacent to this lantibiotic operon is a probable transcriptional regulator and, compatible with these bacterocin systems often forming a quorum-sensing system in conjunction with peptide autoinducers, a small, secreted peptide is also found in this region of ICESpn23FST81. Although this putative bacterocin system has not been functionally investigated, it may provide a competitive advantage against competing strains and species during nasopharyngeal colonization. ICESpn23FST81 carries a putative *pezAT* epsilon toxin-zeta antitoxin system, which may contribute to maintenance of ICESpn23FST81 in pneumococcal daughter cells in a fashion similar to *mosAT* in the STX ICE of *V. cholerae*.

Streptococcus equi subspecies *equi* is the causative agent of strangles, a disease of horses characterized by abscesses in the lymph nodes of the head and neck. Abscess formation can restrict the airway and it is this clinical feature that gives rise to the name strangles. Long-term carriage and shedding results from incomplete drainage of the guttural pouches and/or sinuses following rupture of abscesses formed in the lymph nodes. The related bacterium *Streptococcus equi* subspecies *zooepidemicus*, from which *S. equi* appears to have evolved as a clone, has a wider host range and is (in many cases) not particularly virulent. A comparative analysis of the genome sequences of *S. equi* and *S. zooepidemicus* leads to the suggestion that the acquisition of ICESe2 was a key event in the evolution of *S. equi* from *S. zooepidemicus* and that it contributes to the differing behaviour of these subspecies (Holden *et al.*, 2009). ICESe2 carries the *eqb* locus which encodes a non-ribosomal peptide synthetase system producing a putative siderophore named equibactin which, given the importance of iron acquisition by pathogens in the host, is proposed to play an important role in allowing *S. equi* to cause lymph node abscessation, which in turn is required for subsequent persistence and transmission (Heather *et al.*, 2008).

The association of gene loss with pathogenicity

While much attention focuses on the acquisition of new characteristics through horizontal gene transfer, gene loss also plays a key role in the evolution of pathogenic bacteria. Pseudogenes are coding sequences that have been inactivated by mutations, including non-sense substitutions and frame-shifts, or truncation by deletion or rearrangement. Loss of gene function through pseudogene formation and gene deletion appears to be a hallmark of host-restricted pathogenic bacteria compared with their host-generalist relatives. A good example of this is *S.* Typhi which carries >200 pseudogenes (>4% of coding sequences), the majority of which appear functional in *S.* Typhimurium in which pseudogenes are less frequent (<1% of coding sequences). Interestingly, many of these mutations are shared with *S.* Paratyphi A, which has a life cycle and host range similar to *S.* Typhi (Holt *et al.*, 2009). Such gene loss may be responsible for the narrow human-specific host range of *S.* Typhi and *S.* Paratyphi A, and the loss of function of genes associated with intestinal attachment and persistence is compatible with a life cycle involving invasion of systemic tissues and transmission by excretion via the gall bladder rather than lumenal gut colonization. *Mycobacterium leprae* shows an even more dramatic case of genome decay with less than half of the genome (49.5%) encoding functional genes (Cole *et al.*, 2001). Over 1000 pseudogenes, many of which appear functional in *Mycobacterium tuberculosis*, cover ~27% of the genome and a further 23.5% of the genome does not appear to be coding and may represent gene remnants mutated beyond recognition even as pseudogenes (Cole *et al.*, 2001). As with *S.* Typhi, gene loss in *M. leprae* is associated with a narrow host range. While the examples above deal with large scale gene loss (referred to as genome decay), and there are many more examples of this, including *Yersinia pestis* (Chain *et al.*, 2004) and *Bordetella pertussis* (Parkhill *et al.*, 2003; Preston *et al.*, 2004), more subtle changes can have a significant impact on pathogenicity. Deletion of *cadA*, encoding a lysine decarboxylase, which is present in closely related commensal *E. coli*, is associated with enhanced virulence in *Shigella* spp. and enteroinvasive *E. coli*. Introduction of a wild-type *cadA* gene led to the attenuation of *Shigella flexneri* probably owing to the inhibitory effect on enterotoxin activity of cadaverine, a product of *cadA* (Maurelli *et al.*, 1998). Similarly, loss of an arabinose assimilation operon is associated with increased virulence in *Burkholderia pseudomallei*. This operon is present in the related but non-pathogenic *Burkholderia thailandensis*, and its introduction into *B. pseudomallei*, while permitting growth on arabinose, significantly attenuated the strain in a Syrian hamster infection model (Moore *et al.*, 2004).

Plasmids

The term plasmid, introduced by Joshua Lederberg in 1952 (Lederberg, 1952), refers to mobile, autonomously replicating extra-chromosomal DNA molecules. Many are conjugative plasmids, containing the *tra*-genes required for their transfer to other bacterial cells via conjugation. On the other hand, non-conjugative plasmids are incapable of initiating conjugation and are only transferred with the assistance of conjugative plasmids. Both types may also be spread by transformation or generalized transduction. The medical significance of plasmids was long ago documented with the discovery in 1961 that they were responsible for multi-antibiotic resistance in enteric pathogens (Watanabe and Fukasawa, 1961). Subsequently, the role of plasmids not only in antibiotic resistance, where they do indeed play a significant role, but also in bacterial pathogenicity has become recognized.

The virulence plasmids of *Salmonella*

Certain *S. enterica* serovars, including Typhimurium, Enteriditis, Dublin, Gallinarum, and Choleraesuis carry large (50–90 kb), low copy-number plasmids, which contribute to infection. The presence of these plasmids (termed *Salmonella* plasmid virulence, Spv) increases the virulence of various serovars (Gulig and Doyle, 1993; Gulig *et al.*, 1993, 1998; Matsui *et al.*, 2001) and contributes to the intracellular growth of salmonellae in macrophages (Gulig *et al.*, 1998), and host cell death (Libby *et al.*, 2000; Browne *et al.*, 2002). Furthermore, *spv*-carrying strains dominate among clinical isolates from patients with non-typhoidal bacteraemia, suggesting a role for the virulence plasmids in these infections (Fierer *et al.*, 1992). Although heterogeneous in size and gene content, these virulence plasmids all carry the conserved 7.8-kb region bearing the *spvRABCB* operon which is central to their function in disease pathogenesis. A key role for the *spv* locus was demonstrated by the finding that cloning of this region into a low copy number vector is sufficient to restore the virulence of a plasmid-cured strain (Gulig *et al.*, 1993). More specifically, *spvBC* are the essential functional elements of this locus and are alone sufficient to confer plasmid-mediated virulence (Matsui *et al.*, 2001). *spvB* encodes an ADP-ribosyltransferase which targets host cell actin, destabilizing the cytoskeleton and inducing delayed host cell death (Libby *et al.*, 2000; Lesnick *et al.*, 2001; Browne *et al.*, 2002; Tezcan-Merdol *et al.*, 2001, 2005). This enzymatic activity is required for *spvB* to contribute to virulence (Lesnick *et al.*, 2001). Interestingly, the secretion of SpvB into host cells depends upon the SPI-2 TTSS discussed above (Browne *et al.*, 2002), an example of functional cross-talk and dependence between two distinct types of horizontally acquired determinant of pathogenicity. *spvC* encodes a phosphothreonine lyase active against host mitogen-activated protein kinases and it too requires SPI-2 for delivery into host cells (Mazurkiewicz *et al.*, 2008). There is some evidence that interference by SpvC with host cell signalling down-regulates inflammatory cytokine production by infected cells, which may promote bacterial proliferation (Mazurkiewicz *et al.*, 2008). The functions of SpvA and SpvD are unknown and, while *spvD* contributes to virulence, *spvA* appears dispensable (Roudier *et al.*, 1992). The final gene in the *spv* operon, *spvR*, encodes a transcriptional regulator homologous to the LysR family of regulatory proteins, which drives *spv* gene expression. Compatible with a role in the intracellular growth of salmonellae in macrophages, *spv* gene expression is up-regulated within these cells and under laboratory conditions mimicking this environment (Wilson *et al.*, 1997; Eriksson *et al.*, 2003). In addition to the *spv* locus, *Salmonella* virulence plasmids encode additional virulence-associated genes such as *rck* (resistance to complement) (Hackett *et al.*, 1987), *pef* (plasmid-encoded fimbriae) (Friedrich *et al.*, 1993) or *mig-5* (macrophage-inducible gene) (Valdivia and Falkow, 1997). However, unlike the ubiquitous *spv* locus, the presence of these and other accessory genes depends on the host serovar: for example, while the *S.* Typhimurium and *S.* Enteritidis virulence plasmids contain *spv*, *rck*, *pef* and *mig-5*, only *spv* and *mig-5* are present in the *S.* Gallinarum version.

Integrons

Integrons function as assembly and expression platforms that allow a bacterium to incorporate exogenous DNA into its genome via site-specific recombination and then, subsequently, to express the captured genes (Boucher *et al.*, 2007; Mazel, 2006). Their importance to pathogenicity is highlighted by the fact that they were first identified as being important for antibiotic resistance (Stokes and Hall, 1989). Present in approximately 10% of sequenced bacterial genomes, these diverse elements share three characteristic features: a gene encoding an integrase (*intI*), and a recombination site (*attI*), which together mediate the capture of exogenous DNA, and a promoter that drives transcription of these captured sequences in the absence of their own promoter (Boucher *et al.*, 2007). The process allows the integration of novel genes without potentially deleterious disruption of the core host genome. Multiple capture events are frequent, with the captured genes, referred to as gene cassettes, usually inserted in the same orientation to produce a tandem array of co-transcribed genes. In some cases, large numbers of gene cassettes may be captured leading to the term superintegron: for example, the superintegron of *V. cholerae* serotype O1 biotype El Tor strain N16961 contains at least 215 ORFs and covers approximately 3% of the host genome (Rowe-Magnus *et al.*, 2003). Apart from antibiotic resistance, the biological functions of gene cassettes are largely unknown: most have no known homologues and few have had their functions determined experimentally. However, those that have been studied show that integrons encode for diverse activities, including restriction/methylation systems, lipase, haemagglutinin, polysaccharide biosynthesis and sulphate binding. The importance and role of integrons in bacterial pathogenicity is likely to become more evident as the functions of the gene cassettes they encode are elucidated.

Concluding remarks

As illustrated in this chapter, bacterial determinants of pathogenicity are encoded by a range of different and variable genetic elements. The most important theme that emerges from an understanding of these different genetic elements is that bacterial diversity and changeability are inherent and fundamental to the evolution of pathogenic bacteria. Bacterial evolution can, but does not necessarily, progress by small graded steps. Variation can occur rapidly and dramatically via the acquisition and sharing of genetic elements that can be spread rapidly between a wide range of bacterial species. This rapid variation does not happen to the animals that host pathogenic bacteria, and thus pathogenic bacteria will tend to stay ahead of the game in terms of their ability to infect those hosts and to cause disease. These features make bacterial diseases both so difficult to fight and so fascinating to study.

References

Aanensen, D.M., Mavroidi, A, Bentley, S.D., Reeves, P.R., and Spratt, B.G. (2007). Predicted functions and linkage specificities of the products of the *Streptococcus pneumoniae* capsular biosynthetic loci. J. Bacteriol. *189*, 7856–7876.

Aguiar, S.I., Serrano, I., Pinto, F.R., Melo-Cristino, J., and Ramirez, M. (2008). The presence of the pilus locus is a clonal property among pneumococcal invasive isolates. BMC Microbiol. *8*, 41

Altier, C., (2005). Genetic and environmental control of *Salmonella* invasion. J. Microbiol. *43*, 85–92.

Avery, O.T., Macleod, C.M., and McCarty, M. (1944). Studies on the chemical nature of the substance inducing transformation of pneumococcal types: induction of transformation by a desoxyribonucleic acid fraction isolated from pneumococcus type III.J. Exp. Med. *79*, 137–158.

Aziz, R.K., Edwards, R.A., Taylor, W.W., Low, D.E., McGeer, A. and Kotb, M. (2005). Mosaic prophages with horizontally acquired genes account for the emergence and diversification of the globally disseminated M1T1 clone of *Streptococcus pyogenes*. J. Bacteriol. *187*, 3311–3318.

Aziz, R. K,. and Kotb, M. (2008). Rise and persistence of global M1T1 clone of *Streptococcus pyogenes*. Emerg. Infect. Dis. *14*, 1511–1517.

Bagnoli, F., Moschioni, M., Donati, C., Dimitrovska, V., Ferlenghi, I., Facciotti, C., Muzzi, A., Giusti, F., Emolo, C., Sinisi, A., *et al.* (2008). A second pilus type in *Streptococcus pneumoniae* is prevalent in emerging serotypes and mediates adhesion to host cells. J. Bacteriol. *190*, 5480–5492.

Banks, D.J., Beres, S.B., and Musser, J.M. (2002). The fundamental contribution of phages to GAS evolution, genome diversification and strain emergence. Trends Microbiol. *10*, 515–521.

Barocchi, M.A., Ries, J., Zogaj, X., Hemsley, C., Albiger, B., Kanth, A., Dahlberg, S., Fernebro, J., Moschioni, M., Masignani, V., *et al.* (2006). A pneumococcal pilus influences virulence and host inflammatory responses. Proc. Natl. Acad. Sci. U.S.A. *103*, 2857–2862.

Basset, A., Trzcinski, K., Hermos, C., O'Brien, K.L., Reid, R., Santosham, M., McAdam, A.J., Lipsitch, M., and Malley, R. (2007). Association of the pneumococcal pilus with certain capsular serotypes but not with increased virulence. J. Clin. Microbiol. *45*,1684–1689.

Baumler, A.J. (1997). The record of horizontal gene transfer in *Salmonella*. Trends Microbiol. *5*, 318–322.

Bentley, S.D., Aanensen, D.M., Mavroidi, A., Saunders, D., Rabbinowitsch, E., Collins, M., Donohoe, K., Harris, D., Murphy, L., Quail, M.A., *et al.* (2006). Genetic analysis of the capsular biosynthetic locus from all 90 pneumococcal serotypes. PLoS Genet. *2*, e31.

Beres, S.B., Sylva, G.L., Barbian, K.D., Lei, B., Hoff, J.S., Mammarella, N.D., Liu, M.Y., Smoot, J.C., Porcella, S.F., Parkins, L.D., *et al.* (2002). Genome sequence of a serotype M3 strain of group A *Streptococcus*: phage-encoded toxins, the high-virulence phenotype, and clone emergence. Proc. Natl. Acad. Sci. U.S.A. *99*, 10078–10083.

Beres, S.B., Sylva, G.L., Sturdevant, D.E., Granville, C.N., Liu, M., Ricklefs, S.M., Whitney, A.R., Parkins, L.D., Hoe, N.P., Adams, G.J., *et al.* (2004). Genome-wide molecular dissection of serotype M3 group A *Streptococcus* strains causing two epidemics of invasive infections. Proc. Natl. Acad. Sci. U.S.A. *101*, 11833–11838.

Beres, S.B., Richter, E.W., Nagiec, M.J., Sumby, P., Porcella, S.F., DeLeo, F.R., and Musser, J.M. (2006). Molecular genetic anatomy of inter- and intraserotype variation in the human bacterial pathogen group A *Streptococcus*. Proc. Natl. Acad. Sci. U.S.A. *103*, 7059–7064.

Boucher, Y., Labbate, M., Koenig, J.E., and Stokes, H.W. (2007). Integrons: mobilizable platforms that promote genetic diversity in bacteria. Trends Microbiol. *15*, 301–309.

Browne, S.H., Lesnick, M.L., and Guiney, D.G. (2002). Genetic requirements for *Salmonella*-induced cytopathology in human monocyte-derived macrophages. Infect. Immun. *70*,7126–7135.

Brueggemann, A.B., Griffiths, D.T., Meats, E., Peto, T., Crook, D.W., and Spratt, B.G. (2003). Clonal relationships between invasive and carriage *Streptococcus pneumoniae* and serotype- and clone-specific differences in invasive disease potential. J. Infect. Dis. *187*, 1424–1432.

Brüssow, H., Canchaya, C., and Hardt, W.D. (2004). Phages and the evolution of bacterial pathogens: from genomic rearrangements to lysogenic conversion. Microbiol. Mol. Biol. Rev. *68*, 560–602.

Burrus, V., Marrero, J., and Waldor, M.K. (2006). The current ICE age: biology and evolution of SXT-related integrating conjugative elements. Plasmid 55, 173–183.

Bustamante, V.H., Martinez, L.C., Santana, F.J., Knodler, L.A., Steele-Mortimer, O., and Puente, J.L. (2008). HilD-mediated transcriptional cross-talk between SPI-1 and SPI-2. Proc. Natl. Acad. Sci. U.S.A. *105*, 14591–14596.

Chain, P., Carniel, E., Larimer, F., Lamerdin J., Stoutland, P.O., Regala, W.M., Georgescu, A.M., Vergez, L.M., Land, M., Motin, V.L., *et al.* (2004). Insights into the evolution of *Yersinia pestis* through whole genome comparison with *Yersinia pseudotuberculosis*. Proc. Natl. Acad. Sci. U.S.A. *101*, 13826–13831.

Chaudhuri, R.R., Peters, S.E., Pleasance, S.J., Northen, H., Willers, C., Paterson, G.K., Cone, D.B., Allen, A.G., Owen, P.J., Shalom, G., *et al.* (2009). Comprehensive identification of *Salmonella enterica* serovar Typhimurium genes required for infection of BALB/c mice. PLoS Pathog. *5*, e1000529.

Cirillo, D.M., Valdivia, R.H., Monack, D.M., and Falkow, S. (1998). Macrophage-dependent induction of the *Salmonella* pathogenicity island 2 type III secretion system and its role in intracellular survival. Mol. Microbiol. *30*, 175–188.

Cole, S.T., Eiglmeier, K., Parkhill, J., James, K.D., Thomson, N.R., Wheeler, P.R., Honoré, N., Garnier, T., Churcher, C., Harris, D., *et al.* (2001). Massive gene decay in the leprosy bacillus. Nature *409*, 1007–1011.

Coleman, D., Knights, J., Russell, R., Shanley, D., Birkbeck, T.H., Dougan, G., and Charles, I. (1991). Insertional inactivation of the *Staphylococcus aureus* β-toxin by bacteriophage phi 13 occurs by site- and orientation-specific integration of the ϕ13 genome. Mol. Microbiol. *5*, 933–939.

Croucher, N.J., Walker, D., Romero, P., Lennard, N., Paterson, G.K., Bason, N.C., Mitchell, A.M., Quail, M.A., Andrew, P.W., Parkhill, J., *et al.* (2009). Role of conjugative elements in the evolution of the multidrug-resistant pandemic clone *Streptococcus pneumoniae* (Spain23F) ST81. J. Bacteriol. *191*, 1480–1489.

Crump, J.A., Luby, S.P., and Mintz, E.D. (2004). The global burden of typhoid fever. Bull. WHO *82*, 346–353.

Cundell, D.R., Weiser, J.N., Shen, J., Young, A., and Tuomanen, E.I. (1995). Relationship between colonial morphology and adherence of *Streptococcus pneumoniae*. Infect. Immun. *63*, 757–761.

Deiwick, J., Nikolaus, T., Erdogan, S., and Hensel, M. (1999). Environmental regulation of *Salmonella* pathogenicity island 2 gene expression. Mol. Microbiol. *31*, 1759–1773.

Eichelberg, K., and Galán, J.E. (1999). Differential regulation of *Salmonella typhimurium* type III secreted proteins by pathogenicity island 1 (SPI-1)-encoded transcriptional activators InvF and HilA. Infect. Immun. *67*, 4099–4105.

Eriksson, S., Lucchini, S., Thompson, A., Rhen, M., and Hinton, J.C. (2003). Unravelling the biology of macrophage infection by gene expression profiling of intracellular *Salmonella enterica*. Mol. Microbiol. *47*: 103–118.

Falker, S., Nelson, A.L., Morfeldt, E., Jonas, K., Hultenby, K., Ries, J., Melefors, O., Normark, S., and Henriques-Normark, B. (2008). Sortase-mediated assembly and surface topology of adhesive pneumococcal pili. Mol. Microbiol. *70*, 595–607.

Ferretti, J.J., McShan, W.M., Ajdic, D., Savic, D.J., Savic, G., Lyon, K., Primeaux, C., Sezate, S., Suvorov, A.N., Kenton, S., *et al.* (2001). Complete genome sequence of an M1 strain of *Streptococcus pyogenes*. Proc. Natl. Acad. Sci. USA *98*, 4658–4663.

Fierer, J., Krause, M., Tauxe, R., and Guiney, D. (1992). *Salmonella typhimurium* bacteremia: association with the virulence plasmid. J. Infect. Dis. *166*, 639–642.

Friedrich, M.J., Kinsey, N.E., Vila, J., and Kadner, R.J. (1993). Nucleotide sequence of a 13.9 kb segment of the 90 kb virulence plasmid of *Salmonella typhimurium*: the presence of fimbrial biosynthetic genes. Mol. Microbiol. *8*, 543–558.

Galán, J.E., and Curtiss, R., 3rd. (1989). Cloning and molecular characterization of genes whose products allow *Salmonella typhimurium* to penetrate tissue-culture cells. Proc. Natl. Acad. Sci. U.S.A. *86*, 6383–6387.

Gamage, S.D., Strasser, J.E., Chalk, C.L., and Weiss, A.A. (2003). Nonpathogenic *Escherichia coli* can contribute to the production of Shiga toxin. Infect. Immun. *71*, 3107–3115.

Ginocchio, C.C., Rahn, K., Clarke, R.C., and Galán, J.E. (1997). Naturally occurring deletions in the centisome 63 pathogenicity island of environmental isolates of *Salmonella* spp. Infect. Immun. *65*, 1267–1272.

Granat, S.M., Ollgren, J., Herva, E., Mia, Z., Auranen, K., and Mäkelä, P.H. (2009). Epidemiological evidence for serotype-independent acquired immunity to pneumococcal carriage. J. Infect. Dis. *200*, 99–106.

Griffith, F., (1928). The significance of pneumococcal types. J. Hyg. *27*, 113–159.

Groisman, E.A., and Ochman, H. (1997). How *Salmonella* became a pathogen. Trends Microbiol. *5*, 343–349.

Gulig, P.A., and Doyle, T.J. (1993). The *Salmonella typhimurium* virulence plasmid increases the growth rate of salmonellae in mice. Infect. Immun. *61*, 504–511.

Gulig, P.A., Danbara, H., Guiney, D.G., Lax, A.J., Norel, F., and Rhen, M. (1993). Molecular analysis of *spv* virulence genes of the *Salmonella* virulence plasmids Mol. Microbiol. *7*, 825–830.

Gulig, P.A., Doyle, T.J., Hughes, J.A., and Matsui, H. (1998). Analysis of host cells associated with the Spv-mediated increased intracellular growth rate of *Salmonella typhimurium* in mice. Infect. Immun. *66*, 2471–2485.

Hacker, J., Bender, L., Ott, M., Wingender, J., Lund, B., Marre, R., and Goebel, W. (1990). Deletions

of chromosomal regions coding for fimbriae and hemolysins occur *in vitro* and *in vivo* in various extraintestinal *Escherichia coli* isolates. Microb. Pathog. *8*, 213–225.
Hackett, J., Wyk, P., Reeves, P., and Mathan, V. (1987). Mediation of serum resistance in *Salmonella typhimurium* by an 11-kilodalton polypeptide encoded by the cryptic plasmid. J. Infect. Dis. *155*, 540–549.
Hammerschmidt, S., Wolff, S., Hocke, A., Rosseau, S., Müller, E., and Rohde, M. (2005). Illustration of pneumococcal polysaccharide capsule during adherence and invasion of epithelial cells. Infect. Immun. *73*, 4653–4667.
Hava, D.L., and Camilli, A. (2002). Large-scale identification of serotype 4 *Streptococcus pneumoniae* virulence factors. Mol. Microbiol. *45*, 1389–1406.
Hava, D.L., Hemsley, C.J., and Camilli, A. (2003). Transcriptional regulation in the *Streptococcus pneumoniae rlrA* pathogenicity islet by RlrA. J. Bacteriol. *185*, 413–421.
Heather, Z., Holden, M.T., Steward, K.F., Parkhill, J., Song, L., Challis, G.L., Robinson, C., Davis-Poynter, N., and Waller, A.S. (2008). A novel streptococcal integrative conjugative element involved in iron acquisition. Mol. Microbiol. *70*, 1274–1292.
Hemsley, C., Joyce, E., Hava, D.L., Kawale, A., and Camilli, A. (2003). MgrA, an orthologue of Mga, acts as a transcriptional repressor of the genes within the *rlrA* pathogenicity islet in *Streptococcus pneumoniae*. J. Bacteriol. *185*, 6640–6647.
Hendriksen, W.T., Silva, N., Bootsma, H.J., Blue, C.E., Paterson, G.K., Kerr, A.R., de Jong, A., Kuipers, O.P., Hermans, P.W., and Mitchell, T.J. (2007). Regulation of gene expression in *Streptococcus pneumoniae* by response regulator 09 is strain dependent. J. Bacteriol. *189*, 1382–1389.
Hensel, M., Shea, J.E., Gleeson, C., Jones, M.D., Dalton, E., and Holden, D.W. (1995). Simultaneous identification of bacterial virulence genes by negative selection. Science *269*, 400–403.
Hersh, D., Monack, D.M., Smith, M.R., Ghori, N., Falkow, S., and Zychlinsky, A. (1999). The *Salmonella* invasin SipB induces macrophage apoptosis by binding to caspase-1. Proc. Natl. Acad. Sci. U.S.A. *96*, 2396–2401.
Hilleringmann, M., Giusti, F., Baudner, B.C., Masignani, V., Covacci, A., Rappuoli, R., Barocchi, M.A., and Ferlenghi, I. (2008). Pneumococcal pili are composed of protofilaments exposing adhesive clusters of Rrg A. PLoS Pathog. *4*, e1000026.
Hilleringmann, M., Ringler, P., Muller, S.A., De Angelis, G., Rappuoli, R., Ferlenghi, I., and Engel, A. (2009). Molecular architecture of *Streptococcus pneumoniae* TIGR4 pili. EMBO J. *28*, 3921–3930.
Hindle, Z., Chatfield, S.N., Phillimore, J., Bentley, M., Johnson, J., Cosgrove, C.A., Ghaem-Maghami, M., Sexton, A., Khan, M., Brennan, F.R., *et al.* (2002). Characterization of *Salmonella enterica* derivatives harboring defined *aroC* and *Salmonella* pathogenicity island 2 type III secretion system (*ssaV*) mutations by immunization of healthy volunteers. Infect. Immun. *70*, 3457–3467.
Holden, M.T., Heather, Z., Paillot, R., Steward, K.F., Webb, K., Ainslie, F., Jourdan, T., Bason, N.C., Holroyd, N.E., Mungall, K., *et al.* (2009). Genomic evidence for the evolution of *Streptococcus equi*: host restriction, increased virulence, and genetic exchange with human pathogens. PLoS Pathog. *5*, e1000346.
Holt, K.E., Thomson, N.R., Wain, J., Langridge, G.C., Hasan, R., Bhutta, Z.A., Quail, M.A., Norbertczak, H., Walker, D., Simmonds, M., *et al.* (2009). Pseudogene accumulation in the evolutionary histories of *Salmonella enterica* serovars Paratyphi A and Typhi. BMC Genomics *10*, 36.
Hu, Q., Coburn, B., Deng, W., Li, Y., Shi, X., Lan, Q., Wang, B., Coombes, B.K., and Finlay, B.B. (2008). *Salmonella enterica* serovar Senftenberg human clinical isolates lacking SPI-1. J. Clin. Microbiol. *46*, 1330–1336.
Johnston, J.W., Briles, D.E., Myers, L.E., and Hollingshead, S.K. (2006). Mn^{2+}-dependent regulation of multiple genes in *Streptococcus pneumoniae* through PsaR and the resultant impact on virulence. Infect. Immun. *74*, 1171–1180.
Kerr, A.R., Paterson, G.K., Riboldi-Tunnicliffe, A., and Mitchell, T.J. (2005). Innate immune defense against pneumococcal pneumonia requires pulmonary complement component C3. Infect. Immun. *73*, 4245–4252.
Kim, J.O., and Weiser, J.N. (1998). Association of intrastrain phase variation in quantity of capsular polysaccharide and teichoic acid with the virulence of *Streptococcus pneumoniae*. J. Infect. Dis. *177*, 368–377.
Kim, J.O., Romero-Steiner, S., Sørensen, U.B., Blom, J., Carvalho, M., Barnard, S., Carlone, G., and Weiser, J.N. (1999). Relationship between cell surface carbohydrates and intrastrain variation on opsonophagocytosis of *Streptococcus pneumoniae*. Infect. Immun. *67*: 2327–2333.
King, S.J., Hippe, K.R., Gould, J.M., Bae, D., Peterson, S., Cline, R.T., Fasching, C., Janoff, E.N., and Weiser, J.N. (2004). Phase variable desialylation of host proteins that bind to *Streptococcus pneumoniae in vivo* and protect the airway. Mol. Microbiol. *54*, 159–171.
Kubori, T., Matsushima, Y., Nakamura, D., Uralil, J., Lara-Tejero, M., Sukhan, A., Galán, J.E., and Aizawa, S.I. (1998). Supramolecular structure of the *Salmonella typhimurium* type III protein secretion system. Science *280*, 602–605.
Lederberg, J. (1952). Cell genetics and hereditary symbiosis. Physiol. Rev. *32*, 403–430.
LeMieux, J., Woody, S., and Camilli, A. (2008). Roles of the sortases of *Streptococcus pneumoniae* in assembly of the RlrA pilus. J. Bacteriol. *190*, 6002–6013.
Lesnick, M.L., Reiner, N.E., Fierer, J., and Guiney, D.G. (2001). The *Salmonella spvB* virulence gene encodes an enzyme that ADP-ribosylates actin and destabilizes the cytoskeleton of eukaryotic cells. Mol. Microbiol. *39*, 1464–1470.
Libby, S.J., Lesnick, M., Hasegawa, P., Weidenhammer, E., and Guiney, D.G. (2000). The *Salmonella* virulence plasmid *spv* genes are required for cytopathology in human monocyte-derived macrophages. Cell. Microbiol. *2*, 49–58.

Lin, T.L., Lee, C.Z., Hsieh, P.F., Tsai, S.F., and Wang, J.T. (2008). Characterization of integrative and conjugative element ICEKp1-associated genomic heterogeneity in a *Klebsiella pneumoniae* strain isolated from a primary liver abscess. J. Bacteriol. *190*, 515–526.

Matsui, H., Bacot, C.M., Garlington, W.A., Doyle, T.J., Roberts, S., and Gulig, P.A. (2001). Virulence plasmid-borne *spvB* and *spvC* genes can replace the 90-kilobase plasmid in conferring virulence to *Salmonella enterica* serovar Typhimurium in subcutaneously inoculated mice. J. Bacteriol. *183*, 4652–4658.

Maurelli, A.T., Fernández, R.E., Bloch, C.A., Rode, C.K., and Fasano, A. (1998). 'Black holes' and bacterial pathogenicity: a large genomic deletion that enhances the virulence of *Shigella* spp. and enteroinvasive *Escherichia coli*. Proc. Natl. Acad. Sci. U.S.A. *95*, 3943–3948.

Mavroidi, A., Aanensen, D.M., Godoy, D., Skovsted, I.C., Kaltoft, M.S., Reeves, P.R., Bentley, S.D., and Spratt, B.G. (2007). Genetic relatedness of the *Streptococcus pneumoniae* capsular biosynthetic loci. J. Bacteriol. *189*, 7841–7855.

Mazel, D., (2006). Integrons: agents of bacterial evolution. Nat. Rev. Microbiol. *4*, 608–620.

Mazurkiewicz, P., Thomas, J., Thompson, J.A., Liu, M., Arbibe, L., Sansonetti, P., and Holden, D.W. (2008). SpvC is a *Salmonella* effector with phosphothreonine lyase activity on host mitogen-activated protein kinases. Mol. Microbiol. *67*, 1371–1383.

Medina, E., Paglia, P., Nikolaus, T., Müller, A., Hensel, M., and Guzmán, C.A. (1999) Pathogenicity island 2 mutants of *Salmonella typhimurium* are efficient carriers for heterologous antigens and enable modulation of immune responses. Infect. Immun. *67*, 1093–1099.

Mills, D.M., Bajaj, V., and Lee, C.A. (1995). A 40 Kb chromosomal fragment encoding *Salmonella* Typhimurium invasion genes is absent from the corresponding region of the *Escherichia coli* K-12 chromosome. Mol. Microbiol. *15*, 749–759.

Monack, D.M., Hersh, D., Ghori, N., Bouley, D., Zychlinsky, A., and Falkow, S. (2000). *Salmonella* exploits caspase-1 to colonize Peyer's patches in a murine typhoid model. J. Exp. Med. *192*, 249–258.

Moore, R.A., Reckseidler-Zenteno, S., Kim, H., Nierman, W., Yu, Y., Tuanyok, A., Warawa, J., DeShazer, D., and Woods, D.E. (2004). Contribution of gene loss to the pathogenic evolution of *Burkholderia pseudomallei* and *Burkholderia mallei*. Infect. Immun. 72, 4172–4187.

Morona, J.K., Paton, J.C., Miller, D.C., and Morona, R. (2000). Tyrosine phosphorylation of CpsD negatively regulates capsular polysaccharide biosynthesis in *Streptococcus pneumoniae*. Mol. Microbiol. *35*, 1431–1442.

Moschioni, M., Donati, C., Muzzi, A., Masignani, V., Censini, S., Hanage, W.P., Bishop, C.J., Reis, J.N., Normark, S., Henriques-Normark, B., *et al.* (2008). *Streptococcus pneumoniae* contains 3 *rlrA* pilus variants that are clonally related. J. Infect. Dis. *197*, 888–896.

Musser, J.M., Gray, B.M., Schlievert, P.M., and Pichichero, M.E. (1992). *Streptococcus pyogenes* pharyngitis: characterization of strains by multilocus enzyme genotype, M and T protein serotype, and pyrogenic exotoxin gene probing. J. Clin. Microbiol. *30*, 600–603.

Muzzi, A., Moschioni, M., Covacci, A., Rappuoli, R., and Donati, C. (2008). Pilus operon evolution in *Streptococcus pneumoniae* is driven by positive selection and recombination. PLoS One *3*, e3660.

Nassif, X., Honoré, N., Vasselon, T., Cole, S.T., and Sansonetti, P.J. (1989). Positive control of colanic acid synthesis in *Escherichia coli* by *rmpA* and *rmpB*, two virulence-plasmid genes of *Klebsiella pneumoniae*. Mol. Microbiol. *3*, 1349–1359.

Negrea, A., Bjur, E., Puiac, S., Ygberg, S.E., Aslund, F., and Rhen, M. (2009). Thioredoxin 1 participates in the activity of the *Salmonella enterica* serovar Typhimurium pathogenicity island 2 type III secretion system. J. Bacteriol. *191*, 6918–6927.

Nelson, A.L., Ries, J., Bagnoli, F., Dahlberg, S., Fälker, S., Rounioja, S., Tschöp, J., Morfeldt, E., Ferlenghi, I., Hilleringmann, M., *et al.* (2007a). RrgA is a pilus-associated adhesin in *Streptococcus pneumoniae*. Mol. Microbiol. *66*, 329–340.

Nelson, A.L., Roche, A.M., Gould, J.M., Chim, K., Ratner, A.J., and Weiser, J.N. (2007b). Capsule enhances pneumococcal colonization by limiting mucus-mediated clearance. Infect. Immun. *75*, 83–90.

Ochman, H.,. Soncini, F.C., Solomon, F., and Groisman, E.A. (1996). Identification of a pathogenicity island required for *Salmonella* survival in host cells. Proc. Natl. Acad. Sci. U.S.A. *93*,7800–7804.

Overweg, K., Pericone, C.D., Verhoef, G.G., Weiser, J.N., Meiring, H.D., De Jong, A.P., De Groot, R., and Hermans, P.W. (2000). Differential protein expression in phenotypic variants of *Streptococcus pneumoniae*. Infect. Immun. *68*, 4604–4610.

Parkhill, J., Sebaihia, M., Preston, A., Murphy, L.D., Thomson, N., Harris, D.E., Holden, M.T., Churcher, C.M., Bentley, S.D., Mungall, K.L., *et al.* (2003). Comparative analysis of the genome sequences of *Bordetella pertussis*, *Bordetella parapertussis* and *Bordetella bronchiseptica*. Nat. Genetics *35*, 32–40.

Paterson, G.K., and Mitchell, T.J. (2006). The role of *Streptococcus pneumoniae* sortase A in colonisation and pathogenesis. Microbes Infect. *8*, 145–153.

Paterson, G.K., Blue, C.E., and Mitchell, T.J. (2006). Role of two-component systems in the virulence of *Streptococcus pneumoniae*. J. Med. Microbiol. *55*, 355–363.

Preston, A., Parkhill, J., and Maskell, D.J. (2004). The Bordetellae: lessons from genomics. Nat. Rev. Microbiol. *2*, 379–390.

Rappl, C., Deiwick, J., and Hensel, M. (2003). Acidic pH is required for the functional assembly of the type III secretion system encoded by *Salmonella* pathogenicity island 2. FEMS Microbiol. Lett. *226*,363–372.

Regev-Yochay, G., Lipsitch, M., Basset, A., Rubinstein, E., Dagan, R., Raz, M., and Malley, R. (2009). The pneumococcal pilus predicts the absence of *Staphylococcus aureus* co-colonization in pneumococcal carriers. Clin. Infect. Dis. *48*, 760–763.

Rosch, J.W., Mann, B., Thornton, J., Sublett, J., and Tuomanen, E. (2008). Convergence of regulatory networks on the pilus locus of *Streptococcus pneumoniae*. Infect. Immun. *76*, 3187–3196.

Roudier, C., Fierer, J., and Guiney, D.G. (1992). Characterization of translation termination mutations in the *spv* operon of the *Salmonella* virulence plasmid pSDL2. J. Bacteriol. *174*, 6418–6423.

Rowe-Magnus, D.A., Guerout, A.M., Biskri, L., Bouige, P., and Mazel, D. (2003). Comparative analysis of superintegrons: engineering extensive genetic diversity in the Vibrionaceae. Genome Res. *13*, 428–442.

Saluja, S.K., and Weiser, J.N. (1995) The genetic basis of colony opacity in *Streptococcus pneumoniae*: evidence for the effect of box elements on the frequency of phenotypic variation. Mol. Microbiol. *16*, 215–227.

Shea, J.E., Beuzon, C.R., Gleeson, C., Mundy, R., and Holden, D.W. (1999). Influence of the *Salmonella typhimurium* pathogenicity island 2 type III secretion system on bacterial growth in the mouse. Infect. Immun. *67*, 213–219.

Sitkiewicz, I., Nagiec, M.J., Sumby, P., Butler, S.D., Cywes-Bentley, C., and Musser, J.M. (2006). Emergence of a bacterial clone with enhanced virulence by acquisition of a phage encoding a secreted phospholipase A2. Proc. Natl. Acad. Sci. U.S.A. *103*, 16009–16014.

Smoot, J.C., Barbian, K.D., Van Gompel, J.J., Smoot, L.M., Chaussee, M.S., Sylva, G.L., Sturdevant, D.E., Ricklefs, S.M., Porcella, S.F., Parkins, L.D., *et al.* (2002). Genome sequence and comparative microarray analysis of serotype M18 group A *Streptococcus* strains associated with acute rheumatic fever outbreaks. Proc. Natl. Acad. Sci. U.S.A. *99*, 4668–4673.

Song, X.M., Connor, W., Hokamp, K., Babiuk, L.A., and Potter, A.A. (2009). The growth phase-dependent regulation of the pilus locus genes by two-component system TCS08 in *Streptococcus pneumoniae*. Microb. Pathog. *46*, 28–35.

Stecher, B., Robbiani, R., Walker, A.W., Westendorf, A.M., Barthel, M., Kremer, M., Chaffron, S., Macpherson, A.J., Buer, J., Parkhill, J., *et al.* (2007). *Salmonella enterica* serovar Typhimurium exploits inflammation to compete with the intestinal microbiota. PLoS Biol. *5*, e244.

Stokes, H.W., and Hall, R.M. (1989) A novel family of potentially mobile DNA elements encoding site-specific gene-integration functions: integrons. Mol. Microbiol. 3, 1669–1683.

Sumby, P.,. Barbian, K. D, Gardner, D.J., Whitney, A.R., Welty, D.M., Long, R.D., Bailey, J.R., Parnell, M.J., Hoe, N.P., Adams, G.G., *et al.* (2005a). Extracellular deoxyribonuclease made by group A *Streptococcus* assists pathogenesis by enhancing evasion of the innate immune response. Proc. Natl. Acad. Sci. U.S.A. 102, 1679–1684.

Sumby, P., Porcella, S.F., Madrigal, A.G., Barbian, K.D., Virtaneva, K., Ricklefs, S.M., Sturdevant, D.E., Graham, M.R., Vuopio-Varkila, J., Hoe, N.P., *et al.* (2005b). Evolutionary origin and emergence of a highly successful clone of serotype M1 Group A *Streptococcus* involved multiple horizontal gene transfer events. J. Infect. Dis. *192*, 771–782.

Tezcan-Merdol, D., Engstrand, L., and Rhen, M. (2005). *Salmonella enterica* SpvB-mediated ADP-ribosylation as an activator for host cell actin degradation. Int. J. Med. Microbiol. *295*, 201–212.

Tezcan-Merdol, D., Nyman, T., Lindberg, U., Haag, F., Koch-Nolte, F., and Rhen, M. (2001). Actin is ADP-ribosylated by the *Salmonella enterica* virulence-associated protein SpvB. Mol. Microbiol. *39*, 606–619.

Thijs, I.M., De Keersmaecker, S.C., Fadda, A., Engelen, K., Zhao, H., McClelland, M., Marchal, K., and Vanderleyden, J. (2007). Delineation of the *Salmonella enterica* serovar Typhimurium HilA regulon through genome-wide location and transcript analysis. J. Bacteriol. *189*, 4587–4596.

Valdivia, R.H., and Falkow, S. (1997). Fluorescence-based isolation of bacterial genes expressed within host cells. Science *277*, 2007–2011.

Vandam, J.E., Fleer, A., and Snippe, H. (1990) Immunogenicity and immunochemistry of *Streptococcus pneumoniae* capsular polysaccharides. Antonie Van Leeuwenhoek *58*, 1–47.

van der Velden, A.W., Velasquez, M., and Starnbach, M.N. (2003). *Salmonella* rapidly kill dendritic cells via a caspase-1-dependent mechanism. J. Immunol. *171*, 6742–6749.

Watanabe, T., and Fukasawa, T. (1961). Episome-mediated transfer of drug resistance in Enterobacteriaceae.1. Transfer of resistance factors by conjugation. J. Bacteriol. *81*, 669–678.

Watson, D.A., and Musher, D.M. (1990). Interruption of capsule production in *Streptococcus pneumoniae* serotype3 by insertion of transposon Tn*916*. Infect. Immun. *58*, 3135–3138.

Weiser, J.N., Austrian, R., Sreenivasan, P.K., and Masure, H.R. (1994). Phase variation in pneumococcal opacity: relationship between colonial morphology and nasopharyngeal colonization. Infect. Immun. *62*, 2582–2589.

Weiser, J.N., Bae, D., Epino, H., Gordon, S.B., Kapoor, M., Zenewicz, L.A., and Shchepetov, M. (2001). Changes in availability of oxygen accentuate differences in capsular polysaccharide expression by phenotypic variants and clinical isolates of *Streptococcus pneumoniae*. Infect. Immun. *69*, 5430–5439.

Wilson, J.A., Doyle, T.J., and Gulig, P.A. (1997). Exponential-phase expression of *spvA* of the *Salmonella typhimurium* virulence plasmid: induction in intracellular salts medium and intracellularly in mice and cultured mammalian cells. Microbiology *143*, 3827–3839.

Wozniak, R.A., and Waldor, M.K. (2009). A toxin-antitoxin system promotes the maintenance of an integrative conjugative element. PLoS Genet. *5*, e1000439.

Part II

Bacterial Adhesion to the Cell Surface and Extracellular Matrix of Host Tissues

Fimbrial Adhesins: Adhesive Molecules on a 'Stalk'

4

Hae Joo Kang, Edward N. Baker and Thomas Proft

Abstract

As an important step for the successful and continuous colonization of the host, bacterial pathogens express a variety of specific adhesins on their cell surface, which allows them to interact with receptors on host cells. However, this close interaction might also be detrimental for the bacteria, as it could trigger infiltration and activation of immune cells and eventually lead to phagocytosis. Another obstacle for colonization is electrostatic repulsion due to the negative surface charge on both the bacteria and the host cell. Many bacteria have overcome this problem by expressing adhesins at the tip of a long fibril structure that extends from the bacterial cell surface. These structures are known as fimbriae or pili. Despite their diversity in structure and biogenesis, pili/fimbriae typically consist of a long fibre formed by homopolymerized subunits or pilins, and accessory pilins that often function as adhesins. Some pili are also involved in cell aggregation, biofilm formation, DNA uptake, phage transduction and gliding motility (Table 4.1).

Introduction

Most commensal and pathogenic bacteria express adhesive molecules on the cell surface, which allow specific interaction with receptors on the host cells or with components of the extracellular matrix (ECM). This is an important step for the successful and continuous colonization of the host. However, this close interaction might come at a cost, as it could trigger signal pathways that result in infiltration and activation of immune cells and eventually lead to phagocytosis and bacterial clearance. Another obstacle for colonization is electrostatic repulsion due to the negative surface charge on both the bacteria and the host cell. Many bacteria have overcome this dilemma by expressing adhesins at the tip of a long fibril structure that extends from the bacterial cell surface. These structures, known as pili (singular pilus, Latin for 'hair') or fimbriae (singular fimbria, Latin for 'thread' or 'fibre') provide a 'spacer' between the different cell surfaces and allow cell–cell interaction at a 'safe' distance. The main fibre, also known as the 'fimbrial rod' or 'pilus shaft', is a polymeric structure composed of several hundreds (probably thousands) of small 15–25 kDa subunits or pilins.

Pili are important virulence factors for several pathogens, in particular those that cause infections of the urinary, genital and gastrointestinal tracts. Although often described as adhesive organelles, pili have also been implicated in other functions, such as DNA transfer, phage binding, biofilm formation, cell aggregation, host cell invasion and twitching motility. In this chapter, we will discuss adhesive pili/fimbriae from both Gram-negative and Gram-positive bacteria.

Pili of Gram-negative bacteria

Pili of Gram-negative bacteria were first discovered in the late 1940s as receptors for bacteriophages (Anderson, 1949) and have been studied extensively over several decades. This has been described in a number of excellent reviews and books. Gram-negative pili can be categorized into three different classes according to their structure and assembly mechanisms: (i) pili assembled by the 'chaperone-usher pathway'; (ii) the Type IV pili; and (iii) pili assembled by the extracellular nucleation/precipitation pathway (curli pili).

Chaperone-usher pathway assembled pili

The most extensively characterized pili of this group are the Type I pili, which are found throughout the family of *Enterobacteriaceae,* and the P pili from uropathogenic *Escherichia coli* (UPEC). The adhesive structures are heteropolymers composed of a small number of different protein subunits at various stoichiometries. A flexible fibrillar tip is joined end-to-end to a rigid rod and a single specific adhesive protein, which binds surface carbohydrates on host cells, is located at the distal end of the structure (Sauer *et al.*, 2004; Capitani *et al.*, 2006). Another group of adhesive structures assembled by the chaperone-usher pathway comprises the 'non-pilus adhesins'. Although originally characterized as afimbrial owing to their amorphous or capsule-like morphology at low resolution, several members of this group can in fact be assembled into genuine fimbrial structures. The adhesive structures are generally homopolymers composed of a single protein subunit, like the Afa/Dr family of adhesins of *E. coli* (Anderson *et al.*, 2004; Piatek *et al.*, 2005), and the polymeric F1 capsular antigen of *Yersinia pestis* (Zavialov *et al.*, 2003; Knight, 2007).

During pilus assembly (Fig. 4.1A), the pilus subunits (pilins) are secreted into the periplasmic space via the general secretory pathway and bind to a specific chaperone that assists in protein folding and prevents premature assembly of the subunits. The protein structures of the pilus subunits (pilins) show incomplete immunoglobulin (Ig)-like structures in which the seventh (G) β-strand is missing. The chaperone is composed of two Ig-like domains and the N-terminal domain contains a G1 β-strand that completes the Ig-like structure of the pilin subunit by a mechanism known as 'donor strand complementation'. The pilus subunits polymerize after the chaperone–subunit complex contacts an outer membrane channel-forming protein, called the 'usher' which serves as a platform for pilus assembly. The 'usher' forms a twin-pore secretion complex with a central 2 nm wide β-pore in each subunit (Li *et al.*, 2004), which allows the translocation of chaperone–subunit complexes, but not the assembled pilus rod (Thanassi *et al.*, 1998; Sauer *et al.*, 2000a,b, 2002; Capitani *et al.*, 2006). Pilus assembly is achieved by a mechanism known as 'donor strand exchange'. Each pilus subunit, except the tip adhesin, contains an N-terminal extension of about 15 residues, which can replace the donor strand of the chaperone to complement the fold of the previous pilus subunit (Sauer *et al.*, 2002). The pilus tip adhesin initiates assembly and is the first subunit incorporated into the pilus, based on the higher affinity of the usher for the chaperone–adhesin complex (Munera *et al.*, 2007).

Type I pili

Type I pili are found on most *E. coli* strains and throughout the family of *Enterobacteriaceae.* They are the most prevalent type of pilus in UPEC, where they contribute significantly to bladder infections (cystitis) (Connell *et al.*, 1996). The bladder, a sterile site in healthy individuals, is the primary site of infection in more than 90% of all urinary tract infections (UTIs) (Muhldorfer *et al.*, 2001). Adhesion of UPEC to host cells within the urinary tract enables the bacteria to colonize and to avoid rapid clearance from the flow of urine. The Type I pilus is encoded by the *fim* gene cluster (*fimA-fimH*) and has a composite structure based on a 6.9 nm-thick and 1–2 μm-long helical rod which is formed by a right-handed helical array of 500–3000 copies of the main structural subunit FimA. The rod is connected via FimF to a short stubby 3 nm-wide linear tip fibrillum containing FimG and the specific adhesin FimH (Jones *et al.*, 1995; Hahn *et al.*, 2002).

FimH has a two-domain fold with an N-terminal receptor-binding domain and a C-terminal pilin domain, both showing an immunoglobulin (Ig)-like fold (Choudhury *et al.*, 1999) (Fig. 4.2A). FimH binds to mannose-containing receptors expressed by many types of host cells (Baorto *et al.*, 1997; Zhou *et al.*, 2001). Mannose groups are abundant on the glycan moieties of uroplakins, integral membrane glycoprotein receptors that coat the luminal surface of the bladder epithelium. FimH specifically binds to uroplakin UP1a, but

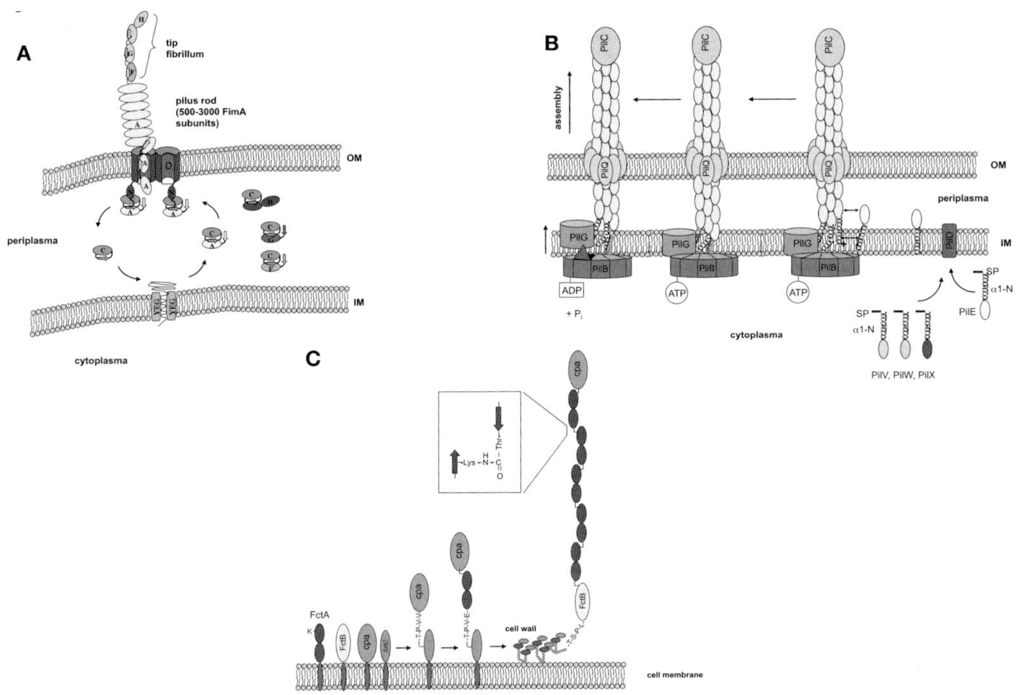

Figure 4.1 Assembly of various pili in Gram-negative and Gram-positive bacteria. (A) The *E. coli* Type I pilus as an example of a chaperone-usher pathway assembled pilus. Pilin subunits are secreted through the inner membrane (IM) into the periplasm by the Sec YEG translocon. The chaperone FimC accelerates subunit folding and stabilizes the proteins by donating a strand to complement the incomplete Ig-like fold of the subunit proteins, a mechanism known as 'donor strand complementation'. The chaperone–subunit complexes are then delivered to the FimD transmembrane assembly platform (usher), which recognizes the binary complexes by an extended N-terminal domain. The FimC chaperone is released back to the periplasm after the donor strand is replaced by an N-terminal extension of another pilin subunit, a mechanism known as 'donor strand exchange'. Although FimD forms a twin–pore complex, only one pore is used for secretion, whereas both pores are used for chaperone-pilin recruitment. The completed pilus is composed of a pilus rod of polymerized FimA subunits and a tip fibrillum with FimF, FimG and the tip adhesin FimH. Adapted from Sauer *et al.*, 2000a. (B) Assembly of the Neisseria GC/MC pilus as an example of a Type IV pilus. The major pilin subunits (PilA) are secreted through the inner membrane (IM) into the periplasm and the signal peptide (SP) is cleaved by the prepilin peptidase PilD. The negative charge of the glutamic acid side chain (E5) in the α1-N region attracts the positively charged N-terminus of the terminal PilA subunit in the growing fibre. Additional interactions between the globular PilA domains enable the subunit to fill the existing gap at the pilus base. The assembly ATPase (PilB), which is associated with the cytoplasmic part of the inner membrane protein PilG, undergoes conformational change during ATP hydrolysis and pushes the pilus filament out of the membrane, providing a gap for the next PilA subunit. Several minor pilins (e.g. PilV, PilW and PilX) also contain the α1-N region and can also be added to the pilus structure. In addition to the assembly ATPase PilB, the GC/MC pili apparatus also contains a structurally related retraction ATPase (PilT) that can remove pilin subunits at the pilus base resulting in 'twitching motility' (not shown). Adapted from Craig and Li, 2008. (C) The GAS serotype M1 pilus as an example of Gram-positive pilus assembly. Pilin proteins are produced as precursor molecules with C-terminal membrane anchors and a cell wall sorting signal (CWSS) such as the LPXTG sequence motif or variations. It is suggested that a pilus polymer grows as pilins are progressively added to the base of the growing pilus. This implies that the fimbrial adhesin Cpa, located at the pilus tip, is incorporated as the first subunit of a pilus. The pilus stalk is a covalent polymer of the major pilin FctA. A specialized sortase (SrtC) cleaves the CWSS motif of FctA between threonine and glycine to generate a covalently linked sortase-pilin intermediate. The sortase-pilin intermediate is then attacked by a nucleophile provided by another pilin in the form of the ε-amino group of a lysine residue (Lys161 for the major pilin Spy0128 of serotype M1). The assembled pilus polymer is finally anchored to the cell wall, again through sortase action; the minor pilin FctB is thought to be incorporated at the base of the pilus with its CWSS linked to the cell wall by a housekeeping sortase. Parts A and B of this figure were reproduced from Proft, T., and Baker, E.N. (2009). Cell. Mol. Life Sci. *66*, 613–635, with permission from the publisher.

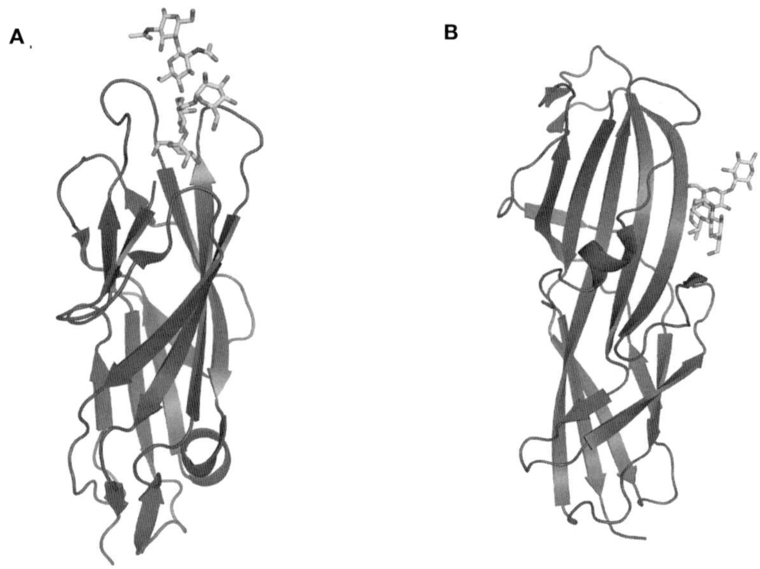

Figure 4.2 Co-crystal structure of fimbrial adhesins with their respective ligand molecules. (A) FimH in complex with oligomannose-3 (PDB code 2VCO). (B) PapG bound to GbO4 glycolipid receptor (PDB code 1J8R).

not to the structurally related UP1b, which is glycosylated differently (Zhou *et al.*, 2001). The basis for this specificity has recently been described by Xie *et al.* (2006), who demonstrated that UPIa presents a high level of terminally exposed mannose residues that are capable of specifically interacting with FimH. In contrast, most terminally exposed glycans of UPIb are non-mannose residues.

Quantitative differences in the adhesiveness of Type I-piliated *E. coli* have been attributed to FimH variants that differ in their receptor-binding domain (Sokurenko *et al.*, 1995). All naturally occurring FimH variants bind tri-mannose receptors, but can vary in their affinity to bind mono-mannose receptors. FimH variants that bind mono-mannose groups with high affinity are preferentially found in UPEC strains, whereas those that bind mono-mannose groups with low affinity are prevalent on commensal *E. coli* strains (Sokurenko *et al.*, 1995, 1998). However, these findings have recently been questioned by Bouckaert *et al.* (2006), who reported that FimH variants from UPEC, enterohaemorrhagic *E. coli* (EHEC) and faecal *E. coli* isolates expressed the same specificities and affinities for high-mannose structures. They also showed that high mannose glycans ending in Manα1-3Manβ1-4GlcNAc are the best FimH receptors and concluded that the carbohydrate expression profiles of the targeted host tissues are stronger determinants of adhesion than FimH variants.

A recent study by Duncan *et al.* (2005) revealed that FimH is not solely responsible for target receptor specificity and that the pilus rod might also play a role. They showed that expression of FimH on a heterologous pilus rod altered the binding of the adhesion, suggesting that the specificity might be modulated through conformational constraints imposed on FimH by the pilus rod.

Infection by Type I pilus-expressing *E. coli* leads to exfoliation of uroplakin-coated cells, which is believed to be a defence mechanism to eliminate infection. However, some bacteria are able to evade the host response by FimH-mediated invasion of bladder epithelial cells. This contributes to the frequent recurrence of UTIs in many patients (Mulvey *et al.*, 1998; Martinez *et al.*, 2000). Once internalized, UPEC rapidly replicates and forms intracellular bacterial communities (IBCs) with biofilm-like properties, a process that requires expression of Type 1 pili and functional FimH by the intracellular bacteria (Wright *et al.*, 2007). It has recently been shown that the FimH receptors for bladder cell invasion differ from the receptors for cell adhesion. FimH specifically recognizes N-linked glycans on α3β1 integrins, which are expressed throughout the urothelium (Eto *et al.*, 2007). This FimH function appears to be independent from mannose binding. In a recent study by Chen *et al.* (2009), a double mutation outside the mannose-binding pocket (A27V, V163A) resulted in a 10,000-fold reduction in mouse bladder colonization and lacked the ability to form IBCs, but had no effect on mannose binding. Bacterial invasion was shown to occur via a cholesterol-and dynamin-dependent phagocytosis-like mechanism that is modulated by calcium levels and clathrin (Eto *et al.*, 2008). A more recent study has shown that interactions of FimH with host receptors, including α3β1 integrin, trigger localized actin rearrangements that lead to internalization of adherent bacteria via a zipper-like mechanism (Dhakal and Mulvey, 2009). FimH binding to bladder cells triggers a signal transduction cascade that results in host actin reorganization, phosphoinositide-3-kinase activation and host protein tyrosine phosphorylation. The small Rho-binding proteins Cdc42 and RhoA are also required for the cell invasion process (Martinez *et al.*, 2000; Martinez and Hultgren, 2002).

Type I pilus expressing *E. coli* are also able to invade mast cells and macrophages after binding of FimH to the GPI-anchored receptor CD48. This process also involves subcellular lipid raft-like structures known as calveolae. It was suggested that this might result in the formation of membrane-bound vacuoles that encapsulate the bacteria and prevent phagocytosis (Shin *et al.*, 2000). A study by Hase *et al.* (2009) has revealed a role for Type I pili in antigen transcytosis across the epithelial barrier into organized lymphoid structures, a process that is mediated by specialized epithelial M cells. They showed that enterobacteria expressing the FimH adhesin specifically bind to glycoprotein 2 (GP2) on the apical surface of M cells, followed by internalization into cytoplasmatic vesicles. Lack of FimH led to defects in trancytosis of Type 1-piliated bacteria and in attenuation of antigen-specific immune responses

Type I pili have also been implicated in biofilm formation. *E. coli* forms biofilms on abiotic surfaces and Type 1 pili are required for initial surface attachment, a process that can be inhibited by mannose (Pratt and Kolter, 1998). Orndorff *et al.* (2004) have shown that aggregation of *E. coli* by secretory IgA (SIgA) is dependent on the pilus structure. Interestingly, this was independent of FimH. The pilus without the FimH adhesin also facilitated SIgA-mediated biofilm formation on polystyrene, although biofilm formation was stronger in the presence of FimH.

Type P pili

P pili are critical virulence factors in the establishment of pyelonephritis by UPEC, mediating recognition of and attachment of the bacteria to tissues of the kidney (Mulvey *et al.*, 2000). The 6.8 nm-wide and several micrometres-long rod is a right-handed helical cylinder composed of repeating pyelonephritis-associated pili (Pap) A proteins with 3.28 subunits per turn. The 2–3 nm-thick tip fibrillum contains the distally located PapG adhesin and the three minor pilus proteins PapE, PapF and PapK (Kuehn *et al.*, 1992; Bullitt and Makowski, 1995). There are three different classes of PapG variants (PapGI, -II, and –III), which differ in their receptor specificity. All receptors bound by the PapG variants contain a common Gal(α1–4)Gal moiety linked to a ceramide group by a β-glucose residue, but differ in the number of *N*-acetylgalactosamine moeties or by the addition of sialic acid residues (Roberts *et al.*, 1994). The receptors are found on human erythrocytes of the P blood group and on uroepithelial cells, and the different binding preferences of the PapG variants seem to contribute to host cell tropism of P-piliated UPEC (Strömberg *et al.*, 1990). The PapG-I variant preferentially binds globotriaosylceramide (GbO3), which is abundant on human uroepithelial cells. PapG-II binds to globoside (GbO4, a glycolipid iso-receptor of the human kidney) (Fig. 4.2B) and is primarily associated with human pyelonephritis and bacteraemia, whereas the PapG-III variant preferentially binds to Forssman antigen (globopentosylceramide, GbO5) and is associated with human cystitis (Strömberg *et al.*, 1991).

The exact contribution of P pili to the pathogenesis of UPEC is not entirely clear. A study by Hagberg *et al.* (1983) showed that an isogenic

P pili mutant was significantly outcompeted by wild-type UPEC in the kidneys of CBA and BALB/c mice during mixed infection. However, another study using a P pilus-defective strain found no significant difference in organism concentration in urine, bladder or kidney between the mutant and wild-type UPEC after transurethral inoculation of CBA mice (Mobley *et al.*, 1993). Given the fact that UPEC can express up to 12 different fimbriae (usually, *E. coli* expresses only one type at a time), it is possible that the loss of P pili-mediated adhesion might be compensated by another fimbrial type.

Afa/Dr adhesin family

The Afa/Dr adhesins are a highly heterogeneous group of homopolymeric adhesive organelles identified in UPEC and diffusely adhering *E. coli* (DAEC). They were originally identified as afimbrial structures, but several members can be assembled into pilus-like structures, which resemble the thin fibrillum at the tip of Type 1 and P pili. Others resemble afimbrial capsules surrounding the bacterial cell (reviewed by Nowicki *et al.*, 2001; Servin, 2005). The structure of the adhesin determines whether the adhesive organelle forms fimbriae (e.g. DraE and DaaE subunits) or an afimbrial structure (e.g. AfaE-III subunit) (Anderson *et al.*, 2004). In the fimbrial structure, many copies of a receptor-binding major subunit (a single-domain adhesin) assemble into thin and flexible fibres.

Most Afa/Dr adhesins (AfaE/DraE) recognize the decay accelerating factor (DAF, CD55) which is expressed on erythrocytes and other tissues, including uroepithelium. DAF is a complement-regulatory membrane protein that protects host tissue from damage by accelerating the decay of C3 and C5 convertases on epithelial cells. In addition, several members of the carcinoembryonic antigen-related cell adhesion molecules (CEACAMs), including carcinoembryonic antigen (CEA), CEACAM-1 and CEACAM-6 are recognized by DraE/AfaE. At least four members of the Dr adhesin family (AfaE-I, AfaE-III, DraE and DaaE) possess two binding sites and bind to both receptors independently (Korotkova *et al.*, 2006). Using a combination of NMR spectroscopy, mutagenesis and biochemical characterization, Korotkova *et al.* have shown that the DraE/CEA interface overlaps with the region for CEA dimerization. They suggest that DraE dissociates CEA dimers before binding to the monomeric CEA receptor (Korotkova *et al.*, 2008a). The DraE subunit (but not the highly homologous AfaE-III) also binds to Type IV collagen. A strong positive selection for amino acid replacements was found in DraE, and variants with decreased sensitivity of DAF binding to high salt were found on isolates associated with UTIs suggesting functional adaptation to their pathogenic niche.

Afa/Dr adhesins are believed to facilitate ascending colonization and chronic infection of the urinary tract and some members are also associated with enteric infection. In addition, they have been shown to facilitate UPEC invasion of uroepithelial cells. Dr-fimbriated *E. coli* cells enter eukaryotic cells via a zipper-like mechanism, which is dependent on dynamic microtubules, lipid rafts, PI3K activation and α5β1 integrin (Kansau *et al.*, 2004; Korotkova *et al.*, 2008b; Guignot *et al.*, 2009). It has been reported that AfaD/DraD has invasin properties by specifically binding to integrin α5β1 (Cota *et al.*, 2006). However, two recent articles question the role of AfaD/DraD in β1 integrin recruitment and bacterial internalization. *E. coli* cells were internalized into CHO cells transfected with DAF or CEACAM receptors in a DraE-mediated, but DraD-independent mechanism. Furthermore, recombinant bacteria expressing DraE were able to adhere and invade primary human bladder cells (Korotkova *et al.*, 2008b; Guignot *et al.*, 2009). A recent study by Diard *et al.* (2006) has shown that Dr fimbriae can be released into the extracellular medium in response to environmental signals, such as temperature and reduced oxygen, but the purpose of this mechanism remains elusive.

S pili

Another pilus type expressed by *E. coli* is the S pilus. S pili are associated with *E. coli* strains that cause sepsis, neonatal meningitis and UTI, including pyelonephritis. The SfaS adhesin interacts with α-sialyl-2,3-galactose receptors on endothelial cells and kidney epithelial cell. Sialic acid residues are found on uroplakin UPIII, which is strongly expressed on the luminal bladder surface suggesting a possible role of S pili in cystitis (Malagolini *et al.*, 2000). Several other pili homologous to S pili, but with different receptor specificity, have been identified and include the F1C pili, which bind to β-GalNAc-1,4β-Gal residues on glycolipids expressed by epithelial and endothelial kidney cells (Khan *et al.*, 2000). The major SfaA pilin also has adhesive properties and binds to endothelial cell glycolipids and plasminogen (reviewed in Mulvey, 2002).

Hif pili

Haemophilus influenzae is a commensal of the upper respiratory tract of humans that can cause diseases such as otitis media and meningitis. Pili are required for *H. influenzae* adherence to respiratory epithelial cells and facilitate colonization of the upper respiratory tract (Gilsdorf *et al.*, 1997). Haemagglutination-positive strains of *H. influenzae* express Hif pili with a diameter of 6–7 nm and a right-handed helical architecture. The main fibre is made up of polymerized HifA subunits and the minor pilins HifD and HifE. HifE is located at the pilus tip and is believed to function as an adhesin, although no specific host cell receptor has been identified. However, HifE was shown to mediate haemagglutination and attachment to human cells (McCrea *et al.*, 1997). The expression of Hif pili is down-regulated soon after nasal infection suggesting that they are important for initial colonization of the human nasopharynx, but not for persistence within the respiratory epithelium and systemic invasion.

Pili expressed by *Proteus mirabilis*

Proteus mirabilis, a common causative agent of cystitis and pyelonephritis, produces several types of pili that are assembled by the chaperone-usher pathway, including the *Proteus mirabilis* fimbriae (PMF), the mannose-resistant/*Proteus*-like (MR/P) fimbriae, the uroepithelial cell adhesin (UCA) and the ambient-temperature fimbriae (ATF) (reviewed in Rocha *et al.*, 2007). Both, PMF and MR/P are believed to play an important role in the colonization of bladder and kidneys, but no receptors have been described yet. A *P. mirabilis* double mutant that lacks functional PMF and MR/P showed reduced *in vitro* binding to uroepithelial cells and significantly diminished colonization of bladder and kidneys in a murine infection model. However, this was not observed with single mutant that lacks either PMF or MR/P suggesting specific and additive roles of these pili in *P. mirabilis* UTI (Zunino *et al.*, 2007). UCA was found to be responsible for the attachment of *P. mirabilis* to uroepithelial cells by recognizing the common carbohydrate sequence GalNAcβ1,4-Gal present in GM1 and asialo-GM2 glycosphingolipids (Lee *et al.*, 2000). ATF own their name to the optimal expression of this fimbria type at 23°C. No association of ATF with pathogenicity in UTI has been described yet.

Type IV pili

Type IV pili are thin (6–8 nm wide), flexible fibres, several micrometres long, which often aggregate laterally to form characteristic bundles. They are found in a large variety of Gram-negative bacteria, including enteropathogenic *E. coli* (EPEC), *Salmonella enterica, Pseudomonas aeruginosa, Legionella pneumophila, Neisseria gonorrhoeae, Neisseria meningitidis,* and *Vibrio cholerae* (reviewed in Strom and Lory, 1993; Craig *et al.*, 2004; Craig and Li, 2008), but have also been discovered in two Gram-positive genera, *Clostridium* and *Ruminococcus* (Rakotoarivonina *et al.*, 2002; Varga *et al.*, 2006), and in cyanobacteria (Bhaya *et al.*, 2000).

In addition to host cell adhesion and biofilm formation, which are also characteristics of many other pili, Type IV pili possess a number of unique features, including DNA uptake during transformation, phage transduction and a flagella-independent form of movement known as twitching motility. These attributes are based on the capacity of Type IV pili to retract, generating a significant mechanical force (reviewed in Mattick, 2002).

Type IV pili are usually homopolymers of a single 15–20 kDa pilin subunit and in some pili possess an adhesive subunit at the pilus tip. In contrast to other pili, Type IV pili biogenesis (Fig. 4.1B) involves a large number of proteins (12 or more) that are conserved among divergent bacterial species. Several of these are core proteins that are required in all Type IV pili systems and include: (i) the major pilin subunit, which is synthesized as precursors (prepilins); (ii) a specific inner-membrane prepilin peptidase that cleaves the N-terminal signal peptide; (iii) a specific ATPase that powers pilus assembly; (iv) an integral inner-membrane protein that recruits the ATPase from the cytoplasma; and (v) an integral outer-membrane secretin that is necessary for the emergence of Type IV pili on the bacterial surface (Wolfgang *et al.*, 2000; Craig and Li, 2008). Retractile pili also require a retraction ATPase for rapid depolymerization of the filament, which confers twitching motility and allows the bacteria to move along semi-solid surfaces, to transduce phages and to take up DNA (reviewed in Burrows, 2005).

The *Neisseria* pilus

Type IV pili are found in two species of the Gram-negative genus *Neisseria*. The gonococcus (GC) pilus is found in *N. gonorrhoeae*, which causes the sexually transmitted disease gonorrhoea, whereas the meningococcal (MC) pilus is found in *N. meningitidis*, the leading cause of fatal sepsis and meningitis. Both pathogens colonize human mucosal surfaces and this is initiated by binding of GC/MC pili to non-ciliated host cells (Swanson, 1973). Bacterial cell interactions are essential for pathogenesis of *Neisseria* and Type IV pili seem to be the only virulence factor that ensures efficient adhesion to the host cell (reviewed in Nassif *et al.*, 1999).

The up to 40 nm-long pilus fibre consists of polymerized subunits of the 17–21 kDa major pilin PilE with the PilC adhesin (110 kDa) located at the pilus tip and also within the bacterial membrane (Rahman *et al.*, 1997). Type IV pili undergo dynamic cycles of assembly, exposure at the bacterial surface and retraction (Morand *et al.*, 2004). Meningococcal pilus-mediated adhesion and retraction contribute to host cell signalling resulting in cytoskeleton rearrangements and the formation of cortical plaques, which are characterized by dense actin polymerization underneath bacterial clusters (Merz *et al.*, 1999).

Purified or recombinant PilC interacts with different human epithelial cell lines, primary epithelial and endothelial cells (Scheuerpflug *et al.*, 1999). There are two highly homologous variants of PilC (PilC1 and PilC2) and either variant is sufficient for pilus assembly in *N. gonorrhoeae* and *N. meningitidis*. However, while either variant is able to promote adhesion of *N. gonorrhoeae* to host cells, expression of the PilC1 variant in *N. meningitidis* results in a non-adhesive, piliated, transformation-competent strain. Therefore, the meningococcal PilC2 protein was considered as a defective variant that was able to promote piliation and transformation competence but not adhesion to host cells (Nassif *et al.*, 1994). However, more recently Morand *et al.* (2009) have demonstrated that, like PilC1, PilC2 is also capable of mediating adhesion to human endometrium-derived ME180 epithelial cells, with formation of cortical plaques and condensation of F-actin. Interestingly, the two PilC variants promoted different effects on the infected host cells. Expression of PilC1 on the meningococcal pilus resulted in a sharp decrease of epidermal growth factor receptor (EGFR) and significant reduction in the motility of infected cells. In contrast, no effect on the EGFR pathway and no change in motility were noticed after PilC2-mediated infection. Only ME180 cells were permissive for PilC2-mediated infection suggesting a discriminative role for PilC in differential adhesion to specific microenvironments and distinct regulation of infected cells.

The human membrane cofactor protein (MCP, CD46) has been reported as a cellular receptor for both GC pili and MC pili (Källström *et al.*, 1997), however this has recently been questioned. Transgenic mice expressing human CD46 are susceptible to meningococcal infection, in particular after intranasal challenge with piliated bacteria, suggesting that human CD46 facilitates pilus-dependent interactions at the epithelial mucosa (Johansson *et al.*, 2003). CD46, which also serves as a receptor for several other pathogens, is a glycoprotein found on all nucleated cells and plays an important role as an inactivator of complement factors C3b and C4b deposited on self-tissue (Riley-Vargas *et al.*, 2004). In an experiment with a human cervical cell line, adherence of piliated gonococci to the cells resulted in up to 80% reduction of CD46 on the cell surface, owing to specific shedding of the protein. This shows that adherence enables the pathogen to manipulate the host cell environment, although the exact role of the CD46 shedding remains elusive (Gill *et al.*, 2003). In addition, phosphorylation of CD46 by the Src family tyrosine kinase c-Yes upon interaction with piliated gonococci has been reported, suggesting that binding of GC pili to CD46 may not simply be a static process but might promote signalling in the host cell that could be important for bacterial virulence (Lee *et al.*, 2002). On the other hand, Tobiason and Seifert (2001) have shown an inverse correlation between gonococcal adherence and surface expression levels of CD46 suggesting that CD46 does not act as a classic receptor for GC pili. The role of CD46 as pilus receptor has also been questioned by Kirchner *et al.* (2005), who reported that binding efficiencies of piliated gonococci as well as purified PilC2 on human epithelial cell lines did not correlate to the level of surface-expressed CD46. Furthermore, piliated gonococci failed to bind CD46-transfected CHO and MDCK cells. While the search for the PilC receptor continues, Kirchner and Meyer (2005) used biochemical analysis to show that the gonococcal PilC2 receptor is a surface protein that is highly expressed on cervix carcinoma-derived epithelial cells. They also found evidence for uptake of PilC2 via receptor-mediated endocytosis and subsequent receptor re-exposure on the cell surface. Specific binding of gonococcal Type IV pili to human complement regulator C4b-binding protein (C4BP) has also been reported, but the molecular mechanism of this interaction remains elusive (Blom *et al.*, 2001).

Another GC/MC pilus component, the PilE subunit-like protein PilV, was demonstrated to be essential for pilus-mediated cell adherence, but dispensable for pilus biogenesis and other pilus-related phenotypes. However, its function in adhesion appears to be indirect by promoting the functional display of the PilC adhesin in the context of the pilus fibre (Winther-Larsen *et al.*, 2001). The GC/MC pilus also binds to erythrocytes causing haemagglutination, and this has been attributed to the major pilin PilE, but not the PilC adhesin (Scheuerpflug *et al.*, 1999).

After the initial pilus-mediated attachment to host tissue, the bacteria form microcolonies and induce the formation of cortical plaques, which are enriched in both components of the cortical cytoskeleton and a subset of integral membrane proteins. Cortical plaque formation depends not only on the presence of the pilus fibre and the PilC adhesin, but also on expression of the PilT protein (Merz *et al.*, 1999). PilT is an inner membrane-associated ATPase that is involved in pilus retraction and is also required for force-dependent pilus elongation enabling the bacteria to modulate interactions with surfaces by controlling the tension on their pili (Maier *et al.*, 2004). Pilus retraction is required for a specialized form of movement called twitching motility that mediates bacterial movement towards the host cell surface for intimate attachment (Pujol *et al.*, 1999; Merz *et al.*, 2000). The PilT-mediated twitching motility is also required for the high level competence for DNA transformation in *Neisseria* (Wolfgang *et al.*, 1998). PilX, an 18-kDa pilin-like protein also plays critical roles in Type IV pili biology. It was demonstrated that, although not required for pilus assembly, PilX is co-localized with the fibre and is necessary for aggregate formation and adhesion to host cells without affecting PilC (Helaine *et al.*, 2007).

The major pilin PilE undergoes extensive antigenic variation. This is mediated by non-reciprocal recombination, in which silent partial pilin genes (*pilS*) are transferred to an active pilin expression locus by a RecA-dependent mechanism (Haas and Meyer, 1986). In addition, the GC/MC pilus undergoes phase variation, which results in a reversible change between piliated and non-piliated bacteria. It was shown that pilus phase variation correlates with an on/off switch in PilC expression suggesting that PilC is also involved in pilus assembly (Jonsson *et al.*, 1991). Non-piliated GC pilus variants are characterized by formation of primarily S-pilin, a polymerization-deficient pilin monomer which was implicated as being crucial for establishment of gonococcal infections (Meyer *et al.*, 1990). The GC pilus structure was solved to 12.5 Å resolution by cryo-electron microscopy and iterative helical space reconstruction (Craig *et al.*, 2006). This showed that the fibre is composed of spiralling three-helix bundles, which form the filament core, and is held together by extensive hydrophobic interactions among the N-terminal α-helices of PilE within the fibre core. The PilE globular domains are more loosely packed on the fibre surface, with protruding hypervariable loops and post-translational modifications that shield conserved functional residues in pronounced grooves resulting in a highly corrugated fibre surface. These grooves are often rich in positively charged residues, which might explain the role of the pilus in DNA uptake.

Pseudomonas aeruginosa pilus

P. aeruginosa is a Gram-negative opportunistic pathogen that is a major cause of nosocomial infections and a leading cause of hospital-acquired pneumonia. It also causes bacteraemia in burn victims and severe persistent respiratory infection in susceptible individuals, in particular cystic fibrosis patients (Mutlu and Wunderink, 2006).

P. aeruginosa produces Pa pili, Type IV pili very similar to the GC/MC pili, which extend as bundles from one or both of the cell poles. These are involved in colonization during infection, twitching motility, biofilm formation, bacteriophage infection, and natural transformation. Fibre diffraction studies on the *P. aeruginosa* strain K and strain O pili (PAK pili and PAO pili) revealed hollow cylinders with five pilin subunits per helical turn, a 52 Å outer diameter, a 12 Å inner diameter and a ~31 Å diameter ring of hydrophobic residues (Folkhard *et al.*, 1981).

Binding to host cells through the *P. aeruginosa* pilus is a tip-associated event that is mediated by the C-terminal region of the 15-kDa pilin structural subunit PilA (Lee *et al.*, 1994). The receptor-binding domain (disulfide-loop region) shows significant sequence variation among different *P. aeruginosa* strains, but this has no effect on receptor recognition (Ramphal *et al.*, 1984). A synthetic peptide with a sequence identical to that of PAK pilin residues 128–144, which contain the predicted cell-binding domain bound to buccal epithelial cells (BECs) and to ciliated tracheal epithelial cells (TECs) (Irvin *et al.*, 1989). The importance of pili in the binding of *P. aeruginosa* to epithelial cell surfaces has been shown by a 90% decrease in the ability of non-piliated strains to bind human A549 Type II pneumocytes (Chi *et al.*, 1991). The Pa pilus is also responsible for binding to DNA (van Schaik *et al.*, 2005) and to abiotic surfaces such as stainless steel, polystyrene and polyvinylchloride (PVC), causing major problems in both the medical and food industries (Giltner *et al.*, 2006). The PilA receptor binding domain peptide (PAK 128–144) bound to steel with high affinity, whereas a single-point mutation (K130I) abolished binding. Interestingly, binding to human buccal epithelial cells was enhanced, suggesting that the C-terminal receptor-binding domain of PilA might have evolved for binding to a variety of surfaces.

The Pa pili were reported to specifically bind the β-GalNAc(1–4)βGal moiety of glycosphingolipids (GSL) asialo-GM1 and asialo-GM2 on epithelial cells (Suh *et al.*, 2001). However, these findings were questioned by Emam *et al.* (2006), who were unable to detect specific binding to GSLs in a panel of 12 clinical and laboratory isolates. Furthermore, binding of *P. aeruginosa* to epithelial cells that lacked expression of GSLs was not affected.

Bundle-forming pilus (BFP) of EPEC

EPEC is a major cause of infant diarrhoea in developing nations (Nataro and Kaper, 1998). In addition to Type I pili (see above), EPEC carry Type IV pili known as the bundle-forming pili (BFP), in recognition of their ability to associate and form intertwined rope-like aggregates. These structures are over 15 μm long and have a polar localization on the cell. BFP-mediated bacterial interactions are responsible for the formation of three-dimensional microcolonies on the surface of epithelia (Girón *et al.*, 1991). This has been referred to as localized adherence (LA) and is an important step in colonization. During adhesion, BFP organize into higher-order cables of increasing diameter and strength (Knutton *et al.*, 1999). Bundlin is the major pilin of BFP and is encoded by the plasmid-borne *bfpA* gene, which is the first cistron of the 14-gene *bfp* operon. Both cable formation and dispersal of aggregates requires the cytoplasmic nucleotide binding protein BfpF (Knutton *et al.*, 1999).

There are conflicting results on the role of BFP as a host cell adhesin, and several different BFP receptors have been proposed. Jagannatha *et al.* (1991) reported EPEC strains that show LA binding to asialo-GM1, asialo-GM2, globoside and lacto-*N*-neotetraose, based on recognition of GalNac-β-1-4 Gal groups. Other groups reported inhibition of LA binding by *N*-acetyl-galactosamine (Scaletsky *et al.*, 1988),

N-acetyl-lactosamine and LewisX (Vanmaele *et al.*, 1999). More recently, recombinant bundlin was shown to inhibit LA to HEp-2 cells. Bundlin also bound to HEp-2 cells and, with millimolar association constants, to synthetic *N*-acetyllactosamine (LacNAc) as shown by nanoelectrospray ionization mass spectrometry (Hyland *et al.*, 2008). Another group showed that purified BFP was able to bind phosphatidylethanolamine (PE), and that this interaction was inhibited by antibodies against bundlin (Barnett Foster *et al.*, 1999). The reason for these conflicting observations probably lies in the variation of the *bfpA* gene. Thus far, nine *bfpA* alleles encoding proteins that are 80% identical have been identified in EPEC isolates from diverse hosts. These are sub-grouped into two categories based on sequence similarity, the α group and the β group (Blank *et al.*, 2000). Humphries *et al.* (2009) have recently shown that only α-bundlins are LacNAc-specific lectins and bind to HEp-2 cells, but not β-bundlins. Conversion of the putative carbohydrate-binding domain, 137-GENNI-141, in α-bundling to that of β-bundlin (136-SPDST-140) resulted in binding deficient BFP. Conversely, the introduction of the α-bundlin domain into β-bundlin resulted in the gain of HEp-2 cell binding. Furthermore, a single mutation (I141T) in the α-bundlin domain altered the binding specificity from LacNAc to LewisX (Humphries *et al.*, 2009).

Toxin co-regulated pilus (TCP) of *Vibrio cholerae*

V. cholerae is the aetiological agent of cholera, a devastating dehydrating diarrhoea caused by toxin production of bacteria colonizing the mucosa of the human small intestine. A critical colonization factor of *V. cholerae* is the toxin-co-regulated pilus (TCP), as TCP-deficient *V. cholerae* mutants are incapable of colonizing the intestines (Tacket *et al.*, 1998).

The TCP structure is assembled as a homopolymer of repeating subunits of TcpA pilin. Like the BFP in EPEC, TCP are polar structures of more than 15 μm in length that self-aggregate, bringing the bacteria together in microcolonies. The primary structure of TcpA is highly conserved among *V. cholerae* serogroups and biotypes shown to be pathogenic to humans, which is in sharp contrast to the antigenic variation seen in other Type IV pili, such as the *Neisseria* pili.

The *tcp* genes that encode the pilin and TCP biogenesis functions are organized as an operon located on the 39-kb *Vibrio* pathogenicity island (VPI) (Karaolis *et al.*, 1998). The acquisition of VPI by *V. cholerae* endows the organism with the ability to express TCP, which acts as a receptor for the cholera toxin gene carrying lysogenic bacteriophage CTXφ (Waldor and Mekalanos, 1996). Although it has been suggested that TCP is involved in epithelial cell adhesion, no specific receptors have been identified and the mechanism by which TCP mediates colonization and whether or not TCP is the only factor required for colonization still remains elusive.

Type IV pili in *Francisella tularensis* (FT pili)

Francisella tularensis, a highly virulent facultative intracellular bacterium, is the aetiological agent of tularaemia, a serious and occasionally fatal disease of humans and animals (reviewed in Ellis *et al.*, 2002). Type IV pili have only recently been discovered by electron microscopy on the surface of the *F. tularensis* live vaccine strain (LVS) (Gil *et al.*, 2004). Disruption of the *pilF* putative pilus assembly ATPase and the *pilT* putative disassembly ATPase genes in LVS was shown to cause a complete loss of pili. The mutants were attenuated for virulence in a mouse model of tularaemia by the intradermal route and were defective for adherence to macrophages, pneumocytes, and hepatocytes, suggesting a role for the *F. tularensis* pili in adhesion and virulence (Chakraborty *et al.*, 2008).

Furthermore, Forslund *et al.* (2006) have shown that disruption of the *pilA* gene results in severe virulence attenuation in mice infected by subcutaneous route and reduced ability to spread from the initial site of infection to the spleen. However, the mutation had no effect on the ability of *F. tularensis* to interact, survive and multiply inside macrophage-like cell lines.

Curli

Curli are proteinaceous coiled fibres found on enteric bacteria, such as *E. coli* and *Salmonella* spp. They are non-branching, β-sheet rich, protease-resistant fibres consisting of repeating subunits of the major pilin CsgA (curlin) and share some biochemical and structural properties with eukaryotic amyloid fibres found in neurodegenerative diseases, such as Parkinson's disease and Alzheimer's disease (Chapman *et al.*, 2002). However, unlike eukaryotic amyloid fibres, curli are produced by a highly regulated process known as the nucleation precipitation pathway, where CsgA is secreted from the bacteria and then nucleated into a fibre by the accessory protein CsgB on the cell surface (Bian and Normark, 1997). Curli have been implicated in a number of biological processes, including biofilm formation, cell aggregation, host cell adhesion and invasion, and as potent inducers of the host's inflammatory response, but their role in pathogenesis is not entirely clear, as it was believed that curli assembly preferentially occurs at temperatures below 30°C (Olsén *et al.*, 1989). However, it is now known that curli expression is strain and condition specific and that many clinical isolates express curli at 37°C. Curli-expressing bacteria adhere to various eukaryotic cell lines, such as uroepithelial cells better than non-curliated strains (Kikuchi *et al.*, 2005). Furthermore, expression of the curli genes in an *E. coli* strain that normally does not express curli resulted in the invasion of human cervical epithelial (HeLa) cells (Gophna *et al.*, 2001). Curli bind to a number of different host proteins. Host colonization is facilitated by binding of curli to extracellular matrix proteins fibronectin (Olsén *et al.*, 1989) and laminin (Olsén *et al.*, 1993). Simultaneous binding to plasminogen and tissue type plasminogen activator (t-PA), resulting in conversion of plasminogen to the active plasmin, has also been reported (Sjöbring *et al.*, 1994). Curliated bacteria might gain an advantage, as plasmin degrades soft tissue, which would allow the bacteria to gain access to deeper tissue. Binding of contact-phase proteins, such as H-kininogen, fibrinogen, and factor XII on the bacterial cell surface slows blood clotting, which could also increase the spread of bacteria to surrounding tissue (Herwald *et al.*, 1998). As a consequence, the pro-inflammatory pathway is activated through the release of bradykinin, a potent inducer of fever, pain and hypotension. The interplay of microbe surface proteins and host contact-phase factors may contribute to the symptoms of sepsis and septic shock. Curli have also been characterized as pathogen-associated molecular patterns (PAMPs) that are recognized by Toll-like receptor 2 (TLR2), resulting in activation of the innate immune response by stimulation of IL-8 expression (Tükel *et al.*, 2005).

Binding of curli to abiotic surfaces has also been reported. Austin *et al.* (1998) have demonstrated that curli allow *Salmonella* Enteriditis to adhere to Teflon and stainless steel and have suggested that the persistence of the bacteria on these surfaces could have implications for the food industry.

Pili of Gram-positive bacteria

Until very recently, pili had been observed in only a few Gram-positive bacterial species, such as *Corynebacterium renale* (Yanagawa and Otsuki, 1970), *Actinomyces naeslundii* (Yeung and Ragsdale, 1997; Yeung *et al.*, 1998) and oral streptococci (Fives-Taylor and Thompson, 1985; Wu *et al.*, 1998). More recently, Ton-That and Schneewind (2003) followed up the work on *Corynebacterium renale* and characterized the pili on *Corynebacterium diphtheriae*. This work showed that *C. diphtheriae* indeed produces long rod-like pili which are composed of an SDS-resistant assembly of pilus subunits. Subsequently, pili have been identified and characterized from several other Gram-positive pathogens, including *Streptococcus pyogenes* (group A *Streptococcus*; GAS) (Mora *et al.*, 2005), *Streptococcus agalactiae* (group B *Streptococcus*; GBS) (Lauer *et al.*, 2005), *Streptococcus pneumoniae* (Barocchi *et al.*, 2006; LeMieux *et al.*, 2006), *Enterococcus faecalis* (Nallapareddy *et al.*, 2006), *Actinomyces naeslundii* (Mishra *et al.*, 2007), and *Bacillus cereus* (Budzik *et al.*, 2007). The covalent assembly of pilus subunits is a common feature of Gram-positive bacterial pili, and is thus fundamentally different from the mechanisms of pilus assembly in Gram-negative bacteria. As will be described in the following sections, the pili of Gram-positive pathogens have been shown to be involved in adherence to host cells and other pathogenic processes.

Structure and assembly of Gram-positive bacterial pili

Gram-positive bacterial pili are composed of pilus subunit proteins that are covalently linked to each other, in striking contrast to Gram-negative bacterial pili which are formed by non-covalent interactions between pilins. The backbone of the pilus is mainly composed of a single type of pilus subunit, the major pilin, and one or two additional types of pilins, the minor (or ancillary) pilins, are also covalently associated with the pilus backbone (Ton-That and Schneewind, 2004; Telford *et al.*, 2006). Each pilus subunit of Gram-positive bacteria contains an LPXTG motif (X for any amino acid) or a variant. This so-called cell wall sorting signal (CWSS) is recognized by sortase enzymes that catalyse the covalent polymerization of pilins and attachment of the pilus fibre to the cell wall (Ton-That and Schneewind, 2004). Gram-positive bacterial pili are encoded by discrete gene clusters that typically contain the genes for pilins, sortases, transcriptional regulators and sometimes transposase-like mobile elements.

The sortase-mediated assembly of Gram-positive bacterial pili (Fig. 4.1C) starts with the secretion of pilin precursors *via* the general secretory Sec pathway. They are exported to the membrane, in which the membrane-anchored sortase cleaves the CWSS between threonine and glycine residues. The resulting sortase-pilin intermediate is subsequently attacked by a nucleophile provided by another pilus subunit in the form of the ε-amino group of a lysine residue, as was first shown in the major pilin SpaA of *C. diphtheriae* where the mutation of this lysine to alanine was shown to abolish pilus polymerization. The sequence around the lysine residue forms a motif that is conserved among major pilins of several other Gram-positive bacterial species such as GBS and *S. pneumoniae* and is referred to as a 'pilin motif'(Ton-That and Schneewind, 2003). The nucleophilic attack was predicted to result in a covalent isopeptide bond between the pilin motif lysine of one major pilin and the CWSS of the next, a linkage that was confirmed by mass spectral analyses of corynebacterial SpaA-pili and *B. cereus* BcpA pili (Budzik *et al.*, 2008a; Kang *et al.*, 2009). A variation to the corynebacterial pilin motif was also found in GAS pilins, although a conserved lysine was used in the same way (Kang *et al.*, 2007).

These covalent pili are quite thin, ranging from 2 to 6 nm in width, and it was thus suggested that the shaft of a pilus would be a single chain of pilins that are covalently linked to each other like beads-on-string (Ton-That and Schneewind, 2004; Kang *et al.*, 2007). This model was confirmed in *S. pneumoniae* where transmission electron microscopy pictures of the *rlrA*-encoded pili clearly showed the head-to-tail arrangement of pilins in the linearly extended pilus structure (Hilleringmann *et al.*, 2009). Crystal structures of the major pilins from GAS, *C. diphtheriae* and *B. cereus* show that in each case these molecules are elongated, consisting of multiple tandem domains with immunoglobulin-like (Ig) folds, such that the pilin motif lysine is at one end and the CWSS at the other, again suggesting the beads-on-string model of pilus polymerization (Kang *et al.*, 2007, 2009; Budzik *et al.*, 2009). These major pilins were also shown to contain internal isopeptide bonds that add stability.

Once assembled, the pilus polymer is then transferred to the cell wall in a sortase-dependent manner, in which the nucleophile is provided by an amino group in the cell wall peptidoglycan cross-bridge, resulting in the linkage between the CWSS of the last pilin in the polymer and the cell wall (Swaminathan *et al.*, 2007; Budzik *et al.*, 2008a; Mandlik *et al.*, 2008). This final step of pilus biogenesis sometimes involves incorporation of a specific minor pilin at the base of a pilus.

There has been much speculation as to how minor pilins are positioned in relation to the major pilin backbone. Early immuno-labelling electron microscopy studies suggested that minor pilins are interspersed along the pilus shaft, leading to the idea that minor pilins may be incorporated into the growing pilus backbone, in the same way as major backbone pilins, or tethered to the pilus backbone laterally, forming branches in the polymerized pilus. Recent experimental results, however, support another, though not necessarily mutually exclusive, model in which minor pilins are attached at the pilus tip and/or at the base; the tip pilins act as pilus-associated adhesins and the basal pilins mediate anchoring of the pilus fibres to the cell wall peptidoglycan, as will be discussed in the following sections.

Fimbrial adhesion in Gram-positive pathogens

The adhesive properties of bacterial pili have been shown to be important in bacterial attachment to the host cell, a critical step towards bacterial colonization. The adhesive function of Gram-positive bacterial pili is mainly attributed to minor pilins, whereas major pilins are the bona fide structural subunits. Whereas minor pilins are not required for the polymerization of extended pili, they are essential for pilus functions such as adherence to host cells and virulence.

Corynebacteria diphtheriae

C. diphtheriae colonizes the human nasopharynx and skin, and causes diphtheria, an acute, contagious disease with various symptoms caused by the highly potent diphtheria toxin (DT) (Hadfield *et al.*, 2000). *C. diphtheriae* has been also known to harbour adhesive properties, independent of DT production (Mattos-Guaraldi *et al.*, 2000), and the recently discovered pili on its surface contribute to bacterial adherence to host cells. The genome of toxigenic *C. diphtheriae* strain NCTC13129 produces three different pili, each the result of sortase-mediated pilin assembly, and each named according to the major pilin, Spa (sortase-mediated pilin assembly) A, D, or H, and encoded by a gene cluster (Ton-That and Schneewind, 2003; Gaspar and Ton-That, 2006; Swierczynski and Ton-That, 2006). All three pili mediate efficient binding to epithelial cells derived from the major sites of corynebacterial infection; the SpaA-type pili preferentially adhere to pharyngeal cells, whereas SpaD- or SpaH-type pili are essential for optimal adherence to laryngeal cells (Mandlik *et al.*, 2007).

The SpaA-type pili have been extensively studied and provided the model system through which the mechanisms of the covalent pilus assembly were first established. The minor pilins of SpaA-type pili appeared to be distributed along the length as well as at the tip (SpaC) and the base (SpaB) of the pilus (Ton-That and Schneewind, 2003). Both minor pilins were essential for optimal binding to pharyngeal epithelial cells and antibodies raised against SpaB or SpaC could block the bacterial adherence, whereas *spaA* deletion did not significantly affect adherence (Mandlik *et al.*, 2007). In the absence of the major pilin, SpaB and SpaC were found anchored to the bacterial surface, and could promote bacterial adherence to host cells similarly to the wild-type pili. Mutants lacking both *spaB* and *spaC* showed a greater reduction in adherence to host cells than the single mutants lacking either of *spaB* or *spaC* alone, indicating that these minor pilins can independently promote corynebacterial binding to pharyngeal cells. Homology searches indicate that SpaC contains a von Willebrand factor A (VWF) domain which is known to interact with extracellular matrix (ECM) proteins and host cell receptors (Mandlik *et al.*, 2007; Ruggeri, 2001). Similar VWF domains are also found in pilus-associated adhesins of other Gram-positive bacteria (see below for more details).

These results indicate that the minor pilins SpaB and SpaC are likely the pilus-associated adhesins of SpaA-type pili, and can remain functional even when anchored to the cell wall in the absence of the extended pilus polymers. SpaB, the smaller of the two minor pilins, was also shown to be required for the cell wall anchoring of SpaA-type pili; it is incorporated in the base of the polymerized pilus shaft and its CWSS is linked to the cell wall by the sortase-catalysed reaction (Mandlik *et al.*, 2008).

Group A *Streptococcus* (GAS; *Streptococcus pyogenes*)

GAS colonizes the throat and skin and causes a range of human infections from sore throat and tonsillitis, to severe, life-threatening, illnesses such as necrotizing fasciitis (flesh-eating disease), acute rheumatic fever and streptococcal toxic shock syndrome (Cunningham, 2000). GAS pilins are encoded in a gene cluster known as the FCT (fibronectin-binding, collagen-binding T-antigen) region, so named because it contains genes for ECM binding proteins and a specific Lancefield T antigen (Bessen and Kalia, 2002); the antisera raised against specific T antigens have been used for serological classification of GAS for decades (Lancefield, 1940; Mora *et al.*, 2005). The pilus components encoded by most FCT regions comprise one major pilin (named FctA for some strains), two minor pilins (FctB and Cpa, collagen-binding protein of group A streptococci), a sortase that specifically catalyses pilus polymerization, and a putative signal peptidase-like protein (SipA or LepA).

The minor pilin Cpa was already known as an adhesin when GAS pili were discovered; Cpa from serotypes M1 and M49 was shown to directly bind to collagen type I, an important ECM protein in the human host (Podbielski *et al.*, 1999; Kreikemeyer *et al.*, 2005). There have been several studies reporting that Cpa is present around the bacterial surface, occasionally extending along the pili, or at the pilus base (Mora *et al.*, 2005; Nakata *et al.*, 2009). However, a more recent electron microscopy study of M3 pili, using monoclonal antibodies, located Cpa at the pilus tip, and this was further supported by the mass spectral analyses of engineered M3 pili and native M1 pili, defining the linkage between Cpa and a succeeding backbone pilin (Quigley *et al.*, 2009; W.D. Smith *et al.*, unpublished). Serotypes M2, M4 and M6 of GAS contain FCT regions with genetic contents and organizations that significantly deviate from other serotypes. The pilins of the M2 serotype are homologous to those of GBS and pneumococcal pili, and the structure of the pneumococcal homologue, RrgA, showed the presence of a VWF domain (Izoré *et al.*, 2010).

GAS pili have been shown to mediate specific adherence to cells prepared or derived from the principal sites of GAS infections, such as human tonsil epithelia, skin keratinocytes and pharyngeal cells (Abbot *et al.*, 2007; Manetti *et al.*, 2007; Crotty Alexander *et al.*, 2010; Smith *et al.*, 2010). Serotype M1 pili were also shown to facilitate bacterial aggregation by binding to soluble gp340, a salivary protein with a possible role in mucosal defence, and it was suggested that the gp340-mediated bacterial aggregation may promote bacterial clearance (Edwards *et al.*, 2008). It was also shown in the T3/13/B serotype that Cpa contributes significantly to GAS virulence in a mouse model of skin infection, even in the absence of the pilus backbone (Lizano *et al.*, 2007). Cpa binding to collagen type I was shown to be responsible for the adherence to HEp-2 cells (derived from human larynx) in serotype M49 (Kreikemeyer *et al.*, 2005). However, for the M1 serotype, pilus-mediated adherence to tonsil and skin cells was not dependent on collagen type I, and the M1 pili did not mediate bacterial adherence to HEp-2 cells (Abbot *et al.*, 2007). Also, different from the T3/13/B serotype, all three pilins of the M1 serotype were shown to be required for production of pili adhesive to human tonsil cells (Abbot *et al.*, 2007). Pili of the M1 serotype were also shown to contribute to aggregation of bacteria in liquid culture, microcolony formation on host cells, and the formation of biofilms (Manetti *et al.*, 2007). Taken together, these results indicate that Cpa is the pilus-associated adhesin of GAS and is located at the pilus tip; however, its collagen-binding activity is not solely responsible for the pilus-mediated adhesion of GAS to host cells. It is also possible that other bacterial surface components assist, complement or mask pilus-mediated adhesion, and there might be serotype-specific functions of GAS pili.

Whereas GBS and pneumococcal pili have been shown to enhance virulence in mouse infection models, expression of pili has been linked to reduction in GAS virulence. For the M1 serotype, it was suggested that pilus expression promotes clearance of GAS by neutrophils and activation of NET (neutrophil extracellular traps) formation, hence resulting in a reduction in virulence (Crotty Alexander *et al.*, 2010). It was suggested for the M49 serotype that the increased virulence observed with blocking the pilus expression could be due to an attenuated bacterial internalization, allowing sustained contact with other virulence factors (Nakata *et al.*, 2009).

The FctB-equivalent in the M1 serotype, Spy0130, was also shown to adhere to human pharyngeal cells, raising the possibility of being an adhesin (Manetti *et al.*, 2007). However, the binding may not be specific, as it was recently shown that binding of GAS pili to human tonsil epithelia was not inhibited by antisera against Spy0130 (W.D. Smith *et al.*, unpublished). Instead, Spy0130 is thought to be incorporated at the base of the pilus and to link the polymer to the cell wall. In support, mass spectral evidence shows that Spy0130 is linked to the CWSS of the major pilin Spy0128, implying a role in anchoring the pilus shaft (C. Linke *et al.*, unpublished).

Streptococcus pneumoniae

S. pneumoniae colonizes the human upper respiratory tract and can cause invasive diseases such as septicaemia, meningitis and pneumonia (Lode, 2009). Some strains of *S. pneumoniae* encode pili from the *rlrA* (Rof-A-like transcriptional regulator) pathogenicity islet (Hava *et al.*, 2003; Barocchi *et al.*, 2006). The *rlrA* islet (also referred to as PI-1, pilus island-1) of the clinical serotype TIGR4 encodes one major pilin, RrgB, and two minor pilins, RrgA and RrgC, along with three sortases SrtC1, SrtC2 and SrtC3 (also known as SrtB, SrtC and SrtD, respectively) (Barocchi *et al.*, 2006; LeMieux *et al.*, 2006; Hilleringmann *et al.*, 2008). Several studies have indicated that the RrgB pilus is polymerized mainly by SrtC1 and also partly by SrtC2, with the minor pilin RrgC being incorporated exclusively by SrtC1, and with the assembled pilus being incorporated into the cell wall by SrtC3 (LeMieux *et al.*, 2006; Falker *et al.*, 2008; Manzano *et al.*, 2008). Earlier attempts to locate the minor pilins in the pilus using antibody labelling and electron microscopy indicated RrgC located at the pilus tip and RrgA mainly at the pilus base and along the pilus shaft (Barocchi *et al.*, 2006; LeMieux *et al.*, 2006). However, a recent, definitive, electron microscopy study using highly specific antibodies with no cross-reactivity showed that the pilus is a single chain of RrgB protein covalently linked head-to-tail and the minor pilins RrgA and RrgC at the tip and base of the pilus, respectively (Hilleringmann *et al.*, 2009).

The *rlrA*-encoded pilus has been shown to be a major virulence factor. Introducing the *rlrA* islet into a *rlrA*-negative strain resulted in pilus expression and corresponding enhancement in adherence to lung epithelial cells (Barocchi *et al.*, 2006). A pilus-expressing pneumococcal isolate was shown to be more virulent than a non-pilated isolate and evoked inflammatory responses that are thought to facilitate bacterial invasion by damaging the mucosal barrier in mouse models (Barocchi *et al.*, 2006). The minor pilin RrgA, the largest of the three pilins, was shown to be responsible for pilus-mediated adherence to human respiratory epithelial cells. An isogenic mutant strain lacking *rrgA* showed significantly diminished pilus-mediated adherence to epithelial cells and complementation of *rrgA* restored adherence (Nelson *et al.*, 2007). Lack of polymerized pilus due to the deletion of the major pilin gene *rrgB* did not affect adherence, implying a role for cell-wall anchored RrgA (Nelson *et al.*, 2007). RrgA was also shown to be essential for the bacterial colonization of the pharyngeal epithelium in mice. The recombinant RrgC protein, the smaller minor pilin, was also shown to be adhesive to epithelial cells (Nelson *et al.*, 2007), although its *in vivo* adhesive role remains to be investigated.

The atomic structure of RrgA provides important insights into the molecular basis of its adhesive functions (Izoré *et al.*, 2010). Like other Gram-positive pilins, RrgA is an elongated molecule, in this case consisting of four domains. Two of its Ig-like domains, D2 and D4, contain internal isopeptide bonds that stabilize the protein like those in the major pilins of other Gram-positive pili. The D3 domain, positioned at the one end of the molecule, contains a region that is structurally homologous to the eukaryotic von Willebrand factor A (VWA)/integrin I fold that was shown to interact with collagen molecules (Cruz *et al.*, 1995; Bienkowska *et al.*, 1997). In addition, the D3 domain contains insertions to the integrin I-like region, which project like arms and form a U-shaped cradle. The authors suggest that the integrin I-like region and the basic U-shaped cradle could potentially be two different recognition sites, the former primarily for binding collagen or other ECM molecules and the latter for binding negatively charged glucosaminoglycans or other acidic ECM molecules (Izoré *et al.*, 2010). The D3 domain is also highly conserved in the RrgA homologues of GBS and GAS M2 serotype, indicating similar adhesive roles for these homologues.

Recently, another pilus-encoding islet was identified in some *S. pneumoniae* strains, referred to as PI-2 (Bagnoli *et al.*, 2008). The major pilin PitB is essential for pilus assembly and also for pilus-mediated adherence to bronchial epithelia, nasopharyngeal cells and laryngeal cells, but these PitB pili showed much lower binding capacity than the *rlrA*-encoded pili. PI-2 does contain another pilin gene *pitA*, but this was shown to be not expressed in all of the isolates studied owing to the presence of a stop codon in some alleles. These results indicate that the PI-2 pilus is composed of only PitB which also acts as an adhesin and a linker to the cell wall (Bagnoli *et al.*, 2008).

Group B *Streptococcus* (GBS; *Streptococcus agalactiae*)

GBS is the leading cause of neonatal sepsis, pneumonia and meningitis, but also affects the elderly, causing invasive infections (Wilkinson, 1978; Schuchat, 1999). Comparative genomic analysis of multiple GBS strains lead to the identification of genetic loci encoding proteins for pilus assembly. PI-1 (pilus island-1) encodes a major pilin, GBS80 and two minor pilins, GBS104 and GBS52 (Lauer *et al.*, 2005; Rosini *et al.*, 2006). The *srtC3-C4* locus in the NEM316 strain (homologous to PI-2a locus in 2603 V/R strain) encodes a major pilin PilB (GBS1477) and two minor pilins, PilA (GBS1478) and PilC (GBS1474) (Dramsi *et al.*, 2006).

PilA was shown to mediate GBS adherence to pulmonary epithelial cells; when pili were removed by deleting the major pilin or sortase genes, the PilA-mediated adhesion was still unaffected (Dramsi *et al.*, 2006). Similarly, it was shown that removing either of the minor pilins, GBS104 (PilA orthologue) or GBS52 (PilC orthologue), significantly affected bacterial adherence to pulmonary cells, whereas the major pilin GBS80 was not required for adhesion (Krishnan *et al.*, 2007). Subsequently, it was shown that pilus-associated PilA is absolutely required for host cell adhesion in the presence of a shear stress caused by a laminar flow (Konto-Ghiorghi *et al.*, 2009). These results indicate that PilA of PI-2a and GBS104 of PI-1 are important for host cell adhesion, and the pilus polymers enable optimal adherence of the adhesins so that bacteria remain attached to host cells in harsh conditions such as in the lung where the mucociliary movement is present. It was also shown that PilA and PilB, but not PilC, are critically involved in biofilm formation on a solid surface (Konto-Ghiorghi *et al.*, 2009). The amino acid sequence of PilA hints at the presence of a VWA domain; this was confirmed by the recent structure of the pneumococcal fimbrial adhesin RrgA (Konto-Ghiorghi *et al.*, 2009; Izoré *et al.*, 2010), which is a PilA homologue. Removing the putative VWA domain from PilA significantly reduced host cell adhesion; it was not required for biofilm formation on an abiotic surface, suggesting that it specifically interacts with host ligands (Konto-Ghiorghi *et al.*, 2009).

PilA in strain NCTC10/84 was also shown to be essential for adherence to human brain microvascular endothelial cells (hBMEC), while the major pilin PilB is dispensable for adherence (Maisey *et al.*, 2007). However, a mutant lacking *pilB* showed a significant reduction in the ability to invade hBMEC, whereas *pilA* deletion did not affect the invasion. In addition, GBS deficient in PilB showed increased susceptibility to phagocytic clearance and a decrease in survival and hence attenuated virulence in the mouse infection model, whereas GBS lacking PilA did not (Maisey *et al.*, 2008). These results highlight that, apart from its obvious role in providing the structural backbone of pilus fibres, the major pilin can also contribute specifically to GBS virulence.

GBS52, the smaller minor pilin of PI-1 pili, was shown to function as an adhesin, as the deletion of *gbs52* significantly reduced bacterial adherence to human pulmonary cells (Krishnan *et al.*, 2007). The atomic structure revealed that GBS52 consists of two domains, both of which are folded into variations of immunoglobulin-like (Ig) folds, with the C-terminal domain of the recombinant protein shown to be capable of binding to pulmonary epithelial cells (Krishnan *et al.*, 2007). However, it was also shown that antibodies against GBS52 do not inhibit the binding of GBS52 to the lung cells (Krishnan *et al.*, 2007). Immuno-labelling of PilC, an orthologue of GBS52, showed it to be distributed mainly at the base of the pilus and occasionally along the pilus shaft, suggesting a role as the basal pilin that links the pilus to the cell wall (Dramsi *et al.*, 2006). It remains to be proven whether GBS52 and PilC indeed play a dual role as the basal pilin and a pilus-associated adhesin.

Enterococcus faecalis

E. faecalis is a major subgroup of enterococci which inhabit the human intestine and can also migrate into the bloodstream. It is a leading cause of hospital-acquired infections, and can cause systemic infections such as urinary tract infection, endocarditis and bacteraemia (Murray, 1990; Edmond *et al.*, 1999). *E. faecalis* produces pili from the *ebp* (endocarditis-and biofilm-associated pili) operon

Table 4.1 Pili/Fimbriae in Gram-negative bacteria

Organism	Pili	Major pilin	Adhesins	Target host cells	Host receptors	Disease
E. coli	Type 1 pili	FimA	FimH	Bladder epithelia, kidney epithelia, erythrocytes, mast cells, neutrophils	Uroplakin UP1a, α3β1 integrins, laminin, CD48, collagen	Cystitis, sepsis, meningitis, UTI, sepsis, meningitis
	S pili	SfaA	SfaS	Bladder epithelia, kidney epithelia, endothelial cells, erythrocytes	Sialic acid residues, plasminogen	
UPEC	P pili	PapA	PapG	Kidney epithelia, erythrocytes	GbO3, GbO4, GbO5	Pyelonephritis
UPEC, DAEC	Dr adhesin	–	DraE	Bladder epithelia, kidney epithelia, erythrocytes	DAF, CEACAMs collagen type IV	Cystitis, pyelonephritis
	Afa adhesin	–	AfaE	Uroepithelia, erythrocytes	DAF, CEACAMs, α5β1-integrins	
EPEC	BFP	BfpA	–	Epithelia	Conflicting reports	Diarrhoea
Enterobacteriaceae	Curli	CsgA	–	Variety	Fibronectin, laminin, plasminogen, t-PA, H-kininogen, factor XII, TLR-2	Wide range
H. influenzae	Hif pili	HifA	HifE	Nasopharyngeal	Unknown	Otitis media, meningitis
P. mirabilis	PMF	PmfA	PmfF	Bladder epithelia, kidney epithelia	Unknown	Cystitis, pyelonephritis
	MR/P	MrpA	MrpG (?)			
	UCA	Unknown	UcaA	Uroepithelia	GM1, asialo-GM2	
	ATF	AtfA		Unknown	Unknown	Not involved
N. gonorrhoeae	GCP	PilE	PilC	Epithelia, endothelia	CD46 (?), C4BP (?)	Gonorrhoea
N. meningitidis	MCP					Meningitis, sepsis
P. aeruginosa	Pa pilus	PilA	–	Epithelia	Asialo-GM1, -GM2	Pneumonia, sepsis
V. cholerae	TCP	TcpA	–	Intestinal epithelia	Unknown	Cholera
F. tularensis	FTP	PilE	–	Pneumocytes, hepatocytes	Unknown	Tularaemia

Pili/Fimbriae in Gram-positive bacteria

Organism	Pili	Major pilin	Minor pilins	Target host cells	Host receptors	Disease
C. diphtheriae	SpaA-type	SpaA	SpaB, SpaC	Pharyngeal epithelia	Unknown	Diphtheria
	SpaD-type	SpaD	SpaE, SpaF	Laryngeal epithelia	Unknown	
	SpaH-type	SpaD	SpaI, SpaG	Laryngeal epithelia	Unknown	
GAS	Serotype M1/M3/M49	Spy0128 (or FctA)	Spy0130 (or FctB), Cpa	Pharyngeal epithelia, tonsil epithelia, skin keratinocytes	Collagen type I, salivary protein gp340	Pharyngitis, impetigo, sepsis, toxic shock
S. pneumoniae	PI-1 (*rlrA* pili)	RrgB	RrgA, RrgC	Lung epithelia, mouse pharyngeal epithelia	Unknown	Sinusitis, pneumonia
	PI-2	PitB	–	Bronchial epithelia, nasopharyngeal epithelia, laryngeal epithelia	Unknown	
GBS	PI-1	GBS80	GBS104, GBS52	Pulmonary epithelia	Unknown	Neonatal sepsis, meningitis
	PI-2a/*srtC3-C4*/*pil* pili	GBS59/PilB (GBS1477)/PilB	GBS67/PilA (GBS1478)/PilA, GBS150/PilC (GBS1474)/PilC	Lung and cervical epithelia, brain microvascular endothelia	Unknown	
E. faecalis	*ebp* pili	EbpC	EbpA, EbpC	Kidney, urinary bladder	Unknown	Endocarditis, UTI, bacteremia
B. cereus	BcpA pili	BcpA	BcpB	Unknown	Unknown	Food poisoning
A. naeslundii	Type I	FimP	FimQ	Tooth enamel	Proline-rich salivary proteins	Dental caries, periodontitis
	Type II	FimA	FimB	Various human cells	Glycoproteins/glycolipids	

which contains three pilin genes, *ebpA*, *ebpB* and *ebpC*, and a sortase gene, *srtC* (Nallapareddy *et al.*, 2006). EbpC was suggested to be the major pilin as antibodies against this pilin labelled the pilus shaft uniformly and its sequence is homologous to the major pilins of other Gram-positive bacteria (Ton-That and Schneewind, 2003, 2004; Nallapareddy *et al.*, 2006). Based on immuno-labelling electron microscopy, EbpA, the larger one of the two minor pilins, appears to be mainly at the pilus base, whereas the location of EbpC could not be identified owing to antibody cross-reactivity (Nallapareddy *et al.*, 2006).

The ability of *E. faecalis* to form biofilm was shown to be associated with endocarditis-causing isolates (Mohamed *et al.*, 2004). The pili encoded by the *ebp* operon are implicated in biofilm production in *E. faecalis*, as deleting any gene coding for the pilins or the sortase significantly affected the initial attachment step of biofilm formation (Nallapareddy *et al.*, 2006). Consistent with this, a non-piliated mutant showed significantly attenuated virulence in a rat endocarditis model. It remains to be seen whether any of the pilins do possess specific adhesive properties. The *ebp* pili were also shown to play significant roles in colonization of kidneys and urinary bladders in a mouse model (Singh *et al.*, 2007).

Actinomyces naeslundii and *Bacillus cereus*

A. naeslundii is an oral pathogen that can cause dental caries and periodontitis. Two distinct types of pili were identified from genospecies 2 strains of this organism (Yeung, 1999). Both pili are encoded by gene clusters, each of which contains two pilin genes and one sortase gene; the major pilin builds the pilus shaft (FimP in type 1 and FimA in type 2), whereas the minor pilin is located at the pilus tip (FimQ and FimB) (Mishra *et al.*, 2007). The apparent absence of a second minor pilin in these pili implies that these pili are anchored to the bacterial surface *via* the linkage between the major pilin and the cell wall peptidoglycan.

Type 1 fimbriae are responsible for bacterial adherence to salivary proteins coating the tooth enamel, whereas type 2 fimbriae mediate binding of the bacteria to the surface of oral streptococci and a wide range of host cells (McIntire *et al.*, 1978; Gibbons *et al.*, 1988; Yeung, 1999). Monoclonal antibodies raised against the major pilins FimA and FimP did not inhibit fimbrial adhesion, an indication that the tip pilins FimB and FimQ are the pilus-associated adhesins (Cisar *et al.*, 1991; Mishra *et al.*, 2007). Supporting this, it was shown that deleting *fimQ* abolished type 1 fimbrial adhesion (Chen *et al.*, 2007).

B. cereus is commonly associated with food-borne illnesses with diarrhoeal and emetic syndromes and can also cause skin infections, bacteraemia and pneumonia (Drobniewski, 1993). Similar to *A. naeslundii*, this pathogen produces pili that comprise two types of subunits. The major pilin BcpA is polymerized by the pilin-specific sortase SrtD to form the pilus backbone, and also can be anchored to the cell wall mediated by the housekeeping sortase SrtA (Budzik *et al.*, 2007, 2008b). The minor pilin BcpB is located at the pilus tip and thus thought to be the fimbrial adhesin (Budzik *et al.*, 2007). The functions and adhesive properties of *B. cereus* pili are currently unknown.

Concluding remarks and future work

Bacterial pili or fimbriae are sophisticated structures that can withstand physical, chemical and enzymatic stresses. A common feature shared by many (if not all) pili is to facilitate adherence to host cells by avoiding electrostatic repulsion, which could result from negative surface charge on both the bacteria and the host cell. Over the last few years, research on bacterial pili has been boosted by the discovery of novel pili structures on a number of pathogenic Gram-positive bacteria, such as streptococci and enterococci.

Fimbrial adherence of Gram-positive bacteria is mainly attributed to minor pilins, which are dispensable for the assembly of extended pilus polymers. Whereas the location of adhesins in some pili remains a contentious issue, several studies support an emerging model in which the adhesins are incorporated at the pilus tip; this model is reminiscent of the *E. coli* type I and type P pili, although these are assembled by entirely different mechanisms.

It is believed that the long pilus shaft allows the pilus-associated adhesins to reach out beyond surface barriers, such as the capsule and the cell wall, for optimal binding to host receptors. However, it has been shown for some species that minor pilins can mediate bacterial adherence to host cells even in the absence of pilus polymers, through anchoring of these adhesins to the cell wall. This brings into question the necessity of producing such extended surface appendages as pili. It is possible that most of the *in vitro* and *in vivo* assays employed thus far are not suitable for testing the functional roles and benefits of long pili (Nelson *et al.*, 2007). Indeed it was shown in GBS that the extended pilus polymer is required for the optimal adherence to the host cells in the presence of a laminar flow that emulates the mucociliary movement in the lung (Konto-Ghiorghi *et al.*, 2009).

The molecular basis of the interactions between pilus-associated adhesins and host receptors is starting to be elucidated. Several adhesive pilins possess regions homologous to the VWF domain, an eukaryotic metal-dependent collagen-binding domain, whose presence has been confirmed in the structure of RrgA (Izoré *et al.*, 2010). Further investigations will be required to identify the host receptors for the fimbrial-associated VWF domain and to determine whether similar structures are responsible for adhesion by the pili of other Gram-positive pili.

References

Abbot, E.L., Smith, W.D., Siou, G.P., Chiriboga, C., Smith, R.J., Wilson, J.A., Hirst, B.H., and Kehoe, M.A. (2007). Pili mediate specific adhesion of *Streptococcus pyogenes* to human tonsil and skin. Cell. Microbiol. *9*, 1822–1833.

Anderson, K.L., Billington, J., Pettigrew, D., Cota, E., Simpson, P., Roversi, P., Chen, H.A., Urvil, P., du Merle, L., Barlow, P.N., *et al.* (2004). An atomic resolution model for assembly, architecture, and function of the Dr adhesins. Mol. Cell *15*, 647–657.

Anderson, T.F. (1949). On the mechanism of adsorption of bacteriophages on host cells. In the Nature of the Bacterial Surface, Miles, A.A., and Pirie, N.W., eds. (Oxford University Press, Oxford, UK), pp.76–95.

Austin, J.W., Sanders, G., Kay, W.W., and Collinson, S.K. (1998). Thin aggregative fimbriae enhance *Salmonella enteritidis* biofilm formation. FEMS Microbiol. Lett. *162*, 295–301.

Bagnoli, F., Moschioni, M., Donati, C., Dimitrovska, V., Ferlenghi, I., Facciotti, C., Muzzi, A., Giusti, F., Emolo, C., Sinisi, A., *et al.* (2008). A second pilus type in *Streptococcus pneumoniae* is prevalent in emerging serotypes and mediates adhesion to host cells. J. Bacteriol. *190*, 5480–5492.

Baorto, D.M., Gao, Z., Malaviya, R., Dustin, M.L., van der Merwe, A., Lublin, D.M., and Abraham, S.N. (1997). Survival of FimH-expressing enterobacteria in macrophages relies on glycolipid traffic. Nature *389*, 636–639.

Barnett Foster, D., Philpott, D., Abul-Milh, M., Huesca, M., Sherman, P.M., and Lingwood, C.A. (1999). Phosphatidylethanolamine recognition promotes enteropathogenic *E. coli* and enterohemorrhagic *E. coli* host cell attachment. Microb. Pathog. *27*, 289–301.

Barocchi, M.A., Ries, J., Zogaj, X., Hemsley, C., Albiger, B., Kanth, A., Dahlberg, S., Fernebro, J., Moschioni, M., Masignani, V., *et al.* (2006). A pneumococcal pilus influences virulence and host inflammatory responses. Proc. Natl. Acad. Sci. U.S.A. *103*, 2857–2862.

Bessen, D.E., and Kalia, A. (2002). Genomic localization of a T serotype locus to a recombinatorial zone encoding extracellular matrix-binding proteins in *Streptococcus pyogenes*. Infect. Immun. *70*, 1159–1167.

Bhaya, D., Bianco, N.R., Bryant, D., and Grossman, A. (2000). Type IV pilus biogenesis and motility in the cyanobacterium *Synechocystis* sp. PCC6803. Mol. Microbiol. *37*, 941–951.

Bian, Z., and Normark, S. (1997). Nucleator function of CsgB for the assembly of adhesive surface organelles in *Escherichia coli*. EMBO J. *16*, 5827–5836.

Bienkowska, J., Cruz, M., Atiemo, A., Handin, R., and Liddington, R. (1997). The von Willebrand factor A3 domain does not contain a metal ion-dependent adhesion site motif. J. Biol. Chem. *272*, 25162–25167.

Blank, T.E., Zhong, H., Bell, A.L., Whittam, T.S., and Donnenberg, M.S. (2000). Molecular variation among type IV pilin (*bfpA*) genes from diverse enteropathogenic *Escherichia coli* strains. Infect. Immun. *68*, 7028–7038.

Blom, A.M., Rytkönen, A., Vasquez, P., Lindahl, G., Dahlbäck, B., and Jonsson, A.B. (2001). A novel interaction between type IV pili of *Neisseria gonorrhoeae* and the human complement regulator C4B-binding protein. J. Immunol. *166*, 6764–6770.

Bouckaert, J., Mackenzie, J., de Paz, J.L., Chipwaza, B., Choudhury, D., Zavialov, A., Mannerstedt, K., Anderson, J., Piérard, D., Wyns, L., *et al.* (2006). The affinity of the FimH fimbrial adhesin is receptor-driven and quasi-independent of *Escherichia coli* pathotypes. Mol. Microbiol. *61*, 1556–1568.

Budzik, J.M., Marraffini, L.A., and Schneewind, O. (2007). Assembly of pili on the surface of *Bacillus cereus* vegetative cells. Mol. Microbiol. *66*, 495–510.

Budzik, J.M., Marraffini, L.A., Souda, P., Whitelegge, J.P., Faull, K.F., and Schneewind, O. (2008a). Amide bonds assemble pili on the surface of bacilli. Proc. Natl. Acad. Sci. U.S.A. *105*, 10215–10220.

Budzik, J.M., Oh, S.Y., and Schneewind, O. (2008b). Cell wall anchor structure of BcpA pili in *Bacillus anthracis*. J. Biol. Chem. *283*, 36676–36686.

Budzik, J.M., Poor, C.B., Faull, K.F., Whitelegge, J.P., He, C., and Schneewind, O. (2009). Intramolecular amide bonds stabilize pili on the surface of bacilli. Proc. Natl. Acad. Sci. U.S.A. *106*, 19992–19997.

Bullitt, E., and Makowski, L. (1995). Structural polymorphism of bacterial adhesion pili. Nature *373*, 164–167.

Burrows, L.L. (2005). Weapons of mass retraction. Mol. Microbiol. *57*, 878–888.

Capitani, G., Eidam, O., Glockshuber, R., and Grütter, M.G. (2006). Structural and functional insights into the assembly of type 1 pili from *Escherichia coli*. Microbes Infect. *8*, 2284–2290.

Chakraborty, S., Monfett, M., Maier, T.M., Benach, J.L., Frank, D.W., and Thanassi, D.G. (2008). Type IV pili in *Francisella tularensis*: roles of *pilF* and *pilT* in fiber assembly, host cell adherence, and virulence. Infect. Immun. *76*, 2852–2861.

Chapman, M.R., Robinson, L.S., Pinkner, J.S., Roth, R., Heuser, J., Hammar, M., Normark, S., and Hultgren, S.J. (2002). Role of *Escherichia coli* curli operons in directing amyloid fiber formation. Science *295*, 851–855.

Chen, P., Cisar, J.O., Hess, S., Ho, J.T., and Leung, K.P. (2007). Amended description of the genes for synthesis of *Actinomyces naeslundii* T14V type 1 fimbriae and associated adhesin. Infect. Immun. *75*, 4181–4185.

Chen, S.L., Hung, C.S., Pinkner, J.S., Walker, J.N., Cusumano, C.K., Li, Z., Bouckaert, J., Gordon, J.I., and Hultgren, S.J. (2009). Positive selection identifies an *in vivo* role for FimH during urinary tract infection in addition to mannose binding. Proc. Natl. Acad. Sci. U.S.A. *106*, 22439–22444.

Chi, E., Mehl, T., Nunn, D., and Lory, S. (1991). Interaction of *Pseudomonas aeruginosa* with A549 pneumocyte cells. Infect. Immun. *59*, 822–828.

Choudhury, D., Thompson, A., Stojanoff, V., Langermann, S., Pinkner, J., Hultgren, S.J., and Knight, S.D. (1999). X-ray structure of the FimC-FimH chaperone–adhesin complex from uropathogenic *Escherichia coli*. Science *285*, 1061–1066.

Cisar, J.O., Barsumian, E.L., Siraganian, R.P., Clark, W.B., Yeung, M.K., Hsu, S.D., Curl, S.H., Vatter, A.E., and Sandberg, A.L. (1991). Immunochemical and functional studies of *Actinomyces viscosus* T14V type 1 fimbriae with monoclonal and polyclonal antibodies directed against the fimbrial subunit. J. Gen. Microbiol. *137*, 1971–1979.

Connell, I., Agace, W., Klemm, P., Schembri, M., Marild, S., and Svanborg, C. (1996). Type 1 fimbrial expression enhances *Escherichia coli* virulence for the urinary tract. Proc. Natl. Acad. Sci. U.S.A. *93*, 9827–9832.

Cota, E., Jones, C., Simpson, P., Altroff, H., Anderson, K.L., du Merle, L., Guignot, J., Servin, A., Le Bouguénec, C., Mardon, H., *et al.* (2006). The solution structure of the invasive tip complex from Afa/Dr fibrils. Mol. Microbiol. *62*, 356–366.

Craig, L., and Li, J. (2008). Type IV pili: paradoxes in form and function. Curr. Opin. Struct. Biol. *18*, 267–277.

Craig, L., Pique, M.E., and Tainer, J.A. (2004). Type IV pilus structure and bacterial pathogenicity. Nat. Rev. Microbiol. *2*, 363–378.

Craig, L., Volkmann, N., Arvai, A.S., Pique, M.E., Yeager, M., Egelman, E.H., and Tainer, J.A. (2006). Type IV pilus structure by cryo-electron microscopy and crystallography: implications for pilus assembly and functions. Mol. Cell *23*, 651–662.

Crotty Alexander, L.E., Maisey, H.C., Timmer, A.M., Rooijakkers, S.H., Gallo, R.L., von Köckritz-Blickwede, M., and Nizet, V. (2010). M1T1 group A streptococcal pili promote epithelial colonization but diminish systemic virulence through neutrophil extracellular entrapment. J. Mol. Med. *88*, 371–381.

Cruz, M.A., Yuan, H., Lee, J.R., Wise, R.J., and Handin, R.I. (1995). Interaction of the von Willebrand factor (vWF) with collagen. Localization of the primary collagen-binding site by analysis of recombinant vWF a domain polypeptides. J. Biol. Chem. *270*, 10822–10827.

Cunningham, M.W. (2000). Pathogenesis of group A streptococcal infections. Clin. Microbiol. Rev. *13*, 470–511.

Dhakal, B.K., and Mulvey, M.A. (2009). Uropathogenic *Escherichia coli* invades host cells via an HDAC6-modulated microtubule-dependent pathway. J. Biol. Chem. *284*, 446–454.

Diard, S., Toribio, A.L., Boum, Y., Vigier, F., Kansau, I., Bouvet, O., and Servin, A. (2006). Environmental signals implicated in Dr fimbriae release by pathogenic *Escherichia coli*. Microbes Infect *8*, 1851–1858.

Dramsi, S., Caliot, E., Bonne, I., Guadagnini, S., Prévost, M.C., Kojadinovic, M., Lalioui, L., Poyart, C., and Trieu-Cuot, P. (2006). Assembly and role of pili in group B streptococci. Mol. Microbiol. *60*, 1401–1413.

Drobniewski, F.A. (1993). *Bacillus cereus* and related species. Clin. Microbiol. Rev. *6*, 324–338.

Duncan, M.J., Mann, E.L., Cohen, M.S., Ofek, I., Sharon, N., and Abraham, S.N. (2005). The distinct binding specificities exhibited by enterobacterial type 1 fimbriae are determined by their fimbrial shafts. J. Biol. Chem. *280*, 37707–37716.

Edmond, M.B., Wallace, S.E., McClish, D.K., Pfaller, M.A., Jones, R.N., and Wenzel, R.P. (1999). Nosocomial bloodstream infections in United States hospitals: a three-year analysis. Clin. Infect. Dis. *29*, 239–244.

Edwards, A.M., Manetti, A.G., Falugi, F., Zingaretti, C., Capo, S., Buccato, S., Bensi, G., Telford, J.L., Margarit, I., and Grandi, G. (2008). Scavenger receptor gp340 aggregates group A streptococci by binding pili. Mol. Microbiol. *68*, 1378–1394.

Ellis, J., Oyston, P.C., Green, M., and Titball, R.W. (2002). Tularemia. Clin. Microbiol. Rev. *15*, 631–646.

Emam, A., Yu, A.R., Park, H.J., Mahfoud, R., Kus, J., Burrows, L.L., and Lingwood, C.A. (2006). Laboratory and clinical *Pseudomonas aeruginosa* strains do not bind glycosphingolipids *in vitro* or during type IV pili-mediated initial host cell attachment. Microbiology *152*, 2789–2799.

Eto, D.S., Jones, T.A., Sundsbak, J.L., and Mulvey, M.A. (2007). Integrin-mediated host cell invasion by type 1-piliated uropathogenic *Escherichia coli*. PLoS Pathog. *3*, e100.

Eto, D.S., Gordon, H.B., Dhakal, B.K., Jones, T.A., and Mulvey, M.A. (2008). Clathrin, AP-2, and the NPXY-binding subset of alternate endocytic adaptors facilitate FimH-mediated bacterial invasion of host cells. Cell. Microbiol. *10*, 2553–2567.

Falker, S., Nelson, A.L., Morfeldt, E., Jonas, K., Hultenby, K., Ries, J., Melefors, O., Normark, S., and Henriques-Normark, B. (2008). Sortase-mediated assembly and surface topology of adhesive pneumococcal pili. Mol. Microbiol. *70*, 595–607.

Fives-Taylor, P.M., and Thompson, D.W. (1985). Surface properties of *Streptococcus sanguis* FW213 mutants non-adherent to saliva-coated hydroxyapatite. Infect. Immun. *47*, 752–759.

Folkhard, W., Marvin, D.A., Watts, T.H., and Paranchych, W. (1981). Structure of polar pili from *Pseudomonas aeruginosa* strains K and O.J. Mol. Biol. *149*, 79–93.

Forslund, A.L., Kuoppa, K., Svensson, K., Salomonsson, E., Johansson, A., Byström, M., Oyston, P.C., Michell, S.L., Titball, R.W., Noppa, L., *et al.* (2006). Direct repeat-mediated deletion of a type IV pilin gene results in major virulence attenuation of *Francisella tularensis*. Mol. Microbiol. *59*, 1818–1830.

Gaspar, A.H., and Ton-That, H. (2006). Assembly of distinct pilus structures on the surface of *Corynebacterium diphtheriae*. J. Bacteriol. *188*, 1526–1533.

Gibbons, R.J., Hay, D.I., Cisar, J.O., and Clark, W.B. (1988). Adsorbed salivary proline-rich protein 1 and statherin: receptors for type 1 fimbriae of *Actinomyces viscosus* T14V-J1 on apatitic surfaces. Infect. Immun. *56*, 2990–2993.

Gil, H., Benach, J.L., and Thanassi, D.G. (2004). Presence of pili on the surface of *Francisella tularensis*. Infect. Immun. *72*, 3042–3047.

Gildsorf, J.R., McCrea, K.W., and Marrs, C.F. (1997). Role of pili in *Haemophilus influenzae* adherence and colonization. Infect. Immun. *65*, 2997–3002.

Gill, D.B., Koomey, M., Cannon, J.G., and Atkinson, J.P. (2003). Down-regulation of CD46 by piliated *Neisseria gonorrhoeae*. J. Exp. Med. *198*, 1313–1322.

Giltner, C.L., van Schaik, E.J., Audette, G.F., Kao, D., Hodges, R.S., Hassett, D.J., and Irvin, R.T. (2006). The *Pseudomonas aeruginosa* type IV pilin receptor binding domain functions as an adhesin for both biotic and abiotic surfaces. Mol. Microbiol. *59*, 1083–1096.

Girón, J.A., Ho, A.S., and Schoolnik, G.K. (1991). An inducible bundle-forming pilus of enteropathogenic *Escherichia coli*. Science *254*, 710–713.

Gophna, U., Barlev, M., Seijffers, R., Oelschlager, T.A., Hacker, J., and Ron, E.Z. (2001). Curli fibers mediate internalization of *Escherichia coli* by eukaryotic cells. Infect. Immun. *69*, 2659–2665.

Guignot, J., Hudault, S., Kansau, I., Chau, I., and Servin, A.L. (2009). Human decay-accelerating factor and CEACAM receptor-mediated internalization and intracellular lifestyle of Afa/Dr diffusely adhering *Escherichia coli* in epithelial cells. Infect. Immun. *77*, 517–531.

Haas, R., and Meyer, T.F. (1986). The repertoire of silent pilus genes in *Neisseria gonorrhoeae*: evidence for gene conversion. Cell *44*, 107–115.

Hadfield, T.L., McEvoy, P., Polotsky, Y., Tzinserling, V.A., and Yakovlev, A.A. (2000). The pathology of diphtheria. J. Infect. Dis. *181* (Suppl. 1), S116–120.

Hagberg, L., Hull, R., Hull, S., Falkow, S., Freter, R., and Svanborg Edén, C. (1983). Contribution of adhesion to bacterial persistence in the mouse urinary tract. Infect. Immun. *40*, 265–272.

Hahn, E., Wild, P., Hermanns, U., Sebbel, P., Glockshuber, R., Häner, M., Taschner, N., Burkhard, P., Aebi, U., and Müller, S.A. (2002). Exploring the 3D molecular architecture of *Escherichia coli* type 1 pili. J. Mol. Biol. *323*, 845–857.

Hase, K., Kawano, K., Nochi, T., Pontes, G.S., Fukuda, S., Ebisawa, M., Kadokura, K., Tobe, T., Fujimura, Y., Kawano, S., *et al.* (2009). Uptake through glycoprotein 2 of FimH(+) bacteria by M cells initiates mucosal immune response. Nature *462*, 226–230.

Hava, D.L., Hemsley, C.J., and Camilli, A. (2003). Transcriptional regulation in the *Streptococcus pneumoniae rlrA* pathogenicity islet by RlrA.J. Bacteriol. *185*, 413–421.

Helaine, S., Dyer, D.H., Nassif, X., Pelicic, V., and Forest, K.T. (2007). 3D structure/function analysis of PilX reveals how minor pilins can modulate the virulence properties of type IV pili. Proc. Natl. Acad. Sci. U.S.A. *104*, 15888–15893.

Herwald, H., Môrgelin, M., Olsén, A., Rhen, M., Dahlbâck, B., Müller-Esterl, W., and Björck, L. (1998). Activation of the contact-phase system on bacterial surfaces-a clue to serious complications in infectious diseases. Nat. Med. *4*, 298–302.

Hilleringmann, M., Giusti, F., Baudner, B.C., Masignani, V., Covacci, A., Rappuoli, R., Barocchi, M.A., and Ferlenghi, I. (2008). Pneumococcal pili are composed of protofilaments exposing adhesive clusters of Rrg A. PLoS Pathog. *4*, e1000026.

Hilleringmann, M., Ringler, P., Müller, S.A., De Angelis, G., Rappuoli, R., Ferlenghi, I., and Engel, A. (2009). Molecular architecture of *Streptococcus pneumoniae* TIGR4 pili. EMBO J. *28*, 3921–3930.

Humphries, R.M., Donnenberg, M.S., Strecker, J., Kitova, E., Klassen, J.S., Cui, L., Griener, T.P., Mulvey, G.L., and Armstrong, G.D. (2009). From alpha to beta: identification of amino acids required for the *N*-acetyllactosamine-specific lectin-like activity of bundlin. Mol. Microbiol. *72*, 859–868.

Hyland, R.M., Sun, J., Griener, T.P., Mulvey, G.L., Klassen, J.S., Donnenberg, M.S., and Armstrong, G.D. (2008). The bundlin pilin protein of enteropathogenic *Escherichia coli* is an *N*-acetyllactosamine-specific lectin. Cell. Microbiol. *10*, 177–187.

Irvin, R.T., Doig, P., Lee, K.K., Sastry, P.A., Paranchych, W., Todd, T., and Hodges, R.S. (1989). Characterization of the *Pseudomonas aeruginosa* pilus adhesin: confirmation that the pilin structural protein subunit contains a human epithelial cell-binding domain. Infect. Immun. *57*, 3720–3726.

Izoré, T., Contreras-Martel, C., El Mortaji, L., Manzano, C., Terrasse, R., Vernet, T., Di Guilmi, A.M., and Dessen, A. (2010). Structural basis of host cell recognition by the pilus adhesin from *Streptococcus pneumoniae.* Structure *18*, 106–115.

Jagannatha, H.M., Sharma, U.K., Ramaseshan, T., Surolia, A., and Balganesh, T.S. (1991). Identification of carbohydrate structures as receptors for localised adherent enteropathogenic *Escherichia coli.* Microb. Pathog. *11*, 259–268.

Johansson, L., Rytkönen, A., Bergman, P., Albiger, B., Källström, H., Hökfelt, T., Agerberth, B., Cattaneo, R., and Jonsson, A.B. (2003). CD46 in meningococcal disease. Science *301*, 373–375.

Jones, C.H., Pinkner, J.S., Roth, R., Heuser, J., Nicholes, A.V., Abraham, S.N., and Hultgren, S.J. (1995). FimH adhesin of type 1 pili is assembled into a fibrillar tip structure in the Enterobacteriaceae. Proc. Natl. Acad. Sci. U.S.A. *92*, 2081–2085.

Jonsson, A.B., Nyberg, G., and Normark, S. (1991). Phase variation of gonococcal pili by frameshift mutation in *pilC*, a novel gene for pilus assembly. EMBO J. *10*, 477–488.

Källström, H., Liszewski, M.K., Atkinson, J.P., and Jonsson, A.B. (1997). Membrane cofactor protein (MCP or CD46) is a cellular pilus receptor for pathogenic *Neisseria*. Mol. Microbiol. *25*, 639–647.

Kang, H.J., Coulibaly, F., Clow, F., Proft, T., and Baker, E.N. (2007). Stabilizing isopeptide bonds revealed in Gram-positive bacterial pilus structure. Science *318*, 1625–1628.

Kang, H.J., Paterson, N.G., Gaspar, A.H., Ton-That, H., and Baker, E.N. (2009). The *Corynebacterium diphtheriae* shaft pilin SpaA is built of tandem Ig-like modules with stabilizing isopeptide and disulfide bonds. Proc. Natl. Acad. Sci. U.S.A. *106*, 16967–16971.

Kansau, I., Berger, C., Hospital, M., Amsellem, R., Nicolas, V., Servin, A.L., and Bernet-Camard, M.F. (2004). Zipper-like internalization of Dr-positive *Escherichia coli* by epithelial cells is preceded by an adhesin-induced mobilization of raft-associated molecules in the initial step of adhesion. Infect. Immun. *72*, 3733–3742.

Karaolis, D.K., Johnson, J.A., Bailey, C.C., Boedeker, E.C., Kaper, J.B., and Reeves, P.R. (1998). A *Vibrio cholerae* pathogenicity island associated with epidemic and pandemic strains. Proc. Natl. Acad. Sci. U.S.A. *95*, 3134–3139.

Khan, A.S., Kniep, B., Oelschlaeger, T.A., Van Die, I., Korhonen, T., and Hacker, J. (2000). Receptor structure for F1C fimbriae of uropathogenic *Escherichia coli*. Infect. Immun. *68*, 3541–3547.

Kikuchi, T., Mizunoe, Y., Takade, A., Naito, S., and Yoshida, S. (2005). Curli fibers are required for development of biofilm architecture in *Escherichia coli* K-12 and enhance bacterial adherence to human uroepithelial cells. Microbiol. Immunol. *49*, 875–884.

Kirchner, M., and Meyer, T.F. (2005). The PilC adhesin of the *Neisseria* type IV pilus-binding specificities and new insights into the nature of the host cell receptor. Mol. Microbiol. *56*, 945–957.

Kirchner, M., Heuer, D., and Meyer, T.F. (2005). CD46-independent binding of neisserial type IV pili and the major pilus adhesin, PilC, to human epithelial cells. Infect. Immun. *73*, 3072–3082.

Knight, S.D. (2007). Structure and assembly of *Yersinia pestis* F1 antigen. Adv. Exp. Med. Biol. *603*, 74–87.

Knutton, S., Shaw, R.K., Anantha, R.P., Donnenberg, M.S., and Zorgani, A.A. (1999). The type IV bundle-forming pilus of enteropathogenic *Escherichia coli* undergoes dramatic alterations in structure associated with bacterial adherence, aggregation and dispersal. Mol. Microbiol. *33*, 499–509.

Konto-Ghiorghi, Y., Mairey, E., Mallet, A., Duménil, G., Caliot, E., Trieu-Cuot, P., and Dramsi, S. (2009). Dual role for pilus in adherence to epithelial cells and biofilm formation in *Streptococcus agalactiae*. PLoS Pathog. *5*, e1000422.

Korotkova, N., Cota, E., Lebedin, Y., Monpouet, S., Guignot, J., Servin, A.L., Matthews, S., and Moseley, S.L. (2006). A subfamily of Dr adhesins of *Escherichia coli* bind independently to decay-accelerating factor and the N-domain of carcinoembryonic antigen. J. Biol. Chem. *281*, 29120–29130.

Korotkova, N., Yang, Y., Le Trong, I., Cota, E., Demeler, B., Marchant, J., Thomas, W.E., Stenkamp, R.E., Moseley, S.L., and Matthews, S. (2008a). Binding of Dr adhesins of *Escherichia coli* to carcinoembryonic antigen triggers receptor dissociation. Mol. Microbiol. *67*, 420–434.

Korotkova, N., Yarova-Yarovaya, Y., Tchesnokova, V., Yazvenko, N., Carl, M.A., Stapleton, A.E., and Moseley, S.L. (2008b). *Escherichia coli* DraE adhesin-associated bacterial internalization by epithelial cells is promoted independently by decay-accelerating factor and carcinoembryonic antigen-related cell adhesion molecule binding and does not require the DraD invasin. Infect. Immun. *76*, 3869–3880.

Kreikemeyer, B., Nakata, M., Oehmcke, S., Gschwendtner, C., Normann, J., and Podbielski, A. (2005). *Streptococcus pyogenes* collagen type I-binding Cpa surface protein. Expression profile, binding characteristics, biological functions, and potential clinical impact. J. Biol. Chem. *280*, 33228–33239.

Krishnan, V., Gaspar, A.H., Ye, N., Mandlik, A., Ton-That, H., and Narayana, S.V. (2007). An IgG-like domain in the minor pilin GBS52 of *Streptococcus agalactiae* mediates lung epithelial cell adhesion. Structure *15*, 893–903.

Kuehn, M.J., Heuser, J., Normark, S., and Hultgren, S.J. (1992). P pili in uropathogenic *E. coli* are composite fibres with distinct fibrillar adhesive tips. Nature *356*, 252–255.

Lancefield, R.C. (1940). Type-specific antigens, M and T, of matt and glossy variants of Group A hemolytic streptococci. J. Exp. Med. *71*, 521–537.

Lauer, P., Rinaudo, C.D., Soriani, M., Margarit, I., Maione, D., Rosini, R., Taddei, A.R., Mora, M., Rappuoli, R., Grandi, G., *et al.* (2005). Genome analysis reveals pili in Group B *Streptococcus*. Science *309*, 105.

Lee, K.K., Sheth, H.B., Wong, W.Y., Sherburne, R., Paranchych, W., Hodges, R.S., Lingwood, C.A., Krivan, H., and Irvin, R.T. (1994). The binding of *Pseudomonas aeruginosa* pili to glycosphingolipids is a tip-associated event involving the C-terminal region of the structural pilin subunit. Mol. Microbiol. *11*, 705–713.

Lee, K.K., Harrison, B.A., Latta, R., and Altman, E. (2000). The binding of *Proteus mirabilis* nonagglutinating fimbriae to ganglio-series asialoglycolipids and lactosyl ceramide. Can. J. Microbiol. *46*, 961–966.

Lee, S.W., Bonnah, R.A., Higashi, D.L., Atkinson, J.P., Milgram, S.L., and So, M. (2002). CD46 is phosphorylated at tyrosine 354 upon infection of epithelial cells by *Neisseria gonorrhoeae*. J. Cell Biol. *156*, 951–957.

LeMieux, J., Hava, D.L., Basset, A., and Camilli, A. (2006). RrgA and RrgB are components of a multisubunit pilus encoded by the *Streptococcus pneumoniae rlrA* pathogenicity islet. Infect. Immun. *74*, 2453–2456.

Li, H., Qian, L., Chen, Z., Thibault, D., Liu, G., Liu, T., and Thanassi, D.G. (2004). The outer membrane usher forms a twin-pore secretion complex. J. Mol. Biol. *344*, 1397–1407.

Lizano, S., Luo, F., and Bessen, D.E. (2007). Role of streptococcal T antigens in superficial skin infection. J. Bacteriol. *189*, 1426–1434.

Lode, H.M. (2009). Clinical impact of antibiotic-resistant Gram-positive pathogens. Clin. Microbiol. Infect. *15*, 212–217.

McCrea, K.W., Watson, W.J., Gilsdorf, J.R., and Marrs, C.F. (1997). Identification of two minor subunits in the pilus of *Haemophilus influenzae*. J. Bacteriol. *179*, 4227–4231.

McIntire, F.C., Vatter, A.E., Baros, J., and Arnold, J. (1978). Mechanism of coaggregation between *Actinomyces viscosus* T14V and *Streptococcus sanguis* 34. Infect. Immun. *21*, 978–988.

Maier, B., Koomey, M., and Sheetz, M.P. (2004). A force-dependent switch reverses type IV pilus retraction. Proc. Natl. Acad. Sci. U.S.A. *101*, 10961–10966.

Maisey, H.C., Hensler, M., Nizet, V., and Doran, K.S. (2007). Group B streptococcal pilus proteins contribute to adherence to and invasion of brain microvascular endothelial cells. J. Bacteriol. *189*, 1464–1467.

Maisey, H.C., Quach, D., Hensler, M.E., Liu, G.Y., Gallo, R.L., Nizet, V., and Doran, K.S. (2008). A group B streptococcal pilus protein promotes phagocyte resistance and systemic virulence. FASEB J. *22*, 1715–1724.

Malagolini, N., Cavallone, D., Wu, X.R., and Serafini-Cessi, F. (2000). Terminal glycosylation of bovine uroplakin III, one of the major integral-membrane glycoproteins of mammalian bladder. Biochim. Biophys. Acta *1475*, 231–237.

Mandlik, A., Swierczynski, A., Das, A., and Ton-That, H. (2007). *Corynebacterium diphtheriae* employs specific minor pilins to target human pharyngeal epithelial cells. Mol. Microbiol. *64*, 111–124.

Mandlik, A., Das, A., and Ton-That, H. (2008). The molecular switch that activates the cell wall anchoring step of pilus assembly in Gram-positive bacteria. Proc. Natl. Acad. Sci. U.S.A. *105*, 14147–14152.

Manetti, A.G., Zingaretti, C., Falugi, F., Capo, S., Bombaci, M., Bagnoli, F., Gambellini, G., Bensi, G., Mora, M., Edwards, A.M., *et al.* (2007). *Streptococcus pyogenes* pili promote pharyngeal cell adhesion and biofilm formation. Mol. Microbiol. *64*, 968–983.

Manzano, C., Contreras-Martel, C., El Mortaji, L., Izoré, T., Fenel, D., Vernet, T., Schoehn, G., Di Guilmi, A.M., and Dessen, A. (2008). Sortase-mediated pilus fiber biogenesis in *Streptococcus pneumoniae*. Structure *16*, 1838–1848.

Martinez, J.J., and Hultgren, S.J. (2002). Requirement of Rho-family GTPases in the invasion of Type 1-piliated uropathogenic *Escherichia coli*. Cell. Microbiol. *4*, 19–28.

Martinez, J.J., Mulvey, M.A., Schilling, J.D., Pinkner, J.S., and Hultgren, S.J. (2000). Type 1 pilus-mediated bacterial invasion of bladder epithelial cells. EMBO J. *19*, 2803–2812.

Mattick, J.S. (2002). Type IV pili and twitching motility. Annu. Rev. Microbiol. *56*, 289–314.

Mattos-Guaraldi, A.L., Duarte Formiga, L.C., and Pereira, G.A. (2000). Cell surface components and adhesion in *Corynebacterium diphtheriae*. Microbes Infect. *2*, 1507–1512.

Merz, A.J., Enns, C.A., and So, M. (1999). Type IV pili of pathogenic Neisseriae elicit cortical plaque formation in epithelial cells. Mol. Microbiol. *32*, 1316–1332.

Merz, A.J., So, M., and Sheetz, M.P. (2000). Pilus retraction powers bacterial twitching motility. Nature *407*, 98–102.

Meyer, T.F., Gibbs, C.P., and Haas, R. (1990). Variation and control of protein expression in *Neisseria*. Annu. Rev. Microbiol. *44*, 451–477.

Mishra, A., Das, A., Cisar, J.O., and Ton-That, H. (2007). Sortase-catalyzed assembly of distinct heteromeric fimbriae in *Actinomyces naeslundii*. J. Bacteriol. *189*, 3156–3165.

Mobley, H.L., Jarvis, K.G., Elwood, J.P., Whittle, D.I., Lockatell, C.V., Russell, R.G., Johnson, D.E., Donnenberg, M.S., and Warren, J.W. (1993). Isogenic P-fimbrial deletion mutants of pyelonephritogenic *Escherichia coli*: the role of alpha Gal(1–4) beta Gal binding in virulence of a wild-type strain. Mol. Microbiol. *10*, 143–155.

Mohamed, J.A., Huang, W., Nallapareddy, S.R., Teng, F., and Murray, B.E. (2004). Influence of origin of isolates, especially endocarditis isolates, and various genes on biofilm formation by *Enterococcus faecalis*. Infect. Immun. *72*, 3658–3663.

Mora, M., Bensi, G., Capo, S., Falugi, F., Zingaretti, C., Manetti, A.G., Maggi, T., Taddei, A.R., Grandi, G., and Telford, J.L. (2005). Group A *Streptococcus* produce pilus-like structures containing protective antigens and Lancefield T antigens. Proc. Natl. Acad. Sci. U.S.A. *102*, 15641–15646.

Morand, P.C., Bille, E., Morelle, S., Eugène, E., Beretti, J.L., Wolfgang, M., Meyer, T.F., Koomey, M., and Nassif, X. (2004). Type IV pilus retraction in pathogenic *Neisseria* is regulated by the PilC proteins. EMBO J. *23*, 2009–2017.

Morand, P.C., Drab, M., Rajalingam, K., Nassif, X., and Meyer, T.F. (2009). *Neisseria meningitidis* differentially controls host cell motility through PilC1 and PilC2 components of type IV Pili. PLoS One *4*, e6834.

Muhldorfer, I., Ziehbur, W., and Hacker, J. (2001). *Escherichia coli* in Urinary Tract Infections. In Molecular Medical Microbiology, Sussman, M., ed. (Academic, Press London), pp. 1739–1748.

Mulvey, M.A. (2002). Adhesion and entry of uropathogenic *Escherichia coli*. Cell. Microbiol. *4*, 257–271.

Mulvey, M.A., Lopez-Boado, Y.S., Wilson, C.L., Roth, R., Parks, W.C., Heuser, J., and Hultgren, S.J. (1998). Induction and evasion of host defenses by type 1-piliated uropathogenic *Escherichia coli*. Science *282*, 1494–1497.

Mulvey, M.A., Schilling, J.D., Martinez, J.J., and Hultgren, S.J. (2000). Bad bugs and beleaguered bladders: interplay between uropathogenic *Escherichia coli* and innate host defenses. Proc. Natl. Acad. Sci. U.S.A. *97*, 8829–8835.

Munera, D., Hultgren, S., and Fernández, L.A. (2007). Recognition of the N-terminal lectin domain of FimH adhesin by the usher FimD is required for type 1 pilus biogenesis. Mol. Microbiol. *64*, 333–346.

Murray, B.E. (1990). The life and times of the *Enterococcus*. Clin. Microbiol. Rev. *3*, 46–65.

Mutlu, G.M., and Wunderink, R.G. (2006). Severe pseudomonal infections. Curr. Opin. Crit. Care *12*, 458–463.

Nakata, M., Köller, T., Moritz, K., Ribardo, D., Jonas, L., McIver, K.S., Sumitomo, T., Terao, Y., Kawabata, S., Podbielski, A., *et al.* (2009). Mode of expression and functional characterization of FCT-3 pilus region-encoded proteins in *Streptococcus pyogenes* serotype M49. Infect. Immun. *77*, 32–44.

Nallapareddy, S.R., Singh, K.V., Sillanpää, J., Garsin, D.A., Höök, M., Erlandsen, S.L., and Murray, B.E. (2006). Endocarditis and biofilm-associated pili of *Enterococcus faecalis*. J. Clin. Invest. *116*, 2799–2807.

Nassif, X., Beretti, J.L., Lowy, J., Stenberg, P., O'Gaora, P., Pfeifer, J., Normark, S., and So, M. (1994). Roles of pilin and PilC in adhesion of *Neisseria meningitidis* to human epithelial and endothelial cells. Proc. Natl. Acad. Sci. U.S.A. *91*, 3769–3773.

Nassif, X., Pujol, C., Morand, P., and Eugène, E. (1999). Interactions of pathogenic *Neisseria* with host cells. Is it possible to assemble the puzzle? Mol. Microbiol. *32*, 1124–1132.

Nataro, J.P., and Kaper, J.B. (1998). Diarrheagenic *Escherichia coli*. Clin. Microbiol. Rev. *11*, 142–201.

Nelson, A.L., Ries, J., Bagnoli, F., Dahlberg, S., Fälker, S., Rounioja, S., Tschöp, J., Morfeldt, E., Ferlenghi, I., Hilleringmann, M., *et al.* (2007). RrgA is a pilus-associated adhesin in *Streptococcus pneumoniae*. Mol. Microbiol. *66*, 329–340.

Nowicki, B., Selvarangan, R., and Nowicki, S. (2001). Family of *Escherichia coli* Dr adhesins: decay-accelerating factor receptor recognition and invasiveness. J. Infect. Dis. *183* (Suppl. 1), S24–27.

Olsén, A., Jonsson, A., and Normark, S. (1989). Fibronectin binding mediated by a novel class of surface organelles on *Escherichia coli*. Nature *338*, 652–655.

Olsén, A., Arnqvist, A., Hammar, M., Sukupolvi, S., and Normark, S. (1993). The RpoS sigma factor relieves H-NS-mediated transcriptional repression of *csgA*, the subunit gene of fibronectin-binding curli in *Escherichia coli*. Mol. Microbiol. *7*, 523–536.

Orndorff, P.E., Devapali, A., Palestrant, S., Wyse, A., Everett, M.L., Bollinger, R.R., and Parker, W. (2004). Immunoglobulin-mediated agglutination of and biofilm formation by *Escherichia coli* K-12 require the type 1 pilus fiber. Infect. Immun. *72*, 1929–1938.

Piatek, R., Zalewska, B., Kolaj, O., Ferens, M., Nowicki, B., and Kur, J. (2005). Molecular aspects of biogenesis of *Escherichia coli* Dr Fimbriae: characterization of DraB–DraE complexes. Infect. Immun. *73*, 135–145.

Podbielski, A., Woischnik, M., Leonard, B.A., and Schmidt, K.H. (1999). Characterization of *nra*, a global negative regulator gene in group A streptococci. Mol. Microbiol. *31*, 1051–1064.

Pratt, L.A., and Kolter, R. (1998). Genetic analysis of *Escherichia coli* biofilm formation: roles of flagella, motility, chemotaxis and type I pili. Mol. Microbiol. *30*, 285–293.

Pujol, C., Eugène, E., Marceau, M., and Nassif, X. (1999). The meningococcal PilT protein is required for induction of intimate attachment to epithelial cells following pilus-mediated adhesion. Proc. Natl. Acad. Sci. U.S.A. *96*, 4017–4022.

Quigley, B.R., Zähner, D., Hatkoff, M., Thanassi, D.G., and Scott, J.R. (2009). Linkage of T3 and Cpa pilins in the *Streptococcus pyogenes* M3 pilus. Mol. Microbiol. *72*, 1379–1394.

Rahman, M., Källström, H., Normark, S., and Jonsson, A.B. (1997). PilC of pathogenic *Neisseria* is associated with the bacterial cell surface. Mol. Microbiol. *25*, 11–25.

Rakotoarivonina, H., Jubelin, G., Hebraud, M., Gaillard-Martinie, B., Forano, E., and Mosoni, P. (2002). Adhesion to cellulose of the Gram-positive bacterium *Ruminococcus albus* involves type IV pili. Microbiology *148*, 1871–1880.

Ramphal, R., Sadoff, J.C., Pyle, M., and Silipigni, J.D. (1984). Role of pili in the adherence of *Pseudomonas aeruginosa* to injured tracheal epithelium. Infect. Immun. *44*, 38–40.

Riley-Vargas, R.C., Gill, D.B., Kemper, C., Liszewski, M.K., and Atkinson, J.P. (2004). CD46: expanding beyond complement regulation. Trends Immunol. *25*, 496–503.

Roberts, J.A., Marklund, B.I., Ilver, D., Haslam, D., Kaack, M.B., Baskin, G., Louis, M., Möllby, R., Winberg, J., and Normark, S. (1994). The Gal(α 1–4)Gal-specific tip adhesin of *Escherichia coli* P-fimbriae is needed for pyelonephritis to occur in the normal urinary tract. Proc. Natl. Acad. Sci. U.S.A. *91*, 11889–11893.

Rocha, S.P., Pelayo, J.S., and Elias, W.P. (2007). Fimbriae of uropathogenic *Proteus mirabilis*. FEMS Immunol. Med. Microbiol. *51*, 1–7.

Rosini, R., Rinaudo, C.D., Soriani, M., Lauer, P., Mora, M., Maione, D., Taddei, A., Santi, I., Ghezzo, C., Brettoni, C., *et al.* (2006). Identification of novel genomic islands coding for antigenic pilus-like structures in *Streptococcus agalactiae*. Mol. Microbiol. *61*, 126–141.

Ruggeri, Z.M. (2001). Structure of von Willebrand factor and its function in platelet adhesion and thrombus formation. Best Pract. Res. Clin. Haematol. *14*, 257–279.

Sauer, F.G., Barnhart, M., Choudhury, D., Knight, S.D., Waksman, G., and Hultgren, S.J. (2000a). Chaperone-assisted pilus assembly and bacterial attachment. Curr. Opin. Struct. Biol. *10*, 548–556.

Sauer, F.G., Mulvey, M.A., Schilling, J.D., Martinez, J.J., and Hultgren, S.J. (2000b). Bacterial pili: molecular mechanisms of pathogenesis. Curr. Opin. Microbiol. *3*, 65–72.

Sauer, F.G., Pinkner, J.S., Waksman, G., and Hultgren, S.J. (2002). Chaperone priming of pilus subunits facilitates a topological transition that drives fiber formation. Cell *111*, 543–551.

Sauer, F.G., Remaut, H., Hultgren, S.J., and Waksman, G. (2004). Fiber assembly by the chaperone-usher pathway. Biochim. Biophys. Acta *1694*, 259–267.

Scaletsky, I.C., Milani, S.R., Trabulsi, L.R., and Travassos, L.R. (1988). Isolation and characterization of the localized adherence factor of enteropathogenic *Escherichia coli*. Infect. Immun. *56*, 2979–2983.

van Schaik, E.J., Giltner, C.L., Audette, G.F., Keizer, D.W., Bautista, D.L., Slupsky, C.M., Sykes, B.D., and Irvin, R.T. (2005). DNA binding: a novel function of *Pseudomonas aeruginosa* type IV pili. J. Bacteriol. *187*, 1455–1464.

Scheuerpflug, I., Rudel, T., Ryll, R., Pandit, J., and Meyer, T.F. (1999). Roles of PilC and PilE proteins in pilus-mediated adherence of *Neisseria gonorrhoeae* and *Neisseria meningitidis* to human erythrocytes and endothelial and epithelial cells. Infect. Immun. *67*, 834–843.

Schuchat, A. (1999). Group B *Streptococcus*. Lancet *353*, 51–56.

Servin, A.L. (2005). Pathogenesis of Afa/Dr diffusely adhering *Escherichia coli*. Clin. Microbiol. Rev. *18*, 264–292.

Shin, J.S., Gao, Z., and Abraham, S.N. (2000). Involvement of cellular caveolae in bacterial entry into mast cells. Science *289*, 785–788.

Singh, K.V., Nallapareddy, S.R., and Murray, B.E. (2007). Importance of the *ebp* (endocarditis- and biofilm-associated pilus) locus in the pathogenesis of *Enterococcus faecalis* ascending urinary tract infection. J. Infect. Dis. *195*, 1671–1677.

Sjöbring, U., Pohl, G., and Olsén, A. (1994). Plasminogen, absorbed by *Escherichia coli* expressing curli or by *Salmonella enteritidis* expressing thin aggregative fimbriae, can be activated by simultaneously captured tissue-type plasminogen activator (t-PA). Mol. Microbiol. *14*, 443–452.

Smith, W.D., Pointon, J.A., Abbot, E., Kang, H.J., Baker, E.N., Hirst, B.H., Wilson, J.A., Banfield, M.J., and Kehoe, M.A. (2010). Roles of minor pilin subunits Spy0125 and Spy0130 in the serotype M1 *Streptococcus pyogenes* strain SF370. J. Bacteriol. *192*, 4651–4659.

Sokurenko, E.V., Courtney, H.S., Maslow, J., Siitonen, A., and Hasty, D.L. (1995). Quantitative differences in adhesiveness of type 1 fimbriated *Escherichia coli* due to structural differences in fimH genes. J. Bacteriol. *177*, 3680–3686.

Sokurenko, E.V., Chesnokova, V., Dykhuizen, D.E., Ofek, I., Wu, X.R., Krogfelt, K.A., Struve, C., Schembri, M.A., and Hasty, D.L. (1998). Pathogenic adaptation of *Escherichia coli* by natural variation of the FimH adhesin. Proc. Natl. Acad. Sci. U.S.A. *95*, 8922–8926.

Strom, M.S., and Lory, S. (1993). Structure–function and biogenesis of the type IV pili. Annu. Rev. Microbiol. *47*, 565–596.

Strömberg, N., Marklund, B.I., Lund, B., Ilver, D., Hamers, A., Gaastra, W., Karlsson, K.A., and Normark, S. (1990). Host-specificity of uropathogenic *Escherichia coli* depends on differences in binding specificity to Galα 1-4Gal-containing isoreceptors. EMBO J. *9*, 2001–2010.

Strömberg, N., Nyholm, P.G., Pascher, I., and Normark, S. (1991). Saccharide orientation at the cell surface affects glycolipid receptor function. Proc. Natl. Acad. Sci. U.S.A. *88*, 9340–9344.

Suh, J.Y., Spyracopoulos, L., Keizer, D.W., Irvin, R.T., and Sykes, B.D. (2001). Backbone dynamics of receptor binding and antigenic regions of a *Pseudomonas aeruginosa* pilin monomer. Biochemistry *40*, 3985–3995.

Swaminathan, A., Mandlik, A., Swierczynski, A., Gaspar, A., Das, A., and Ton-That, H. (2007). Housekeeping sortase facilitates the cell wall anchoring of pilus polymers in *Corynebacterium diphtheriae*. Mol. Microbiol. *66*, 961–974.

Swanson, J. (1973). Studies on gonococcus infection. IV. Pili: their role in attachment of gonococci to tissue culture cells. J. Exp. Med. *137*, 571–589.

Swierczynski, A., and Ton-That, H. (2006). Type III pilus of corynebacteria: Pilus length is determined by the level of its major pilin subunit. J. Bacteriol. *188*, 6318–6325.

Tacket, C.O., Taylor, R.K., Losonsky, G., Lim, Y., Nataro, J.P., Kaper, J.B., and Levine, M.M. (1998). Investigation of the roles of toxin-coregulated pili and mannose-sensitive hemagglutinin pili in the pathogenesis of *Vibrio cholerae* O139 infection. Infect. Immun. *66*, 692–695.

Telford, J.L., Barocchi, M.A., Margarit, I., Rappuoli, R., and Grandi, G. (2006). Pili in Gram-positive pathogens. Nat. Rev. Microbiol. *4*, 509–519.

Thanassi, D.G., Saulino, E.T., Lombardo, M.J., Roth, R., Heuser, J., and Hultgren, S.J. (1998). The PapC usher forms an oligomeric channel: implications for pilus biogenesis across the outer membrane. Proc. Natl. Acad. Sci. U.S.A. *95*, 3146–3151.

Tobiason, D.M., and Seifert, H.S. (2001). Inverse relationship between pilus-mediated gonococcal adherence and surface expression of the pilus receptor, CD46. Microbiology *147*, 2333–2340.

Ton-That, H., and Schneewind, O. (2003). Assembly of pili on the surface of *Corynebacterium diphtheriae*. Mol. Microbiol. *50*, 1429–1438.

Ton-That, H., and Schneewind, O. (2004). Assembly of pili in Gram-positive bacteria. Trends Microbiol. *12*, 228–234.

Tükel, C., Raffatellu, M., Humphries, A.D., Wilson, R.P., Andrews-Polymenis, H.L., Gull, T., Figueiredo, J.F., Wong, M.H., Michelsen, K.S., Akçelik, M., *et al.* (2005). CsgA is a pathogen-associated molecular pattern of *Salmonella enterica* serotype Typhimurium that is recognized by Toll-like receptor 2. Mol. Microbiol. *58*, 289–304.

Vanmaele, R.P., Heerze, L.D., and Armstrong, G.D. (1999). Role of lactosyl glycan sequences in inhibiting enteropathogenic *Escherichia coli* attachment. Infect. Immun. *67*, 3302–3307.

Varga, J.J., Nguyen, V., O'Brien, D.K., Rodgers, K., Walker, R.A., and Melville, S.B. (2006). Type IV pili-dependent gliding motility in the Gram-positive pathogen *Clostridium perfringens* and other Clostridia. Mol. Microbiol. *62*, 680–694.

Waldor, M.K., and Mekalanos, J.J. (1996). Lysogenic conversion by a filamentous phage encoding cholera toxin. Science *272*, 1910–1914.

Wilkinson, H.W. (1978). Group B streptococcal infection in humans. Annu. Rev. Microbiol. *32*, 41–57.

Winther-Larsen, H.C., Hegge, F.T., Wolfgang, M., Hayes, S.F., van Putten, J.P., and Koomey, M. (2001). *Neisseria gonorrhoeae* PilV, a type IV pilus-associated protein essential to human epithelial cell adherence. Proc. Natl. Acad. Sci. U.S.A. *98*, 15276–15281.

Wolfgang, M., Lauer, P., Park, H.S., Brossay, L., Hébert, J., and Koomey, M. (1998). PilT mutations lead to simultaneous defects in competence for natural transformation and twitching motility in piliated *Neisseria gonorrhoeae*. Mol. Microbiol. *29*, 321–330.

Wolfgang, M., van Putten, J.P., Hayes, S.F., Dorward, D., and Koomey, M. (2000). Components and dynamics of fiber formation define a ubiquitous biogenesis pathway for bacterial pili. EMBO J. *19*, 6408–6418.

Wright, K.J., Seed, P.C., and Hultgren, S.J. (2007). Development of intracellular bacterial communities of uropathogenic *Escherichia coli* depends on type 1 pili. Cell. Microbiol. *9*, 2230–2241.

Wu, H., Mintz, K.P., Ladha, M., and Fives-Taylor, P.M. (1998). Isolation and characterization of Fap1, a fimbriae-associated adhesin of *Streptococcus parasanguis* FW213. Mol. Microbiol. *28*, 487–500.

Xie, B., Zhou, G., Chan, S.Y., Shapiro, E., Kong, X.P., Wu, X.R., Sun, T.T., and Costello, C.E. (2006). Distinct glycan structures of uroplakins Ia and Ib: structural basis for the selective binding of FimH adhesin to uroplakin Ia. J. Biol. Chem. *281*, 14644–14653.

Yanagawa, R., and Otsuki, K. (1970). Some properties of the pili of *Corynebacterium renale*. J. Bacteriol. *101*, 1063–1069.

Yeung, M.K. (1999). Molecular and genetic analyses of *Actinomyces* spp. Crit. Rev. Oral. Biol. Med. *10*, 120–138.

Yeung, M.K., and Ragsdale, P.A. (1997). Synthesis and function of *Actinomyces naeslundii* T14V type 1 fimbriae require the expression of additional fimbria-associated genes. Infect. Immun. *65*, 2629–2639.

Yeung, M.K., Donkersloot, J.A., Cisar, J.O., and Ragsdale, P.A. (1998). Identification of a gene involved in assembly of *Actinomyces naeslundii* T14V type 2 fimbriae. Infect. Immun. *66*, 1482–1491.

Zavialov, A.V., Berglund, J., Pudney, A.F., Fooks, L.J., Ibrahim, T.M., MacIntyre, S., and Knight, S.D. (2003). Structure and biogenesis of the capsular F1 antigen from *Yersinia pestis*: preserved folding energy drives fiber formation. Cell *113*, 587–596.

Zhou, G., Mo, W.J., Sebbel, P., Min, G., Neubert, T.A., Glockshuber, R., Wu, X.R., Sun, T.T., and Kong, X.P. (2001). Uroplakin Ia is the urothelial receptor for uropathogenic *Escherichia coli*: evidence from *in vitro* FimH binding. J. Cell Sci. *114*, 4095–4103.

Zunino, P., Sosa, V., Schlapp, G., Allen, A.G., Preston, A., and Maskell, D.J. (2007). Mannose-resistant *Proteus*-like and *P. mirabilis* fimbriae have specific and additive roles in *P. mirabilis* urinary tract infections. FEMS Immunol. Med. Microbiol. *51*, 125–133.

Non-pilus (Non-fimbrial) Adhesins

5

Amanda J. Sheets and Joseph W. St. Geme, III

Abstract

Adherence to a biological surface allows bacteria to persist and spread within the host and represents an essential early step in the pathogenesis of most bacterial diseases. Bacteria produce a variety of pilus and non-pilus adhesive structures that mediate specific adherence to host tissues. Among non-pilus adhesive structures, most can be classified according to the mechanism of secretion or the mechanism of anchoring to the bacterial surface. The majority of non-pilus adhesins are proteins, but other structures such as lipopolysaccharide and lipoteichoic acid also have adhesive function. This chapter summarizes the classes of bacterial non-pilus adhesins and highlights the roles of prototype adhesins in the context of disease pathogenesis. Elucidation of conserved mechanisms of secretion and anchoring of adhesins may facilitate the development of novel therapeutic agents that combat infectious diseases by effectively disrupting adherence and initial interactions with the host.

Bacterial adherence to host tissues

In a normal host, the primary surfaces to which bacteria adhere include the skin and mucosa. Owing to the presence of layers of dead, keratinized epidermal cells, normal intact skin is relatively resistant to bacterial infection. However, infections oftentimes occur when there is disruption of the skin barrier, for example by lacerations, abrasions, punctures, chronic wounds, or burns. Once the barrier is opened, adherent microorganisms (such as *Streptococcus pyogenes, Staphylococcus* spp., and *Propionibacterium* spp.) can easily penetrate the skin and invade the body. Mucosal epithelial cell surfaces include oral, respiratory, gastric, intestinal, and genitourinary mucosa. Upon entering the respiratory or genital tract, bacteria initially interact with mucus, an aqueous solution containing glycoproteins (mucins) and various salts, and are largely eliminated by mucociliary clearance due to the action of ciliated epithelial cells. Non-adherent bacteria are removed from the gastrointestinal tract and the urinary tract by peristalsis and excretion. To persist on mucosal surfaces, organisms must adhere to epithelial cells and often form discrete microcolonies. Heavy colonization at initial sites (upper respiratory tract, lower genital tract) facilitates contiguous spread of organisms to normally sterile sites such as the lower respiratory tract, sinuses, middle ear, and upper genital tract.

Bacteria that disseminate to other sites within the host often adhere to endothelial cells, serum components, or components of connective tissue extracellular matrix (ECM), which normally serves as a substrate for the adhesion of host cells. A number of bacterial pathogens exploit the host cell adhesion system to attach to host tissues. ECM components that support bacterial adherence include fibronectin, collagen, fibrinogen/fibrin, elastin, vitronectin, laminin, as well as heparin sulfate-containing proteoglycans. Bacterial proteins responsible for adhesion to ECM molecules are termed **m**icrobial **s**urface **c**omponents **r**ecognizing **a**dhesive **m**atrix **m**olecules (MSCRAMMs).

The process of adhesion is largely achieved

by highly specific ligand–receptor interactions. A variety of molecules can function as receptors, including proteins, glycoproteins, and glycolipids. The presence of complementary receptors on specific cell types in part determines the tissue tropism associated with a bacterial species. Some bacteria have a broad tissue tropism and can infect many types of cells, tissues, and hosts. Other bacteria may infect primarily a single tissue owing to limited receptor expression or may be uniquely host-restricted owing to genetic polymorphisms and structural differences associated with key receptors.

Bacteria produce a variety of adhesive structures that facilitate colonization and other aspects of the pathogenic process. In contrast to adhesive pili, non-pilus adhesins are monomeric or oligomeric and are anchored directly to the outer membrane, generally protruding only a short distance from the bacterial cell surface. Consequently, most non-pilus adhesins are not visible using standard electron microscopy and are associated with intimate contact between the bacterium and the host cell or target substrate. Non-pilus adhesins that appear as fibre-like structures are usually shorter and thinner than polymeric pili. While there is a great variety of non-pilus adhesins, some common themes have emerged relating to the process of surface localization and anchoring as well as the means of binding to host tissues. For the purpose of consistency, in this chapter the term non-pilus will be used instead of non-fimbrial or afimbrial.

Gram-negative bacterial non-pilus adhesins

Surface proteins expressed by Gram-negative bacteria are generally classified into families based upon sequence similarities within conserved structural and functional domains, which reflect distinct mechanisms of secretion across the outer membrane. Therefore, this section has been organized according to the mechanism of secretion underlying the surface localization of several prototype Gram-negative bacterial non-pilus adhesins. Most non-pilus adhesins are secreted by the type V secretion pathway and are classified as autotransporters or two-partner secretion systems (Fig. 5.1). Autotransporter proteins are further divided into two subfamilies called classical monomeric autotransporters (Va) and trimeric autotransporters (Vc), and two-partner secretion systems are sometimes referred to as unlinked autotransporters (Vb). In addition, a number of integral outer membrane proteins with porin-like structural features and lipopolysaccharide molecules from selected organisms have been shown to enhance bacterial adherence to host surfaces. Selected non-pilus adhesins are highlighted in Table 5.1.

Autotransporters

Autotransporter proteins share conserved structural and functional domains, including an N-terminal signal sequence, an internal secreted passenger domain, and a C-terminal membrane anchor domain (Fig. 5.1A and 5.1B). These proteins were originally designated autotransporters based on the belief that their primary structure contains all the information necessary for transport from the periplasm to the bacterial surface (Henderson *et al.*, 1998; Henderson and Nataro, 2001; Wells *et al.*, 2007). However, recent studies suggest that the Bam complex (also known as the YaeT or Omp85 complex) is required for integration of all outer membrane proteins into the outer membrane, including autotransporters (Voulhoux *et al.*, 2003; Bernstein, 2007).

The first step in the process of autotransporter secretion involves traversal of the cytoplasmic membrane. The N-terminal signal peptide directs export across the cytoplasmic membrane via the general secretory pathway. A number of autotransporters contain an atypical signal peptide with a conserved extension of the n-domain followed by a typical hydrophobic h-domain and a typical signal peptidase I cleavage site (Henderson *et al.*, 1998; St Geme and Cutter, 2000). Once the protein is in the periplasm, the C-terminal domain inserts into the outer membrane and facilitates surface localization of the passenger domain (Fig. 5.1A and 5.1B). The membrane anchor β-domain of classical, monomeric autotransporters is formed by 12 amphipathic β-strands that align antiparallel to each other to create a β-barrel with a central hydrophilic channel. A transmembrane α-helical linker region occludes this pore and connects with

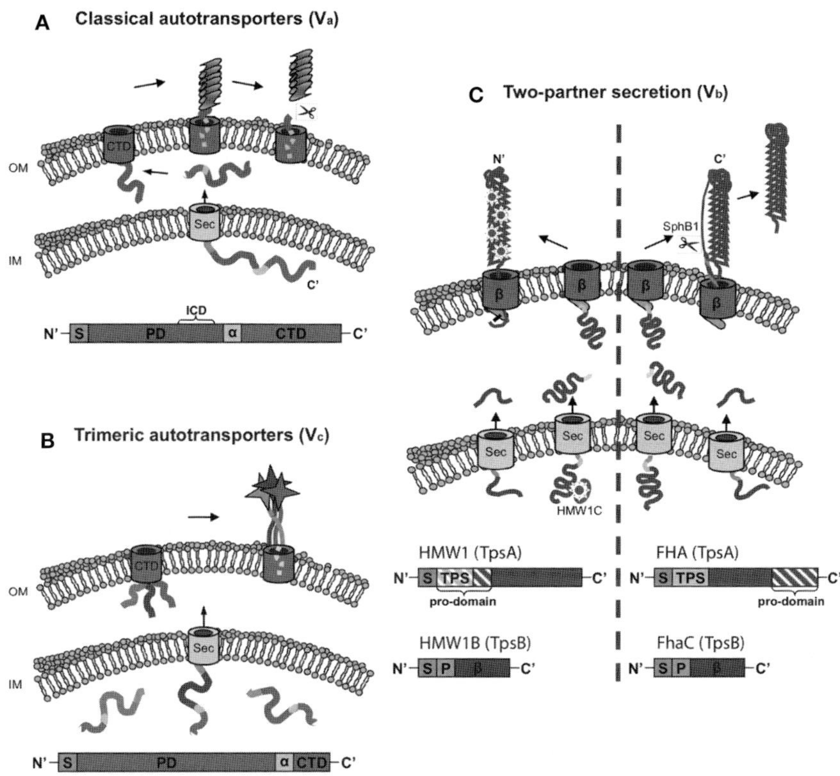

Figure 5.1 Gram-negative non-pilus adhesins. (A) Classical autotransporters. An N-terminal signal peptide (S) directs translocation of the protein across the cytoplasmic membrane and is cleaved. A C-terminal β-domain (CTD) targets the autotransporter to the outer membrane, forms a β-barrel pore, and is occluded by a transmembrane α-helical linker region (α) that connects with the extracellular passenger domain (PD). A number of classical autotransporter adhesins have an intramolecular chaperone domain (ICD) that is important for folding of the passenger domain, which mediates binding to host receptors. In virtually all conventional autotransporters, the passenger domain is cleaved from the translocator domain and is either tethered non-covalently to the bacterial surface or released extracellularly. (B) Trimeric autotransporters. The C-terminal domain from each monomer contributes four β-strands which trimerize to form a 12-stranded β-barrel pore. All known trimeric autotransporters harbour adhesive activity in the passenger domain, lack an apparent intramolecular chaperone domain, form heat-resistant oligomers, and remain uncleaved at the cell surface. Oftentimes these proteins form short, thin surface fibres that are visible by transmission electron microscopy. (C) Two-partner secretion. The extracellular effector TpsA proteins contain four functional domains: the signal peptide (S), the two-partner secretion (TPS) domain, the pro-domain (striped), and the extracellular effector domain (solid). The outer membrane translocator TpsB proteins contain three functional domains: the signal peptide (S), the periplasmic domain (P), and the β-barrel domain (β). N-terminal signal peptides direct translocation of TpsA and TpsB across the cytoplasmic membrane by the Sec secretion system. Each TpsB protein inserts into the outer membrane and forms a β-barrel. The TPS domain targets the TpsA pro-protein to the cognate TpsB protein, interacting with the periplasmic domain and initiating translocation across the outer membrane to the extracellular environment where it folds into a β-helix or β-solenoid. The prototype adhesins of this family, *H. influenzae* HMW1 and *B. pertussis* FHA, exhibit slight differences with respect to TPS sequence and orientation of the mature protein and may represent two distinct TPS subfamilies. Secretion of HMW1: The N-terminal TPS-containing pro-domain is cleaved and released extracellularly, generating mature HMW1. The C-terminus undergoes disulfide bond formation, which serves to anchor mature HMW1 to HMW1B. Surface anchoring of HMW1 also requires glycosylation, which is mediated by the HMW1C glycosyltransferase. Secretion of FHA: A C-terminal pro-domain is cleaved at the cell surface by the serine protease SphB1 and other unidentified factors, releasing a significant amount of mature FHA into the extracellular environment. A colour version of this figure is located in the plate section at the back of the book.

Table 5.1 Selected examples of Gram-negative bacterial adhesins

Category	Organism	Adhesin	Target cells/receptor/substrate	Host/disease
Classical autotransporters	*Bordetella* spp.	Pertactin	Mammalian ciliated respiratory epithelial cells, integrins	Mammals (human, rodent) – RTIs
	Bordetella spp.	BrkA	Human lung fibroblasts, epithelial cells	Mammals (human, rodent) – RTIs
	Bordetella pertussis	TcfA	Mouse trachea	Mammals (human) – whooping cough
	Chlamydia spp.	POMPs	Genital epithelial cells	Mammals – genital infections, urethritis, spontaneous abortion, enteritis, pneumonia, conjunctivitis
	Escherichia coli (EHEC, ETEC, UPEC)	Antigen 43	Mammalian epithelial cells, self-associates	Mammals (human) – gastroenteritis, sepsis, UTIs
	E. coli (diarrheagenic)	AIDA-I	Mammalian epithelial cells, *N*-glycosylated membrane protein, gp119, self-associates	Mammals (human) – gastroenteritis
	E. coli (ETEC)	TibA	Intestinal and colonic epithelial cells, self-associates	Mammals (human) – gastroenteritis
	Haemophilus influenzae	Hap	Respiratory epithelial cells, collagen IV, laminin, fibronectin, self-associates	Mammals (human) – RTIs, sepsis, meningitis
	Helicobacter pylori	AlpA	Gastric tissue	Mammals (human) – gastritis, gastric ulcers
	Helicobacter pylori	BabA	Gastric epithelial cells, Lewis b (α-1,3/4-difucosylated) blood group antigen	Mammals (human) – gastritis, gastric ulcers
Trimeric autotransporters	*Bartonella henselae*	BadA	Endothelial and epithelial cells, collagen IV, fibronectin	Mammals (human, cat) – bacillary angiomatosis, peliosis hepatis, cat scratch disease
	Haemophilus influenzae	Hia	Respiratory epithelial cells	Mammals (human) – RTIs, sepsis, meningitis
	Haemophilus influenzae	Hsf	Respiratory epithelial cells	Mammals (human) – RTIs, sepsis, meningitis
	Haemophilus cryptic genospecies	Cha	Genital and respiratory epithelial cells, self-associates	Mammals (human) – female genital and neonatal infections
	Moraxella catarrhalis	Hag	Respiratory epithelial cells, collagen IV	Mammals (human) – RTIs
	Moraxella catarrhalis	UspA1	Respiratory epithelial cells, CEACAM1, fibronectin, laminin	Mammals (human) – RTIs
	Moraxella catarrhalis	UspA2	Fibronectin, laminin, vitronectin	Mammals (human) – RTIs
	Moraxella catarrhalis	UspA2h	Respiratory epithelial cells	Mammals (human) – RTIs
	Neisseria meningitidis	NadA	Respiratory epithelial cells	Mammals (human) – meningitis, sepsis

	Neisseria meningitidis	NhhA	Genital and respiratory epithelial cells, laminin, heparin sulphate	Mammals (human) – meningitis, sepsis
	Yersinia spp.	YadA	Human epidermoid cells, β1 integrins, collagen type I/III/IV/V/XI, laminin, fibronectin, self-associates	Mammal (human, rodent) – enteritis, mesenteric lymphadenitis, reactive arthritis
Two-partner secretion	*Bordetella pertussis*	FHA	Respiratory epithelial cells, macrophages, CR3 integrin, heparan sulfate, sulfatide (via galactose residues)	Mammals (human) – RTIs
	E. coli (ETEC)	EtpA	Intestinal epithelial cells	Mammals (human) – gastroenteritis
	Erwinia chrysanthemi	HecA	Leaf epidermal cells	Plants – necrosis, soft-rot
	Haemophilus influenzae	HMW1/ HMW2	Respiratory epithelial cells, unknown glycoprotein receptor	Mammals (human) – RTIs, sepsis, meningitis
	Moraxella catarrhalis	MchA1/ MchA2	Human bronchial epithelial cells	Mammals (human) – RTIs
OMPs/Porins	*Neisseria meningitidis*	Opc	Heparan sulphate glycosaminoglycans, vitronectin, fibronectin	Mammals (human) – meningitis, sepsis
	Neisseria spp.	Opa	CD66 proteins/CEACAM receptors	Mammals (human) – meningitis, sepsis, gonorrhoea
	Neisseria gonorrhoeae	Opa50	Heparin sulphate proteoglycans, vitronectin	Mammals (human) – sepsis, gonorrhoea
	Haemophilus influenzae	P2	Sialic acid-containing oligosaccharides, nasopharyngeal mucin	mammals (human) – RTIs, sepsis, meningitis
	Haemophilus influenzae	P5	CEACAM1, oropharyngeal cells and nasopharyngeal mucin	Mammals (human) – RTIs, sepsis, meningitis
	Moraxella catarrhalis	OmpCD	human lung cells	mammals (human) – RTIs
	Vibrio cholerae	OmpU	Intestinal epithelial cells, fibronectin, RGD tripeptide	Mammals (human) – gastroenteritis (cholera)
	Chlamydia trachomatis	OmcB	Glycosaminoglycan (GAG)-like receptors, heparan sulfate-like GAG	Mammals – genital and neonatal infections, urethritis, trachoma

RTI, respiratory tract infection; UTI, urinary tract infection.

the extracellular passenger domain (Oomen *et al.*, 2004). In the case of trimeric autotransporters, a shorter C-terminal domain from each monomer contributes an α-helical linker and four β-strands which trimerize in the outer membrane to form a 12-stranded β-barrel pore (Roggenkamp *et al.*, 2003; Surana *et al.*, 2004a; Ackermann *et al.*, 2008), analogous to the anchor domain of monomeric autotransporters. Deletion of the α-helical linker region disrupts surface localization of the passenger domain in both conventional and trimeric autotransporters, suggesting that the α-helix and the β-barrel regions together function as the 'minimal translocation unit' (Maurer *et al.*, 1999; Oliver *et al.*, 2003a). The mechanism by which the passenger domain is translocated across the outer membrane is unknown. According to one model, the unfolded passenger domain is extruded through the hydrophilic channel of the membrane anchor in a hairpin conformation, since it is unlikely that the large, fully folded passenger domain can pass through the ~2-nm β-barrel pore (Oomen *et al.*, 2004; Jong *et al.*, 2007; Ieva and Bernstein, 2009). An alternative model is that the β-domain functions primarily to target and anchor the protein and that an external factor translocates the folded passenger domain across the outer membrane (Skillman *et al.*, 2005).

While the C-terminal β-domains associated with surface localization of the passenger proteins are highly homologous among autotransporters, the extracellular passenger domains exhibit considerable sequence variation (Henderson *et al.*, 2004). Consequently, autotransporter passenger domains confer a wide range of virulence-associated effector functions (sometimes in combination), including adhesion, invasion, protease activity, cytotoxicity, and serum resistance (Henderson *et al.*, 1998; Henderson and Nataro, 2001; Cotter *et al.*, 2005). Some also mediate bacterial aggregation and are classified as self-associating autotransporters (SAATs) (Benz and Schmidt, 1992; Klemm *et al.*, 2004, 2006; Sherlock *et al.*, 2004, 2005).

Classical autotransporters

Although autotransporter passenger domains vary considerably in sequence and size, almost all are predicted to form a β-helical structure (Junker *et al.*, 2006). Recent studies with the *Bordetella* BrkA adhesin have also revealed the presence of a conserved ~100-residue intramolecular chaperone domain (Pfam002475) that is important for extracellular folding of the passenger domain and is present in a number of conventional autotransporters (Oliver *et al.*, 2003b; Surana *et al.*, 2004a). Biochemical analysis of BrkA suggests that partial folding of the passenger domain takes place in the periplasm and that additional folding occurs simultaneously with translocation (Hodak and Jacob-Dubuisson, 2007). In the absence of this region, the BrkA passenger domain is translocated across the outer membrane but remains in a protease-sensitive, improperly folded conformation (Oliver *et al.*, 2003b).

Upon surface presentation, the passenger domain in virtually all conventional autotransporters is cleaved from the translocator domain and is either tethered non-covalently to the bacterial surface or released extracellularly. Many passenger domains contain a protease subdomain that mediates release by autocatalytic processing (Henderson *et al.*, 1998). In other cases, an outer membrane protease is involved in processing. The *Haemophilus influenzae* Hap autotransporter represents an example of an adhesive autotransporter protein that undergoes autoproteolysis owing to the activity of a serine protease catalytic site, resulting in extracellular release of the passenger domain. The C-terminal portion of the secreted Hap passenger domain mediates bacterial adherence to cultured human respiratory cells, bacterial adherence to laminin, fibronectin, and collagen IV, and bacterial aggregation (St Geme, 2002; Fink *et al.*, 2003). Interestingly, secretory leucocyte protease inhibitor, a serine protease inhibitor that is present in respiratory secretions, is upregulated during infection and is capable of blocking Hap autoproteolytic activity (Fink and St Geme, 2003), suggesting that the local host environment may influence the proteolytic and adhesive activities of this multifunctional protein.

Escherichia coli self-associating autotransporter proteins Antigen 43 (Ag43), AIDA-1, and TibA appear to belong to an autotransporter subfamily based on amino acid similarities and the presence of repetitive amino acid motifs (Owen *et al.*, 1996; Henderson and Nataro, 2001). Although the roles of the repeated amino acid sequence motifs have not been investigated, repetitive elements are often characteristic of proteins with adherence functions (Henderson and Nataro, 2001). Ag43 (widely conserved among enteropathogenic, enterohaemorrhagic, and uropathogenic *E. coli* strains) confers low level adhesion to certain mammalian cells and mediates bacterial aggregation, which contributes to biofilm formation and presumably enhances colonization of the mammalian intestine (Caffrey and Owen, 1989; Owen *et al.*, 1996; Henderson and Owen, 1999). AIDA-1 mediates aggregation and the diffuse adherence phenotype associated with diarrheagenic *E. coli* strains (Benz and Schmidt, 1992). The TibA adhesin/invasin is expressed by enterotoxigenic *E. coli* (ETEC) and mediates adherence and invasion in assays with intestinal epithelial cells (Elsinghorst and Weitz, 1994; Lindenthal and Elsinghorst, 2001). Beyond mediating bacterial aggregation and initiating biofilm formation through self-association, Ag43, TibA, and AIDA-I can also interact with each other via heterologous interactions, resulting in mixed bacterial aggregates (Klemm *et al.*, 2006). Interestingly, independent adherence assays using unmodified Ag43, AIDA-I, and TibA with various human epithelial cells have demonstrated that glycosylation of each of these proteins is essential for adherence (Benz and Schmidt, 2001; Lindenthal and Elsinghorst, 2001; Sherlock *et al.*, 2006).

Adhesive autotransporters often contain an Arg-Gly Asp (RGD) tripeptide, a motif that is associated in some systems with binding to β-integrin molecules on the plasma membrane of mammalian cells (D'Souza *et al.*, 1991; Henderson *et al.*, 1998). Examples include Ag43, the *Bordetella* BrkA and pertactin proteins, and the *Bordetella pertussis* TcfA (Leininger *et al.*, 1991, 1992; Finn and Stevens, 1995; Owen *et al.*, 1996). One of the loops protruding from the pertactin β-helix contains the RGD tripeptide (Emsley *et al.*, 1996). RGD-containing peptides derived from pertactin inhibit binding of purified pertactin to HeLa cells and entry of *B. pertussis* into HeLa cells, supporting a role for the RGD motif in adhesive activity (Leininger *et al.*, 1992). Whether the RGD motifs in other autotransporters play a role in adhesive activity awaits further investigation.

Trimeric autotransporters

Trimeric autotransporters (sometimes referred to as the oligomeric coiled-coil adhesin (Oca) family) represent a rapidly growing subfamily of autotransporters. The prototypic members of this subfamily are the *Yersinia enterocolitica* YadA adhesin and the *H. influenzae* Hia/Hsf adhesins. The multifaceted YadA protein promotes adherence to epithelial cells and ECM proteins and also mediates bacterial aggregation and serum resistance (El Tahir and Skurnik, 2001). During infection, the YadA adhesin aids in binding of *Yersinia* to intestinal M cells (Vazquez-Torres and Fang, 2000). The Hia adhesin is expressed by ~25% of non-typeable *H. influenzae* strains, and the Hsf adhesin is ubiquitous among encapsulated (sero-typeable) strains of *H. influenzae* (St Geme *et al.*, 1998). Hia and Hsf are homologues and have the same binding specificity, mediating attachment to the same battery of epithelial cell lines (St Geme *et al.*, 1996). Hsf is larger than Hia and forms short, thin surface fibres that are visible by transmission electron microscopy. All trimeric autotransporters that have been studied to date form heat-resistant oligomers and contain a passenger domain that remains uncleaved (Elkins *et al.*, 2000; Sandt and Hill, 2000; Ray *et al.*, 2002; Riess *et al.*, 2004). Studies with Hia have demonstrated that trimerization of the passenger domain is required for protein stability and adhesive activity (Cotter *et al.*, 2006).

Trimeric autotransporters exhibit a highly repetitive modular architecture, with a passenger domain that contains a variety of structural motifs categorized into head, connector, and stalk domains (Szczesny and Lupas, 2008). The head domain of YadA is composed of a cluster of Hep/Hag degenerate repeats (NSVAIGXXS), a β-roll-forming sequence element that is named for the *Moraxella catarrhalis* haemagglutinin Hag adhesin (Pearson *et al.*, 2002). This repetitive sequence motif is now frequently referred to as a Ylhead (for YadA-like head) domain. Other head domains include: HIN(1,2) domains (two Ylhead insert motifs, defined by GYDP and FxG motifs), Trp-rings (β-meander), GIN motifs (β-prism), and FxG motifs (unknown structure) (Szczesny and Lupas, 2008). Connector motifs allow for the transition from one type of structure to another.

Most common among these connectors is the helical 'neck' or HIM motif (Pfam05662), which frequently connects a Ylhead domain to a coiled-coil stalk domain. Examination of the amino acid sequences among numerous predicted trimeric autotransporter adhesins reveals wide conservation of these structural building blocks in different arrangements. Owing to the repetitive nature of these sequence motifs, they likely represent recombination hotspots that facilitate antigenic variation and adaptation to the host environment.

Trimeric autotransporter adhesins oftentimes bind several ECM components in addition to specific host cell receptors. However, despite the presence of highly conserved structural motifs, binding domains of trimeric autotransporters vary greatly in their composition. The Ylhead repeats are critical to host cell and collagen IV binding activity of both the YadA and the *Bartonella henselae* BadA head domains, but are unnecessary for adherence by the *M. catarrhalis* Hag protein (Bullard *et al.*, 2007; Kaiser *et al.*, 2008). The BadA head domain also contains a critical Trp-ring/GIN motif that is adjacent to the Ylhead repeats. Characterization of Hia binding activity has demonstrated two distinct binding domains, both composed of neighbouring GIN/neck/Trp-ring motifs. These two Hia binding domains interact with the same proteinaceous host cell receptor with differing affinities. The lower affinity binding domain is located towards the N-terminus of the 1098-residue protein, most distal from the bacterial surface (Laarmann *et al.*, 2002). One hypothesis is that this distal binding domain mediates initial attachment to host cells and is then displaced by the higher affinity binding domain via a ratcheting effect, resulting in more intimate and stable bacterial attachment to host cells. Crystal structure information from several trimeric autotransporter adhesins has shown that the adhesive domains consistently have a threefold symmetry (Cotter *et al.*, 2005), with three identical binding surfaces and the potential for multivalent interactions with host cells.

Location of the binding domains at the tip of the protein distal from the bacterial surface is a common theme among trimeric autotransporters, presumably because this position makes them easily accessible to host cell receptor structures (Nummelin *et al.*, 2004; Kaiser *et al.*, 2008; Yu *et al.*, 2008). The presence of a stalk domain further enhances accessibility of an adhesive head domain. The *Yersinia* YadA (Koretke *et al.*, 2006), the *Neisseria meningitidis* NadA (Comanducci *et al.*, 2002), and the *B. henselae* BadA (Kaiser *et al.*, 2008) adhesins exemplify the classic head-stalk-anchor organization of trimeric autotransporter adhesins, although the stalk lengths vary dramatically among these proteins. Owing to the presence of stalk domains, trimeric autotransporters are often visible by conventional negative staining electron microscopy as thin fibres. One example is YadA, which has been described as having a lollipop appearance (Hoiczyk *et al.*, 2000; Szczesny *et al.*, 2008). The sequence composition of left-handed coiled-coiled stalk domains varies from protein to protein but is usually characterized by repetitive sequences with heptad periodicity. The seven positions are denoted a-g, with small hydrophobic residues (commonly alanine) preferentially occurring in the first (a) and fourth (d) positions. In recent studies of the *Haemophilus* cryptic genospecies Cha adhesin, it was observed that variation in tandem identical 28-residue repeats comprising a predicted stalk domain both alters the length of Cha surface fibres and modulates the adhesive activity of the protein. Specifically, increased repeat copy number results in decreased bacterial attachment to genital and respiratory epithelial cells (Sheets and St. Geme, 2011). It is possible that the presence of a contracted stalk domain facilitates more intimate bacterial attachment to host cells than is possible with long, filamentous projections.

Two-partner secretion systems

Numerous important human, animal, and plant pathogens secrete adhesins and other virulence factors by means of the two-partner secretion (TPS) pathway (Henderson *et al.*, 2004; Jacob-Dubuisson *et al.*, 2004; Aoki *et al.*, 2005). TPS systems are conceptually similar to autotransporters, except that the passenger and the membrane anchor domains are synthesized as separate proteins. Extracellular secretion of the exoprotein (TpsA) is dependent on a cognate, channel-forming outer membrane β-barrel protein (TpsB). Both the TpsA protein and the TpsB protein contain an

N-terminal signal peptide that mediates export across the cytoplasmic membrane via the Sec translocon. A distinctive N-proximal module in TpsA (called a TPS domain or secretion domain) is essential for secretion across the outer membrane and has been shown to interact directly with the associated TpsB translocator (Schönherr *et al.*, 1993; Surana *et al.*, 2004b; Hodak *et al.*, 2006).

The best-studied TPS systems with adhesive activity are filamentous haemagglutinin (FHA)/FhaC of *B. pertussis* and *Bordetella bronchiseptica*, and HMW1/HMW1B and HMW2/HMW2B of *H. influenzae* (Fig. 5.1C). FHA forms filamentous structures on the bacterial surface and mediates bacterial attachment to ciliated respiratory epithelial cells, a key process in overcoming mucociliary clearance in the upper respiratory tract. Although FHA is not the only adhesin involved in *Bordetella* colonization, elimination of FHA results in a significant reduction in the ability of the organism to colonize the trachea in the mouse colonization model, and antibodies against FHA provide protection against mouse and human infection (Kimura *et al.*, 1990). As a result, this highly immunogenic protein is a primary component of acellular pertussis vaccines.

The HMW1 and HMW2 adhesins are expressed by ~75% of non-typeable *H. influenzae* strains and mediate attachment to respiratory epithelial cells (St Geme *et al.*, 1993, 1998). The pathogenesis of *H. influenzae* disease begins with colonization of the nasopharynx. Contiguous spread within the respiratory tract leads to infections associated with the middle ear, the sinuses, the conjunctiva, or the lungs. The presence of high antibody titres against HMW1/HMW2 in sera from patients recovering from non-typable *H. influenzae* disease indicates that the HMW1/HMW2 adhesins are expressed *in vivo* (presumably to aid in colonization of the nasopharynx) and are a prime target of the serum antibody response (Barenkamp and Bodor, 1990). Studies examining paired nasopharyngeal and middle ear isolates of *H. influenzae* from children with acute otitis media revealed decreased HMW1/HMW2 expression associated with migration to the middle ear (Dawid *et al.*, 1999). This association supports the idea that modulated HMW expression enables the organism to balance the need to colonize the respiratory tract early during infection against the ability to evade the immune response at later points in the infection process.

TpsB proteins belong to the Omp85 family and contain a conserved N-terminal domain (designated POTRA motifs for polypeptide-transport-associated), which is presumed to be involved in protein–protein interactions such as translocator/substrate recognition. Consistent with this model, partial deletion of the HMW1B periplasmic domain with a predicted POTRA domain abolished secretion of HMW1 without affecting channel formation (Surana *et al.*, 2006). Using overlay and co-immunoprecipitation assays, Hodak and colleagues have shown that a non-native conformation of the FHA TPS domain is specifically recognized by the periplasmic domain of FhaC, while the folded domain is not (Hodak *et al.*, 2006). Upon surface localization, FHA folds into a conformation rich in β-helices (Kajava 2001). The crystal structure of the N-terminus of the mature FHA protein indicates a folded diameter of 4 nm, which is incompatible with secretion through a 1–2-nm TpsB (FhaC) translocation channel in a native, folded state (Clantin *et al.*, 2004). Taken together, this information suggests that complete folding of FHA (and presumably other TpsA substrates) occurs at the cell surface, during or after translocation. Interestingly, a number of point mutations in the TPS domain of FHA disrupt secretion without affecting FHA/FhaC interactions, suggesting an additional function of this domain such as nucleation of folding or stabilization of the native TpsA structure. It has been proposed that the beta-helical structure of the TPS domain of FHA functions as a scaffold and initiates folding of the β-helical motifs that form the central domain of the adhesin. Sequence alignment of numerous TPS domains has suggested that TpsA proteins fall into two distinct subfamilies, HMW1/HMW2-like and FHA-like. This information has led to the hypothesis that the complex mechanism by which TPS domains interact with TpsB partners likely differs among TPS subfamilies and may play a role in determining the mechanism of secretion and the ultimate orientation of the TpsA protein on the cell surface.

The TpsA exoprotein often is synthesized with a pro-domain at either the N-terminus or

C-terminus that is required for surface presentation of the mature TpsA but is removed during translocation. HMW1 is synthesized with an N-terminal 68 amino acid signal peptide followed by a 373 amino acid pro-domain (including the TPS domain) that is cleaved by an unknown protease either in the periplasm or in the extracellular environment. FHA is synthesized as a 367 kDa pre-pro-protein (called FhaB) and undergoes cleavage of the 71 amino acid N-terminal signal peptide and the C-terminal 1225 amino acid pro-domain, generating a mature protein that is ~240 kDa (Coutte *et al.*, 2001; Kajava *et al.*, 2001; Chevalier *et al.*, 2004; Mazar and Cotter, 2006). Although the mechanism is unclear, the large C-terminal pro-domain seems to control secretion-dependent folding of FHA and in particular the globular C-domain (Julio *et al.*, 2009). Cleavage of the pro-domain is mediated at least in part by SphB1, a surface-exposed autotransporter and putative serine protease (Coutte *et al.*, 2001). Recent evidence indicates that other factors are also capable of cleaving the FHA pro-domain. While some mature FHA is detectable on the cell surface, most is released into the extracellular milieu under standard *in vitro* conditions.

Surface retention of HMW1 requires the C-terminal 20 amino acids, which form a disulfide bond and result in occlusion of the HMW1B pore, perhaps by forming a globular domain in the HMW1B pore or on the periplasmic side of the pore (Buscher *et al.*, 2006). A number of other HMW1 subfamily members have two cysteine residues near their C-terminus, suggesting a conserved mechanism of anchoring these proteins within the cognate translocator proteins. Surface anchoring of HMW1 also requires glycosylation, which is mediated by the HMW1C glycosyltransferase (Grass *et al.*, 2003). In the absence of glycosylation, HMW1 is partially degraded and is freely released from the surface of the organism, resulting in reduced adherence.

The mature FHA protein contains several features implicated in binding to host cell receptors on respiratory epithelial cells. A heparin-binding site associated with haemagglutination activity has been shown to be important for recognizing sulphated polysaccharides on non-ciliated epithelial cells (Hannah *et al.*, 1994). A carbohydrate-recognition domain (CRD) located near the centre of the mature protein interacts with galactose residues on a sulfated glycolipid called sulfatide, which is very common on the surface of ciliated cells (Prasad *et al.*, 1993). A nearby RGD (Arg-Gly Asp) motif has been implicated in binding to integrins on epithelial cells and monocytes to activate host signalling pathways (Relman *et al.*, 1989; Ishibashi *et al.*, 1994; Ishibashi and Nishikawa, 2002, 2003). Recent work by Julio *et al.* demonstrated that a globular domain formed by the C-terminal ~500 amino acids (named MCD, for mature C-terminal domain), which is located distally from the cell surface, mediates attachment to epithelial and macrophage-like cells *in vitro* and is required for colonization of the rat respiratory tract (Mazar and Cotter, 2006; Julio *et al.*, 2009). Taken together, studies with *B. pertussis* FHA suggest that distinct binding domains within this protein allow it to function as an adhesin when surface-associated (by mediating attachment to respiratory epithelia) and as an immunomodulator when released from the cell surface (by interacting with integrin receptors on the surface of epithelial cells and macrophages).

OMPs/porins

A group of Gram-negative bacterial proteins called porins form β-barrel channels through the lipid bilayer of the outer membrane. The porins are designed to allow passage of hydrophilic molecules and exclusion of larger or harmful compounds. Interestingly, a variety of porin-like integral outer membrane proteins can also contribute to host–pathogen interactions (Table 5.1) (Holm *et al.*, 2004). The outer membrane protein CD (OMPCD) in *M. catarrhalis* mediates binding to A549 human lung cells (Holm *et al.*, 2004) and middle ear mucin (Reddy *et al.*, 1997). The P5 protein in *H. influenzae* is also a major outer membrane protein with adhesive activity. This protein shares approximately 50% similarity with the *E. coli* OmpA β-barrel protein and has been shown to enhance adherence to oropharyngeal cells as well as nasopharyngeal mucin and to mediate attachment to CHO and HeLa cells via CEACAM1 (Reddy *et al.*, 1996; Hill *et al.*, 2001; Bakaletz *et al.*, 2005; Bookwalter *et al.*, 2008). The P2 protein in *H. influenzae* is another outer membrane porin with adhesive activity and specifically binds to nasopharyngeal mucin via sialic acid-containing oligosaccharides (Reddy *et al.*, 1996). As targets of the immune response, both P2 and P5 undergo frequent size and sequence variation during the course of chronic infection, especially in putative surface-exposed loop regions, enabling the organism to evade clearance by protective antibodies (Groeneveld *et al.*, 1988; Forbes *et al.*, 1992; Duim *et al.*, 1997; Smith-Vaughan *et al.*, 1997; Webb and Cripps, 1998).

Work with pathogenic *N. meningitidis* and *Neisseria gonorrhoeae* suggests that pili mediate initial attachment to mucosal epithelia and that colony opacity-associated (Opa) proteins, an antigenically distinct Opc protein, and other surface structures mediate more intimate attachment. Tight adherence mediated by the Opa and Opc integral outer membrane proteins leads to invasion and transcytosis of epithelial cells. The *Neisseria* Opc protein is a ten-stranded β-barrel molecule with five surface-exposed loops and is relatively invariant. The Opa proteins are eight-stranded transmembrane β-barrel structures with four surface-exposed loops, and a single *N. gonorrhoeae* strain can express as many as 11 distinct Opa variants encoded by chromosomally unlinked phase variable alleles (Heckels, 1981; Stern and Meyer, 1987; Bhat *et al.*, 1991; Kupsch *et al.*, 1993). Antigenic diversity is generated by both intergenic and interstrain recombination between *opa* genes, allowing bacteria to escape antibody-mediated immunity. Differential expression of Opa variants can potentially alter receptor specificity and thus influence the bacterial tropism for epithelium, endothelium, and phagocytic cells. Recovery of Opa^+ bacteria in studies with human volunteers inoculated with Opa variants has demonstrated that Opa expression is selected for during urogenital, cervical, or rectal infection (Swanson *et al.*, 1988; Jerse *et al.*, 1994).

The majority of *Neisseria* Opa variant proteins bind to CD66 proteins, which represent a subset of the carcinoembryonic antigen (CEA) receptor family of the immunoglobulin (Ig) superfamily. Binding of Opa to CD66 receptors mediates transcytosis of *N. gonorrhoeae* across polarized T84 epithelial cell monolayers (Wang *et al.*, 1998). The presence of CD66 on cervical and endometrial epithelium (Prall *et al.*, 1996) suggests that gonococcal dissemination from these sites may be mediated by Opa–CD66 interactions (Dehio *et al.*, 1998a). The Opa_{50} variant expressed by *N. gonorrhoeae* and the Opc protein of meningococci have been shown to bind cell surface-associated heparin sulphate proteoglycans (HSPGs) (Chen *et al.*, 1995; van Putten and Paul, 1995; de Vries *et al.*, 1998), which consist of long repeating heparan sulphate glycosaminoglycans (HS-GAGs) that are linked to various core proteins expressed on the surface of mammalian cells (Carey, 1997). Binding of HSPGs leads to bacterial entry into some epithelial cell lines and is dependent on the activation of phosphatidylcholine-dependent phospholipase C and acidic sphingomyelinase (Grassmé *et al.*, 1997). Opa_{50}-mediated bacterial entry into other cell lines is dependent on co-ligation of HSPGs with secondary receptors such as integrins via interactions with RGD-containing ECM proteins. In particular, Opa_{50} interacts with the serum-derived ECM protein vitronectin, which contains an RGD sequence and interacts with several integrin receptors. Vitronectin-dependent invasion by Opa50-expressing gonococci into HeLa cells occurs in an α_V integrin-dependent manner (Duensing and van Putten, 1997; Gómez-Duarte *et al.*, 1997; Dehio *et al.*, 1998b). After gonococcal transcytosis across the mucosal barrier in the genital tract, binding to the ECM via vitronectin also presumably enhances colonization of the basolateral surface of urethral or vaginal epithelium (Dehio *et al.*, 1998a). The Opc protein of meningococci has also been shown to recruit vitronectin from serum to facilitate interactions with $\alpha_V\beta_3$ integrins on the apical side of endothelial cells (Virji *et al.*, 1994, 1995). Beyond binding to vitronectin, Opa_{50}-expressing bacteria can bind to fibronectin. This interaction facilitates bacterial attachment and entry in assays with HEp-2 cells via co-ligation of HSPGs and the fibronectin-integrin receptor $\alpha_5\beta_1$ (van Putten *et al.*, 1998). Since HSPGs are typically expressed basolaterally (Carrey, 1997; Gómez-Duarte *et al.*, 1997), it has been suggested that HSPG/Opa-dependent uptake and transmigration across epithelial cells may provide a mechanism by which bacteria recolonize the apical surface and reseed the urethral lumen, facilitating transmission to a new host (Dehio *et al.*, 1998a).

Non-pilus adhesins exported via the FGL-chaperone/usher pathway

A number of adhesins assembled by the chaperone/usher pathway were originally identified as atypical fimbrial, afimbrial, or non-fimbrial structures. Initial morphological descriptions based on negative staining electron microscopy suggested that these structures are not polymeric despite being associated with gene clusters homologous to those of chaperone/usher assembled pili. The distinct chaperones (designated FGL chaperones) associated with this subfamily of atypical adhesins exhibit slight structural differences compared with the chaperones involved in assembly of pili. Most significantly, the FGL chaperones contain a longer amino acid loop between their G1 β-strand (the strand donated during donor strand complementation to stabilize pilus subunits) and the previous strand. As a result, more contacts are made between the chaperone and the adhesive subunit, potentially contributing to the production of atypical non-pilus structures (Hung and Hultgren, 1998; Zavialov *et al.*, 2003). Members of the Afa/Dr family of adhesins associated with uropathogenic and diarrhoea-associated *E. coli* and envelope antigen F1 (Caf1) expressed by *Yersinia pestis* are the best-characterized examples of these atypical adhesive organelles (Zavialov *et al.*, 2003, 2005; Zavialov and Knight, 2007). AfaE-III forms an afimbrial sheath at the bacterial surface and contains subunits that are assembled end-to-end by donor-strand complementation into polymeric fibres. Interestingly, AfaE-III appears as both ~2-nm thin flexible fibres and much thicker fibres, which are proposed to be a consequence of tighter subunit interactions or coiling of fibres (Anderson *et al.*, 2004). Several adhesin fibres such as Afa-III, Afa-VII, Afa-VIII, Dr-II, NFA-1, and Caf1 can aggregate to form amorphous masses or capsules (Zavialov *et al.*, 2007). Within the subfamily of FGL-chaperone/usher assembled adhesins, it seems that the final architecture is determined by the adhesive subunit, as highlighted by a chimeric construct in which the *draE* gene (fimbrial) and the *afaE-3* gene (afimbrial) were switched. In contrast to FGS-chaperone assembled pili, FGL-chaperone assembled adhesins lack a specialized adhesive subunit either at the distal end or the edge of a tip fibrillum and instead are polyadhesive, reflecting adhesive activity in the major structural subunits. As a consequence, these adhesins promote polyvalent binding to host cells, aggregation of receptors, and triggering of downstream signalling events.

Lipopolysaccharide

Lipopolysaccharide (LPS) is a major component of the outer membrane of Gram-negative bacteria and has been shown in several bacterial species to be involved in adherence to epithelial cells, mucus, and the ECM component laminin (Cohen *et al.*, 1985; Gupta *et al.*, 1994; Valkonen *et al.*, 1994; Jacques, 1996). LPS is composed of a lipid A moiety that is embedded in the outer membrane and a hydrophilic polysaccharide that extends outward from the cell surface. The heteropolysaccharide is composed of the core oligosaccharide and the variable O-antigen, a polymer of repeating glycosyl residues. The core oligosaccharide contains a unique octose 2-keto-3-deoxyoctulosonic acid (KDO) linked to a triheptose backbone, which in turn is linked to an outer region composed of a variety of hexoses. *N. gonorrhoeae, M. catarrhalis* and *H. influenzae* are examples of pathogens that elaborate a rough LPS, which lacks repeating O-antigen and is referred to as short-chain LPS or lipo-oligosaccharide (LOS) (Flesher and Insel, 1978; Hood *et al.*, 2004).

Bélanger and colleagues first proposed that LPS acts as an adhesin and showed that high molecular weight LPS contributes to *Actinobacillus pleuropneumoniae* (the causative agent of porcine fibrinohaemorrhagic necrotizing pleuropneumonia) adherence to porcine respiratory tract cells (Bélanger *et al.*, 1990). Later studies demonstrated that purified detoxified LPS (polysaccharides free of lipid A) and monoclonal antibodies directed against the O-antigen of LPS inhibit *A. pleuropneumoniae* adherence (Paradis *et al.*, 1994; Jacques, 1996). Other studies have reported a relationship between LPS and enteric bacterial colonization using animal models. LPS mutants of *Salmonella* Typhimurium and *Klebsiella pneumoniae* have been found to be less efficient colonizers of chicken gut (Camprubi *et al.*, 1993; Craven, 1994). Work with *N. meningitidis* and *H. influenzae* has suggested that LPS/LOS is also important for the pathogenicity of these pathogens, both facilitating adherence to epithelial cells and promoting destruction of the mucosal barrier. Interestingly, variants of *H. influenzae* expressing high molecular weight LOS have been found to colonize the nasopharynx of infant rats much more efficiently than variants producing low molecular weight LOS (Weiser, 1993). Further analysis has demonstrated that expression of phosphorylcholine (ChoP) on *Haemophilus* LOS contributes to colonization of the nasopharynx in rats (Weiser *et al.*, 1998). $ChoP^+$ LOS glycoforms have also been shown to mediate bacterial attachment and invasion of human bronchial epithelial cells via interactions with the platelet activating factor (PAF) receptor (Swords *et al.*, 2000). LPS of *Pseudomonas aeruginosa* has been reported to bind to the glycolipid asialo-GM1 on host cells as well as to mucin and laminin (Gupta *et al.*, 1994; Jacques, 1996).

Two gonococcal LOS receptors have been identified in studies using HepG2 hepatoma cells as a tool. The oligosaccharide region of LOS binds to an uncharacterized 70 kDa protein (p70), and sialylated LOS binds the major hepatic asialoglycoprotein receptor (ASGP-R) (Porat *et al.*, 1995). ASGP-R normally functions to remove aged compounds (glycoproteins and glycolipids) from the circulation but has also been identified on the surface of male urogenital cells and sperm, suggesting that LOS plays a role in gonococcal colonization of the urogenital tract and in dissemination of gonococci between sexual partners (Harvey *et al.*, 2000).

Gram-positive bacterial non-pilus adhesins

Gram-positive bacteria are distinguished from Gram-negative bacteria by a thicker peptidoglycan layer and the lack of an outer membrane and have developed distinct mechanisms to export and anchor adhesive structures on their surface. Adhesive proteins are either covalently linked to the cell wall, lipidated and embedded in the plasma membrane, or non-covalently associated with teichoic acids or lipoteichoic acids (Fig. 5.2A). In some cases, membrane-bound lipoteichoic acid has also been demonstrated to play a role in bacterial adherence. The mechanism of surface-localization for a number adhesins belonging to a novel class of proteins designated 'anchorless' is as yet unknown. The following text reviews some well-studied prototype Gram-positive bacterial adhesins, highlighting their roles as virulence factors, their domain organization, and their mechanism of interaction with host substrates.

Cell-wall anchored adhesins (LPXTG)

The most common means by which Gram-positive organisms attach proteins to their cell wall is by a process termed sorting (Navarre and Schneewind, 1999). Proteins attached by sorting contain an LPXTG amino acid motif preceding a C-terminal hydrophobic domain and a positively charged tail. After secretion, the hydrophobic domain retains the protein in the plasma membrane, and the positively charged tail remains in the cytoplasm. A sortase enzyme then cleaves the LPXTG motif between the threonine and glycine residues and covalently links the carboxyl group of the threonine to a free amino group of the branch peptide within the growing peptidoglycan cell wall (Schneewind *et al.*, 1995; Ton-That *et al.*, 1997; Navarre and Schneewind, 1999). A proline-rich stretch of amino acids is commonly found on the N-terminal side of the LPXTG motif and is believed to span the peptidoglycan layer. While cell-wall-anchored Gram-positive bacterial adhesins exhibit conserved C-terminal domains, significant variety is observed among N-proximal domains, which are associated with distinct binding activities. Examples are discussed below.

M proteins

S. pyogenes is the causative agent of group A streptococcal infections, which typically begin in the posterior pharynx or skin and range from superficial, localized skin infections (impetigo) and pharyngitis (strep throat) to more invasive cellulitis and potentially life-threatening necrotizing fasciitis. Group A streptococci produce a hypervariable LPXTG-sorted adhesin called M protein that influences the predilection for specific disease states and serves as the basis

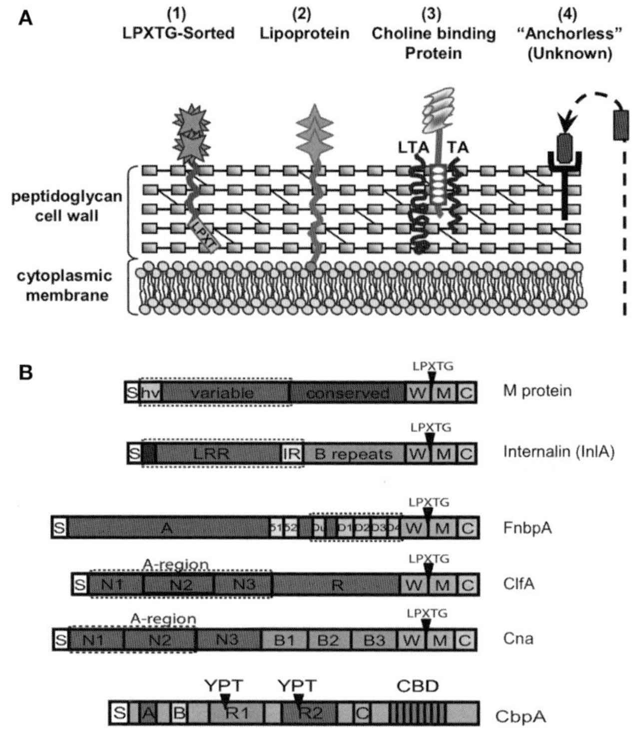

Figure 5.2 Gram-positive non-pilus adhesins. (A) Surface protein anchoring mechanisms. (i) The majority of Gram-positive bacterial adhesins are anchored to the cell wall by a sortase. Each of these proteins contains an N-terminal signal peptide (S), which directs secretion across the cytoplasmic membrane. The C-terminus of these adhesins contains features typical of surface-anchored proteins, a proline-rich wall-spanning region (W), a hydrophobic membrane-spanning (M), and a positively charged tail (C). An LPXTG motif near the C-terminus is recognized by sortase, which covalently links the protein to the peptidoglycan layer of the cell wall. (2) Lipoproteins are anchored in the cytoplasmic membrane. Pre-prolipoproteins contain a lipobox sequence motif located in the C-terminal part of the leader peptide, with a conserved cysteine serving as the target amino acid for lipidation. (3) Members of the choline-binding protein (Cbp) family of proteins are non-covalently bound to a phosphorylcholine moiety on cell wall teichoic acids and membrane-bound lipoteichoic acids via a choline-binding repeat domain. (4) Anchorless proteins are secreted and re-associate with the bacterial cell surface by an as-yet-unknown mechanism. (B) Domain organization of prototype adhesins: M protein of group A streptococci (*S. pyogenes*) contains an N-proximal variable domain with a hypervariable tip (hv) and a C-terminal conserved domain. The variable domain is associated with adhesive activity and mediates binding to a variety of substrates. Internalin (InlA) of *L. monocytogenes* contains two repeat regions, the leucine-rich repeats (LRR) and the B repeats, which are connected by the inter-repeat region. Structure studies indicate that the E-cadherin-binding LRR domain forms a bowed tube with a calcium-binding cap and that the IR domain forms an Ig-like fold. *S. aureus* ECM-binding adhesins (MSCRAMMs): FnbpA of *S. aureus* is a prototype fibronectin-binding adhesin and is characterized by N-proximal non-repetitive sequence containing two almost identical 30 amino acid repeats of unknown function (δ1, δ2) and five fibronectin-binding units (Du-D4). ClfA of *S. aureus* is a plasma clumping factor that binds fibrinogen. The ligand-binding A-region contains three subdomains (N1, N2 and N3) and is followed by unique dipeptide repeat region R consisting of mainly aspartic acid and serine residues. Cna of *S. aureus* contains an N-terminal collagen-binding A-region and a 187-residue motif repeated 1–3 times (B1, B2, B3), which varies from strain to strain. For each protein, adhesive regions are highlighted by dashed boxes. *S. pneumoniae* choline-binding protein CbpA: Leucine zipper motifs cluster within five domains: A, B, and C form coiled-coil dimers, and each R-domain adopts a unique 3-α-helix, antiparallel raft-like structure. The R-domains bind the secretory IgA receptor (pIgR), enabling pneumocci to adhere to and invade respiratory epithelial cells. The locations of critical YPT motifs are noted. The choline-binding repeats (CBD) associate non-covalently with phosphocholine on cell-wall teichoic acids and membrane-bound lipoteichoic acids. A colour version of this figure is located in the plate section at the back of the book.

for serotyping isolates (along with T antigen) (Navarre and Schneewind, 1999; Cunningham, 2000). M proteins are displayed as a coiled-coil dimer on the bacterial surface and are composed of an N-terminal variable domain with a hypervariable tip and a C-terminal conserved domain (Fig. 5.2B) (Cunningham, 2000; Courtney *et al.*, 2002). Although repeat regions within the conserved domain mediate attachment to complement regulatory factor CD46 on keratinocytes, most binding interactions are mediated by the variable N-terminal region (Cunningham, 2000; Courtney *et al.*, 2002). Because of variation in the binding region, group A streptococci interact with a variety of host cell receptors depending on their M serotype. M proteins also recognize a variety of substrates and ECM components, including albumin, fibrinogen, fibronectin, kininogen, and plasminogen (Cunningham, 2000; Courtney *et al.*, 2002).

Internalins

Listeria monocytogenes is a food-borne pathogen that can cause gastroenteritis, septicaemia, and meningitis and enters mammalian cells by inducing internalization. This process is dependent on the production of a sortase-anchored adhesive molecule called internalin (InlA) that mediates attachment to and invasion of intestinal epithelial cells via binding to host cell E-cadherin (Braun and Cossart, 2000; Cabanes *et al.*, 2002; Lecuit *et al.*, 1997). In polarized epithelial cells, E-cadherin is located at adherens junctions that seal epithelial cells together. InlA/E–cadherin interactions also facilitate traversal of the human maternofetal barrier by *L. monocytogenes* (Lecuit *et al.*, 2004). Co-production of a related adhesive molecule called internalin B (InlB) is necessary to promote *Listeria* entry into several other cell types, including enterocytes, hepatocytes, fibroblasts, endothelial cells, and epithelial cells such as Vero, HeLa, CHO, or Hep-2 cells (Lecuit *et al.*, 1997). InlB uses the HGF–SF growth factor receptor as a cellular ligand, also known as the c-met proto-oncogene or Met (Shen *et al.*, 2000). Activation of the Met tyrosine kinase receptor results in actin cytoskeleton rearrangements, membrane ruffling, and bacterial invasion. Since Met co-localizes with E-cadherin at cell–cell junctions (Kamei *et al.*, 1999), it has been proposed that released InlB may open the junctions and facilitate interactions between internalin and E-cadherin (Cossart, 2001).

InlA and InlB belong a large family of virulence-associated listerial proteins (InlA-J) that is characterized by an N-terminal cap domain and a conserved leucine-rich repeat (LRR) domain with 22-amino acid repeats. Some members of the family also contain an inter-repeat region (IR) domain and varying C-terminal repeats (Fig. 5.2B) (Braun and Cossart, 2000; Cabanes *et al.*, 2002). Most contain a C-terminal LPXTG sorting motif and are thus covalently anchored to the cell wall. However, InlB loosely associates with lipoteichoic acids in the cell wall through its C-terminal GW modules (repeats in which the first residues are Gly and Trp). A small subfamily of internalins do not contain any anchoring motifs and are predicted to be secreted.

Functional analyses with InlA have shown that the LRR domain recognizes the first of five extracellular immunoglobulin-like domains of human E-cadherin (hEC1) and that the IR region is important for proper folding of the molecule (Braun and Cossart, 2000; Schubert *et al.*, 2002). The first crystal structures determined were the cap/LRR-domain of InlB (Marino *et al.*, 1999) and the cap/LRR/IR-domain of InlB and InlH (Schubert *et al.*, 2001), revealing a hydrophilic N-terminal calcium-binding cap followed by a curved solenoid formed by the LRR-domain. Each LRR repeat contains a β strand-loop-helix-loop motif with the β strand nearly parallel to the helix. The β strands form an eight-stranded, parallel β sheet that represents the concave face of the curved LRR domain. The concave face contains a cluster of surface-exposed hydrophobic residues and a cluster of negatively charged residues that most likely mediate interactions with the host receptors (Marino *et al.*, 1999). Additionally, the N-terminal cap region contains two half calcium-binding sites, which Marino *et al.* proposed could assist in recognition of certain ligands by bridging protein–protein interactions (Marino *et al.*, 1999). The IR region folds into an immunoglobulin (Ig)-like domain at the base of the molecule and is thought to cap the LRR curve. This general structure was expected to be similar to that of the LRR domain in InlA in spite of distinct receptor specificities. Later, crystal structures of the functional domain of InlA alone and complexed with the N-terminal domain of human E-cadherin (hEC1) confirmed that the LRR-domain of InlA surrounds and specifically recognizes hEC1 (Schubert *et al.*, 2002). The InlA IR region may also be important for heterophilic interactions with the Ig fold of E-cadherin (Schubert *et al.*, 2001).

Interestingly, mice are resistant to infection by *L. monocytogenes*, reflecting the fact that InlA does not interact with mouse E-cadherin or adhere to murine epithelia. The specificity of InlA for human E-cadherin is determined by residue 16, which is a proline in hEC1 and a glutamic acid in mEC1. Replacement of glutamic acid with proline in transfected cells expressing mEC1(E16P) renders them susceptible to invasion of InlA-coated beads (Lecuit *et al.*, 1999). Conservation of this proline residue is a major determinant of human susceptibility to *L. monocytongenes* infection, demonstrating how slight sequence and/or structural differences can alter adhesin/receptor interactions and thus influence host and tissue tropism.

Staphylococcus aureus ECM-binding adhesins (MSCRAMMs): FnbpA, ClfA, and Cna

Several well-studied, cell wall-anchored adhesins of staphylococci represent prototype MSCRAMMs of Gram-positive bacteria. Studies suggest that these ECM-binding proteins typify conserved domain organizations and mechanisms of ligand recognition. *S. aureus* is a major human pathogen known to cause a broad spectrum of diseases, including native and prosthetic valve endocarditis. Fibronectin is a dimeric glycoprotein found in the ECM and body fluids of animals and also in the conditioning layer that coats many implanted medical devices, such as heart valves and catheters. Fibrinogen plays an important role in the wound healing process and in platelet thrombus formation and is found at high concentrations in blood plasma. Fibrinogen is also the most abundant host protein in endothelial lesions and the major blood protein deposited on implanted biomaterials. Expression of fibronectin-binding and fibrinogen-binding proteins thus modulates the ability of *S. aureus* to colonize vascular tissue and implanted medical devices.

Binding of *S. aureus* to fibronectin is mediated by two fibronectin-binding proteins called FnbpA and FnbpB (Flock *et al.*, 1987; Fröman *et al.*, 1987; Jonsson *et al.*, 1991). These proteins enable the bacteria to interact with cellular integrins and to trigger internalization into host cells. The prototype FnbpA protein is characterized by a N-proximal non-repetitive sequence containing two almost identical 30 amino acid repeats (δ) of unknown function and five fibronectin-binding units (Du-D4) located adjacent to the proline-rich wall-spanning region (Fi. 5.2B). The ~45 residue binding units are tandemly repeated three times (D1, D2, D3) and are followed by a partial repeat unit (D4). A fifth, slightly divergent, unit (Du) is located approximately 100 amino acids N-terminal to D1. The overall arrangement of fibronectin-binding units is also found in FnbpB and in the streptococcal proteins SfbI/protein F1, SfbII, Sof22, protein F2, FnbA, FnbB, and FNZ, albeit with slight variations (Hanski and Caparon, 1992; McGavin *et al.*, 1993; Lindgren *et al.*, 1994; Talay *et al.*, 1994; Kreikemeyer *et al.*, 1995; Jaffe *et al.*, 1996; Ozeri *et al.*, 1996). 6x-His-tagged recombinant proteins derived from the ligand-binding sites of FnbpA, FnbA/FnbB, and SfbI cross-inhibit fibronectin-binding by *S. aureus*, *S. pyogenes*, and *Streptococcus dysgalactiae*, suggesting that different MSCRAMMs bind to overlapping regions in fibronectin (F1 modules) by similar mechanisms. Extensive sequence alignments of ligand-binding regions has suggested that a conserved core sequence, ED(T/S)(9 or 10 Xs) GG(3 or 4 Xs)(I/V)DF, is important for fibronectin-binding activity (McGavin *et al.*, 1991, 1993).

S. aureus has long been known to form clumps in the presence of blood plasma *in vitro*, reflecting the presence of surface proteins that bind to fibrinogen (Hawiger *et al.*, 1982). Using transposon mutagenesis, McDevitt and colleagues identified the *S. aureus* fibrinogen adhesin, designated ClfA for clumping factor (McDevitt *et al.*, 1994). ClfA has been shown to be an important virulence factor in rabbit and rat models of infective endocarditis and in a mouse model of septic arthritis. *S. aureus* expresses other fibrinogen-binding proteins as well, most notably coagulase, which activates the coagulase pathway by binding both fibrinogen and prothrombin (Hemker *et al.*, 1975; Kawabata *et al.*, 1985; Bodén and Flock, 1989). However, ClfA is considered to be the major staphylococcal fibrinogen-binding MSCRAMM, as evidenced by the observation that coagulase-negative mutants can still bind to fibrinogen-coated surfaces and exhibit a clumping phenotype in the presence of plasma (McDevitt *et al.*, 1994). ClfA contains features of typical Gram-positive surface proteins, including a signal peptide and a C-terminal LPXTG motif. The N-proximal ligand-binding A-region contains three subdomains called N1, N2 and N3 (Perkins *et al.*, 2001; Ponnuraj *et al.*, 2003). ClfA also contains a unique dipeptide repeat region R consisting of mainly aspartic acid and serine repeated 154 times (Fig. 5.2B). Structural analysis indicates that ClfA interacts with the C-terminal region of the fibrinogen (Fγ) gamma-chain in a manner similar to the fibrinogen-binding SdrG protein of *Staphylococcus epidermidis* (Ganesh *et al.*, 2008).

Collagenous proteins are major components of the ECM and frequently represent a major target substrate for bacteria. The collagen-binding MSCRAMM of *S. aureus* is called Cna and is the prototype member of this family (Switalski *et al.*, 1989). The Cna protein contains an N-terminal collagen-binding A-region (Patti *et al.*, 1993), a 187-amino acid motif repeated three times (B1, B2, B3), a 64-amino acid lysine/proline rich region, a hydrophobic membrane-spanning, and a positively charged C-terminal tail (Fig. 5.2B). However, despite the presence of an LPKTG motif, protein-sorting studies have demonstrated that this protein is predominantly anchored to the staphylococcal cell membrane and not sorted to the cell wall owing to shorter spacing between the sorting motif and charged tail (Schneewind *et al.*, 1993).

Cna has been shown to be a virulence factor in many different animal models of staphylococcal infections, including arthritis, endocarditis, osteomyelitis, mastitis, and keratitis, implying that the ability to interact with collagen plays a fundamental role in the pathogenesis of disease (Patti *et al.*, 1993, 1994; Hienz *et al.*, 1996; Mamo *et al.*, 2000; Rhem *et al.*, 2000; Elasri *et al.*, 2002). Examination of *S. aureus* strains isolated from patients with septic arthritis has revealed broad conservation of the *cna* gene, and expression of Cna is both necessary and sufficient for *S. aureus* adherence to cartilage, an important step in the pathogenesis of septic arthritis (Switalski *et al.*, 1993). Interestingly, *S. aureus* strains contain different forms of the *cna* gene with one, two, or three B repeats, resulting in strain-dependent size variation. However, strains expressing the Cna size variants and region B deletion mutants do not differ in collagen-binding capacity, raising questions regarding the functional significance of the B repeats (Hartford *et al.*, 1999).

By generating chimeric proteins containing various fragments of the Cna A-region, Patti and colleagues further localized the minimal collagen-binding domain to a central segment defined by residues 151–318 (Patti *et al.*, 1993), and crystal structures of this domain revealed a novel variant of the IgG-fold (Symersky *et al.*, 1997; Zong *et al.*, 2005). Based on Cna amino acid sequence similarities with *S. aureus* fibrinogen-binding proteins, Zong *et al.* (2005) proposed that the A-region of Cna is similarly composed of three subdomains: N1 (residues 31–140), N2 (residues 141–344), and N3 (residues 345–531). Later studies demonstrated that full collagen binding activity requires both the N1 and N2 putative subdomains. Crystal structure analysis of Cna N1N2 apo-protein and N1N2 complexed with a synthetic collagen-like triple helical peptide led Zong *et al.* (2005) to propose that ligand recognition involves a multistep binding mechanism where these two subdomains cooperate to wrap around and 'hug' the rope-like structure of the collagen fibre.

Lipoproteins

Bacterial lipoproteins are a functionally heterogeneous class of proteins that have an N-terminal lipid moiety that serves to anchor these proteins in the plasma membrane. They are synthesized as pre-prolipoproteins and undergo post-translational modifications. Lipoprotein signal peptide features are typical of signal sequences and consist of a positively charged N-region (usually containing lysine and/or arginine), a central hydrophobic H-region, and a cleavage C-region. The lipobox motif is located in the C-terminal part of the leader peptide and consists of four amino acids

[LVI/ASTVI/GAS/C] (Babu *et al.*, 2006). The conserved cysteine residue in this lipid modification recognition motif is lipidated after the precursor lipoprotein is translocated in a Sec-dependent manner across the plasma membrane, although recent evidence suggests that a few lipoproteins may also be translocated by the twin-arginine translocation (Tat) system (McDonough *et al.*, 2005). In Gram-positive bacteria, lipidation depends on the activity of two integral-membrane enzymes, prolipoprotein diacylglyceryl transferase (Lgt) and signal peptidase II (LspA). Lgt adds a diacylglyerol residue to the thiol group of the conserved cysteine within the lipobox. A number of studies have identified adhesins that have been directly demonstrated to be lipoproteins or are predicted to be lipoproteins based on sequence analyses.

The LraI (for 'lipoprotein receptor antigen') family of surface-associated lipoproteins is involved in adherence by oral streptococci (Jenkinson, 1994). *Streptococcus sanguinis* is a commensal of the human oral cavity (found particularly in dental plaque) that is capable of causing acute bacterial endocarditis and produces a 34.7-kDa lipoprotein SsaB that interacts with saliva and salivary receptors (Ganeshkumar *et al.*, 1988, 1991, 1993). SsaB has also been implicated in mediating coaggregation with *Actinomyces naeslundii*, another commensal member of the human oral flora (Kolenbrander *et al.*, 1993). *Streptococcus gordonii* is also a commensal in the oral cavity and produces SarA and ScaA, two multifunctional lipoproteins that aid in colonization of the oral cavity, aggregation, and coaggregation of organisms (Jenkinson and Easingwood, 1990; Jenkinson, 1992; Kolenbrander *et al.*, 1994; Jenkinson *et al.*, 1995). SarA also plays an essential role in uptake of hexa- and heptapeptides as a source of nutrients to the organism (Jenkinson *et al.*, 1996).

PsaA is an immunogenic lipoprotein that is broadly conserved among different *Streptococcus pneumoniae* capsular types and was originally hypothesized to function as an adhesin owing to sequence similarities with SsaB and the FimA adhesin from *S. parasanguis* (Sampson *et al.*, 1994). Further investigation has confirmed that PsaA plays an adhesive role in pneumococcal colonization of the nasopharynx (Berry and Paton, 1996; McAllister *et al.*, 2004). Srivastava *et al.* (2000) and Romero-Steiner *et al.* (2006) identified two peptides that inhibit PsaA adherence to Detroit 562 nasopharyngeal cells. Immunization with these peptides significantly reduces pneumococcal nasopharyngeal colonization in mice, underscoring the role of PsaA and these putative binding domains *in vivo*. Recent studies by Anderton *et al.* (2007) demonstrated that E-cadherin is the eukaryotic cellular receptor for PsaA.

Interestingly, these highly homologous streptococcal proteins (SsaB, ScaA, FimA, and PsaA) all belong to ABC-type transport protein complexes. Other ABC-type transporters that mediate bacterial adherence to host cells have been identified in *Bacillus anthracis* (Read *et al.*, 2002) and *Shigella flexneri* 2a (Jin *et al.*, 2002), and comparative genomic sequencing suggests the presence of such dual-function lipoproteins in various organisms, including *Listeria* spp., *Lactobacillus* spp., *Lactococcus* spp., and *Leuconostoc* spp. (Rajam *et al.*, 2008).

Choline-binding proteins

Pneumococci attach phosphocholine on cell wall teichoic acids and membrane-bound lipoteichoic acids (Fischer, 2000). Members of the choline-binding protein (Cbp) family of proteins are non-covalently bound to this phosphorylcholine moiety via a choline-binding repeat domain (CBD). The CBD motif adopts a unique β-solenoid structure, with choline groups binding between β-hairpin steps of the staircase-like structure (Fernández-Tornero *et al.*, 2001). CbpA, also called SpsA, is one such choline-binding protein that functions as an adhesin and plays a critical role in colonization of the nasopharynx by *S. pneumoniae* (Rosenow *et al.*, 1997). CbpA enables pneumococci to adhere to and invade human cells by binding to the polymeric immunoglobulin receptor (pIgR). This receptor normally functions to transport secretory IgA from the basolateral surface to the apical surface of nasopharyngeal cells (Zhang *et al.*, 2000). CbpA binds to the ectodomain of pIgR via two 'R' domains to hijack this endocytosis machinery to translocate pneumococci across the nasopharyngeal epithelial cells into the bloodstream (Zhang *et al.*, 2000; Luo *et al.*, 2005).

CBD motifs have been found among surface proteins of other bacteria, and in all cases these proteins have a characteristic modular architecture, with the C-terminal CBD region being responsible for attachment to the pathogen and N-proximal modules being responsible for biological function. The N-terminal modules differ widely among choline-binding proteins, reflecting different biological functions ranging from bacterial autolysis (LytA) and glucosaminidase activity (LytB) to adhesion to host cells (CbpA). The CbpA N-terminus contains repeated leucine zipper motifs (Landschulz *et al.*, 1988) that cluster within five domains termed A, B, R1, R2 and C (Fig. 5.2B). Leucine zipper motifs generally mediate intra- or intermolecular α-helix/α-helix interactions (Lupas, 1996). Domains A, B, and C are 21–25 amino acids in length and are predicted to form coiled-coil dimers. The 110-amino-acid R domains (R1 and R2 are 78% identical) (Zhang *et al.*, 2000) comprise 12 imperfect copies of the leucine zipper heptad motif and adopt a unique 3-α-helix, antiparallel raft-like structure. A tyrosine fork near the loop between helices 1 and 2 of each R domain contains a conserved YPT motif that plays a key role in binding to pIgR. These structural features appear to be conserved among CbpA proteins from most pneumococcal strains, thus underscoring their importance in CbpA binding to pIgR.

Anchorless adhesins

Recently, a few Gram-positive bacterial cell-surface proteins with adhesive and invasive functions have been described that lack an apparent signal peptide, lipobox, CBD, or LPXTG motif. Because the mechanism of secretion and surface location of these proteins has not yet been elucidated, they are described as 'anchorless.' *S. pneumoniae* secretes at least one such adhesin, which Holmes *et al.* named PavA (Holmes *et al.*, 2001). Despite the absence of cell-localization signal sequences, PavA localizes to the pneumococcal cell surface, mediates binding to immobilized fibronectin, and is essential for pneumococcal virulence in the mouse infection models of septicaemia (Holmes *et al.*, 2001) and meningitis (Pracht *et al.*, 2005). PavA-fibronectin binding is heparin sensitive, suggesting that PavA binds to the C-terminal heparin-binding domain of fibronectin (Holmes *et al.*, 2001).

The Fbp54 protein expressed by *S. pyogenes* shows significant homology to PavA, lacks apparent anchor motifs, and binds soluble fibronectin (Courtney *et al.*, 1994). The binding domain of Fbp54 has been localized to the N-terminal 89 residues of the protein (Courtney *et al.*, 1996), a region that can inhibit Fbp54-mediated adherence to human buccal epithelial cells. Fbp54–fibronectin interactions do not appear to play a role in group A streptococcal adhesion to HEp-2 tissue culture cells, suggesting that Fbp54 directs adherence to specific cell types (Courtney *et al.*, 1996). Analysis of sera from patients with post-streptococcal glomerulonephritis and acute rheumatic fever indicates that Fbp54 is expressed *in vivo* and is immunogenic in the human host (Courtney *et al.*, 1996).

Other streptococcal surface proteins that lack sequence motifs associated with Sec-dependent export and cell-surface linkage include group A streptococcal enolase (SEN) and pneumococcal enolase (Eno). These proteins bind to and activate plasminogen, which may facilitate bacterial penetration of the basement membrane (Pancholi and Fischetti, 1998; Bergmann *et al.*, 2001; Derbise *et al.*, 2004; Ge *et al.*, 2004; Kinnby *et al.*, 2008). Plasminogen can be converted following cleavage to the serine protease plasmin, which in turn can degrade ECM proteins. Thus, Eno-binding activity has been proposed to facilitate streptococcal tissue invasion and dissemination through epithelial barriers. Eno has also been reported to play a role in *Streptococcus mutans* binding to salivary mucin MG2, which is thought to be a first step in bacterial attachment to a tooth surface coated with salivary pellicle or saliva-coated plaque (Ge *et al.*, 2004).

It has been hypothesized that bacteria utilize a novel, conserved mechanism for export and surface retention of anchorless proteins such as PavA, Fbp54, SEN, and Eno. However, elucidation of the precise mechanism by which these proteins are secreted and re-associate with the bacterial cell surface awaits further investigation. It should be noted that anchorless proteins that execute biological functions other than adherence and invasion have also been described (Chhatwal, 2002).

Lipoteichoic acid

While bacterial adherence is mediated principally by surface proteins, lipoteichoic acid (LTA) has been shown to play a role in bacterial interactions with host tissue components. LTA is composed of long chains of ribitol phosphate anchored to the lipid bilayer by a glyceride and is a major component of the cell wall of Gram-positive bacteria. Exogenous LTA has been shown to varying degrees to inhibit adherence by oral streptococci to fibronectin-coated hydroxyapatite beads (Hogg and Manning, 1988), providing circumstantial evidence for the involvement of LTA in the adhesion of this group of bacteria. More recently, two groups have demonstrated that LTA binds to an N-terminal fragment of fibronectin (Courtney *et al.*, 1988; Wagner *et al.*, 1988). In addition, LTA is involved in the attachment of group A streptococci to buccal cells and certain pharyngeal epithelial cell lines (Courtney *et al.*, 1992). It has been suggested that initial, weak interactions between group A streptococci and certain host tissues are mediated by LTA, which facilitates more intimate and high–affinity interactions between M proteins and specific host receptors (Courtney *et al.*, 1992, 2002; Hasty *et al.*, 1992; Cunningham, 2000).

Concluding remarks

Most commensal and pathogenic bacteria express a repertoire of surface-associated adhesive structures that mediate specific interactions with receptors on host cells or with components of the ECM to avoid clearance from the host. Since successful colonization is a key step in bacterial pathogenesis, bacterial adhesins represent key virulence factors. As highlighted in this chapter, bacterial non-pilus adhesins are secreted and surface anchored by a limited set of conserved mechanisms and display diverse structural domains involved in specific receptor recognition. Interestingly, a number of non-pilus adhesins appear to be multifunctional virulence factors and have been associated with enzymatic functions, bacterial aggregation, host cell invasion, serum resistance, and host immune response modulation.

Engagement of an adhesin with a host-cell receptor frequently leads to activation of signalling pathways in both the bacterium and the host cell, resulting in numerous potential downstream events. Adherent bacteria may remain extracellular, form microcolonies or biofilms, induce internalization into non-phagocytic cells, or transcytose cell layers. Bacterial attachment may also cause localized tissue destruction, which may promote transmission of the organism or enhance bacterial colonization, in part due to exposure of the underlying ECM. Cellular invasion helps bacteria to find nutrients, escape host defence mechanisms, and invade tissues surrounding the initial site of colonization. Organisms may then disseminate systemically through the lymphatic or circulatory system to reach locations beyond the initial site of adherence.

Since microbial adhesion to host tissues is a critical initial event in the pathogenesis of most infections, blocking initial host/pathogen interactions and preventing bacterial colonization should prevent progression to disease. Considering this idea, bacterial adhesins that mediate initial interactions with the host are attractive targets for new antimicrobials and have potential as vaccine components. Because bacterial surface structures are frequently associated with antigenic or phase variation as a means to evade antibody-mediated immunity, therapeutic agents that specifically target conserved factors involved in the secretion, assembly, or anchoring of adhesins represent ideal candidates for treating disease.

References

Ackermann, N., Tiller, M., Anding, G., Roggenkamp, A., and Heesemann, J. (2008). Contribution of trimeric autotransporter C-terminal domains of oligomeric coiled-coil adhesin (Oca) family members YadA, UspA1, EibA, and Hia to translocation of the YadA passenger domain and virulence of *Yersinia enterocolitica*. J. Bacteriol. *190*, 5031–5043.

Anderson, K.L., Billington, J., Pettigrew, D., Cota, E., Simpson, P., Roversi, P., Chen, H.A., Urvil, P., du Merle, L., Barlow, P.N., *et al.* (2004). An atomic resolution model for assembly, architecture, and function of the Dr adhesins. Mol. Cell. *15*, 647–657.

Anderton, J.M., Rajam, G., Romero-Steiner, S., Summer, S., Kowalczyk, A.P., Carlone, G.M., Sampson, J.S., and Ades, E.W. (2007). E-cadherin is a receptor for the common protein pneumococcal surface adhesin A (PsaA) of *Streptococcus pneumoniae*. Microb. Pathog. *42*, 225–236.

Aoki, S.K., Pamma, R., Hernday, A.D., Bickham, J.E., Braaten, B.A., and Low, D.A. (2005).

Contact-dependent inhibition of growth in *Escherichia coli*. Science *309*, 1245–1248.

Babu, M.M., Priya, M.L., Selvan, A.T., Madera, M., Gough, J., Aravind, L., and Sankaran, K. (2006). A database of bacterial lipoproteins (DOLOP) with functional assignments to predicted lipoproteins. J. Bacteriol. *188*, 2761–2773.

Bakaletz, L.O., Baker, B.D., Jurcisek, J.A., Harrison, A., Novotny, L.A., Bookwalter, J.E., Mungur, R., and Munson, R.S., Jr. (2005). Demonstration of Type IV pilus expression and a twitching phenotype by *Haemophilus influenzae*. Infect. Immun. *73*, 1635–1643.

Barenkamp, S.J., and Bodor, F.F. (1990). Development of serum bactericidal activity following nontypable *Haemophilus influenzae* acute otitis media. Pediatr. Infect. Dis. J *9*, 333–339.

Bélanger, M., Dubreuil, D., Harel, J., Girard, C., and Jacques, M. (1990). Role of lipopolysaccharides in adherence of *Actinobacillus pleuropneumoniae* to porcine tracheal rings. Infect. Immun. *58*, 3523–3530.

Benz, I., and Schmidt, M.A. (1992). AIDA-I, the adhesin involved in diffuse adherence of the diarrhoeagenic *Escherichia coli* strain 2787 (O126:H27), is synthesized via a precursor molecule. Mol. Microbiol. *6*, 1539–1546.

Benz, I., and Schmidt, M.A. (2001). Glycosylation with heptose residues mediated by the *aah* gene product is essential for adherence of the AIDA-I adhesin. Mol. Microbiol. *40*, 1403–1413.

Bergmann, S., Rohde, M., Chhatwal, G.S., and Hammerschmidt, S. (2001). alpha-Enolase of *Streptococcus pneumoniae* is a plasmin(ogen)-binding protein displayed on the bacterial cell surface. Mol. Microbiol. *40*, 1273–1287.

Bernstein, H.D. (2007). Are bacterial 'autotransporters' really transporters? Trends Microbiol *15*, 441–447.

Berry, A.M., and Paton, J.C. (1996). Sequence heterogeneity of PsaA, a 37-kilodalton putative adhesin essential for virulence of *Streptococcus pneumoniae*. Infect. Immun. *64*, 5255–5262.

Bhat, K.S., Gibbs, C.P., Barrera, O., Morrison, S.G., Jähnig, F., Stern, A., Kupsch, E.M., Meyer, T.F., and Swanson, J. (1991). The opacity proteins of *Neisseria gonorrhoeae* strain MS11 are encoded by a family of 11 complete genes. Mol. Microbiol. *5*, 1889–1901.

Bodén, M.K., and Flock, J.I. (1989). Fibrinogen-binding protein/clumping factor from *Staphylococcus aureus*. Infect. Immun. *57*, 2358–2363.

Bookwalter, J.E., Jurcisek, J.A., Gray-Owen, S.D., Fernandez, S., McGillivary, G., and Bakaletz, L.O. (2008). A carcinoembryonic antigen-related cell adhesion molecule 1 homologue plays a pivotal role in nontypeable *Haemophilus influenzae* colonization of the chinchilla nasopharynx via the outer membrane protein P5-homologous adhesin. Infect. Immun. *76*, 48–55.

Braun, L., and Cossart, P. (2000). Interactions between *Listeria monocytogenes* and host mammalian cells. Microbes Infect. *2*, 803–811.

Bullard, B., Lipski, S., and Lafontaine, E.R. (2007). Regions important for the adhesin activity of *Moraxella catarrhalis* Hag. BMC Microbiol. *7*, 65.

Buscher, A.Z., Grass, S., Heuser, J., Roth, R., and St Geme, J.W., 3rd (2006). Surface anchoring of a bacterial adhesin secreted by the two-partner secretion pathway. Mol. Microbiol. *61*, 470–483.

Cabanes, D., Dehoux, P., Dussurget, O., Frangeul, L., and Cossart, P. (2002). Surface proteins and the pathogenic potential of *Listeria monocytogenes*. Trends Microbiol. *10*, 238–245.

Caffrey, P., and Owen, P. (1989). Purification and N-terminal sequence of the alpha subunit of antigen 43, a unique protein complex associated with the outer membrane of *Escherichia coli*. J. Bacteriol. *171*, 3634–3640.

Camprubi, S., Merino, S., Benedi, V.J., and Tomas, J.M. (1993). The role of the O-antigen lipopolysaccharide and capsule on an experimental *Klebsiella pneumoniae* infection of the rat urinary tract. FEMS Microbiol. Lett. *111*, 9–13.

Carey, D.J. (1997). Syndecans: multifunctional cell-surface co-receptors. Biochem. J. *327)*, 1–16.

Chen, T., Swanson, J., Wilson, J., and Belland, R.J. (1995). Heparin protects Opa $^+$ *Neisseria gonorrhoeae* from the bactericidal action of normal human serum. Infect. Immun. *63*, 1790–1795.

Chevalier, N., Moser, M., Koch, H.G., Schimz, K.L., Willery, E., Locht, C., Jacob-Dubuisson, F., and Müller, M. (2004). Membrane targeting of a bacterial virulence factor harbouring an extended signal peptide. J. Mol. Microbiol. Biotechnol. *8*, 7–18.

Chhatwal, G.S. (2002). Anchorless adhesins and invasins of Gram-positive bacteria: a new class of virulence factors. Trends Microbiol. *10*, 205–208.

Clantin, B., Hodak, H., Willery, E., Locht, C., Jacob-Dubuisson, F., and Villeret, V. (2004). The crystal structure of filamentous hemagglutinin secretion domain and its implications for the two-partner secretion pathway. Proc. Natl. Acad. Sci. U.S.A. *101*, 6194–6199.

Cohen, P.S., Arruda, J.C., Williams, T.J., and Laux, D.C. (1985). Adhesion of a human fecal *Escherichia coli* strain to mouse colonic mucus. Infect. Immun. *48*, 139–145.

Comanducci, M., Bambini, S., Brunelli, B., Adu-Bobie, J., Aricò, B., Capecchi, B., Giuliani, M.M., Masignani, V., Santini, L., Savino, S., *et al.* (2002). NadA, a novel vaccine candidate of *Neisseria meningitidis*. J. Exp. Med. *195*, 1445–1454.

Cossart, P. (2001). Met, the HGF-SF receptor: another receptor for *Listeria monocytogenes*. Trends Microbiol. *9*, 105–107.

Cotter, S.E., Surana, N.K., and St Geme, J.W., 3rd (2005). Trimeric autotransporters: a distinct subfamily of autotransporter proteins. Trends Microbiol. *13*, 199–205.

Cotter, S.E., Surana, N.K., Grass, S., and St Geme, J.W., 3rd (2006). Trimeric autotransporters require trimerization of the passenger domain for stability and adhesive activity. J. Bacteriol. *188*, 5400–5407.

Courtney, H.S., Stanislawski, L., Ofek, I., Simpson, W.A., Hasty, D.L., and Beachey, E.H. (1988). Localization of a lipoteichoic acid binding site to a 24-kilodalton NH2-terminal fragment of fibronectin. Rev. Infect. Dis. *10* (Suppl. 2), S360–362.

Courtney, H.S., von Hunolstein, C., Dale, J.B., Bronze, M.S., Beachey, E.H., and Hasty, D.L. (1992). Lipoteichoic acid and M protein: dual adhesins of group A streptococci. Microb. Pathog. *12*, 199–208.

Courtney, H.S., Li, Y., Dale, J.B., and Hasty, D.L. (1994). Cloning, sequencing, and expression of a fibronectin/fibrinogen-binding protein from group A streptococci. Infect. Immun. *62*, 3937–3946.

Courtney, H.S., Dale, J.B., and Hasty, D.I. (1996). Differential effects of the streptococcal fibronectin-binding protein, FBP54, on adhesion of group A streptococci to human buccal cells and HEp-2 tissue culture cells. Infect. Immun. *64*, 2415–2419.

Courtney, H.S., Hasty, D.L., and Dale, J.B. (2002). Molecular mechanisms of adhesion, colonization, and invasion of group A streptococci. Ann. Med. *34*, 77–87.

Coutte, L., Antoine, R., Drobecq, H., Locht, C., and Jacob-Dubuisson, F. (2001). Subtilisin-like autotransporter serves as maturation protease in a bacterial secretion pathway. EMBO J. *20*, 5040–5048.

Craven, S.E. (1994). Altered colonizing ability for the ceca of broiler chicks by lipopolysaccharide-deficient mutants of *Salmonella typhimurium*. Avian Dis. *38*, 401–408.

Cunningham, M.W. (2000). Pathogenesis of group A streptococcal infections. Clin. Microbiol. Rev. *13*, 470–511.

D'Souza, S.E., Ginsberg, M.H., and Plow, E.F. (1991). Arginyl-glycyl-aspartic acid (RGD): a cell adhesion motif. Trends Biochem. Sci. *16*, 246–250.

Dawid, S., Barenkamp, S.J., and St Geme, J.W., 3rd (1999). Variation in expression of the *Haemophilus influenzae* HMW adhesins: a prokaryotic system reminiscent of eukaryotes. Proc. Natl. Acad. Sci. U.S.A. *96*, 1077–1082.

Dehio, C., Gray-Owen, S.D., and Meyer, T.F. (1998a). The role of neisserial Opa proteins in interactions with host cells. Trends Microbiol. *6*, 489–495.

Dehio, M., Gómez-Duarte, O.G., Dehio, C., and Meyer, T.F. (1998b). Vitronectin-dependent invasion of epithelial cells by *Neisseria gonorrhoeae* involves alpha(v) integrin receptors. FEBS Lett. *424*, 84–88.

Derbise, A., Song, Y.P., Parikh, S., Fischetti, V.A., and Pancholi, V. (2004). Role of the C-terminal lysine residues of streptococcal surface enolase in Glu- and Lys-plasminogen-binding activities of group A streptococci. Infect. Immun. *72*, 94–105.

Duensing, T.D., and van Putten, J.P. (1997). Vitronectin mediates internalization of *Neisseria gonorrhoeae* by Chinese hamster ovary cells. Infect. Immun. *65*, 964–970.

Duim, B., Bowler, L.D., Eijk, P.P., Jansen, H.M., Dankert, J., and van Alphen, L. (1997). Molecular variation in the major outer membrane protein P5 gene of nonencapsulated *Haemophilus influenzae* during chronic infections. Infect. Immun. *65*, 1351–1356.

El Tahir, Y., and Skurnik, M. (2001). YadA, the multifaceted *Yersinia* adhesin. Int. J. Med. Microbiol. *291*, 209–218.

Elasri, M.O., Thomas, J.R., Skinner, R.A., Blevins, J.S., Beenken, K.E., Nelson, C.L., and Smeltzer, M.S. (2002). *Staphylococcus aureus* collagen adhesin contributes to the pathogenesis of osteomyelitis. Bone *30*, 275–280.

Elkins, C., Morrow, K.J., Jr., and Olsen, B. (2000). Serum resistance in *Haemophilus ducreyi* requires outer membrane protein DsrA. Infect. Immun. *68*, 1608–1619.

Elsinghorst, E.A., and Weitz, J.A. (1994). Epithelial cell invasion and adherence directed by the enterotoxigenic *Escherichia coli tib* locus is associated with a 104-kilodalton outer membrane protein. Infect. Immun. *62*, 3463–3471.

Emsley, P., Charles, I.G., Fairweather, N.F., and Isaacs, N.W. (1996). Structure of *Bordetella pertussis* virulence factor P.69 pertactin. Nature *381*, 90–92.

Fernández-Tornero, C., López, R., Garcia, E., Giménez-Gallego, G., and Romero, A. (2001). A novel solenoid fold in the cell wall anchoring domain of the pneumococcal virulence factor LytA. Nat. Struct. Biol. *8*, 1020–1024.

Fink, D.L., and St Geme, J.W., 3rd (2003). Chromosomal expression of the *Haemophilus influenzae* Hap autotransporter allows fine-tuned regulation of adhesive potential via inhibition of intermolecular autoproteolysis. J. Bacteriol. *185*, 1608–1615.

Fink, D.L., Buscher, A.Z., Green, B., Fernsten, P., and St Geme, J.W., 3rd (2003). The *Haemophilus influenzae* Hap autotransporter mediates microcolony formation and adherence to epithelial cells and extracellular matrix via binding regions in the C-terminal end of the passenger domain. Cell. Microbiol. *5*, 175–186.

Finn, T.M., and Stevens, L.A. (1995). Tracheal colonization factor: a *Bordetella pertussis* secreted virulence determinant. Mol. Microbiol. *16*, 625–634.

Fischer, W. (2000). Phosphocholine of pneumococcal teichoic acids: role in bacterial physiology and pneumococcal infection. Res. Microbiol. *151*, 421–427.

Flesher, A.R., and Insel, R.A. (1978). Characterization of lipopolysaccharide of *Haemophilus influenzae*. J. Infect. Dis. *138*, 719–730.

Flock, J.I., Fröman, G., Jönsson, K., Guss, B., Signäs, C., Nilsson, B., Raucci, G., Höök, M., Wadström, T., and Lindberg, M. (1987). Cloning and expression of the gene for a fibronectin-binding protein from *Staphylococcus aureus*. EMBO J. *6*, 2351–2357.

Forbes, K.J., Bruce, K.D., Ball, A., and Pennington, T.H. (1992). Variation in length and sequence of porin (*ompP2*) alleles of non-capsulate *Haemophilus influenzae*. Mol. Microbiol. *6*, 2107–2112.

Fröman, G., Switalski, L.M., Speziale, P., and Höök, M. (1987). Isolation and characterization of a fibronectin receptor from *Staphylococcus aureus*. J. Biol. Chem. *262*, 6564–6571.

Ganesh, V.K., Rivera, J.J., Smeds, E., Ko, Y.P., Bowden, M.G., Wann, E.R., Gurusiddappa, S., Fitzgerald, J.R., and Höök, M. (2008). A structural model of the

Staphylococcus aureus ClfA–fibrinogen interaction opens new avenues for the design of anti-staphylococcal therapeutics. PLoS Pathog. *4*, e1000226.

Ganeshkumar, N., Song, M., and McBride, B.C. (1988). Cloning of a *Streptococcus sanguis* adhesin which mediates binding to saliva-coated hydroxyapatite. Infect. Immun. *56*, 1150–1157.

Ganeshkumar, N., Hannam, P.M., Kolenbrander, P.E., and McBride, B.C. (1991). Nucleotide sequence of a gene coding for a saliva-binding protein (SsaB) from *Streptococcus sanguis* 12 and possible role of the protein in coaggregation with *Actinomyces*. Infect. Immun. *59*, 1093–1099.

Ganeshkumar, N., Arora, N., and Kolenbrander, P.E. (1993). Saliva-binding protein (SsaB) from *Streptococcus sanguis* 12 is a lipoprotein. J. Bacteriol. *175*, 572–574.

Ge, J., Catt, D.M., and Gregory, R.L. (2004). *Streptococcus mutans* surface alpha-enolase binds salivary mucin MG2 and human plasminogen. Infect. Immun. *72*, 6748–6752.

Gómez-Duarte, O.G., Dehio, M., Guzmán, C.A., Chhatwal, G.S., Dehio, C., and Meyer, T.F. (1997). Binding of vitronectin to Opa-expressing *Neisseria gonorrhoeae* mediates invasion of HeLa cells. Infect. Immun. *65*, 3857–3866.

Grass, S., Buscher, A.Z., Swords, W.E., Apicella, M.A., Barenkamp, S.J., Ozchlewski, N., and St Geme, J.W., 3rd (2003). The *Haemophilus influenzae* HMW1 adhesin is glycosylated in a process that requires HMW1C and phosphoglucomutase, an enzyme involved in lipooligosaccharide biosynthesis. Mol. Microbiol. *48*, 737–751.

Grassmé, H., Gulbins, E., Brenner, B., Ferlinz, K., Sandhoff, K., Harzer, K., Lang, F., and Meyer, T.F. (1997). Acidic sphingomyelinase mediates entry of *N. gonorrhoeae* into nonphagocytic cells. Cell *91*, 605–615.

Groeneveld, K., van Alphen, L., Eijk, P.P., Jansen, H.M., and Zanen, H.C. (1988). Changes in outer membrane proteins of nontypable *Haemophilus influenzae* in patients with chronic obstructive pulmonary disease. J. Infect. Dis. *158*, 360–365.

Gupta, S.K., Berk, R.S., Masinick, S., and Hazlett, L.D. (1994). Pili and lipopolysaccharide of *Pseudomonas aeruginosa* bind to the glycolipid asialo GM1. Infect. Immun. *62*, 4572–4579.

Hannah, J.H., Menozzi, F.D., Renauld, G., Locht, C., and Brennan, M.J. (1994). Sulfated glycoconjugate receptors for the *Bordetella pertussis* adhesin filamentous hemagglutinin (FHA) and mapping of the heparin-binding domain on FHA. Infect. Immun. *62*, 5010–5019.

Hanski, E., and Caparon, M. (1992). Protein F, a fibronectin-binding protein, is an adhesin of the group A streptococcus *Streptococcus pyogenes*. Proc. Natl. Acad. Sci. U.S.A. *89*, 6172–6176.

Hartford, O., McDevitt, D., and Foster, T.J. (1999). Matrix-binding proteins of *Staphylococcus aureus*: functional analysis of mutant and hybrid molecules. Microbiology *145*, 2497–2505.

Harvey, H.A., Porat, N., Campbell, C.A., Jennings, M., Gibson, B.W., Phillips, N.J., Apicella, M.A., and Blake, M.S. (2000). Gonococcal lipooligosaccharide is a ligand for the asialoglycoprotein receptor on human sperm. Mol. Microbiol. *36*, 1059–1070.

Hasty, D.L., Ofek, I., Courtney, H.S., and Doyle, R.J. (1992). Multiple adhesins of streptococci. Infect. Immun. *60*, 2147–2152.

Hawiger, J., Timmons, S., Strong, D.D., Cottrell, B.A., Riley, M., and Doolittle, R.F. (1982). Identification of a region of human fibrinogen interacting with staphylococcal clumping factor. Biochemistry *21*, 1407–1413.

Heckels, J.E. (1981). Structural comparison of *Neisseria gonorrhoeae* outer membrane proteins. J. Bacteriol. *145*, 736–742.

Hemker, H.C., Bas, B.M., and Muller, A.D. (1975). Activation of a pro-enzyme by a stoichiometric reaction with another protein. The reaction between prothrombin and staphylocoagulase. Biochim. Biophys. Acta *379*, 180–188.

Henderson, I.R., and Nataro, J.P. (2001). Virulence functions of autotransporter proteins. Infect. Immun. *69*, 1231–1243.

Henderson, I.R., and Owen, P. (1999). The major phase-variable outer membrane protein of *Escherichia coli* structurally resembles the immunoglobulin A1 protease class of exported protein and is regulated by a novel mechanism involving Dam and oxyR. J. Bacteriol. *181*, 2132–2141.

Henderson, I.R., Navarro-Garcia, F., and Nataro, J.P. (1998). The great escape: structure and function of the autotransporter proteins. Trends Microbiol. *6*, 370–378.

Henderson, I.R., Navarro-Garcia, F., Desvaux, M., Fernandez, R.C., and Ala'Aldeen, D. (2004). Type V protein secretion pathway: the autotransporter story. Microbiol. Mol. Biol. Rev. *68*, 692–744.

Hienz, S.A., Schennings, T., Heimdahl, A., and Flock, J.I. (1996). Collagen binding of *Staphylococcus aureus* is a virulence factor in experimental endocarditis. J. Infect. Dis. *174*, 83–88.

Hill, D.J., Toleman, M.A., Evans, D.J., Villullas, S., Van Alphen, L., and Virji, M. (2001). The variable P5 proteins of typeable and non-typeable *Haemophilus influenzae* target human CEACAM1. Mol. Microbiol. *39*, 850–862.

Hodak, H., Clantin, B., Willery, E., Villeret, V., Locht, C., and Jacob-Dubuisson, F. (2006). Secretion signal of the filamentous haemagglutinin, a model two-partner secretion substrate. Mol. Microbiol. *61*, 368–382.

Hodak, H., and Jacob-Dubuisson, F. (2007). Current challenges in autotransport and two-partner protein secretion pathways. Res. Microbiol. *158*, 631–637.

Hogg, S.D., and Manning, J.E. (1988). Inhibition of adhesion of viridans streptococci to fibronectin-coated hydroxyapatite beads by lipoteichoic acid. J. Appl. Bacteriol. *65*, 483–489.

Hoiczyk, E., Roggenkamp, A., Reichenbecher, M., Lupas, A., and Heesemann, J. (2000). Structure and sequence analysis of *Yersinia* YadA and *Moraxella* UspAs reveal a novel class of adhesins. EMBO J. *19*, 5989–5999.

Holm, M.M., Vanlerberg, S.L., Foley, I.M., Sledjeski, D.D., and Lafontaine, E.R. (2004). The *Moraxella catarrhalis* porin-like outer membrane protein CD is an adhesin for human lung cells. Infect. Immun. *72*, 1906–1913.

Holmes, A.R., McNab, R., Millsap, K.W., Rohde, M., Hammerschmidt, S., Mawdsley, J.L., and Jenkinson, H.F. (2001). The *pavA* gene of *Streptococcus pneumoniae* encodes a fibronectin-binding protein that is essential for virulence. Mol. Microbiol. *41*, 1395–1408.

Hood, D.W., Randle, G., Cox, A.D., Makepeace, K., Li, J., Schweda, E.K., Richards, J.C., and Moxon, E.R. (2004). Biosynthesis of cryptic lipopolysaccharide glycoforms in *Haemophilus influenzae* involves a mechanism similar to that required for O-antigen synthesis. J. Bacteriol. *186*, 7429–7439.

Hung, D.L., and Hultgren, S.J. (1998). Pilus biogenesis via the chaperone/usher pathway: an integration of structure and function. J. Struct. Biol. *124*, 201–220.

Ieva, R., and Bernstein, H.D. (2009). Interaction of an autotransporter passenger domain with BamA during its translocation across the bacterial outer membrane. Proc. Natl. Acad. Sci. U.S.A. *106*, 19120–19125.

Ishibashi, Y., Claus, S., and Relman, D.A. (1994). *Bordetella pertussis* filamentous hemagglutinin interacts with a leukocyte signal transduction complex and stimulates bacterial adherence to monocyte CR3 (CD11b/CD18). J. Exp. Med. *180*, 1225–1233.

Ishibashi, Y., and Nishikawa, A. (2002). *Bordetella pertussis* infection of human respiratory epithelial cells up-regulates intercellular adhesion molecule-1 expression: role of filamentous hemagglutinin and pertussis toxin. Microb. Pathog. *33*, 115–125.

Ishibashi, Y., and Nishikawa, A. (2003). Role of nuclear factor-kappa B in the regulation of intercellular adhesion molecule 1 after infection of human bronchial epithelial cells by *Bordetella pertussis*. Microb. Pathog. *35*, 169–177.

Jacob-Dubuisson, F., Fernandez, R., and Coutte, L. (2004). Protein secretion through autotransporter and two-partner pathways. Biochim. Biophys. Acta *1694*, 235–257.

Jacques, M. (1996). Role of lipo-oligosaccharides and lipopolysaccharides in bacterial adherence. Trends Microbiol. *4*, 408–409.

Jaffe, J., Natanson-Yaron, S., Caparon, M.G., and Hanski, E. (1996). Protein F2, a novel fibronectin-binding protein from *Streptococcus pyogenes*, possesses two binding domains. Mol. Microbiol. *21*, 373–384.

Jenkinson, H.F. (1992). Adherence, coaggregation, and hydrophobicity of *Streptococcus gordonii* associated with expression of cell surface lipoproteins. Infect Immun *60*, 1225–1228.

Jenkinson, H.F. (1994). Adherence and accumulation of oral streptococci. Trends Microbiol. *2*, 209–212.

Jenkinson, H.F., and Easingwood, R.A. (1990). Insertional inactivation of the gene encoding a 76-kilodalton cell surface polypeptide in *Streptococcus gordonii* Challis has a pleiotropic effect on cell surface composition and properties. Infect. Immun. *58*, 3689–3697.

Jenkinson, H.F., McNab, R., Loach, D.M., and Tannock, G.W. (1995). Lipoprotein receptors in oral streptococci. Dev. Biol. Stand. *85*, 333–341.

Jenkinson, H.F., Baker, R.A., and Tannock, G.W. (1996). A binding-lipoprotein-dependent oligopeptide transport system in *Streptococcus gordonii* essential for uptake of hexa- and heptapeptides. J. Bacteriol. *178*, 68–77.

Jerse, A.E., Cohen, M.S., Drown, P.M., Whicker, L.G., Isbey, S.F., Seifert, H.S., and Cannon, J.G. (1994). Multiple gonococcal opacity proteins are expressed during experimental urethral infection in the male. J. Exp. Med. *179*, 911–920.

Jin, Q., Yuan, Z., Xu, J., Wang, Y., Shen, Y., Lu, W., Wang, J., Liu, H., Yang, J., Yang, F., *et al.* (2002). Genome sequence of *Shigella flexneri* 2a: insights into pathogenicity through comparison with genomes of Escherichia coli K12 and O157. Nucleic Acids Res. *30*, 4432–4441.

Jong, W.S., ten Hagen-Jongman, C.M., den Blaauwen, T., Slotboom, D.J., Tame, J.R., Wickström, D., de Gier, J.W., Otto, B.R., and Luirink, J. (2007). Limited tolerance towards folded elements during secretion of the autotransporter Hbp. Mol. Microbiol. *63*, 1524–1536.

Jonsson, K., Signäs, C., Müller, H.P., and Lindberg, M. (1991). Two different genes encode fibronectin binding proteins in *Staphylococcus aureus*. The complete nucleotide sequence and characterization of the second gene. Eur. J. Biochem. *202*, 1041–1048.

Julio, S.M., Inatsuka, C.S., Mazar, J., Dieterich, C., Relman, D.A., and Cotter, P.A. (2009). Natural-host animal models indicate functional interchangeability between the filamentous haemagglutinins of *Bordetella pertussis* and *Bordetella bronchiseptica* and reveal a role for the mature C-terminal domain, but not the RGD motif, during infection. Mol. Microbiol. *71*, 1574–1590.

Junker, M., Schuster, C.C., McDonnell, A.V., Sorg, K.A., Finn, M.C., Berger, B., and Clark, P.L. (2006). Pertactin beta-helix folding mechanism suggests common themes for the secretion and folding of autotransporter proteins. Proc. Natl. Acad. Sci. U.S.A. *103*, 4918–4923.

Kaiser, P.O., Riess, T., Wagner, C.L., Linke, D., Lupas, A.N., Schwarz, H., Raddatz, G., Schäfer, A., and Kempf, V.A. (2008). The head of *Bartonella* adhesin A is crucial for host cell interaction of *Bartonella henselae*. Cell. Microbiol. *10*, 2223–2234.

Kajava, A.V., Cheng, N., Cleaver, R., Kessel, M., Simon, M.N., Willery, E., Jacob-Dubuisson, F., Locht, C., and Steven, A.C. (2001). Beta-helix model for the filamentous haemagglutinin adhesin of *Bordetella pertussis* and related bacterial secretory proteins. Mol. Microbiol. *42*, 279–292.

Kamei, T., Matozaki, T., Sakisaka, T., Kodama, A., Yokoyama, S., Peng, Y.F., Nakano, K., Takaishi, K., and Takai, Y. (1999). Coendocytosis of cadherin and c-Met coupled to disruption of cell-cell adhesion in MDCK cells--regulation by Rho, Rac and Rab small G proteins. Oncogene *18*, 6776–6784.

Kawabata, S., Morita, T., Iwanaga, S., and Igarashi, H. (1985). Enzymatic properties of staphylothrombin, an active molecular complex formed between staphylocoagulase and human prothrombin. J. Biochem. *98*, 1603–1614.

Kimura, A., Mountzouros, K.T., Relman, D.A., Falkow, S., and Cowell, J.L. (1990). *Bordetella pertussis* filamentous hemagglutinin: evaluation as a protective antigen and colonization factor in a mouse respiratory infection model. Infect. Immun. *58*, 7–16.

Kinnby, B., Booth, N.A., and Svensater, G. (2008). Plasminogen binding by oral streptococci from dental plaque and inflammatory lesions. Microbiology *154*, 924–931.

Klemm, P., Hjerrild, L., Gjermansen, M., and Schembri, M.A. (2004). Structure–function analysis of the self-recognizing Antigen 43 autotransporter protein from *Escherichia coli*. Mol. Microbiol. *51*, 283–296.

Klemm, P., Vejborg, R.M., and Sherlock, O. (2006). Self-associating autotransporters, SAATs: functional and structural similarities. Int. J. Med. Microbiol. *296*, 187–195.

Kolenbrander, P.E., Ganeshkumar, N., Cassels, F.J., and Hughes, C.V. (1993). Coaggregation: specific adherence among human oral plaque bacteria. FASEB J. *7*, 406–413.

Kolenbrander, P.E., Andersen, R.N., and Ganeshkumar, N. (1994). Nucleotide sequence of the *Streptococcus gordonii* PK488 coaggregation adhesin gene, *scaA*, and ATP-binding cassette. Infect. Immun. *62*, 4469–4480.

Koretke, K.K., Szczesny, P., Gruber, M., and Lupas, A.N. (2006). Model structure of the prototypical non-fimbrial adhesin YadA of *Yersinia enterocolitica*. J. Struct. Biol. *155*, 154–161.

Kreikemeyer, B., Talay, S.R., and Chhatwal, G.S. (1995). Characterization of a novel fibronectin-binding surface protein in group A streptococci. Mol. Microbiol. *17*, 137–145.

Kupsch, E.M., Knepper, B., Kuroki, T., Heuer, I., and Meyer, T.F. (1993). Variable opacity (Opa) outer membrane proteins account for the cell tropisms displayed by *Neisseria gonorrhoeae* for human leukocytes and epithelial cells. EMBO J. *12*, 641–650.

Laarmann, S., Cutter, D., Juehne, T., Barenkamp, S.J., and St Geme, J.W. (2002). The *Haemophilus influenzae* Hia autotransporter harbours two adhesive pockets that reside in the passenger domain and recognize the same host cell receptor. Mol. Microbiol. *46*, 731–743.

Landschulz, W.H., Johnson, P.F., and McKnight, S.L. (1988). The leucine zipper: a hypothetical structure common to a new class of DNA binding proteins. Science *240*, 1759–1764.

Lecuit, M., Dramsi, S., Gottardi, C., Fedor-Chaiken, M., Gumbiner, B., and Cossart, P. (1999). A single amino acid in E-cadherin responsible for host specificity towards the human pathogen *Listeria monocytogenes*. EMBO J. *18*, 3956–3963.

Lecuit, M., Nelson, D.M., Smith, S.D., Khun, H., Huerre, M., Vacher-Lavenu, M.C., Gordon, J.I., and Cossart, P. (2004). Targeting and crossing of the human maternofetal barrier by *Listeria monocytogenes*: role of internalin interaction with trophoblast E-cadherin. Proc. Natl. Acad. Sci. U.S.A. *101*, 6152–6157.

Lecuit, M., Ohayon, H., Braun, L., Mengaud, J., and Cossart, P. (1997). Internalin of *Listeria monocytogenes* with an intact leucine-rich repeat region is sufficient to promote internalization. Infect. Immun. *65*, 5309–5319.

Leininger, E., Roberts, M., Kenimer, J.G., Charles, I.G., Fairweather, N., Novotny, P., and Brennan, M.J. (1991). Pertactin, an Arg-Gly Asp-containing *Bordetella pertussis* surface protein that promotes adherence of mammalian cells. Proc. Natl. Acad. Sci. U.S.A. *88*, 345–349.

Leininger, E., Ewanowich, C.A., Bhargava, A., Peppler, M.S., Kenimer, J.G., and Brennan, M.J. (1992). Comparative roles of the Arg-Gly Asp sequence present in the *Bordetella pertussis* adhesins pertactin and filamentous hemagglutinin. Infect. Immun. *60*, 2380–2385.

Lindenthal, C., and Elsinghorst, E.A. (2001). Enterotoxigenic *Escherichia coli* TibA glycoprotein adheres to human intestine epithelial cells. Infect. Immun. *69*, 52–57.

Lindgren, P.E., Signäs, C., Rantamäki, L., and Lindberg, M. (1994). A fibronectin-binding protein from *Streptococcus equisimilis*: characterization of the gene and identification of the binding domain. Vet. Microbiol. *41*, 235–247.

Luo, R., Mann, B., Tuomanen, E., and Kriwacki, R.W. (2005). NMR assignment of the R2 domain of pneumococcal choline binding protein A (CbpA). J. Biomol. NMR *32*, 93.

Lupas, A. (1996). Coiled coils: new structures and new functions. Trends Biochem. Sci. *21*, 375–382.

McAllister, L.J., Tseng, H.J., Ogunniyi, A.D., Jennings, M.P., McEwan, A.G., and Paton, J.C. (2004). Molecular analysis of the *psa* permease complex of *Streptococcus pneumoniae*. Mol. Microbiol. *53*, 889–901.

McDevitt, D., Francois, P., Vaudaux, P., and Foster, T.J. (1994). Molecular characterization of the clumping factor (fibrinogen receptor) of *Staphylococcus aureus*. Mol. Microbiol. *11*, 237–248.

McDonough, J.A., Hacker, K.E., Flores, A.R., Pavelka, M.S., Jr., and Braunstein, M. (2005). The twin-arginine translocation pathway of *Mycobacterium smegmatis* is functional and required for the export of mycobacterial beta-lactamases. J. Bacteriol. *187*, 7667–7679.

McGavin, M.J., Raucci, G., Gurusiddappa, S., and Höök, M. (1991). Fibronectin binding determinants of the *Staphylococcus aureus* fibronectin receptor. J. Biol. Chem. *266*, 8343–8347.

McGavin, M.J., Gurusiddappa, S., Lindgren, P.E., Lindberg, M., Raucci, G., and Höök, M. (1993). Fibronectin receptors from *Streptococcus dysgalactiae* and *Staphylococcus aureus*. Involvement of conserved residues in ligand binding. J. Biol. Chem. *268*, 23946–23953.

Mamo, W., Fröman, G., and Müller, H.P. (2000). Protection induced in mice vaccinated with recombinant collagen-binding protein (CnBP) and alpha-toxoid against intramammary infection with *Staphylococcus aureus*. Microbiol. Immunol. *44*, 381–384.

Marino, M., Braun, L., Cossart, P., and Ghosh, P. (1999). Structure of the lnlB leucine-rich repeats, a domain that triggers host cell invasion by the bacterial pathogen *L. monocytogenes*. Mol. Cell *4*, 1063–1072.

Maurer, J., Jose, J., and Meyer, T.F. (1999). Characterization of the essential transport function of the AIDA-I autotransporter and evidence supporting structural predictions. J. Bacteriol. *181*, 7014–7020.

Mazar, J., and Cotter, P.A. (2006). Topology and maturation of filamentous haemagglutinin suggest a new model for two-partner secretion. Mol. Microbiol. *62*, 641–654.

Navarre, W.W., and Schneewind, O. (1999). Surface proteins of Gram-positive bacteria and mechanisms of their targeting to the cell wall envelope. Microbiol. Mol. Biol. Rev. *63*, 174–229.

Nummelin, H., Merckel, M.C., Leo, J.C., Lankinen, H., Skurnik, M., and Goldman, A. (2004). The *Yersinia* adhesin YadA collagen-binding domain structure is a novel left-handed parallel beta-roll. EMBO J. *23*, 701–711.

Oliver, D.C., Huang, G., and Fernandez, R.C. (2003a). Identification of secretion determinants of the *Bordetella pertussis* BrkA autotransporter. J. Bacteriol. *185*, 489–495.

Oliver, D.C., Huang, G., Nodel, E., Pleasance, S., and Fernandez, R.C. (2003b). A conserved region within the *Bordetella pertussis* autotransporter BrkA is necessary for folding of its passenger domain. Mol. Microbiol. *47*, 1367–1383.

Oomen, C.J., van Ulsen, P., van Gelder, P., Feijen, M., Tommassen, J., and Gros, P. (2004). Structure of the translocator domain of a bacterial autotransporter. EMBO J. *23*, 1257–1266.

Owen, P., Meehan, M., de Loughry-Doherty, H., and Henderson, I. (1996). Phase-variable outer membrane proteins in *Escherichia coli*. FEMS Immunol. Med. Microbiol. *16*, 63–76.

Ozeri, V., Tovi, A., Burstein, I., Natanson-Yaron, S., Caparon, M.G., Yamada, K.M., Akiyama, S.K., Vlodavsky, I., and Hanski, E. (1996). A two-domain mechanism for group A streptococcal adherence through protein F to the extracellular matrix. EMBO J. *15*, 989–998.

Pancholi, V., and Fischetti, V.A. (1998). alpha-enolase, a novel strong plasmin(ogen) binding protein on the surface of pathogenic streptococci. J. Biol. Chem. *273*, 14503–14515.

Paradis, S.E., Dubreuil, D., Rioux, S., Gottschalk, M., and Jacques, M. (1994). High-molecular-mass lipopolysaccharides are involved in *Actinobacillus pleuropneumoniae* adherence to porcine respiratory tract cells. Infect. Immun. *62*, 3311–3319.

Patti, J.M., Boles, J.O., and Höök, M. (1993). Identification and biochemical characterization of the ligand binding domain of the collagen adhesin from *Staphylococcus aureus*. Biochemistry 32, 11428–11435.

Patti, J.M., Bremell, T., Krajewska-Pietrasik, D., Abdelnour, A., Tarkowski, A., Rydén, C., and Höök, M. (1994). The *Staphylococcus aureus* collagen adhesin is a virulence determinant in experimental septic arthritis. Infect. Immun. *62*, 152–161.

Pearson, M.M., Lafontaine, E.R., Wagner, N.J., St Geme, J.W., 3rd, and Hansen, E.J. (2002). A *hag* mutant of *Moraxella catarrhalis* strain O35E is deficient in hemagglutination, autoagglutination, and immunoglobulin D-binding activities. Infect. Immun. *70*, 4523–4533.

Perkins, S., Walsh, E.J., Deivanayagam, C.C., Narayana, S.V., Foster, T.J., and Höök, M. (2001). Structural organization of the fibrinogen-binding region of the clumping factor B MSCRAMM of *Staphylococcus aureus*. J. Biol. Chem. *276*, 44721–44728.

Ponnuraj, K., Bowden, M.G., Davis, S., Gurusiddappa, S., Moore, D., Choe, D., Xu, Y., Höök, M., and Narayana, S.V. (2003). A 'dock, lock, and latch' structural model for a staphylococcal adhesin binding to fibrinogen. Cell *115*, 217–228.

Porat, N., Apicella, M.A., and Blake, M.S. (1995). A lipooligosaccharide-binding site on HepG2 cells similar to the gonococcal opacity-associated surface protein Opa. Infect. Immun. *63*, 2164–2172.

Pracht, D., Elm, C., Gerber, J., Bergmann, S., Rohde, M., Seiler, M., Kim, K.S., Jenkinson, H.F., Nau, R., and Hammerschmidt, S. (2005). PavA of *Streptococcus pneumoniae* modulates adherence, invasion, and meningeal inflammation. Infect. Immun. *73*, 2680–2689.

Prall, F., Nollau, P., Neumaier, M., Haubeck, H.D., Drzeniek, Z., Helmchen, U., Löning, T., and Wagener, C. (1996). CD66a (BGP), an adhesion molecule of the carcinoembryonic antigen family, is expressed in epithelium, endothelium, and myeloid cells in a wide range of normal human tissues. J. Histochem. Cytochem. *44*, 35–41.

Prasad, S.M., Yin, Y., Rodzinski, E., Tuomanen, E.I., and Masure, H.R. (1993). Identification of a carbohydrate recognition domain in filamentous hemagglutinin from *Bordetella pertussis*. Infect. Immun. *61*, 2780–2785.

van Putten, J.P., and Paul, S.M. (1995). Binding of syndecan-like cell surface proteoglycan receptors is required for *Neisseria gonorrhoeae* entry into human mucosal cells. EMBO J. *14*, 2144–2154.

van Putten, J.P., Duensing, T.D., and Cole, R.L. (1998). Entry of OpaA $^+$ gonococci into HEp-2 cells requires concerted action of glycosaminoglycans, fibronectin and integrin receptors. Mol. Microbiol. *29*, 369–379.

Rajam, G., Anderton, J.M., Carlone, G.M., Sampson, J.S., and Ades, E.W. (2008). Pneumococcal surface adhesin A (PsaA): a review. Crit. Rev. Microbiol. *34*, 131–142.

Ray, S.K., Rajeshwari, R., Sharma, Y., and Sonti, R.V. (2002). A high-molecular-weight outer membrane protein of *Xanthomonas oryzae* pv. *oryzae* exhibits similarity to non-fimbrial adhesins of animal pathogenic bacteria and is required for optimum virulence. Mol. Microbiol. *46*, 637–647.

Read, T.D., Salzberg, S.L., Pop, M., Shumway, M., Umayam, L., Jiang, L., Holtzapple, E., Busch, J.D., Smith, K.L., Schupp, J.M., *et al.* (2002). Comparative genome sequencing for discovery of novel polymorphisms in *Bacillus anthracis*. Science *296*, 2028–2033.

Reddy, M.S., Bernstein, J.M., Murphy, T.F., and Faden, H.S. (1996). Binding between outer membrane proteins of nontypeable *Haemophilus influenzae* and human nasopharyngeal mucin. Infect. Immun. *64*, 1477–1479.

Reddy, M.S., Murphy, T.F., Faden, H.S., and Bernstein, J.M. (1997). Middle ear mucin glycoprotein: purification and interaction with nontypable *Haemophilus influenzae* and *Moraxella catarrhalis*. Otolaryngol. Head Neck Surg. *116*, 175–180.

Relman, D.A., Domenighini, M., Tuomanen, E., Rappuoli, R., and Falkow, S. (1989). Filamentous hemagglutinin of *Bordetella pertussis*: nucleotide sequence and crucial role in adherence. Proc. Natl. Acad. Sci. U.S.A. *86*, 2637–2641.

Rhem, M.N., Lech, E.M., Patti, J.M., McDevitt, D., Höök, M., Jones, D.B., and Wilhelmus, K.R. (2000). The collagen-binding adhesin is a virulence factor in *Staphylococcus aureus* keratitis. Infect. Immun. *68*, 3776–3779.

Riess, T., Andersson, S.G., Lupas, A., Schaller, M., Schäfer, A., Kyme, P., Martin, J., Wälzlein, J.H., Ehehalt, U., Lindroos, H., *et al.* (2004). *Bartonella* adhesin a mediates a proangiogenic host cell response. J. Exp. Med. *200*, 1267–1278.

Roggenkamp, A., Ackermann, N., Jacobi, C.A., Truelzsch, K., Hoffmann, H., and Heesemann, J. (2003). Molecular analysis of transport and oligomerization of the *Yersinia enterocolitica* adhesin YadA.J. Bacteriol. *185*, 3735–3744.

Romero-Steiner, S., Caba, J., Rajam, G., Langley, T., Floyd, A., Johnson, S.E., Sampson, J.S., Carlone, G.M., and Ades, E. (2006). Adherence of recombinant pneumococcal surface adhesin A (rPsaA)-coated particles to human nasopharyngeal epithelial cells for the evaluation of anti-PsaA functional antibodies. Vaccine *24*, 3224–3231.

Rosenow, C., Ryan, P., Weiser, J.N., Johnson, S., Fontan, P., Ortqvist, A., and Masure, H.R. (1997). Contribution of novel choline-binding proteins to adherence, colonization and immunogenicity of *Streptococcus pneumoniae*. Mol. Microbiol. *25*, 819–829.

Sampson, J.S., O'Connor, S.P., Stinson, A.R., Tharpe, J.A., and Russell, H. (1994). Cloning and nucleotide sequence analysis of *psaA*, the *Streptococcus pneumoniae* gene encoding a 37-kilodalton protein homologous to previously reported *Streptococcus* sp. adhesins. Infect. Immun. *62*, 319–324.

Sandt, C.H., and Hill, C.W. (2000). Four different genes responsible for nonimmune immunoglobulin-binding activities within a single strain of *Escherichia coli*. Infect. Immun. *68*, 2205–2214.

Schneewind, O., Mihaylova-Petkov, D., and Model, P. (1993). Cell wall sorting signals in surface proteins of Gram-positive bacteria. EMBO J. *12*, 4803–4811.

Schneewind, O., Fowler, A., and Faull, K.F. (1995). Structure of the cell wall anchor of surface proteins in *Staphylococcus aureus*. Science *268*, 103–106.

Schönherr, R., Tsolis, R., Focareta, T., and Braun, V. (1993). Amino acid replacements in the *Serratia marcescens* haemolysin ShlA define sites involved in activation and secretion. Mol. Microbiol. *9*, 1229–1237.

Schubert, W.D., Göbel, G., Diepholz, M., Darji, A., Kloer, D., Hain, T., Chakraborty, T., Wehland, J., Domann, E., and Heinz, D.W. (2001). Internalins from the human pathogen *Listeria monocytogenes* combine three distinct folds into a contiguous internalin domain. J. Mol. Biol. *312*, 783–794.

Schubert, W.D., Urbanke, C., Ziehm, T., Beier, V., Machner, M.P., Domann, E., Wehland, J., Chakraborty, T., and Heinz, D.W. (2002). Structure of internalin, a major invasion protein of *Listeria monocytogenes*, in complex with its human receptor E-cadherin. Cell *111*, 825–836.

Sheets, A.J., and St. Geme, J.W., 3rd. (2011). Adhesive activity of the *Haemophilus* cryptic genospecies Cha autotransporter is modulated by a novel mechanism involving variation in peptide repeats. J. Bacteriol. *193*, 329–339.

Shen, Y., Naujokas, M., Park, M., and Ireton, K. (2000). InIB-dependent internalization of *Listeria* is mediated by the Met receptor tyrosine kinase. Cell *103*, 501–510.

Sherlock, O., Schembri, M.A., Reisner, A., and Klemm, P. (2004). Novel roles for the AIDA adhesin from diarrheagenic *Escherichia coli*: cell aggregation and biofilm formation. J. Bacteriol. *186*, 8058–8065.

Sherlock, O., Vejborg, R.M., and Klemm, P. (2005). The TibA adhesin/invasin from enterotoxigenic *Escherichia coli* is self recognizing and induces bacterial aggregation and biofilm formation. Infect. Immun. *73*, 1954–1963.

Sherlock, O., Dobrindt, U., Jensen, J.B., Munk Vejborg, R., and Klemm, P. (2006). Glycosylation of the self-recognizing *Escherichia coli* Ag43 autotransporter protein. J. Bacteriol. *188*, 1798–1807.

Skillman, K.M., Barnard, T.J., Peterson, J.H., Ghirlando, R., and Bernstein, H.D. (2005). Efficient secretion of a folded protein domain by a monomeric bacterial autotransporter. Mol. Microbiol. *58*, 945–958.

Smith-Vaughan, H.C., Sriprakash, K.S., Mathews, J.D., and Kemp, D.J. (1997). Nonencapsulated *Haemophilus influenzae* in Aboriginal infants with otitis media: prolonged carriage of P2 porin variants and evidence for horizontal P2 gene transfer. Infect. Immun. *65*, 1468–1474.

Srivastava, N., Zeiler, J.L., Smithson, S.L., Carlone, G.M., Ades, E.W., Sampson, J.S., Johnson, S.E., Kieber-Emmons, T., and Westerink, M.A. (2000). Selection of an immunogenic and protective epitope of the PsaA protein of *Streptococcus pneumoniae* using a phage display library. Hybridoma *19*, 23–31.

St Geme, J.W., 3rd (2002). Molecular and cellular determinants of non-typeable *Haemophilus influenzae* adherence and invasion. Cell. Microbiol. *4*, 191–200.

St Geme, J.W., 3rd, and Cutter, D. (2000). The *Haemophilus influenzae* Hia adhesin is an autotransporter protein that remains uncleaved at the C terminus and fully cell associated. J. Bacteriol. *182*, 6005–6013.

St Geme, J.W., 3rd, Falkow, S., and Barenkamp, S.J. (1993). High-molecular-weight proteins of nontypable *Haemophilus influenzae* mediate attachment to human epithelial cells. Proc. Natl. Acad. Sci. U.S.A. *90*, 2875–2879.

St Geme, J.W., 3rd, Cutter, D., and Barenkamp, S.J. (1996). Characterization of the genetic locus encoding *Haemophilus influenzae* type b surface fibrils. J. Bacteriol. *178*, 6281–6287.

St Geme, J.W., 3rd, Kumar, V.V., Cutter, D., and Barenkamp, S.J. (1998). Prevalence and distribution of the *hmw* and *hia* genes and the HMW and Hia adhesins among genetically diverse strains of non-typeable *Haemophilus influenzae*. Infect. Immun. *66*, 364–368.

Stern, A., and Meyer, T.F. (1987). Common mechanism controlling phase and antigenic variation in pathogenic neisseriae. Mol. Microbiol. *1*, 5–12.

Surana, N.K., Cutter, D., Barenkamp, S.J., and St Geme, J.W., 3rd (2004a). The *Haemophilus influenzae* Hia autotransporter contains an unusually short trimeric translocator domain. J. Biol. Chem. *279*, 14679–14685.

Surana, N.K., Grass, S., Hardy, G.G., Li, H., Thanassi, D.G., and Geme, J.W., 3rd (2004b). Evidence for conservation of architecture and physical properties of Omp85-like proteins throughout evolution. Proc. Natl. Acad. Sci. U.S.A. *101*, 14497–14502.

Surana, N.K., Buscher, A.Z., Hardy, G.G., Grass, S., Kehl-Fie, T., and St Geme, J.W., 3rd (2006). Translocator proteins in the two-partner secretion family have multiple domains. J. Biol. Chem. *281*, 18051–18058.

Swanson, J., Barrera, O., Sola, J., and Boslego, J. (1988). Expression of outer membrane protein II by gonococci in experimental gonorrhea. J. Exp. Med. *168*, 2121–2129.

Switalski, L.M., Speziale, P., and Höök, M. (1989). Isolation and characterization of a putative collagen receptor from *Staphylococcus aureus* strain Cowan 1. J. Biol. Chem. *264*, 21080–21086.

Switalski, L.M., Patti, J.M., Butcher, W., Gristina, A.G., Speziale, P., and Höök, M. (1993). A collagen receptor on *Staphylococcus aureus* strains isolated from patients with septic arthritis mediates adhesion to cartilage. Mol. Microbiol. *7*, 99–107.

Swords, W.E., Buscher, B.A., Ver Steeg Ii, K., Preston, A., Nichols, W.A., Weiser, J.N., Gibson, B.W., and Apicella, M.A. (2000). Non-typeable *Haemophilus influenzae* adhere to and invade human bronchial epithelial cells via an interaction of lipooligosaccharide with the PAF receptor. Mol. Microbiol. *37*, 13–27.

Symersky, J., Patti, J.M., Carson, M., House-Pompeo, K., Teale, M., Moore, D., Jin, L., Schneider, A., DeLucas, L.J., Höök, M., *et al.* (1997). Structure of the collagen-binding domain from a *Staphylococcus aureus* adhesin. Nat. Struct. Biol. *4*, 833–838.

Szczesny, P., and Lupas, A. (2008). Domain annotation of trimeric autotransporter adhesins--daTAA. Bioinformatics *24*, 1251–1256.

Szczesny, P., Linke, D., Ursinus, A., Bar, K., Schwarz, H., Riess, T.M., Kempf, V.A., Lupas, A.N., Martin, J., and Zeth, K. (2008). Structure of the head of the *Bartonella* adhesin BadA. PLoS Pathog. *4*, e1000119.

Talay, S.R., Valentin-Weigand, P., Timmis, K.N., and Chhatwal, G.S. (1994). Domain structure and conserved epitopes of Sfb protein, the fibronectin-binding adhesin of *Streptococcus pyogenes*. Mol. Microbiol. *13*, 531–539.

Ton-That, H., Faull, K.F., and Schneewind, O. (1997). Anchor structure of staphylococcal surface proteins. A branched peptide that links the carboxyl terminus of proteins to the cell wall. J. Biol. Chem. *272*, 22285–22292.

Valkonen, K.H., Wadström, T., and Moran, A.P. (1994). Interaction of lipopolysaccharides of *Helicobacter pylori* with basement membrane protein laminin. Infect. Immun. *62*, 3640–3648.

Vazquez-Torres, A., and Fang, F.C. (2000). Cellular routes of invasion by enteropathogens. Curr. Opin. Microbiol. *3*, 54–59.

Virji, M., Makepeace, K., and Moxon, E.R. (1994). Distinct mechanisms of interactions of Opc-expressing meningococci at apical and basolateral surfaces of human endothelial cells; the role of integrins in apical interactions. Mol. Microbiol. *14*, 173–184.

Virji, M., Makepeace, K., Peak, I.R., Ferguson, D.J., Jennings, M.P., and Moxon, E.R. (1995). Opc- and pilus-dependent interactions of meningococci with human endothelial cells: molecular mechanisms and modulation by surface polysaccharides. Mol. Microbiol. *18*, 741–754.

Voulhoux, R., Bos, M.P., Geurtsen, J., Mols, M., and Tommassen, J. (2003). Role of a highly conserved bacterial protein in outer membrane protein assembly. Science *299*, 262–265.

de Vries, F.P., Cole, R., Dankert, J., Frosch, M., and van Putten, J.P. (1998). *Neisseria meningitidis* producing the Opc adhesin binds epithelial cell proteoglycan receptors. Mol. Microbiol. *27*, 1203–1212.

Wagner, B., Schmidt, K.H., Wagner, M., and Wadström, T. (1988). Localization and characterization of fibronectin-binding to group A streptococci. An electron microscopic study using protein-gold-complexes. Zentralbl. Bakteriol. Mikrobiol. Hyg. A *269*, 479–491.

Wang, J., Gray-Owen, S.D., Knorre, A., Meyer, T.F., and Dehio, C. (1998). Opa binding to cellular CD66 receptors mediates the transcellular traversal of *Neisseria gonorrhoeae* across polarized T84 epithelial cell monolayers. Mol. Microbiol. *30*, 657–671.

Webb, D.C., and Cripps, A.W. (1998). Secondary structure and molecular analysis of interstrain variability in the P5 outer-membrane protein of non-typable *Haemophilus influenzae* isolated from diverse anatomical sites. J. Med. Microbiol. *47*, 1059–1067.

Weiser, J.N. (1993). Relationship between colony morphology and the life cycle of *Haemophilus influenzae*: the contribution of lipopolysaccharide phase variation to pathogenesis. J. Infect. Dis. *168*, 672–680.

Weiser, J.N., Pan, N., McGowan, K.L., Musher, D., Martin, A., and Richards, J. (1998). Phosphorylcholine on the lipopolysaccharide of *Haemophilus influenzae* contributes to persistence in the respiratory tract and sensitivity to serum killing mediated by C-reactive protein. J. Exp. Med. *187*, 631–640.

Wells, T.J., Tree, J.J., Ulett, G.C., and Schembri, M.A. (2007). Autotransporter proteins: novel targets at the bacterial cell surface. FEMS Microbiol. Lett. *274*, 163–172.

Yu, C., Ruiz, T., Lenox, C., and Mintz, K.P. (2008). Functional mapping of an oligomeric autotransporter adhesin of *Aggregatibacter actinomycetemcomitans*. J. Bacteriol. *190*, 3098–3109.

Zavialov, A.V., and Knight, S.D. (2007). A novel self-capping mechanism controls aggregation of periplasmic chaperone Caf1M. Mol. Microbiol. *64*, 153–164.

Zavialov, A.V., Berglund, J., Pudney, A.F., Fooks, L.J., Ibrahim, T.M., MacIntyre, S., and Knight, S.D. (2003). Structure and biogenesis of the capsular F1 antigen from *Yersinia pestis*: preserved folding energy drives fiber formation. Cell *113*, 587–596.

Zavialov, A.V., Tischenko, V.M., Fooks, L.J., Brandsdal, B.O., Aqvist, J., Zav'yalov, V.P., Macintyre, S., and Knight, S.D. (2005). Resolving the energy paradox of chaperone/usher-mediated fibre assembly. Biochem J. *389*, 685–694.

Zavialov, A., Zav'yalova, G., Korpela, T., and Zav'yalov, V. (2007). FGL chaperone-assembled fimbrial polyadhesins: anti-immune armament of Gram-negative bacterial pathogens. FEMS Microbiol. Rev. *31*, 478–514.

Zhang, J.R., Mostov, K.E., Lamm, M.E., Nanno, M., Shimida, S., Ohwaki, M., and Tuomanen, E. (2000). The polymeric immunoglobulin receptor translocates pneumococci across human nasopharyngeal epithelial cells. Cell *102*, 827–837.

Zong, Y., Xu, Y., Liang, X., Keene, D.R., Hook, A., Gurusiddappa, S., Höök, M., and Narayana, S.V. (2005). A 'Collagen Hug' model for *Staphylococcus aureus* CNA binding to collagen. EMBO J. *24*, 4224–4236.

Biofilms: The Secret Story of Microbial Communities

6

Christophe S. Bernard, Caroline Giraud, Jennifer Spagnolo and Sophie de Bentzmann

Abstract

This chapter is dedicated to a particular phase of the bacterial cell cycle known as the biofilm, in which single-celled individuals gather together to form a sedentary but dynamic community with a complex structure, displaying spatial and functional heterogeneity. In response to the perception of environmental signals by sensing systems, appropriate responses are triggered, leading to biofilm formation. This process involves various molecular systems (described in detail here for Gram-negative and Gram-positive bacteria) enabling bacteria to identify appropriate surfaces on which to anchor themselves, to stick to those surfaces and to each other, to construct multicellular communities several hundreds of micrometers thick and to detach from the community. These molecular systems are used antagonistically or synergistically, depending on the microenvironment confronting the bacterium. The biofilm microbial community is a unique, highly competitive and crowded environment facilitating microevolutionary processes and horizontal gene transfer between distantly related microorganisms. It is governed by social rules, based on the production and use of 'public' goods, with actors and recipients. Biofilms constitute a unique shield against external aggressions, including drug treatment and immune reactions. Biofilm-associated infections in humans have therefore generated major problems for the diagnosis and treatment of disease. Improvements in our understanding of biofilms have led to innovative research aiming to interfere with the process of biofilm formation.

Introduction

Over the last 30 years, microbiologists have begun to study a particular phase of the bacterial cell cycle known as the biofilm formation. Almost all bacterial species can live in biofilms, but interest in biofilms has rocketed recently, because the emergence of new biofilm-associated infectious diseases in humans has resulted in considerable problems for diagnosis and treatment. Biofilms are a particular problem for human health, not only because they may cause several infectious diseases, but also because they may be associated with many inert surfaces, including those of medical devices for internal or external use. However, their presence is probably underestimated, principally due to the need for *in vivo* diagnosis (Lynch and Robertson, 2008).

Early reports described biofilm formation as an aspect of microbial physiology (Henrici, 1933), but recent advances in genetics, genomics, global -omics approaches and visualization techniques (Pamp *et al.*, 2009) have provided a considerable body of data relating to many aspects of the bacterial process of biofilm formation. New areas of research have also emerged, such as biofomics (Azevedo *et al.*, 2009), an approach which brings insight into how biofilm formation ability is affected by a number of environmental, physiological and mutational factors and how these factors interplay between themselves. Sociobiological aspects (Foster *et al.*, 2007) of this microbial community represent also new research axes.

The notion of a biofilm is based on single-celled individuals gathering together to form a sedentary community with a complex structure, displaying spatial and functional heterogeneity (Parsek and Tolker-Nielsen, 2008). For some time, biofilms were considered to be a developmental process (Danese *et al.*, 2001), during which a specific genetic programme was switched on (Lazazzera, 2005; Moorthy and Watnick, 2005; An and Parsek; 2007), with coordinated patterns of gene expression and protein production. However, this view was recently called into question (Monds and O'Toole, 2009). Indeed, the process is dynamic, with motility, for example, not simply involved in the initial attachment step. The biofilm is not a rigid structure, consisting instead of functional entities, in which the cells display specific adaptations to a biofilm lifestyle (O'Toole *et al.*, 2000; Nadell *et al.*, 2009). Global -omics approaches have had a considerable impact in rapidly increasing our knowledge about the biofilm lifestyle (Azevedo *et al.*, 2009). However, heterogeneity and the composition of the biofilm in terms of specific subpopulations have been largely neglected or overlooked in transcriptome analyses, owing to the small amounts of RNA present in slow-growing cells for example as compared with total RNA (An and Parsek, 2007). Furthermore, many biofilm-dependent regulatory mechanisms are post-transcriptional, for example those governed by c-di-GMP, and cannot be evaluated by such approaches.

The biofilm lifestyle is highly dependent on the bacterial species and environmental conditions, so it is difficult to make generalizations for all bacterial species. However, common properties have been defined, such as the need for a substrate and 'preconditioning surfaces', the specialization of subpopulations (known as 'division of labour') (Aparna and Yadav, 2008), the production of a hydrated matrix shaping the community and the division of this life cycle into stages.

Genetics and genomics have, over the last ten years, provided us with considerable amounts of data concerning the molecular actors involved in this structuring of the community. In this chapter, we will consider these actors for the Gram-positive and Gram-negative bacteria associated with human infections. The multicellular structure of the biofilm makes it possible for the bacteria concerned to undergo dormancy and hibernation, enabling them to survive and to disseminate their genomes. It may therefore be considered as a step in the bacterial cell cycle. Biofilms also display unique properties, such as multidrug tolerance, and resistance to both opsonization and phagocytosis, which may protect them, enabling them to survive in hostile environmental conditions and to resist selective pressures (Weitao, 2009). This community is also unique in that it brings together different species (bacteria, fungi or yeasts) in a structure in which they can cooperate, rather than compete. The biofilm thus constitutes a microbial society, with its own sets of social rules and patterns of behaviour, including altruism and cooperation, both of which favour the success of the group (Shapiro, 1998; Parsek and Greenberg, 2005), on the one hand, and competition, on the other. Certain subpopulations may display specialization. All these patterns of behaviour are orchestrated by communication, which may be chemical or genetic in nature (e.g. exchange of plasmids) (Weigel *et al.*, 2007). The biofilm thus constitutes a unique way to stabilize interactions between species, inducing marked changes in the symbiotic relationships between them and affecting the function of the microbial community (Hansen *et al.*, 2007). Regardless of whether the biofilm can truly be considered as a multicellular community based on task-sharing behaviour, this cellular arrangement creates chemical and metabolite gradients and heterogeneity in oxygen availability. As such, it is a potentially stressful environment for aerobic species, necessitating adaptation to oxygen availability (Xu *et al.*, 1998). Starvation is an important trigger of stress responses (Stanley *et al.*, 2003) and is associated with changes in metabolic and catabolic pathways and with signs of membrane stress (Beloin *et al.*, 2004; Cerca *et al.*, 2005). Stressful conditions in biofilms represent a unique way to generate genetic diversity and to drive evolution. The emergence of new subpopulations – such as small-colony variants (genetic diversification), persisters and cells tolerant to imposed constraints – represents a new challenge for microbiologists, who will probably need to develop an integral, holistic view of the community if they are to decipher this bacterial lifestyle.

Real biofilms are far more complex than those studied *in vitro,* owing to the underestimation of environmental parameters or the numbers of species controlling biofilm formation. However, several possible solutions for counteracting biofilms in biofilm-associated human infections have been proposed, based on the large body of data collected to date.

The 'developmental' process and its associated specific genetic programme

Stages

Biofilms, like other animal communities, form gradually over time. Changes in their form and function play a key role in the life cycle of the microorganisms, and this has led to suggestions that the time-dependent evolution of the microcommunity equates to a developmental process (O'Toole *et al.*, 2000). However, this developmental model has been called into question, because it is based principally on phenomenological descriptions of changes in the form and function of biofilms, rather than a detailed understanding of the causal basis of biofilm formation and its molecular basis (Monds and O'Toole, 2009).

Whatever the bacterial species involved and the complexity of the resulting biofilm, biofilm formation is a dynamic process highly dependent on environmental signals. Biofilms pass through a four-stage universal growth cycle illustrated in Fig. 6.1, consisting of initiation, maturation, maintenance, and dissolution phases, regardless of the phenotype of the constituent microorganisms (O'Toole *et al.*, 2000; Sauer *et al.*, 2002).

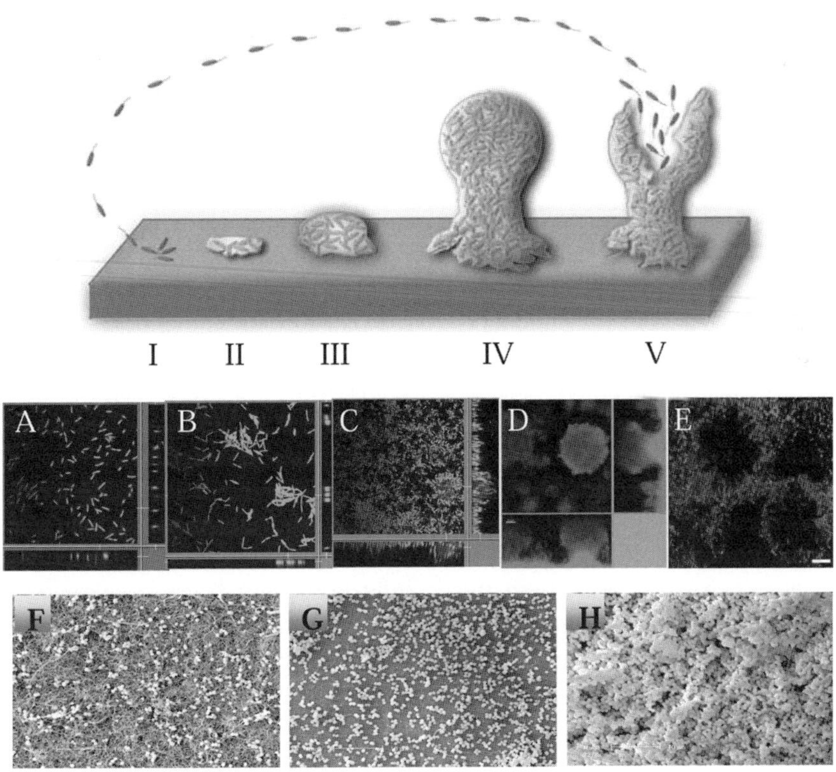

Figure 6.1 Temporal evolution of biofilm. Schematization of the four-stage universal growth cycle of a biofilm with common characteristics (that includes initiation (I), maturation (II and III), maintenance (IV), and dissolution (V)). Underneath are presented those steps in *P. aeruginosa* either labelled with DAPI (A-C), chromosomal GFP (D) (personal data) or LIVE/DEAD BacLight kit (E) (Boles *et al.*, 2005) observed with confocal microscopy and in *S. aureus* (F-H) in scanning electron microscopy (personal data).

Initiation

Contact is established during the initiation stage, which involves cell-surface interactions and the development of microcolonies through cell–cell interactions. This phase is very rapid and is highly dependent on environmental signals, the roughness of the surfaces to which the bacteria become attached and hydrodynamics (Goller and Romeo, 2008). This stage comprises a reversible step, during which the bacteria may detach from the surface, followed by an irreversible step. During the initiation stage, the bacteria are growing exponentially. Motility and chemotaxis are important because they bring the bacteria into contact with surfaces suitable for colonization. However, non-motile bacteria also form biofilms. Several elements, such as conjugative pili, CU pili, curli and LPS in Gram-negative bacteria, and MSCRAMMs (microbial surface components recognizing adhesive matrix molecules) in Gram-positive bacteria, are involved in this phase (O'Toole *et al.*, 2000). However, a combination of these factors, rather than a single molecular determinant, is generally involved.

Maturation and maintenance

The hallmark of a sessile lifestyle is the production of a complex matrix encasing the bacteria. Extracellular matrices (ECM) are as different as the bacteria producing them, but they undoubtedly play an essential role in structuring and maintaining multicellular communities (Branda *et al.*, 2005), with the final architecture of the community depending largely on the nature of the matrix produced. This maturation is accompanied by a three-dimensional (3D) expansion of the community, involving several different molecules, such as adhesins from the AT (autotransporters) and TPS (two partner secretion) families and molecules involved in initiation. The highly hydrated and glycosylated polymer in which the bacteria are embedded at this stage is generated by exopolysaccharide biosynthesis. This process is principally dependent on cell density, except in *Vibrio cholerae*, in which exopolysaccharide (VPS) synthesis is thought to be induced at low cell density and also involved in the initiation stage.

Dispersion

Dispersion is the stage at which some of the bacterial cells detach and move away from the biofilm, reverting to a planktonic lifestyle, thereby promoting dissemination from an initial site of infection to other sites. This stage was largely ignored in the past, but has been studied extensively more recently.

Mechanisms other than the production of dispersion-promoting molecules may be involved in dispersion. For example, in *V. cholerae*, virulence gene expression and biofilm formation occur at low cell density, in the absence of autoinducers, and these phenotypes are repressed at high cell density, when the signal molecules have accumulated (Xavier and Foster, 2007). Quorum-sensing (QS) mutants form thicker biofilms, and QS signals repress biofilm formation by repressing *vps* gene expression. The dispersion of individual cells from the biofilm enables these cells to switch off QS-mediated repression of the cholera toxin and toxin-coregulated pilus genes required for colonization (Zhu and Mekalanos, 2003).

Horizontal gene transfer mediated by transformation and conjugation has been linked to the promotion of biofilm structuration. By contrast, bacteriophage acquisition has been linked to dispersion, particularly in *Pseudomonas fluorescens* biofilms infected with the phiIBB-PF7A phage, and the integration of phages into the biofilm matrix and their entrapment at the surface may also be beneficial (Sillankorva *et al.*, 2008). Phage acquisition, resulting in lysogeny, has been shown to alter the capacity of *Bacillus anthracis* to sporulate, produce exopolysaccharide, form biofilms, and survive in the long term in the soil, the positive effects of the phages depending strongly on their nature (Schuch and Fischetti, 2009). The influence of prophages has recently been investigated in the wild-type (WT) PAO1 strain of *Pseudomonas aeruginosa* carrying the filamentous Pf4 prophage (Rice *et al.*, 2009). Cell death occurs in WT biofilms, whereas biofilms formed by the Pf4 phage-deficient mutant of *P. aeruginosa* do not develop hollow centres or display the cell death typical of the process of WT *P. aeruginosa* PAO1 biofilm differentiation. These findings highlight the correlation between the conversion of the Pf4 phage into a superinfective form, and the appearance of variants in the dispersing population. Prophages may also have a direct effect on the shape of the biofilm. This has been elegantly demonstrated in *P. aeruginosa*, in which the Pf4 phage has been shown to be involved not only in the formation of hollow colonies and cell death within microcolonies, but also in microcolony expansion and the control of microlony size and stability (Rice *et al.*, 2009). In addition to phage-mediated cell death in *P. aeruginosa* biofilms (Webb *et al.*, 2003), the phage-mediated appearance of highly infectious small-colony variants has also been described (Rice *et al.*, 2009). The genes of the CP4–57 prophage of the BW25113 strain of *Escherichia coli* clearly act to reduce early biofilm formation (Wang *et al.*, 2009). Furthermore, prophages can excise naturally from *E. coli* biofilms, particularly during the early stages of biofilm formation, causing the remaining prophage-deficient bacteria to switch to a phenotype of enhanced motility, slower growth and greater biofilm dispersal.

As an external signal triggering dispersion, NO also induces the transition from a biofilm to a planktonic lifestyle in *P. aeruginosa*. It acts via the secondary messenger c-di-GMP network, through the post-translational activation of phosphodiesterase activity rather than by transcriptional control (Barraud *et al.*, 2009).

Specific genetic programme

Biofilm formation is highly dependent on regulatory networks governing the switch between planktonic and sedentary lifestyles. These networks integrate environmental signals through adequate sensing systems triggering appropriate responses and QS (a multicellular response) resulting in the production of autoinducers, a type of small, diffusible molecules, and other molecules (c-di-GMP, ppGpp and cAMP) (Senadheera and Cvitkovitch, 2008; Jonas *et al.*, 2009; de Kievit, 2009; Pesavento and Hengge, 2009). The stepwise formation of the biofilm, like developmental processes, involves the switching on of a specific genetic programme, leading to coordinated patterns of gene expression and protein production. The development of DNA microarray technology and its use for analysis of global gene expression profiles during the various stages of bacterial biofilm development have made it possible to identify the genes required for biofilm formation and to determine the environmental signals regulating biofilm formation, the extent to which biofilm cells are different from other cells and the extent to which biofilm formation is a developmental process (Lazazzera, 2005). However, in DNA microarray studies, particularly those determining the gene expression profile in bacterial biofilms, no information is available concerning the number of cells in the population differentially expressing a gene and the location of these cells. There is a clear need for standardization of genome-scale microarray-based transcriptional profiling experiments, with the use of uniform experimental designs to resolve the spatial and functional heterogeneity in the biofilm community in particular (An and Parsek, 2007). Multiple strains should also be studied, to avoid strain-specific bias in the interpretation of results, as almost no overlap has been found between the regulons identified in different studies (Nagarajan *et al.*, 2009). This is particularly important because hydrodynamics, nutrient load and intracellular carbon flux all have a major impact, presumably by altering the expression of cellular traits essential for bacterial adaptation during the various stages of biofilm formation (Kjelleberg and Molin, 2002).

Despite these difficulties, DNA microarray analyses have suggested that bacteria may display a common response upon biofilm formation, with the repression of flagellar genes and the overexpression of genes for adhesion (encoding surface appendages, such as pili, or secreted polysaccharides) and ribosomal proteins, probably due to an increase in the growth rate of at least some of the cells in the biofilm (Lazazzera, 2005). Almost all profiling studies carried out have indicated significant metabolic changes in biofilm communities in response to changes in oxygen and nutrient availability, and stress response signatures resulting principally from starvation (An and Parsek, 2007). DNA microarray analyses have also helped to identify the transcription factors affecting biofilms and have indicated that there may be no biofilm-specific genes, suggesting that biofilm formation is not a developmental process (Lazazzera, 2005). Instead, DNA microarray data suggest that biofilms may display a unique pattern of gene expression, in which certain subsets of genes may be expressed independently in biofilms or in planktonic conditions, but are expressed simultaneously only in biofilms (Lazazzera, 2005; Moorthy and Watnick, 2005; An and Parsek, 2007). This is consistent with the reprogramming of gene expression in biofilms resulting from changes in multiple environmental physicochemical conditions: bacteria within biofilms encounter conditions of higher osmolarity, poorer oxygen supply and higher cell density (Prigent-Combaret and Lejeune, 1999; Moorthy and Watnick 2005).

Global transcriptomic approaches have shown that prophage genes are among the most strongly induced genes during biofilm development in Gram-negative bacteria, such as *P. aeruginosa* (Whiteley *et al.*, 2001) and *E. coli* (Domka *et al.*, 2006, 2007; Herzberg *et al.*, 2006; García-Contreras *et al.*, 2008) and in Gram-positive bacteria (Stanley *et al.*, 2003; Webb *et al.*, 2003; Ren *et al.*, 2004). Phage induction may be important for biofilms (Whiteley *et al.*, 2001).

The molecules involved

Our understanding of the molecular basis of bacterial biofilm development has benefited from improvements in genetics and genomics and from the development of visualization techniques, making it possible to unravel the processes involved in biofilm development, physiology and adaptation. A plethora of systems enabling bacteria to identify and anchor themselves onto appropriate surfaces, to stick to each other, to construct multicellular communities several hundreds of micrometres thick and to detach from the community has been identified and characterized in many biofilm-forming bacteria. It is not possible to identify general molecular profiles for a given bacterial species, because some genes are important for biofilm formation under both static and dynamic conditions, whereas others are important only under dynamic biofilm conditions (Ramsey and Whiteley, 2004). However, throughout the bacterial kingdom, these genes can be separated into those encoding appendages consisting of oligomerized subunits responsible for motility (Type IV pili designated TFP, flagella) or with other functions (fimbriae, other types of pili, curli), ECM, and surface adhesins or other secreted elements illustrated in Fig. 6.2. The molecular machines responsible for assembling or secreting these systems are, of course, highly dependent on the simple or double membranes of Gram-positive and Gram-negative bacteria, respectively. Their role in biofilm initiation and structuring is also highly dependent on environmental conditions and the surfaces encountered by the bacteria (Fig. 6.3). Each bacterium has its own adhesion toolkit, containing a number of molecules, different for each species, that may be used antagonistically or synergistically, depending on the situation with which the bacterium is faced.

Appendages responsible for motility

Motile biofilm-forming species are generally equipped with flagella and/or TFP. By using these appendages in active motility towards surfaces, bacteria are able to overcome the hydrodynamic and electrostatic forces to which they are subjected.

Flagellum

The bacterial flagellum is an extracellular appendage assembled by a complex self-assembling nanomachine allowing bacteria to move in their environment. It is composed of three substructures: the basal body, the filament acting as the propeller, which is formed by the oligomerization of flagellin and the addition of a cap protein that acts as an adhesin, and the hook, which connects the basal body and the filament. The basal body anchors the structure in the cell envelope and contains the motor, which consists of a rotor connected to the hook and filament, and a non-rotating part, or stator, which conducts coupling ions and is responsible for energy conversion.

Swimming motility and coordinated swarming motility, both of which are driven by flagella (Samatey *et al.*, 2001; Chevance and Hugues, 2008; Jarrell and McBride, 2008), drive the movement of bacteria in liquid and semi-liquid media, respectively. Both types of motility enable bacteria to form communities. The rotation of single or multiple flagella drives swimming motility in fluid environments and swarming along wetted surfaces. Swarming motility has been associated

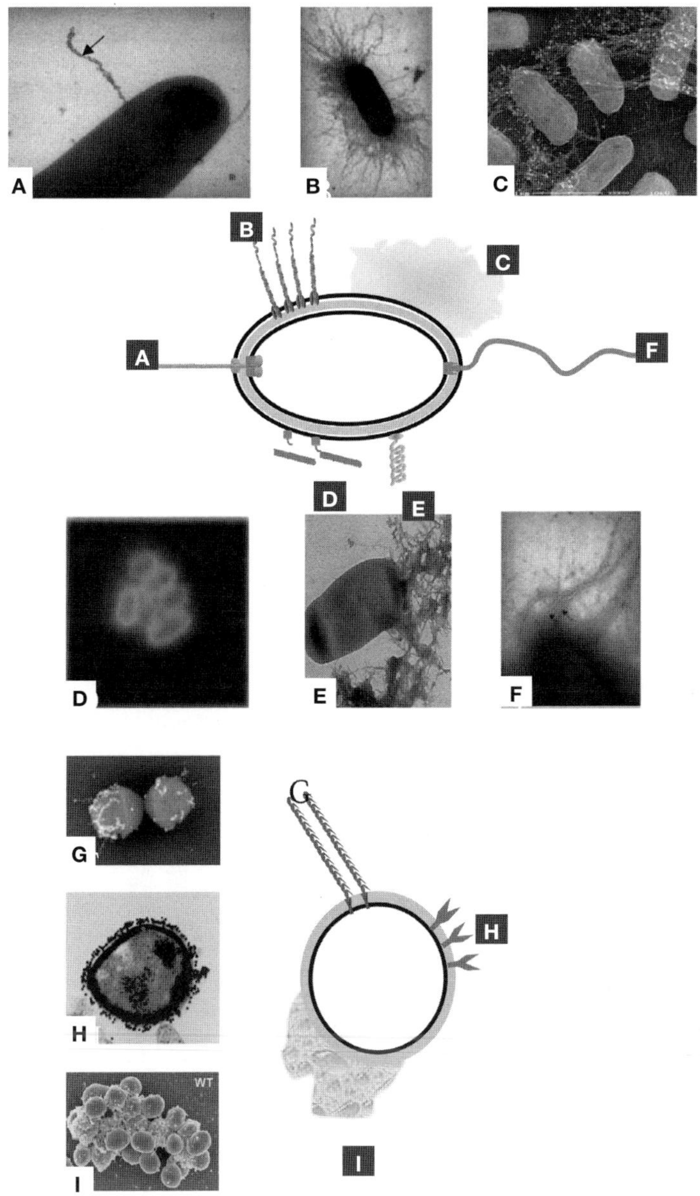

Figure 6.2 Examples of structures and molecules recovered in Gram-negative (upper panel) and Gram-positive bacteria (lower panel) that have been described to play a role in biofilm formation. Upper panel (A-F) (A) TFPb pili from the Flp subfamily assembled at the surface of *P. aeruginosa* visualized in transmission electron microscopy coupled to immunolabelling (de Bentzmann *et al.*, 2006). (B) Pili of CU family assembled at the surface of *P. aeruginosa* visualized in transmisson electron microscopy (personal data). (C) As an example of matrix production, polar cellulose secretion in *E. coli* visualized in scanning electron microscopy coupled to immunogold labelling (Lequeré and Ghigo, 2009). (D) Ag43, an AT protein revealed immunolocalization at the surface of *E. coli* by immunofluorescence (Henderson *et al.*, 1997). (E) Curli structures visualized in tranmission electron microscopy in *E. coli* (Epstein *et al.*, 2009). (F) Multiple flagella at the surface of *P. aeruginosa* vizualized in transmission electron microscopy (Giraud *et al.*, 2010). Lower panel (G-I) (G) Sortase-assembled pili assembled at the surface of *Streptococcus agalactiae* viewed in scanning electron and transmission electron microscopies coupled to immunolabelling detection (Dramsi *et al.*, 2006). (H) MSCRAMM and cell surface–associated proteins at the surface of *Staphylococcus aureus* (personal data) detected with immunolabelling in transmission electron microscopy. (I) PIA visualization on the *S. epidermidis* surface viewed in scanning electron microscopy (Vuong *et al.*, 2004).

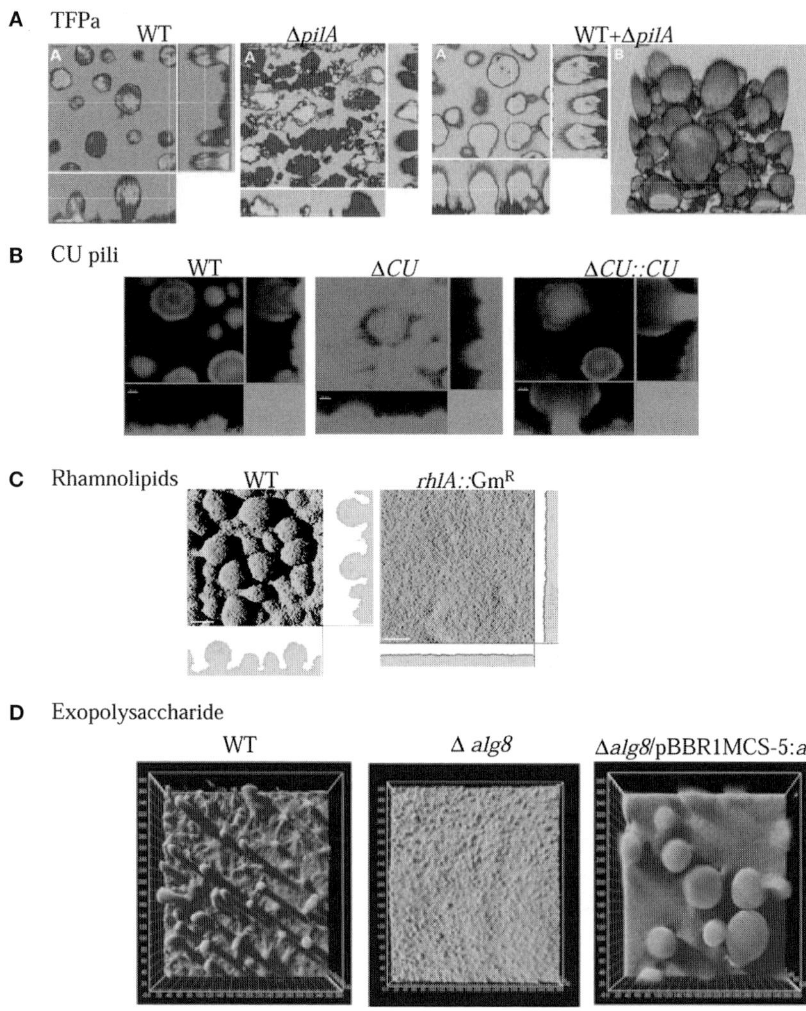

Figure 6.3 Examples of contribution of several molecular actors involved in bacterial biofilm development and structuration viewed in confocal microscopy: (A) Contribution of TFPa pili exemplified in *P. aeruginosa*, WT and Δ*pilA* each being tagged with CFP and YFP populations separately (I and II) or mixed (III and IV) with Δ*pilA* tagged with CFP and WT tagged with YFP (Klausen *et al.*, 2003a). (B) Contribution of CU pili in *P. aeruginosa* biofilm development, WT, Δ*CU* (deletion mutant) and Δ*CU::CU* (deletion mutant with cis-complementation) being tagged with chromosomal GFP (personal data). (C) Contribution of rhamnolipids in *P. aeruginosa* biofilm development, WT and *rhlA*::Gm^R (insertional mutant) being tagged with YFP (Pamp and Tolker-Nielsen, 2007). (D) Contribution of exopolysaccharide alginate (alg) in *P. aeruginosa* biofilm development, WT, Δ*alg8* and Δ*alg8*/pBBR1MCS-5*alg8* being tagged with LIVE/DEAD BacLight kit (Hay *et al.*, 2009).

with the presence of a larger number of flagella (Harshey and Matsuyama, 1994; Köhler *et al.*, 2000), and is mostly, but not exclusively, associated with chemotaxis, which subsequently induces reversal of the flagellar motor (Verstraeten *et al.*, 2008). However, under certain conditions, flagella are dispensable for biofilm formation.

Flagella were initially shown to play a critical role in biofilm development in studies aiming to identify the genetic determinants of biofilm formation. Many studies have identified mutants with defects in the initiation of biofilm formation on an abiotic surface that were affected in flagellum-mediated motility. Such mutants have been identified for *P. aeruginosa* (O'Toole and Kolter, 1998) and *E. coli* (Pratt and Kolter, 1998), with the best biofilm-formers (strains ATCC 25,404 and MG1655) displaying the highest motility, whereas the worst biofilm former (strain BW25113) is motility-impaired (Wood *et al.*, 2006). The contribution of motility to biofilm formation has also been shown to differ between strains in *V. cholerae* (Watnick and Kolter, 1999; Watnick *et al.*, 2001). However, nutrient availability may affect the involvement of the flagella in biofilm formation. The role of the flagellum in initial attachment and biofilm structuring (O'Toole and Kolter, 1998) depends on environmental conditions, as flagella play no role in attachment during *P. aeruginosa* biofilm formation in citrate minimal medium (Klausen *et al.*, 2003b).

In *P. aeruginosa* PA14, the opposite patterns of flagellum biosynthesis (Tart *et al.*, 2006) and motility-dependent phenotype (Caiazza *et al.*, 2007), together with exopolysaccharide synthesis, determine whether the bacteria form biofilms or adopt swarming behaviour. A similar pattern of opposing effects of flagellum-dependent motility and exopolysaccharide production has also been observed in *V. cholerae* O139 (Watnick *et al.*, 2001). This strongly suggests that flagella are involved in the initial adhesion stage of biofilm development (O'Toole and Kolter, 1998), but cease to be required once the biofilm reaches the microcolony formation stage (Whiteley *et al.*, 2001). This role of flagella is clearly dispensable in *E. coli* strains, in particular, in the presence of other robust systems, such as curli (Prigent-Combaret *et al.*, 2000), conjugative pili (Ghigo, 2001; Reisner *et al.*, 2003), exopolysaccharide PGA (Wang *et al.*, 2004) and aggregative adhesion fimbriae for enteroaggregative *E. coli* strains (Sheikh *et al.*, 2001). The contribution to biofilms of motor reversal, leading to swarming behaviour, has been studied in *P. aeruginosa*. Mutations affecting the MotAB or the MotCD stator abolish biofilm formation in both static and flow conditions, consistent with an important role for these stators in early biofilm formation (Toutain *et al.*, 2007). Both MotAB and MotCD are involved in the initial polar attachment of the bacterial cell to the surface (reversible attachment). The MotAB stator is also involved in subsequent adhesion, leading to irreversible attachment, whereas the MotCD stator is involved in swarming. A high rate of flagellar motor reversals in high-viscosity conditions coupled with a hyperswarming phenotype in a *sadB* mutant is associated with defective biofilm formation (Caiazza *et al.*, 2007). SadB-dependent control of flagellar motor reversals on reversible polar attachment to a solid surface might decrease rotation about the cell pole, thereby increasing the time of interaction between the bacterium and its substrate and promoting biofilm formation. *E. coli* strains locked in the 'tumbling' chemotaxis mode by mutation are less able to attach to an abiotic surface than the wild type or mutants locked in the 'running' mode. This suggests that a combination of clockwise and counter-clockwise flagellar rotations and the coupling of these rotations are important in cell attachment (McClaine and Ford, 2002).

In terms of the initial contribution to attachment, once anchorage has occurred, motility genes, including flagellum genes in particular, have been reported to be repressed (Prigent-Combaret and Lejeune, 1999; Whiteley *et al.*, 2001) in *E. coli* and *P. aeruginosa*. In *Vibrio parahaemolyticus*, in which polar and lateral flagella are responsible for swimming and swarming motilities, respectively (Merino *et al.*, 2006), the polar flagellum is important for motility in liquid media, but, after host attachment, lateral flagella are induced, linking the bacteria to the surface and contributing to microcolony formation through firm bacterial adhesion (Belas and Colwell, 1982; Kirov *et al.*, 2002). However, biofilm development in polar flagellar mutants is arrested at the monolayer stage and blocked before three-dimensional expansion (Enos-Berlage *et al.*, 2005), suggesting a role for polar flagellum-dependent motility in later stages of biofilm development in *V. parahaemolyticus*. Similar findings have been reported for *P. aeruginosa* (Klausen *et al.*, 2003a). However, the possibility that disruption of the polar flagellar system in *V. cholerae* affects the regulation of other gene systems related to biofilm development, such as those encoding lateral flagella or extracellular polysaccharide, cannot be ruled out (Watnick *et al.*, 2001).

In addition to playing a role in biofilm initiation, the flagellum and flagellum-dependent motility enable bacteria to leave the community. Nutrient-induced biofilm dispersion has been associated with an increase in *fliC* expression and a corresponding decrease in *pilA* expression in dispersed cells (Sauer *et al.*, 2004). During the development of a *P. aeruginosa* biofilm in rich medium, local dispersion – known as seeding dispersal – is observed, in the form of the hollowing out of some microcolonies, with a wall-forming subpopulation of non-motile cells present on the outside of the microcolonies and a rapidly moving motile subpopulation present within the microcolonies (Sauer *et al.*, 2002; Purevdorj-Gage *et al.*, 2005).

Retractile pili

Type IV pili (TFP) are formed by the oligomerization of type IV pilins by a dedicated machine. The resulting long, flexible organelles emanate from the surfaces of many Gram-negative bacteria and some Gram-positive bacteria and therefore represent the most widespread organ of bacterial attachment (Mattick, 2002). There are two subtypes of Type IV pilins, Type IVa and IVb pilins, which differ in the length of their leader peptides, with Type IVa (TFPa) pilins having a short leader peptide and Type IVb (TFPb) pilins having a longer leader peptide (Craig and Li, 2008; Giraud *et al.*, 2010). Both subtypes are associated with twitching motility, mediated by retraction of these organelles (Craig and Li, 2008; Pelicic, 2008). This movement is colonial in nature and usually involves cell-cell contact. Assembly and retraction are made possible by the presence of two independent ATPases. This property is common to TFPa and TFPb, except for the Flp TFPb subfamily, in which TadA, the only ATPase present, is dedicated to pilus elongation (see the next section). Twitching motility, which can be defined as bacterial translocation over solid or semi-solid surfaces, involves the extension, tethering and subsequent retraction of TFPa. This movement has been well characterized for TFPa, and can bring cells together into complex structures, such as biofilms, under conditions of nutrient depletion. It may promote rapid exploratory movements of colonies over new surfaces under nutrient-rich conditions. These pili are found in many human pathogens and their contribution to multicellular behaviour has been studied in detail in *Neisseria gonorrhoeae, Neisseria meningitidis, P. aeruginosa,* enterohaemorrhagic *E. coli* (Xicohtencatl-Cortes *et al.*, 2009), *Francisella tularensis* (Proft and Baker, 2009), *Legionella pneumophilia* (Coil and Anné, 2009), *V. parahaemolyticus* (Shime-Hattori *et al.*, 2006) and *Haemophilus influenzae* (Jurcisek *et al.*, 2007). However, under certain conditions, TFP are not required for biofilm formation (Klausen *et al.*, 2003a). TFP have both adhesive and motile properties, and the relative contributions of these different properties to bacterial biofilm initiation and structuration are difficult to separate clearly.

Role of TFP adhesion

TFP seem to attach to inert surfaces by relatively non-specific adhesion at their tip (Skerker and Berg, 2001). The contribution of TFPa to adhesion and structuring of the biofilm has been extensively studied in *P. aeruginosa*. The C-terminal *P. aeruginosa* PilA receptor-binding domain (residues 128–144 of the PilA protein) exposed at the tip of TFPa mediates direct binding to both biotic [through recognition of βGalNAc(1–4)βGal, the minimal structural carbohydrate receptor sequence of asialo-GM1 and asialo-GM2 glycolipids] and abiotic surfaces (Lee *et al.*, 1994; Sheth *et al.*, 1994; Giltner *et al.*, 2006). A post-translational modification, such as glycosylation of the PilA pilin in *P. aeruginosa* 1244, greatly reduces the hydrophobicity of these fibres, but does not alter the structure of the biofilm formed by this strain (Smedley *et al.*, 2005). A non-typeable *H. influenzae* (NHTI) *pilA* mutant forms a heterogeneous layer of biomass about half as thick as that formed by the parental isolate. This suggests a role for NTHI TFP in stabilizing the structure of the biofilm, possibly through interbacterial associations (Jurcisek *et al.*, 2007), which may be facilitated through the shedding of TFP within the biofilm matrix, followed by binding to double-stranded DNA (Jurcisek and Bakaletz, 2007). DNA binding to the intact Type IVa pilus structure of *P. aeruginosa* also strongly suggests that TFP-DNA binding function plays a role in biofilm formation (van Schaik *et al.*, 2005).

Moraxella catarrhalis pili are essential for biofilm maturation, as biofilm formation has been shown to be delayed in a *pilA* mutant, resulting in a less robust biofilm phenotype. This finding highlights a possible direct role for TFP production in the regulation of biofilm structure and depth, although the loss of TFP does not completely abolish the formation of biofilms by *M. catarrhalis* (Luke *et al.*, 2007).

V. parahaemolyticus contains two types of TFP: ChiRP and MSHA pili. The MSHA pili are involved in the attachment of *V. parahaemolyticus* to the support, whereas ChiRP pili contribute to bacterium–bacterium interactions during biofilm formation. These two types of pilus are both produced during biofilm formation, but seem to have different functions in different steps (Shime-Hattori *et al.*, 2006).

The TFPb bundle pili mediate the autoaggregation of cells to form discrete three-dimensional microcolonies. Regardless of TFP-dependent motility, a discrete segment within the D-region of the *V. cholerae* TcpA TFPb pilin has been identified as involved in pilus–pilus interactions (Kirn *et al.*, 2000). This is also the case for another TFPb from EPEC known to facilitate autoaggregation and microcolony formation (Girón *et al.*, 1991; Bieber *et al.*, 1998; Knutton *et al.*, 1999). The shufflon protein in the TFPb of *Salmonella enterica* serovar Typhi contributes to aggregate dispersion and further dissemination by inhibiting the bacterial self-association mediated by bundled TFPb (Morris *et al.*, 2003).

Role of TFP motility

Klausen *et al.* (2003a) showed that TFPa are not required for the initial attachment of *P. aeruginosa*, in confocal microscopy studies and COMSTAT image analysis in flow chambers. However, it has been shown in surface plasmon resonance (SPR) studies that *P. aeruginosa* mutants either lacking pili (*pilA*) or carrying an excess of non-retractable pili (*pilT*) have binding properties different from those of the wild type (Jenkins *et al.*, 2005). The possible role of TFP in later stages of biofilm development has also been studied. Mushroom-shaped structures are formed in wild-type *P. aeruginosa* biofilms. The stalk is formed by the proliferation of bacteria with downregulated twitching motility, and the cap is formed by bacteria that climb the microcolony stalks using TFPa, aggregating at the top (Klausen *et al.*, 2003b). However, there is also evidence to suggest that TFPa are involved in the formation of mature multicellular structures in *P. aeruginosa* biofilms and that this process involves TFP-dependent adhesion rather than TFP-dependent motility (Barken *et al.*, 2008). In *N. meningitidis*, the TFP-dependent twitching motility mediated by PilX is correlated with the formation of microcolonies within the biofilm in unencapsulated meningococci (Lappann *et al.*, 2006), whereas no such correlation is observed for autoaggregation.

In *Clostridium perfringens*, a Gram-positive bacterium, strains with impaired motility due to mutations in genes encoding TFP-associated functions generate biofilms with significantly lower bacterial densities and lower levels of intercellular matrix material than wild-type strains. However, pili are shed from the bacterial surfaces within the biofilm (Varga *et al.*, 2008).

TFP-dependent motility may also play a role in multispecies microbial associations. In mixed biofilms formed by *P. aeruginosa* and *Agrobacterium tumefasciens*, *P. aeruginosa* tends to dominate the biofilm, 'blanketing' or burying immature *A. tumefasciens* microcolonies via TFPa-dependent and flagellum-dependent motilities. *A. tumefasciens* uses its flagellar motility to emigrate from coculture biofilms (An *et al.*, 2006).

Appendages not involved in motility

FLP TFPb

As described above, the Type IVb pilin family includes an additional, clearly monophyletic Flp prepilin subfamily that is present in many bacterial and archaeal species (Tomich *et al.*, 2007), including human pathogens such as *Aggregatibacter actinomycetemcomitans* (previously *Actinobacillus actinomycetemcomitans*), *Yersinia* spp., *Bordetella pertussis*, *P. aeruginosa*, *Corynebacterium diphtheriae* and *Mycobacterium tuberculosis* (Planet *et al.*, 2003; Tomich *et al.*, 2007). Flp pili, including those of *A. actinomycetemcomitans*, are assembled as bundles of long thin fibrils (Kachlany *et al.*, 2000, 2001; de Bentzmann *et al.*, 2006). Flp retraction has not been described, and no retraction ATPase has yet been identified. This TFPb Flp is linked to the *tad* (tight adherence) locus and was initially identified in *A. actinomycetemcomitans* clinical isolates of which were found to form rough colonies on solid medium and extremely tenacious biofilms in liquid medium (Rosan *et al.*, 1988), in a non-specific manner (Kachlany *et al.*, 2000). Non-adherent mutants generated in a transposon library of *A. actinomycetemcomitans* CU1000N (Kachlany *et al.*, 2001; Planet *et al.*, 2003) or by directed mutagenesis (Perez *et al.*, 2006) are defective in rough colony formation, autoaggregation, adhesion, the production of Flp pili and the formation of a tenacious biofilm. In *P. aeruginosa*, the Flp pili are required for biofilm formation, particularly during the clustering process occurring early in microcolony formation, on abiotic and epithelial cell surfaces, regardless of the motility, of any type, associated with TFPa (de Bentzmann *et al.*, 2006; Bernard *et al.*, 2009). However, a putative post-translational modification of the Flp pilin affects specific bacterium-host tissue interactions (Bernard *et al.*, 2009). In *Haemophilus ducreyi*, inactivation of any one of several genes in the *tad* gene cluster greatly decreases the ability of the bacteria to attach to plastic and to human foreskin fibroblasts (HFF), and results in a lack of microcolony formation in cocultures with HFF (Nika *et al.*, 2002). The expression of these genes is therefore required for full virulence of *H. ducreyi* in humans (Spinola *et al.*, 2003).

Chaperone-usher (CU) pili

The Chaperone-usher (CU) pathway is responsible for the synthesis of linear multisubunit pili/fimbriae (Fronzes *et al.*, 2008) in Gram-negative bacteria. It is found principally in the Enterobacteriaceae, in a number of *Pseudomonas, Haemophilus, Bordetella, Xylella, Burkholderia, Acinetobacter* and *Ralstonia* species (Sauer *et al.*, 2004) and in *Yersinia pestis* (Felek *et al.*, 2007). Non redundant multiple copies of CU clusters are generally present in a given genome, highlighting the diversity of roles and the particular functions these systems may play in bacterial attachment and biofilm formation. CU pili are thought to be responsible for specific host attachment and/or the evasion of host responses (Wright *et al.*, 2007; Zavialov and Knight, 2007), but there is increasing evidence to suggest that they are also involved in non-specific binding to abiotic surfaces (Pratt and Kolter, 1998; Vallet *et al.*, 2001). This is the case for *E. coli* Type 1 fimbriae, in which both the major fimbrial subunit FimA and the tip adhesin FimH are critical for the initial attachment of *E. coli* (Pratt and Kolter, 1998; Beloin *et al.*, 2004). However, because the assembly of CU in *E. coli* occurs with the chaperone-adhesin (FimD-FimH) complex targeted first to the usher and then opens the translocation pore (Saulino *et al.*, 1998; Ng *et al.*, 2004; Remaut *et al.*, 2008), it remains difficult to elucidate the contribution of the tip adhesin FimH to biofilm formation. Type 3 CU genes similar to those present in *Klebsiella pneumoniae* have recently been identified in uropathogenic *E. coli* recovered from patients with catheter-associated urinary tract infection (CAUTI), and the products of these genes have been found to influence biofilm formation by the CAUTI *E. coli* strain MS2027 (Ong *et al.*, 2008). The UPEC strain CFT073 genome contains ten gene loci sharing sequence identity with the CU class of fimbrial genes. However, attention has recently been focused on one of these clusters, cluster F9. F9 fimbriae mediate strong biofilm formation in an *E. coli* K-12 background devoid of Type 1 fimbriae, in the absence of haemagglutination or properties associated with cell adhesion (Ulett *et al.*, 2007). In *P. aeruginosa*, the CU pathway was initially detected in transposon mutants derived from a *P. aeruginosa* PAK strain with no TFPa (PAKNP) that displayed defective biofilm formation (Vallet *et al.*, 2001). Genome mining has also identified several other CU clusters (Giraud *et al.*, 2010), the *cupB* and *cupC* (Vallet *et al.*, 2001), *cupF* and *cupE* (Filloux *et al.*, 2004a) clusters in the PAO1 genome sequence, and an additional cluster, the *cupD* cluster, in the PAP-I pathogenicity island of the PA14 and PA7 genome sequences (He *et al.*, 2004; Mikkelsen *et al.*, 2009; Nicastro *et al.*, 2009). The CupA system is thought to be required for the stable attachment of bacteria to abiotic surfaces (Vallet *et al.*, 2001). It therefore has an impact on biofilm structure independently of, and at an earlier stage than TFPa. The CupA system is involved in the establishment and physical structure of the exopolysaccharide

Pel-dependent pellicle (Friedman and Kolter, 2004a). Overproduction of this system results in the auto-aggregative SCV (small colony variant) morphotype (Häussler, 2004) observed in strains infecting cystic fibrosis (CF) patients (D'Argenio *et al.*, 2002; Meissner *et al.*, 2007). This morphotype forms biofilms particularly efficiently. The CupB and C systems cooperate in cell–cell interactions and microcolony formation during the biofilm maturation process (Kulasekara *et al.*, 2005; Ruer *et al.*, 2007, 2008). The PvrR-dependent abolition of the auto-aggregative rough SCV of *P. aeruginosa* strain PA14 suggests possible effects on the CupD system (Drenkard and Ausubel, 2002; Häussler, 2004). An additional CU cluster in *P. aeruginosa* (Giraud *et al.*, 2010), *cupE* (Giraud *et al.*, 2011), is very similar to a recently identified *csu* cluster present in *Acinetobacter baumannii* (Tomaras *et al.*, 2003; de Breij *et al.*, 2009). This *csu* locus is required for pilus formation, concomitant attachment to plastic surfaces and the subsequent formation of *A. baumannii* biofilms (Tomaras *et al.*, 2003). Further evidence of the involvement of CU pili in the community lifestyle is provided by *K. pneumoniae*. The lack of Type 3 appendages on the surface of *K. pneumoniae* is associated with poor biofilm formation *in vitro* early and later in development, with the tip adhesin MrkD not required for efficient biofilm formation on plastic surfaces (Langstraat *et al.*, 2001). Indeed, MrkD is essential for attachment and subsequent biofilm formation on surfaces coated with host-derived proteins (Jagnow and Clegg, 2003). MrkF, a minor structural component of *K. pneumoniae* Type 3 fimbriae, appears to be intermittently incorporated into recombinant Type 3 fimbriae, resulting in a shortening of fimbrial length and an increase in autoaggregation and biofilm formation (Huang *et al.*, 2009).

Curli

Curli are the major protein components of a complex ECM produced by many Enterobacteriaceae (Barnhart and Chapman, 2006). They are produced by a dedicated machine, which secretes CsgA, a major structural protein, in an unpolymerized form and CsgB, which is responsible for nucleation to generate a fibre in the extracellular environment.

The contribution of adhesive curli to initial cell-surface interactions and subsequent cell–cell interactions have been described for *E. coli* K-12 (Vidal *et al.*, 1998), and *Salmonella* Typhimurium (Jonas *et al.*, 2007). Curli are important for the attachment phase, in the initial stages of biofilm development. In *E. coli* K-12, strains with curli form a 3D mature biofilm, with characteristic water channels and pillars of bacteria, whereas curli-deficient strains form flat biofilms (Kikuchi *et al.*, 2005). Bacterial curli make an important individual contribution to the initiation and development of biofilms, but curli and cellulose may function synergistically, at least in entero-haemorragic and enteropathogenic *E. coli* strains, thereby favouring host colonization or biofilm development (Saldaña *et al.*, 2009). By contrast, in *E. coli* K-12 MG1655, cellulose seems to interfere with the adhesion mediated by curli (Gualdi *et al.*, 2008).

Conjugative pili

Natural conjugative plasmids promote biofilm development (Ghigo, 2001; Reisner *et al.*, 2006). Conjugative pili play an important role in promoting initial adhesion events and biofilm development in *E. coli* biofilms, and this effect is not restricted to a specific incompatibility group or specific type of pilus (Ghigo, 2001). It has been suggested that transient derepression of conjugative pilus synthesis in transconjugants and their immediate descendants, after an initial horizontal transfer between a small subpopulation of donor bacteria and the surface-attached recipient population, increases the rate of transfer by attaching donor and recipient cells before other factors intervene to promote biofilm cohesion. It has been suggested that conjugative pili may themselves act as cell adhesins, connecting cells and stabilizing biofilm structures in hydrodynamic biofilm systems (Molin and Tolker-Nielsen, 2003). Consistent with this notion, it has been reported that plasmids for which transfer is repressed may induce biofilm formation when the bacterial population present in the biofilm also includes plasmid-free recipient cells (Ghigo, 2001). It has also been suggested that the presence of derepressed IncF plasmids contributes to biofilm maturation and differentiation (Reisner *et al.*, 2003). Furthermore, various conjugative plasmids have been shown to contribute to the shaping of the biofilm (Folkesson *et al.*, 2008).

Pili from Gram-positive bacteria

Unlike Gram-negative bacteria, Gram-positive bacteria assemble pili by a mechanism involving a transpeptidase called sortase, which cross-links individual pilin monomers, ultimately joining the resulting covalent polymer to the cell-wall peptidoglycan (Mandlik *et al.*, 2008; Proft and Baker, 2009). The biological importance of these surface pili in Gram-positive bacteria has only recently been investigated. In group A streptococci (GAS, *Streptococcus pyogenes*), pilus-negative mutants lacking the genes encoding the pilus backbone structural protein or the sortase C1 form cellular aggregates much less efficiently in liquid culture and cannot produce the typical three-dimensional layer with bacterial microcolonies embedded in a matrix of carbohydrate polymers (Manetti *et al.*, 2007). In *Enterococcus faecalis*, the Ebp proteins assembled in Ebp pili at surface of the bacterium are important for biofilm production in the OG1RF strain (Nallapareddy *et al.*, 2006).

The extracellular matrix

The structural integrity of biofilms is critically dependent on the ECM produced by their constituent cells. ECMs are potentially as diverse as the biofilms producing them and make a significant contribution to community organization by embedding the bacteria (Branda *et al.*, 2005). The ECM consists of exopolysaccharide polymers, proteins, nucleic acids, lipids, nutrients and metabolites in a highly hydrated hydrogel (97% water). The genomes of many bacterial species living in biofilms include loci encoding components of the exopolysaccharide biosynthesis pathway and export machines. Exopolysaccharides fill the intercellular space in microbial aggregates and determine the structure and architecture of the biofilm matrix, by stabilizing the 3D structure of the biofilm (Watnick and Kolter, 2000).

Secreted exopolysaccharides

Different bacterial species produce different polymers and a given bacterial species may produce several different exopolysaccharides (Friedman and Kolter, 2004b). Many of these molecules have only recently been identified, their production previously being masked by that of other molecules, or because the conditions in which they are produced and the regulatory networks controlling their expression have only recently been discovered. They have mostly been classified in terms of the acidic, neutral or basic nature of the exopolysaccharide produced. According to this classification, acidic polysaccharides, such as the alginate of *P. aeruginosa* and the K antigens and colanic acid of *E. coli*, may be considered dispensable for biofilm formation and may even be considered to have a negative effect (Hanna *et al.*, 2003; Wozniak *et al.*, 2003; Schembri *et al.*, 2004). Biofilm formation is instead driven by neutral and basic exopolysaccharides, such as the β-1,6-*N*-acetyl-D-glucosamine exopolysaccharides recovered from staphylococci, *E. coli*, *Y. pestis* and actinobacilli (Branda *et al.*, 2005), PEL and PSL from *P. aeruginosa*, the cellulose recovered from many eubacteria (Lasa, 2006) and D-glucans from *Streptococcus mutans* (Munro *et al.*, 1995).

Whereas the same exopolysaccharides are used by diverse bacterial species to produce the biofilm matrix, it is particularly interesting to note that a single bacterial species has the capacity to produce several different extracellular polysaccharides. For example, three exopolysaccharides have been identified in *E. coli* biofilms: β-1,6-*N*-acetyl-D-glucosamine polymer (PGA), colonic acid and cellulose (Beloin *et al.*, 2008). Similarly, *P. aeruginosa* produces at least three exopolysaccharides: PEL, PSL and alginate (Giraud *et al.*, 2010). PGA (also known as PIA) is also produced by Gram-positive bacteria such as *Staphylococcus aureus* (a derivative named PNAG) and *Staphylococcus epidermidis*. It is thus widespread in bacterial biofilms (Otto, 2008). However, PNAG plays fundamentally different structural roles in *S. aureus* and *S. epidermidis* biofilms. This difference is illustrated by the ability of the PNAG-degrading enzyme dispersin B to detach preformed biofilms produced by *S. epidermidis*, whereas this enzyme has no effect on the attachment of preformed *S.*

aureus biofilms (Izano *et al.*, 2008). Deacetylation of polyacetylglucosamine, the PIA component of the exopolysaccharide matrix of *S. epidermidis*, generates a positively charged polymer and is essential for the biological function of this molecule in biofilm formation. PIA is the only exopolysaccharide known to be deacetylated. This deacetylation is mediated by the IcaB protein and leads to biofilm formation, and the non-cationic nature of non-deacetylated polyacetylglucosamine in an isogenic *icaB* mutant strain has been associated with the inability of PIA to attach to the bacterial cell surface and, therefore implicating IcaB in PIA-dependent biofilm formation by *S. epidermidis* (Vuong *et al.*, 2004). The O-acetylation of mannuronate residues of alginate in *P. aeruginosa* also conditions the structure of the biofilm. The acetylation-defective mutant produces a thinner and more homogeneous biofilm, owing to the lower viscosity of the extracellular material produced and weaker exopolymer interactions (Nivens *et al.*, 2001; Tielen *et al.*, 2005). The nature of the exopolysaccharides produced differs between bacterial species. For example, in *V. cholerae* O1 El Tor A1552, the VPS exopolysaccharide produced is a mixture of glucose and galactose (Yildiz and Schoolnik, 1999) with *N*-acetyl glucosamine, mannose and xylose present as minor components, whereas CPSA from *V. parahaemolyticus* is an equal mixture of fucose, galactose and *N*-acetylglucosamine (Enos-Berlage and McCarter, 2000; Enos-Berlage *et al.*, 2005). Most exopolysaccharides have a positive effect on biofilm initiation and/or structuration or associated phenotypes, such as rough colony formation or pellicle formation. However, a number of bacterial exopolysaccharides have been reported to have no effect or, even, a negative effect on biofilms. For example, in *Vibrio fisheri*, the biofilm-dependent and pellicle phenotypes have been attributed to the *syp* locus responsible for cellulose production, but not to the *vps* locus responsible for VPS exopolysaccharide production, which is specific to *V. fisheri* and is not found in *V. cholerae* (Darnell *et al.*, 2008).

The colonic acid polymer produced by *E. coli* may be tightly associated with the cell surface or released into the extracellular medium. It has been reported to mask the Ag43 and AIDA-I adhesins, revealing negative effects of capsular polysaccharides on other actors in biofilm structuration. This phenomenon is common within a range of Gram-negative bacteria including *K. pneumoniae*, *Enterobacter cloacae*, *S. liquefaciens*, *Burkholderia cepacia*, and *P. aeruginosa* (Schembri *et al.*, 2004).

Cell-anchored polysaccharides

In addition to secreted polysaccharides, many bacteria produce cell-anchored polysaccharides. These polysaccharides include capsule polysaccharides and the lipopolysaccharides present in Gram-negative bacteria. Owing to their proximity to the bacterial cell surface, capsule polysaccharides are often described as forming a shield covering other cell-surface adhesins, with negative effects on biofilm formation (Joseph and Wright, 2004; Enos-Berlage *et al.*, 2005). However, although the capsule of *K. pneumoniae* contributes to the correct spatial distribution of adhering bacteria on the substrate and modulates bacterial aggregation, thus acting as a shield, it also actively shapes the typical three-dimensional organization of the mature *K. pneumoniae* biofilm (Balestrino *et al.*, 2008). *V. cholerae* O139 contains a locus with genes for CPS and lipopolysaccharide (LPS) O antigen biosynthesis, which also plays a negative role in biofilm formation (Kierek and Watnick, 2003). Loss of the genes encoding components of the O antigen biosynthesis pathway in *V. parahaemolyticus* increases attachment (Enos-Berlage *et al.*, 2005). Several lines of evidence suggest that lipopolysaccharide (LPS) may play a role in *E. coli* biofilm formation. The production of a truncated form of LPS is linked to the abolition of biofilm formation in an uropathogenic strain (Beloin *et al.*, 2006). Furthermore, the oxygen deprivation-dependent overproduction of LPS results in an increase in the overall negative electrical charge on the cell envelope, increasing the repulsion between the surface of the cell and negatively charged surfaces, thereby further decreasing biofilm formation in *E. coli* W3100 (Landini and Zehnder, 2002). Decreasing LPS production has also been shown to facilitate biofilm formation (Beloin *et al.*, 2006). Remarkable findings concerning the role of capsule polysaccharides have also been reported for the uropathogenic strain CFT073 and all *E. coli* strains with group II capsules. These strains release a soluble polysaccharide into their environment, which induces changes in the physicochemical properties of the surface, preventing a wide range of Gram-positive and Gram-negative bacteria from forming a biofilm; this soluble polysaccharide may therefore be considered a potential bacterial interference molecule for preventing adhesion (Valle *et al.*, 2006).

The nature of the exopolysaccharides produced conditions the shape of the biofilm formed. This has been demonstrated for *Pseudomonas putida*, a bacterial species producing multiple exopolysaccharides. Alginate, a mannuronic-guluronic acid-containing exopolysaccharide produced in conditions of water limitation, is associated with the production of a thicker biofilm, with a thicker exopolysaccharide layer at the interface with the air, covering a smaller surface area than the biofilm formed by the isogenic *algD* mutant (Chang *et al.*, 2007). This exopolysaccharide seems to play a greater role than the others produced by *P. putida* in limiting the dehydration stress of the bacterial community. Thus, alginate-producing species may have a competitive advantage in environments in which water is limiting.

DNA

Extracellular DNA (eDNA) has been recovered from many biofilms. The release of eDNA into the biofilm is probably triggered by the lysis of a subpopulation of bacteria. Alternatively, it may occur through a lysis-independent mechanism involving the release of small vesicles, in conjunction with exopolysaccharide biosynthesis (Whitchurch *et al.*, 2002) or through a specific mechanism. Exogenous H_2O_2 has recently been shown to induce the release of DNA from *Streptococcus sanguinis* and *Streptococcus gordonii* with functionally intact walls and in which the ribosomes are retained over time (Kreth *et al.*, 2009). It has been suggested that the release of DNA in the absence of detectable autolysis is an adaptation to the competitive oral biofilm environment, facilitating exchanges of genetic material between competent cells in conditions in which autolysis would create open spaces for competitors to invade. The association of membrane vesicles with DNA within biofilms certainly affects the interaction between matrix components (Schooling *et al.*, 2009). Furthermore, actively produced eDNA has been reported to function as a spatial structure, forming a filamentous network, in an aquatic bacterium F8 (Böckelmann *et al.*, 2006). This structure may protect newly formed biofilms in this species. It seems to be required for biofilm initiation in several other species, including *P. aeruginosa* (Whitchurch *et al.*, 2002). During the two-step process of biofilm formation in *E. faecalis*, the initial attachment phase is DNA-independent, and the autolysin-dependent release of DNA during the accumulative stage is crucial for the growth, maturation, and structural stability of *E. faecalis* biofilms (Guiton *et al.*, 2009). In *S. aureus*, eDNA plays a structural role in all stages of biofilm development, in a gene-dependent manner involving bacterial programmed cell death and *cidA*-mediated lysis (Rice *et al.*, 2007), whereas the staphylococcal thermonuclease actively degrades the eDNA, possibly to promote biofilm dispersal (Mann *et al.*, 2009).

Other matrix components

P. aeruginosa rapidly kills neutrophils, resulting in the release of F-actin and neutrophilic DNA. These released components form a mesh to which *P. aeruginosa* cells can specifically attach by sticking to the F-actin, thereby improving *P. aeruginosa* biofilm formation (Walker *et al.*, 2005).

Secreted factors

Several different types of molecules are secreted, including adhesins, which are associated with the surface, and proteins transported into the ECM by an active mechanism. The adhesins of the Type V secretion pathway of Gram-negative bacteria and MSCRAMM from Gram-positive bacteria, in particular, have been implicated in adhesion and the development and structuration of biofilms.

Type V adhesins: autotransporters (ATs) and TPS

AT are proteins secreted via the Type Va secretion system (also referred to as the AT pathway) in Gram-negative bacteria (Desvaux *et al.*, 2004; Henderson *et al.*, 2004; Wells *et al.*, 2007). AT proteins are characterized by the presence of three distinct domains: an N-terminal signal sequence that mediates export of the protein across the cytoplasmic membrane, a surface-located 'mature' protein (the passenger or α-domain) and a carboxy-terminal domain (the translocation or β-domain) that facilitates secretion of the passenger domain through the outer membrane. The passenger domain, which may be an adhesin, is secreted as a large monomer. Ag43 (for antigen 43), which has been identified in most commensal and pathogenic *E. coli* and is distributed over the entire surface of the cell (Henderson *et al.*, 1997), mediates cell–cell association. The autoaggregation displayed by Ag43 is also characteristic of several other AT proteins, including AIDA-I, TibA from *E. coli* and Hap from *H. influenzae* (Hendrixson and St Geme, 1998; Sherlock *et al.*, 2004, 2005). Ag43 is thus a critical enhancer for microcolony formation and biofilm formation (Danese *et al.*, 2000; Kjaergaard *et al.*, 2000). A considerable and increasing number of other adhesins from the AT subfamily have been shown to play a role in biofilm formation in *E. coli* (Torres *et al.*, 2002; Sherlock *et al.*, 2004, 2005; Roux *et al.*, 2005; Valle *et al.*, 2008; Wells *et al.*, 2008, 2009; Herold *et al.*, 2009) and *Y. pestis* (Felek *et al.*, 2008). Much of the diverse adhesin repertoire of *E. coli* remains to be explored (Beloin *et al.*, 2008).

The Type Vb secretion system is TPS system consisting of the TpsA and TpsB proteins. The transporter, TpsB, is embedded in the outer membrane and serves as a dedicated transporter for TpsA. Many TpsA proteins, such as HMW1, HMW2, FHA, LspA1 and LspA2, from *H. influenzae, B. pertussis,* and *H. ducreyi*, respectively (Jacob-Dubuisson *et al.*, 2004) act as adhesins. HrpA, the TpsA of *N. meningitidis*, has recently been reported to be involved in maintaining biofilm structure, possibly through interactions between bacteria and independently of any involvement in interactions between bacteria and eukaryotic cells (Neil and Apicella, 2009). FHA from *Bordetella bronchiseptica* plays a key role in the formation of the strong biofilm phenotype during the Bvg^+ and Bvg^i phases, but not during Bvg^- phase. This action is strongly associated with fimbriae and probably promotes attachment to surfaces (Irie *et al.*, 2004).

P. aeruginosa has at least six identified TpsA substrates (Filloux *et al.*, 2004b). No clear role has been identified for the TpsA6/CupB5 protein in *P. aeruginosa* biofilm formation, although several of the characteristics of this protein are consistent with a role in adhesion and structuring of the biofilm (Ruer *et al.*, 2008; Giraud *et al.*, 2010). Potential involvement in *P. aeruginosa* adhesion or biofilm formation and structuring has not been evaluated for any of the other TPS cargoes.

MSCRAMM adhesins

Microbial surface components recognizing adhesive matrix molecules (MSCRAMM adhesins) are surface adhesins of Gram-positive bacteria that specifically recognize human ECM proteins. These proteins are linked to the peptidoglycan in a covalent sortase-dependent manner or non-covalently and have an exposed binding domain that may display affinity for several matrix molecules. These molecules have been extensively described in *S. aureus* (Clarke and Foster, 2006). Owing to their specificity of binding to the ECM, the role of these molecules in biofilm formation by Gram-positive bacteria is probably limited to the initial stages of biofilm formation, up to the ECM preconditioning of surfaces (Otto, 2008). However, in *E. faecium*, the SgrA LPXTG protein and EcbA, two novel surface-exposed adhesins, are not only involved in adhesion to components of the ECM, but also play specific roles in biofilm formation. SgrA mediates the binding of *E. faecium* cells to its ligands in the ECM, and is involved in biofilm formation (Hendrickx *et al.*, 2009).

Surface-associated proteins

Bap from *S. aureus*, Esp from *E. faecalis* and BapA from *Salmonella enterica* ser. Enteritidis, three archetypal biofilm-associated proteins (BAP), have been recovered from the surface of Gram-negative or Gram-positive bacteria (Latasa *et al.*, 2006). These proteins have several structural features in common, including the presence of a core region consisting of 13 successive repeats of 86 amino acids predicted to fold into a seven-strand sandwich forming an immunoglobulin-like fold involved in cell adhesion. Bap is involved not only in the primary attachment step, but also, together with the polysaccharide PIA/PNAG, in cell-to cell aggregation and, thus, biofilm maturation (Cucarella *et al.*, 2001). Through its homophilic interactions, Bap hinders interactions between bacterial MSCRAMM receptors and host proteins (Cucarella *et al.*, 2002).

Type VI secretion

The production of components of the recently described Type VI secretion system (T6SS), including the hexameric Hcp nanotube-forming protein (Mougous *et al.*, 2006; Ballister *et al.*, 2008) in particular, has been reported to be systematically associated with the development of a biofilm community lifestyle in *Burkholderia cenocepacia* K56-2 (Aubert *et al.*, 2008) for example. This association is supported by transcriptional (Wagner *et al.*, 2004) and translational (Sauer *et al.*, 2002; Waite *et al.*, 2006; Anderson and O'Toole, 2008) evidence in *P. aeruginosa*. Hcp proteins have also been shown to play a role in cohesive biofilm development in *V. parahaemolyticus* (Enos-Berlage *et al.*, 2005) and in enteroaggregative *E. coli* (Aschtgen *et al.*, 2008). These findings suggest that the T6SS machine may be involved in the adhesive and/or cohesive properties of biofilms, either directly, owing to its own adhesive properties, or through the secretion of an unidentified molecule that binds the community. The systematic expression of t6ss clusters with other elements indispensable for biofilm development (e.g. cable pili, adhesins) suggests that the products of these genes may cooperate in the structuring of the biofilm, or the existence of a coregulation with CU fimbriae (de Pace *et al.*, 2010).

Soluble factors

The extraction of soluble factors from the complex insoluble matrix surrounding bacterial communities is not easy. This section thus deals with an emerging field of research that has not yet been approached in many bacterial species.

Lectins are multivalent soluble proteins that bind specific glycans and are specific to pseudomonads. The role of lectins depends strongly on their distribution. Most of the activity of these proteins is stored within the cells and is released by cell lysis (Glick and Garber, 1983). However, small but significant fractions of activity reach the cytoplasmic membrane, the periplasmic space and the outer membrane, through an unknown process (Glick and Garber, 1983; Tielker *et al.*, 2005). *P. aeruginosa* lectins increase bacterial cell aggregation and are therefore probably involved in biofilm structuring. Their multivalent nature makes it possible to develop a network of polysaccharides (Imberty *et al.*, 2004). PA-IL has been shown to affect the area covered by a biofilm on various substrates, in both static and dynamic assays, and to be mostly involved in maturation, rather than in initial attachment to the substrate (Diggle *et al.*, 2006). PA-IIL has also been implicated in the coverage of surfaces by biofilms (Tielker *et al.*, 2005), and it has recently been shown that multivalent fucosyl dendrimers competing with fucose-PA-IIL ligand can disperse *P. aeruginosa* biofilms (Johansson *et al.*, 2008). It has been suggested that this *P. aeruginosa* lectin-dependent cell–cell interaction may connect bacteria of different species within the community.

Molecules with catalytic activity against exopolysaccharides promote dispersion. Dispersin, DspB, an *N*-acetyl-glucosaminidase that degrades the *N*-acetylglucosamine-containing extracellular polysaccharide of biofilms formed by the periodontal pathogen *A. actinomycetemcomitans*, causes the detachment of *A. actinomycetemcomitans* cells and the dispersion of the biofilm (Kaplan *et al.*, 2003). It also rapidly and efficiently removes *S. epidermidis* from surfaces (Kaplan *et al.*, 2004). The *P. aeruginosa* neuraminidase appears to be important for biofilm production. However, its precise role remains unclear, particularly because of the linking of its activity to the cleavage of 2,3-linked sialic acids from glycoconjugates (Soong *et al.*, 2006).

In *P. aeruginosa*, the AT EstA, which is an esterase rather than an adhesin, acts through rhamnolipid production and cellular motility: swimming, twitching and swarming (Wilhelm *et al.*, 2007). Thus, *estA* mutants have a biofilm phenotype different from that of the wild type. The biofilms formed by *estA* mutants cover a significantly smaller surface area and consist of large aggregates of cells separated by regions of extremely low cell density.

Proteases, such as the gingipains in *Porphyromonas gingivalis,* clearly participate in the development of the biofilm and its dispersion

(Kuboniwa *et al.*, 2009). Two of these cell-associated cysteine proteinases, specific for arginine (Rgp) and lysine (Kgp), have been implicated in *P. gingivalis* biofilm development. Kgp prevents microcolony expansion, whereas Rgp mediates transverse enlargement and restricts longitudinal extension (Kubowina *et al.*, 2009). A loss of proteolytic activity has been associated with a significant enhancement of biofilm formation and a strongly attenuated detachment phenotype in *S. aureus*, whereas an increase in serine protease levels has been detected in effluents from detaching biofilms (Boles and Horswill, 2008).

Toxin–antitoxin (TA) pairs consist of a stable toxin and a labile antitoxin encoded by pairs of genes found on the chromosomes of many bacteria and on low-copy number plasmids. It has been suggested that toxin–antitoxin pairs are involved in defence against phages (Pecota and Wood, 1996). However, TA pairs also seem to have other effects on biofilm formation, as 'genomic junk', in growth rate control, programmed cell death and persister formation (Magnuson, 2007). TA pairs are widespread among bacteria, being found in both Gram-positive and Gram-negative species. TA-mediated programmed cell death and the subsequent regulated lysis of bacterial cells provide enough genomic DNA to ensure that the biofilm remains stable (Bayles, 2007; Wood, 2009). The biofilms produced by the *S. aureus cidA* mutant are less compact and less adherent than those of the wild type, because of the smaller amounts of DNA released by these cells rather than differences in initial attachment (Rice *et al.*, 2007). An equivalent mechanism has been identified in *S. epidermidis*, in which AtlE-mediated cell lysis promotes surface attachment and stabilizes established biofilms (Qin *et al.*, 2007). In *E. coli* MG1655, the five TA systems control biofilm formation. The deletion of all five TA systems impairs early stages of biofilm production by repressing the generation of Type 1 fimbriae, but increases the amount of mature biofilm by decreasing biofilm dispersal (Kim *et al.*, 2009). The other TA pair in *E. coli*, HhA-TomB, promotes biofilm dispersion through controlled cell lysis, again illustrating the particular role of these systems in *E. coli* biofilm physiology (García-Contreras *et al.*, 2008). The TA systems that fine-tune the regulation of programmed cell death are clearly involved in biofilm formation and dispersion.

Non-protein factors

In addition to acting as a chemical signal driving the coordinate response of the biofilm population (Hooshangi and Bentley, 2008; Senadheera and Cvitkovitch, 2008), acyl-HSLs in single-species biofilms have also been shown to play an important role in cementing the cells together. In *P. aeruginosa*, acyl-homoserine lactones (HSLs) define the separations between bacterial pillars in the three-dimensional structure of the biofilm (Davies *et al.*, 1998). *P. aeruginosa* mutants that do not produce acyl-HSL, form biofilms with closely packed cells and are easily disrupted by sodium dodecyl sulfate. Thus, extracellular signals enforce the zoning laws in single-species biofilms.

Growing *P. aeruginosa* cells produce *cis*-2-decenoic acid, an organic compound capable of inducing the dispersion of established biofilms and inhibiting biofilm development. This molecule has also been shown to induce the dispersion of biofilms formed by *E. coli*, *K. pneumoniae*, *Proteus mirabilis*, *S. pyogenes*, *Bacillus subtilis*, *S. aureus*, and the yeast *Candida albicans* (Davies and Marques, 2009).

Rhamnolipids are amphipatic molecules produced by *P. aeruginosa* (Soberón-Chávez *et al.*, 2005), and other pseudomonads. They function as biosurfactants, interfering with cell–cell and cell-substrate interactions by disrupting reversible adhesion, irreversible attachment and influencing the development of biofilm structure. Rhamnolipids play several roles in biofilm development and shaping, despite being synthesized in biofilms only between stalk formation and capping of the mushroom-like structures (Lequette and Greenberg, 2005). They are thought to be critical for the initiation and maintenance of biofilm architecture, including, in particular, the development of 3D structures. They are also important for dipersion. Biosurfactants promote microcolony formation during the initial phase and facilitate subsequent migration-dependent structural development (Pamp and Tolker-Nielsen, 2007). Rhamnolipids also play a key role in keeping channels open, by affecting cell–cell interactions and the attachment of bacterial cells to surfaces at late stages of biofilm development (Davey *et al.*, 2003). The exogenous addition of rhamnolipids and the induction of rhamnolipid gene expression promote detachment in wild-type *P. aeruginosa* biofilms (Boles *et al.*, 2005). The detachment function of rhamnolipids results in a central hollowing pattern. These molecules may act directly on the biofilm matrix, disrupting and solubilizing components and possibly even incorporating the matrix into micelles. They must also disrupt cell surface structures that act as adhesion factors. Interestingly, rhamnolipids secreted by *P. aeruginosa* also disrupt *B. bronchiseptica* biofilms (Irie *et al.*, 2005) and impede further *P. fluorescens* and *E. coli* establishment (Davey *et al.*, 2003).

Biofilm diversity

The concept of 'biofilm diversity' takes into account both the selection of subpopulations with different phenotypic traits, such as metabolic activity, antibiotic tolerance and spore formation, in biofilms formed by a single clonal population and the various bacterial species present in multispecies biofilms. In both situations, there is a convergence towards the selection of the fittest subpopulations.

Genotypic evolution

In various bacterial species, biofilm growth can lead to the generation of genetic diversity in microbial populations. This population diversity results from the occurrence of double-stranded DNA breaks due to endogenous oxidative stress in some cells within biofilms. The repair of these breaks by a mutagenic mechanism involving recombinatorial DNA repair genes then generates genetic variants. The mutations produced promote the adaptation of biofilm communities to changing conditions, in addition to generating diversity (Boles and Singh, 2008). Biofilms are highly competitive and crowded environments in which the nutrient supply is limited and microevolutionary processes, such as mutation selection, occur (Conibear *et al.*, 2009). These processes are important for the growth of the biofilm and for the management of oxidative stress. Indeed, oxidative stress is undoubtedly the cause of many mutations in bacterial biofilms (Allegrucci and Sauer, 2008; Boles and Singh, 2008). The mutation frequency of bacterial cells in biofilms is six to 100 times higher than that in planktonic cells, and the *in situ* mutation rate is heterogeneous within biofilms, with mutations preferentially occurring in forming microcolonies at an early stage of clonal development (Conibear *et al.*, 2009). Thus, in *P. aeruginosa* at least, microcolonies constitute specific foci of genetic adaptation and evolution, leading to the selection of subpopulations with a greater ability to acquire antibiotic resistance (Blázquez, 2003; Maciá *et al.*, 2005), the horizontal acquisition of exogenous DNA (Townsend *et al.*, 2003) and the dispersal of new genetic variants. The cells selected are dividing sessile cells, rather than planktonic cells, highlighting the occurrence of specific DNA-damage stresses at this particular stage of biofilm development.

The small colony variant (SCV) phenotype is a typical phenotype, probably resulting from adaptive mutations and selection (Häussler, 2004), observed in biofilm-associated infections. It is associated with high levels of antibiotic resistance, autoaggregative behaviour and an enhanced ability to form biofilms. The SCV phenotype has been described in several bacterial species, including *P. aeruginosa* (von Götz *et al.*, 2004), *E. coli* (Roggenkamp *et al.*, 1998) and *S. aureus* (Chatterjee *et al.*, 2007). However, the mechanisms influencing SCV occurrence and reversion remain unknown. Certain features of the SCV phenotype, such as slow growth and metabolic changes, favour its appearance and maintenance in biofilms. SCV selection also confers resistance to other bacteria, as reported for *S. aureus* SCV selection, which results in resistance to the exotoxins secreted by *Pseudomonas* (Biswas *et al.*, 2009). The SCV phenotype and phage production are linked, because Pf4 mediates phenotypic variation in the *P. aeruginosa* biofilm (Webb *et al.*, 2004) and the absence of phage Pf4 from the PAO1 *P. aeruginosa* genome abolishes SCV shedding during dispersion (Rice *et al.*, 2009), consistent with a role for phages in the generation of genetic diversity in the biofilm (Whiteley *et al.*, 2001). However, the SCV variants of the PA14 strain formed under static conditions are not linked to Pf5 phage activity (Mooij *et al.*, 2007).

Genetic exchange

Biofilms containing distantly related microorganisms are ideal environments for horizontal gene transfer because they sustain high densities of bacteria and high levels of metabolic activity even in the harshest environments; they therefore provide a unique environment for microbial evolution (Ochman *et al.*, 2000). They provide a stable physical environment for cell-cell contact (Sørensen *et al.*, 2005), facilitating DNA transfer from donor to recipient cells by conjugation. The presence of eDNA in the biofilm also provides cells with a unique means of acquiring DNA through transformation, if the cells in the biofilm are naturally competent. This is the case for *S. mutans* (Senadheera and Cvitkovitch, 2008), the etiologic agent of dental caries in humans. Biofilm-grown *S. mutans* cells display transformation rates 10 to 600 times higher than those of free-living *S. mutans* cells. Their transformation is peptide pheromone-dependent and involves activation of the competence *comCDE locus*. This system functions optimally when the cells are living in actively growing biofilms. Biofilm environments provide optimal conditions for QS systems, as shown for the induction of genetic competence (Li *et al.*, 2001). Conversely, the competence system also plays a role in determining the ability of *S. mutans* to initiate biofilm formation (Li *et al.*, 2002).

The high bacterial density in biofilms favours plasmid dispersal by conjugation, which requires intimate cell-cell contact via conjugative pili; the conjugation mechanism may itself stimulate biofilm development (Molin and Tolker-Nielsen, 2003). High rates of plasmid transfer occur in biofilm populations, regardless of the concentration of nutrients in the medium, if a strain of *Alcaligenes eutrophus* is used as the recipient and the RK2 plasmid derivative from *E. coli* is used as the donor strain (Hausner and Wuertz, 1999). When the donor *E. coli* BJ4 R1*drd*19 and recipient *E. coli* BJ4 populations are present at fixed spatial locations, as in a biofilm, transfer occurs very rapidly in the initial phase, but no further transfer is detected thereafter, despite the persistence of a large recipient population (Licht *et al.*, 1999). This suggests that the fixed position of either donor or recipient cells influences accessibility and that transconjugants form a 'layer' a few bacteria thick, which remains on top of the recipient microcolonies attached to the glass surface, preventing transfer of the plasmid to the lower layers of recipients (Christensen *et al.*, 1998). This situation occurs particularly in cases in which the donor strain encounters an already structured recipient cell biofilm. However, motility, including the swimming motility of certain cells in mixed-species biofilms formed from different *Pseudomonas* spp., might help to disseminate plasmid transfer to other locations in the community (Tolker-Nielsen *et al.*, 2000). Moreover, direct biofilm shaping influences the rate of transfer by transformation and conjugation. During conjugation, open channels and pores allow donor cells to penetrate more deeply into the biofilm, increasing the chances of cells coming into contact and, therefore, of genes being transferred (Hendrickx *et al.*, 2003; Wuertz *et al.*, 2004).

Physiological heterogeneity

Bacterial cells living in a biofilm are subject to a remarkably heterogeneous microenvironment, resulting in considerable physiological heterogeneity of the embedded population as illustrated in Fig. 6.4. The balance between metabolic activities and diffusion processes creates gradients of nutrients, waste and diffusible molecules. The diverse microenvironments thus create increased chemical, physiological and genetic heterogeneity over both space and time (Stewart and Franklin, 2008). This has been observed in *S. aureus* biofilms, in which cells exist in at least four distinct states: aerobic growth, fermentative growth, dead and dormant (Rani *et al.*, 2007). A particular subset of cells forming the major interior domains of *Shewanella oneidensis* biofilms and carrying out biofilm-specific metabolism generates energy without actively growing (Teal *et al.*, 2006). This physiological pattern is often associated with gene expression activity and has been used to decipher the antibiotic tolerance and sensitivity of bacterial biofilms (de Kievit *et al.*, 2001; Bagge *et al.*, 2004). The development of new technologies for investigating physiological activities at the cellular level that cannot be detected by global -omics methods has opened up new possibilities for investigating the physiological heterogeneity within biofilms. However, it is difficult to observe

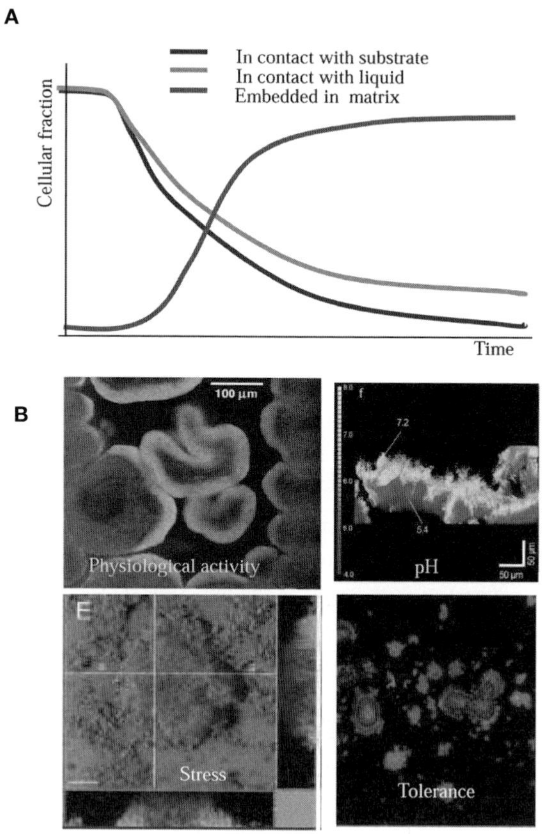

Figure 6.4 Heterogeneity generated within bacterial biofilms. (A) Scheme inspired from Sporrmann (2008) describing the increasing proportion of the microbial community embedded in matrix while cells in contact with substrate or liquid are decreasing with time during this bacterial life cycle. (B) Examples of metabolic (Werner *et al.*, 2004), pH (Hidalgo *et al.*, 2009), stress (Ito *et al.*, 2009) and tolerance to antimicrobial substances (Folkesson *et al.*, 2008) within the microcommunity observed with two-colour markers. A colour version of this figure is available in the plate section at the back of the book.

multiples responses in different or identical cells in these conditions.

Multiple species

The biofilm is also a multispecies community amenable to observation. However, the interactions between different species within the biofilm have only recently begun to be unravelled and these multispecies communities have often been considered to be a simple sum of their parts. Nonetheless, the studies carried out to date tend to support the holistic view that the whole is much more than a sum of its parts (Kuramitsu *et al.*, 2007). Studies of biofilms in the oral cavity were the first to investigate these aspects and revealed a fascinating universe of interspecies interactions in biofilms. The multispecies microbial community present in dental plaque has emerged as a unique example of the complex relationships driving the evolution of mixed communities of this type. Competition for nutrients occurs, together with clear synergistic and mutualistic interactions for the metabolism of nutrients, the production of products for use by the entire community, biofilm shaping, the degradation of complex molecules and protection against toxic compounds (Moons *et al.*, 2009). In this particular environment, one species may affect the fitness or metabolic activity of another, resulting in continual alternation between episodes of 'war' and 'peace'. This aspect, leading to the evolution of the community (Hansen *et al.*, 2007), is a very new area of research, but paves the way to a fascinating microcosm.

Biofilm: a shield against aggressions

Multifactorial antimicrobial resistance and tolerance

Multidrug resistance, more than any other property of biofilms, provides a clear demonstration that population behaviour is not the sum of the contributions of single cells.

It is now widely accepted that life in a sedentary community confers a unique type of resistance, known as biofilm-associated antimicrobial resistance, on bacteria. This resistance is multifactorial and differs between species. We focus here on resistance to antimicrobial molecules, but these ideas can easily be extended to multimetal resistance and tolerance (Harrison *et al.*, 2007). These mechanisms of tolerance and resistance to antimicrobial drugs or heavy metals are a multifactorial phenomenon reflecting the unique variability of the physiological state of bacteria within biofilms.

Biofilms are unique multicellular constructions of bacteria from one or several species, in which horizontal genetic transfer may occur easily, thus facilitating the acquisition of resistance genes. The bacteria within biofilms are embedded in a matrix of polysaccharides (EPS) that they produce themselves. This matrix limits antibiotic diffusion. The association within the biofilm of molecules of various types, including exopolysaccharides and DNA, results in a physical barrier to the diffusion of antimicrobial agents. However, many studies have surprisingly shown that the penetration of antibiotics is not limited in bacterial biofilms. For example, fluoroquinolones diffuse rapidly within *P. aeruginosa* (Vrany *et al.*, 1997) and *K. pneumoniae* (Anderl *et al.*, 2000) biofilms, tetracycline diffuses rapidly in *E. coli* biofilms (Stone *et al.*, 2002) and vancomycin diffuses rapidly in *S. epidermidis* biofilms (Darouiche *et al.*, 1994). Aminoglycosides are the only molecules for which poor penetration has been reported, in biofilms of an alginate-producing strain of *P. aeruginosa* (Shigeta *et al.*, 1997). As exopolysaccharide matrices differ considerably between, and even within species, the limited diffusion of antimicrobial drugs within bacterial biofilms has almost certainly been underestimated.

Biofilms impose unique environmental stresses or starvation conditions, resulting in the regulation of specific drug resistance-associated genes. These conditions may favour the emergence of dormant cells called persisters (Anderson and O'Toole, 2008). Persisters are dormant cells in a particular physiological state with low levels of translation but a unique gene expression profile (Shah *et al.*, 2006) in which genes encoding metabolic proteins are switched off and operons encoding toxin–antitoxin pairs are switched on, these genes probably contributing to dormancy. Persisters account for about 1% of the cells in the stationary growth phase in planktonic and biofilm cultures. They are multidrug-tolerant cells that emerge randomly in the stationary phase, in the absence of growth. They constitute a subpopulation of altruistic cells ensuring kin survival and propagation of the genome. Persisters are not specific to the biofilm lifestyle, as they have also been identified in planktonic cultures (Spoering and Lewis, 2001). However, these cells embedded in the matrix enable the biofilm to escape immune attack, recovering from dormancy when the threat subsides (Lewis, 2005, 2007). However, persister cells, which are resistant to killing by antibiotics and can survive drug treatment, account for only a small proportion of the biofilm population (Lewis, 2001). Indeed, when dispersed mechanically, most biofilm cells seem to be as susceptible to inhibitors as bacteria growing in a planktonic suspension. This strongly suggests that a number of cells are drug-tolerant because of their particular physiological state in the biofilm, due, for example, to nutrient and oxygen limitation. The molecular basis of persister physiology remains unclear, although significant transcriptional or proteomic differences between planktonic and biofilm bacteria have been identified (Whiteley *et al.*, 2001; Sauer *et al.*, 2002; Bagge *et al.*, 2004). It remains unclear whether persister cells are present before drug exposure or whether they evolve in response to drug exposure.

Some resistance mechanisms may be stronger in biofilms than in other environments, given that some efflux pumps have been shown to be more efficient in *P. aeruginosa* (Gillis *et al.*, 2005) and *E. coli* (Hancock and Klemm, 2007) biofilms. However, this mechanism is not universal, and some efflux pump inhibitors can reduce, or even abolish *E. coli* biofilm formation (Kvist *et al.*, 2008). The observed resistance of *E. coli* biofilms may be attributed to the formation of ampicillin-resistant subpopulations in the deeper layers of mature biofilms but not in young colony biofilms and to biofilm-specific phenotypes, such as slow growth and the induction of *rpoS*-mediated stress responses, in subpopulations of cells (Ito *et al.*, 2009).

A novel mechanism of biofilm-associated antibiotic resistance has recently been described, in which, in the presence of DNA, the highly anionic polymer acting as a cation chelator in the ECM, creates a localized cation-limited environment that is detected by *P. aeruginosa,* leading to the induction of LPS modification genes and resistance to antimicrobial drugs, such as cationic antimicrobial peptides and aminoglycosides (Mulcahy *et al.*, 2008). Another interesting biofilm-specific resistance mechanism has also been identified in this bacterium. The *ndvB*-dependent production of glucans in the biofilm leads to aminoglycosides being sequestered in the periplasm, preventing them from reaching their ribosomal targets; this mechanism is unique to *P. aeruginosa* biofilms (Mah *et al.*, 2003).

In multimicrobial biofilms containing *C. albicans* and *S. aureus,* like that occurring on the surface of indwelling medical devices, differences in phenotype between polymicrobial and monomicrobial biofilms were reported. Compared with monomicrobial biofilms, the resistance of *S. aureus* to vancomycin is higher in the polymicrobial biofilm, requires viable *C. albicans* and is mediated in part by the *C. albicans* matrix (Harriott and Noverr, 2009), although *C. albicans* growth and sensitivity to amphotericin B are not altered in the polymicrobial biofilm.

Immune failure

In microbes infecting eukaryotic hosts, a biofilm lifestyle may enable bacteria to evade the host response, as illustrated in Fig. 6.5. However, the efficacy of this approach depends on the bacterial species considered.

For *S. epidermidis,* biofilm cells are more resistant to opsonization than their planktonic counterparts. This resistance, mediated by an otherwise protective antibody, is due not to a biofilm-specific phenotype but to the presence of high levels of antigen (poly *N*-acetylglucosamine) within the biofilm, preventing bacterial opsonization by the antibody (Cerca *et al.*, 2006).

The alginate-producing biofilm of *P. aeruginosa* resists killing by macrophages in the presence of interferon gamma, whereas biofilms that do not produce alginate are killed by phagocytes (Leid *et al.*, 2005). *In vitro* biofilms of the *P. aeruginosa rhlA* mutant grown in a continuous-culture flow-cell system induce the necrosis of only a few polymorphonuclear leucocytes (PMNs), by contrast to what is observed with biofilms formed by the wild-type bacterium (van Gennip *et al.*, 2009). The presence of PMNs induces an aggressive response in the *P. aeruginosa* biofilm, involving the production of a QS-dependent cocktail of toxic components (Bjarnsholt *et al.*, 2005), including rhamnolipids, which stick to the biofilm, forming a deadly shield around the biofilm cells. As a consequence, approaching PMNs are killed on contact and the bacteria persist (Alhede *et al.*, 2009), probably, as previously reported, because PMNs are immobilized on *P. aeruginosa* biofilms (Jesaitis *et al.*, 2003). Furthermore, biofilm formation induces changes in the structure of LPS in *P. aeruginosa,* in terms of the levels of both lipid A and the polysaccharide moieties, strengthening the inflammatory response (TNF-α and IL-6) of human monocytes. No such response is observed in LPS-deficient Gram-positive bacteria, such as *S. aureus* (Ciornei *et al.*, 2010). The phagocytosis of the *P. aeruginosa* biofilm is blocked, but it has been reported that, in conditions mimicking physiological shearing, leucocytes attach to maturing or fully matured *S. aureus* biofilm, penetrating the biofilm and producing cytokines (Leid *et al.*, 2002). A recent report showed that PMNs move on intact *Staphylococcus* biofilms, and that biofilm-free areas appear at the sites vacated by the PMNs. The PMNs thus seem to take up the *S. aureus* biofilm as they move along. When *S. aureus*

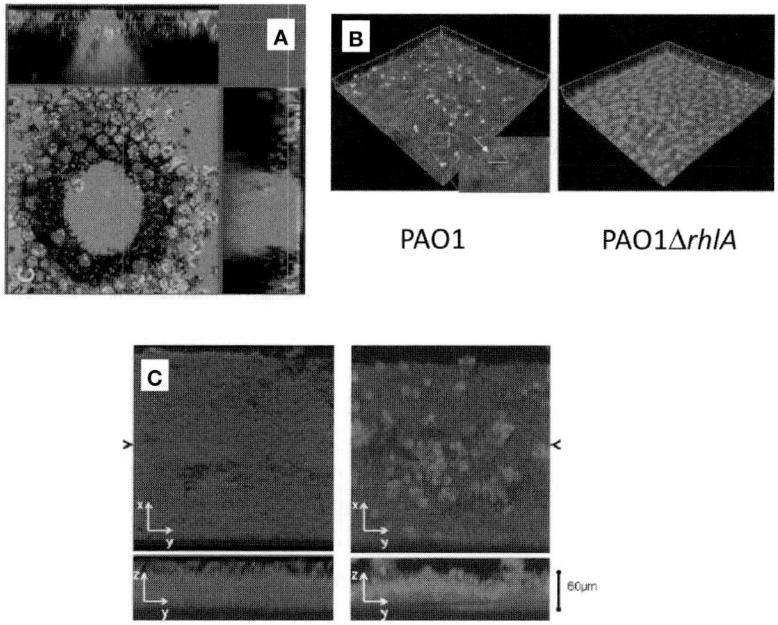

Figure 6.5 Biofilm against host immune defence. (A) Confocal microscopic observation of GFP-tagged *P. aeruginosa* and PMNs stained with cyto62. Note that intact PMNs are never identified with the top layer of the biofilm (Alhede *et al.*, 2009). (B) Confocal microscopic observation of *P. aeruginosa* GFP-tagged inducing necrosis of PMNs stained with ethidium bromide (red) whereas biofilm of the *rhlA* mutant does not (Van Gennip *et al.*, 2009). (C) SytoBC-labelled *S. aureus* biofilms are rapidly infiltrated by PMNs which move as a front, and turn yellow, the latter indicative again of biofilm uptake (biofilms in red). PMNs are labelled with antiCD66b and Cy3 conjugated antibody (Günther *et al.*, 2009). A colour version of this figure is located in the plate section at the back of the book.

biofilm fragments are used, the PMN migrate towards the biofilm, disrupting and eventually destroying the biofilm. The PMN-sensitivity of *S. aureus* biofilms clearly depends on the stage of maturation of biofilms, with young, immature biofilms seeming to be more sensitive than mature biofilms, probably because of changes in the architecture and composition of the extracellular polymer matrix over time (Günther *et al.*, 2009). The same is true of *Mycoplasma* biofilms, which display heterogeneous resistance to antimicrobial agents and the innate immune system as a function of the location of the cells in the biofilm. Resistance to complement and gramicidin is limited to the tower structures, in which cell density is highest and polysaccharides are abundant (Simmons and Dybvig, 2007). The intracellular communities formed by uropathogenic *E. coli* (UPEC) in superficial umbrella cells make use of filamentation mediated by SulA to resist killing by neutrophils (Justice *et al.*, 2006). The mechanism of exopolysaccharide modification – the deacetylation of polyacetylglucosamine, a component of the exopolysaccharide matrix produced by biofilm forming *S. epidermidis* – mediated by the IcaB protein is involved in biofilm formation and indispensable for resistance to phagocytosis by human neutrophils and resistance to antimicrobial peptides (Vuong *et al.*, 2004).

There is evidence to suggest that interactions between community members mediate prokaryotic resistance to host innate immunity. The oral pathogen *A. actinomycetemcomitans* resides in microaerophilic niches in gingival crevices, as part of a complex multispecies community including several streptococci, such as *S. gordonii*. This non-pathogenic species of *Streptococcus* converts sugars into hydrogen peroxide and lactic acid. In mixed communities, *A. actinomycetemcomitans* has evolved mechanisms for resisting the action of these compounds. It also uses lactic acid preferentially as a carbon source (Brown and Whiteley, 2007) and hydrogen peroxide to trigger a specific, targeted genetic programme. This programme involves the upregulation of genes encoding catalases for the transformation of hydrogen peroxide and adhesins enabling the bacteria to resist host immunity (Ramsey and Whiteley, 2009). This is a remarkable example of cross-feeding and cross-induced protection against killing by the alternative complement component of the innate immune system in a multispecies biofilm. The metabolites produced by *S. gordonii* effectively 'warn' *A. actinomycetemcomitans,* enabling this bacterium to implement its resistance mechanisms.

Sociobiology

Despite the difficulties involved in trying to define characteristics common to biofilms that apply to the entire bacterial kingdom and the various environmental conditions in which biofilms develop, the ubiquitous nature of biofilms strongly suggests that living in communities is a key element of bacterial ecology and evolution. Studies of the social behaviour of bacteria within biofilms have thus received particular attention recently. It was thought that communication between individuals was essential to coordinate the behaviour of the whole community in this complex multicellular environment. However, it now seems that cooperation and communication are not the only social connections between the bacteria within a biofilm. Biofilms are now seen as communities with social rules, based on the production and use of public goods, with actors and recipients. Different types of social behaviour can therefore be classified as a function of the benefits that the community derives from each individual, which can be assessed by measuring fitness (West *et al.*, 2006; Diggle *et al.*, 2007; Foster and Xavier, 2007; Foster *et al.*, 2007; Dunny *et al.*, 2008; Nadell *et al.*, 2009). In this classification, cooperation is defined as beneficial to the recipient. Cooperation may be subdivided into mutualism, which is also beneficial to the actor, and altruism, which is detrimental to the actor. Competition has a negative effect on the recipient, and may be beneficial, neutral or detrimental to the actor. The microbial society living in the biofilm may therefore be governed by social rules or patterns of behaviour, including cooperative behaviour, such as mutualism and altruism, competitive behaviour and specialization, involving various types of communication based on chemical or genetic signals (Weigel *et al.*, 2007). This is illustrated in Fig. 6.6.

Cooperation results in the success of the group (Parsek and Greenberg, 2005; Shapiro, 1998), and perfect cooperation in biofilms is more likely if they contain only one strain (high level of genetic relatedness). Genetically identical cells, as found in most multicellular organisms, do not have evolutionary conflicts of interest. Their predicted behaviour is therefore simply that which is optimal for the group. However, social cooperation between species may also be essential for survival, and this requires collective action.

Altruism may be defined as acting for the benefit of the community or producing a resource of use to the community to one's own detriment (West *et al.*, 2006). Altruistic behaviour in biofilms tends to be observed when all the individuals present are derived from a common ancestor, whereas the exploitation of certain individuals by individuals of other bacterial species rapidly tends to develop into competition (Nadell *et al.*, 2009). Cell death during biofilm development is thought to be caused by an individual cell destroying itself in a suicide mechanism to ensure the continued survival of the population as a whole, through the release of eDNA, a phenomenon that has been described for *S. aureus* (Bayles, 2007; Rice *et al.*, 2007) and *S. mutans* (Perry *et al.*, 2009). Subpopulations of *E. faecalis* producing an immunity protein commit fratricide against a subpopulation of cells not producing this protein, and this mechanism also generates eDNA to the general benefit of the entire community (Thomas *et al.*, 2009). However, the biofilm lifestyle may also be a way of stabilizing interactions between distantly related species, causing marked changes in their symbiotic nature and affecting community function (Hansen *et al.*, 2007). This is particularly true for the dental plaque biofilm, in which the periodontopathogen *P. gingivalis* may be an integral

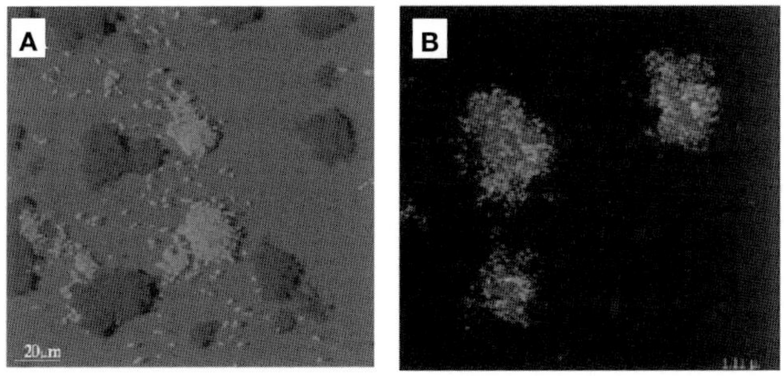

Figure 6.6 Confocal microscopic observation of an example of commensal relationship between *Burkholderia* sp. LB400 hybridized with a CY3-labelled (dark grey) probe and *Pseudomonas* sp. B13 hybridized with a FITC-labelled (pale grey) probe. In citrate growth conditions (A), consortium consisted mostly of separate microcolonies of *Burkholderia* sp. LB400 and *Pseudomonas* sp. B13, whereas grown on low concentrations of 3-chlorobiphenyl (B), multispecies community consists of thin microcolonies of *Burkholderia* sp. LB400, which are mostly often covered with *Pseudomonas* sp. B13 bacteria (Nielsen *et al.*, 2000).

component of mutualistic multispecies oral biofilm communities at all stages of dental plaque development (Periasamy and Kolenbrander, 2009a and 2009b).

Diffusible secreted molecules, such as extracellular proteases, iron-chelating siderophores and autoinducers (Griffin *et al.*, 2004; D'Argenio *et al.*, 2007; Diggle *et al.*, 2007), and extracellular polymeric substances (Hammer and Bassler, 2003; Xavier and Foster, 2007) are commodities useful to the entire community that are costly to produce. The secretion of extracellular polymers by a cell enables it to act altruistically, driving its descendants into a more oxygen-rich environment. It also provides a strong competitive advantage at the scale of the cell lineage, by overcoming neighbouring non-producing cells (Xavier and Foster, 2007). This particular example highlights the subtle switch between cooperation and competition. Thus, subpopulations of cells may turn off (or, more rarely, on) the production of a particular factor (Hammer and Bassler, 2003), usually by mutation, whilst benefitting from the production of this factor by their wild-type counterparts. This situation generally results in the outcompetition of the mutant strain in mixed populations.

In all cooperative groups, there are always some individuals that receive great benefit from the group whilst investing little or nothing. These individuals may be regarded as obligate social parasites relying on others for their survival, whereas the other individuals may be considered to be either altruistic or victims. The biofilm community is not exempt from such social parasites and the cheaters of the bacterial world are individuals emerging in conditions of starvation or high levels of QS activity. Cell-to-cell communication, known as QS, is an important cooperative mechanism for controlling multicellular behaviour. However, the emergence of QS mutants, highlights the possible existence of other types of behaviour, such as various types of cheating. Cheaters have been identified in various bacterial species, including *Myxococcus xanthus* and *P. aeruginosa* (Fiegna *et al.*, 2006; Diggle *et al.*, 2007; Sandoz *et al.*, 2007) and have been reported to benefit from the community whilst making little or no contribution themselves. They may therefore be seen as a parasitic subpopulation with a greater fitness or a growth advantage over other individuals in the community (Dunny *et al.*, 2008). There seems to be a balance between cheaters and altruists in the *M. xanthus* development cycle in the absence of evolutionary change or extinction events provoked by the cheaters (Fiegna and Velicer, 2003). However, this population of cheaters may become cooperative if there are enough cheaters to have a detrimental effect on the population as a whole, as observed in *M. xanthus* (Fiegna *et al.*, 2006) and *P. aeruginosa* (Sandoz *et al.*, 2007). The identification of *P. aeruginosa lasR* mutants in CF airways, probably arising as a consequence of the biofilm lifestyle, provides further evidence for the existence of cheating behaviour among bacteria involved in biofilm-associated infections. Finally, there is also evidence of cannibalistic behaviour in *B. subtilis* biofilms. Cannibalism-associated toxins produced by a self-protected population with a resistance mechanism, induce the mass death of unprotected cells, leading to a release of nutrients that the survivors then use for matrix production (López *et al.*, 2009).

These examples demonstrate the need for microbiological studies of biofilms to include this social dimension, to ensure that this particular bacterial lifestyle cycle is fully deciphered.

Conclusions

As only specific original aspects of biofilms have been dealt with in this chapter, the reader should bear in mind that this particular bacterial lifecycle is governed by a complex regulatory network that is currently poorly understood but integrates environmental signals (Stanley and Lazazzera, 2004), QS (Kirisits and Parsek, 2006) and ci-di-GMP signalling (Jonas *et al.*, 2009). As this bacterial lifestyle may be associated with human infectious diseases, several lines of research are currently focusing on the possibility of interfering with biofilm initiation, structuration or communication, and promoting dispersion. All new information concerning the functioning of biofilms may potentially lead to strategies for dismantling this microbial community (Hunter, 2008). Much effort has focused on compounds interfering with QS autoinducers (Estrela *et al*, 2009), molecules enhancing dispersion, such as NO, capsules or dispersin-like molecules and, more recently, phages (Lu and Collins, 2007). Increasing efforts are also being dedicated to molecules interfering with adhesive structures and to the development of new surfaces for internal or external medical devices (Dwyer, 2008). High-resolution automated confocal laser scanning microscopy has also been used to identify new genes involved in *P. aeruginosa* biofilm formation (Müsken *et al.*, 2009). This work indicates that increasing the resolution of biofilm examinations will probably increase the number of bacterial genes identified as involved in this process, this number probably being underestimated on rough examinations of phenotypes. The incorporation of these technical developments into research efforts should advance our knowledge of this fascinating area of the microbial world.

Acknowledgements

SdB's work is supported by the French cystic fibrosis foundation (VLM), the foundation Bettencourt-Schueller and CNRS institutional and ANR grants: ERA-NET ADHRES 27481, PCV-ANR 27628, ANR Jeunes Chercheurs JC09-473544, Europathogenomics 2005–2010. CG and JS are supported by the French research and technology ministry.

References

Alhede, M., Bjarnsholt, T., Jensen, P.Ø., Phipps, R.K., Moser, C., Christophersen, L., Christensen, L.D., van Gennip, M., Parsek, M., Høiby, N., *et al.* (2009). *Pseudomonas aeruginosa* recognizes and responds aggressively to the presence of polymorphonuclear leukocytes. Microbiology *155*, 3500–3508.

Allegrucci, M., and Sauer, K. (2008). Formation of *Streptococcus pneumoniae* non-phase-variable colony variants is due to increased mutation frequency present under biofilm growth conditions. J. Bacteriol. *190*, 6330–6339.

An, D., and Parsek, M.R. (2007). The promise and peril of transcriptional profiling in biofilm communities. Curr. Opin. Microbiol. *10*, 292–296.

An, D., Danhorn, T., Fuqua, C., and Parsek, M.R. (2006). Quorum sensing and motility mediate interactions between *Pseudomonas aeruginosa* and *Agrobacterium tumefaciens* in biofilm cocultures. Proc. Natl. Acad. Sci. U.S.A. *103*, 3828–3833.

Anderl, J.N., Franklin, M.J., and Stewart, P.S. (2000). Role of antibiotic penetration limitation in *Klebsiella pneumoniae* biofilm resistance to ampicillin and ciprofloxacin. Antimicrob. Agents Chemother. *44*, 1818–1824.

Anderson, G.G., and O'Toole, G.A. (2008). Innate and induced resistance mechanisms of bacterial biofilms. Curr. Top Microbiol. Immunol. *22*, 85–105.

Aparna, M.S., and Yadav, S. (2008). Biofilms: microbes and disease. Braz. J. Infect. Dis. *12*, 526–530.

Aschtgen, M.S., Bernard, C.S., de Bentzmann, S., Lloubès, R., and Cascales, E. (2008). SciN is an outer membrane lipoprotein required for type VI secretion in enteroaggregative *Escherichia coli*. J. Bacteriol. *190*, 7523–7531.

Aubert, D.F., Flannagan, R.S., and Valvano, M.A. (2008). A novel sensor kinase-response regulator hybrid controls biofilm formation and type VI secretion system activity in *Burkholderia cenocepacia*. Infect. Immun. *76*, 1979–1991.

Azevedo, N.F., Lopes, S.P., Keevil, C.W., Pereira, M.O., and Vieira, M.J. (2009). Time to 'go large' on biofilm research: advantages of an omics approach. Biotechnol. Lett. *31*, 477–485.

Bagge, N., Hentzer, M., Andersen, J.B., Ciofu, O., Givskov, M., and Høiby, N. (2004). Dynamics and spatial distribution of beta-lactamase expression in *Pseudomonas aeruginosa* biofilms. Antimicrob. Agents Chemother. *48*, 1168–1174.

Balestrino, D., Ghigo, J.M., Charbonnel, N., Haagensen, J.A., and Forestier, C. (2008). The characterization of functions involved in the establishment and maturation of *Klebsiella pneumoniae in vitro* biofilm reveals dual roles for surface exopolysaccharides. Environ. Microbiol. *10*, 685–701.

Ballister, E.R., Lai, A.H., Zuckermann, R.N., Cheng, Y., and Mougous, J.D. (2008). *In vitro* self-assembly of tailorable nanotubes from a simple protein building block. Proc. Natl. Acad. Sci. U.S.A. *105*, 3733–3738.

Barken, K.B., Pamp, S.J., Yang, L., Gjermansen, M., Bertrand, J.J., Klausen, M., Givskov, M., Whitchurch, C.B., Engel, J.N., and Tolker-Nielsen, T. (2008). Roles of type IV pili, flagellum-mediated motility and extracellular DNA in the formation of mature multicellular structures in *Pseudomonas aeruginosa* biofilms. Environ. Microbiol. *10*, 2331–2343.

Barnhart, M. M, and Chapman, M.R. (2006). Curli biogenesis and function. Annu. Rev. Microbiol. *60*, 131–147.

Barraud, N., Schleheck, D., Klebensberger, J., Webb, J.S., Hassett, D.J., Rice, S.A., and Kjelleberg, S. (2009). Nitric oxide signaling in *Pseudomonas aeruginosa* biofilms mediates phosphodiesterase activity, decreased cyclic di-GMP levels, and enhanced dispersal. J. Bacteriol. *191*, 7333–7342.

Bayles, K.W. (2007). The biological role of death and lysis in biofilm development. Nat. Rev. Microbiol. *5*, 721–726.

Belas, M.R., and Colwell, R.R. (1982). Scanning electron microscope observation of the swarming phenomenon of *Vibrio parahaemolyticus*. J. Bacteriol. *150*, 956–959.

Beloin, C., Michaelis, K., Lindner, K., Landini, P., Hacker, J., Ghigo, J.M., and Dobrindt, U. (2006). The transcriptional antiterminator RfaH represses biofilm formation in *Escherichia coli*. J. Bacteriol. *188*, 1316–1331.

Beloin, C., Roux, A., and Ghigo, J.M. (2008). *Escherichia coli* biofilms. Curr. Top. Microbiol. Immunol. *322*, 249–289.

Beloin, C., Valle. J., Latour-Lambert, P., Faure, P., Kzreminski, M., Balestrino, D., Haagensen, J.A., Molin, S., Prensier, G., Arbeille, B., *et al.* (2004). Global impact of mature biofilm lifestyle on *Escherichia coli* K-12 gene expression. Mol. Microbiol. *51*, 659–674.

de Bentzmann, S., Aurouze, M., Ball, G., and Filloux, A. (2006). FppA, a novel *Pseudomonas aeruginosa* prepilin peptidase involved in assembly of type IVb pili. J. Bacteriol. *188*, 4851–4860.

Bernard, C.S., Bordi, C., Termine, E., Filloux, A., and de Bentzmann, S. (2009). Organization and PprB-dependent control of the *Pseudomonas aeruginosa tad* locus, involved in Flp pilus biology. J. Bacteriol. *191*, 1961–1973.

Bieber, D., Ramer, S.W., Wu, C.Y., Murray, W.J., Tobe, T., Fernandez, R., and Schoolnik, G.K. (1998). Type IV pili, transient bacterial aggregates, and virulence of enteropathogenic *Escherichia coli*. Science *280*, 2114–2118.

Biswas, L., Biswas, R., Schlag, M., Bertram, R., and Götz, F. (2009). Small-colony variant selection as a survival strategy for *Staphylococcus aureus* in the presence of *Pseudomonas aeruginosa*. Appl. Environ. Microbiol. *75*, 6910–6912.

Bjarnsholt, T., Jensen, P.Ø., Burmølle, M., Hentzer, M., Haagensen, J.A., Hougen, H.P., Calum, H., Madsen, K.G., Moser, C., Molin, S., *et al.* (2005). *Pseudomonas aeruginosa* tolerance to tobramycin, hydrogen peroxide and polymorphonuclear leukocytes is quorum-sensing dependent. Microbiology *151*, 373–383.

Blázquez, J. (2003). Hypermutation as a factor contributing to the acquisition of antimicrobial resistance. Clin. Infect. Dis. *37*, 1201–1209.

Böckelmann, U., Janke, A., Kuhn, R., Neu, T.R., Wecke, J., Lawrence, J.R., and Szewzyk, U. (2006). Bacterial extracellular DNA forming a defined network-like structure. FEMS Microbiol. Lett. *262*, 31–38.

Boles, B.R., and Horswill, A.R. (2008). Agr-mediated dispersal of *Staphylococcus aureus* biofilms. PLoS Pathog. *4*, e1000052.

Boles, B.R., and Singh, P.K. (2008). Endogenous oxidative stress produces diversity and adaptability in biofilm communities. Proc. Natl. Acad. Sci. U.S.A. *105*, 12503–12508.

Boles, B.R., Thoendel, M., and Singh, P.K. (2005). Rhamnolipids mediate detachment of *Pseudomonas aeruginosa* from biofilms. Mol. Microbiol. *57*, 1210–1223.

Branda, S.S., Vik, S., Friedman, L., and Kolter, R. (2005). Biofilms: the matrix revisited. Trends Microbiol. *13*, 20–26.

de Breij, A., Gaddy, J., van der Meer, J., Koning, R., Koster, A., van den Broek, P., Actis, L., Nibbering, P., and Dijkshoorn, L. (2009). CsuA/BABCDE-dependent pili are not involved in the adherence of *Acinetobacter baumannii* ATCC19606(T) to human airway epithelial cells and their inflammatory response. Res. Microbiol. *160*, 213–218.

Brown, S.A., and Whiteley, M. (2007). A novel exclusion mechanism for carbon resource partitioning in *Aggregatibacter actinomycetemcomitans*. J. Bacteriol. *189*, 6407–6414.

Caiazza, N.C., Merritt, J.H., Brothers, K.M., and O'Toole, G.A. (2007). Inverse regulation of biofilm formation and swarming motility by *Pseudomonas aeruginosa* PA14. J. Bacteriol. *189*, 3603–3612.

Cerca, N., Pier, G.B., Vilanova, M., Oliveira, R., and Azeredo, J. (2005). Quantitative analysis of adhesion and biofilm formation on hydrophilic and hydrophobic surfaces of clinical isolates of *Staphylococcus epidermidis*. Res. Microbiol. *156*, 506–514.

Cerca, N., Jefferson, K.K., Oliveira, R., Pier, G.B., and Azeredo, J. (2006). Comparative antibody-mediated phagocytosis of *Staphylococcus epidermidis* cells grown in a biofilm or in the planktonic state. Infect. Immun. *74*, 4849–4855.

Chang, W.S., van de Mortel, M., Nielsen, L., Nino de Guzman, G., Li, X., and Halverson, L.J. (2007). Alginate production by *Pseudomonas putida* creates a hydrated microenvironment and contributes to biofilm architecture and stress tolerance under water-limiting conditions. J. Bacteriol. *189*, 8290–8299.

Chatterjee, I., Herrmann, M., Proctor, R.A., Peters, G., and Kahl, B.C. (2007). Enhanced post-stationary-phase survival of a clinical thymidine-dependent small-colony variant of *Staphylococcus aureus* results from lack of a functional tricarboxylic acid cycle. J. Bacteriol. *189*, 2936–2940.

Chevance, F.F., and Hughes, K.T. (2008). Coordinating assembly of a bacterial macromolecular machine. Nat. Rev. Microbiol. *6*, 455–465.

Christensen, B.B., Sternberg, C., Andersen, J.B., Eberl, L., Moller, S., Givskov, M., and Molin, S. (1998). Establishment of new genetic traits in a microbial biofilm community. Appl. Environ. Microbiol. *64*, 2247–2255.

Ciornei, C.D., Novikov, A., Beloin, C., Fitting, C., Caroff, M., Ghigo, J.M., Cavaillon, J.M., and Adib-Conquy, M. (2010). Biofilm-forming *Pseudomonas aeruginosa* bacteria undergo lipopolysaccharide structural modifications and induce enhanced inflammatory cytokine response in human monocytes. Innate Immun. *16*, 288–301.

Clarke, S.R., and Foster, S.J. (2006). Surface adhesins of *Staphylococcus aureus*. Adv. Microb. Physiol. *51*, 187–224.

Coil, D.A., and Anné, J. (2009). Twitching motility in *Legionella pneumophila*. FEMS Microbiol. Lett. *293*, 271–277.

Conibear, T.C., Collins, S.L., and Webb, J.S. (2009). Role of mutation in *Pseudomonas aeruginosa* biofilm development. PLoS One *4*, e6289.

Craig, L., and Li, J. (2008). Type IV pili: paradoxes in form and function. Curr. Opin. Struct. Biol. *18*, 267–277.

Cucarella, C., Solano, C., Valle, J., Amorena, B., Lasa, I., and Penadés, J.R. (2001). Bap, a *Staphylococcus aureus* surface protein involved in biofilm formation. J. Bacteriol. *183*, 2888–2896.

Cucarella, C., Tormo, M.A., Knecht, E., Amorena, B., Lasa, I., Foster, T.J.,and Penadés, J.R. (2002). Expression of the biofilm-associated protein interferes with host protein receptors of *Staphylococcus aureus* and alters the infective process. Infect. Immun. *70*, 3180–3186.

Danese, P.N., Pratt, L.A., Dove, S.L., and Kolter, R. (2000). The outer membrane protein, antigen 43, mediates cell-to-cell interactions within *Escherichia coli* biofilms. Mol. Microbiol. *37*, 424–432.

Danese, P.N., Pratt, L.A., and Kolter, R. (2001). Biofilm formation as a developmental process. Methods Enzymol. *336*, 19–26.

D'Argenio, D.A., Calfee, M.W., Rainey, P.B., and Pesci, E.C. (2002). Autolysis and autoaggregation in *Pseudomonas aeruginosa* colony morphology mutants. J. Bacteriol. *184*, 6481–6489.

D'Argenio, D.A., Wu, M., Hoffman, L.R., Kulasekara, H.D., Déziel, E., Smith, E.E., Nguyen, H., Ernst, R.K., Larson Freeman, T.J., Spencer, D.H., *et al.* (2007). Growth phenotypes of *Pseudomonas aeruginosa lasR* mutants adapted to the airways of cystic fibrosis patients. Mol. Microbiol *64*, 512–533.

Darnell, C.L., Hussa, E.A., and Visick, K.L. (2008). The putative hybrid sensor kinase SypF coordinates biofilm formation in *Vibrio fischeri* by acting upstream of two response regulators, SypG and VpsR. J. Bacteriol. *190*, 4941–4950.

Darouiche, R.O., Dhir, A., Miller, A.J., Landon, G.C., Raad, I.I., and Musher, D.M. (1994). Vancomycin penetration into biofilm covering infected prostheses and effect on bacteria. J. Infect. Dis. *170*, 720–723.

Davey, M.E., Caiazza, N.C., and O'Toole, G.A. (2003). Rhamnolipid surfactant production affects biofilm architecture in *Pseudomonas aeruginosa* PAO1. J. Bacteriol. *185*, 1027–1036.

Davies, D.G., and Marques, C.N. (2009). A fatty acid messenger is responsible for inducing dispersion in microbial biofilms. J. Bacteriol. *191*, 1393–1403.

Davies, D.G., Parsek, M.R., Pearson, J.P., Iglewski, B.H., Costerton, J.W., and Greenberg, E.P. (1998). The involvement of cell-to-cell signals in the development of a bacterial biofilm. Science *280*, 295–298.

Desvaux, M., Parham, N.J., and Henderson, I.R. (2004). The autotransporter secretion system. Res. Microbiol. *155*, 53–60.

Diggle, S.P., Stacey, R.E., Dodd, C., Cámara, M., Williams, P., and Winzer, K. (2006). The galactophilic lectin, LecA, contributes to biofilm development in *Pseudomonas aeruginosa*. Environ. Microbiol. *8*, 1095–1104.

Diggle, S.P., Griffin, A.S., Campbell, G.S., and West, S.A. (2007). Cooperation and conflict in quorum-sensing bacterial populations. Nature *450*, 411–414.

Domka, J., Lee, and J., Wood, T.K. (2006). YliH (BssR) and YceP (BssS) regulate *Escherichia coli* K-12 biofilm formation by influencing cell signaling. Appl. Environ. Microbiol. *72*, 2449–2459.

Domka, J., Lee, J., Bansal, T., and Wood, T.K. (2007). Temporal gene-expression in *Escherichia coli* K-12 biofilms. Environ. Microbiol. *9*, 332–346.

Dramsi, S., Caliot, E., Bonne, I., Guadagnini, S., Prévost, M.C., Kojadinovic, M., Lalioui, L., Poyart, C., and Trieu-Cuot, P. (2006). Assembly and role of pili in group B streptococci. Mol. Microbiol. *60*, 1401–1413.

Drenkard, E., and Ausubel, F.M. (2002). *Pseudomonas* biofilm formation and antibiotic resistance are linked to phenotypic variation. Nature *416*, 740–743.

Dunny, G.M., Brickman, T.J., and Dworkin, M. (2008). Multicellular behavior in bacteria: communication, cooperation, competition and cheating. Bioessays. *30*, 296–298.

Dwyer, A. (2008). Surface-treated catheters-a review. Semin. Dial. *21*, 542–546.

Enos-Berlage, J.L., and McCarter, L.L. (2000). Relation of capsular polysaccharide production and colonial cell organization to colony morphology in *Vibrio parahaemolyticus*. J. Bacteriol. *182*, 5513–5520.

Enos-Berlage, J.L., Guvener, Z.T., Keenan, C.E., and McCarter, L.L. (2005). Genetic determinants of biofilm development of opaque and translucent *Vibrio parahaemolyticus*. Mol. Microbiol. *55*, 1160–1182.

Epstein, E.A., Reizian, M.A., and Chapman, M.R. (2009). Spatial clustering of the curlin secretion lipoprotein requires curli fiber assembly. J. Bacteriol. *191*, 608–615.

Estrela, A.B., Heck, M.G., and Abraham, W.R. (2009). Novel approaches to control biofilm infections. Curr. Med. Chem. *16*, 1512–1530.

Felek, S., Runco, L.M., Thanassi, D.G., and Krukonis, E.S. (2007). Characterization of six novel chaperone/usher systems in *Yersinia pestis*. Adv. Exp. Med. Biol. *603*, 97–105.

Felek, S., Lawrenz, M.B., and Krukonis, E.S. (2008). The *Yersinia pestis* autotransporter YapC mediates host cell binding, autoaggregation and biofilm formation. Microbiology *154*, 1802–1812.

Fiegna, F., and Velicer, G.J. (2003). Competitive fates of bacterial social parasites: persistence and self-induced extinction of *Myxococcus xanthus* cheaters. Proc. Biol. Sci. *270*, 1527–1534.

Fiegna, F., Yu, Y.T., Kadam, S.V., and Velicer, G.J. (2006). Evolution of an obligate social cheater to a superior cooperator. Nature *441*, 310–314.

Filloux, A., de Bentzmann, S., Aurouze, M., Lazdunski, A., and Vallet, I. (2004a). Fimbrial genes in *Pseudomonas aeruginosa* and *Pseudomonas putida*. In *Pseudomonas*, Genomics, Life Style and Molecular Architecture, vol. 1, Ramos, J.-L., ed. (Kluwer Academic/Plenum Publishers, New York), pp. 721–748.

Filloux, A., Bleves, S., van Ulsen, P., and Tommassen, J. (2004b). Protein secretion mechanisms in *Pseudomonas*. In *Pseudomonas*, Genomics, Life Style and Molecular Architecture, vol. 1, Ramos, J.-L., ed. (Kluwer Academic/Plenum Publishers, New York), pp. 749–791.

Folkesson, A., Haagensen, J.A., Zampaloni, C., Sternberg, C., and Molin, S. (2008). Biofilm induced tolerance towards antimicrobial peptides. Plos One *3*, e1891.

Foster, K.R., and Xavier, J.B. (2007). Cooperation: bridging ecology and sociobiology. Curr. Biol. *17*, R319–321.

Foster, K.R., Parkinson, K., and Thompson, C.R. (2007). What can microbial genetics teach sociobiology? Trends Genet. *23*, 74–80.

Friedman, L., and Kolter, R. (2004a). Genes involved in matrix formation in *Pseudomonas aeruginosa* PA14 biofilms. Mol. Microbiol. *51*, 675–690.

Friedman, L, and Kolter R. (2004b). Two genetic loci produce distinct carbohydrate-rich structural components of the *Pseudomonas aeruginosa* biofilm matrix. J. Bacteriol. *186*, 4457–4465.

Fronzes, R, Remaut, H, and Waksman, G. (2008). Architectures and biogenesis of non-flagellar protein appendages in Gram-negative bacteria. EMBO J. *27*, 2271–2280.

García-Contreras, R., Zhang, X.S., Kim, Y., and Wood, T.K. (2008). Protein translation and cell death: the role of rare tRNAs in biofilm formation and in activating dormant phage killer genes. PLoS One *3*, e2394.

van Gennip, M., Christensen, L.D., Alhede, M., Phipps, R., Jensen, P.Ø., Christophersen, L., Pamp, S.J., Moser, C., Mikkelsen, P.J., Koh, A.Y., *et al.* (2009). Inactivation of the *rhlA* gene in *Pseudomonas aeruginosa* prevents rhamnolipid production, disabling the protection against polymorphonuclear leukocytes. APMIS *117*, 537–546.

Ghigo, J.M. (2001). Natural conjugative plasmids induce bacterial biofilm development. Nature *412*, 442–445.

Gillis, R.J., White, K.G., Choi, K.H., Wagner, V.E., Schweizer, H.P., and Iglewski, B.H. (2005). Molecular basis of azithromycin-resistant *Pseudomonas aeruginosa* biofilms. Antimicrob. Agents Chemother. *49*, 3858–37.86

Giltner, C.L., van Schaik, E.J., Audette, G.F., Kao, D., Hodges, R.S., Hassett, D.J., and Irvin, R.T. (2006). The *Pseudomonas aeruginosa* type IV pilin receptor binding domain functions as an adhesin for both biotic and abiotic surfaces. Mol. Microbiol. *59*, 1083–1096.

Giraud, C., Bernard, C., Ruer, S., and de Bentzmann, S. (2010). Biological 'glue' and 'Velcro': Molecular tools for adhesion and biofilm formation in the hairy and gluey bug *Pseudomonas aeruginosa*. Environ. Microbiol. Rep. *2*, 343–358.

Giraud, C., Bernard, C.S., Calderon, V., Yang, L., Filloux, A., Molin, S., Fichant, G., Bordi, C., and de Bentzmann, S. (2011). The PprA-PprB two component system activates CuPE, the first non-archetypal *Pseudomonas aeruginosa* chaperon-usher pathway system assembling fimbriae. Environ. Microbiol. *13*, 666–683.

Girón, J.A., Ho, A.S., and Schoolnik, G.K. (1991). An inducible bundle-forming pilus of enteropathogenic *Escherichia coli*. Science *254*, 710–713.

Glick J, and Garber N. (1983). The intracellular localization of *Pseudomonas aeruginosa* lectins. J. Gen. Microbiol. *129*, 3085–3090.

Goller, C.C., and Romeo, T. (2008). Environmental influences on biofilm development. Curr. Top Microbiol. Immunol. *322*, 37–66.

von Götz, F., Häussler, S., Jordan, D., Saravanamuthu, S.S., Wehmhöner, D., Strüssmann, A., Lauber, J., Attree, I., Buer, J., Tümmler, B., *et al.* (2004). Expression analysis of a highly adherent and cytotoxic small colony variant of *Pseudomonas aeruginosa* isolated from a lung of a patient with cystic fibrosis. J. Bacteriol. *186*, 3837–3847.

Griffin, A.S., West, S.A., and Buckling, A. (2004). Cooperation and competition in pathogenic bacteria. Nature *430*, 1024–1027.

Gualdi, L., Tagliabue, L., Bertagnoli, S., Ieranò, T., De Castro, C., and Landini, P. (2008). Cellulose modulates biofilm formation by counteracting curli-mediated colonization of solid surfaces in *Escherichia coli*. Microbiology *154*, 2017–2024.

Guiton, P.S., Hung, C.S., Kline, K.A., Roth, R., Kau, A.L., Hayes, E., Heuser, J., Dodson, K.W., Caparon, M.G., and Hultgren, S.J. (2009). Contribution of autolysin and sortase A during *Enterococcus faecalis* DNA-dependent biofilm development. Infect. Immun. *77*, 3626–3638.

Günther, F., Wabnitz, G.H., Stroh, P., Prior, B., Obst, U., Samstag, Y., Wagner, C., and Hänsch, G.M. (2009). Host defence against *Staphylococcus aureus* biofilms infection: phagocytosis of biofilms by polymorphonuclear neutrophils (PMN). Mol. Immunol. *46*, 1805–1813.

Hammer, B.K., and Bassler, B.L. (2003). Quorum sensing controls biofilm formation in *Vibrio cholerae*. Mol. Microbiol. *50*, 101–104. Erratum in: Mol. Microbiol. (2004). *51*,1521.

Hancock, V., and Klemm, P. (2007). Global gene expression profiling of asymptomatic bacteriuria *Escherichia coli* during biofilm growth in human urine. Infect. Immun. *75*, 966–976.

Hanna, A., Berg, M., Stout, V., and Razatos, A. (2003). Role of capsular colanic acid in adhesion of uropathogenic *Escherichia coli*. Appl. Environ. Microbiol. *69*, 4474–4481.

Hansen, S.K., Rainey, P.B., Haagensen, J.A., and Molin. S. (2007). Evolution of species interactions in a biofilm community. Nature *445*, 533–536.

Harriott, M.M., and Noverr, M.C. (2009). *Candida albicans* and *Staphylococcus aureus* form polymicrobial biofilms: effects on antimicrobial resistance. Antimicrob. Agents Chemother. *53*, 3914–3922.

Harrison, J.J., Ceri, H., and Turner, R.J. (2007). Multimetal resistance and tolerance in microbial biofilms. Nat. Rev. Microbiol. *5*, 928–938.

Harshey, R.M., and Matsuyama, T. (1994). Dimorphic transition in *Escherichia coli* and *Salmonella typhimurium*: surface-induced differentiation into hyperflagellate swarmer cells. Proc. Natl. Acad. Sci. U.S.A. *91*, 8631–8635.

Hausner, M., and Wuertz, S. (1999). High rates of conjugation in bacterial biofilms as determined by quantitative *in situ* analysis. Appl. Environ. Microbiol. *65*, 3710–3713.

Häussler, S. (2004). Biofilm formation by the small colony variant phenotype of *Pseudomonas aeruginosa*. Environ. Microbiol. *6*, 546–551.

Hay, I.D., Gatland, K., Campisano, A., Jordens, J.Z., and Rehm, B.H. (2009). Impact of alginate overproduction on attachment and biofilm architecture of a supermucoid *Pseudomonas aeruginosa* strain. Appl. Environ. Microbiol. *75*, 6022–6025.

He, J., Baldini, R.L., Déziel, E., Saucier, M., Zhang, Q., Liberati, N.T., Lee, D., Urbach, J., Goodman, H.M., and Rahme, L.G. (2004). The broad host range pathogen *Pseudomonas aeruginosa* strain PA14 carries two pathogenicity islands harboring plant and animal virulence genes. Proc. Natl. Acad. Sci. U.S.A. *101*, 2530–2535.

Henderson, I.R., Meehan, M., and Owen, P. (1997). Antigen 43, a phase-variable bipartite outer membrane protein, determines colony morphology and autoaggregation in *Escherichia coli* K-12. FEMS Microbiol. Lett. *149*, 115–120.

Henderson, I.R., Navarro-Garcia, F., Desvaux, M., Fernandez, R.C, and Ala'Aldeen, D. (2004). Type V protein secretion pathway: the autotransporter story. Microbiol. Mol. Biol. Rev. *68*, 692–744.

Hendrickx, L., Hausner, M., and Wuertz, S. (2003). Natural genetic transformation in monoculture *Acinetobacter* sp. strain BD413 biofilms. Appl. Environ. Microbiol. *69*, 1721–1727.

Hendrickx, A.P., van Luit-Asbroek, M., Schapendonk, C.M., van Wamel, W.J., Braat, J.C., Wijnands, L.M., Bonten, M.J., and Willems, R.J. (2009). SgrA, a nidogen-binding LPXTG surface adhesin implicated in biofilm formation, and EcbA, a collagen binding MSCRAMM, are two novel adhesins of hospital-acquired *Enterococcus faecium*. Infect. Immun. *77*, 5097–5106.

Hendrixson, D.R., and St Geme, J.W., 3rd. (1998). The *Haemophilus influenzae* Hap serine protease promotes adherence and microcolony formation, potentiated by a soluble host protein. Mol. Cell 2, 841–850.

Henrici, A.T. (1933). Studies of freshwater bacteria. I. A direct microscopic technique. J. Bacteriol. *25*, 277–287.

Herold, S., Paton, J.C., and Paton, A.W. (2009). Sab, a novel autotransporter of locus of enterocyte effacement-negative Shiga-toxigenic *Escherichia coli* O113:H21, contributes to adherence and biofilm formation. Infect. Immun. *77*, 3234–3243.

Herzberg, M., Kaye, I.K., Peti, W., and Wood, T.K. (2006). YdgG (TqsA) controls biofilm formation in *Escherichia coli* K-12 through autoinducer 2 transport. J. Bacteriol. *188*, 587–598.

Hidalgo, G., Burns, A., Herz, E., Hay, A.G., Houston, P.L., Wiesner, U., and Lion, L.W. (2009). Functional tomographic fluorescence imaging of pH microenvironments in microbial biofilms by use of silica nanoparticle sensors. Appl. Environ. Microbiol. *75*, 7426–7435.

Hooshangi, S., and Bentley, W.E. (2008). From unicellular properties to multicellular behavior: bacteria quorum sensing circuitry and applications. Curr. Opin. Biotechnol. *19*, 550–555.

Huang, Y.J., Liao, H.W., Wu, C.C., and Peng, H.L. (2009). MrkF is a component of type 3 fimbriae in *Klebsiella pneumoniae*. Res. Microbiol. *160*, 71–79.

Hunter, P. (2008). The mob response. The importance of biofilm research for combating chronic diseases and tackling contamination. EMBO Rep. *9*, 314–317.

Imberty, A., Wimmerová, M., Mitchell, E.P., and Gilboa-Garber, N. (2004). Structures of the lectins from *Pseudomonas aeruginosa*: insight into the molecular basis for host glycan recognition. Microbes Infect. *6*, 221–228.

Irie, Y., Mattoo, S., and Yuk, M.H. (2004). The Bvg virulence control system regulates biofilm formation in *Bordetella bronchiseptica*. J. Bacteriol. *186*, 5692–5698.

Irie, Y., O'toole, G.A., and Yuk, M.H. (2005). *Pseudomonas aeruginosa* rhamnolipids disperse *Bordetella bronchiseptica* biofilms. FEMS Microbiol. Lett. *250*, 237–243.

Ito, A., May, T., Taniuchi, A., Kawata, K., and Okabe, S. (2009). Localized expression profiles of *rpoS* in *Escherichia coli* biofilms. Biotechnol. Bioeng. *103*, 975–983.

Izano, E.A., Amarante, M.A., Kher, W.B., and Kaplan, J.B. (2008). Differential roles of poly *N*-acetylglucosamine surface polysaccharide and extracellular DNA in *Staphylococcus aureus* and *Staphylococcus epidermidis* biofilms. Appl. Environ. Microbiol. *74*, 470–476.

Jacob-Dubuisson, F., Fernandez, R., and Coutte, L. (2004). Protein secretion through autotransporter and two-partner pathways. Biochim. Biophys. Acta *1694*, 235–257.

Jagnow, J., and Clegg, S. (2003). *Klebsiella pneumoniae* MrkD-mediated biofilm formation on extracellular matrix- and collagen-coated surfaces. Microbiology. *149*, 2397–2405.

Jarrell, K.F., and McBride, M.J. (2008). The surprisingly diverse ways that prokaryotes move. Nat. Rev. Microbiol. *6*, 466–476.

Jenkins, AT, Buckling, A., McGhee, M., and Ffrench-Constant, R.H. (2005). Surface plasmon resonance shows that type IV pili are important in surface attachment by *Pseudomonas aeruginosa*. J.R. Soc. Interface *2*, 255–259.

Jesaitis, A.J., Franklin, M.J., Berglund, D., Sasaki, M., Lord, C.I., Bleazard, J.B., Duffy, J.E., Beyenal, H., and Lewandowski, Z. (2003). Compromised host defense on *Pseudomonas aeruginosa* biofilms: characterization of neutrophil and biofilm interactions. J. Immunol. *171*, 4329–4339.

Johansson, E.M., Crusz, S.A., Kolomiets, E., Buts, L., Kadam, R.U., Cacciarini, M., Bartels, K.M., Diggle, S.P., Cámara, M., Williams, P., *et al.* (2008). Inhibition and dispersion of *Pseudomonas aeruginosa* biofilms by glycopeptide dendrimers targeting the fucose-specific lectin LecB. Chem. Biol. *15*, 1249–1257.

Jonas, K., Tomenius, H., Kader, A., Normark, S., Römling, U., Belova, L.M., and Melefors, O. (2007). Roles of curli, cellulose and BapA in *Salmonella* biofilm morphology studied by atomic force microscopy. BMC Microbiol. *7*, 70.

Jonas, K., Melefors, O., and Römling, U. (2009). Regulation of c-di-GMP metabolism in biofilms. Future Microbiol. *4*, 341–358.

Joseph, L.A., and Wright, A.C. (2004). Expression of *Vibrio vulnificus* capsular polysaccharide inhibits biofilm formation. J. Bacteriol. *186*, 889–893.

Jurcisek, J.A., and Bakaletz, L.O. (2007). Biofilms formed by nontypeable *Haemophilus influenzae in vivo* contain both double-stranded DNA and type IV pilin protein. J. Bacteriol. *189*, 3868–3875.

Jurcisek, J.A., Bookwalter, J.E., Baker, B.D., Fernandez, S., Novotny, L.A., Munson, R.S. Jr, and Bakaletz, L.O. (2007). The PilA protein of non-typeable *Haemophilus influenzae* plays a role in biofilm formation, adherence to epithelial cells and colonization of the mammalian upper respiratory tract. Mol. Microbiol. *65*, 1288–1299.

Justice, S.S., Hunstad, D.A., Seed, P.C., and Hultgren, S.J. (2006). Filamentation by *Escherichia coli* subverts innate defenses during urinary tract infection. Proc. Natl. Acad. Sci. U.S.A. *103*, 19884–19889.

Kachlany, S.C., Planet, P.J., Bhattacharjee, M.K., Kollia, E., DeSalle, R., Fine, D.H., and Figurski, D.H. (2000). Non-specific adherence by *Actinobacillus actinomycetemcomitans* requires genes widespread in bacteria and archaea. J. Bacteriol. *182*, 6169–6176.

Kachlany, S.C., Planet, P.J., DeSalle, R., Fine, D.H., and Figurski, D.H. (2001). Genes for tight adherence of *Actinobacillus actinomycetemcomitans*: from plaque to plague to pond scum. Trends Microbiol. *9*, 429–437.

Kaplan, J.B., Ragunath, C., Ramasubbu, N., and Fine, D.H. (2003). Detachment of *Actinobacillus actinomycetemcomitans* biofilm cells by an endogenous β-hexosaminidase activity. J. Bacteriol. *185*, 4693–4698.

Kaplan, J.B., Ragunath, C., Velliyagounder, K., Fine, D.H., and Ramasubbu, N. (2004). Enzymatic detachment of *Staphylococcus epidermidis* biofilms. Antimicrob. Agents Chemother. *48*, 2633–2636.

Kierek, K., and Watnick, P.I. (2003). The *Vibrio cholerae* O139 O-antigen polysaccharide is essential for Ca^{2+}-dependent biofilm development in sea water. Proc. Natl. Acad. Sci. U.S.A. *100*, 14357–14362.

de Kievit, T.R. (2009). Quorum sensing in *Pseudomonas aeruginosa* biofilms. Environ. Microbiol. *11*, 279–288.

de Kievit, T.R., Parkins, M.D., Gillis, R.J., Srikumar, R., Ceri, H., Poole, K., Iglewski, B.H., and Storey, D.G. (2001). Multidrug efflux pumps: expression patterns and contribution to antibiotic resistance in *Pseudomonas aeruginosa* biofilms. Antimicrob. Agents Chemother. *45*, 1761–1770.

Kikuchi, T., Mizunoe, Y., Takade, A., Naito, S., and Yoshida, S. (2005). Curli fibers are required for development of biofilm architecture in *Escherichia coli* K-12 and enhance bacterial adherence to human uroepithelial cells. Microbiol. Immunol. *49*, 875–884.

Kim, Y., Wang, X., Ma, Q., Zhang, X.S., and Wood, T.K. (2009). Toxin-antitoxin systems in *Escherichia coli* influence biofilm formation through YjgK (TabA) and fimbriae. J. Bacteriol. *191*, 1258–1267.

Kirisits, M.J., and Parsek, M.R. (2006). Does *Pseudomonas aeruginosa* use intercellular signalling to build biofilm communities? Cell. Microbiol. *8*, 1841–1849.

Kirn, T.J., Lafferty, M.J., Sandoe, C.M., and Taylor, R.K. (2000). Delineation of pilin domains required for bacterial association into microcolonies and intestinal colonization by *Vibrio cholerae*. Mol. Microbiol. *35*, 896–910.

Kirov, S.M., Tassell, B.C., Semmler, A.B., O'Donovan, L.A., Rabaan, A.A., and Shaw, J.G. (2002). Lateral flagella and swarming motility in *Aeromonas* species. J. Bacteriol. *184*, 547–555.

Kjaergaard, K., Schembri, M.A., Hasman, H., and Klemm, P. (2000). Antigen 43 from *Escherichia coli* induces inter- and intraspecies cell aggregation and changes in colony morphology of *Pseudomonas fluorescens*. J. Bacteriol. *182*, 4789–4796.

Kjelleberg, S., and Molin, S. (2002). Is there a role for quorum sensing signals in bacterial biofilms? Curr. Opin. Microbiol. *5*, 254–258.

Klausen, M., Aaes-Jørgensen, A., Molin, S., and Tolker-Nielsen, T. (2003a). Involvement of bacterial migration in the development of complex multicellular structures in *Pseudomonas aeruginosa* biofilms. Mol. Microbiol. *50*, 61–68.

Klausen, M., Heydorn, A., Ragas, P., Lambertsen, L., Aaes-Jørgensen, A., Molin, S., and Tolker-Nielsen, T. (2003b). Biofilm formation by *Pseudomonas aeruginosa* wild type, flagella and type IV pili mutants. Mol. Microbiol. *48*, 1511–1524.

Knutton, S., Shaw, R.K., Anantha, R.P., Donnenberg, M.S., and Zorgani, A.A. (1999). The type IV bundle-forming pilus of enteropathogenic *Escherichia coli* undergoes dramatic alterations in structure associated with bacterial adherence, aggregation and dispersal. Mol. Microbiol. 33, 499–509.

Köhler, T., Curty, L.K., Barja, F., van Delden, C., and Pechère, J.C. (2000). Swarming of *Pseudomonas aeruginosa* is dependent on cell-to-cell signaling and requires flagella and pili. J. Bacteriol. *182*, 5990–5996.

Kreth, J., Vu, H., Zhang, Y., and Herzberg, M.C. (2009). Characterization of hydrogen peroxide-induced DNA release by *Streptococcus sanguinis* and *Streptococcus gordonii*. J. Bacteriol. *191*, 6281–6291.

Kuboniwa, M., Amano, A., Hashino, E., Yamamoto, Y., Inaba, H., Hamada, N., Nakayama, K., Tribble, G.D., Lamont, R.J., and Shizukuishi, S. (2009). Distinct roles of long/short fimbriae and gingipains in homotypic biofilm development by *Porphyromonas gingivalis*. BMC Microbiol. 9, 105.

Kulasekara, H.D., Ventre, I., Kulasekara, B.R., Lazdunski, A., Filloux, A., and Lory, S. (2005). A novel two-component system controls the expression of *Pseudomonas aeruginosa* fimbrial cup genes. Mol. Microbiol. *55*, 368–380.

Kuramitsu, H.K., He, X., Lux, R., Anderson, M.H., and Shi, W. (2007). Interspecies interactions within oral microbial communities. Microbiol. Mol. Biol. Rev. *71*, 653–670.

Kvist, M., Hancock,V., and Klemm, P. (2008). Inactivation of efflux pumps abolishes bacterial biofilm formation. Appl. Environ. Microbiol. *74*, 7376–7382.

Landini, P., and Zehnder, A.J. (2002). The global regulatory *hns* gene negatively affects adhesion to solid surfaces by anaerobically grown *Escherichia coli* by modulating expression of flagellar genes and lipopolysaccharide production. J. Bacteriol. *184*, 1522–1529.

Langstraat, J., Bohse, M., and Clegg, S. (2001). Type 3 fimbrial shaft (MrkA) of *Klebsiella pneumoniae*, but not the fimbrial adhesin (MrkD), facilitates biofilm formation. Infect. Immun. *69*, 5805–5812.

Lappann, M., Haagensen, J.A, Claus, H., Vogel, U., and Molin, S. (2006). Meningococcal biofilm formation: structure, development and phenotypes in a standardized continuous flow system. Mol. Microbiol. *62*, 1292–1309.

Lasa, I. (2006). Towards the identification of the common features of bacterial biofilm development. Int. Microbiol. *9*, 21–28.

Latasa, C., Solano, C., Penadés, J.R., and Lasa, I. (2006). Biofilm-associated proteins. C.R. Biol. *329*, 849–857.

Lazazzera, B.A. (2005). Lessons from DNA microarray analysis: the gene expression profile of biofilms. Curr. Opin. Microbiol. *8*, 222–227.

Lee, K.K., Sheth, H.B., Wong, W.Y., Sherburne, R., Paranchych, W., Hodges, R.S., Lingwood, C.A., Krivan, H., and Irvin, R.T. (1994). The binding of *Pseudomonas aeruginosa* pili to glycosphingolipids is a tip-associated event involving the C-terminal region of the structural pilin subunit. Mol. Microbiol. *11*, 705–713.

Leid, J.G., Shirtliff, M.E., Costerton, J.W., and Stoodley, A.P. (2002). Human leukocytes adhere to, penetrate, and respond to *Staphylococcus aureus* biofilms. Infect. Immun. *70*, 6339–6345.

Leid, J.G., Willson, C.J., Shirtliff, M.E., Hassett, D.J., Parsek, M.R., and Jeffers, A.K. (2005). The exopolysaccharide alginate protects *Pseudomonas aeruginosa* biofilm bacteria from IFN-γ-mediated macrophage killing. J. Immunol. *175*, 7512–7518.

Le Quéré, B., and Ghigo, J.M. (2009). BcsQ is an essential component of the *Escherichia coli* cellulose biosynthesis apparatus that localizes at the bacterial cell pole. Mol. Microbiol. *72*, 724–740.

Lequette, Y., and Greenberg, E.P. (2005). Timing and localization of rhamnolipid synthesis gene expression in *Pseudomonas aeruginosa* biofilms. J. Bacteriol. *187*, 37–44.

Lewis, K. (2001). Riddle of biofilm resistance. Antimicrob. Agents Chemother. *45*, 999–1007.

Lewis, K. (2005). Persister cells and the riddle of biofilm survival. Biochemistry (Mosc.) *70*, 267–274.

Lewis, K. (2007). Persister cells, dormancy and infectious disease. Nat. Rev. Microbiol. *5*, 48–56.

Li, Y.H., Lau, P.C., Lee, J.H., Ellen, R.P., and Cvitkovitch, D.G. (2001). Natural genetic transformation of *Streptococcus mutans* growing in biofilms. J. Bacteriol. *183*, 897–908.

Li, Y.H., Tang, N., Aspiras, M.B., Lau, P.C., Lee, J.H., Ellen, R.P., and Cvitkovitch, D.G. (2002). A quorum-sensing signaling system essential for genetic competence in *Streptococcus mutans* is involved in biofilm formation. J. Bacteriol. *184*, 2699–2708.

Licht, T.R., Christensen, B.B., Krogfelt, K.A., and Molin, S. (1999). Plasmid transfer in the animal intestine and other dynamic bacterial populations: the role of community structure and environment. Microbiology *145*, 2615–2622.

López, D., Vlamakis, H., Losick, R., and Kolter, R. (2009). Cannibalism enhances biofilm development in *Bacillus subtilis*. Mol. Microbiol. *74*, 609–618.

Lu, T.K., and Collins, J.J. (2007). Dispersing biofilms with engineered enzymatic bacteriophage. Proc. Natl. Acad. Sci. U.S.A. *104*, 11197–11202.

Luke, N.R., Jurcisek, J.A., Bakaletz, L.O., and Campagnari, A.A. (2007). Contribution of *Moraxella catarrhalis* type IV pili to nasopharyngeal colonization and biofilm formation. Infect. Immun. *75*, 5559–5564.

Lynch, A.S., and Robertson, G.T. (2008). Bacterial and fungal biofilm infections. Annu. Rev. Med. *59*, 415–428.

Maciá, M.D., Blanquer, D., Togores, B., Sauleda, J., Pérez, J.L., and Oliver, A. (2005). Hypermutation is a key factor in development of multiple-antimicrobial resistance in *Pseudomonas aeruginosa* strains causing chronic lung infections. Antimicrob. Agents Chemother. *49*, 3382–3386.

McClaine, J.W., and Ford, R.M. (2002). Reversal of flagellar rotation is important in initial attachment of *Escherichia coli* to glass in a dynamic system with high- and low-ionic-strength buffers. Appl. Environ. Microbiol. *68*, 1280–1289.

Magnuson, R.D. (2007). Hypothetical functions of toxin-antitoxin systems. J. Bacteriol. *189*, 6089–6092.

Mah, T.F., Pitts, B., Pellock, B., Walker, G.C., Stewart, P.S., and O'Toole, G.A. (2003). A genetic basis for *Pseudomonas aeruginosa* biofilm antibiotic resistance. Nature *426*, 306–310.

Mandlik, A., Das, A., and Ton-That, H. (2008). The molecular switch that activates the cell wall anchoring step of pilus assembly in Gram-positive bacteria. Proc. Natl. Acad. Sci. U.S.A. *105*, 14147–14152.

Manetti, A.G., Zingaretti, C., Falugi, F., Capo, S., Bombaci, M., Bagnoli, F., Gambellini, G., Bensi, G., Mora, M., Edwards, A.M., *et al.* (2007). *Streptococcus pyogenes* pili promote pharyngeal cell adhesion and biofilm formation. Mol. Microbiol. *64*, 968–983.

Mann, E.E., Rice, K.C., Boles, B.R., Endres, J.L., Ranjit, D., Chandramohan, L., Tsang, L.H., Smeltzer, M.S., Horswill, A.R., and Bayles, K.W. (2009). Modulation of eDNA release and degradation affects *Staphylococcus aureus* biofilm maturation. PloS One *4*, e5822.

Mattick, J.S. (2002). Type IV pili and twitching motility. Annu. Rev. Microbiol. *56*, 289–314.

Meissner, A., Wild, V., Simm, R., Rohde, M., Erck, C., Bredenbruch, F., Morr, M., Römling, U., and Häussler, S. (2007). *Pseudomonas aeruginosa cupA*-encoded fimbriae expression is regulated by a GGDEF and EAL domain-dependent modulation of the intracellular level of cyclic diguanylate. Environ. Microbiol. *9*, 2475–2485.

Merino, S., Shaw, J.G., and Tomás, J.M. (2006). Bacterial lateral flagella: an inducible flagella system. FEMS Microbiol. Lett. *263*, 127–135.

Mikkelsen, H., Ball, G., Giraud, C., and Filloux, A. (2009). Expression of *Pseudomonas aeruginosa cupD* fimbrial genes is antagonistically controlled by RcsB and the EAL-containing PvrR response regulators. PloS One *4*, e6018.

Molin, S., and Tolker-Nielsen, T. (2003). Gene transfer occurs with enhanced efficiency in biofilms and induces enhanced stabilisation of the biofilm structure. Curr. Opin. Biotechnol. *14*, 255–261.

Monds, R.D., and O'Toole, G.A. (2009). The developmental model of microbial biofilms: ten years of a paradigm up for review. Trends Microbiol. *17*, 73–87.

Mooij, M.J., Drenkard, E., Llamas, M.A., Vandenbroucke-Grauls, C.M., Savelkoul, P.H., Ausubel, F.M., and Bitter, W. (2007). Characterization of the integrated filamentous phage Pf5 and its involvement in small-colony formation. Microbiology *153*, 1790–1798.

Moons, P., Michiels, C.W., and Aertsen, A. (2009). Bacterial interactions in biofilms. Crit. Rev. Microbiol. *35*, 157–168.

Moorthy, S., and Watnick, P.I. (2005). Identification of novel stage-specific genetic requirements through whole genome transcription profiling of *Vibrio cholerae* biofilm development. Mol. Microbiol. *57*, 1623–1635.

Morris, C., Yip, C.M., Tsui, I.S., Wong, D.K., and Hackett, J. (2003). The shufflon of *Salmonella enterica* serovar Typhi regulates type IVB pilus-mediated bacterial self-association. Infect. Immun. *71*, 1141–1146.

Mougous, J.D., Cuff, M.E., Raunser, S., Shen, A., Zhou, M., Gifford, C.A., Goodman, A.L., Joachimiak, G., Ordoñez, C.L., Lory, S., *et al.* (2006). A virulence locus of *Pseudomonas aeruginosa* encodes a protein secretion apparatus. Science *312*, 1526–1530.

Mulcahy, H., Charron-Mazenod, L., and Lewenza, S. (2008). Extracellular DNA chelates cations and induces antibiotic resistance in *Pseudomonas aeruginosa* biofilms. PLoS Pathog. *4*, e1000213.

Munro, C.L., Michalek, S.M., and Macrina, F.L. (1995). Sucrose-derived exopolymers have site-dependent roles in *Streptococcus mutans*-promoted dental decay. FEMS Microbiol. Lett. *128*, 327–332.

Müsken, M., Di Fiore, S., Dötsch, A., Fischer, R., and Häussler, S. (2010). Genetic determinants of *Pseudomonas aeruginosa* biofilm establishment. Microbiology *156*, 431–441.

Nadell, C.D., Xavier, J.B., and Foster, K.R. (2009). The sociobiology of biofilms. FEMS Microbiol. Rev. *33*, 206–224.

Nagarajan, V., Smeltzer, M.S., and Elasri, M.O. (2009). Genome-scale transcriptional profiling in *Staphylococcus aureus*: bringing order out of chaos. FEMS Microbiol. Lett. *295*, 204–210.

Nallapareddy, S.R., Singh, K.V., Sillanpää, J., Garsin, D.A., Höök, M., Erlandsen, S.L., and Murray, B.E. (2006). Endocarditis and biofilm-associated pili of *Enterococcus faecalis*. J. Clin. Invest. *116*, 2799–2807.

Neil, R.B., and Apicella, M.A. (2009). Role of HrpA in biofilm formation of *Neisseria meningitidis* and regulation of the *hrpBAS* transcripts. Infect. Immun. *77*, 2285–2293.

Ng, T.W., Akman, L., Osisami, M., and Thanassi, D.G. (2004). The usher N terminus is the initial targeting site for chaperone-subunit complexes and participates in subsequent pilus biogenesis events. J. Bacteriol. *186*, 5321–5331.

Nicastro, G.G., Boechat, A.L., Abe, C.M., Kaihami, G.H., and Baldini, R.L. (2009). *Pseudomonas aeruginosa* PA14 *cupD* transcription is activated by the RcsB response regulator, but repressed by its putative cognate sensor RcsC. FEMS Microbiol. Lett. *301*, 115–123.

Nielsen, A.T., Tolker-Nielsen, T., Barken, K.B., and Molin, (2000). S. Role of commensal relationships on the spatial structure of a surface-attached microbial consortium. Environ. Microbiol. *2*, 59–68.

Nika, J.R., Latimer, J.L., Ward, C.K., Blick, R.J., Wagner, N.J., Cope, L.D., Mahairas, G.G., Munson, R.S. Jr, and Hansen, E.J. (2002). *Haemophilus ducreyi* requires the *flp* gene cluster for microcolony formation *in vitro*. Infect. Immun. *70*, 2965–2975.

Nivens, D.E., Ohman, D.E., Williams. J., and Franklin, M.J. (2001). Role of alginate and its O acetylation in formation of *Pseudomonas aeruginosa* microcolonies and biofilms. J. Bacteriol. *183*, 1047–1057.

Ochman, H., Lawrence, J.G., and Groisman, E.A. (2000). Lateral gene transfer and the nature of bacterial innovation. Nature *405*, 299–304.

Ong, C.L., Ulett, G.C., Mabbett, A.N., Beatson, S.A., Webb, R.I., Monaghan, W., Nimmo, G.R., Looke, D.F., McEwan, A.G., and Schembri, M.A. (2008). Identification of type 3 fimbriae in uropathogenic *Escherichia coli* reveals a role in biofilm formation. J. Bacteriol. *190*, 1054–1063.

O'Toole, G.A., and Kolter, R. (1998). Flagellar and twitching motility are necessary for *Pseudomonas aeruginosa* biofilm development. Mol. Microbiol. *30*, 295–304.

O'Toole, G., Kaplan, H.B., and Kolter, R. (2000). Biofilm formation as microbial development. Annu. Rev. Microbiol. *54*, 49–79.

Otto, M. (2008). Staphylococcal biofilms. Curr. Top. Microbiol. Immunol. *322*, 207–228.

de Pace, F., Nakazato, G., Pacheco, A., de Paiva, J.B., Sperandio, V., and de Silveira, W.D. (2010). The type VI secretion system plays a role in type 1 fimbria expression and pathogenesis of an avian pathogenic *Escherichia coli* strain. Infect. Immun. *78*, 4990–4998.

Pamp, S.J., and Tolker-Nielsen, T. (2007). Multiple roles of biosurfactants in structural biofilm development by *Pseudomonas aeruginosa*. J. Bacteriol. *189*, 2531–2539.

Pamp, S.J., Sternberg, C., and Tolker-Nielsen, T. (2009). Insight into the microbial multicellular lifestyle via flow-cell technology and confocal microscopy. Cytometry A *75*, 90–103.

Parsek, M.R., and Greenberg, E.P. (2005). Sociomicrobiology: the connections between quorum sensing and biofilms. Trends Microbiol. *13*, 27–33.

Parsek, M.R., and Tolker-Nielsen, T. (2008). Pattern formation in *Pseudomonas aeruginosa* biofilms. Curr. Opin. Microbiol. *11*, 560–566.

Pecota, D.C., and Wood, T.K. (1996). Exclusion of T4 phage by the *hok/sok* killer locus from plasmid R1. J. Bacteriol. *178*, 2044–2050.

Pelicic, V. (2008). Type IV pili: *e pluribus unum*? Mol. Microbiol. *68*, 827–837.

Perez, BA, Planet, P.J., Kachlany, S.C., Tomich, M., Fine, D.H., and Figurski, D.H. (2006). Genetic analysis of the requirement for *flp-2, tadV*, and *rcpB* in *Actinobacillus actinomycetemcomitans* biofilm formation. J. Bacteriol. *188*, 6361–6375.

Periasamy, S., and Kolenbrander, P.E. (2009a). *Aggregatibacter actinomycetemcomitans* builds mutualistic biofilm communities with *Fusobacterium nucleatum* and *Veillonella* species in saliva. Infect. Immun. *77*, 3542–3551.

Periasamy, S., and Kolenbrander, P.E. (2009b). Mutualistic biofilm communities develop with *Porphyromonas gingivalis* and initial, early, and late colonizers of enamel. J. Bacteriol. *191*, 6804–6811.

Perry, J.A., Cvitkovitch, D.G., and Lévesque, C.M. (2009). Cell death in *Streptococcus mutans* biofilms: a link between CSP and extracellular DNA. FEMS Microbiol. Lett. *299*, 261–266.

Pesavento, C., and Hengge, R. (2009). Bacterial nucleotide-based second messengers. Curr. Opin. Microbiol. *12*, 170–176.

Planet, P.J., Kachlany, S.C., Fine, D.H., DeSalle, R., and Figurski, D.H. (2003). The widespread colonization island of *Actinobacillus actinomycetemcomitans*. Nat. Genet. *34*, 193–198.

Pratt, L.A., and Kolter, R. (1998). Genetic analysis of *Escherichia coli* biofilm formation: roles of flagella, motility, chemotaxis and type I pili. Mol. Microbiol. *30*, 285–293.

Prigent-Combaret, C., and Lejeune, P. (1999). Monitoring gene expression in biofilms. Methods Enzymol. *310*, 56–79.

Prigent-Combaret, C., Prensier, G., Le Thi, T.T., Vidal, O., Lejeune, P., and Dorel, C. (2000). Developmental pathway for biofilm formation in curli-producing *Escherichia coli* strains: role of flagella, curli and colanic acid. Environ. Microbiol. *2*, 450–464.

Proft, T., and Baker, E.N. (2009). Pili in Gram-negative and Gram-positive bacteria – structure, assembly and their role in disease. Cell. Mol. Life Sci. *66*, 613–635.

Purevdorj-Gage, B., Costerton, W.J., and Stoodley, P. (2005). Phenotypic differentiation and seeding dispersal in non-mucoid and mucoid *Pseudomonas aeruginosa* biofilms. Microbiology *151*, 1569–1576.

Qin, Z., Ou, Y., Yang, L., Zhu, Y., Tolker-Nielsen, T., Molin, S., and Qu, D. (2007). Role of autolysin-mediated DNA release in biofilm formation of *Staphylococcus epidermidis*. Microbiology *153*, 2083–2092.

Ramsey, M.M., and Whiteley, M. (2004). *Pseudomonas aeruginosa* attachment and biofilm development in dynamic environments. Mol. Microbiol. *53*, 1075–1087.

Ramsey, M.M., and Whiteley, M. (2009). Polymicrobial interactions stimulate resistance to host innate immunity through metabolite perception. Proc. Natl. Acad. Sci. U.S.A. 106, 1578–1583.

Rani, S.A., Pitts, B., Beyenal, H., Veluchamy, R.A., Lewandowski, Z., Davison, W.M., Buckingham-Meyer, K., and Stewart, P.S. (2007). Spatial patterns of DNA replication, protein synthesis, and oxygen concentration within bacterial biofilms reveal diverse physiological states. J. Bacteriol. *189*, 4223–4233.

Reisner, A., Haagensen, J.A., Schembri, M.A., Zechner, E.L., and Molin, S. (2003). Development and maturation of *Escherichia coli* K-12 biofilms. Mol. Microbiol. *48*, 933–946.

Reisner, A., Höller, BM., Molin, S., and Zechner, E.L. (2006). Synergistic effects in mixed *Escherichia coli* biofilms: conjugative plasmid transfer drives biofilm expansion. J. Bacteriol. *188*, 3582–3588.

Remaut, H., Tang, C., Henderson, N.S., Pinkner, J.S., Wang, T., Hultgren, S.J., Thanassi, D.G., Waksman, G., and Li, H. (2008). Fiber formation across the bacterial outer membrane by the chaperone/usher pathway. Cell *133*, 640–652.

Ren, D., Bedzyk, L.A., Setlow, P., Thomas, S.M., Ye, R.W., and Wood, T.K. (2004). Gene expression in *Bacillus subtilis* surface biofilms with and without sporulation and the importance of *yveR* for biofilm maintenance. Biotechnol. Bioeng. *86*, 344–364.

Rice, K.C., Mann, E.E., Endres, J.L., Weiss, E.C., Cassat, J.E., Smeltzer, M.S., and Bayles, K,W. (2007). The cidA murein hydrolase regulator contributes to DNA release and biofilm development in *Staphylococcus aureus*. Proc. Natl. Acad. Sci. U.S.A. *104*, 8113–8118.

Rice, S.A., Tan, C.H., Mikkelsen, P.J., Kung, V., Woo, J., Tay, M., Hauser, A., McDougald, D., Webb, J.S., and Kjelleberg, S. (2009). The biofilm life cycle and virulence of *Pseudomonas aeruginosa* are dependent on a filamentous prophage. ISME J. *3*, 271–282.

Roggenkamp, A., Sing, A., Hornef, M., Brunner, U., Autenrieth, I.B., and Heesemann, J. (1998). Chronic prosthetic hip infection caused by a small-colony variant of *Escherichia coli*. J. Clin. Microbiol. *36*, 2530–2534.

Rosan, B., Slots, J., Lamont, R.J., Listgarten, M.A., and Nelson, G.M. (1988). *Actinobacillus actinomycetemcomitans* fimbriae. Oral Microbiol. Immunol. *3*, 58–63.

Roux, A., Beloin, C., and Ghigo, J.M. (2005). Combined inactivation and expression strategy to study gene function under physiological conditions: application to identification of new *Escherichia coli* adhesins. J. Bacteriol. *187*, 1001–1013.

Ruer, S., Stender, S., Filloux, A., and de Bentzmann, S. (2007). Assembly of fimbrial structures in *Pseudomonas aeruginosa*: functionality and specificity of chaperone-usher machineries. J. Bacteriol. *189*, 3547–3555.

Ruer, S., Ball, G., Filloux, A., and de Bentzmann, S. (2008). The 'P-usher', a novel protein transporter involved in fimbrial assembly and TpsA secretion. EMBO J. *27*, 2669–2680.

Saldaña, Z., Xicohtencatl-Cortes, J., Avelino, F., Phillips, A.D., Kaper, J.B., Puente, J.L., and Girón, J.A. (2009). Synergistic role of curli and cellulose in cell adherence and biofilm formation of attaching and effacing *Escherichia coli* and identification of Fis as a negative regulator of curli. Environ. Microbiol. *11*, 992–1006.

Samatey, F.A., Imada, K., Nagashima, S., Vonderviszt, F., Kumasaka, T., Yamamoto, M., and Namba, K. (2001). Structure of the bacterial flagellar protofilament and implications for a switch for supercoiling. Nature *410*, 331–337.

Sandoz, K.M., Mitzimberg, S.M., and Schuster, M. (2007). Social cheating in *Pseudomonas aeruginosa* quorum sensing. Proc. Natl. Acad. Sci. U.S.A. *104*, 15876–15881.

Sauer, K., Camper, A.K., Ehrlich, G.D., Costerton, J.W., and Davies, D.G. (2002). *Pseudomonas aeruginosa* displays multiple phenotypes during development as a biofilm. J. Bacteriol. *184*, 1140–1154.

Sauer, K., Cullen, M.C., Rickard, A.H., Zeef, L.A., Davies, D.G., and Gilbert, P. (2004). Characterization of nutrient-induced dispersion in *Pseudomonas aeruginosa* PAO1 biofilm. J. Bacteriol. *186*, 7312–7326.

Saulino, E.T., Thanassi, D.G., Pinkner, J.S., and Hultgren, S.J. (1998). Ramifications of kinetic partitioning on usher-mediated pilus biogenesis. EMBO J. *17*, 2177–2185

van Schaik, E.J., Giltner, C.L., Audette, G.F., Keizer, D.W., Bautista, D.L., Slupsky, C.M., Sykes, B.D., and Irvin, R.T. (2005). DNA binding: a novel function of *Pseudomonas aeruginosa* type IV pili. J. Bacteriol. *187*, 1455–1464.

Schembri, M.A., Dalsgaard, D., and Klemm, P. (2004). Capsule shields the function of short bacterial adhesins. J. Bacteriol. *186*, 1249–1257.

Schooling, S.R., Hubley, A., and Beveridge, T.J. (2009). Interactions of DNA with biofilm-derived membrane vesicles. J. Bacteriol. *191*, 4097–4102.

Schuch, R., and Fischetti, V.A. (2009). The secret life of the anthrax agent *Bacillus anthracis*: bacteriophage-mediated ecological adaptations. PLoS One *4*, e6532.

Senadheera, D., and Cvitkovitch, D.G. (2008). Quorum sensing and biofilm formation by *Streptococcus mutans*. Adv. Exp. Med. Biol. *631*, 178–188.

Shah, D., Zhang, Z., Khodursky, A., Kaldalu, N., Kurg, K., and Lewis, K. (2006). Persisters: a distinct physiological state of *E. coli*. BMC Microbiol. *6*, 53.

Shapiro, J.A. (1998). Thinking about bacterial populations as multicellular organisms. Annu. Rev. Microbiol. *52*, 81–104.

Sheikh, J., Hicks, S., Dall'Agnol, M., Phillips, A.D., and Nataro, J.P. (2001). Roles for Fis and YafK in biofilm formation by enteroaggregative *Escherichia coli*. Mol. Microbiol. *41*, 983–997.

Sheth, H.B., Lee, K.K., Wong, W.Y., Srivastava, G., Hindsgaul, O., Hodges, R.S., Paranchych, W., and Irvin, R.T. (1994). The pili of *Pseudomonas aeruginosa* strains PAK and PAO bind specifically to the carbohydrate sequence βGalNAc(1–4)βGal found in glycosphingolipids asialo-GM1 and asialo-GM2. Mol. Microbiol. *11*, 715–723.

Sherlock, O., Schembri, M.A., Reisner, A., and Klemm, P. (2004). Novel roles for the AIDA adhesin from diarrheagenic *Escherichia coli*: cell aggregation and biofilm formation. J. Bacteriol. *186*, 8058–8065.

Sherlock, O., Vejborg, R.M., and Klemm, P. (2005). The TibA adhesin/invasin from enterotoxigenic *Escherichia coli* is self recognizing and induces bacterial aggregation and biofilm formation. Infect. Immun. *73*, 1954–1963.

Shigeta, M., Tanaka, G., Komatsuzawa, H., Sugai, M., Suginaka, H., and Usui, T. (1997). Permeation of antimicrobial agents through *Pseudomonas aeruginosa* biofilms: a simple method. Chemotherapy *43*, 340–345.

Shime-Hattori, A., Iida, T., Arita, M., Park, K.S., Kodama, T., and Honda, T. (2006). Two type IV pili of *Vibrio parahaemolyticus* play different roles in biofilm formation. FEMS Microbiol. Lett. *264*, 89–97.

Sillankorva, S., Neubauer, P., and Azeredo, J. (2008). *Pseudomonas fluorescens* biofilms subjected to phage phiIBB-PF7A. BMC Biotechnol. *8*, 79.

Simmons, W.L., and Dybvig, K. (2007). Biofilms protect *Mycoplasma pulmonis* cells from lytic effects of complement and gramicidin. Infect. Immun. *75*, 3696–3699.

Skerker, J.M., and Berg, H.C. (2001). Direct observation of extension and retraction of type IV pili. Proc. Natl. Acad. Sci. U.S.A. *98*, 6901–6904.

Smedley, J.G., 3rd, Jewell, E., Roguskie, J., Horzempa, J., Syboldt, A., Stolz, D.B., and Castric, P. (2005). Influence of pilin glycosylation on *Pseudomonas aeruginosa* 1244 pilus function. Infect. Immun. *73*, 7922–7931.

Soberón-Chávez, G., Lépine, F., and Déziel, E. (2005). Production of rhamnolipids by *Pseudomonas aeruginosa*. Appl. Microbiol. Biotechnol. *68*, 718–725.

Soong, G., Muir, A., Gomez, M.I., Waks, J., Reddy, B., Planet, P., Singh, P.K., Kaneko, Y., Wolfgang, M.C., Hsiao, Y.S., *et al.* (2006). Bacterial neuraminidase facilitates mucosal infection by participating in biofilm production. J. Clin. Invest. *116*, 2297–2305.

Sørensen, S.J., Bailey, M., Hansen, L.H., Kroer, N., and Wuertz, S. (2005). Studying plasmid horizontal transfer *in situ*: a critical review. Nat. Rev. Microbiol. *3*, 700–710.

Spinola, S.M., Fortney, K.R., Katz, B.P., Latimer, J.L., Mock, J.R., Vakevainen, M., and Hansen, E.J. (2003). *Haemophilus ducreyi* requires an intact *flp* gene cluster for virulence in humans. Infect. Immun. *71*, 7178–7182.

Spoering, A.L., and Lewis, K. (2001). Biofilms and planktonic cells of *Pseudomonas aeruginosa* have similar resistance to killing by antimicrobials. J. Bacteriol. *183*, 6746–6751.

Spormann, A.M. (2008). Physiology of microbes in biofilms. Curr. Top Microbiol. Immunol. *322*, 17–36.

Stanley, N.R., and Lazazzera, B.A. (2004). Environmental signals and regulatory pathways that influence biofilm formation. Mol. Microbiol. *52*, 917–924.

Stanley, N.R., Britton, R.A., Grossman, A.D., and Lazazzera, B.A. (2003). Identification of catabolite repression as a physiological regulator of biofilm formation by *Bacillus subtilis* by use of DNA microarrays. J. Bacteriol. *185*, 1951–1957.

Stewart, P.S., and Franklin, M.J. (2008). Physiological heterogeneity in biofilms. Nat. Rev. Microbiol. *6*, 199–210.

Stone, G., Wood, P., Dixon, L., Keyhan, M., and Matin, A. (2002). Tetracycline rapidly reaches all the constituent cells of uropathogenic *Escherichia coli* biofilms. Antimicrob. Agents Chemother. *46*, 2458–2461.

Tart, A.H., Blanks, M.J., and Wozniak, D.J. (2006). The AlgT-dependent transcriptional regulator AmrZ (AlgZ) inhibits flagellum biosynthesis in mucoid, nonmotile *Pseudomonas aeruginosa* cystic fibrosis isolates. J. Bacteriol. *188*, 6483–6489.

Teal, T.K., Lies, D.P., Wold, B.J., and Newman, D.K. (2006). Spatiometabolic stratification of *Shewanella oneidensis* biofilms. Appl. Environ. Microbiol. *72*, 7324–7330.

Thomas, V.C., Hiromasa, Y., Harms, N., Thurlow, L., Tomich, J., and Hancock, L.E. (2009). A fratricidal mechanism is responsible for eDNA release and contributes to biofilm development of *Enterococcus faecalis*. Mol. Microbiol. 72, 1022–1036.

Tielen, P., Strathmann, M., Jaeger, K.E, Flemming, H.C., and Wingender, J. (2005). Alginate acetylation influences initial surface colonization by mucoid *Pseudomonas aeruginosa*. Microbiol. Res. *160*, 165–176.

Tielker, D., Hacker, S., Loris, R., Strathmann, M., Wingender, J., Wilhelm, S., Rosenau, F., and Jaeger, K.E. (2005). *Pseudomonas aeruginosa* lectin LecB is located in the outer membrane and is involved in biofilm formation. Microbiology *151*, 1313–1323.

Tolker-Nielsen, T., Brinch, U.C., Ragas, P.C., Andersen, J.B., Jacobsen, C.S., and Molin, S. (2000). Development and dynamics of *Pseudomonas* sp. biofilms. J. Bacteriol. *182*, 6482–6489.

Tomaras, A.P., Dorsey, C.W., Edelmann, R.E., and Actis, L.A. (2003). Attachment to and biofilm formation on abiotic surfaces by *Acinetobacter baumannii*: involvement of a novel chaperone-usher pili assembly system. Microbiology *149*, 3473–3484.

Tomich, M., Planet, P.J., and Figurski, D.H. (2007). The *tad* locus: postcards from the widespread colonization Island. Nat. Rev. Microbiol. *5*, 363–375.

Torres, A.G., Perna, N.T., Burland, V., Ruknudin, A., Blattner, F.R., and Kaper, J.B. (2002). Characterization of Cah, a calcium-binding and heat-extractable autotransporter protein of enterohaemorrhagic *Escherichia coli*. Mol. Microbiol. *45*, 951–966.

Toutain, C.M., Caizza, N.C., Zegans, M.E., and O'Toole, G.A. (2007). Roles for flagellar stators in biofilm formation by *Pseudomonas aeruginosa*. Res. Microbiol. *158*, 471–477.

Townsend, J.P., Nielsen, K.M., Fisher, D.S., and Hartl, D.L. (2003). Horizontal acquisition of divergent chromosomal DNA in bacteria: effects of mutator phenotypes. Genetics *164*, 13–21. Erratum in: Genetics (2003). *164*, 1241.

Ulett, G.C., Mabbett, A.N., Fung, K.C., Webb, R.I., and Schembri, M.A. (2007). The role of F9 fimbriae of uropathogenic *Escherichia coli* in biofilm formation. Microbiology *153*, 2321–2331.

Valle, J., Da Re, S., Henry, N., Fontaine, T., Balestrino, D., Latour-Lambert, P., and Ghigo, J.M. (2006). Broad-spectrum biofilm inhibition by a secreted bacterial polysaccharide. Proc. Natl. Acad. Sci. USA *103*, 12558–12563.

Valle, J., Mabbett, A.N., Ulett, G.C., Toledo-Arana, A., Wecker, K., Totsika, M., Schembri, M.A., Ghigo, J.M., and Beloin, C. (2008). UpaG, a new member of the trimeric autotransporter family of adhesins in uropathogenic *Escherichia coli*. J. Bacteriol. *190*, 4147–4161.

Vallet, I., Olson, J.W., Lory, S., Lazdunski, A., and Filloux, A. (2001). The chaperone/usher pathways of *Pseudomonas aeruginosa*: identification of fimbrial gene clusters (*cup*) and their involvement in biofilm formation. Proc. Natl. Acad. Sci. U.S.A. *98*, 6911–6916.

Varga, J.J., Therit, B., and Melville, S.B. (2008). Type IV pili and the CcpA protein are needed for maximal biofilm formation by the Gram-positive anaerobic pathogen *Clostridium perfringens*. Infect. Immun. *76*, 4944–4951.

Verstraeten, N., Braeken, K., Debkumari, B., Fauvart, M., Fransaer, J., Vermant, J., and Michiels, J. (2008). Living on a surface: swarming and biofilm formation. Trends Microbiol. *16*, 496–506.

Vidal, O., Longin, R., Prigent-Combaret, C., Dorel, C., Hooreman, M., and Lejeune, P. (1998). Isolation of an *Escherichia coli* K-12 mutant strain able to form biofilms on inert surfaces: involvement of a new *ompR* allele that increases curli expression. J. Bacteriol. *180*, 2442–2449.

Vrany, J.D., Stewart, P.S., and Suci, P.A. (1997). Comparison of recalcitrance to ciprofloxacin and levofloxacin exhibited by *Pseudomonas aeruginosa* biofilms displaying rapid-transport characteristics. Antimicrob. Agents Chemother. *41*, 1352–1358.

Vuong, C., Kocianova, S., Voyich, J.M., Yao, Y., Fischer, E.R., DeLeo, F.R., and Otto, M. (2004). A crucial role for exopolysaccharide modification in bacterial biofilm formation, immune evasion, and virulence. J. Biol. Chem. *279*, 54881–54886.

Wagner, V.E., Gillis, R.J., and Iglewski, B.H. (2004). Transcriptome analysis of quorum-sensing regulation and virulence factor expression in *Pseudomonas aeruginosa*. Vaccine *22*, S15–20.

Waite, R.D., Paccanaro, A., Papakonstantinopoulou, A., Hurst, J.M., Saqi, M., Littler, E., and Curtis, M.A. (2006). Clustering of *Pseudomonas aeruginosa* transcriptomes from planktonic cultures, developing and mature biofilms reveals distinct expression profiles. BMC Genomics *7*, 162.

Walker, T.S., Tomlin, K.L., Worthen, G.S., Poch, K.R., Lieber, J.G., Saavedra, M.T., Fessler, M.B., Malcolm, K.C., Vasil, M.L., and Nick, J.A. (2005). Enhanced *Pseudomonas aeruginosa* biofilm development mediated by human neutrophils. Infect. Immun. *73*, 3693–3701.

Wang, X., Preston, J.F. 3rd, and Romeo, T. (2004). The *pgaABCD* locus of *Escherichia coli* promotes the synthesis of a polysaccharide adhesin required for biofilm formation. J. Bacteriol. *186*, 2724–2734.

Wang, X., Kim, Y., and Wood, T.K. (2009). Control and benefits of CP4–57 prophage excision in *Escherichia coli* biofilms. ISME J. *3*, 1164–1179.

Watnick, P.I., and Kolter, R. (1999). Steps in the development of a *Vibrio cholerae* El Tor biofilm. Mol. Microbiol. *34*, 586–595.

Watnick, P., and Kolter, R. (2000). Biofilm, city of microbes. J. Bacteriol. *182*, 2675–2679.

Watnick, P.I., Lauriano, C.M., Klose, K.E., Croal, L., and Kolter, R. (2001). The absence of a flagellum leads to altered colony morphology, biofilm development and virulence in *Vibrio cholerae* O139. Mol. Microbiol. *39*, 223–235.

Webb, J.S., Thompson, L.S., James, S., Charlton, T., Tolker-Nielsen, T., Koch, B., Givskov, M., and Kjelleberg, S. (2003). Cell death in *Pseudomonas aeruginosa* biofilm development. J. Bacteriol. *185*, 4585–4592.

Webb, J.S., Lau, M., and Kjelleberg, S. (2004). Bacteriophage and phenotypic variation in *Pseudomonas aeruginosa* biofilm development. J. Bacteriol. *186*, 8066–8073.

Weigel, L.M., Donlan, R.M., Shin, D.H., Jensen, B., Clark, N.C., McDougal, L.K., Zhu, W., Musser, K.A., Thompson, J., Kohlerschmidt, D., *et al.* (2007). High-level vancomycin-resistant *Staphylococcus aureus* isolates associated with a polymicrobial biofilm. Antimicrob. Agents Chemother. *51*, 231–238.

Weitao, T. (2009). Multicellularity of a unicellular organism in response to DNA replication stress. Res. Microbiol. *160*, 87–88.

Wells, T.J., Tree, J.J., Ulett, G.C., and Schembri, M.A. (2007). Autotransporter proteins: novel targets at the bacterial cell surface. FEMS Microbiol. Lett. *274*, 163–172.

Wells, T.J., Sherlock, O., Rivas, L., Mahajan, A., Beatson, S.A., Torpdahl, M., Webb, R.I., Allsopp, L.P., Gobius, K.S., Gally, D.L.,*et al.* (2008). EhaA is a novel autotransporter protein of enterohemorrhagic *Escherichia coli* O157:H7 that contributes to adhesion and biofilm formation. Environ. Microbiol. *10*, 589–604.

Wells, T.J., McNeilly, T.N., Totsika, M., Mahajan, A., Gally, D.L., and Schembri, M.A. (2009). The *Escherichia coli* O157:H7 EhaB autotransporter protein binds to laminin and collagen I and induces a serum IgA response in O157:H7 challenged cattle. Environ. Microbiol. *11*, 1803–1814.

Werner, E., Roe, F., Bugnicourt, A., Franklin, M.J., Heydorn, A., Molin, S., Pitts, B., and Stewart, P.S. (2004). Stratified growth in *Pseudomonas aeruginosa* biofilms. Appl. Environ. Microbiol. *70*, 6188–6196.

West, S.A., Griffin, A.S., Gardner, A., and Diggle, S.P. (2006). Social evolution theory for microorganisms. Nat. Rev. Microbiol. *4*, 597–607.

Whitchurch, C.B., Tolker-Nielsen, T., Ragas, P.C., and Mattick, J.S. (2002). Extracellular DNA required for bacterial biofilm formation. Science *295*, 1487.

Whiteley, M., Bangera, M.G., Bumgarner, R.E., Parsek, M.R., Teitzel, G.M., Lory, S., and Greenberg, E.P. (2001). Gene expression in *Pseudomonas aeruginosa* biofilms. Nature *413*, 860–864.

Wilhelm, S., Gdynia, A., Tielen, P., Rosenau, F., and Jaeger, K.E. (2007). The autotransporter esterase EstA of *Pseudomonas aeruginosa* is required for rhamnolipid production, cell motility, and biofilm formation. J. Bacteriol. *189*, 6695–6703.

Wood, T.K. (2009). Insights on *Escherichia coli* biofilm formation and inhibition from whole-transcriptome profiling. Environ. Microbiol. *11*, 1–15.

Wood, T.K., González Barrios, A.F., Herzberg, M., and Lee, J. (2006). Motility influences biofilm architecture in *Escherichia coli*. Appl. Microbiol. Biotechnol. *72*, 361–367.

Wozniak, D.J., Wyckoff, T.J., Starkey, M., Keyser, R., Azadi, P., O'Toole, G.A., and Parsek, M.R. (2003). Alginate is not a significant component of the extracellular polysaccharide matrix of PA14 and PAO1 *Pseudomonas aeruginosa* biofilms. Proc. Natl. Acad. Sci. U.S.A. *100*, 7907–7912.

Wright, K.J., Seed, P.C., and Hultgren, S.J. (2007). Development of intracellular bacterial communities of uropathogenic *Escherichia coli* depends on type 1 pili. Cell. Microbiol. *9*, 2230–2241.

Wuertz, S., Okabe, S., and Hausner, M. (2004). Microbial communities and their interactions in biofilm systems: an overview. Water Sci. Technol. *49*, 327–336.

Xavier, J.B., and Foster, K.R. (2007). Cooperation and conflict in microbial biofilms. Proc. Natl. Acad.Sci. U.S.A. 104, 876–881.

Xicohtencatl-Cortes, J., Monteiro-Neto, V., Saldaña, Z., Ledesma, M.A., Puente, J.L., and Girón, J.A. (2009). The type 4 pili of enterohemorrhagic *Escherichia coli* O157:H7 are multipurpose structures with pathogenic attributes. J. Bacteriol. *191*, 411–421.

Xu, K.D., Stewart, P.S., Xia, F., Huang, C.T., and McFeters, G.A. (1998). Spatial physiological heterogeneity in *Pseudomonas aeruginosa* biofilm is determined by oxygen availability. Appl. Environ. Microbiol. *64*, 4035–4039.

Yildiz, F.H., and Schoolnik, G.K. (1999). *Vibrio cholerae* O1 El Tor: identification of a gene cluster required for the rugose colony type, exopolysaccharide production, chlorine resistance, and biofilm formation. Proc. Natl. Acad. Sci. U.S.A. *96*, 4028–4033.

Zavialov, A.V., and Knight, S.D. (2007). A novel self-capping mechanism controls aggregation of periplasmic chaperone Caf1M. Mol. Microbiol. *64*, 153–164.

Zhu, J., and Mekalanos, J.J. (2003). Quorum sensing-dependent biofilms enhance colonization in *Vibrio cholerae*. Dev. Cell 5, 647–656.

Part III

Poisoning the Host by Toxins

Toxins Damaging Cellular Membranes: Paradigms and Molecular Features

7

Joseph E. Alouf

Abstract

The repertoire of the bacterial cytolytic pore-forming protein toxins (PTFs) comprises ca. 86 identified members produced by both Gram-positive and -negative bacteria. The essential functional feature of these cytolysins is their capacity to provoke the formation of hydrophilic pores in the cytoplasmic membranes of target eukaryotic cells. This process results from the binding of the proteins on the cell surface, followed by their oligomerization, which leads to the insertion of the oligomers into the membrane and formation of protein-lined channels. This insertion provokes the impairment of the osmotic balance of the cell and subsequent cytolysis. The classification and molecular aspects of a number of important PTFs are described, as well as the pathophysiological features of some of these cytolysins and their relation to human diseases.

Introduction

Among the repertoire of the ca. 350 bacterial protein toxins identified so far (Alouf and Popoff, 2006), about 120 (35% of the whole members of the repertoire) constitute the superfamily of the so-called membrane-damaging toxins (MDTs). These toxins, also known as cytolysins or cytolytic toxins, are characterized by their property to damage or disrupt the integrity of the cytoplasmic phospholipid bilayer membrane of appropriate human and/or animal cells. The impairment of membrane integrity causes osmotic imbalance reflected by cell swelling due to water influx and dissipation of electrochemical gradients, which may lead to cell lysis and death (van der Goot, 2001; Alouf *et al.*, 2006). The disruption of the cytoplasmic membrane provokes the release of intracytoplasmic components outside the cell (e.g. haemoglobin by MDT-damaged erythrocytes). Moreover, the damage of the cytoplasmic membrane may be followed by that of the membranes surrounding intracytoplasmic organelles (e.g. lysosomes, mitochondria, endoplasmic reticulum, various bodies or granules) and thereby the release of their internal contents, particularly certain ions (Ca^{++}, enzymes or pharmacological effectors, such as serotonin, cytokines, inflammatory factors) (Launay and Alouf, 1979; Bhakdi and Tranum-Jensen, 1984; Goebel and Kreft 1997).

Many MDTs have been shown to be lethal for experimental animals and to damage the organs and tissues of the host. Moreover, several lines of experimental evidence and clinical observations demonstrated (particularly by genetic deletions) that many toxins contribute to the virulence of the bacteria and play a pivotal role in bacterial pathogenesis in humans and animals. A wide range of information may be found in recent textbooks (Burns *et al.*, 2001; van der Goot, 2001; Alouf and Popoff, 2006).

Historical background

Historically, the lytic actions of most MDTs were first identified (some of them at the turn of the twentieth century) by virtue of their lytic action *in vitro* on erythrocyte (human or from other animal species) suspensions in saline (haemolysis) or the formation of haemolysis zones around bacterial colonies growing on blood agar plates. For this reason, the toxins were first called 'haemolysins'. This approach is still used for the phenotypic characterization of haemolytic bacteria and, consequently, for the relevant haemolytic toxins. However, it became clear from an early stage, that many if not all bacterial haemolysins acted as damaging agents on a variety of other eukaryotic cells and tissues, including tumoral cells. Moreover, many MDTs caused tissue damage or death when injected into experimental animals, thereby justifying their status as 'toxins'. Interestingly, a number of MDTs failed to lyse erythrocytes, while they damaged other eukaryotic cells, particularly leucocytes (leucocidins), produced by a number of bacterial species (*Pasteurella* spp., *Bordetella* spp., *Pseudomonas* spp., *Escherichia coli, Staphylococcus* spp.). The repeats in toxin (RTX) cytolysins from Gram-negative bacteria are typical examples of leucocidins (Welch, 2001; Oxhamre and Richter-Dahlfors, 2003). For this reason, Alan Bernheimer (1970), a pioneer in the field of membrane-damaging toxins, proposed the terms 'cytotoxin' or 'cytolytic toxin' in an attempt to describe more accurately the range of the biological activities of MDTs.

A new period in the study of bacterial cytolysins started at the turn of the 1970s. It concerned the molecular mechanism of actions of these proteins on various eukaryotic cells or cell lines, isolated cell membranes (ghosts) and artificial model membrane systems, such as phospholipid films and liposomes investigated by a variety of biochemical, physical and electron microscopical techniques. A considerable body of literature is available for this important field of bacterial toxins. The wealth of biochemical, cellular, genetic and structural knowledge on MDTs accumulated in the past 15 years is illustrated by four major achievements:

i the discovery of more than 40 novel cytolysins and the finding that many of them could be grouped into specific families.
ii the cloning and sequencing of the structural genes of ca. 70 MDTs, many of them being members of the newly defined pathogenicity islands (Kaper and Hacker, 1999).
iii the outstanding progress in the elucidation of the mechanisms of action of many MDTs investigated by the tools and concepts of cell biology, as described below in the case of the pore-forming cholesterol-dependent cytolysins (Heuck *et al.*, 2003) and certain members oft the RTX family (Welch, 2001; Sotomayor *et al.*, 2010).
iv the establishment by X-ray crystallography to determine the three-dimensional structure of many MDTs in the past 12 years (Lesieur *et al.* 1997; Rossjohn *et al.*, 1997; Ramachandran *et al.*, 2004; Parker and Feil, 2005; Alouf and Popoff, 2006).

Functional classification of the MDTS

The progress in the study of the mechanisms of membrane damage of target cells by this wide group of proteins allowed investigators to distinguish three main classes of MDTs (Bernheimer and Rudy 1986):

i The toxins which enzymatically hydrolyse the phospholipids constituting the bilayer membrane of eukaryotic cells. The majority of this class of toxins exhibits phospholipase C activity, some of them being zinc metallophospholipases or they display sphingomyelinase activity and in some cases phospholipase D activity. *Clostridium perfringens* α-toxin (phospholipase C), *Staphylococcus aureus* β-toxin (sphingomyelinase C) and *Vibrio damselae* haemolysin (phospholipase D) are among the most important toxins of this class.

From an historical point of view, the first bacterial protein toxin whose molecular mechanism of action was identified was the *C. perfringens* (*welchii*) α-toxin (Macfarlane and Knight, 1941).

ii Toxins exhibiting a detergent-like (surfactant) activity resulting in membrane 'solubilization' and/or partial insertion into the hydrophobic regions of target membranes. This appears the case for the 26-aminoacid δ and δ–like toxins of *S. aureus*, *Staphylococcus haemolyticus* and *Staphylococcus lugdunensis*, *Bacillus subtilis* surfactin, probably *Streptococcus pyogenes* streptolysin S and the heat-stable haemolysin from *Pseudomonas aeruginosa*.

iii The family of the pore-forming cytolytic toxins (PFTs) which create channels (pores) through the cytoplasmic membrane (7–9 nm) of target cells. The three families comprise 26, 10 and ca. 86 proteins, respectively.

In the present review, we will briefly describe the major molecular, functional and certain pathophysiological features of the family of PFTs.

The concept of pore-forming toxins

This concept concerns a general process of cell membrane disorganization elicited by a great number of proteins of prokaryotic origin (bacterial toxins) or from eukaryotic organisms (complement, perforin). The essential feature of this process is the formation of hydrophilic pores (channels) in the cytoplasmic membrane of target cells, resulting from the insertion of foreign animal, plant or bacterial proteins into the membrane, followed by the impairment of cell permeability and thereby cell destruction (lysis).

Historically, the first proposal that proteins may create pores in cell membranes concerns the mechanism of erythrocyte lysis by insertion of the C_5–C_9 components ('membrane complex attack') of plasma complement described by Manfred Mayer (1972) and later on by Bhakdi and Tranum-Jansen (1984). This process was rapidly extended to an increasing number of cytolytic bacterial protein toxins since the 1980s. Several lines of experimental evidence led to postulate the following scheme: the toxins are released by the bacteria as water -soluble monomeric proteins and then undergo a series of remarkable changes upon interacting with target cell membranes. Toxin binding to the membranes facilitates the concentration of the monomers and their transition to form non-covalently-associated oligomers. Toxin molecule oligomerization leads to an amphipathic state (β-barrel profile), insertion into the membrane and the formation of protein-lined pores of various sizes depending on the toxin involved. In these pores, the hydrophobic side of the toxin molecule is exposed to the membrane acyl chains and the hydrophilic sides line up the pore (Shatursky *et al.*, 1999; Ramachandran *et al.*, 2004; Valeva *et al.*, 2005). Toxin receptors on membranes may be either specific or non-specific. Cholesterol has been shown to be specific for the sulphydryl-activated toxins (cholesterol-binding toxins) (Alouf, 2000). Interestingly, aerolysin and pertussis toxin share a common receptor-binding domain (Rossjohn *et al.*, 1997).

Visualization of membrane-associated oligomers of PFTs

Since the pioneering electron microscopic studies of Dourmashkin and Ross (1966) on erythrocyte damage by complement and streptolysin O, several investigations on PFTs have been reported in this respect. These authors and many others (Bernheimer and Rudy, 1986; Bhakdi and Tranum-Jensen 1988, 1991) observed the formation of ring- and arc-shaped structures by electron microscopy of various cholesterol-binding toxins, particularly streptolysin O, pneumolysin and perfringolysin O. The recent detailed studies on these toxins revealed that the oligomeric structures on damaged membranes contained ca. 50 subunits and two concentric rings of streptolysin O monomers were observed on erythrocytes treated with this toxin (Sekiya *et al.*, 1996; Palmer, 2001).

Sizes of the toxin-induced pores

The pores elicited by the toxins investigated to date can differ widely in size. The smallest pores formed in mammalian cell membranes have been sized to around 1–2 nm in effective diameter. This is the case of the *S. aureus* α-toxin, the *E. coli* α-haemolysin, the *Aeromonas* aerolysin and the *Vibrio* El Tor haemolysin (Bakhdi *et al.*, 1996). In contrast, the cholesterol-binding toxins streptolysin O, pneumolysin and perfringolysin O create holes up to 25–30 nm in diameter (Sekiya *et al.*, 1996).

Pore-forming toxins: classification and molecular characteristics

The majority of PFTs are simple single-chain proteins secreted by the bacteria (extracellular toxins) into the surrounding environment. However, some of them may remain associated at the surface of these microorganisms (e.g. cell-bound form of *S. pyogenes* streptolysin S, *Bordetella pertussis* adenylate cyclase-haemolysin) or remain located in the cytoplasm (intracytoplasmic toxins, e.g. pneumolysin from *Streptococcus pneumoniae)*. The molecular weights of PFTs range from ca. 2.5 kDa for the *S. aureus* δ-toxin up to 177 kDa for the *B. pertussis* adenylate cyclase. However, many PFTs show a broad diversity of molecular topology, complex gene structures, as well as architectural and functional originality (Alouf *et al.*, 2006).

The repertoire of the pore-forming toxins

The cloning and sequencing of the majority of the structural genes of PFTs over the past 20 years and the purification of a great number of these toxins, as well as the determination of the three–dimensional structure of many of them and the progress in our understanding of the mechanism of action of these cytolysins allowed investigators to establish a rational classification of most PFTs into 'congeneric' and 'individualistic' toxins, as initially suggested by Bernheimer and Rudy (1986). The former category comprises those PFTs that share close structural and/or similar physico-chemical and functional lytic properties and mechanisms of action, while the latter concerns PFTs exhibiting an appreciable degree of individuality or uniqueness.

Individualistic pore-forming toxins

This group comprises *S. aureus* α-toxin, *Aeromonas* aerolysin, *Enterococcus faecalis* and *Bacillus cereus* bi-and tri-component toxins, *Legionella pneumophila* legiolysin, *P. aeruginosa* leucotoxin, *Clostridium septicum* α-toxin. Interestingly, the *S. aureus* α-toxin and the bi-component leucocidins, which are distant in sequence, are similar in structure. This family of cytotoxins is beyond the scope of the present review. We refer the reader to the books published by van der Goot (2001), Burns *et al.* (2003), Alouf and Popoff (2006).

Congeneric pore-forming toxins

Most members of PFTs belong to this category of cytolysins. They are produced by either Gram-positive or -negative bacteria. The main features of some congeneric PFTs are hereafter briefly summarized.

Cytolytic toxins from Gram-positive bacteria

These cytolysins comprise a large number of families, including cholesterol-dependent cytolysins (CDC).

The family of the cholesterol-dependent cytolysins

This family is a prototype of simple single-gene encoded pore-forming toxins. These 50- to 60-kDa single chain proteins, formerly known as the SH-activated toxins, constitute the largest group of structurally related bacterial toxins (Palmer, 2001; Giddings *et al.*, 2003; Heuck *et al.*, 2003; Alouf *et al.*, 2006; Farrand *et al.*, 2010). These toxins are a group of β-barrel pore-forming proteins produced and expressed as water-soluble monomers. They are highly related at the primary, secondary and tertiary structural levels. CDCs constitute the largest family of PFTs. The first CDCs (perfringolysin O and streptolysin O) were described over a century ago (Alouf, 2000). Since then, ca. 30 members have been identified and many of these toxins function as primary or ancillary pathogenesis factors.. They are produced by eight genera (27 species):

1. *Streptococcus* (eight species: *S. pyogenes*, *S. equisimilis*, *S. canis*, *S. pneumoniae*, *S. suis*, *S. intermedius*, *S. mitis*, S. sp. 597);
2. *Bacillus* (three species: *B. cereus*, *B. thuringiensis*, *B. anthracis*);
3. *Brevibacillus* (one species: *B. latersporus*);
4. *Paenibacillus* (one species: *P. alvei*);
5. *Clostridium* (nine species: *C. tetani*, *C. botulinum*, *C. perfringens*, *C. septicum*, *C. histolyticum*, *C. novyi*, *C. chauvoei*, *C. bifermentans*, *C. sordellii*);

6 *Listeria* (three species: *L. monocytogenes, L. ivanovii, L. seeligeri*);
7 *Arcanobacterium* (one species: *A. pyogenes*);
8 *Gardnerella* (one species: *G. vaginalis*).

General characteristics of CDCs

These toxins are secreted into the extracellular medium, except pneumolysin, which remains intracytoplasmic (absence of a signal peptide) and is only released upon bacterial lysis. Among the CDC-producing species only *L. monocytogenes* and *L. ivanovii* are intracellular pathogens, which grow and release their toxins in the phagocytic cells.

CDCs share many features. They are antigenically related, lethal to animals and highly lytic towards eukaryotic cells, including erythrocytes. Their lytic and lethal properties are suppressed by sulfhydryl-group blocking agents and reversibly restored by thiols or other reducing agents (Alouf, 2000). These properties are irreversibly suppressed by very low concentrations of cholesterol and other 3β-hydroxysterols. Membrane cholesterol is thought to be the toxin-binding site at the surface of target eukaryotic cells. Toxin molecules bind as monomers to the membrane surface with subsequent oligomerization (ca. 50 monomers) into arc- and ring-shaped structures surrounding large pores generated by this process (Sekiya *et al.*, 1996; Tweten *et al.*, 2001). Interestingly, the eukaryotic sea anemone *Metridium senile* produces a cholesterol-binding cytolysin sharing several biological properties of bacterial CDCs, including the formation of arc- and ring-shaped structures at the surface of target cells (Bernheimer *et al.*, 1979).

The structural genes of the toxins cloned and sequenced to date are all chromosomal. The primary structure of the proteins deduced from the nucleotide sequences shows obvious similarities, particularly in the carboxy-terminal part, and a characteristic common consensus sequence (undecapeptide) containing a unique Cys residue (ECTGLAWEWWR); except for *Arcanobacterium pyolysin* and *intermidilysin*, which do not contain this residue long supposed absolutely necessary for biological activity. Indeed, the genetic replacement of the Cys residue in the undecapeptide by other amino acids demonstrated that this residue is not *per se* essential for toxin function. Other residues in the undecapeptide have been mutagenized, particularly the Trp (W) residues. One of them appeared critical for lytic activity. The elucidation of the 3D structure of perfringolysin O provided important information on the structure–activity relationship of this CDC, which appeared to comprise four domains (Rossjohn *et al.*, 1997; Lesieur *et al.*, 1997). The Trp-rich region in the undecapeptide within domain 4 is assumed to conformationally adapt to cholesterol.

Oligomerization of the CDCs

After binding to the target cells, the soluble, monomeric forms of CDC molecules oligomerize on the membrane surface forming a ring-like prepore complex, and finally insert a large β-barrel into the membrane (about 250 Å in diameter). Formation of such a large transmembrane structure requires multiple and coordinated conformational changes. The presence of cholesterol in the target membrane is absolutely required for pore formation and therefore, it was long thought that cholesterol was the cellular receptor for these toxins. However, not all the CDCs need cholesterol for binding. Intermedilysin, secreted by *Streptoccocus intermedius*, only binds to membranes containing a protein receptor, but forms pores only if the membrane contains sufficient cholesterol. In contrast, perfringolysin O, secreted by *C. perfringens*, only binds to membranes containing substantial amounts of cholesterol. The mechanisms by which cholesterol regulates the cytolytic activity of the CDCs are not understood at the molecular level. The C-terminus of perfringolysin O is involved in cholesterol recognition, and changes in the conformation of the loops located at the distal tip of this domain affect the toxin–membrane interactions. At the same time, the distribution of cholesterol in the membrane can modulate toxin binding. Recent studies support the concept that there is a dynamic interplay between the cholesterol-binding domain of the CDCs and the excess of cholesterol molecules in the target membrane (Farrand *et al.*, 2010).

Essential role of cholesterol

Cholesterol dependence of the CDC-induced toxicity mechanism is a hallmark of these toxins, yet the identity of the CDC-cholesterol recognition motif has remained elusive. A detailed analysis of membrane interactive structures at the tip of perfringolysin O (PFO) domain 4 reveals that a Thr–Leu pair mediates CDC recognition of, and binding to membrane cholesterol. This motif is conserved in all known CDCs, and conservative changes in its sequence or others are not well tolerated. Thus, the Thr–Leu pair constitutes a common structural basis for mediating CDC-cholesterol recognition and binding.

Atypical CDCs

Vaginolysin (VLY) is unusual among CDCs in that it exhibits exquisite host specificity (Gelber *et al.*, 2008). Binding the human complement regulatory CD59, and not its homologues in other animals, is required for VLY mediated erythrocyte cytolysis. This mechanism is shared by one other member of the CDC family, intermedilysin from *S. intermedius*. In addition to causing erythrocyte lysis, VLY activates the conserved epithelial p38 mitogen-activated protein kinase pathway and induces interleukin-8 production by human epithelial cells. Transfection of human CD59 into non-susceptible cells renders them sensitive to VLY mediated lysis. In addition, a single amino acid substitution in the VLY undecapeptide (P480W) generates a toxoid that does not form pores (Gelber *et al.*, 2008).

Cytolytic toxins of *S. aureus*

This family (Guillet *et al.*, 2004; Prévost *et al.*, 2006) comprises two groups of cytolytic virulence factors: the α-helix α-toxin and the family of the bi-component pore-forming staphylococcal leucocidins. We will review here the family of these bi-component toxins. The leucocidins consist of six classes of S proteins and five classes of F proteins with 30 possible binary combinations. The γ-toxin of this group is also haemolytic. These toxins target the membrane of human macrophages, neutrophils and monocytes into which they insert and form pores. Among these toxins, the so-called Panton-Valentine leucocidin corresponds to the bi-component LukS-PV- LukF-PV, respectively. Both proteins share 30% sequence identity and display highly similar 3D structures. They are organized in three domains: a central β sandwich, a loosely structured rim domain and the stem domain composed of three anti-parallel β-strands. The first step of the pore formation is the interaction of LukS-PV with an unidentified compound of the membrane of the target cell. This triggers the recruitment of LukF-PV to the membrane and a hetero-octameric alternated prepore associates at the cell surface. Interactions of the proteins with the membranes occur at the rim domain. Pore formation results from the extension of the stem domains of each protein into the membrane as a β-hairpin, leading to a 16-strand beta-barrel (Maveyraud *et al.*, 2009).

Streptolysin S

Group A *Streptococcus* (GAS) is an important human pathogen that causes pharyngitis and invasive infections, including necrotizing faciitis. Streptolysin S (SLS) is the cytolytic factor that creates the zone of β-haemolysis surrounding GAS colonies grown on blood agar. The discovery of a potential genetic determinant involved in SLS production, *sagA*, encoding a small peptide of 53 amino acids has been reported (Betschel *et al.*, 1998). Using transposon mutagenesis, chromosomal walking steps, and data from the GAS genome sequencing project (www.genome.ou.edu/strep.html), the authors identified a contiguous nine-gene locus (*sagA* to *sagI*) involved in SLS production. The *sag* locus is conserved among GAS strains, regardless of M protein type. Mutagenesis of each gene in the *sag* operon resulted in an SLS-negative phenotype, whereas targeted mutations (i) upstream of the *sagA* promoter and (ii) downstream of a terminator sequence after *sagI* did not affect SLS production, establishing the functional boundaries of the operon. A Rho-independent terminator sequence between *sagA* and *sagB* appears to regulate the amount of *sagA* transcript produced versus transcript for the entire operon. Reintroduction of the nine-gene *sag* locus on a plasmid vector restored SLS activity to the non-haemolytic *sagA* knockout mutant. Finally, heterologous expression of the intact *sag* operon conferred the SLS β-haemolytic phenotype to the non-haemolytic *Lactococcus lactis*. Therefore, gene products of the GAS *sag* operon are both necessary and sufficient for SLS production. Sequence similarities of the *sag* operon gene products suggest that SLS is related to the bacteriocin family of microbial toxins. The β-haemolytic group G *Streptococcus* chromosome also contains a homologue of the nine- gene group A *sag* operon encoding β-haemolysin SLS (Humar *et al.*, 2002). Interestingly, *Streptococcus iniae*, a major emerging zoonotic fish pathogen was also reported to cause human infections (Facklam *et al.*, 2005; Locke *et al.*, 2007; Baiano and Barnes, 2009). This microorganism was shown to damage erythrocytes, neutrophils, lymphocytes and some tissue cultures through the release of a cytolysin analogous to SLS, sharing 73% sequence similarity with the *S. pyogenes* SLS. The number of genes and their genomic order were identical in the various streptococci investigated.

The possible role played by SLS of GAS in the pathophysiology of streptococcal infections and in post-streptococcal sequelae has been discussed by Ginsburg (1999). The following properties of SLS justify its definition as a virulence factor: (1) its presence on the *Streptococcus* surface in a cell-bound form, (2) its continuous and prolonged synthesis by resting streptococci, (3) its non-immunogenicity, (4) its extractability by serum proteins (albumin, α-lipoprotein), (5) its ability to be transferred directly to target cells, while being protected from inhibitory agents in the milieu of inflammation, (6) its capacity to generate holes in the membrane phospholipids in a large variety of mammalian cells, (7) its ability to synergize with oxidants, proteolytic enzymes and with additional host-derived pro-inflammatory agonists, and (8) its absence in streptococcal mutants associated with a lower pathogenicity for animals. Because tissue damage in streptococcal and post-streptococcal sequelae might be the end result of a distinct synergism between streptococcal and host-derived pro-inflammatory agonists, it is proposed that only cocktails of anti-inflammatory agents, including distinct inhibitors of SLS (phospholipids), γ globulin, inhibitors of reactive oxygen species, proteinases, cationic proteins, cytokines, will be effective in inhibiting the multiple synergistic interactions which lead to fasciitis, myositis and the flesh-eating syndromes, and often develop into sepsis, septic shock and multiple organ failure. The creation of mutants deficient in SLS and in proteases will help to shed light on the specific role played by SLS in the virulence of group A haemolytic streptococci.

CAMP factor

The pore-forming toxin CAMP factor from group B streptococci (Podbielski *et al.*, 1994) and *S. iniae* (Baiano and Barns, 2009) synergistically act with sphingomyelinase-producing *S. aureus* to produce a distinct arrow -shaped area of complete lysis of erythrocytes in sheep-blood agar (Nizet *et al.*, 2000).

Pore-forming families of toxins from Gram-negative bacteria

The family of the multigenic encoded RTX toxins comprises at least 19 members produced by 15 species from *Escherichia*, *Proteus*, *Morganella*, *Pasteurella*, *Mannheimia*, *Actinobacillus*, *Vibrio*, *Bordetella*, *Moraxella* and *Enterobacter*. These toxins behave as leucotoxins and/or haemolysins (Welch, 2001; Oxhamre and Richter-Dahlfors, 2003). RTX motifs are non-apeptide sequences, which are able to fold into an idiosyncratic structure, called the parallel β-roll, in the presence of calcium. The adenylate cyclase toxin (CyaA) produced by *B. pertussis*, the causative agent of whooping cough, is one of the best characterized RTX cytolysins. CyaA contains a C-terminal receptor domain (RD) that mediates toxin binding to the eukaryotic cell receptor. The receptor-binding domain is composed of about 40 RTX motifs organized in five successive blocks (I to V). The RTX blocks are separated by non-RTX flanking regions of variable lengths. It has been shown that block V, with its N- and C-terminal flanking regions, constitutes an autonomous subdomain required for the toxicity of CyaA. It was also shown that the RTX polypeptides, in the absence of calcium, exhibit the hallmarks of intrinsically disordered proteins (Sotomayor *et al.*, 2010).

The family of the bi-component *Serratia* and *Proteus* cytolysins

These toxins comprise two components separately inactive, which become haemolytic and cytolytic in association. They are produced by *Serratia marcescens*, *Proteus mirabilis* and *Proteus vulgaris*, as well as by *Haemophilus ducreyi* and the fish pathogen *Edwardsiella tarda*.

The family of haemolysins of *Vibrio cholerae* and other *Vibrio* species

These toxins are divided into three groups: the haemolysins related to the El Tor pore-forming haemolysin from *V. cholerae* Ol biotype El Tor, the haemolysins related to *Vibrio parahaemolyticus* thermostable direct haemolysins and other haemolysins.

Biological and pharmacological consequences of the membrane damaging effects of pore-forming and other cytolysins

Toxin-producing pathogenic bacteria, frequently subvert host cell functions. In this respect, PFTs and other cytolytic toxins play a pivotal role in the perturbation of the homeostasis of the organism, which may lead directly or indirectly to the pathogenesis of acute and chronic diseases. The subversion of essential functions in the host concerns particularly humoral and cellular adaptive immunity, as well as innate immunity, cytokine network dysregulation, perturbation of signal transduction pathways and of the expression of certain adhesion molecules.

PFTs and other cytolysins as virulence factors

Cell membrane disruption or damage may have important consequences besides the lytic effect per se. It can facilitate bacterial spread through the tissues of the host in concert with non-lytic enzymes released by the invading bacteria, such as hyaluronidases, collagenases and other tissue damaging effectors (certain phospholipases), which contribute to degrade cellular surfaces or matrices.

On the other hand, both cytolytic toxins and certain enzymes (proteases, NADases, etc.) allow the invading bacteria to acquire essential nutrients from the damaged cells, particularly intracellular iron for their growth. Similarly, certain cytokines (e.g. IL-2) released by the damaged cells may also help bacterial development. These effects create favourable conditions to the bacteria to establish appropriate survival niches inside the cell, such as those created via the action of listeriolysin O (Goebel and Kreft, 1997). In this respect, many PFTs and other cytolysins behave as virulence factors of the corresponding bacteria. This aspect of toxin behaviour is well documented in the case of *S. aureus* α-toxin (Bhakdi *et al.*, 1989; Bhakdi and Tranum-Jensen, 1991), certain RTX leucocidins, *E. faecalis* haemolysin, listeriolysin O, pneumolysin, and streptolysin O, *Arcanobacterium pyogenes* pyolysin. Certain membrane-damaging phospholipases are also virulence factors. On the other hand, cell apoptosis has been reported to be elicited by certain toxins, such as listeriolysin O and aerolysin.

Consequences of the biological effects of sublytic concentrations of PFTs on target cells

The most important pathogenic effects of PFTs and other cytolysins do not primarily concern the damage of host's erythocytes (haemolysis) as initially thought (except perhaps for *C. perfringens* α-haemolysin). It rather concerns the damage inflicted to immune system cells, endothelial and epithelial cells, particularly hepatocytes and many internal tissues. As a matter of fact, these effects take place in most cases at sublytic concentrations, which may be more representative of toxin concentrations *in vivo*. Many effectors are released by toxin-damaged cells (particularly monocytes, polymophonuclear neutrophils, dendritic cells, NK cells, lymphocytes).

Concluding remarks

Our knowledge on PFTs and, more generally, on other classes of membrane-damaging toxins has progressed considerably over the past decade concerning their structure, genomic aspects and structure–activity relationships. Remarkable developments were also made concerning their biological effects at the subcellular levels and their pathophysiological properties. The involvement of some of them in the pathogenesis of various diseases is currently well established. However, the role of many other PFTs in acute and chronic infections remains to be elucidated.

References

Alouf, J.E. (2000). Cholesterol-binding cytolytic protein toxins. Int. J. Med. Microbiol. *290*, 351–356.

Alouf, J.E., and Popoff, M.R. (2006). The Comprehensive Sourcebook of Bacterial Protein Toxins. (Academic Press/Elsevier, London, New York).

Alouf, J.E., Billington, S.J., and Jost, B.H. (2006). Repertoire and general features of the family of cholesterol-dependent cytolysins. In The Comprehensive Sourcebook of Bacterial Protein Toxins, Alouf, J.E., and Popoff, M.R., eds. (New York: Academic Press/Elsevier, London), pp. 643–658.

Baiano, J.C., and Barnes, A.C. (2009). Towards control of *Streptococcus iniae*. Emerg. Infect. Dis. *15*, 1891–1896.

Bhakdi, S., and Tranum-Jensen, J (1984). Mechanism of complement cytolysis and the concept of channel forming proteins. Philos. Trans. R. Soc. Lond. B Biol. Sci. *306*, 311–324.

Bhakdi, S., and Tranum-Jensen, J. (1988). Damage to cell membranes by pore-forming bacterial cytolysins. Prog. Allergy *40*, 1–43.

Bhakdi, S., and Tranum-Jensen, J. (1991). Complement lysis: a hole is a hole. Immunol. Today *12*, 318–320.

Bhakdi, S., Muhly, M., Korom, S., and Hugo, F. (1989). Release of interleukin-1β associated with potent cytocidal action of staphylococcal alpha-toxin on human monocytes. Infect. Immun. *57*, 3512–3519.

Bhakdi, S., Bayley, H., Valeva, A., Walev, I., Walker, B., Kehoe, M., and Palmer, M. (1996). Staphylococcal alpha-toxin, streptolysin O, and *Escherichia coli* hemolysin: prototypes of pore-forming bacterial cytolysins. Arch. Microbiol. *165*, 73–79.

Bernheimer, A.W. (1970). Cytolytic toxins of bacteria. In Microbial Toxins, vol. 1, Ajil, S.J., Montie, T.C., and Kadis, S., eds. (Academic Press, New York), pp. 183–212.

Bernheimer, A.W., and Rudy, B. (1986). Interactions between membranes and cytolytic peptides. Biochim. Biophys. Acta *864*, 123–141.

Bernheimer, A.W., Avigad, L.S., and Kim, K. (1979). Comparison of metridiolysin from the sea anemone with thiol-activated cytolysins from bacteria. Toxicon *17*, 69–75.

Betschel, S.D., Borgia, S.M., Barg, N.L., Low, D.E., and De Azavedo, J.C. (1998). Reduced virulence of group A streptococcal Tn*916* mutants that do not produce streptolysin S. Infect. Immun. *66*, 1671–1679.

Burns, D.L., Barbieri, J.T., Iglewski, B.H., and Rappuoli, R. (2003). Bacterial Protein Toxins (ASM Press, Washington DC).

Dourmashkin, R.R., and Rosse, W.F. (1966). Morphologic changes in the membranes of red blood cells undergoing hemolysis. Am. J. Med. *41*, 699–710.

Facklam, R, Elliot, J., Shewmaker, L., and Reingold, A. (2005). Identification and characterization of sporadic isolates of *Streptococcus iniae* isolated from humans. J. Clin. Microbiol. *43*, 933–937.

Farrand, A.J., LaChapelle, S., Hotze, E.M., Johnson, A.E., and Tweten, R.K. (2010). Only two amino acids are essential for cytolytic recognition of cholesterol at the membrane surface. Proc. Natl. Acad. Sci. U.S.A. *107*, 4341–4346.

Gelber, S.E., Aguilar, J.L., Lewis, K.L., and Ratner, A.J. (2008). Functional and phylogenetic characterization of vaginolysin, the human-specific cytolysin from *Gardnerella vaginalis*. J. Bacteriol. *190*, 3896–3903.

Giddings, K.S., Johnson, A.E., and Tweten, R.K. (2003). Redefining cholesterol's role in the mechanism of the cholesterol-dependent cytolysins. Proc. Natl. Acad. Sci. U.S.A. *100*, 11315–11320.

Ginsburg, I. (1999). Is streptolysin S of group A streptococci a virulence factor? APMIS *107*, 1051–1059.

Goebel, W., and Kreft, J. (1997). Cytolysins and the intracellular life of bacteria. Trends Microbiol. *5*, 86–88.

van der Goot, F.G. (2001). Pore-forming Toxins, Current Topics in Microbiology and Immunology, (Springer-Verlag, Berlin Heidelberg).

Guillet, V., Roblin, P., Werner, S., Coraiola, M., Menestrina, G., Monteil, H., Prévost, G., and Mourey, L. (2004). Crystal structure of leucotoxin S component: new insight into the staphylococcal β-barrel pore-forming toxins. J. Biol. Chem. 279, 41028–41037.

Heuck, A.P., Tweten, R.K., and Johnson, A.E. (2003). Assembly and topography of the prepore complex in cholesterol-dependent cytolysins. J. Biol. Chem. *278*, 31218–31225.

Humar, D., Datta, V., Bast, D.J., Beall, B., De Azavedo, J.C., and Nizet, V. (2002). Streptolysin S and necrotising infections produced by group G *Streptococcus*. Lancet *359*, 124–129.

Kaper, J.B., and Hacker, J. (1999). Pathogenicity Islands and Other Mobile Virulence Elements (Washington DC: ASM Press).

Launay, J.M., and Alouf, J.E. (1979). Biochemical and ultrastructural study of the disruption of blood platelets by streptolysin O. Biochim. Biophys. Acta *556*, 278–291.

Lesieur, C., Vécsey-Semjén, B., Abrami, L., Fivaz, M., and van der Goot, G.F. (1997). Membrane insertion: The strategies of toxins (review). Mol. Membr. Biol. *14*, 45–64.

Locke, J.B., Colvin, K.M., Varki, N.,Vicknair, M.R., Nizet, V., and Buchanan, J.T. (2007). *Streptococcus iniae* beta-hemolysin streptolysin S is a virulence factor in fish infection. Dis. Aquat. Org. *76*, 17–26.

Macfarlane, M.G., and Knight B.C. (1941). The biochemistry of bacterial toxins. The lecithinase activity of *Cl. welchii* toxins. Biochem. J. *35*, 884–902.

Mayer, M.M. (1972). Mechanism of cytolysis by complement. Proc. Natl. Acad. Sci. U.S.A. *69*, 2954–2958.

Maveyraud *et al.* (2009). 14th European Workshop on Bacterial Protein Toxins, Obernai, France, 27 June–2 July, p. 73.

Nizet, V., Beall, B., Bast, D.J., Datta, V., Kilburn, L., Low, D.E., and De Azavedo, J.C. (2000). Genetic locus for streptolysin S production by group A *Streptococcus*. Infect Immun. *68*, 4245–4254.

Oxhamre, C., and Richter-Dahlfors, A. (2003). Membrane-damaging toxins: family of RTX toxins. In Bacterial Protein Toxins, Burns, D.L., Barbieri, J.T., Iglewski, B.H., and Rappuoli, R., eds. (ASM Press, Washington DC), pp. 203–214.

Palmer, M. (2001). The family of thiol-activated, cholesterol-binding cytolysins. Toxicon *39*, 1681–1689.

Parker, M.W., and Feil, S.C. (2005). Pore-forming protein toxins: from structure to function. Prog. Biophys. Mol. Biol. *88*, 91–142.

Podbielski, A, Blankenstein, O., and Lütticken, R. (1994). Molecular characterization of the *cfb* gene encoding group B streptococcal CAMP factor. Med. Microb. Immunol. *183*, 239–256.

Prévost, G., Mourey, L., Colin, D.A., Monteil, H., Dalla Serra M., and Menestrina, G. (2006). Alpha-helix and beta-barrel pore-forming toxins (leucocidins, alpha-, gamma-, and delta-cytolysins) of *Staphylococcus aureus*. In The Comprehensive Sourcebook of Bacterial Protein Toxins, Alouf, J.E., and Popoff, M.R., eds. (Academic Press/Elsevier, London, New York), pp. 590–607.

Ramachandran, R., Tweten, R.K., and Johnson, A.E. (2004). Membrane-dependent conformational changes initiate cholesterol-dependent cytolysin oligomerization and intersubunit beta-strand alignment. Nat. Struct. Mol. Biol. *11*, 697–705.

Rossjohn, J., Feil, S.C., McKinstry, W.J., Tweten, R.K., and Parker M.W. (1997). Structure of a cholesterol-binding, thiol-activated cytolysin and a model of its membrane form. Cell *89*, 685–692.

Shatursky, O., Heuck, A.P., Shapard, L.A., Rossjohn, J., Parker, M.W., Johnson, A.E., and Tweten, R. (1999). The mechanism of membrane insertion for a cholesterol-dependent cytolysin: a novel paradigm for pore-forming toxins. Cell *99*, 293–299.

Sekiya, K., Danbara, H., Yase, K., and Futaesaku, Y. (1996). Electron microscopic evaluation of a two-step theory of pore formation by streptolysin O.J. Bacteriol. *178*, 6998–7002.

Sotomayor Pérez, A.C., Karst, J.C., Davi, M., Guijarro, J.I., Ladant, D., and Chenal, A. (2010). Characterization of the regions involved in the calcium-induced folding of the intrinsically disordered RTX motifs from the *Bordetella pertussis* adenylate cyclase toxin. J. Mol. Biol. *397*, 534–549.

Tweten, R.K., Parker, M.W., and Johnson, A.E. (2001). The cholesterol-dependent cytolysins. Curr. Top. Microbiol. Immunol. *257*, 15–33.

Valeva, A., Walev, I., Boukhallouk, F., Wassenaar, T.M., Heinz, N., Hedderich, J., Lautwein, S., Möcking, M., Weis, S., Zitzer, A., and Bhakdi, S. (2005). Identification of the membrane penetrating domain of *Vibrio cholerae* cytolysin as a β-barrel structure. Mol. Microbiol. *57*, 124–131.

Welch, R.A. (2001). RTX toxin structure and function: a story of numerous anomalies and few analogies in toxin biology. In Pore-forming Toxins, Current Topics in Microbiology and Immunology, van der Goot, F.G., ed. (Springer-Verlag, Berlin Heidelberg), pp. 85–111.

Toxins Acting on Intracellular Targets: Only Foes or Also Friends?

8

Teresa Frisan, Riccardo Guidi and Lina Guerra

Abstract

Bacteria possess an arsenal of virulence factors that allow them to colonize, invade and replicate within hostile niches, such as immunocompetent individuals. Bacterial toxins are among the most sophisticated virulence factors. They are highly specific for their target. Specificity is defined at the level of the target cell (presence of the receptor) and at the level of the substrate, since most of them are enzymes and require specific protein–protein interactions to exert their toxic activity. Some of them are also very powerful, and are among the most toxic compounds known to date. To gain access to their targets within the host cell, these molecules have exploited all the possible cellular routes of internalization: from direct translocation through the plasma membrane, to retrograde transport through the Golgi complex and the endoplasmic reticulum. Identification of their mode of action, their receptors and internalization routes has been instrumental to understand the role of toxins in the pathogenesis of bacteria-induced diseases, and has allowed production of prophylactic vaccines. Last but not least, bacterial toxins have been and are still used as tools in biomedical research to characterize basic cellular processes.

General overview

Many bacteria produce diffusible molecules, mostly proteins, which can enter mammalian cells, target specific functions, alter cellular signals, and can finally lead to cell death. Since these toxins are secreted by the producing bacteria and can act at remote distances, they are often defined as 'exotoxins'. Toxins are part of the arsenal of the bacterial virulence factors that favour colonization of the host, evasion of the immune system, entry into the host cells and acquisition of nutrients. In this chapter we will discuss the mode of action of exotoxins that target intracellular components, their contribution to bacterial pathogenicity and the beneficial use in medical treatments. It will be clear that bacterial exotoxins cannot only be regarded as foes, but also as friends.

Most bacterial exotoxins are (i) highly specific, since most of them are enzymes and require direct protein–protein interactions with the cellular target, and (ii) highly toxic. Indeed, some of these molecules are the most powerful human poisons. The *Clostridium botulinum* toxin has a lethal dose 50% (LD_{50}) of roughly 0.005–0.05 μg/kg, while the LD_{50} of the very well known poison strychnine is approximately 10 mg/kg. Owing to their high toxicity, several toxins have been studied as biological weapons. As recently as September 2001, spores of *Bacillus anthracis*, which produce the anthrax toxin, were spread by mail delivery via the US postal office. This resulted in twenty-two confirmed or suspect cases of anthrax infection. Eleven of these were inhalational cases, of whom five died; 11 were cutaneous cases (seven confirmed, four suspected) (Schwartz, 2009). Fig. 8.1 summarizes the location and the mode of action of the most well characterized exotoxins. This information is expanded in Table 8.1, which lists the exotoxins, the producing bacterium, the toxin activity, the cellular target, and the receptor, when identified.

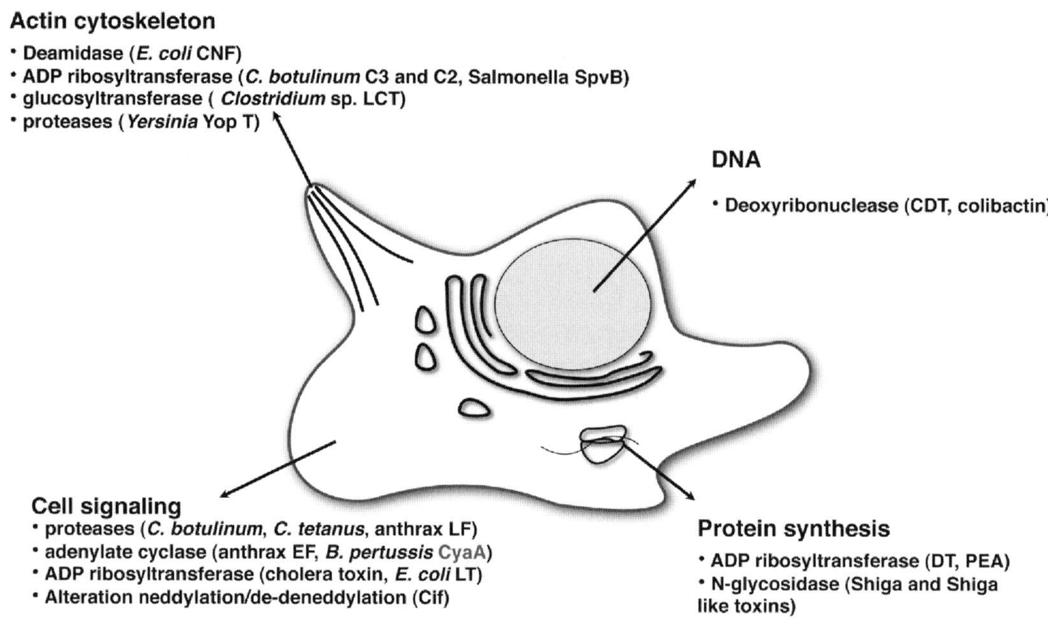

Figure 8.1 Schematic representation of the best characterized bacterial exotoxins and their intracellular targets. DT, diphtheria toxin; PEA, *P. aeruginosa* exotoxin A; CDT, cytolethal distending toxin; LF, lethal factor; EF, oedema factor; CNF, cytotoxic necrotizing factor; LT, labile toxin; CIF, cycle inhibiting factor.

Toxin receptors and the concept of A–B toxins

The initial step in intoxication is the binding of the exotoxin to the target cell. For most of the exotoxins identified, this process involves the recognition of a specific receptor on the surface of the host cells. The nature of the receptors varies depending on the toxin, from glycosphingolipids, such as the GM1 ganglioside which binds to cholera toxin, to surface glycoproteins, such as the tumour endothelial marker 8 (TEM8) and the capillary morphogenesis protein 2 (CMG2) which are the receptors for the anthrax toxin (Bradley *et al.*, 2001; Scobie *et al.*, 2003). Identification of a toxin receptor still remains a major challenge, and only few receptors have been identified, possibly owing to the limited number of molecules present on the target cells or limitation in the technical development (Table 8.1). However, identification of the toxin receptor still represents essential information because it allows a full understanding on how the toxin accesses the target cells and can lead to development of therapeutic intervention to block toxin entry and consequently intoxication.

The expression of the receptor is what defines the tissue tropism and the species specificity of an exotoxin. Some toxins have a very restricted tissue tropism, such as the botulinum toxin (BoNT), known as neurotoxin since it binds to molecules presents on neuronal cells: the G1b-series gangliosides and the membrane protein synaptotagmin (Caleo and Schiavo, 2009). The simultaneous interaction of the botulinum toxin with the ganglioside and the protein receptor represents a high affinity binding, required for further internalization of the toxin (Montecucco, 1986).

Toxins that act within the cytosolic or nuclear compartments require crossing cellular membranes upon receptor binding. Most of these toxins are defined as type III toxins or A-B toxins, because they consist of two parts: the component that binds the cell receptor and mediates internalization (named 'B', for binding), and the component that possesses the enzymatic activity and exerts the toxic effect once internalized (named 'A', for active). It follows that the A subunit is not able to intoxicate the target cells, unless coupled to its B component. The structure of an

Table 8.1 Summary of the best-characterized exotoxins acting intracellularly

Toxin	Bacterium	Activity	Target	Receptor
Anthrax toxins	*B. anthracis*	Metalloproteases (LF) Adenylate cyclase (EF)	MAP2K ATP	TEM8 and CMG2
BoNT	*C. botulinum*	Zn-metalloprotease	Synaptic vesicle components	Synaptotagmin (BoNT B and G)
TeNT	*C. tetanii*	Zn-metalloprotease	Synaptic vesicle components	–
Bot C2	*C. botulinum*	ADP-ribosyl transferase	Actin	–
Bot C3	*C. botulinum*	ADP-ribosyl transferase	Small Rho GTPases	–
CNF, DNT	*E. coli*	Deaminase	Small Rho GTPases	LRP (CNF)
CT, LT	*V. cholerae*, *E. coli*	ADP-ribosyl transferase	G-proteins	GM1 and GD1b (only LT)
CDT	Gram-negative	Deoxyribonuclease	DNA	–
Colibactin	*E. coli*	Deoxyribonuclease	DNA	–
DT	*C. diphtheriae*	ADP-ribosyl transferase	EF-2	EGFR-like and CD9
PEA	*P. aeruginosa*	ADP-ribosyl transferase	EF-2	–
LCT (CGT)	*Clostridium* spp.	Glucosyl transferase	Small GTPases (Rho, Ras)	–
CyaA	*B. pertussis*	Adenylate cyclase	ATP	CD11b/CD18
SptB	*S. enterica*	GAP	Small Rho GTPases	–
SopE and SopE2	*S. enterica*	GEF	Small Rho GTPases	–
Stx, SLTs	*S. dysenteriae*, *E. coli*	N-glycosidase	28S rRNA	Gb3
YopE	*Yersinia* spp.	GAP	Small Rho GTPases	–
YopT	*Yersinia* spp.	Protease	Small Rho GTPases	–
YpkA/YopO	*Yersinia* spp.	GDI	Small Rho GTPases	–

A–B toxin can be quite heterogeneous, as illustrated in Fig. 8.2.

A–B linked by disulfide bond

The diphtheria toxin (DT) is secreted as a single-chain polypeptide by the producing bacterium. Trypsin-like proteases cleave this protein into two fragments (A and B) that remain associated by the formation of covalent disulfide bonds. The B subunit contains the receptor-binding and the translocation domains, required to mediate the internalization of the enzymatically active A fragment which ADP-ribosylates a specific amino acid of the elongation factor (EF) 2, thereby inhibiting protein synthesis (Collier, 1975; Choe *et al.*, 1992). The *Pseudomonas aeruginosa* exotoxin A (PEA) is similar in structure and function to DT (Iglewski *et al.*, 1977; Hwang *et al.*, 1987). The two clostridial tetanus and botulinum toxins are also produced as a single polypeptide, which is then cleaved into the A and B fragments, linked by a disulfide bond. Once internalized into the target cell, the A fragment – a zinc metalloprotease – cleaves key molecules involved in exocytosis of neurotransmitter vesicles, blocking the delivery of specific synaptic signals (Binz and Rummel, 2009).

A–B in a single polypeptide

The cytotoxic necrotizing factor (CNF) is produced by certain *Escherichia coli* strains as a single polypeptide, which contains a putative N-terminal receptor-binding domain (the B domain), a central hydrophobic translocation domain, and a C-terminal active domain (Lemichez *et al.*, 1997); the latter catalyses the deamination of a glutamine residue on the actin-cytoskeleton regulating small GTPases of the Rho family, leading to constitutive activation of these molecules and enhanced actin polymerization (Flatau *et al.*, 1997; Schmidt *et al.*, 1997).

Binary A–B

In the binary toxins, the A and B subunits are produced as two individual non-linked proteins. The B subunit is cleaved upon binding to the cellular receptor, and subsequently oligomerizes to form a homoheptamer that serves as recruitment site for the active (A) subunit. Anthrax toxin from *B. anthracis*, C2 toxin from *C. botulinum*, and the *Clostridium perfringens* iota toxin are among the best-studied A–B binary toxins (Collier and Young, 2003; Barth and Stiles, 2008). Interestingly, the active components of these toxins have

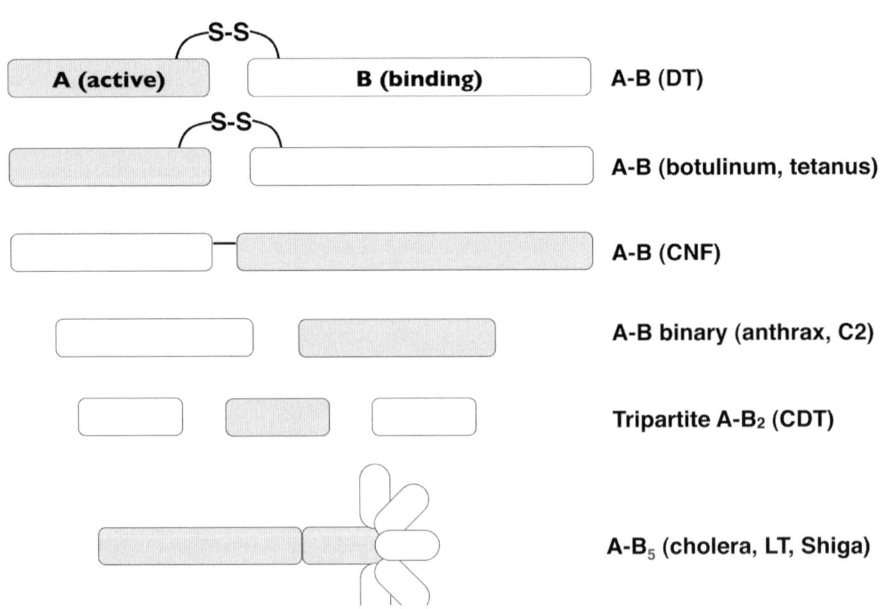

Figure 8.2 Organization of the active (A) and binding (B) domains in the different A–B toxins. DT, diphtheria toxin; CDT, cytolethal distending toxin; CNF, cytotoxic necrotizing factor; LT, labile toxin.

various enzymatic activities: while the C2 and iota toxin are ADP-ribosyltransferases targeting actin, the anthrax toxin is characterized by two distinct A subunits which bind mutually exclusive to the heptameric B component: the oedema factor (EF, an adenylate cyclase) (Leppla, 1982) and the lethal factor (LF, a metalloprotease) (Klimpel *et al.*, 1994).

Tripartite A–B

The family of the cytolethal distending toxins (CDTs) is the only example so far of tripartite A–B bacterial toxins, where the active subunit requires non-covalent association with two other proteins, CdtA and CdC, in order to bind the surface of the target cells, prior to internalization (Nesic *et al.*, 2004; Nesic and Stebbins, 2005). Therefore, CDT may be regarded as AB_2 toxin.

$A–B_5$

The $A–B_5$ toxins share similar structures, but target distinct intracellular molecules and therefore, the effect of the intoxication varies. These toxins consist of five B subunits non-covalently linked to the C-terminus of the single A subunit. This class comprises cholera toxin, the *E. coli* heat labile toxin (LT), pertussis toxin from *Bordetella pertussis*, Shiga toxin from *Shigella dysenteriae*, and Shiga-like toxins 1 and 2 produced by certain *E. coli* strains (Merritt and Hol, 1995; Lacy and Stevens, 1998). The A subunit of the cholera, the *E. coli* LT, and pertussis toxins are ADP-ribosyltransferases, and constitutively activate the α-subunit of trimeric G proteins. The A subunit of Shiga and Shiga-like toxins acts as N-glycosidases, causing the removal of a specific adenine from the 28S ribosomal RNA of the 60S subunit. This depurination results in inhibition of the cellular protein synthesis (Endo *et al.*, 1988).

Toxin internalization

Bacterial toxins that act within the cell need to cross cellular membranes. Interestingly, several key routes of intracellular trafficking have been discovered by studying how bacterial toxins get access to the intracellular environment. Some toxins, such as cholera, diphtheria and Shiga toxins are used as markers for specific endocytosis processes. Therefore, these molecules have been valuable tools to explore cellular processes in biomedical research. In this chapter we will briefly review the main mechanisms of bacterial toxins internalization, which are summarized in Fig. 8.3.

Direct delivery from the plasma membrane

The *B. pertussis* adenylate cyclase (CyaA) is the only toxin known so far to cross directly the plasma membrane of the target cell (Rogel and Hanski, 1992). The toxin is a bimodular molecule containing an adenylate cyclase (AC) domain, the A domain, which binds calmodulin and catalyses the conversion of ATP to cAMP, and an RTX (repeat in toxin) haemolytic domain responsible for binding and translocation of the AC (Glaser *et al.*, 1988). The RTX domain of CyaA binds with high affinity the $\alpha_M\beta_2$ integrin CD11b/CD18 present on macrophages, neutrophils and dendritic cells (Guermonprez *et al.*, 2001); however, this toxin can also exert its effect on a large variety of other cells that are CD11b/CD18 negative, indicating that binding to the host cells is mediated by additional factors (Hanski, 1989). CyaA is synthesized as an inactive precursor, which is converted to the active toxin by a post-translational palmitoylation of the lysine residues Lys_{860} and Lys_{983} mediated by the bacterial CyaC acyltransferase, co-expressed with CyaA (Barry *et al.*, 1991; Hackett *et al.*, 1994). The entry/translocation of CyaA into target cells is strictly dependent on the presence of calcium ions, and the RTX domain is supposed to be directly involved in this process, as it harbours the main low affinity Ca^{2+}-binding sites. The exact mechanism involved in the translocation of the AC domain across the membrane remains still poorly understood. Once within the intracellular environment, the catalytic active domain of CyaA is cleaved and released into the cytosol where, in the presence of calmodulin, it catalyses the uncontrolled conversion of ATP to cAMP (Fig. 8.3). The intoxication results in an alteration of cell signalling and inhibition of the bactericidal functions of professional phagocytes, possibly contributing to the escape of the pathogen from some of the key effectors of the innate (macrophages and neutrophils) and adaptive immunity (dendritic cells) (Boyd *et al.*, 2005).

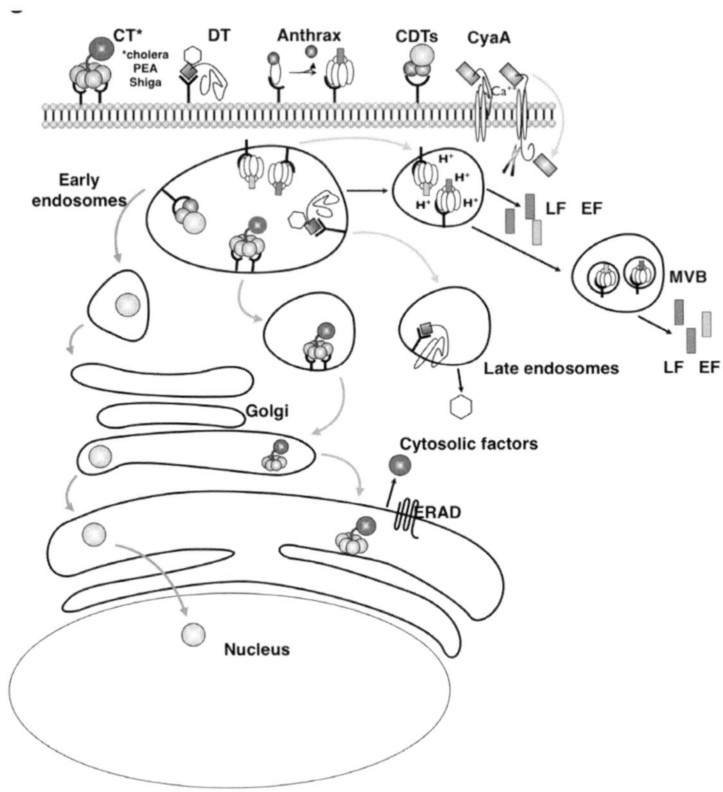

Figure 8.3 Summary of the internalization routes hijacked by bacterial exotoxins. CT, cholera toxin; DT, diphtheria toxin; PEA, *P. aeruginosa* exotoxin A; CDT, cytolethal distending toxin; LF, lethal factor; EF, oedema factor; MBV, multivesicular bodies; ERAD, ER-associated degradation pathway; MB, multivesicular body.

Translocation from endosomes

DT binds to a heparin-binding epidermal growth factor-like precursor and the molecule CD9, via the C-terminal domain (B domain). The toxin is then internalized in the early endosomal compartment via clathrin-coated vesicles. Historically, DT was the first toxin for which translocation was shown to be dependent on an acid environment, indicating that translocation into the cytosol occurs from endosomal compartments (Sandvig and Olsnes, 1980). Within the endosomes, DT is cleaved by furin into two polypetides (A and B), which remain linked by disulfide bonds. The low endosomal pH triggers a conformational change of the translocation domain, located in the central part of the molecule, which allows it to insert into the vesicle membrane and to form an 18–22 Å channel. The active (A) domain of DT is possibly translocated into the cytosol in a partially unfolded state through the channel where, upon re-folding, it exerts the toxic effect (ADP-ribosylation of EF2 and inhibition of protein synthesis) (Fig. 8.3). It still remains to clarify whether translocation occurs spontaneously or requires cellular factors (Falnes and Sandvig, 2000).

Anthrax toxin is also translocated from the endosomal compartment; however, translocation of this toxin seems to be more complex than that described for DT (van der Goot and Young, 2009; Young and Collier, 2007). The 83-kDa B subunit of anthrax toxin (known as the protective antigen, PA_{83}), binds to two receptors: the tumour endothelial marker (TEM) 8 and the capillary morphogenesis protein (CMG) 2, and it is cleaved by furin-like proteolytic activity into a 63-kDa molecule (PA_{63}) that undergoes oligomerization to form a homoheptamer. The pre-pore $(PA_{63})_7$ can bind either of the two active (A) subunits:

the lethal factor (LF) or the oedema factor (EF) to form a non-covalently linked A–B toxin. The receptor–toxin complexes accumulate into microdomains of the membranes, known as lipid rafts, where the cytosolic part of the receptor undergoes ubiquitination, and endocytosis can be triggered. It is still unclear whether anthrax toxin is internalized via clathrin-coated vesicles, since a direct role of this molecule has not been shown yet; however, many of the accessory molecules involved in clathrin-mediated endocytosis, such as dynamin and eps15, are required for toxin internalization. The toxin–receptor complexes are trafficked to the early endosomes, and upon acidification, the PA heptamer is inserted into the membrane, and the partially unfolded EF and LF can be translocated into the cytosolic compartment where they can exert their toxic activity: constitutive production of cAMP (EF, adenylate cyclase) and cleavage of the MAP kinase kinases (LF, metalloprotease) (Fig. 8.3). A complication in the translocation of anthrax toxin derives from the observation that only part of the receptor–toxin complexes are translocated from the early endosomes, since a proportion is found in the intraluminal vesicles (ILV) of the late endosomes, known also as multivesicular bodies (MBVs). Little is known on how the A subunits get access to the cytosol from the ILVs. The most reasonable mechanism is a back fusion of the ILVs with the limiting membrane of the MVBs (Fig. 8.3). Other toxins translocated from the acidic endosomal compartments are: the *E. coli* cytotoxic necrotizing factor (CNF) 1, the *C. botulinum* C2, the clostridial botulinum and tetanus neurotoxins, and the *Clostridium difficile* toxin B (Sandvig and van Deurs, 2005).

Retrograde transport

A group of bacterial toxins reaches their cellular targets trafficking from the endosomal compartment via the Golgi apparatus and the endoplasmic reticulum (ER), before accessing the cytosol or the nuclear compartments (Fig. 8.3). In this paragraph, internalization of Shiga toxin will be used as paradigm of this family of toxins (Sandvig *et al.*, 1992; Sandvig and van Deurs, 2000). The CDTs will also be discussed, since differently from other retrogradely transported toxins which act in the cytosol, they need to access the nuclear compartment.

Shiga toxin (Stx) is an A–B_5 toxin, and its internalization is initiated by binding of the pentameric B subunit to the specific receptor, the glycosphingolipid Gb3 (known also as CD77). This binding triggers a series of signalling events, such as activation of the tyrosine kinase Syk, which induce clathrin phosphorylation and uptake of Stx into clathrin-coated vesicles. The internalized Stx travels to the early endosomes and is then sorted to the *trans* Golgi network (TGN) and Golgi apparatus and further transported to the ER. Some evidence indicates that translocation across the ER membrane requires the Sec61 translocation complex. Sec61 is used to retro-translocate misfolded proteins from the ER to the cytosol, where they can be degraded by cellular proteases, notably the ubiquitin proteasome system. Retro-translocation of StxA via Sec61 involves interaction with two distinct ER chaperones, BiP and HEDJ/ERdj3. How Stx escapes the proteasome degradation is still unclear. The ER-to-cytosol translocation of Stx is also associated with a furin-mediated cleavage of the A subunit, which is nicked, leading to release of the major A_1 fragment that targets the 28S ribosomal RNA, inactivates the ribosomal function, and consequently inhibits protein synthesis.

Other bacterial toxins, such as cholera toxin (CT), the *E. coli* LT, and *P. aeruginosa* A toxin (PEA) are internalized via a similar route (Lord *et al.*, 2005). They are possibly endocytosed using different mechanisms, e.g. caveolae-mediated (CT and LT) or clathrin-dependent (PEA) endocytosis (Sandvig and van Deurs, 2005). The same toxin can also enter via different routes (CT) (Chinnapen *et al.*, 2007). However, trafficking via the early endosomes seems to be a common pathway, independently of early endocytic processes. These toxins further share the requirement of the Sec61 translocon to cross the ER membrane (Sandvig and van Deurs, 2005).

The A subunit of the CDT family is structurally and functionally homologous to the mammalian DNAse I and targets the cellular DNA (Lara-Tejero and Galán, 2000; Elwell *et al.*, 2001). Intoxication with CDT causes DNA damage and activates the classical cellular responses to genotoxic stress, leading to cell cycle arrest or apoptosis (Smith and Bayles, 2006). Similarly to Stx, CT, LT, and PEA, CDT is retrogradely transported from the plasma membrane via the Golgi apparatus to the ER (Guerra *et al.*, 2005). However, exit from the ER does not require the same translocation system as the other mentioned toxins, and biochemical evidences failed to detect the A subunit in the cytosolic compartment (Guerra *et al.*, 2005, 2008a). Based on these findings, it is believed that CDT is directly translocated from the ER to the nucleus where it can exert the toxic activity.

Role of toxins in bacterial pathogenesis

In this section, we will discuss in more details the mechanism of action of some of the best-characterized bacterial exotoxins. Special attention will be devoted to their role in bacterial pathogenesis. In relation to pathogenesis, we can distinguish situations where the toxin is the main cause of the disease (toxin diseases, such as tetanus, botulism, haemolytic–uraemic syndrome, anthrax, and diphtheria) and situations in which the toxin contributes to the disease.

Toxin disease

Clostridial neurotoxins

This family includes the tetanus (TeNT) and the botulinum (BoNT) toxins, the causative agents of tetanus (spastic paralysis) and botulism (flaccid paralysis), respectively (Goonetilleke and Harris, 2004). One TeNT and seven BoNTs (from A to G) have been described and, as discussed in section B, they share a common structure: a heavy (H_C, the B domain) and a light chain (L_C, the A domain) linked by a disulfide bond and non-covalent interactions. The catalytically active domain of TeNT and BoNT is a zinc metalloprotease specific for the SNARE proteins, which regulate the fusion of the synaptic vesicles with the plasma membrane. The result of BoNT or TeNT intoxication is inhibition of the neurotransmitter exocytosis, and consequent inactivation of the neuronal function. The final effect, spastic versus flaccid paralysis, depends on the different neuronal target: TeNT acts at the level of the inhibitory interneurons located in the spinal cord, and impairs release of the inhibitory transmitters, glycine and gamma-aminobutyric acid (GABA), while BoNT acts at the level of the neuromuscular junction (NMJ), and inhibits the release of acetylcholine (Lalli *et al.*, 2003a) and (Fig. 8.4).

The neurotropism of BoNT and TeNT depends on the specificity of the B domain for molecules expressed on the synaptic membrane. Initial binding of these toxins occurs via polysialogangliosides, and a double-receptor model has been proposed (Montecucco, 1986), where the oligosaccharide portion of the polysialogangliosides captures BoNT and TeNT, leading to an enrichment of the toxins on the plasma membrane, facilitating the interaction with the cognate protein receptor. The latter has been identified only for a subset of clostridial neurotoxins: the BoNT/B and BoNT/G bind to synaptotagmins I and II, while BoNT/A binds to the synaptic vesicle protein SV2. Ganglioside G1b mediates the initial binding of TeNT to the plasma membrane of the motor neuron. However, up to date there is no clear identification of the neuronal protein receptor (Binz and Rummel, 2009; Caleo and Schiavo, 2009). Once bound to the plasma membrane of the target cells, BoNT is internalized via clathrin-mediated endocytosis, and acidification of the endosomal vesicles induces insertion of the B domain (H_C) into the membrane, leading to translocation of the A domain (L_C) into the cytosol at the NMJ, where it cleaves the target SNAREs (Breidenbach and Brunger, 2005).

Intoxication with BoNT is known as botulism, which can occur in a variety of forms depending on the entry route of the bacterial spores/toxin. Food-borne botulism occurs after ingestion of food contaminated with *C. botulinum,* which has already germinated from the spore form and has produced the toxin. In this case, the disease arises as a consequence of an intoxication, and is not due to colonization of the host by the bacterium (Goonetilleke and Harris, 2004). Intestinal botulism, instead, occurs when bacterial spores are ingested and germinate in the intestinal tract, producing the toxin *in loco* (Goonetilleke and Harris, 2004). This is the most common type of botulism in infants, and the main cause is accidental ingestion of soil or contaminated honey. The higher frequency of intestinal botulism in infants may be due to the fact that the intestinal

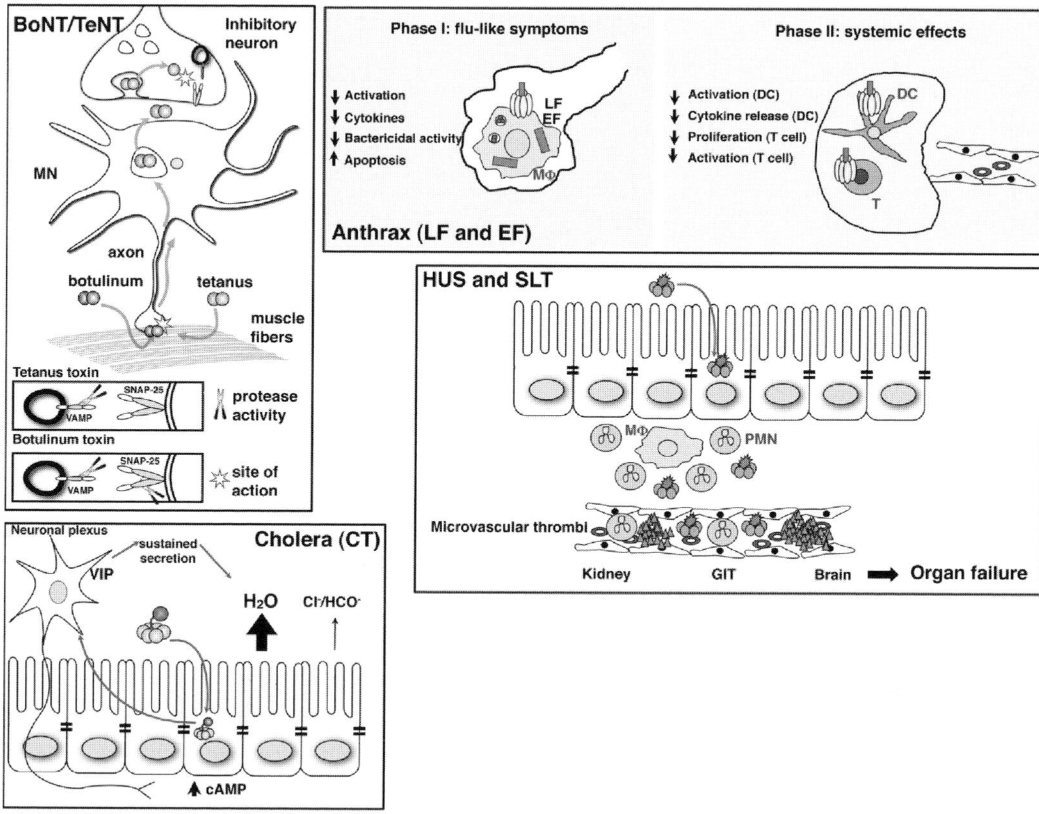

Figure 8.4 Toxin diseases. The different panels represent the role of *Clostridium* neurotoxins (BoNT and TeNT), cholera toxin (CT), Shiga-like toxin (SLT) and anthrax (lethal factor, LF, and oedema factor, EF) in disease pathogenesis. MN, motor neuron; VIP, vasointestinal peptide; HUS, haemolytic uremic syndrome; GIT, gastro-intestinal tract; MF, macrophages; DC dendritic cells; T, T lymphocytes; PMN, neutrophils.

tract of newborns is still not fully colonized by the intestinal protective microflora, and therefore represents a suitable environment for germination of the ingested *C. botulinum* spores. Based on this, honey should not be given to infants younger than 12 months. Wound botulism depends on contamination of deep penetrating wounds with bacterial spores. In conditions when blood supply is reduced or interrupted owing to extensive tissue damage, the spores find suitable conditions for germination (Goonetilleke and Harris, 2004). All forms of botulism produce similar effects: initial involvement of the cranial nerves, leading to blurred vision, dysphonia or dysphagia, followed by acute symmetric descending flaccid paralysis that may lead to respiratory failure. Any case of descending flaccid paralysis, without fever, should be considered a possible intoxication by BoNT. The incubation period and the onset of the disease depend on the amount of toxin absorbed. In food-borne botulism, this can vary from 2 h to 36 h, and up to a week (Goonetilleke and Harris, 2004). The diagnosis is predominantly clinical, and it can be confirmed by detection of the toxin or the bacteria in specimens, such as serum, in the very early stages of infection/intoxication. The treatment is mainly supportive, patients are usually treated in intensive care unit, and in case of respiratory failure, intubation and ventilation should be applied (Goonetilleke and Harris, 2004).

TeNT is internalized at the level of the membrane of the motor neuron (MN) via clathrin-dependent endocytosis (Roux *et al.*, 2005). Vesicles containing the toxin are retrogradely transported along the axon to the soma of the MN, using a microtubule-based (dynein) and a F-actin-based motor (myosin Va) (Lalli *et al.*, 2003b). These vesicles present a neutral pH during the transport, contrary to the classical endosomal vesicles, which undergo rapid acidification upon internalization (Bohnert and Schiavo, 2005). This prevents the release of the L_C of TeNT into the cytosol of the MN, and the toxin can reach the final destination: the inhibitory interneurons of the spinal cord. How TeNT is translocated from the soma of the MN to the afferent inhibitory terminals, where it cleaves VAMP/synaptobrevin, is not understood.

Tetanus arises by contamination of deep wounds or burns with soil, manure or rusty metals where *C. tetani* spores are present (Brook, 2008). The death toll in developing countries due to neonatal tetanus, as a consequence of unhygienic childbirth, is reported to be around 200,000 cases per year (http://www.who.int/immunization_monitoring/diseases/tetanus/en/). The clinical pattern of tetanus is characterized by initial neck stiffness, sore throat and difficulties in opening the mouth, followed by muscular spasms, which can lead to laryngeal contraction, as well as chest wall compliance, resulting in compromised respiratory function (Brook, 2008). The incubation period (time from inoculation to the first symptoms) is around 7–10 days, and the period of onset (from first symptoms to the start of the spasms) is 1–7 days. The diagnosis is clinical, as there is no specific confirmatory test (Brook, 2008). The management of the disease is immediate antibiotic therapy by metronidazole, and passive vaccination with anti-tetanus immunoglobulin can reduce the severity and the duration of the illness. Muscular spasms are usually treated by administration of sedatives (e.g. benzodiazepines, which enhance the GABA activity). If muscle spasms persist, in spite of sedation, muscle relaxants may be required (Brook, 2008). Development of tetanus can be prophylactically prevented by active immunization with inactivated TeNT (also known as tetanus toxoid) (Brook, 2008).

Cholera toxin

Cholera toxin (CT) is produced and secreted by *Vibrio cholerae*, a Gram-negative rod-shaped bacterium. As previously discussed, CT is closely related in structure and function to LT produced by some strains of enterotoxigenic *E coli* (ETEC) (Sánchez and Holmgren, 2005). The B subunit (also known as L, for light, or CTB/LTB) forms a circular homopentamer, which is non-covalently associated with the A subunit (also known as H, for heavy, or CTA/LTA) (Sánchez and Holmgren, 2008). The CTA subunit is cleaved by a *V. cholerae* protease in two fragments (CTA1 and CTA2) that remain linked by a disulfide bond. The enzymatic ADP-ribosylating activity of CTA resides within the CTA1 fragment, while the CTA2 is required for interaction with the CTB pentamer (Sánchez and Holmgren, 2008). Both CT and LT bind to the same receptor: the ganglioside GM1, although LT can also bind to GD1b (Holmgren *et al.*, 1985). Upon binding, internalization can occur via clathrin-dependent or caveolae-dependent endocytosis. Independently of the initial steps, the toxin reach the early endosome compartment and from there, it is retrogradely transported to the ER similarly to Stx (Chinnapen *et al.*, 2007). The cellular target of both CT and LT is the enzyme adenylate cyclase (AC). ADP-ribosylation of regulatory protein Gs causes constitutive AC activation, resulting in an uncontrolled production of cAMP (Sánchez and Holmgren, 2008).

Infection with *V. cholerae* or LT-producing ETEC causes secretory diarrhoea, although cholera is usually more severe than the ETEC-induced disease, possibly due to the much higher efficacy of secretion of CT from the producing bacterium (Sánchez and Holmgren, 2005). How does CT cause secretory diarrhoea? The constitutive production of cAMP in the epithelial cells of the upper part of the small intestine causes an imbalance in the electrolyte flux in the intestinal epithelial cells: increased luminal secretion of chloride ion (Cl^-) via the cystic fibrosis trans-membrane conductance regulator (CFTR), which in turn causes a passive flow of bicarbonate ($HCOO^-$). This unbalance produces a decreased Na^+ uptake, leading to osmotic secretion of water from the enterocytes (Burleigh and Banks, 2007; Sánchez and Holmgren, 2008). The CT-induced ion flux into the intestinal lumen produces a massive loss of water in the order of 500–1000 ml/h (Sánchez and Holmgren, 2008) (Fig. 8.4). The altered ion transport is further potentiated by the fact that CT, but not LT, also stimulates the enterochromaffin cells to release serotonin, which in turn induces secretion of the vasointestinal peptide (VIP) from the local neuronal submucosal plexus (Turvill *et al.*, 1998) (Fig. 8.4). The evidence for this role of CT comes mainly from *in vivo* experiments aimed at blocking, either pharmacologically or by local denervation, the secretion of VIP, leading to reduced water loss upon CT intoxication.

Proper and prompt treatment of cholera can save even severely ill patients. The goal of the treatment is to restore lost fluids and salts by administration of rehydration salts solution. To date, three oral cholera vaccines are available. One is based on recombinant CTB and inactivated *V. cholerae* cell wall (WCs) of different serotypes (trade name Dukoral). A variant of the Dukoral vaccine containing no recombinant B-subunit has been produced and tested in Vietnam, and licensed only in this country. The CVD 103-HgR vaccine (trade name Orochol®) contains an attenuated live oral genetically modified *V. cholerae* O1 strain (CVD 103-HgR).

Shiga-like toxins (SLT) and the haemolytic uremic syndrome

Shiga-like toxins (SLT), also known as verotoxins, are produced by certain strains of enterohaemorrhagic *E. coli* (EHEC), such as the O157:H7 strain (Karmali *et al.*, 1985). Infection with these strains can trigger haemolytic uraemic syndrome (HUS), which is the most common cause of renal failure in childhood. HUS is a thrombotic disorder, characterized by microvascular thrombi, haemolytic anaemia, thromobocytopenia and acute renal failure (Palermo *et al.*, 2009).

This toxin family was first described in 1977 by its ability to kill Vero cells (African green monkey kidney cells), hence the original name of verotoxin (Konowalchuk *et al.*, 1977). Two major subfamilies of SLT have been identified: Stx1 and Stx2 (O'Loughlin and Robins-Browne, 2001). As previously described, these toxins are $A–B_5$ toxins that contain an A subunit, composed of two fragments (A_1 and A_2, linked by a disulfide bond), bound to a homopentameric structure formed by the B subunit. The receptor is the globotriasyolceramide (Gb3), and upon internalization and retrograde transport to the ER, the A subunit is cleaved into two fragments. The active A_1 fragment is released into the cytosol, where it catalyses N-glycosidation of the 28S ribosomal RNA (Sandvig and van Deurs, 2000; O'Loughlin and Robins-Browne, 2001). The result of the intoxication is inhibition of protein synthesis, and subsequent induction of cell death.

How can humans be infected with SLT-producing EHEC? Infected bovine species are the main reservoir of these *E. coli* strains. Consumption of products from unpasteurized milk or undercooked beef is the primary cause of infection. Upon infection, the bacteria colonize the intestinal mucosa and invade the epithelium causing extensive necrosis, and subsequent bloody diarrhoea and dehydration. Usually, EHEC do not give rise to bacteraemia, but remain localized within the intestinal mucosa where they produce and secrete SLT upon lysis (O'Loughlin and Robins-Browne, 2001). 10–15% of patients infected with EHEC can develop HUS, and this possibly occurs when a sufficient amount of the toxin is translocated through the intestinal epithelium and enters the systemic circulation.

Endothelial cells are the primary targets of SLT. Intoxicated cells not only undergo cell death, upon inhibition of protein synthesis, but also activate an array of pro-inflammatory mediators, such as interleukin (IL)-1β and tumour necrosis factor (TNF)-α, which increase the thrombocytic effect of SLT and enhance its toxicity. The damage to endothelial cells and secretion of pro-inflammatory mediators may induce expression of adhesion molecules, such as ICAM1 and VCAM1, leading to enhanced adhesion and activation of leucocytes and a reduced blood flow through the inflamed vessels. Activated leucocytes start to release their products (e.g. degradative enzymes and reactive oxygen species), exacerbating the damage of the endothelium and providing a potent stimulus for platelet activation and formation of thrombi. Human kidneys express high levels of Gb3 in both the cortex and the medulla, and this explains why this organ is frequently affected by SLT intoxication.

Diagnosis is established by detecting the bacterium in the stool of patients, and commercially available kits include direct detection of SLT. Early detection of the EHEC is a key factor for administering the proper therapeutic support, since all the therapies available to date have a very narrow window for application: 2 days from the beginning of the *E coli* O157:H7-associated diarrhoea (Palermo *et al.*, 2009). The treatment of HUS is mainly supportive and includes control of fluid and electrolyte imbalance, use of dialysis and blood transfusion if required. Antibiotic treatment should be avoided during the acute phase of HUS, since this may enhance bacterial lysis and toxin secretion (Palermo *et al.*, 2009).

Anthrax toxins

Anthrax is a disease caused by infection with spores from *B. anthracis*. Infection of humans is an accident, since anthrax is mainly a disease of animals (herbivores, such as sheep and cattle). The clinical manifestation and the mortality rate depend on the route of infection: the skin form of the disease is the least dangerous, with a mortality rate ranging between 5% and 20%, followed by gastrointestinal anthrax (25–60%), while pulmonary anthrax has the highest mortality rate (close to 100%), if not treated in time (http://www.who.int/vaccine_research/diseases/zoonotic/en/index1.html). Cutaneous anthrax presents as a small pruritic papule that develops within a week into a vesicle usually on an exposed part of the body such as the face, the neck or arm. The vesicle eventually ruptures, revealing a depressed ulcer crater that develops into a black eschar (http://www.who.int/vaccine_research/diseases/zoonotic/en/index1.html). Gastrointestinal anthrax develops within 2–5 days following ingestion of contaminated meat, with nausea, vomiting, fever, abdominal pain and diarrhoea (http://www.who.int/vaccine_research/diseases/zoonotic/en/index1.html). Pulmonary anthrax develops after inhalation of *B. anthracis* spores, which can be easily disseminated, and that is why anthrax has been considered as a weapon in biological warfare since World War I (Schwartz, 2009). As recently as 2001, letters containing anthrax spores were delivered via the US mail system. At least 22 people developed anthrax infection; among these cases, 11 exhibited the pulmonary form of the disease, and five of them died (http://jama.ama-assn.org/cgi/content/full/287/1/34).

The anthrax toxin receptors and its complex internalization pathway have already been described. In this paragraph, we will discuss the symptoms of pulmonary anthrax and how they are related to the activity of the two anthrax toxins: (i) the lethal factor (LF), a zinc metalloprotease that cleaves the MAP kinase kinases (Klimpel *et al.*, 1994), inhibiting their activation, and (ii) the oedema factor (EF), a calmodulin-dependent adenylate cyclase (Leppla, 1982), which induces uncontrolled production of intracellular cAMP (Fig. 8.4).

Pulmonary anthrax starts with non-specific flu-like symptoms, such as fever, chills, cough, myalgia and malaise. Subsequently, patients develop mediastinal lymphadenitis and pleural effusion, respiratory failure and shock. This biphasic course depends on the initial recognition of the spores by the innate immunity effectors, such as macrophages (flu-like symptoms). This stage can be controlled by antibiotic therapy. The cardiac and respiratory failure depends on the dissemination of the bacteria, which leads to a systemic production of the toxin. In this phase of the disease, antibiotic therapy is not efficient since the illness is mainly caused by the toxic activity of LF and EF (Golden *et al.*, 2009).

Why are the spores and the germinating bacteria not cleared by the host immune defences? Once the spores reach the pulmonary alveoli, they are phagocytosed by resident macrophages and dendritic cells (DCs). The interplay between the bactericidal activity of these professional phagocytes and the ability of the spore to evade killing and germinate is crucial for the development of anthrax. Therefore, it is not surprising that LF and EF present in the spores and produced by the germinating bacteria cause an immune paralysis of the macrophages (Tournier *et al.*, 2009). LF shuts off the bactericidal activity of macrophages, down-regulates production of pro-inflammatory cytokines (such as IL-8, TNF-α), reduces the recruitment of cellular and molecular effectors of the immune system (e.g. neutrophils and the complement system), and induces apoptosis of macrophages. Similarly, the increased levels of cAMP, caused by EF, induce constitutive activation of protein kinase A (PKA), reducing cell motility and inhibiting secretion of pro-inflammatory cytokines (Fig. 8.4). LF and EF not only paralyse the activation and function of key cells of the innate immunity, but also prevent activation of the adaptive immunity, using two different strategies: (i) they inhibit activation of DCs, the professional antigen-presenting cells required for activation of T lymphocytes; (ii) they directly impair T cell activation and proliferation. To exert this effect, the toxins need to be delivered to the regional lymph nodes, via the lymphatic vessels, and from the draining lymphnodes LF and EF can reach the blood stream (Fig. 8.4), causing pulmonary and cardiac failure. Concerning the systemic effects of the anthrax toxin, it is well established from animal experiments that the toxin is central in the pathogenesis of the disease. However, the exact mechanisms on how LT and EF contribute to the cardiovascular and respiratory failure are still not clear.

The definitive diagnosis of anthrax is made by the isolation of *B. anthracis* from sputum in pulmonary anthrax and from vomitus, faeces and ascites in intestinal anthrax. The management of anthrax-associated shock and vascular collapse depends on antimicrobial and supportive therapy. It is noteworthy to remember that antibiotic treatment does not neutralize or prevent the biological effects of the toxins, once they have been produced and released in the body. The therapeutic effect of post-exposure vaccination is limited owing to the rapid onset of the disease and the immunosuppressive activity of LF and EF (Golden *et al.*, 2009). Several experimental approaches have been undertaken to induce toxin neutralization, such as: production of neutralizing antibody against the B domain of the toxin to be passively administered in case of *B. anthracis* infection, inhibitors of furin, which mediate the cleavage of the PA subunit, and allows its oligomerization and pore formation, inhibitors of LF, development of anthrax decoy receptors (Golden *et al.*, 2009). The optimal preventive measure is vaccination. Anthrax vaccines are available for both animals and humans. However, in humans, their use has been confined to high-risk groups, such as occupationally exposed workers and military personnel. The human anthrax vaccines are subunit vaccines that consist of supernatants from toxigenic strains of *B. anthracis* inactivated with formaldehyde (AVA/BioThrax; Bioport) (Brey, 2005). To develop and maintain protective immunity in humans, these vaccines must be administered subcutaneously six times over 18 months, followed by yearly booster injections (Pittman *et al.*, 2002; Singer *et al.*, 2008).

Diphtheria

Diphtheria is caused by infection with toxigenic strains of the air-borne pathogen *Corynebacterium diphtheriae*. DT is an A-B toxin, which is internalized by clathrin-dependent endocytosis and is translocated into the cytosol upon acidification of the early endosomes. DT ADP-ribosylates EF2 and inhibits protein synthesis, causing rapid cell death.

The infection is limited to the upper respiratory tract, and does not induce bacteraemia (Hadfield *et al.*, 2000). However, the toxin produced within the infection foci can be systemically disseminated causing lesions of the heart and (to a lesser extent) nerves. At the site of infection, the mucosa of the respiratory tract presents ulcerations, due to the effect of the toxin, and formation of an obstructive inflammatory pseudomembrane (Hadfield *et al.*, 2000). In severe cases, infection may terminate with acute cardiac failure. About three-quarters of patients with severe disease develop neuropathy, and the incidence of neurological complications is related directly to the severity of respiratory symptoms (Hadfield *et al.*, 2000). Treatment consists of immediate administration of diphtheria antitoxin and antibiotics.

From an historical perspective, the German physician Emil von Behring developed an antitoxin in the 1890s, and was awarded the first Nobel Prize in Physiology and Medicine. William H. Park and Anna Wessels Williams, and Emile Roux and Auguste Chaillou also independently developed diphtheria antitoxin in the 1890s (http://en.wikipedia.org/wiki/Diphtheria). Diphtheria vaccines are based on diphtheria toxoid, a modified bacterial toxin that induces protective antitoxin. Diphtheria toxoid combined with tetanus and pertussis vaccines (DTwP) has been part of the WHO Expanded Programme on Immunization (EPI) since 1974, and during the period 1980–2000, the total number of reported diphtheria cases was reduced by >90% (http://www.scielosp.org/scielo.php?script=sci_arttext&pid=S0042-96862009000700015&lng=en&nrm=iso&tlng=en).

Toxins that contribute to the disease

Toxins targeting the Rho GTPases

Before discussing the bacterial toxins that act on the small GTPases of the Rho family, we will briefly describe the role of these proteins and how their activity is tightly regulated in physiological conditions. The Rho GTPases belong to the superfamily of the Ras proteins. To date more than 20 different mammalian Rho GTPases have been identified. However, our discussion will focus on the three best characterized members: RhoA, Rac1 and Cdc42 (Jaffe and Hall, 2005). These proteins have been initially described as key regulators of the actin cytoskeleton dynamics, such as formation of actin stress fibres (RhoA), lamellipodia (Rac1), filopodia (Cdc42), and organization of focal adhesion contacts (RhoA, Rac and Cdc42). It is now well recognized that the small GTPases of the Rho family not only regulate actin cytoskeleton dynamics, but also secretion, production of reactive oxygen species, regulation of transcription, cell polarity, cell cycle progression, apoptosis and cell survival (Jaffe and Hall, 2005).

As their name indicates, these proteins bind GTP in their active state, while they are GDP-bound when inactive. In order to regulate cellular processes, the small GTPases need to interact with their down-stream effectors. This interaction is temporary and spatially regulated: only the GTP-bound molecules directly interact with the cognate effectors, and activation is associated with isoprenylation, which recruits the active GTP-bound form in proximity of the plasma membrane (Jaffe and Hall, 2005). As schematically illustrated in Fig. 8.5, the inactive GDP-bound GTPase is associated with guanine nucleotide dissociation inhibitor (GDI). Upon specific stimuli, guanine nucleotide exchange factors (GEFs) catalyse the GDP to GTP exchange. The process is controlled by activation of the intrinsic GTPase activity, a process facilitated by GTPase-activating proteins (GAPs).

Based on the pleiotropic effects of the small GTPases of the Rho family, it is not surprising that these proteins have been targeted by bacteria. Indeed, by regulating the actin cytoskeleton dynamics, a bacterium can promote its own internalization into non-professional phagocytic cells in order to replicate and evade the host immune response (*Salmonella enterica*) (Ly and Casanova, 2007), or can prevent phagocytosis and consequent destruction from macrophages (*Yersinia* spp.) (Dube, 2009). Furthermore, alteration of actin cytoskeleton dynamics may favour bacterial colonization and invasion through the epithelial barriers (via disruption of cell junctions), or bacterial immune evasion by inhibiting leucocyte migration (macrophages and neutrophils) and activation of T lymphocytes (Aktories and Barbieri, 2005). In the following paragraphs we will discuss the mechanism of action of some toxins inhibiting or activating the small Rho GTPases and their contribution to pathogenesis.

Large clostridial toxins

The large clostridial toxins (LCT, also known as clostridial glucosylating toxins) are produced by several *Clostridium* species, such as *C. difficile, Clostridium sordellii* and *Clostridium novyi* (Jank and Aktories, 2008). The best characterized of this family are the *C. difficile* toxins A and B. They are produced as a single polypeptide, which contains three domains: the catalytically active N-terminal domain, a central translocation domain, and a C-terminal receptor-binding domain. Upon binding to a still not clearly identified cellular receptor, the toxins are internalized into early endosomes. The low pH of the endosomal environment induces a conformational change of the translocation domain, allowing the toxin to insert into the membrane and to form a pore, through which the catalytic domain is translocated. It has been recently demonstrated that only the 543 amino acids of the N-terminal part of the toxin are able to reach the cytosol, owing to activation of an auto-proteolytic process. The toxins catalyse monoglucosylation on a specific threonine residue of the small GTPases of the Rho family, using UDP-glucose as the donor substrate. The glusosylated form the Rho protein can still switch from the GDP-bound to the GTP-bound state, but it is no longer able to interact with its downstream effectors (Fig. 8.5).

C. difficile is the causative agent pseudomembranous colitis, an inflammatory condition of the

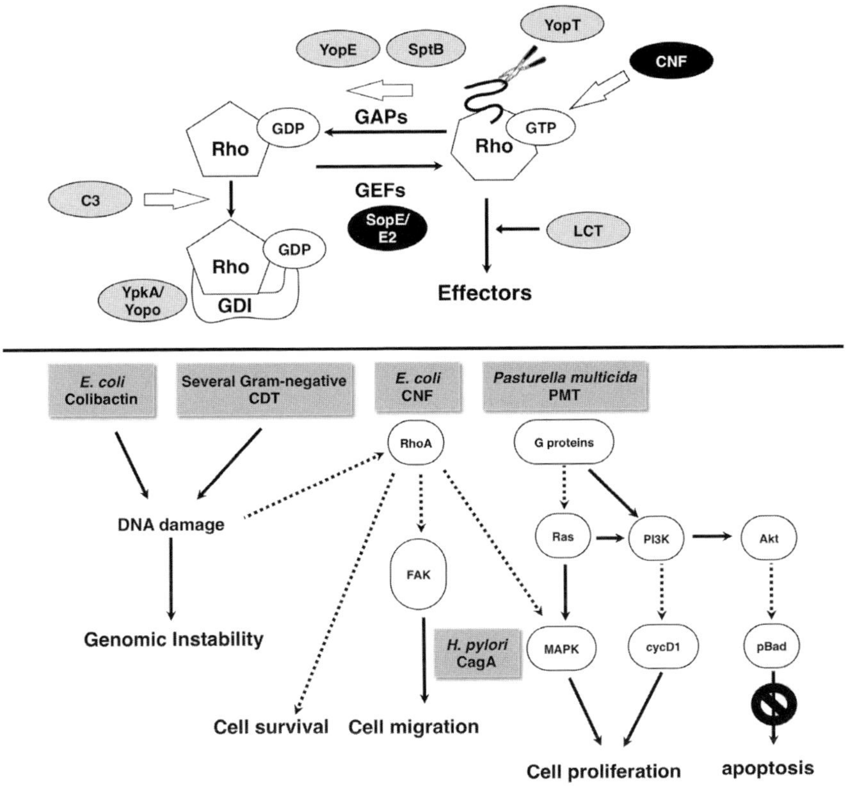

Figure 8.5 Toxins associated with bacterial-induced diseases. (A) Toxins that interfere with activation of the small GTPases of Rho family. GDI, guanine nucleotide dissociation inhibitor; GAP, GTPase-activating proteins; GEF, guanine nucleotide exchange factors. (B) Toxins that interfere with cell cycle progression. The figure summarizes the signal transduction pathways and the effects of intoxication with cyclomodulins that may promote carcinogenesis. CNF, cytotoxic necrotizing factor.

colon that occurs in some people who received antibiotics therapy. The most common and the initial sign of pseudomembranous colitis is watery, and sometimes bloody, diarrhoea (Rupnik *et al.*, 2009). The pathogenicity of *C. difficile* is mediated by TcdA and TcdB, therefore this should be considered as a toxin-induced disease. It is included in this section because of the activity of the *C. difficile* toxins on the small Rho GTPase. TcdA and TcdB damage the human colonic mucosa, inducing tissue damage and alteration of the intestinal barrier (Thelestam and Chaves-Olarte, 2000; Aktories and Just, 2005; Rupnik *et al.*, 2009). Toxins A and B also induce a potent inflammatory response, characterized by massive infiltration of neutrophils, release of pro-inflammatory mediators, enhancement of the tissue destruction and impairment of enterocyte functions (Rupnik *et al.*, 2009).

Yersinia spp. Yops effectors

The *Yersinia* genus includes three human pathogens. The most infamous is *Yersinia pestis*, which causes bubonic plague when transmitted by the bite of a flea, and pneumonic plague when acquired through aerosol transmission. *Yersinia enterocolitica* and *Yersinia pseudotuberculosis*, instead, are transmitted via consumption of contaminated food or water (Pujol and Bliska, 2005). Infection with these enteropathogenic species causes a broad range of gastrointestinal diseases, from enteritis to mesenteric lymphadenitis. Infection starts at the level of the small intestine, but rapidly the bacteria invade the intestinal mucosa passing via the M cells, specialized cells of the follicle-associated epithelia that function in antigen uptake. After invasion of the intestinal epithelium, the bacteria replicate within the gut-associated lymphoid tissue, known as Peyer's patches. The bacteria also spread to mesenteric lymph nodes, resulting in mesenteric lymphadenitis (Pujol and Bliska, 2005).

The ability of these pathogens to successfully invade the intestinal mucosa and evade the innate immune responses depends on several virulence factors, including six Yop effectors (YopE, YopH, YpkA/YopO, YopM, YopJ/P and YopT), which are directly injected into the cytosol of host cells via a specialized needle-like structure, known as type III secretion system (T3SS) (Galán and Wolf-Watz, 2006; Matsumoto and Young, 2009). Three of these T3SS effectors are toxins that inhibit the activation of the Rho GTPases (Matsumoto and Young, 2009). YopE is a Rho GTPase-activating protein (RhoGAP) that has been shown to stimulate GTP hydrolysis, thereby interfering with actin depolymerization mediated by the Rho family of proteins. YpkA/YopO contains a C-terminal domain that interacts with RhoA and Rac1 to mimic RhoGDI, thereby preventing nucleotide exchange and activation. YopT acts as a cysteine protease that removes the lipid modification of RhoA, resulting in protein displacement from the plasma membrane and inhibition of its signalling function. The function of these three effectors is summarized in Fig. 8.5.

S. enterica SopE, SopB and SptB

Differently from *Yersinia*, which is mainly an extracellular pathogen and requires escape from phagocytosis, *S. enterica* is a facultative intracellular bacterium, and its virulence depends on survival and replication within professional phagocytes (macrophages). *Salmonella* also promotes its own uptake by non-professional phagocytic epithelial cells, as a mean to invade the intestinal mucosa. Infection with *Salmonella* occurs via ingestion of contaminated food, and in the majority of cases, it causes a self-limiting enteritis. However, in some cases, the bacterium disseminates systemically to colonize the liver and spleen, and gives rise to a more serious condition, known as typhoid fever (Ibarra and Steele-Mortimer, 2009).

Successful infection depends on the ability of the bacteria to gain entry into host cells. To achieve this, *Salmonella* injects an array of bacterial effectors into the host cytoplasm via one of the two T3SSs (SPI-1) (Galán and Wolf-Watz, 2006; Ly and Casanova, 2007). At least 13 proteins have been identified whose secretion is mediated by the SPI-1 T3SS, and six of them have been shown to regulate directly or indirectly the actin cytoskeleton dynamics, triggering a dramatic reorganization of the actin cytoskeleton that engulfs bacteria. In this paragraph, we will focus only on the effectors that directly act in the Rho GTPases: SopE, SopE2, and SptP. SopE and SopE2 act as GEFs for Rac1 and Cdc42, and catalyse nucleotide exchange from GDP to GTP (Fig. 8.5). After bacterial entry, the host cell cytoskeleton largely returns to its basal state, and this is mediated by the third *S. enterica* T3SS effector SptB, which is a RhoGAP (Fig. 8.5).

The *E. coli* cytotoxic necrotizing factor

The cytotoxic necrotizing factor (CNF) is a family of bacterial toxins produced by *Escherichia* and *Yersinia* (CNF1, CNF2, CNF3 and CNFy) (Lemonnier *et al.*, 2007). CNF1 is synthesized as a single-chain polypeptide, organized in three functional domains: the N-terminal receptor-binding domain, which has been shown to bind to the laminin receptor LRP on the surface of the host cell, a central domain that mediates the translocation of the active domain (located at the C-terminal end). Upon binding to the cellular receptor, CNF is internalized into early endosomes via clathrin-independent endocytosis, and subsequently it reaches late endosomes, where the acid condition results in translocation of the catalytic domain (Lemonnier *et al.*, 2007). The detail of this process and which part of the molecule gains access to the cytosol is still poorly characterized. CNF1 is a deaminase and catalyses the deamination of the glutamine in position 63 of RhoA. Rac1 and Cdc42 are also targets for the CNF1-mediated deamination (Flatau *et al.*, 1997; Schmidt *et al.*, 1997). This results in inhibition of the intrinsic GTPase activity of the small GTPases, which thereby remain constitutively activated (Fig. 8.5).

CNF1 is present in one third of the uropathogenic *E. coli* (UPEC). UPEC preferentially colonizes the bladder and causes cystitis, but can also ascend through the ureters into the kidneys, causing a more severe condition, pyelonephritis (Wiles *et al.*, 2008). Upon bacterial colonization of the urinary tract epithelium, the host reacts by activating a strong pro-inflammatory response, leading to neutrophil infiltration, exfoliation of infected bladder epithelial cells, and the generation of reactive oxygen species and other antimicrobial compounds (Svanborg *et al.*, 2006). Functional studies in animal models suggest that expression of this toxin contributes to the characteristic inflammatory response and enhances tissue damage caused by UPEC (Wiles *et al.*, 2008). Indeed, this toxin triggers a potent secretion of pro-inflammatory cytokines, including IL-8, a chemoattractant for neutrophils. CNF can also contribute to bacterial invasion by inducing disassembly of cellular junctions (Mills *et al.*, 2000). Alteration of the actin cytoskeleton dynamics can also stimulate plasma membrane ruffling and filopodia formation, resulting in increased host-cell motility and bacterial uptake (Doye *et al.*, 2002). It is common consensus that these intracellular bacteria can help to establish a chronic infection, and cause recurrent cystitis (Dhakal *et al.*, 2008).

Cyclomodulins

This section is dedicated to bacterial toxins known as cyclomodulins that interfere with the regulation of the cell cycle in eukaryotic cells (Oswald *et al.*, 2005). The possible contribution of these toxins in cancer development will also be discussed. Therefore, we will briefly review the current understanding on how chronic inflammation and bacterial infection can lead to tumour development. Some of the most well characterized cyclomodulins and their potential contribution to carcinogenesis are reported in Fig. 8.5.

Carcinogenesis is a complex process that requires six essential alterations of normal cell physiology: self-sufficiency in growth signals, insensitivity to growth inhibition, evasion of apoptosis, immortalization-sustained angiogenesis, and tissue invasion and metastasis (Hanahan and Weinberg, 2000). To date there is solid evidence that links chronic inflammation, either induced by bacterial infection (*Helicobacter pylori*) or tissue damage (e.g. hepatitis), with cancer development (Moss and Blaser, 2005; Karin *et al.*, 2006; Lin and Karin, 2007). Several mechanisms can contribute to the alterations indicated by Hanahan and Weinberg (2000): (i) constitutive activation of transcription factors can upregulate expression of genes that have anti-apoptotic and proliferative activity, such as Bcl-X_L, cIAP, cFLIP, c-Myc and cyclin D, NF-κB being the most prominent one (Karin and Greten, 2005); (ii) production of reactive oxygen species that can cause DNA damage and therefore lead to genomic instability (Hussain *et al.*, 2003); (iii) bacterial toxins that directly promote cell proliferation (PMT), enhance inflammatory and anti-apoptotic signals (CNF), cause DNA damage (CDT and colibactin) (Lax, 2005; Oswald *et al.*, 2005).

Cyclomodulins promoting cell proliferation

The cytotoxic necrotizing factor (CNF) secreted from *E. coli* and *Y. pseudotuberculosis* and the dermonecrotic toxin (DNT) from *Bordetella* spp. are deaminases and transglutaminases, respectively (Lemonnier *et al.*, 2007). As discussed previously, these activities induce a constitutive activation of the small GTPases of the Rho family: RhoA, Rac1 and Cdc42. CNF1 activates several proliferating and survival signals, ranging from promotion of G1/S transition and induction of DNA replication to activation of NFκB, enhanced expression of adhesion molecules (integrins, focal adhesion kinase, cadherins and catenins) that prevent anoikis (cell death induced by cell detachment from the extracellular matrix) (Travaglione *et al.*, 2008) (Fig. 8.5). The CNF-induced alteration of the actin cytoskeleton dynamics can further promote cell junction disruption and enhance motility, a phenomenon linked to invasiveness and metastasis (Travaglione *et al.*, 2008).

PMT is a protein toxin from *Pasteurella multocida*, a bacterium that can cause disease in a wide range of animal species, including avian fowl cholera, bovine haemorrhagic septicaemia, enzoonotic pneumonia and swine atrophic rhinitis. This toxin is the only factor identified to date to be responsible for the clinical and pathological signs of atrophic rhinitis (Harper *et al.*, 2006). In cell biology, PMT is known to act on various types of cell lines as a highly potent mitogen, causing anchorage-independent growth, especially in some fibroblastic cells, because of its effect on multiple signalling transducers, such as the small GTPase RhoA, the MAP kinase ERK, the STAT proteins, and tyrosine phosphorylation of focal adhesion kinases and paxillin. These effects are mainly induced by the PMT-dependent activation of members of two heterotrimeric G protein families: G_q and $G_{12/13}$ (Preuss *et al.*, 2009) (Fig. 8.5).

The Gram-negative bacterium *H. pylori* colonizes the human gastric mucosa and is one of the most successful bacterial pathogens, since 50% of the world population carries *H. pylori*. Infection causes chronic gastritis and more severe gastric disease in 10–15% of the cases, such as gastric adenocarcinoma and mucosa-associated lymphoid tissue (MALT) lymphoma (Backert and Selbach, 2008). Today, *H. pylori* is the only bacterium classified as human carcinogen by the International Agency for Research on Cancer (IARC) and the World Health Organization (WHO). This bacterium produces one well-characterized cyclomodulin, the Cytotoxin-associated antigen A (CagA), encoded by the *cagA* gene located within the *cag*-pathogenicity island (Backert and Selbach, 2008). CagA is injected into the cytosol of the host cells via a needle-like structure known as the type IV secretion system (T4SS), and it is phosphorylated on tyrosine residues ($CagA^{PY}$) by members of two cellular kinase families: the Src family kinases (SFKs), and the Abl kinases. $CagA^{PY}$ mainly signals to the actin cytoskeleton, induces disruption of tight and adherent junctions, loss of cell polarity, and pro-inflammatory and mitogenic responses. CagA can also induce indirect activation of transcription factors, which have been shown to play a key role in carcinogenesis: NF-κB and β-catenin/LEF, leading to upregulation of mitogenic genes, such as c-Myc and cyclin D1. In line with these observations, CagA-positive *H. pylori* strains have been reported to affect the cell cycle during infection by stimulating cyclin D1 expression, G1-S phase progression and host cell survival (Backert and Selbach, 2008). Very recently, the oncogenic potential of CagA was demonstrated in a transgenic mouse model: CagA expression induces abnormal proliferation of gastric epithelial cells and haematopoietic cells, followed by the development of gastrointestinal carcinomas and leukemias/lymphomas (Ohnishi *et al.*, 2008).

Cyclomodulins inhibiting cell proliferation

The CDTs are produced by a variety of Gram-negative bacteria, such as *E. coli, Aggregatibacter actinomycetemcomitans, Haemophilus ducreyi, S. dysenteriae, Campylobacter* spp. and *Helicobacter* spp., and *S. enterica* (Smith and Bayles, 2006). The active holotoxin is a tripartite complex, formed by the CdtA, CdtB and CdtC subunits. The CdtB component is the active subunit, that shares structural and functional homology with mammalian deoxyribonuclease I (DNase I). The CdtA and CdtC subunits are ricin-like lectin domains, involved in cellular internalization of the toxin (Nesic *et al.*, 2004; Nesic and Stebbins, 2005). Upon internalization and translocation into the nuclear compartment, CDT causes DNA double strand breaks and activates the cellular DNA damage checkpoint responses that induce cell cycle arrest in the G1 or G2 phases of the cell cycle (Cortes-Bratti *et al.*, 2001). In adherent cells, CDT intoxication is associated with formation of actin stress fibres. This effect is regulated by the activation of the small GTPase RhoA, and promotes cell survival of intoxicated cells. The DNA damage-dependent RhoA signalling diverges into two different effector cascades: one dependent on the RhoA kinases ROCKI and ROCKII, which controls the formation of actin stress fibres; and one regulated by the mitogen-activated protein kinase (MAPK) p38 and its downstream target MAPK-activated protein kinase 2, which promotes cell survival (Guerra *et al.*, 2008b). An unusual hybrid non-ribosomal peptide-polyketide compound, named colibactin, is produced by some strains of *E. coli*, and intoxication induces DNA double-strand breaks and DNA damage checkpoint responses, similarly to those observed in CDT-intoxicated cells (Nougayrède *et al.*, 2006).

The cycle inhibiting factor (Cif) is a cyclomodulin injected into eukaryotic cells by T3SS of enteropathogenic and enterohaemorrhagic *E. coli* (EPEC and EHEC). Intoxication induces cell cycle arrest at the G_2/M and G_1/S phase transitions and, at least in HeLa cells, reorganization of the actin network (Marchès *et al.*, 2003; Taieb *et al.*, 2006). Differently from CDT or colibactin, Cif does not cause DNA damage, but it directly interacts with neddylated cullin-RING ubiquitin ligase complexes (CRLs), and consequently inhibits proteasomal-mediated degradation of the cyclin-dependent kinase inhibitors (CKIs) $p21^{waf1/cip1}$ and $p27^{kip1}$, which actively participate in the control of cell cycle progression (Samba-Louaka *et al.*, 2008; Jubelin *et al.*, 2010).

Mycobacterium ulcerans is a causative agent of Buruli ulcer, a debilitating skin disease found in Australia and West Africa. The disease is characterized by persistent, severe necrotic lesions of the skin and underlying fat, and lack of an acute inflammatory response. *M. ulcerans* secretes a polyketide-derived macrolide, named mycolactone, which induces cell rounding, inhibits protein synthesis and causes cell cycle arrest in G0/G1 and eventually cell death (George *et al.*, 1999).

Beneficial role of bacterial toxins

Immunotoxins

The possibility to physically separate the active domain from a receptor-binding domain of a bacterial toxin made it possible to develop chimeric molecules carrying toxin moieties that can specifically target tumour cells (Fig. 8.6). The concept of the 'magic bullet' using toxin compounds coupled to antibodies to kill transformed cells, which express or over-express tumour markers, was developed by Paul Ehrlich in the early 1900s (Strebhardt and Ullrich, 2008). However, technical issues have considerably slowed the effective production of immunotoxins.

To date the choice of the binding moieties are either growth factor receptors (e.g. IL2 receptor) or single chain variable fragments, which contain

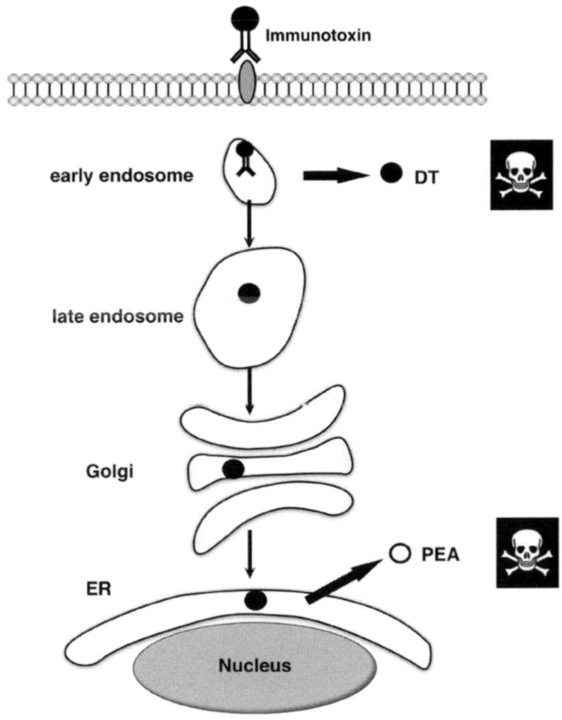

Figure 8.6 Schematic representation of immunotoxins where the B moiety of an A–B toxin is replaced with an antibody or a single variable fragment directed against antigens exclusively expressed or over-expressed in tumour cells. Growth factor receptor has also been used as B moieties. DT, diphtheria toxin; PEA, *P. aeruginosa* exotoxin A.

the antibody variable region of the light and heavy chains separated by a flexible linker. The bacterial A domain mostly used is the one derived from the *P. aeruginosa* PEA or from DT, both inhibit protein synthesis by ADP-ribosylating EF-2 (Pastan *et al.*, 2007; Kreitman, 2009). To improve specificity, the toxin components have been mutated to avoid recognition from neutralizing antibodies that may be present in the patients. The shorter versions recently used include PE38, composed of amino acids 253 to 364 and 381 to 613 of PEA, and DT388 or $DTAB_{389}$, containing the first 388 amino acids of DT. Table 8.2 summarizes the immunotoxins that are presently tested in clinical trials. Denileukin diftitox has been approved by the Food and Drug Administration for the treatment of advanced cutaneous T-cell lymphomas.

Development of an immunotoxin poses quite a number of challenges: stimulation of neutralizing antibodies in patients after the first single cycle of treatment, and toxicity, mainly vascular leak syndrome, characterized by hypotension, peripheral oedema, and hypoalbuminaemia (Kreitman, 2009). This type of toxicity is not unexpected since a cytotoxic protein must traverse endothelial cells to exit the blood vessels. Small recombinant fusion toxins that contain either growth factor or Fv fragments as ligands are the most successful immunotoxins developed so far. The most sensitive diseases appear to be hematologic malignancies. Future development will need to address combinations of immunotoxins with other anticancer therapies in order to overcome problems of tumour penetration, toxicity, and immunogenicity (Kreitman, 2009).

Botulinum toxin

BoNT/A has a wide variety of clinical applications, a part from aesthetic use, which are often related to abnormal muscle contractures (de Maio, 2008). The modern BoNT-based clinical treatment was pioneered by Alan B. Scott and Edward J. Schantz in the early 1970s, when the toxin was used to correct strabismus (Scott, 1980). The toxin was then employed successfully to treat various forms of dystonia and other forms of contractions of smooth and skeletal muscles, as well as to conditions of glandular hypersecretion and pain. BoNT/A is found in three commercial preparations: Botox® (Allergan, Inc., Irvine, California, USA), Dysport® (Ipsen Limited, Berkshire UK) and Xeomin® (Merz Pharmaceuticals GmbH, Frankfurt am Main, Germany). The only approved BoNT type B formulation is Myobloc® (US)/ Neurobloc® (outside US) (de Maio, 2008). BoNT is the treatment of choice for a form of cervical dystonia, known as spasmodic torticolli, characterized by involuntary contraction of muscles in the neck and shoulder. Almost 90% of patients respond positively to the BoNT treatment (de Maio, 2008). In case of spasticity, the target of the treatment with BoNT is reduction of tonus, pain relief, prevention of contractures and functional improvements (Purkiss *et al.*, 2000). Owing to its anti-spasm effects, BoNT has been also use to alleviate certain symptoms associated with Parkinson's disease (Shah *et al.*, 2005). Hemifacial spasm and facial asymmetry, commonly caused by vascular compression of the facial nerve at the root exit zone, is also reduced by targeted BoNT injections (de Maio, 2008).

Bacterial toxins as tools in cell biology research

Last by not least, bacterial toxins have been and are still instrumental in cell biology to characterize cellular processes (Schiavo and van der Goot, 2001). It was by studying Shiga toxin that retrograde transport from endosome to the ER was initially discovered. Toxins can be used as markers for specific membrane microdomains, such as lipid rafts, since they bind receptors, which are enriched in these fractions (e.g. CT), or for probing different endocytic routes. Studies on TeNT and how it is transported retrogradely through the axon has made it possible to study also trafficking of neurotropic growth factor receptors (e.g. p75 and TrkA). Toxins such as the *B. pertussis* CyaA and the *E. coli* LT have been used as vectors to deliver proteins within the cells (e.g. delivery of antigen for induction of specific immune responses).

Concluding remarks

Evolution has shaped the development of very specific and sophisticated toxins, which allow bacteria to modulate the host cellular responses

Table 8.2 Immunotoxins

Name	Antigen	Toxin	Disease	Clinical efficacy	Side-effects	Trial
DAB398IL2, Denileukin diftitox (Ontak)	IL2R	DT	CTCL, CLL, NHL	–	VLS	FDA approved
LMB-2	CD25	PE	HCL, CLL, ATL, NHL	35% PR, 5% CR	Fever, elevated transaminases	Phase II
BL22	CD22	PE	HCL, CLL, NHL	19% PR, 61% CR	Reversible HUS	Phase II
LMB-1, LMB-7, LMB-9	Lewis Y	PE	Adenocarcinoma	Not reported yet	Renal and gastric toxicity	Phase I
SS1P	Mesothelin	PE	Mesothelioma Ovarian carcinoma	50% stabilization disease	Pleuritis	Planned phase II
	IL13R	PE	Glioblastomas	–	–	Phase I
	EGFR	PE	Glioblastomas	–	–	Phase I
HA22	CD22	PE	HCL	–	–	Phase I
DT388-GM-CSF, DTGM	GM-CSFR	DT	AML	31% CR, 62% PR	–	–
DT388-IL3	IL3R	DT	AML, MDS	40% PR, 40% CR	–	Phase I

ATL, adult T-cell leukaemia; AML, acute myelogenous leukaemia; CLL, chronic lymphocytic leukaemia; CR, complete remission; CTCL, cutaneous T-cell lymphoma; FDA, Food and Drug Administration; HCL, hairy cell leukaemia; HUS haemolytic uremic syndrome; MDS, myelodysplastic syndrome; NHL, non-Hodgkin lymphoma; PR, partial response; VLS, vascular leak syndrome.

and establish a successful colonization within their specific niches. The study of bacterial toxins has helped to understand the pathogenicity of bacteria-induced diseases, has allowed the development of efficient vaccines to prevent or reduce the severity of disease, has contributed to the design of drugs with clinical efficacy, and has provided tools to assist biomedical research. The discovery of toxins with novel modes of action (e.g. genotoxins) demonstrates that, in spite of more than 100 years of research, this field still reserves a lot of surprises.

Acknowledgements

This work has been supported by the Swedish Research Council, the Swedish Cancer Society, the Åke Wiberg Foundation, and the Karolinska Institutet. TF is supported by the Swedish Cancer Society.

References

Aktories, K., and Barbieri, J.T. (2005). Bacterial cytotoxins: targeting eukaryotic switches. Nat. Rev. Microbiol. *3*, 397–410.

Aktories, K., and Just, I. (2005). Clostridial Rho-inhibiting protein toxins. Curr. Top. Microbiol. Immunol. *291*, 113–145.

Anthrax. (2001). Update: Investigations of Bioterrorism-related Anthrax – Connecticut, 2001 (WHO). Available at: http://www.who.int/vaccine_research/diseases/zoonotic/en/index1.html.<Please provide further details, e.g. article name, publication details, access dates>

Anthrax_attach. (2011). Vaccine research and development. (WHO). Available at: http://jama.ama-assn.org/cgi/content/full/287/1/34.

Backert, S., and Selbach, M. (2008). Role of type IV secretion in *Helicobacter pylori* pathogenesis. Cell. Microbiol. *10*, 1573–1581.

Barry, E.M., Weiss, A.A., Ehrmann, I.E., Gray, M.C., Hewlett, E.L., and Goodwin, M.S. (1991). *Bordetella pertussis* adenylate cyclase toxin and hemolytic activities require a second gene, *cyaC*, for activation. J. Bacteriol. *173*, 720–726.

Barth, H., and Stiles, B.G. (2008). Binary actin-ADP-ribosylating toxins and their use as molecular Trojan horses for drug delivery into eukaryotic cells. Curr. Med. Chem. *15*, 459–469.

Binz, T., and Rummel, A. (2009). Cell entry strategy of clostridial neurotoxins. J. Neurochem. *109*, 1584–1595.

Bohnert, S., and Schiavo, G. (2005). Tetanus toxin is transported in a novel neuronal compartment characterized by a specialized pH regulation. J. Biol. Chem. *280*, 42336–42344.

Boyd, A.P., Ross, P.J., Conroy, H., Mahon, N., Lavelle, E.C., and Mills, K.H. (2005). *Bordetella pertussis* adenylate cyclase toxin modulates innate and adaptive immune responses: distinct roles for acylation and enzymatic activity in immunomodulation and cell death. J. Immunol. *175*, 730–738.

Bradley, K.A., Mogridge, J., Mourez, M., Collier, R.J., and Young, J.A. (2001). Identification of the cellular receptor for anthrax toxin. Nature *414*, 225–229.

Breidenbach, M.A., and Brunger, A.T. (2005). New insights into clostridial neurotoxin–SNARE interactions. Trends Mol. Med. *11*, 377–381.

Brey, R.N. (2005). Molecular basis for improved anthrax vaccines. Adv. Drug Deliv. Rev. *57*, 1266–1292.

Brook, I. (2008). Current concepts in the management of *Clostridium tetani* infection. Expert Rev. Anti Infect. Ther. *6*, 327–336.

Burleigh, D.E., and Banks, M.R. (2007). Stimulation of intestinal secretion by vasoactive intestinal peptide and cholera toxin. Auton. Neurosci. *133*, 64–75.

Caleo, M., and Schiavo, G. (2009). Central effects of tetanus and botulinum neurotoxins. Toxicon *54*, 593–599.

Chinnapen, D.J., Chinnapen, H., Saslowsky, D., and Lencer, W.I. (2007). Rafting with cholera toxin: endocytosis and trafficking from plasma membrane to ER. FEMS Microbiol. Lett. *266*, 129–137.

Choe, S., Bennett, M.J., Fujii, G., Curmi, P.M., Kantardjieff, K.A., Collier, R.J., and Eisenberg, D. (1992). The crystal structure of diphtheria toxin. Nature *357*, 216–222.

Cholera. (2011). Cholera vaccines. (WHO). Available at: http://www.who.int/topics/cholera/vaccines/current/en/index.html.

Collier, R.J. (1975). Diphtheria toxin: mode of action and structure. Bacteriol. Rev. *39*, 54–85.

Collier, R.J., and Young, J.A. (2003). Anthrax toxin. Annu. Rev. Cell. Dev. Biol. *19*, 45–70.

Cortes-Bratti, X., Karlsson, C., Lagergård, T., Thelestam, M., and Frisan, T. (2001). The *Haemophilus ducreyi* cytolethal distending toxin induces cell cycle arrest and apoptosis via the DNA damage checkpoint pathways. J. Biol. Chem. *276*, 5296–5302.

Dhakal, B.K., Kulesus, R.R., and Mulvey, M.A. (2008). Mechanisms and consequences of bladder cell invasion by uropathogenic *Escherichia coli*. Eur. J. Clin. Invest. *38* (Suppl. 2), 2–11.

Diphtheria. (2011). WHO and UNICEF estimates of national infant immunization coverage: methods and processes. Available at: http://en.wikipedia.org/wiki/Diphtheria.

Doye, A., Mettouchi, A., Bossis, G., Clément, R., Buisson-Touati, C., Flatau, G., Gagnoux, L., Piechaczyk, M., Boquet, P., and Lemichez, E. (2002). CNF1 exploits the ubiquitin-proteasome machinery to restrict Rho GTPase activation for bacterial host cell invasion. Cell *111*, 553–564.

Dube, P. (2009). Interaction of *Yersinia* with the gut: mechanisms of pathogenesis and immune evasion. Curr. Top. Microbiol. Immunol. *337*, 61–91.

Elwell, C., Chao, K., Patel, K., and Dreyfus, L. (2001). *Escherichia coli* CdtB mediates cytolethal distending toxin cell cycle arrest. Infect. Immun. *69*, 3418–3422.

Endo, Y., Tsurugi, K., Yutsudo, T., Takeda, Y., Ogasawara, T., and Igarashi, K. (1988). Site of action of a Vero toxin (VT2) from *Escherichia coli* O157:H7 and of Shiga toxin on eukaryotic ribosomes. RNA N-glycosidase activity of the toxins. Eur. J. Biochem. *171*, 45–50.

Falnes, P.O., and Sandvig, K. (2000). Penetration of protein toxins into cells. Curr. Opin. Cell Biol. *12*, 407–413.

Flatau, G., Lemichez, E., Gauthier, M., Chardin, P., Paris, S., Fiorentini, C., and Boquet, P. (1997). Toxin-induced activation of the G protein p21 Rho by deamidation of glutamine. Nature *387*, 729–733.

Galán, J.E., and Wolf-Watz, H. (2006). Protein delivery into eukaryotic cells by type III secretion machines. Nature *444*, 567–573.

George, K.M., Chatterjee, D., Gunawardana, G., Welty, D., Hayman, J., Lee, R., and Small, P.L. (1999). Mycolactone: a polyketide toxin from *Mycobacterium ulcerans* required for virulence. Science *283*, 854–857.

Glaser, P., Sakamoto, H., Bellalou, J., Ullmann, A., and Danchin, A. (1988). Secretion of cyclolysin, the calmodulin-sensitive adenylate cyclase-haemolysin bifunctional protein of *Bordetella pertussis*. EMBO J. *7*, 3997–4004.

Golden, H.B., Watson, L.E., Lal, H., Verma, S.K., Foster, D.M., Kuo, S.R., Sharma, A., Frankel, A., and Dostal, D.E. (2009). Anthrax toxin: pathologic effects on the cardiovascular system. Front. Biosci. *14*, 2335–2357.

Goonetilleke, A., and Harris, J.B. (2004). Clostridial neurotoxins. J. Neurol. Neurosurg. Psychiatry *75* (Suppl. 3), 35–39.

van der Goot, G., and Young, J.A. (2009). Receptors of anthrax toxin and cell entry. Mol. Aspects Med. *30*, 406–412.

Guermonprez, P., Khelef, N., Blouin, E., Rieu, P., Ricciardi-Castagnoli, P., Guiso, N., Ladant, D., and Leclerc, C. (2001). The adenylate cyclase toxin of *Bordetella pertussis* binds to target cells via the $\alpha_M\beta_2$integrin (CD11b/CD18). J. Exp. Med. *193*, 1035–1044.

Guerra, L., Teter, K., Lilley, B.N., Stenerlöw, B., Holmes, R.K., Ploegh, H.L., Sandvig, K., Thelestam, M., and Frisan, T. (2005). Cellular internalization of cytolethal distending toxin: a new end to a known pathway. Cell. Microbiol. *7*, 921–934.

Guerra, L., Nemec, K.N., Massey, S., Tatulian, S.A., Thelestam, M., Frisan, T., and Teter, K. (2008a). A novel mode of translocation for cytolethal distending toxin. Biochim. Biophys. Acta *1793*, 489–495.

Guerra, L., Carr, H.S., Richter-Dahlfors, A., Masucci, M.G., Thelestam, M., Frost, J.A., and Frisan, T. (2008b). A bacterial cytotoxin identifies the RhoA exchange factor Net1 as a key effector in the response to DNA damage. PLoS One 3, e2254.

Hackett, M., Guo, L., Shabanowitz, J., Hunt, D.F., and Hewlett, E.L. (1994). Internal lysine palmitoylation in adenylate cyclase toxin from *Bordetella pertussis*. Science *266*, 433–435.

Hadfield, T.L., McEvoy, P., Polotsky, Y., Tzinserling, V.A., and Yakovlev, A.A. (2000). The pathology of diphtheria. J. Infect. Dis. *181* (Suppl. 1.), S116–120.

Hanahan, D., and Weinberg, R.A. (2000). The hallmarks of cancer. Cell *100*, 57–70.

Hanski, E. (1989). Invasive adenylate cyclase toxin of *Bordetella pertussis*. Trends Biochem. Sci. *14*, 459–463.

Harper, M., Boyce, J.D., and Adler, B. (2006). *Pasteurella multocida* pathogenesis: 125 years after Pasteur. FEMS Microbiol. Lett. *265*, 1–10.

Holmgren, J., Lindblad, M., Fredman, P., Svennerholm, L., and Myrvold, H. (1985). Comparison of receptors for cholera and *Escherichia coli* enterotoxins in human intestine. Gastroenterology *89*, 27–35.

Hussain, S.P., Hofseth, L.J., and Harris, C.C. (2003). Radical causes of cancer. Nat. Rev. Cancer *3*, 276–285.

Hwang, J., Fitzgerald, D.J., Adhya, S., and Pastan, I. (1987). Functional domains of *Pseudomonas* exotoxin identified by deletion analysis of the gene expressed in *E. coli*. Cell *48*, 129–136.

IARC. (2011). Section of Infections–Infections and Cancer Biology Group. Available at: http://www.iarc.fr/en/research-groups/ICB/index.php.

Ibarra, J.A., and Steele-Mortimer, O. (2009). *Salmonella*-the ultimate insider. *Salmonella* virulence factors that modulate intracellular survival. Cell. Microbiol. *11*, 1579–1586.

Iglewski, B.H., Liu, P.V., and Kabat, D. (1977). Mechanism of action of *Pseudomonas aeruginosa* exotoxin Aiadenosine diphosphate-ribosylation of mammalian elongation factor 2 *in vitro* and *in vivo*. Infect. Immun. *15*, 138–144.

Jaffe, A.B., and Hall, A. (2005). Rho GTPases: biochemistry and biology. Annu. Rev. Cell Dev. Biol. *21*, 247–269.

Jank, T., and Aktories, K. (2008). Structure and mode of action of clostridial glucosylating toxins: the ABCD model. Trends Microbiol. *16*, 222–229.

Jubelin, G., Taieb, F., Duda, D., Hsu, Y., Samba-Louaka, A., Nobe, R., Penary, M., Watrin, C., Nougayrède, J.P., Schulman, B.A., *et al.* (2010). Pathogenic bacteria target NEDD8-conjugated cullins to hijack host-cell signaling pathways. PLoS Pathog. *6*, e1001128.

Karin, M., and Greten, F.R. (2005). NF-κB: linking inflammation and immunity to cancer development and progression. Nat. Rev. Immunol. *5*, 749–759.

Karin, M., Lawrence, T., and Nizet, V. (2006). Innate immunity gone awry: linking microbial infections to chronic inflammation and cancer. Cell *124*, 823–835.

Karmali, M.A., Petric, M., Lim, C., Fleming, P.C., Arbus, G.S., and Lior, H. (1985). The association between idiopathic hemolytic uremic syndrome and infection by verotoxin-producing *Escherichia coli*. J. Infect. Dis. *151*, 775–782.

Klimpel, K.R., Arora, N., and Leppla, S.H. (1994). Anthrax toxin lethal factor contains a zinc metalloprotease consensus sequence which is required for lethal toxin activity. Mol. Microbiol. *13*, 1093–1100.

Konowalchuk, J., Speirs, J.I., and Stavric, S. (1977). Vero response to a cytotoxin of *Escherichia coli*. Infect. Immun. *18*, 775–779.

Kreitman, R.J. (2009). Recombinant immunotoxins containing truncated bacterial toxins for the treatment of hematologic malignancies. BioDrugs *23*, 1–13.

Lacy, D.B., and Stevens, R.C. (1998). Unraveling the structures and modes of action of bacterial toxins. Curr. Opin. Struct. Biol. *8*, 778–784.

Lalli, G., Bohnert, S., Deinhardt, K., Verastegui, C., and Schiavo, G. (2003a). The journey of tetanus and botulinum neurotoxins in neurons. Trends Microbiol. *11*, 431–437.

Lalli, G., Gschmeissner, S., and Schiavo, G. (2003b). Myosin Va and microtubule-based motors are required for fast axonal retrograde transport of tetanus toxin in motor neurons. J. Cell Sci. *116*, 4639–4650.

Lara-Tejero, M., and Galán, J.E. (2000). A bacterial toxin that controls cell cycle progression as a deoxyribonuclease I-like protein. Science *290*, 354–357.

Lax, A.J. (2005). Opinion: bacterial toxins and cancer – a case to answer? Nat. Rev. Microbiol. *3*, 343–349.

Lemichez, E., Flatau, G., Bruzzone, M., Boquet, P., and Gauthier, M. (1997). Molecular localization of the *Escherichia coli* cytotoxic necrotizing factor CNF1 cell-binding and catalytic domains. Mol. Microbiol. *24*, 1061–1070.

Lemonnier, M., Landraud, L., and Lemichez, E. (2007). Rho GTPase-activating bacterial toxins: from bacterial virulence regulation to eukaryotic cell biology. FEMS Microbiol. Rev. *31*, 515–534.

Leppla, S.H. (1982). Anthrax toxin edema factor: a bacterial adenylate cyclase that increases cyclic AMP concentrations of eukaryotic cells. Proc. Natl. Acad. Sci. U.S.A. *79*, 3162–3166.

Lin, W.W., and Karin, M. (2007). A cytokine-mediated link between innate immunity, inflammation, and cancer. J. Clin. Invest. *117*, 1175–1183.

Lord, J.M., Roberts, L.M., and Lencer, W.I. (2005). Entry of protein toxins into mammalian cells by crossing the endoplasmic reticulum membrane: co-opting basic mechanisms of endoplasmic reticulum-associated degradation. Curr. Top. Microbiol. Immunol. *300*, 149–168.

Ly, K.T., and Casanova, J.E. (2007). Mechanisms of *Salmonella* entry into host cells. Cell. Microbiol. *9*, 2103–2111.

de Maio, M. (2008). Therapeutic uses of botulinum toxin: from facial palsy to autonomic disorders. Expert Opin. Biol. Ther. *8*, 791–798.

Marchès, O., Ledger, T.N., Boury, M., Ohara, M., Tu, X., Goffaux, F., Mainil, J., Rosenshine, I., Sugai, M., De Rycke, J., *et al.* (2003). Enteropathogenic and enterohaemorrhagic *Escherichia coli* deliver a novel effector called Cif, which blocks cell cycle G2/M transition. Mol. Microbiol. *50*, 1553–1567.

Matsumoto, H., and Young, G.M. (2009). Translocated effectors of *Yersinia*. Curr. Opin. Microbiol. *12*, 94–100.

Merritt, E.A., and Hol, W.G. (1995). AB5 toxins. Curr. Opin. Struct. Biol. *5*, 165–171.

Mills, M., Meysick, K.C., and O'Brien, A.D. (2000). Cytotoxic necrotizing factor type 1 of uropathogenic *Escherichia coli* kills cultured human uroepithelial 5637 cells by an apoptotic mechanism. Infect. Immun. *68*, 5869–5880.

Montecucco, C. (1986). How do tetanus and botulinum toxins bind to neuronal membranes? Trends Biochem. Sci. *11*, 314–317.

Moss, S.F., and Blaser, M.J. (2005). Mechanisms of disease: inflammation and the origins of cancer. Nat. Clin. Pract. Oncol. *2*, 90–97.

Nesic, D., and Stebbins, C.E. (2005). Mechanisms of assembly and cellular interactions for the bacterial genotoxin CDT. PLoS Pathog. *1*, e28.

Nesic, D., Hsu, Y., and Stebbins, C.E. (2004). Assembly and function of a bacterial genotoxin. Nature *429*, 429–433.

Nougayrède, J.P., Homburg, S., Taieb, F., Boury, M., Brzuszkiewicz, E., Gottschalk, G., Buchrieser, C., Hacker, J., Dobrindt, U., and Oswald, E. (2006). *Escherichia coli* induces DNA double-strand breaks in eukaryotic cells. Science *313*, 848–851.

O'Loughlin, E.V., and Robins-Browne, R.M. (2001). Effect of Shiga toxin and Shiga-like toxins on eukaryotic cells. Microbes Infect. *3*, 493–507.

Ohnishi, N., Yuasa, H., Tanaka, S., Sawa, H., Miura, M., Matsui, A., Higashi, H., Musashi, M., Iwabuchi, K., Suzuki, M., *et al.* (2008). Transgenic expression of *Helicobacter pylori* CagA induces gastrointestinal and hematopoietic neoplasms in mouse. Proc. Natl. Acad. Sci. U.S.A. *105*, 1003–1008.

Oswald, E., Nougayrède, J.P., Taieb, F., and Sugai, M. (2005). Bacterial toxins that modulate host cell-cycle progression. Curr. Opin. Microbiol. *8*, 83–91.

Palermo, M.S., Exeni, R.A., and Fernández, G.C. (2009). Hemolytic uremic syndrome: pathogenesis and update of interventions. Expert Rev. Anti Infect. Ther. *7*, 697–707.

Pastan, I., Hassan, R., FitzGerald, D.J., and Kreitman, R.J. (2007). Immunotoxin treatment of cancer. Annu. Rev. Med. *58*, 221–237.

Pittman, P.R., Hack, D., Mangiafico, J., Gibbs, P., McKee, K.T., Jr., Friedlander, A.M., and Sjogren, M.H. (2002). Antibody response to a delayed booster dose of anthrax vaccine and botulinum toxoid. Vaccine *20*, 2107–2115.

Preuss, I., Kurig, B., Nürnberg, B., Orth, J.H., and Aktories, K. (2009). *Pasteurella multocida* toxin activates Gβγ dimers of heterotrimeric G proteins. Cell. Signal *21*, 551–558.

Pujol, C., and Bliska, J.B. (2005). Turning *Yersinia* pathogenesis outside in: subversion of macrophage function by intracellular yersiniae. Clin. Immunol. *114*, 216–226.

Purkiss, J., Welch, M., Doward, S., and Foster, K. (2000). Capsaicin-stimulated release of substance P from cultured dorsal root ganglion neurons: involvement of two distinct mechanisms. Biochem. Pharmacol. *59*, 1403–1406.

Rogel, A., and Hanski, E. (1992). Distinct steps in the penetration of adenylate cyclase toxin of *Bordetella pertussis* into sheep erythrocytes. Translocation of the toxin across the membrane. J. Biol. Chem. *267*, 22599–22605.

Roux, S., Colasante, C., Saint Cloment, C., Barbier, J., Curie, T., Girard, E., Molgó, J., and Brûlet, P. (2005). Internalization of a GFP-tetanus toxin C-terminal fragment fusion protein at mature mouse neuromuscular junctions. Mol. Cell. Neurosci. *30*, 572–582.

Rupnik, M., Wilcox, M.H., and Gerding, D.N. (2009). *Clostridium difficile* infection: new developments in epidemiology and pathogenesis. Nat. Rev. Microbiol. *7*, 526–536.

Samba-Louaka, A., Nougayrède, J.P., Watrin, C., Jubelin, G., Oswald, E., and Taieb, F. (2008). Bacterial cyclomodulin Cif blocks the host cell cycle by stabilizing the cyclin-dependent kinase inhibitors p21 and p27. Cell. Microbiol. *10*, 2496–2508.

Sánchez, J., and Holmgren, J. (2005). Virulence factors, pathogenesis and vaccine protection in cholera and ETEC diarrhea. Curr. Opin. Immunol. *17*, 388–398.

Sánchez, J., and Holmgren, J. (2008). Cholera toxin structure, gene regulation and pathophysiological and immunological aspects. Cell. Mol. Life Sci. *65*, 1347–1360.

Sandvig, K., and van Deurs, B. (2000). Entry of ricin and Shiga toxin into cells: molecular mechanisms and medical perspectives. EMBO J. *19*, 5943–5950.

Sandvig, K., and van Deurs, B. (2005). Delivery into cells: lessons learned from plant and bacterial toxins. Gene Ther. *12*, 865–872.

Sandvig, K., and Olsnes, S. (1980). Diphtheria toxin entry into cells is facilitated by low pH.J. Cell Biol. *87*, 828–832.

Sandvig, K., Garred, O., Prydz, K., Kozlov, J.V., Hansen, S.H., and van Deurs, B. (1992). Retrograde transport of endocytosed Shiga toxin to the endoplasmic reticulum. Nature *358*, 510–512.

Schiavo, G., and van der Goot, F.G. (2001). The bacterial toxin toolkit. Nat. Rev. Mol. Cell Biol. *2*, 530–537.

Schmidt, G., Sehr, P., Wilm, M., Selzer, J., Mann, M., and Aktories, K. (1997). Gln 63 of Rho is deamidated by *Escherichia coli* cytotoxic necrotizing factor-1. Nature *387*, 725–729.

Schwartz, M. (2009). Dr. Jekyll and Mr. Hyde: a short history of anthrax. Mol. Aspects Med. *30*, 347–355.

Scobie, H.M., Rainey, G.J., Bradley, K.A., and Young, J.A. (2003). Human capillary morphogenesis protein 2 functions as an anthrax toxin receptor. Proc. Natl. Acad. Sci. U.S.A. *100*, 5170–5174.

Scott, A.B. (1980). Botulinum toxin injection into extraocular muscles as an alternative to strabismus surgery. Ophthalmology *87*, 1044–1049.

Shah, S.N., Hornyak, J., and Urquhart, A.G. (2005). Flexion contracture after total knee arthroplasty in a patient with Parkinson's disease: successful treatment with botulinum toxin type A.J. Arthroplasty *20*, 1078–1080.

Singer, D.E., Schneerson, R., Bautista, C.T., Rubertone, M.V., Robbins, J.B., and Taylor, D.N. (2008). Serum IgG antibody response to the protective antigen (PA) of *Bacillus anthracis* induced by anthrax vaccine adsorbed (AVA) among U.S. military personnel. Vaccine *26*, 869–873.

Smith, J.L., and Bayles, D.O. (2006). The contribution of cytolethal distending toxin to bacterial pathogenesis. Crit.Rev. Microbiol. *32*, 227–248.

Strebhardt, K., and Ullrich, A. (2008). Paul Ehrlich's magic bullet concept: 100 years of progress. Nat. Rev. Cancer *8*, 473–480.

Svanborg, C., Bergsten, G., Fischer, H., Godaly, G., Gustafsson, M., Karpman, D., Lundstedt, A.C., Ragnarsdottir, B., Svensson, M., and Wullt, B. (2006). Uropathogenic *Escherichia coli* as a model of host–parasite interaction. Curr. Opin. Microbiol. *9*, 33–39.

Taieb, F., Nougayrède, J.P., Watrin, C., Samba-Louaka, A., and Oswald, E. (2006). *Escherichia coli* cyclomodulin Cif induces G2 arrest of the host cell cycle without activation of the DNA-damage checkpoint-signalling pathway. Cell. Microbiol. *8*, 1910–1921.

Tetanus. http://www.who.int/immunization_monitoring/diseases/tetanus/en

Thelestam, M., and Chaves-Olarte, E. (2000). Cytotoxic effects of the *Clostridium difficile* toxins. Curr. Top. Microbiol. Immunol. *250*, 85–96.

Tournier, J.N., Rossi Paccani, S., Quesnel-Hellmann, A., and Baldari, C.T. (2009). Anthrax toxins: a weapon to systematically dismantle the host immune defenses. Mol. Aspects Med. *30*, 456–466.

Toxin_LD50. http://www.absoluteastronomy.com/topics/LD50.

Travaglione, S., Fabbri, A., and Fiorentini, C. (2008). The Rho-activating CNF1 toxin from pathogenic *E. coli*: a risk factor for human cancer development? Infect. Agent Cancer 3, 4.

Turvill, J.L., Mourad, F.H., and Farthing, M.J. (1998). Crucial role for 5-HT in cholera toxin but not *Escherichia coli* heat-labile enterotoxin-intestinal secretion in rats. Gastroenterology *115*, 883–890.

WHO_Diphtheria. http://www.scielosp.org/scielo.php?script=sci_arttext&pid=S0042-96862009000700015&lng=en&nrm=iso&tlng=en.

Wiles, T.J., Kulesus, R.R., and Mulvey, M.A. (2008). Origins and virulence mechanisms of uropathogenic *Escherichia coli*. Exp. Mol. Pathol. *85*, 11–19.

Young, J.A., and Collier, R.J. (2007). Anthrax toxin: receptor binding, internalization, pore formation, and translocation. Annu. Rev. Biochem. *76*, 243–265.

Part IV

Cellular Invasion by Bacterial Pathogens

Mechanisms of Bacterial Entry into Host Cells

9

Kevin Moreau and Frank Lafont

Abstract

Most invading bacteria enter the host cell by using either a triggered or a zippered mechanism. The former depends on membrane ruffles induced by injection of bacteria-derived effectors into the eukaryotic cell. A hallmark of the latter is a 'sliding' of the bacteria into the cell through a clathrin-mediated structure, which is distinct from the pits in conventional clathrin-mediated endocytosis. Bacteria hijacking either of these mechanisms can also take advantage of signalling platforms activated within specialized membrane domains (lipid rafts). At the entry site, activated signalling pathways regulate the fate of the invading microorganism. Bacteria may then replicate in either cytoplasmic or vacuolar niches. Alternatively, the host immune system can deal with the infection and target the pathogen for elimination via several degradation pathways (notably including autophagy).

Introduction

To establish and maintain a successful infection, pathogenic bacteria have evolved a variety of strategies for invading the host and avoiding or resisting the innate immune response (Cossart and Sansonetti, 2004). Based on their ability to respond to these challenges, bacteria can replicate in a specific compartment called a 'niche'. In contrast to bacterial intoxications (which depend on the secretion of toxins), infections require the adhesion of bacteria to the host cell. Host–pathogen interactions are specific and occur either over the cell body or on very thin filopodia protruding from the cell (referred to as 'nanotubes' when observed *in vitro*) (Rustom *et al.*, 2004; Sowinski *et al.*, 2008). These interactions may involve high-affinity or low-affinity receptors. Pathogenic bacteria express a huge diversity of surface proteins for binding to cell receptors (Alonso and Garcia-del Portillo, 2004). The interactions between bacterial proteins and cell receptors trigger a transduction signal (such as phosphorylation) which leads to the recruitment of adaptor and host effector proteins involved in the remodelling of actin cytoskeleton machinery. This is followed by cellular internalization of the bacterium. This mechanism of bacterial entry involves a change in the plasma membrane topology, engulfment of the pathogen and the formation of bacteria-containing vacuoles (Cossart and Sansonetti, 2004) (Figs. 9.1 and 9.2). Pathogens can modulate the cytoskeleton directly by injecting bacterial effector proteins via dedicated secretory systems. These effector molecules cause massive cytoskeletal changes, which result in the formation of a macropinocytic pocket and then internalization of the bacteria via a triggering process (Cossart and Sansonetti, 2004).

The zipper mechanism

The zipper mechanism of host cell invasion can be divided into three steps. The first step (bacterial adhesion to the cell surface) is actin-independent and involves interactions between a bacterial ligand and a receptor on the host cell surface. The transient signalling induced by this ligand–receptor interaction initiates the second

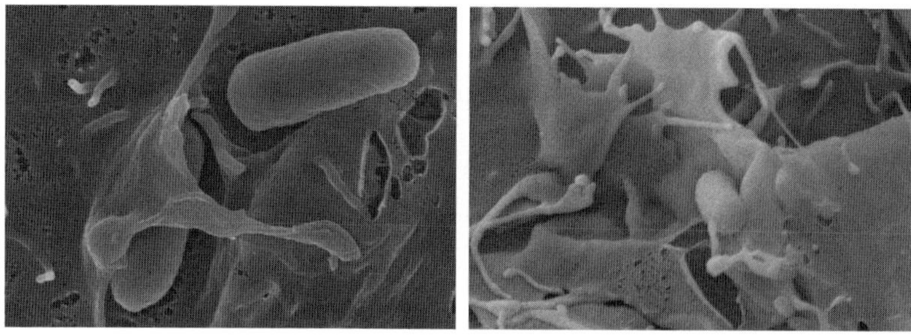

Figure 9.1 The zipper and trigger mechanisms of bacterial entry. Scanning electron micrographs illustrate how *Yersinia* (zipper model; left panel) and Shigella (trigger model; right panel) enter the host cell.

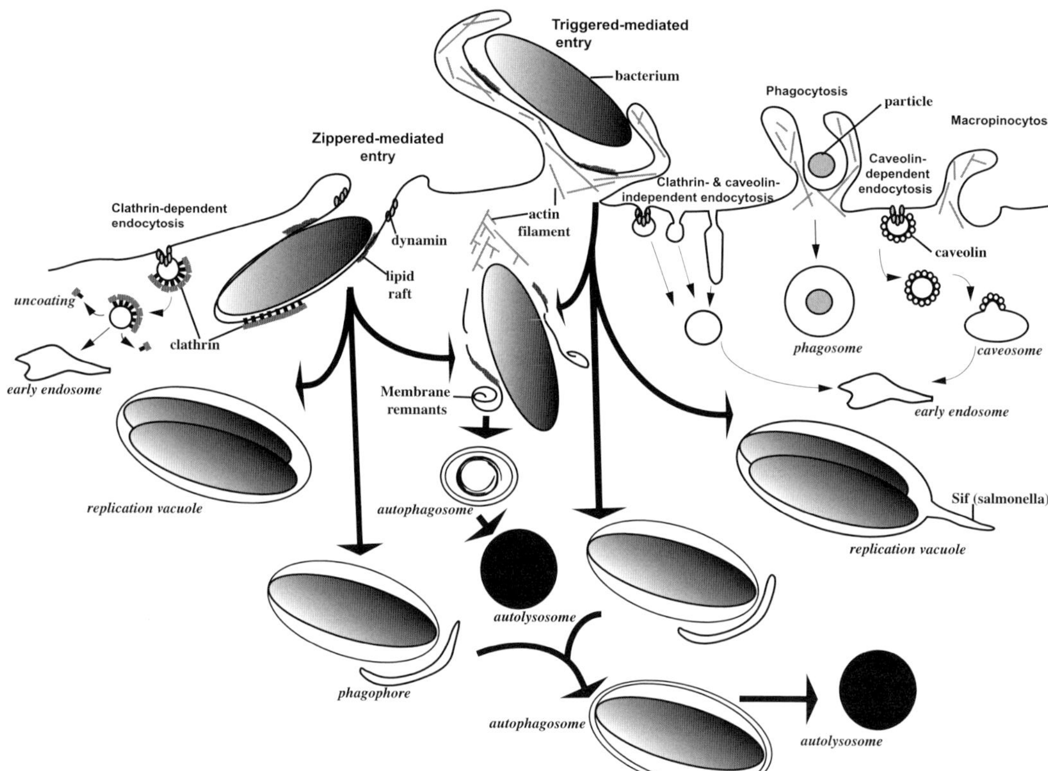

Figure 9.2 Cell entry pathways. Illustration of the several modes of entry for solutes and molecules on the one hand and particles and bacteria on the other.

step – phagocytic cup formation, with actin polymerization and membrane extension around the bacteria. The third and last step corresponds to phagocytic cup closure and internalization of the bacteria. This zipper invasion mechanism is used by *Yersinia* species, in particular. *Yersinia pseudotuberculosis* express two bacterial adhesins (invasin and YadA) which bind to the β1 integrin receptors in the host's extracellular matrix (Isberg and Barnes, 2001). Invasin binds β1 integrins through a superdomain on the C-terminal part of the protein, which triggers integrin clustering and activation (Leong *et al.*, 1990). The cytoplasmic tail of the β1 chain (which usually interacts with the cytoskeleton at the focal adhesion complex) is required for the bacterial entry (Gustavsson *et al.*, 2002). It is worth noting that mutations, which disrupt the cytoplasmic tail's NPxY sequence, prevent integrin-mediated bacterial uptake. Surprisingly, other mutations in the β1 chain, which reduce cytoskeletal binding can result in enhanced bacterial uptake (Tran Van Nhieu *et al.*, 1996). This observation can be explained by an increase in receptor mobility at the plasma membrane, because of the loss of interactions with the cytoskeleton. This increased mobility raises the probability of interactions with bacterial adhesins. Integrin activation leads to phosphorylation of the tyrosine kinases involved in actin remodelling and then to internalization of the bacterium. The focal adhesion kinase (FAK) is probably involved in signal transduction prompted by integrin clustering, since this tyrosine kinase binds to the integrin β1 chain. Over-expression of a dominant interfering form of FAK decreases invasin-dependent bacterial entry (Alrutz and Isberg, 1998). Invasin-dependent bacterial internalization also requires the activation of small GTPases like Rac1 (a master regulator of actin polymerization) and ADP-ribosylation factor 6 (Arf6, involved in membrane trafficking) (Wong and Isberg, 2003). These two proteins induce the activation of phosphatidylinositol 4-phosphate 5-kinase (PIP5K); this kinase produces phosphatidylinositol 4,5-biphosphate (an important signalling molecule for regulating vesicle exo- and endocytosis and the accompanying actin cytoskeletal rearrangements) at the plasma membrane (Wong and Isberg, 2003; Pizarro-Cerdá and Cossart, 2004).

Listeria monocytogenes is a Gram-positive bacterium responsible for intestinal disorders, meningitis and spontaneous abortions (Dussurget *et al.*, 2004). These various symptoms result from the bacterium's ability to cross the intestinal, blood–brain and placental barriers. Like the *Yersinia* species, *Listeria* uses a zipper mechanism for invasion of the host cell. The *Listeria* adhesins internalin A and internalin B (InlA and InlB) participate in the entry mechanism. Internalin A binds to human E-cadherin (Lecuit *et al.*, 1999), a protein involved in cell junction integrity (Yap *et al.*, 1997). Interaction between InlA and E-cadherin leads to the recruitment of ß-and α-catenins at the bacterial entry site, which in turn bind to actin and trigger its polymerization (Lecuit *et al.*, 2000). Actin polymerization is also dependent on Rac, cortactin and the Arp2/3 complex. Myosin VIIA and its ligand vezatin have been implicated in InlA-dependent internalization (Sousa *et al.*, 2004). Internalin B-dependent internalization requires interaction with the hepatocyte growth factor (HGF) receptor Met (Shen *et al.*, 2000). The InlB–Met interaction is species-specific, since InlB can bind only the human form of Met (Khelef *et al.*, 2006). Activation of Met by InlB induces the recruitment of a few adaptor proteins, such as Gab1, Cbl and Shc, which in turn leads to the activation of the phosphoinositide-3 kinase (PI3K) at the plasma membrane (Ireton *et al.*, 1996). The PI3K activates Rac, which triggers actin polymerization (as described above). Actin reorganization is also modulated upstream by cofilin (involved in actin depolymerization of pre-existing filaments) and the cofilin-modulating protein LIM-kinase (Bierne *et al.*, 2001). It is also noteworthy that clathrin (the hallmark component in clathrin-dependent endocytosis) has been implicated in the entry mechanism; this will be described in more detail below, in a specific part of this chapter. In essence, it is currently considered that the zipper invasion mechanism is clathrin-dependent and that the trigger invasion mechanism is not.

The trigger mechanism

The trigger mechanism of host cell invasion is used by Gram-negative bacteria, such as *Shigella flexneri* and *Salmonella* spp. (Cossart and Sansonetti, 2004). One key feature of the trigger mechanism is massive reorganization of the actin cytoskeleton at the plasma membrane (Cossart and Sansonetti, 2004). Bacterial adhesion to the host cell is mediated by a type III secretion system (T3SS). The bacteria use the T3SS to inject bacterial effectors, which specifically target cytosol proteins involved in actin remodelling. The *Salmonella* T3SS is encoded by the chromosomal pathogenicity island (PAI) SP-1, whereas that of *Shigella* is encoded by a plasmid PAI. Two proteins encoded by T3SS (SipB/C in *Salmonella* and IpaB/C in *Shigella*) induce pore formation and allow the delivery of effectors into the cytosol (Galán, 2001; Sansonetti, 2001). Pore formation is highly regulated and involves the multifunctional protein IpaD in *Shigella*. IpaD is located at the tip of the T3SS and controls IpaB/C exposure at the point of contact between the bacterium and the cell (Espina *et al.*, 2006). Bacterial interaction with the host cell occurs in many successive steps. Prior to the interaction with the host, the bacterial effector proteins in the cytosol are associated with specific, degradation-preventing chaperones (Parsot *et al.*, 2003). When bacteria adhere to the host-cell surface, the secretion signal is activated. This step is probably coordinated by specific phosphatases at the base of the T3SS and leads to effector injection (Blocker *et al.*, 2003). Interaction between IpaB and the hyaluronic acid receptor CD44 is thought to be an important step in adhesion, activation of the secretion machinery and insertion of IpaB/C into the plasma membrane (Skoudy *et al.*, 2000). It has been suggested that interactions between IpaB/IpaC/IpaD and integrin receptors are also important for internalization (Watarai *et al.*, 1996). After the injection of bacterial effectors into the cytosol, massive cytoskeleton reorganization leads to the formation of filopodia around the bacterium. These filopodia are referred to as membrane ruffles and appear to have the same operating mechanism in both *Shigella* and *Salmonella*. The required actin nucleation is induced by IpaC in *Shigella* and SipC in *Salmonella*; in both cases, nucleation is mediated by the cytosolic, C-terminal domain of the respective protein (Hayward and Koronakis, 1999; Tran Van Nhieu *et al.*, 1999). In *Salmonella*, actin filament extension occurs at the bacterial entry site and is dependent on the SopE effectors (SopE1 and SopE2) mimicking an exchange factor for Cdc42 and Rac1 (Hardt *et al.*, 1998), whereas the SipA effector binds to and stabilizes actin filaments (Zhou *et al.*, 1999). Once bacteria have been engulfed in the macropinocytic compartment, actin is depolymerized so that internalization can be completed. The *Salmonella* effector protein SptP (secreted by the T3SS) possesses both a tyrosine phosphatase activity (for MAPK) and a GAP activity (for Cdc42 and Rac). It acts as a SopE antagonist and therefore blocks actin polymerization. So that these processes are coordinated in a timely manner, SopE is degraded more rapidly by the proteasome than SptP is (Kubori and Galán, 2003). The *Shigella* effector IpaA (also secreted by the T3SS) binds to the N-terminal domain of vinculin, a protein involved in actin depolymerization (Bourdet-Sicard *et al.*, 1999).

Endocytosis mechanisms involved in bacterial entry

The differences between the zipper and trigger mechanisms are primarily related to morphological characteristics, as a result of the specific molecular features involved. From a mechanistic viewpoint, the two invasion processes involve the hijacking of different endocytosis pathways. The plasma membrane is a highly dynamic compartment, which separates the inside of the cell from the outside environment. Although ions and low molecular weight compounds (such as amino acids and sugars) can cross the plasma membrane through pores, larger molecules require more complex mechanisms for internalization via a process called endocytosis. Different kinds of endocytosis occur in the cell: phagocytosis, macropinocytosis, clathrin-dependent endocytosis, caveolin-dependent endocytosis and both clathrin- and caveolin-independent endocytosis (Conner and Schmid, 2003). Phagocytosis is restricted to 'professional' phagocytes (such as neutrophils, macrophages and dendritic cells), whereas the other endocytosis modalities are ubiquitous.

Clathrin-dependent endocytosis

Clathrin-mediated endocytosis is a major mechanism for protein internalization at the plasma membrane (Fig. 9.2). Clathrin is the main component of clathrin-coated vesicles. As a result of *in vitro* experiments on clathrin triskelion formation, it was suggested that clathrin-mediated endocytosis could not internalize molecules larger than 120 nm (Heuser, 1980; Keen, 1987; McMahon, 1999). In view of this size-based model, it was long considered that bacteria could not be internalized via clathrin-mediated endocytosis. However, following the description of large, flat clathrin coats in the endosomal compartment, this view had to be reconsidered (Raiborg *et al.*, 2006). It has now been proven that bacteria using the zipper mechanism (and not the trigger mechanism) require clathrin for their internalization (Cossart and Sansonetti, 2004; Pizarro-Cerdá and Cossart, 2006). This finding was in accordance with the known role of actin in the entry mechanism (Engqvist-Goldstein and Drubin, 2003).

Interaction between *L. monocytogenes* InlB and Met induces receptor internalization via a clathrin-dependent mechanism (Veiga and Cossart, 2005). Clathrin, dynamin and ubiquitin ligases are recruited to the bacterial entry site and are key factors in bacterial internalization (Veiga and Cossart, 2005). Similar recruitment was also observed in *Escherichia coli* expressing the *Rickettsia* outer membrane protein B gene (Chan *et al.*, 2009). Microscopy findings and experimental data obtained after pharmacological inhibition of clathrin-mediated endocytosis suggest that clathrin-mediated endocytosis has a role in the internalization of many bacteria, including *Escherichia coli* (Oelschlaeger *et al.*, 1994; Jouve *et al.*, 1997; Green and Brown, 2006; Nazli *et al.*, 2006), *Staphylococcus aureus* (Almeida *et al.*, 1996; Ellington *et al.*, 1999), *Streptococcus pneumoniae* (Cundell *et al.*, 1995; Radin *et al.*, 2005), *Streptococcus dysgalactiae* (Calvinho and Oliver, 1998), *Ehrlichia risticii* (Messick and Rikihisa, 1993), *Brucella abortus* (Detilleux *et al.*, 1991), *Klebsiella pneumoniae* (Oelschlaeger and Tall, 1997), *Legionella dumoffii* (Maruta *et al.*, 1998), *Campylobacter jejuni* (Oelschlaeger *et al.*, 1993), *Citrobacter freundii* (Oelschlaeger *et al.*, 1993), uropathogenic *E. coli* (UPEC; Eto *et al.*, 2008) and *Porphyromonas gingivalis* (Boisvert and Duncan, 2008). However, the latter case remains to be confirmed, since conflicting results have been reported using different cell models and approaches (Furuta *et al.*, 2009).

Y. pseudotuberculosis invades the host cell following interaction between bacterial adhesins (YadA and invasin) and integrin receptors. Expression of the invasin gene in a non-pathogenic strain of *E. coli* enables the latter (*E. coli inv*) to invade non-phagocytic cells (Isberg and Falkow, 1985). Electron micrographs of *E. coli inv* revealed clathrin-coated pits (CCPs) at the bacterial entry site (Tran Van Nhieu *et al.*, 1996). Inhibition of clathrin-mediated endocytosis (either by potassium depletion or by micro-injection of an antibody against the AP-2 adaptor) also inhibits bacterial entry (Tran Van Nhieu *et al.*, 1996). Markers of the clathrin-mediated endocytosis pathway (such as the transferrin receptor) have been observed in the vacuole containing *E. coli inv* (Lafont *et al.*, 2002). These data indicate that clathrin-mediated endocytosis has a role in bacterial internalization. This discovery also raises new questions in cell biology, since the data suggest that the clathrin coat, which forms around internalized bacteria, differs from conventional CCPs. This is illustrated by the recruitment of clathrin by enteropathogenic *E. coli*, which remain outside the cell but bind actin-dependent pedestals (the formation of which requires clathrin) (Veiga *et al.*, 2007).

Neisseria gonorrhoeae binds to the asialoglycoprotein receptor, which is endocytosed in a clathrin-dependent process. Furthermore, monodansylcadaverin, a known inhibitor of clathrin-mediated endocytosis, impairs the entry of *N. gonorrhoeae* into epithelial cells (Harvey *et al.*, 2001), although contradictory results have been reported (Grassmé *et al.*, 1996). Ultrastructural electron imaging studies have confirmed the presence of clathrin at the entry site of *Chlamydia* spp. (Hodinka and Wyrick, 1986; Hodinka *et al.*, 1988; Wyrick *et al.*, 1989). However, monodansylcadaverin treatment and the expression of dominant inhibitory mutants of the clathrin-mediated pathway, such as dynamin I K44A and Eps15 delta95–295, did not argue in favour of clathrin involvement (Ward and Murray, 1984; Boleti *et al.*, 1999). Hence, the results of experiments examining the entry of *Chlamydia* spp. and *Neisseria* suggest that these pathogens could use several different internalization pathways – some of which are clathrin-dependent.

Lipid raft-dependent endocytosis

Over the last two decades, an increasing body of evidence has suggested that plasma membrane components are dynamically partitioned into specific microdomains, rather than being randomly distributed (Simons and Ikonen, 1997). Of the various microdomains characterized to date, lipid rafts have emerged as important clustering devices for cell signalling (Simons and Ikonen, 1997). Even though the size, biophysical properties, stability and exact composition of these units have yet to be fully characterized, lipid rafts are now thought to play a major role in the membrane-associated signalling associated with several cell physiological processes and, indeed, a number of disorders, such as Alzheimer's disease, Parkinson's disease and prion and other infectious diseases (Simons and Ikonen, 1997; Brown and London, 2000; Simons and Toomre, 2000; Anderson and Jacobson, 2002; Edidin, 2003; Mañes *et al.*, 2003; Lafont *et al.*, 2004; Lafont and van der Goot, 2005; Michel and Bakovic, 2007). A hallmark of lipid rafts is the high proportion of cholesterol and glycosphingolipids, in association with glycosylphosphatitylinositol-anchored proteins (GPi-APs). In biochemical terms, the rafts' particular lipid composition renders them resistant to solubilization by detergents (Lingwood and Simons, 2010). When used with appropriate controls, this property of detergent resistance can be utilized to identify proteins in collapsed microdomains (Schuck *et al.*, 2003). Membrane cholesterol extraction with methyl-β-cyclodextrin (MBCD) perturbs the binding of target proteins to detergent-resistant membranes and thus represents one way to study protein partitioning in lipid rafts and functionally destabilize it. Combined with these relatively indiscriminate assays, detailed microscopy analysis can provide information of binding between lipid raft markers and target proteins (Owen *et al.*, 2009).

In microbiology, lipid rafts were studied with respect to the internalization of bacterial toxins. Aerolysin (a pore-forming toxin produced by *Aeromonas hydrophila*) interacts with lipid rafts via a mechanism involving toxin heptamerization (Abrami and van Der Goot, 1999). Other subsequent examples described the impact of the lipid environment on the function of pore-forming toxins (Abrami *et al.*, 2000; Lafont *et al.*, 2004). Certain toxins have also been seen to associate with lipid rafts via the induction of receptor clustering. For instance, cholera toxin binds to the GM1 ganglioside in a pentameric arrangement (Lencer *et al.*, 1999). With respect to bacterial internalization, lipid rafts have been implicated in the entry mechanism of *N. gonorrhoeae* (Grassmé *et al.*, 1997), pathogenic strains of *E. coli* (Baorto *et al.*, 1997), *Pseudomonas aeruginosa* (Grassmé *et al.*, 2003), *S. flexneri* (Lafont *et al.*, 2002), *L. monocytogenes* (Seveau *et al.*, 2004), *Chlamydia* spp. (Stuart *et al.*, 2003), *C. jejuni* (Wooldridge *et al.*, 1996), *B. abortus* (Kim *et al.*, 2002), and *Mycobacterium* spp. (Gatfield and Pieters, 2000) (Table 9.1).

N. gonorrhoeae was one of the first bacteria found to hijack lipids and membrane domains in order to infect human epithelial cells. The initial research showed that *N. gonorrhoeae* activates sphingomyelinase a few minutes after infection of epithelial cells; this leads to sphingomyelin hydrolysis and the release of ceramides, which form membrane platforms for bacterial internalization (Grassmé *et al.*, 1997). Further studies have found that *N. gonorrhoeae* infection induces lipid raft clustering and membrane macrodomain formation (Schmitter *et al.*, 2004). These macrodomains serve as platforms for recruiting and clustering specific host receptors involved in *N. gonorrhoeae* entry, such as CEACAM3 and Rac (Schmitter *et al.*, 2004).

As with *N. gonorrhoeae*, the infection of epithelial cells by *P. aeruginosa* leads to lipid raft clustering. Bacterial infection also induces the activation of sphingomyelinase and the formation of ceramide-rich membrane platforms (Grassmé *et al.*, 2003). Functional perturbation of the lipid rafts by MBCD treatment prevents bacterial internalization (Grassmé *et al.*, 2000, 2003; Kowalski and Pier, 2004).

During the infection of epithelial cells by *S. flexneri*, bacterial effectors are injected through a T3SS; this leads to massive reorganization of the cytoskeleton and then bacterial internalization (as described above). The IpaB–CD44 interaction is dependent on the receptor diffusion to lipid rafts (Lafont *et al.*, 2002). This interaction is probably transient, since IpaB can also bind directly to cholesterol- and sphingolipid-enriched liposomes (van der Goot *et al.*, 2004) and cholesterol itself, as does *Salmonella* SipB (Hayward *et al.*, 2005). The IpaB-CD44 binding triggers actin remodelling via a Rho-dependent and raft-dependent mechanism and may enable subsequent interaction between IpaB/IpaC and the integrin receptors (Watarai *et al.*, 1996; Demali *et al.*, 2006).

During *L. monocytogenes* infection, lipid raft markers are recruited to the bacterial entry site (Seveau *et al.*, 2004). Moreover, the E-cadherin

Table 9.1 Bacteria and bacterial toxins which bind to lipid rafts during host cell entry, together with a list of the molecules known to be involved in these processes

Bacteria or bacterial effectors	Molecules localized in lipid rafts at bacterial entry site
Toxins	
LPS	CD14, TLR4, CD55, GDF5, CXCR4, Hsp70 and 90, GM1
Tetanus toxin	GT1 and/or GD1b, GPI-AP
Botulinum toxin	GD1b
Cholera toxin	GM1
Thermolabile enterotoxin	GD1
Shiga toxin	Gb3, GM1
Aerolysin, toxin α of *Clostridium*	GPI-anchored proteins (AP)
Cry 1A endotoxin	Aminopeptidase N
Bin toxin	Cpm1
SLO, PFO, LLO	Cholesterol
Lysenin	Sphingomyelin
Cytolysin of *Vibrio cholerae*	Sphingolipids, cholesterol
Anthrax toxin	ATR
VacA	Receptor-like protein tyrosine phosphatase α & β, sphingomyelin
Bacteria	
Brucella spp.	GM1, GPI-AP (CD48)
C. jejuni	?
C. trachomatis	Caveolin, GM1
E. coli (FimH+)	CD48, caveolin-1, GM1
E. coli (Afa/Dr+)	CD55, CD66, α5β1
Ehrlichia chaffeensis, *Anaplasma phagocytophilum*	GPI-AP, HM1, caveolin-1, PLC-γ2
F. tularensis	Caveolin
M. tuberculosis, *M. bovis*, *M. kansasii*	CR3, CD55, CD66b, CD16b, CD14
P. gingivalis	Caveolin
Salmonella Typhimurium	CD55
S. flexneri	CD44, GPI-AP, and sphingolipids
Y. pseudotuberculosis	?

molecules involved in internalin A-mediated bacterial internalization were found exclusively in a lipid raft-enriched fraction (Seveau *et al.*, 2004). The localization of E-cadherin within lipid rafts is essential for initial interaction with internalin A and thus for triggering of the bacterial entry mechanism (Seveau *et al.*, 2004). In contrast, the initial InlB–Met interaction does not depend on partitioning of the latter into rafts, whereas rafts perturbation has been shown to delay internalization and to affect downstream signalling (Seveau *et al.*, 2004). *L. monocytogenes* is one typical example of a bacterium which uses both clathrin and lipid rafts in the invasion mechanism, as do other bacteria exploiting a receptor-mediated zipper mechanism (Veiga *et al.*, 2007). The fact that rafts can be recruited within clathrin domains has been also nicely demonstrated in studies investigating bacterial toxin internalization (Abrami *et al.*, 2003, 2010). *B. abortus* infects macrophages in a lipid raft-dependent manner. The structure which forms during bacterial internalization is similar to that observed in the trigger mechanism and contains lipid raft markers, such as GM1 (Kim *et al.*, 2002). Bacterial internalization is inhibited by cholesterol extraction with MBCD or cholesterol clustering with filipin.

Mycobacterium spp. are able to replicate inside macrophages. Work by Gatfield and Pieters (2000) has shown that intact lipid rafts are required for bacterial internalization. MBCD treatment inhibits the invasion of macrophages by *Mycobacterium tuberculosis* and *Mycobacterium bovis* (Gatfield and Pieters, 2000). Complement receptor 3 (CR3) has been identified as a *Mycobacterium kansasii* receptor involved in lipid raft-dependent internalization (Peyron *et al.*, 2000). It is associated with a GPI-AP and is preferentially located within lipid rafts. Treatment with anti-CR3 antibodies blocks bacterial entry and perturbation of lipid rafts redistributes CR3 to a non-raft compartment (Peyron *et al.*, 2000). Cholesterol depletion does not inhibit CR3 activity but impairs association between cholesterol, CR3 and GPI-APs, which thus inhibits bacterial entry (Zaas *et al.*, 2005a).

Chlamydia trachomatis is an intracellular bacterium, which invades both phagocytic and non-phagocytic cells. The destabilization of functional lipid rafts (as described above) decreases bacterial entry (Stuart *et al.*, 2003). However, there are some strain-dependent variations: *C. trachomatis* serovar A invades host cells independently of lipid rafts, in contrast to *Chlamydia pneumoniae* and *Chlamydia psittaci* (Stuart *et al.*, 2003). Internalization of *C. trachomatis* serovar A is clathrin-dependent (Stuart *et al.*, 2003). These disparities may be explained by the involvement of different receptors.

C. jejuni infects intestinal cells in a lipid raft-dependent manner, since MBCD treatment inhibits bacterial entry. However, the internalization receptor has not yet been identified (Wooldridge *et al.*, 1996).

Host cell adhesion of pathogenic *E. coli* depends on the FimH bacterial adhesin. The latter has a mannose-binding domain and binds to the CD48 receptor (a GPI-AP), which induces bacterial internalization (Shin *et al.*, 2000). The functional role of lipid rafts has been demonstrated by using MBCD as an inhibitor of bacterial entry. It is interesting to note that opsonised bacteria are internalized in a clathrin-dependent process and traffic to the lysosomal compartment where they are degraded. This observation suggests that pathogenic bacteria have evolved strategies for avoiding lysosomal degradation by hijacking lipid rafts.

Caveolae-dependent endocytosis

Caveolae were first identified by electron microscopy and are defined as 60- to 80-nm-wide pits in the plasma membrane (Palade, 1953) (Fig. 9.2). These structures have been implicated in endocytosis, transcytosis, calcium signalling and many other signal transduction events (Lisanti *et al.*, 1995; Kurzchalia and Parton, 1999; Parton and Richards, 2003). Caveolae and their main membrane protein components (the caveolins) have also been linked to disease states; mutations in the caveolins have been found in breast cancer and limb girdle muscular dystrophy. It has been estimated that a single caveolar structure contains 144 molecules of caveolin (Pelkmans and Zerial, 2005). Caveola formation requires the association between caveolins and lipid rafts (Parton and Simons, 2007). Cholesterol depletion with MBCD disorganizes the caveolar structure and the caveolae are enriched in sphingolipids such as the gangliosides GM1 and GM3 (Parton and Simons, 2007). During plasma membrane fractionation, caveolae are found in the lipid raft fraction (Simons and Vaz, 2004; Hancock, 2006). Of the various molecules involved in caveola-dependent endocytosis, a few proteins, such as dynamin, the Src kinases, kinase C protein and actin, are known to play major roles in caveola internalization (Oh *et al.*, 1998; Henley *et al.*, 1999; Pelkmans and Helenius, 2002; Pelkmans *et al.*, 2002; Sharma *et al.*, 2004; Yao *et al.*, 2005). In essence, caveolae correspond to specialized, morphologically distinct sphingolipid-cholesterol microdomains which are stabilized by caveolin proteins (Parton and Simons, 2007).

Caveolae are also exploited by pathogens, both as direct portals for endocytic entry and for facilitating the entry of pathogens which are larger than a single caveolae, as discussed below (Parton and Richards, 2003). As in clathrin-mediated endocytosis, theoretical models of the caveolae are not compatible with the internalization of bacteria, since it is thought that only molecules smaller than 80 nm can be internalized via these structures. The initial reports on caveolae-mediated pathogen internalization were published for simian virus 40 and poliovirus (Anderson *et al.*, 1996; Pelkmans *et al.*, 2001). In 2004, data obtained with *E. coli* FimH indicated that bacterial internalization in bladder epithelial cells is dependent on caveolin-1 (Duncan *et al.*, 2004). It was suggested that caveolin-1 is important for the formation of the lipid raft platforms where bacteria are internalized (Duncan *et al.*, 2004). Soon afterwards, another report showed that *P. aeruginosa* internalization in epithelial cells is impaired by caveolin-1 or caveolin-2 knockdown with RNA interference techniques (Zaas *et al.*, 2005b). Furthermore, the lipid raft-dependent tyrosine phosphorylation of caveolin-2 appeared to be a critical regulator of *P. aeruginosa* invasion (Zaas *et al.*, 2005b). Caveolin-1 knockdown also decreases the internalization of *P. gingivalis* in KB cells (derived from HeLa cells) (Tamai *et al.*, 2005). This involvement of caveolin-1 has also been extended to *Francisella tularensis*, since a live vaccine strain recruits cholesterol-rich lipid domains with caveolin-1 for successful entry into macrophages (Tamilselvam and Daefler, 2008).

Autophagy: news and views

Autophagy ('self-digestion') is a cellular pathway involved in protein and organelle degradation. It has an astonishing number of links to pathophysiological functions in humans. For example, autophagy dysfunction is associated with cancer, neurodegeneration, ageing and microbial infection (Kirkegaard *et al.*, 2004; Levine and Deretic, 2007). In recent years, it has been shown that autophagy is involved in the host innate immune response against pathogens (discussed in Chapters 10 and 11).

Autophagosome formation occurs through dynamic rearrangements of cellular membrane structures initiated by the emergence of an

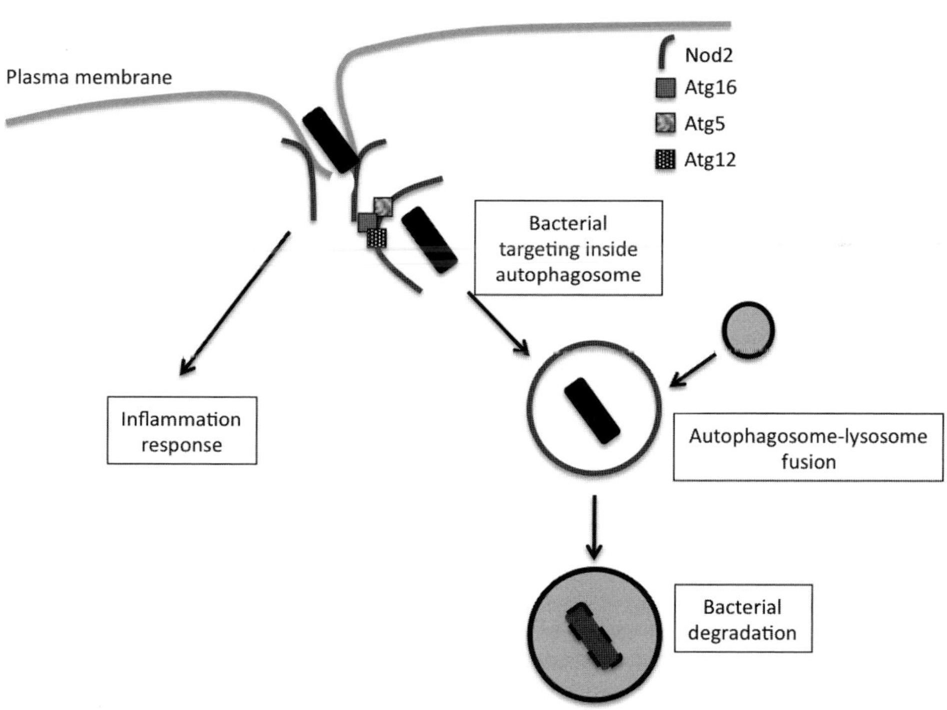

Figure 9.3 Nod proteins are involved in the autophagic host cell response to pathogen invasion. The pathogen-associated molecular pattern activation of Nods allows further recruitment of Atg16L1, which triggers the autophagic response. Nods can also modulate the NF-κB pathways involved in inflammatory response and survival. Adapted from Travassos *et al.* (2010).

isolation membrane that starts to enwrap portions of the cytoplasm and forms a phagophore (also called a pre-autophagic structure) (Xie and Klionsky, 2007). The identity of the membrane donor compartment is still under debate and all intracellular membrane compartments have so far been involved. If specificity of the donor membrane occurs, the detailed mechanism awaits demonstration whilst on the other hand specificity in function of the cargo appears as a consensus. Herein we refer to the autophagy targeting of bacteria, named xenophagy. The phagophore membranes extend and fuse to create a double-membrane autophagosome. More than thirty proteins (the Atg proteins) have been found to regulate autophagy (Xie and Klionsky, 2007). Some contribute to two separate conjugation reactions required for autophagosome formation. The first involves the conjugation of Atg12 and Atg5. The Atg12–Atg5 conjugate interacts non-covalently with Atg16L1 to form a complex, which is essential for phagophore expansion. The Atg16L1–Atg12–Atg5 complexes locate to the phagophore and dissociate once the autophagosome has fully formed. A second conjugation reaction (involving Atg8/LC3) is involved in the completion of autophagosome formation (Xie and Klionsky, 2007).

Although autophagy has been implicated in regulating the fate of internalized bacteria (Deretic and Levine, 2009) and the signalling associated with remnant vacuolar membranes for cytoplasmically replicating bacteria (Dupont *et al.*, 2009), recent data have revealed the molecular links between autophagy and bacterial entry *per se*. The peptidoglycan sensors Nod1 and Nod2 trigger the recruitment of Atg16L1 to the site of bacterial entry, where it in turn triggers bacterial autophagy (Cooney *et al.*, 2010; Travassos *et al.*, 2010) (Fig. 9.3). In addition, these recent studies demonstrate that the most common Crohn's disease-associated Nod2 mutant is defective in recruiting Atg16L1 to the plasma membrane. This impairment has a profound effect on the autophagic response induced by intracellular bacterial infection. Taken together, these reports suggest that autophagy has a role in both bacterial internalization and trafficking inside autophagosomes. The recruitment of Atg16L1 to the plasma membrane at the bacterial entry site is a new interesting feature that will probably reveal novel aspects of the relationship between bacterial endocytosis and autophagosome formation.

Acknowledgements

The authors gratefully acknowledge the receipt of grants-in-aid from the *Fondation pour la Recherche Médicale* and the French Ministry of Research (*Chaire d'Excellence*). We wish to thank Nicolas Barois from the BioImaging Centre Lille-Nord de France and MICPaL facility and the PFMU from the Pasteur Insitute (Paris) for the EM data.

References

Abrami, L., and van Der Goot, F.G. (1999). Plasma membrane microdomains act as concentration platforms to facilitate intoxication by aerolysin. J. Cell Biol. *147*, 175–184.

Abrami, L., Fivaz, M., and van der Goot, F.G. (2000). Surface dynamics of aerolysin on the plasma membrane of living cells. Int. J. Med. Microbiol. *290*, 363–367.

Abrami, L., Liu, S., Cosson, P., Leppla, S.H., and van der Goot, F.G. (2003). Anthrax toxin triggers endocytosis of its receptor via a lipid raft-mediated clathrin-dependent process. J. Cell Biol. *160*, 321–328.

Abrami, L., Bischofberger, M., Kunz, B., Groux, R., and van der Goot, F.G. (2010). Endocytosis of the anthrax toxin is mediated by clathrin, actin and unconventional adaptors. PLoS Pathog. *6*, e1000792.

Almeida, R.A., Matthews, K.R., Cifrian, E., Guidry, A.J., and Oliver, S.P. (1996). *Staphylococcus aureus* invasion of bovine mammary epithelial cells. J. Dairy Sci. *79*, 1021–1026.

Alonso, A., and Garcia-del Portillo, F. (2004). Hijacking of eukaryotic functions by intracellular bacterial pathogens. Int. Microbiol. *7*, 181–191.

Alrutz, M.A., and Isberg, R.R. (1998). Involvement of focal adhesion kinase in invasin-mediated uptake. Proc. Natl. Acad. Sci. U S A *95*, 13658–13663.

Anderson, H.A., Chen, Y., and Norkin, L.C. (1996). Bound simian virus 40 translocates to caveolin-enriched membrane domains, and its entry is inhibited by drugs that selectively disrupt caveolae. Mol. Biol. Cell. *7*, 1825–1834.

Anderson, R.G., and Jacobson, K. (2002). A role for lipid shells in targeting proteins to caveolae, rafts, and other lipid domains. Science *296*, 1821–1825.

Baorto, D.M., Gao, Z., Malaviya, R., Dustin, M.L., van der Merwe, A., Lublin, D.M., and Abraham, S.N. (1997). Survival of FimH-expressing enterobacteria in macrophages relies on glycolipid traffic. Nature *389*, 636–639.

Bierne, H., Gouin, E., Roux, P., Caroni, P., Yin, H.L., and Cossart, P. (2001). A role for cofilin and LIM kinase in *Listeria*-induced phagocytosis. J. Cell Biol. *155*, 101–112.

Blocker, A., Komoriya, K., and Aizawa, S. (2003). Type III secretion systems and bacterial flagella: insights into their function from structural similarities. Proc. Natl. Acad. Sci. U.S.A. *100*, 3027–3030.

Boisvert, H., and Duncan, M.J. (2008). Clathrin-dependent entry of a gingipain adhesin peptide and *Porphyromonas gingivalis* into host cells. Cell. Microbiol. *10*, 2538–2552.

Boleti, H., Benmerah, A., Ojcius, D.M., Cerf-Bensussan, N., and Dautry-Varsat, A. (1999). *Chlamydia* infection of epithelial cells expressing dynamin and Eps15 mutants: clathrin-independent entry into cells and dynamin-dependent productive growth. J. Cell Sci. *112*, 1487–1496.

Bourdet-Sicard, R., Rüdiger, M., Jockusch, B.M., Gounon, P., Sansonetti, P.J., and Tran Van Nhieu, G. (1999). Binding of the *Shigella* protein IpaA to vinculin induces F-actin depolymerization. EMBO J. *18*, 5853–5862.

Brown, D.A., and London, E. (2000). Structure and function of sphingolipid-and cholesterol-rich membrane rafts. J. Biol. Chem. *275*, 17221–17224.

Calvinho, L.F., and Oliver, S.P. (1998). Characterization of mechanisms involved in uptake of *Streptococcus dysgalactiae* by bovine mammary epithelial cells. Vet. Microbiol. *63*, 261–274.

Chan, Y., Cardwell, M., Hermanas, T., Uchiyama, T., and Martinez, J. (2009). Rickettsial outer-membrane protein B (rOmpB) mediates bacterial invasion through Ku70 in an actin, c-Cbl, clathrin and caveolin 2-dependent manner. Cell. Microbiol. *11*, 629–644.

Conner, S.D., and Schmid, S.L. (2003). Regulated portals of entry into the cell. Nature *422*, 37–44.

Cooney, R., Baker, J., Brain, O., Danis, B., Pichulik, T., Allan, P., Ferguson, D.J., Campbell, B.J., Jewell, D., and Simmons, A. (2010). NOD2 stimulation induces autophagy in dendritic cells influencing bacterial handling and antigen presentation. Nat. Med. *16*, 90–97.

Cossart, P., and Sansonetti, P.J. (2004). Bacterial invasion: the paradigms of enteroinvasive pathogens. Science *304*, 242–248.

Cundell, D.R., Gerard, N.P., Gerard, C., Idanpaan-Heikkila, I., and Tuomanen, E.I. (1995). *Streptococcus pneumoniae* anchor to activated human cells by the receptor for platelet-activating factor. Nature *377*, 435–438.

Demali, K.A., Jue, A.L., and Burridge, K. (2006). IpaA targets β1 integrins and Rho to promote actin cytoskeleton rearrangements necessary for *Shigella* entry. J. Biol. Chem. *281*, 39534–39541.

Deretic, V., and Levine, B. (2009). Autophagy, immunity, and microbial adaptations. Cell Host Microbe *5*, 527–549.

Detilleux, P.G., Deyoe, B.L., and Cheville, N.F. (1991). Effect of endocytic and metabolic inhibitors on the internalization and intracellular growth of *Brucella abortus* in Vero cells. Am. J. Vet. Res. *52*, 1658–1664.

Duncan, M.J., Li, G., Shin, J.S., Carson, J.L., and Abraham, S.N. (2004). Bacterial penetration of bladder epithelium through lipid rafts. J. Biol. Chem. *279*, 18944–18951.

Dupont, N., Lacas-Gervais, S., Bertout, J., Paz, I., Frèche, B., Tran Van Nhieu, G., van der Goot, F.G., Sansonetti, P.J., and Lafont, F. (2009). *Shigella* phagocytic vacuolar membrane remnants participate in the cellular response to pathogen invasion and are regulated by autophagy. Cell Host Microbe *6*, 137–149.

Dussurget, O., Pizarro-Cerdá, J., and Cossart, P. (2004). Molecular determinants of *Listeria monocytogenes* virulence. Annu. Rev. Microbiol. *58*, 587–610.

Edidin, M. (2003). The state of lipid rafts: from model membranes to cells. Annu. Rev. Biophys. Biomol. Struct. *32*, 257–283.

Ellington, J.K., Reilly, S.S., Ramp, W.K., Smeltzer, M.S., Kellam, J.F., and Hudson, M.C. (1999). Mechanisms of *Staphylococcus aureus* invasion of cultured osteoblasts. Microb. Pathog. *26*, 317–323.

Engqvist-Goldstein, A.E., and Drubin, D.G. (2003). Actin assembly and endocytosis: from yeast to mammals. Annu. Rev. Cell. Dev. Biol. *19*, 287–332.

Espina, M., Olive, A.J., Kenjale, R., Moore, D.S., Ausar, S.F., Kaminski, R.W., Oaks, E.V., Middaugh, C.R., Picking, W.D., and Picking, W.L. (2006). IpaD localizes to the tip of the type III secretion system needle of *Shigella flexneri*. Infect. Immun. *74*, 4391–4400.

Eto, D.S., Gordon, H.B., Dhakal, B.K., Jones, T.A., and Mulvey, M.A (2008). Clathrin, AP-2, and the NPXY-binding subset of alternate endocytic adaptors facilitate FimH-mediated bacterial invasion of host cells. Cell. Microbiol. *10*, 2553–2567.

Furuta, N., Tsuda, K., Omori, H., Yoshimori, T., Yoshimura, F., and Amano, A. (2009). *Porphyromonas gingivalis* outer membrane vesicles enter human epithelial cells via an endocytic pathway and are sorted to lysosomal compartments. Infect. Immun. *77*, 4187–4196.

Galán, J.E. (2001). *Salmonella* interactions with host cells: type III secretion at work. Annu. Rev. Cell. Dev. Biol. *17*, 53–86.

Gatfield, J., and Pieters, J. (2000). Essential role for cholesterol in entry of mycobacteria into macrophages. Science *288*, 1647–1650.

van der Goot, F.G., Tran Van Nhieu, G., Allaoui, A., Sansonetti, P., and Lafont, F. (2004). Rafts can trigger contact-mediated secretion of bacterial effectors via a lipid-based mechanism. J. Biol. Chem. *279*, 47792–47798.

Grassmé, H.U., Ireland, R.M., and van Putten, J.P. (1996). Gonococcal opacity protein promotes bacterial entry-associated rearrangements of the epithelial cell actin cytoskeleton. Infect. Immun. *64*, 1621–1630.

Grassmé, H., Gulbins, E., Brenner, B., Ferlinz, K., Sandhoff, K., Harzer, K., Lang, F., and Meyer, T.F. (1997). Acidic sphingomyelinase mediates entry of *N. gonorrhoeae* into nonphagocytic cells. Cell *91*, 605–615.

Grassmé, H., Kirschnek, S., Riethmueller, J., Riehle, A., von Kürthy, G., Lang, F., Weller, M., and Gulbins, E. (2000). CD95/CD95 ligand interactions on epithelial cells in host defense to *Pseudomonas aeruginosa*. Science *290*, 527–530.

Grassmé, H., Jendrossek, V., Riehle, A., von Kürthy, G., Berger, J., Schwarz, H., Weller, M., Kolesnick, R., and Gulbins, E. (2003). Host defense against *Pseudomonas aeruginosa* requires ceramide-rich membrane rafts. Nat. Med. *9*, 322–330.

Green, B.T., and Brown, D.R. (2006). Differential effects of clathrin and actin inhibitors on internalization of *Escherichia coli* and *Salmonella choleraesuis* in porcine jejunal Peyer's patches. Vet. Microbiol. *113*, 117–122.

Gustavsson, A., Armulik, A., Brakebusch, C., Fässler, R., Johansson, S., and Fällman, M. (2002). Role of the β1-integrin cytoplasmic tail in mediating invasin-promoted internalization of *Yersinia*. J. Cell Sci. *115*, 2669–2678.

Hancock, J.F. (2006). Lipid rafts: contentious only from simplistic standpoints. Nat. Rev. Mol. Cell. Biol. *7*, 456–462.

Hardt, W.D., Chen, L.M., Schuebel, K.E., Bustelo, X.R., and Galán, J.E. (1998). *S. typhimurium* encodes an activator of Rho GTPases that induces membrane ruffling and nuclear responses in host cells. Cell *93*, 815–826.

Harvey, H.A., Jennings, M.P., Campbell, C.A., Williams, R., and Apicella, M.A. (2001). Receptor-mediated endocytosis of *Neisseria gonorrhoeae* into primary human urethral epithelial cells: the role of the asialoglycoprotein receptor. Mol. Microbiol. *42*, 659–672.

Hayward, R.D., and Koronakis, V. (1999). Direct nucleation and bundling of actin by the SipC protein of invasive *Salmonella*. EMBO J. *18*, 4926–4934.

Hayward, R.D., Cain, R.J., McGhie, E.J., Phillips, N., Garner, M.J., and Koronakis, V. (2005). Cholesterol binding by the bacterial type III translocon is essential for virulence effector delivery into mammalian cells. Mol. Microbiol. *56*, 590–603.

Henley, J.R., Cao, H., and McNiven, M.A. (1999). Participation of dynamin in the biogenesis of cytoplasmic vesicles. FASEB J. *13* (Suppl. 2), S243–247.

Heuser, J. (1980). Three-dimensional visualization of coated vesicle formation in fibroblasts. J. Cell Biol. *84*, 560–583.

Hodinka, R.L., and Wyrick, P.B. (1986). Ultrastructural study of mode of entry of *Chlamydia psittaci* into L-929 cells. Infect. Immun. *54*, 855–863.

Hodinka, R.L., Davis, C.H., Choong, J., and Wyrick, P.B. (1988). Ultrastructural study of endocytosis of *Chlamydia trachomatis* by McCoy cells. Infect; Immun. *56*, 1456–1463.

Ireton, K., Payrastre, B., Chap, H., Ogawa, W., Sakaue, H., Kasuga, M., and Cossart, P. (1996). A role for phosphoinositide 3-kinase in bacterial invasion. Science *274*, 780–782.

Isberg, R.R., and Barnes, P. (2001). Subversion of integrins by enteropathogenic *Yersinia*. J. Cell Sci. *114*, 21–28.

Isberg, R.R., and Falkow, S. (1985). A single genetic locus encoded by *Yersinia pseudotuberculosis* permits invasion of cultured animal cells by *Escherichia coli* K-12. Nature *317*, 262–264.

Jouve, M., Garcia, M.I., Courcoux, P., Labigne, A., Gounon, P., and Le Bouguénec, C. (1997). Adhesion to and invasion of HeLa cells by pathogenic *Escherichia coli* carrying the *afa-3* gene cluster are mediated by the AfaE and AfaD proteins, respectively. Infect. Immun. *65*, 4082–4089.

Keen, J.H. (1987). Clathrin assembly proteins: affinity purification and a model for coat assembly. J. Cell Biol. *105*, 1989–1998.

Khelef, N., Lecuit, M., Bierne, H., and Cossart, P. (2006). Species specificity of the *Listeria monocytogenes* InlB protein. Cell. Microbiol. *8*, 457–470.

Kim, S., Watarai, M., Makino, S., and Shirahata, T. (2002). Membrane sorting during swimming internalization of *Brucella* is required for phagosome trafficking decisions. Microb. Pathog. *33*, 225–237.

Kirkegaard, K., Taylor, M.P., and Jackson, W.T. (2004). Cellular autophagy: surrender, avoidance and subversion by microorganisms. Nat. Rev. Microbiol. *2*, 301–314.

Kowalski, M.P., and Pier, G.B. (2004). Localization of cystic fibrosis transmembrane conductance regulator to lipid rafts of epithelial cells is required for *Pseudomonas aeruginosa*-induced cellular activation. J. Immunol. *172*, 418–425.

Kubori, T., and Galán, J.E. (2003). Temporal regulation of *Salmonella* virulence effector function by proteasome-dependent protein degradation. Cell *115*, 333–342.

Kurzchalia, T.V., and Parton, R.G. (1999). Membrane microdomains and caveolae. Curr Opin Cell Biol *11*, 424–431.

Lafont, F., and van der Goot, F.G. (2005). Bacterial invasion via lipid rafts. Cell. Microbiol. *7*, 613–620.

Lafont, F., Tran Van Nhieu, G., Hanada, K., Sansonetti, P., and van der Goot, F.G. (2002). Initial steps of *Shigella* infection depend on the cholesterol/sphingolipid raft-mediated CD44–IpaB interaction. EMBO J. *21*, 4449–4457.

Lafont, F., Abrami, L., and van der Goot, F.G. (2004). Bacterial subversion of lipid rafts. Curr. Opin. Microbiol. *7*, 4–10.

Lecuit, M., Dramsi, S., Gottardi, C., Fedor-Chaiken, M., Gumbiner, B., and Cossart, P. (1999). A single amino acid in E-cadherin responsible for host specificity towards the human pathogen *Listeria monocytogenes*. EMBO J. *18*, 3956–3963.

Lecuit, M., Hurme, R., Pizarro-Cerdá, J., Ohayon, H., Geiger, B., and Cossart, P. (2000). A role for α-and β-catenins in bacterial uptake. Proc. Natl. Acad. Sci. U.S.A. *97*, 10008–10013.

Lencer, W.I., Hirst, T.R., and Holmes, R.K. (1999). Membrane traffic and the cellular uptake of cholera toxin. Biochim. Biophys. Acta *1450*, 177–190.

Leong, J.M., Fournier, R.S., and Isberg, R.R. (1990). Identification of the integrin binding domain of the *Yersinia pseudotuberculosis* invasin protein. EMBO J. *9*, 1979–1989.

Levine, B., and Deretic, V. (2007). Unveiling the roles of autophagy in innate and adaptive immunity. Nat. Rev. Immunol. *7*, 767–777.

Lingwood, D., and Simons, K. (2010). Lipid rafts as a membrane-organizing principle. Science *327*, 46–50.

Lisanti, M.P., Tang, Z., Scherer, P.E., Kübler, E., Koleske, A.J., and Sargiacomo, M. (1995). Caveolae, transmembrane signalling and cellular transformation. Mol. Membr. Biol. *12*, 121–124.

McMahon, H.T. (1999). Endocytosis: an assembly protein for clathrin cages. Curr. Biol. *9*, R332–335.

Mañes, S., Ana Lacalle, R., Gómez-Moutón, C., and Martinez-A, C. (2003). From rafts to crafts: membrane asymmetry in moving cells. Trends Immunol. *24*, 320–326.

Maruta, K., Ogawa, M., Miyamoto, H., Izu, K., and Yoshida, S.I. (1998). Entry and intracellular localization of *Legionella dumoffii* in Vero cells. Microb. Pathog. *24*, 65–73.

Messick, J.B., and Rikihisa, Y. (1993). Characterization of *Ehrlichia risticii* binding, internalization, and proliferation in host cells by flow cytometry. Infect. Immun. *61*, 3803–3810.

Michel, V., and Bakovic, M. (2007). Lipid rafts in health and disease. Biol. Cell *99*, 129–140.

Nazli, A., Wang, A., Steen, O., Prescott, D., Lu, J., Perdue, M.H., Söderholm, J.D., Sherman, P.M., and McKay, D.M. (2006). Enterocyte cytoskeleton changes are crucial for enhanced translocation of nonpathogenic *Escherichia coli* across metabolically stressed gut epithelia. Infect. Immun. *74*, 192–201.

Oelschlaeger, T.A., and Tall, B.D. (1997). Invasion of cultured human epithelial cells by *Klebsiella pneumoniae* isolated from the urinary tract. Infect. Immun. *65*, 2950–2958.

Oelschlaeger, T.A., Guerry, P., and Kopecko, D.J. (1993). Unusual microtubule-dependent endocytosis mechanisms triggered by *Campylobacter jejuni* and *Citrobacter freundii*. Proc. Natl. Acad. Sci. U.S.A. *90*, 6884–6888.

Oelschlaeger, T.A., Barrett, T.J., and Kopecko, D.J. (1994). Some structures and processes of human epithelial cells involved in uptake of enterohemorrhagic *Escherichia coli* O157:H7 strains. Infect. Immun. *62*, 5142–5150.

Oh, P., McIntosh, D.P., and Schnitzer, J.E. (1998). Dynamin at the neck of caveolae mediates their budding to form transport vesicles by GTP-driven fission from the plasma membrane of endothelium. J. Cell Biol. *141*, 101–114.

Owen, D.M., Williamson, D., Rentero, C., and Gaus, K. (2009). Quantitative microscopy: protein dynamics and membrane organisation. Traffic *10*, 962–971.

Palade, G.E. (1953). An electron microscope study of the mitochondrial structure. J. Histochem. Cytochem. *1*, 188–211.

Parsot, C., Hamiaux, C., and Page, A.L. (2003). The various and varying roles of specific chaperones in type III secretion systems. Curr. Opin. Microbiol. *6*, 7–14.

Parton, R.G., and Richards, A.A. (2003). Lipid rafts and caveolae as portals for endocytosis: new insights and common mechanisms. Traffic *4*, 724–738.

Parton, R., and Simons, K. (2007). The multiple faces of caveolae. Nat. Rev. Mol. Cell. Biol. *8*, 185–194.

Pelkmans, L., and Helenius, A. (2002). Endocytosis via caveolae. Traffic *3*, 311–320.

Pelkmans, L., and Zerial, M. (2005). Kinase-regulated quantal assemblies and kiss-and-run recycling of caveolae. Nature *436*, 128–133.

Pelkmans, L., Kartenbeck, J., and Helenius, A. (2001). Caveolar endocytosis of simian virus 40 reveals a new two-step vesicular-transport pathway to the ER. Nat. Cell Biol. *3*, 473–483.

Pelkmans, L., Püntener, D., and Helenius, A. (2002). Local actin polymerization and dynamin recruitment in SV40-induced internalization of caveolae. Science *296*, 535–539.

Peyron, P., Bordier, C., N'Diaye, E.N., and Maridonneau-Parini, I. (2000). Nonopsonic phagocytosis of *Mycobacterium kansasii* by human neutrophils depends on cholesterol and is mediated by CR3 associated with glycosylphosphatidylinositol-anchored proteins. J. Immunol. *165*, 5186–5191.

Pizarro-Cerdá, J., and Cossart, P. (2004). Subversion of phosphoinositide metabolism by intracellular bacterial pathogens. Nat. Cell Biol. *6*, 1026–1033.

Pizarro-Cerdá, J., and Cossart, P. (2006). Bacterial adhesion and entry into host cells. Cell *124*, 715–727.

Radin, J.N., Orihuela, C.J., Murti, G., Guglielmo, C., Murray, P.J., and Tuomanen, E.I. (2005). β-Arrestin 1 participates in platelet-activating factor receptor-mediated endocytosis of *Streptococcus pneumoniae*. Infect. Immun. *73*, 7827–7835.

Raiborg, C., Wesche, J., Malerod, L., and Stenmark, H. (2006). Flat clathrin coats on endosomes mediate degradative protein sorting by scaffolding Hrs in dynamic microdomains. J. Cell Sci. *119*, 2414–2424.

Rustom, A., Saffrich, R., Markovic, I., Walther, P., and Gerdes, H.H. (2004). Nanotubular highways for intercellular organelle transport. Science *303*, 1007–1010.

Sansonetti, P.J. (2001). Rupture, invasion and inflammatory destruction of the intestinal barrier by *Shigella*, making sense of prokaryote–eukaryote cross-talks. FEMS Microbiol. Rev. *25*, 3–14.

Schmitter, T., Agerer, F., Peterson, L., Munzner, P., and Hauck, C.R. (2004). Granulocyte CEACAM3 is a phagocytic receptor of the innate immune system that mediates recognition and elimination of human-specific pathogens. J. Exp. Med. *199*, 35–46.

Schuck, S., Honsho, M., Ekroos, K., Shevchenko, A., and Simons, K. (2003). Resistance of cell membranes to different detergents. Proc. Natl. Acad. Sci. U.S.A. *100*, 5795–5800.

Seveau, S., Bierne, H., Giroux, S., Prévost, M.C., and Cossart, P. (2004). Role of lipid rafts in E-cadherin- and HGF-R/Met-mediated entry of *Listeria monocytogenes* into host cells. J. Cell Biol. *166*, 743–753.

Sharma, D.K., Brown, J.C., Choudhury, A., Peterson, T.E., Holicky, E., Marks, D.L., Simari, R., Parton, R.G., and Pagano, R.E. (2004). Selective stimulation of caveolar endocytosis by glycosphingolipids and cholesterol. Mol. Biol. Cell *15*, 3114–3122.

Shen, Y., Naujokas, M., Park, M., and Ireton, K. (2000). InIB-dependent internalization of *Listeria* is mediated by the Met receptor tyrosine kinase. Cell *103*, 501–510.

Shin, J.S., Gao, Z., and Abraham, S.N. (2000). Involvement of cellular caveolae in bacterial entry into mast cells. Science *289*, 785–788.

Simons, K., and Ikonen, E. (1997). Functional rafts in cell membranes. Nature *387*, 569–572.

Simons, K., and Toomre, D. (2000). Lipid rafts and signal transduction. Nat. Rev. Mol. Cell. Biol. *1*, 31–39.

Simons, K., and Vaz, W.L. (2004). Model systems, lipid rafts, and cell membranes. Annu. Rev. Biophys. Biomol. Struct. *33*, 269–295.

Skoudy, A., Mounier, J., Aruffo, A., Ohayon, H., Gounon, P., Sansonetti, P., and Tran Van Nhieu, G. (2000). CD44 binds to the *Shigella* IpaB protein and participates in bacterial invasion of epithelial cells. Cell. Microbiol. *2*, 19–33.

Sousa, S., Cabanes, D., El-Amraoui, A., Petit, C., Lecuit, M., and Cossart, P. (2004). Unconventional myosin VIIa and vezatin, two proteins crucial for *Listeria* entry into epithelial cells. J. Cell Sci. *117*, 2121–2130.

Sowinski, S., Jolly, C., Berninghausen, O., Purbhoo, M.A., Chauveau, A., Köhler, K., Oddos, S., Eissmann, P., Brodsky, F.M., Hopkins, C., *et al.* (2008). Membrane nanotubes physically connect T cells over long distances presenting a novel route for HIV-1 transmission. Nat. Cell Biol. *10*, 211–219.

Stuart, E.S., Webley, W.C., and Norkin, L.C. (2003). Lipid rafts, caveolae, caveolin-1, and entry by *Chlamydiae* into host cells. Exp. Cell. Res. *287*, 67–78.

Tamai, R., Asai, Y., and Ogawa, T. (2005). Requirement for intercellular adhesion molecule 1 and caveolae in invasion of human oral epithelial cells by *Porphyromonas gingivalis*. Infect. Immun. *73*, 6290–6298.

Tamilselvam, B., and Daefler, S. (2008). *Francisella* targets cholesterol-rich host cell membrane domains for entry into macrophages. J. Immunol. *180*, 8262–8271.

Tran Van Nhieu, G., Caron, E., Hall, A., and Sansonetti, P.J. (1999). IpaC induces actin polymerization and filopodia formation during *Shigella* entry into epithelial cells. EMBO J. *18*, 3249–3262.

Tran Van Nhieu, G., Krukonis, E.S., Reszka, A.A., Horwitz, A.F., and Isberg, R.R. (1996). Mutations in the cytoplasmic domain of the integrin β1 chain indicate a role for endocytosis factors in bacterial internalization. J. Biol. Chem. *271*, 7665–7672.

Travassos, L., Carneiro, L., Ramjeet, M., Hussey, S., Kim, Y., Magalhães, J., Yuan, L., Soares, F., Chea, E., Le Bourhis, L., *et al.* (2010). Nod1 and Nod2 direct autophagy by recruiting ATG16L1 to the plasma membrane at the site of bacterial entry. Nat. Immunol. *11*, 55–62.

Veiga, E., and Cossart, P. (2005). *Listeria* hijacks the clathrin-dependent endocytic machinery to invade mammalian cells. Nat. Cell Biol. *7*, 894–900.

Veiga, E., Guttman, J.A., Bonazzi, M., Boucrot, E., Toledo-Arana, A., Lin, A.E., Enninga, J., Pizarro-Cerda, J., Finlay, B.B., Kirchhausen, T., *et al.* (2007). Invasive and adherent bacterial pathogens co-Opt host clathrin for infection. Cell Host Microbe *2*, 340–351.

Ward, M.E., and Murray, A. (1984). Control mechanisms governing the infectivity of *Chlamydia trachomatis* for HeLa cells: mechanisms of endocytosis. J. Gen. Microbiol. *130* 1765–1780.

Watarai, M., Funato, S., and Sasakawa, C. (1996). Interaction of Ipa proteins of *Shigella flexneri* with alpha5beta1 integrin promotes entry of the bacteria into mammalian cells. J. Exp. Med. *183*, 991–999.

Wong, K.W., and Isberg, R.R. (2003). Arf6 and phosphoinositol-4-phosphate-5-kinase activities permit bypass of the Rac1 requirement for β1 integrin-mediated bacterial uptake. J. Exp. Med. *198*, 603–614.

Wooldridge, K.G., Williams, P.H., and Ketley, J.M. (1996). Host signal transduction and endocytosis of *Campylobacter jejuni*. Microb. Pathog. *21*, 299–305.

Wyrick, P.B., Choong, J., Davis, C.H., Knight, S.T., Royal, M.O., Maslow, A.S., and Bagnell, C.R. (1989). Entry of genital *Chlamydia trachomatis* into polarized human epithelial cells. Infect. Immun. *57* 2378–2389.

Xie, Z., and Klionsky, D.J. (2007). Autophagosome formation: core machinery and adaptations. Nat. Cell Biol. *9*, 1102–1109.

Yao, Q., Chen, J., Cao, H., Orth, J.D., McCaffery, J.M., Stan, R.V., and McNiven, M.A. (2005). Caveolin-1 interacts directly with dynamin-2. J. Mol. Biol. *348*, 491–501.

Yap, A.S., Brieher, W.M., and Gumbiner, B.M. (1997). Molecular and functional analysis of cadherin-based adherens junctions. Annu. Rev. Cell. Dev. Biol. *13*, 119–146.

Zaas, D.W., Duncan, M., Rae Wright, J., and Abraham, S.N. (2005a). The role of lipid rafts in the pathogenesis of bacterial infections. Biochim. Biophys. Acta *1746*, 305–313.

Zaas, D.W., Duncan, M.J., Li, G., Wright, J.R., and Abraham, S.N. (2005b). *Pseudomonas* invasion of type I pneumocytes is dependent on the expression and phosphorylation of caveolin-2. J. Biol. Chem. *280*, 4864–4872.

Zhou, D., Mooseker, M.S., and Galán, J.E. (1999). An invasion-associated *Salmonella* protein modulates the actin-bundling activity of plastin. Proc. Natl. Acad. Sci. U.S.A. *96*, 10176–10181.

The Bacterial Life in a Vacuole

10

Ana Rita Furtado and Agathe Subtil

Abstract

Several intracellular bacteria survive and multiply inside membrane-bounded compartments called vacuoles. While this lifestyle offers several advantages against host cell defences, it also imposes important constraints. Pathogens have developed independent strategies to survive within their host, and as a consequence, each 'vacuole' is unique to a given bacterium in its composition and behaviour. However, common requirements and strategies emerge. We describe how intracellular pathogens interact with the host trafficking pathways to transform their vacuole into a favourable niche, and undermine the host defence mechanisms against intruders. This is mainly achieved through the action of bacterial proteins translocated out of the vacuoles into the host cytoplasm. An increasing number of these 'effector' proteins are currently being identified, and the functions of a few of them start to be understood.

Introduction: the pros and cons of life in a vacuole

Unlike viruses, which can fuse with the plasma membrane to enter host cells, bacteria only have one way to penetrate into a cell while preserving its integrity: by engulfment through the invagination and closure of the plasma membrane. Therefore, the very first moment of bacterial intracellular life inevitably starts within a vacuole. Then comes the 'choice': to develop inside this compartment, or to get access to the cytosol. Both strategies have been adopted through evolution, because each confers advantages to the microbes against host cell defences. In each case, these advantages come at a price (Table 10.1). This chapter describes the bacterial properties selected through evolution to accommodate an intravacuolar lifestyle. They will be illustrated by examples taken from the most common intracellular pathogens: *Mycobacterium*, *Salmonella*, *Chlamydia*, *Brucella*, *Coxiella* and *Legionella*. These pathogens are phylogenetically very diverse, and have evolved different molecular mechanisms to adapt to their host, which is reflected by the diversity in the composition of their respective vacuoles. However, all intravacuolar bacteria meet identical challenges: the need to avoid destruction by host cell defence mechanisms, restricted access to the host cell resources, necessity to expand the vacuolar volume to accompany bacterial multiplication, need to ensure host survival until replication has successfully taken place and then to make a way out. Therefore, in spite of their diversity, intravacuolar bacteria share many properties, which constitute the different sections of this chapter. Since the vacuole is in permanent contact with other intracellular compartments of the host cell, the first section gives an overview of these compartments and of the mechanisms that regulate exchanges between them. Finally, the diversity in pathogens is reflected by the variations in the terminologies to designate the vacuoles: 'parasitophorous vacuole', 'inclusion', 'phagosome'. For convenience, we will use 'vacuole' in all cases.

Table 10.1 The pros and cons of life in a vacuole to escape host defence mechanisms (vs. life in the cytosol)

Points to consider	Pros	Cons
Vacuole acidification and fusion with lysosomes	–	Need to prevent fusion with lysosomes or to develop resistance to the degradative abilities of this compartment
Access to nutrients	–	Cytosolic nutrients are not directly accessible: need to develop transport mechanisms and to increase the surface of exchange by expansion of the vacuole membrane
Microenvironment	Possibility to create a favourable environment different from the cytosol	–
Innate immunity	Many of the pathogenic patterns recognized by innate immunity remain hidden because the receptors are cytosolic or in other compartments	–
Adaptive immunity	Different arms of the adaptive immunity ensure that an immune response is raised against all antigens, whether present in the cytosol or in intracellular compartments	
Autophagy	Protection against direct ubiquitination of the pathogen, followed by autophagy	Intracellular compartment are also subjected to autophagy, so a need remains to counteract this host defence mechanism, or use it for dissemination

Intracellular compartments and regulation of membrane traffic

The eukaryotic cell is compartmentalized in several structures delimited by lipid bilayers, the organelles. Organelles have distinct lipid and protein compositions, which enable them to perform distinct functions within the cell. However, they are highly interconnected and constantly exchange material by means of vesicles or tubules that transit from one compartment to another. In spite of this continuous membrane flow, compartments preserve their biochemical identity by very efficient sorting processes that ensure that only resident proteins and lipids are retained in a given organelle. The vacuole in which a pathogen resides constitutes an organelle of its own, also capable of exchanging material with other intracellular compartments. In fact, intracellular development is only possible because pathogens take control of the interactions between the vacuole and the intracellular trafficking of the host. To understand how this is achieved, we will briefly introduce the different compartments of the cell, and the mechanisms that regulate their interactions (Fig. 10.1).

The endocytic pathway

The endocytic pathway is composed of several subcompartments, which correspond to distinct steps in the endocytic process. These compartments have been defined by the use of 'markers': protein or lipid components that are found almost exclusively in one type of compartment. After internalization into eukaryotic cells, cell surface proteins and lipids, together with solutes, first appear in peripheral early endosomes, sometimes referred to as sorting endosomes (the classical markers for early endosomes are the early endosomal antigen 1 EEA1, and the small GTPases Rab4 and Rab5). From this compartment, sorting occurs: the majority of the membrane constituents are recycled back to the plasma membrane, at least in part via recycling endosomes (which, as sorting endosomes, contain recycling receptors such as the transferrin receptor, and Rab35 and Rab11), while most solute and some membrane components are transported to late endosomes (classical markers for late endosomes are Rab7, mannose-6 phosphate receptors, tetraspan proteins such as CD63 and lysosomal glycoproteins). Late endosomes

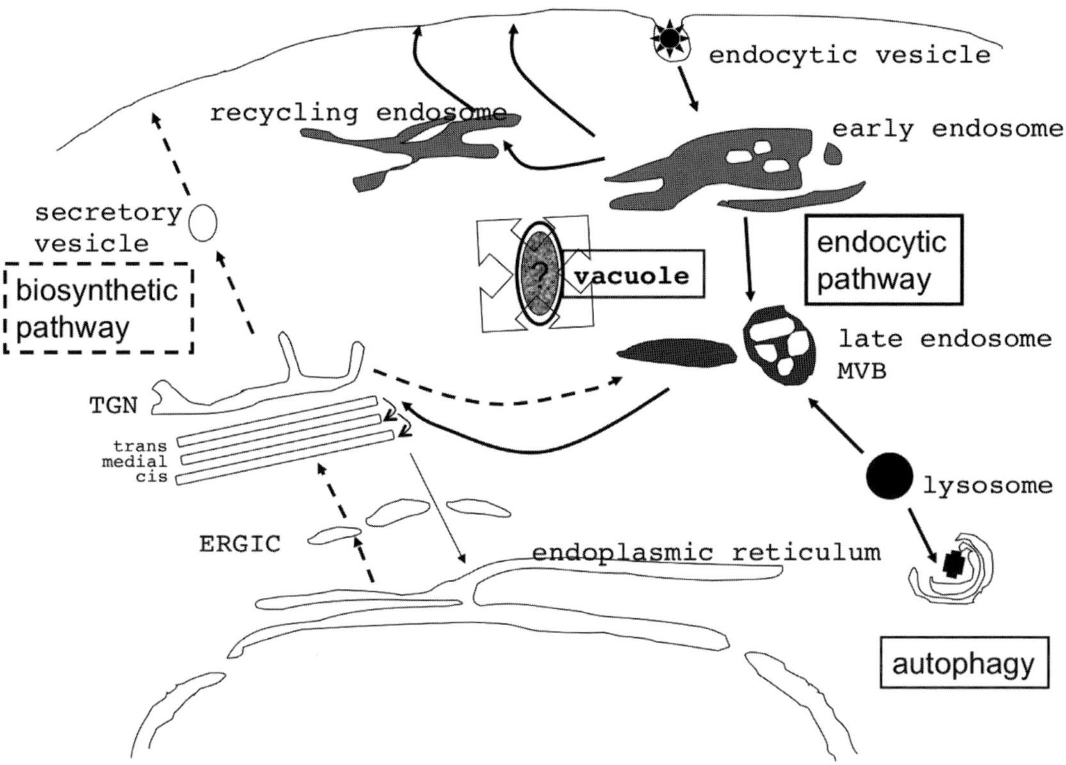

Figure 10.1 Intracellular compartments with which the bacteria-containing vacuole might interact. As for other external particles or nutrients (star), vacuole formation takes bacteria to the endocytic route. The endocytic pathway (solid arrows) involves a succession of compartments of different shape, content and intraluminal pH. The default pathway ends up with the fusion of the vacuole with lysosomes, which contain degradative hydrolases. However, bacteria have acquired different strategies to escape this fate (see text). The endocytic pathway is connected to the biosynthetic pathway (dotted arrows), that is followed by soluble proteins and membrane proteins cotranslationally translocated into the endoplasmic reticulum (ER). Following protein folding, they exit the ER in vesicles that fuse with the ER–Golgi intermediate compartment (ERGIC). The proteins then sequentially pass through the Golgi apparatus and arrive in the *trans* Golgi network (TGN), from where they are sorted to different destinations. Many proteins are retrieved back to the ER or early Golgi stacks by retrograde transport. Finally, degradation of part of the intracellular material, including pathogens, can occur via autophagy, which also involves fusion with lysosomes.

eventually fuse with lysosomes, which contain hydrolytic enzymes.

Endocytic compartments come in many different shapes. Early endosomes consist of tubular and cisternal elements associated with vesicular regions of 0.3–0.5 μm in diameter and are located at the cellular periphery. The tubular elements resemble recycling endosomes: they do not contain internal membranes and are involved in rapid recycling back to the cell surface. In contrast, the vesicular ones accumulate membrane invaginations, and are often called multivesicular bodies (MVBs) or endosomal carrier vesicles (ECVs). The internal pH of early endosomes is slightly acidic (pH 6.0–6.2), because of the presence of the vacuolar proton pump (Arai *et al.*, 1993). Recycling endosomes mainly consist of networks of very thin tubules (50–60 nm in diameter and up to several μm in length), often concentrated around the microtubule organizing centre. The luminal pH of this compartment is slightly higher than that of sorting endosomes (pH 6.4).

The degradation pathway

During endocytosis, recycled receptors and lipids are very rapidly segregated into the tubular portion of endosomes, while most ligands and some membrane proteins targeted to degradation accumulate in the vesicular part. The content of these multivesicular compartments is delivered to late endosomes, which are generally believed to provide the last sorting station before lysosomes. They exhibit a complex morphological organization with cisternal, tubular and multivesicular regions, and their internal membranes accumulate the unique phospholipid lysobisphosphatidic acid (LBPA) (Kobayashi *et al.*, 1998). Some proteins distribute preferentially in the internal portion of the membranes rather than in the periphery of the compartment, although the mechanisms for this partition are not clear.

The terminal compartment of the degradation pathway is the lysosome, which contains more than 45 different hydrolases including lipases, lysozyme and proteases. They are typically spherical, with an acidic luminal pH (4.6–5.0) generated by the vacuolar proton pump and typically distributed throughout the cytoplasm, often with a high concentration in the perinuclear region. By electron microscopy, lysosomes can be distinguished from late endosomes because they do not contain internal membranes and are more electron dense. However, both compartments share the same abundant proteins, the lysosomal glycoproteins or lamps, and contain lysosomal enzymes. Surprisingly, while lysosomes clearly perform different functions, only the absence of some proteins (such as mannose-6-phosphate receptors or the small GTPases Rab7 and Rab9) differentiates them from late endosomes.

Autophagy

Autophagy is a mechanism by which cells sequester defective organelles and cytoplasmic material from the rest of the cytoplasm in a double-membrane vesicle, called an autophagosome. The outer membrane of the autophagosome eventually fuses with a lysosome to form an autolysosome or autophagolysosome, whose contents is degraded via acidic lysosomal hydrolases. Autophagy is a tightly regulated process that plays a normal part in cell homeostasis, helping to maintain a balance between the synthesis, degradation, and subsequent recycling of cellular products (Klionsky, 2007). It is also an important killing mechanism, shown to be active against a variety of pathogens including *Mycobacterium*, *Legionella* and *Salmonella* (Gutierrez *et al.*, 2004; Amer and Swanson, 2005; Birmingham *et al.*, 2006). Markers of autophagosome formation include class III phosphoinositide 3-kinase and autophagy-related genes (Atg), in particular the ubiquitin-like protein Atg8, known as LC3 in mammalian cells (He and Klionsky, 2009). In recent years, ubiquitylation has emerged as a signal for targeting various types of cargo, ranging from protein aggregates to membrane-bounded organelles and microbes, to the autophagocytic route (Kirkin *et al.*, 2009; Lerena *et al.*, 2010).

The biosynthetic pathway

The endocytic pathway is highly connected with the biosynthetic pathway taken up by soluble proteins destined for secretion, and by membrane proteins. Upon biosynthesis, these proteins are translocated in the endoplasmic reticulum (ER), a very abundant tubular network that expands throughout the cell. Protein folding occurs in the ER, and some modifications such as N-linked glycosylation take place in this compartment. Once correctly folded, proteins are transported through the Golgi apparatus, a complex organelle composed of membrane-bounded stacks known as cisternae, generally located in proximity to the microtubule organizing centre. Transport between the ER and the Golgi apparatus is bi-directional. The anterograde pathway (ER to Golgi) involves the Coat protein complex II (COPII), a multi-subunit protein complex responsible for the formation of membrane vesicles at the ER (Fromme *et al.*, 2008). Conversely, COPI proteins mark the vesicles to be brought back from the Golgi to the RER, through the retrograde pathway.

Proteins enter the Golgi apparatus through the *cis*-face and then travel through the different cisternae. The Golgi apparatus contains many enzymes, which selectively modify the cargo proteins. For example, addition of a mannose-6 phosphate tag to acid hydrolase precursor proteins, that are eventually destined for incorporation into lysosomes, occurs in the *cis*-Golgi. Proteins progress through the stacks until they reach the *trans* Golgi network (TGN). This area of the Golgi is the point at which proteins are sorted and shipped to their intended destinations, which can be the plasma membrane, secretory vesicles, or other intracellular compartments of the endocytic pathway, in particular late endosomes (Bard and Malhotra, 2006). In polarized cells, an additional level of complexity is reached because proteins destined for the apical surface are sorted from those destined for the basolateral membrane.

Coat components

Exchanges between intracellular compartments are accomplished by vesicles and tubules. The formation of a shuttle compartment opens the possibility of selecting the cargo while excluding other components, which is the basis of intracellular sorting. Selection is ensured by the interaction of membrane proteins with coat components that surround the nascent vesicle. This process has been studied in detail in the case of clathrin-coated endocytosis, which allows internalization at the plasma membrane of many membrane receptors and ligands. These membrane proteins contain tyrosine or dileucine-based motifs in their cytoplasmic tail, which interact with an adaptor complex called AP-2. This complex in turn recruits clathrin and various accessory molecules and enzymes that are necessary for clathrin-coated vesicle formation (Sorkin, 2004). Since the discovery of AP-2 which operates at the plasma membrane, three other adaptors complexes (AP-1, 3 and 4) have been shown to operate at intracellular locations (Robinson, 2004; Baust *et al.*, 2008). In addition, as mentioned earlier, vesicles that travel between the ER and the Golgi apparatus are also coated, in the anterograde (COPII) and retrograde (COPI) directions. COPI is also involved in early endosome maturation (Gu and Gruenberg, 1999), and has many points in common with clathrin coats (McMahon and Mills, 2004).

Specificity of fusion between compartments

Microscopic observations in living cells demonstrate that intracellular compartments are highly dynamic structures, constantly subjected to membrane fission and fusion. Membrane fusion between two compartments of the same nature (for example, two early endosomes) is called homotypic fusion, while heterotypic fusion involves compartments of different composition. Fusion between membranes is tightly controlled, and two families of proteins emerge for their central role in this process: the SNAREs (soluble N-ethylene maleimide-sensitive factor attachment protein receptors) and the Rab proteins. Both families of proteins, together with accessory proteins, act in concert to ensure the specificity of membrane fusion (Wickner and Schekman, 2008).

SNARE proteins play a critical role both in membrane docking and in fusion (Söllner, 2003; Jahn and Scheller, 2006; Südhof and Rothman, 2009). They are anchored in two lipid bilayers and associate in complexes involving four molecules that associate in a helical bundle, via a common SNARE-domain. Complex formation is needed for the fusion of the two lipid bilayers. SNAREs can also have inhibitory roles in membrane fusion, by substituting for or binding to a subunit of a fusogenic SNARE bundle to form a non-fusogenic complex (Varlamov *et al.*, 2004). Upon vesicle fusion, all SNAREs are embedded in the same membrane and after dissociation of the SNARE complex, the different chains are recycled back to their cognate compartment to be used in a new round of membrane fusion.

Rab proteins are small GTPases that can anchor to the cytoplasmic surface of specific organelles. In humans, there are more than 60 members of the Rab family that are localized to distinct intracellular membranes. When bound to the nucleotide GTP, Rab proteins are able to interact with specific effectors. Hydrolysis of GTP to GDP, facilitated by a GTPase activating protein (GAP), leads to inactivation, until a guanine nucleotide exchange factor (GEF) allows recovery of the active GTP-bound state. Rab GTPases play a central role in ensuring the specificity of membrane fusion events. They control membrane identity and vesicle budding, uncoating, motility and fusion through the recruitment of effector proteins, such as sorting adaptors, tethering factors, kinases, phosphatases and motors. Crosstalk between multiple Rab GTPases through shared effectors or through effectors that recruit selective Rab activators, ensures the spatiotemporal regulation of vesicle traffic (Grosshans *et al.*, 2006; Stenmark, 2009).

Considering the central role played by Rabs in regulating intracellular trafficking, it is not surprising that many intracellular pathogens modulate Rab recruitment to define their replication compartment (Brumell and Scidmore, 2007). Recruitment of SNARE proteins to the vacuolar membrane appears to be less common, although it might not have been investigated extensively. Unlike Rab recruitment to the vacuole, SNARE acquisition by the vacuolar membrane requires a membrane fusion event to take place, because SNARE proteins remain membrane-associated. The ER-associated SNARE protein Sec22b was shown to be delivered to *Legionella*-containing vacuoles, together with Rab1 (Kagan *et al.*, 2004). Another way to use the SNARE machinery would be to mimic the SNARE domain, and then engage in formation of SNARE complexes with host molecules. This mimicry mechanism has been adopted by *Chlamydia*, which express SNARE-like proteins at the surface of their vacuole, and recruit several host SNAREs in its vicinity (Delevoye *et al.*, 2008).

Lipids

Although our understanding of the lipid composition of the different organelles is still very poor compared with their characterization at the protein level, it is clear that lipids are not homogenously distributed between the different compartments, and that within a compartment, different membrane domains of distinct lipid content, coexist (van Meer *et al.*, 2008). The 'raft hypothesis' postulated more that ten years ago that long largely saturated sphingolipids can pack together more than phospholipids, which are rich in shorter and unsaturated acyl-chains (Simons and Ikonen, 1997; Lingwood and Simons, 2010). This difference in structure is at the basis of the formation of distinct domains in membranes, which has been documented at the plasma membrane, but also in intracellular compartments. Raft domains have many functional implications, including signal transduction and intracellular sorting. They are important for the entry of several pathogens, as discussed in Chapter 9 by Moreau and Lafont. It is likely that bacteria-containing vacuoles also contain subdomains of specific lipid composition, which may be relevant for signalling or for interactions with other compartments, but this has not yet been documented. In contrast, the ability of intracellular pathogens to modulate and exploit phosphoinositide (PI) glycerolipid of the vacuole has been illustrated in many cases (Weber *et al.*, 2009). PIs are key regulators of eukaryotic signal transduction, cytoskeleton architecture and membrane dynamics. They are used by bacteria to anchor some of their effector proteins, such as the effector SidC from *Legionella pneumophila*, which specifically binds to PI(4)P (Weber *et al.*, 2006). Some bacterial effector proteins themselves were shown to have PI phosphatase activity (Norris *et al.*, 1998). Modulation of the PI content of the vacuole also participates in the control exerted by the pathogen on its maturation. This is best illustrated with *Mycobacterium tuberculosis*, which uses several complementary strategies to deplete its vacuolar membrane from PI(3)P (Weber *et al.*, 2009). Finally, it is quite remarkable that many pathogen-containing vacuoles accumulate cholesterol. It probably influences exchanges between the vacuole and intracellular compartments, and increases the vacuole stability.

Adaptation to life in a vacuole

The normal fate after cell invasion

Successful pathogens are, by definition, able to cause disease in spite of the defences of the host. To understand how this is achieved, we will describe the different lines of defence of the host against intruders, which are sufficient to eradicate the very large majority of microbes. Pathogens usually have a tropism for macrophages and/or epithelial cells. Since these two cell types differ in their ability to fight intrusion, we will describe these two situations separately.

Phagocytic cells

The ability to internalize large particles, a process named phagocytosis, allows animal cells to eliminate foreign material, dying cells or intruders such as bacteria. It is mostly used by professional phagocytes (neutrophils, monocytes, macrophages, dendritic cells and mast cells), which present specific receptors on their surface that recognize the material to eliminate. Within minutes, the phagosome merges with either a lysosome or a granule to form a phagolysosome. The particle is then subjected to a formidable battery of killing mechanisms. The main mechanism for destruction of phagocytosed material comes from the activity of hydrolases that are delivered to the phagosome lumen by fusion with the lysosomes. All these enzymes are acid hydrolases that require an acidic environment for optimal activity, and the pH of phagolysosomes is maintained acidic through the activity of a proton pump (Arai *et al.*, 1993). In addition, small cationic antimicrobial peptides have been shown in some cases to play a role in innate resistance against intracellular pathogens, by acting on bacteria sequestered within vacuolar compartments (Radtke and O'Riordan, 2006). Recently, a novel antibacterial activity for lysosomes has been described. In macrophages infected with *M. tuberculosis*, ubiquitin-derived peptides contained inside lysosomes appear to carry at least part of the antimycobacterial activity of this organelle (Purdy and Russell, 2007). For the vast majority of microbes, internalization and exposure to the acidic, hydrolytically active environment of the phagosome is sufficient to bring about their demise. In addition, activation of macrophages leads to increased microbicidal activity through the expression of the inducible nitric oxide synthase (iNOS) and components of the nicotinamide adenine dinucleotide phosphate (NADPH) oxidase complex, which release reactive oxygen and nitrogen intermediates into the nascent phagosome (a process called respiratory or oxidative burst) (Bedard *et al.*, 2007). Finally, phagocytes are equipped with a full range of Toll-like receptors (TLRs) and other pattern-recognition receptors capable of recognizing and inducing a preliminary, inflammatory response against a microbial presence (Rohde *et al.*, 2007b).

In the longer term, phagosomes are also central for the building of cellular immune responses to infection, because this is the place where some bacterial antigenic peptides are loaded on major histocompatibility complex (MHC) molecules of class II. In some cases, phagosomes fuse with the ER in which MHC class I molecules transit, and as a result, peptide loading on MHC-class I molecules can also occur in the phagosome (Guermonprez *et al.*, 2003). Therefore, phagosomes are not only important in the immediate innate response to bacterial invasion: they also constitute an important platform in building cellular immunity against these pathogens (Ramachandra *et al.*, 2009).

Non-professional phagocytic cells

Phagocytes are better equipped to internalize and eliminate microorganisms than other cell types. However, the basic molecules that sustain phagocytosis are present in non-phagocytic cells, and bacteria have developed the ability to penetrate into many cell types, especially into epithelial cells, which are most exposed to intruders. After entry, the compartment containing the microorganism normally matures along the endocytic pathway, acidifies, and eventually fuses with lysosomes, leading to the degradation of its content. While NADPH oxidase activity has been widely characterized in phagocytes, it is noteworthy that it is also found in non-phagocytic cell types, including epithelial cells, in which ROS production upon bacterial infection has also been documented (Cheng *et al.*, 2001; Bánfi *et al.*, 2003). In addition, epithelial cells express MHC class II molecules, enabling them to present bacterial peptides and therefore to participate in cellular immunity during infection.

Making the vacuole a suitable environment

To survive the defence mechanisms described above, intracellular pathogens have developed two strategies: to rapidly get out of the vacuole or to make this compartment suitable for survival and proliferation. Here, we will describe some of the strategies acquired by different pathogens to counteract the killing abilities of the host, while remaining within the vacuole. Although these strategies are diverse, they all rely on the fact that shortly after internalization, bacteria take control of the interactions between the vacuolar membrane and other intracellular compartments, rapidly building a niche that resists host killing mechanisms.

Interactions between the vacuole and intracellular compartments have taken three main paths during evolution. Bacteria such as *Mycobacterium* and *Salmonella* restrict their interactions with the endocytic pathway, each to a different degree, and as a result reside within compartments that do not fuse with lysosomes, thereby escaping degradative enzymes. Fully opposite is the strategy opted by *Coxiella*, which allows the maturation of the vacuole in a compartment that eventually fuses with lysosomes. Finally, some bacteria establish intimate links with intracellular organelles out of the endocytic pathway, from which they can obtain shelter and/or nutrients. This is the case of *Legionella* and *Brucella*, which develop in an ER-derived environment, and to some extent of *Chlamydia*, which develops tightly controlled interactions with both the endocytic and the biosynthetic pathways.

Arrest in phagosome maturation: the example of *M. tuberculosis* in resting macrophages

In this paragraph, we will illustrate how bacteria restrict their interactions with the endocytic pathway with the example of *M. tuberculosis*. For other pathogens such as *Salmonella*, *Chlamydia* or *Ehrlichia*, we recommend recent reviews (Garciá-del Portillo *et al.*, 2008; Steele-Mortimer, 2008; Kumar and Valdivia, 2009; Wakeel *et al.*, 2010).

Fusion of an intracellular organelle with the vacuole results in the appearance of its membrane constituents in the vacuolar membrane, while its luminal content accesses the lumen of the vacuole. Therefore, microscopic observations using markers of different intracellular compartments were made to describe the interactions between vacuoles and the host components. Shortly after internalization, phagosomes containing *M. tuberculosis* are decorated with Rab5 and are accessible to transferrin, two indications of their ability to fuse with early endosomal compartments (Clemens and Horwitz, 1996). From that stage, phagosome maturation appears to be blocked, and the luminal pH remains only slightly acidic (6.3) (Sturgill-Koszycki *et al.*, 1994). This relatively high pH seems essential for bacterial replication, as mutants that are defective in their ability to prevent phagosome acidification fail to replicate (Pethe *et al.*, 2004).

The mechanisms by which *M. tuberculosis* prevents maturation of the phagosomes are not yet fully understood and recent progress in the understanding of *Mycobacterium* secretion machineries will probably shed light on this aspect. It has been known for a long time that it requires an active process, because killed bacteria end up in lysosomes (Armstrong and Hart, 1971). Cell wall constituents have been implicated, as well as bacterial enzymes (Rohde *et al.*, 2007b). However, the way these luminal signals are transmitted to the external leaflet of the vacuole, which probably controls membrane fusion events, is not clear. As mentioned above, several bacterial proteins regulate the content of PI glycerolipids in the *M. tuberculosis* vacuole, underlining the importance of these lipids in defining a suitable niche for this pathogen (Weber *et al.*, 2009).

Life in an acidic environment: the solutions from *Coxiella*

Coxiella infect primarily mononuclear phagocytes. Rather than escaping the endolysosomal pathway, these obligate intracellular pathogens direct the formation of a parasitophorous vacuole, which gradually acquires markers of early and late endocytic compartments, as well as markers of autophagosomes, before fusing with lysosomes (Voth and Heinzen, 2007). By doing that, these bacteria are not just passively following the default pathway: bacterial protein synthesis is required to sustain interactions with late endocytic compartments and autophagosomes (Romano *et al.*, 2007).

Coxiella is the only known example of a bacterial pathogen that replicates throughout its infectious cycle in a vacuole that has the composition of lysosomes. How the bacteria resist the inhospitable environment of the phagolysosome is not understood. It might result from specific features of the surface of these bacteria (Voth and Heinzen, 2007). Interestingly, infection of resting macrophages fail to elicit an oxidative burst, suggesting that *Coxiella* might escape this cellular defence mechanism by inhibiting the activity of the NADPH oxidase (Baca *et al.*, 1994), a feature also reported in the case of *Erhlichia* infection in phagocytic cells (Rikihisa,

2006) and *Chlamydia* infection in epithelial cells (Boncompain *et al.*, 2010). The mechanisms by which *Coxiella* modulate maturation of the vacuole are completely unknown, mostly because the lack of a system to genetically manipulate the microorganism has limited investigations at the molecular level. However, it is highly likely that the Dot/Icm type IV secretion system harboured by the *Coxiella* genome is used by the bacteria to deliver effectors in the host cells, as it does in the case of *Legionella* (see below). Heterologous secretion of potential *Coxiella* effector proteins has recently been possible in *Legionella* (Pan *et al.*, 2008), a methodology that should permit the identification of some of the molecular mechanisms by which *Coxiella* subverts its host (Voth and Heinzen, 2009).

Hijacking compartments outside the endocytic pathway

Legionella and *Brucella* are the best known examples of pathogens which not only restrict interaction with the endocytic pathway, but have co-opted another intracellular compartment, the ER. Within the first 5 min of infection, the *Legionella*-containing vacuole diverges from the endocytic pathway, becoming surrounded by ER-derived vesicles, mitochondria and ribosomes. Bacterial proteins secreted by the Dot/Icm type IV secretion system are involved in this process (Franco *et al.*, 2009; Isberg *et al.*, 2009). For example, the bacterial protein RalF acts as a GEF that localizes to the *Legionella*-containing vacuole, recruiting and activating Arf1, the small GTPase involved in COPI-mediated Golgi to ER retrograde transport (Nagai *et al.*, 2002). In addition, at least three other bacterial proteins contribute to the recruitment of Rab1, which is involved in ER to Golgi traffic, to the vacuolar membrane (Franco *et al.*, 2009).

Not surprisingly, *Brucella's* way to the ER environment takes another route, implicating molecular mechanisms that are different from those acquired by *Legionella* (Roop *et al.*, 2009). Nonetheless, in both cases, the ER-derived niche is the most favourable to intracellular replication, maybe because it offers an optimal access to metabolites synthesized or translocated in this compartment.

Interactions with the host

Secretion mechanisms

One of the main necessities of intravacuolar life is to be able to exert actions beyond the vacuolar membrane, in the host cytoplasm. This is mainly accomplished using secretion systems dedicated to the export of proteins across one (Gram-positive bacteria) or two (Gram-negative bacteria) bacterial membranes, and across the vacuolar membrane. In Gram-negative bacteria, the best understood example of such a secretion system is the type III secretion (TTS) machinery, which consists of a multiprotein complex of more than 20 proteins that delivers bacterial proteins into eukaryotic cells (Galán and Wolf-Watz, 2006). Genes coding for this apparatus are often clustered in a pathogenicity island, either on a plasmid or on the bacterial chromosome. Many pathogenic Gram-negative bacteria possess TTS systems, including plant pathogens, and often TTS occurs from an extracellular location, at the host cell surface, and can either trigger or prevent uptake. For pathogens living within a vacuole, such as *Chlamydia* or *Salmonella*, TTS is also active after internalization. Interestingly, *Salmonella* has acquired two pathogenicity islands, coding for two secretion machineries, expressed either at the entry phase (SPI-1) or, upon uptake, from an intracellular location (SPI-2). In contrast, *Chlamydia* possess only one set of genes, although the secretion apparatus is active both at the entry phase and later in development. How temporal expression and translocation of effectors is regulated is not known. *Chlamydia*, with only about 1000 genes, has undergone a drastic reduction of its genome and relies on the host for many essential functions. Remarkably, the number of type III secreted proteins is estimated to be over 100, illustrating how central this secretion system is in exploiting host resources. Noticeably, TTS is not the only means by which *Chlamydia* translocates proteins into the host. For example the protease CPAF (Zhong *et al.*, 2001), which exerts several functions in the host cytosol, is secreted by a different mechanism, which is still undetermined (Heuer *et al.*, 2003).

The second best known secretion machineries are of type IV, present for example in *Coxiella*,

Erhlichia, *Brucella* and *Legionella*. These secretion systems have evolved to transfer both DNA and protein between cells. They have been characterized in depth in the case of the Dot/Icm secretion system of *Legionella*, which contains more than 20 proteins (Marra *et al.*, 1992). Mutations in many of their genes lead to defective recruitment of ER-derived material to the *Legionella*-containing vacuole, and to rapid acquisition of late endosomal markers. Identification of proteins secreted by this machinery is ongoing, using different strategies such as bioinformatics, search for secretion signals or for proteins that disrupt cellular functions. Even though more than 100 candidates have been identified, the function of only a few is known (Isberg *et al.*, 2009). Regarding the regulation of secretion, one recent study indicates that phagocytosis is required to trigger effector delivery (Charpentier *et al.*, 2009).

Even less is known concerning the type VI secretion machinery that has been recognized as a novel secretion system only recently (Pukatzki *et al.*, 2009). Regarding intracellular pathogens that reside within a vacuole, the implication of type VI secretion in virulence was reported in the case of *Salmonella* Typhimurium (Parsons and Heffron, 2005). Finally, the qualification of type VII secretion was recently proposed for a newly identified secretion mechanism in Gram-positive bacteria (Abdallah *et al.*, 2007). In *M. tuberculosis*, this secretion mechanism, also called ESX, plays an essential role in virulence. Whether this secretion system forms injectisomes that can directly translocate proteins across the phagosome membrane, or plays a role in the escape out of the phagosome and subsequent life in the cytosol remain open questions (Simeone *et al.*, 2009).

Import of nutrients

Like all eukaryotic membranes, the membrane of the vacuole is not permeable to ions and solutes (Heinzen and Hackstadt, 1997). This property enables the bacteria to create a favourable environment in terms of pH and ionic composition. However, it also means that to utilize host cell resources, bacteria have to possess not only import mechanisms, as found in cytosolic bacteria, but also transport mechanisms to get these resources across the vacuolar membrane. Conversely, the limitation of nutrient access to the vacuole constitutes a powerful antibacterial weapon for the host. In this line, Nramp is a transporter for divalent cations present at the phagosomal membrane, which restricts microbial access to essential micronutrients, such as iron and manganese (Jabado *et al.*, 2000; Cellier *et al.*, 2007). Expression of a Nramp analogue by *M. tuberculosis* may compete for the same divalent cations and constitute a survival strategy (Agranoff *et al.*, 1999).

Our knowledge on putative transport mechanisms in vacuolar membranes is extremely limited. Fusion of the vacuolear membrane with vesicles might be a way to supply nutrients to the vacuolar lumen. Fusion of ER-derived vesicles with the *Brucella* vacuole was proposed to supply the bacteria with iron contained in haem and targeted to the ER for degradation by the haem oxygenase (Roop *et al.*, 2009). Translocation of essential molecules across the vacuolar membrane can also result from the activity of transporters. These can be of host origin, such as the vacuolar proton pump, but can potentially also be of bacterial origin. For example, chlamydiae translocate more than 60 different proteins to their vacuolar membrane: their functions are largely unknown, and it is entirely possible that some of them are involved in the import of nutrients from the cytosol into the vacuole.

Recently, a very intriguing finding was the observation that lipid droplets, which are vesicles of significant size filled with neutral lipids, are imported inside the *Chlamydia*-containing vacuole (Cocchiaro *et al.*, 2008). Translocation of lipid droplets into the chlamydial inclusion lumen represents one potential source of neutral lipids for the bacteria. The mechanism by which this translocation occurs is unknown and seems to involve invagination of the vacuolar membrane.

Expansion of the vacuole

Life in a vacuole takes different shapes. Some bacteria, such as *Coxiella* or *Chlamydia*, divide within a single compartment, which grows to occupy most of the cytosol (Fig. 10.2). In others, bacterial division is accompanied with fission events of the vacuolar membrane, so that only one or few bacteria are contained within an individual vacuole. In this case, the ratio of membrane area per bacterium

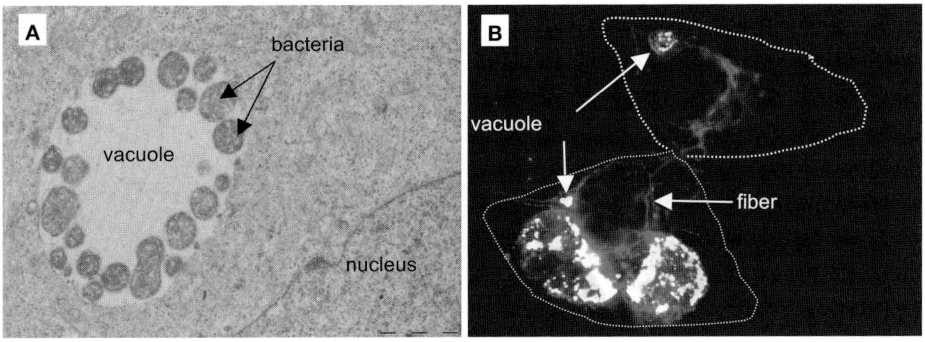

Figure 10.2 *C. trachomatis*-infected cells. (A) *C. trachomatis* grow in a single vacuole that counts up to several hundred bacteria at the end of the infectious cycle. During the multiplication phase, bacteria are mostly found in contact with the vacuole membrane. (B) Cell shape is represented by dotted lines, while vacuole membranes (grey shade) and bacteria (white) were labelled with specific antibodies and observed by immunofluorescence microscopy. This technique allows visualization of fibres emanating from the vacuoles and spreading throughout the cytosol, an observation not possible in the fine sections used for electron microscopy (A). (Pictures reproduced with permission from N. Cayet (A) and A. Subtil (B), Institut Pasteur.

is more important than when multiple bacteria are contained within a single vacuole. *Chlamydia*- or *Salmonella*-containing vacuoles make membrane extensions throughout the cytoplasm, whose function is not known (Fig. 10.2). One hypothesis is that these 'filaments' increase the vacuole surface and facilitate exchanges with the host. Another interesting hypothesis proposed in the case of *Chlamydia* is that membrane extensions facilitate the formation of secondary vacuoles that are inherited upon cell division (Suchland *et al.*, 2005). In that respect, again different strategies exist, as *Chlamydia* vacuoles can divide between daughter cells, while *Coxiella* vacuoles cannot (Roman *et al.*, 1986).

Expansion of the vacuolar membrane requires import of lipids from the host, although incorporation of bacterial lipids is not excluded. The mature *Coxiella*-containing vacuole occupies nearly the entire cytoplasm, and its growth is sustained by an increase in cholesterol uptake and biosynthesis (Voth and Heinzen, 2007). Filipin staining indicates that the vacuolar membrane has approximately the same cholesterol content as the cholesterol-rich plasma membrane, a feature which confers a high resistance of the vacuole to mechanical disruption, despite its spaciousness (Howe and Heinzen, 2006). *Chlamydia* also relies on host-cell lipids for growth of the vacuolar membrane. Eukaryotic-derived phospholipids, sphingomyelin, and cholesterol accumulate in chlamydiae, suggesting that these host-derived constituents traverse the vacuolar membrane, with subsequent incorporation into the bacterium (Hackstadt *et al.*, 1995; Wylie *et al.*, 1997). To do this, the *Chlamydia*-containing vacuole intersects at least two trafficking pathways: export from the Golgi apparatus and traffic of multivesicular bodies (Robertson *et al.*, 2009). The main role of the fragmentation of the Golgi apparatus observed upon *Chlamydia* infection might be to enhance acquisition of Golgi-derived trafficking (Heuer *et al.*, 2009).

Interactions with the host cytoskeleton

The cytoskeleton is a cellular scaffold contained within the cytoplasm and is made out of three networks: actin filaments, intermediate filaments and the microtubule network. It is a dynamic structure that maintains cell shape, protects the cell against mechanical stress, enables cellular motion, and plays important roles in both intracellular trafficking and cellular division. These properties make the cytoskeleton an essential target of bacterial activity, to enable invagination of the plasma membrane and entry, but also further on, during intracellular life.

The cytoskeleton is used for positioning the organelles in the cell, and for their motion.

Therefore, interception of intracellular trafficking by the vacuole can be facilitated by the position adopted by the vacuole along the intracellular trafficking routes. Both *Chlamydia* and *Salmonella* vacuoles are located in proximity of the microtubule organization centre, which might be a strategic choice, close to compartments, such as the Golgi apparatus and endosomal recycling compartments. The cytoskeleton is not only important for bringing material to the vacuole: it is used to position the vacuole in the cell and can also take part in shaping the vacuole, as illustrated further by those two pathogens. *Chlamydia* uses the microtubule network to move its vacuole towards the microtubule organizing centre in the first hours after entry, using the host motor protein dynein (Grieshaber *et al.*, 2003). Later, the scaffolding created by the cytoskeleton could limit vacuole growth. It has been proposed that disruption of cytokeratin-8 by the chlamydial protease CPAF would allow vacuole expansion (Dong *et al.*, 2004). Recently, it was shown that the *Chlamydia* vacuole is encased in a dynamic scaffold of actin and intermediate filaments. It was proposed that such a dynamic network might stabilize the vacuole while minimizing the exposure of the inclusion content to cytoplasmic innate immune surveillance pathways (Kumar and Valdivia, 2008).

Several hours after uptake into different host cell types, *Salmonella* induces the formation of an F-actin meshwork around bacterial vacuoles (Méresse *et al.*, 2001). The role of this actin 'nest' is not clear, but it appears to be needed to establish an intracellular niche that supports bacterial growth. Another remarkable feature of *Salmonella*-containing vacuoles is the appearance of tubules named Sifs (*Salmonella*-induced filaments), which are sustained by the microtubule network. Mutants lacking the TTS effector SifA do not form Sifs and eventually escape in the vacuole (Beuzón *et al.*, 2000). While the formation of the actin nest and of Sifs is clearly dependent on TTS, the exact role of the effectors implicated and their molecular targets remain unclear (Steele-Mortimer, 2008).

Probing the environment

Homeostasis in the eukaryotic host ensures a rather stable environment to intracellular bacteria. Nonetheless, parameters such as nutrient access or host cell defence activity can vary throughout development within the host. For all intracellular bacteria, replication is usually delayed for some time after entry (between hours and days, in the case of *Coxiella* development), probably until the intravacuolar niche has become able to sustain multiplication. The mechanisms developed by bacteria to sense their environment are mostly unknown.

In several cases, intravacuolar pH contributes to the regulation of bacterial gene expression. For example, when acidification of the *Brucella*-containing vacuole is blocked early after internalization, brucellae will not initiate intracellular replication. The low pH induces expression of genes that encode the type IV secretion system, which are themselves necessary to turn the vacuole into a hospitable niche (Boschiroli *et al.*, 2002). The intravacuolar pH was also shown to be an essential parameter for the regulation of gene expression during *Mycobacterium* infection (Rohde *et al.*, 2007c). In this study, a high-resolution profile of *M. tuberculosis* invasion-linked gene expression was generated by assaying the transcriptome at 20 min intervals up to 2 h post-infection. Transcriptional responses were detected within minutes of phagocytosis, including gene subsets with distinct temporal profiles. Pharmacological manipulation of phagosomal pH and *in vitro* acid stress studies revealed that vacuole acidification is an important trigger for differential gene expression. Interestingly, there are marked species-specific differences in the responses of *M. tuberculosis* and *Mycobacterium bovis* BCG to intraphagosomal cues.

One strategy employed by many bacteria for the coordinated expression of factors influencing interactions with their environment is quorum sensing, a cell density-dependent system for intracellular communication, allowing coordinated gene regulation within a bacterial population. Two such transcriptional regulators have been identified in *Brucella*, which appear not to be redundant in function, although both play a role in the regulation of the expression of type IV and flagellar biosynthesis secretion genes (Delrue *et al.*, 2005; Rambow-Larsen *et al.*, 2008, 2009).

How *Chlamydia* senses the intracellular environment is a particularly intriguing question in view of its unique developmental cycle. The infectious form, the elementary body (EB), is not able to replicate and shows very little, if any, metabolic activity. Shortly after internalization, metabolic activity is detected in the EBs, and they change in morphology into the replicative form, the reticulate body (RB). Contact of EBs with the plasma membrane is probably the initial trigger for entering in the 'activation' programme. Secretion of one TTS effector protein, TARP, stored in the bacteria, occurs at the entry step (Clifton *et al.*, 2004). Whether activation of TTS is sufficient to trigger the beginning of bacterial gene expression that is observed a few hours after entry, or whether additional parameters of the nascent vacuole are sensed to engage into the differentiation programme is unknown. Later in the cycle, RBs convert into EBs, and again it is not known whether this occurs in response to a change in the intracellular or intravacuolar environment. This differentiation is not synchronous for all the bacteria contained within the same vacuole, an observation that has led to a model proposing that the loss of contact with the vacuolar membrane, which does not affect all the bacteria simultaneously, is the triggering event of RB to EB differentiation (Hoare *et al.*, 2008). The question of how chlamydiae sense the environment is particularly important because under some conditions, chlamydiae are able to enter into a 'persistence' phase, with very little replication. Under this state, the bacteria are refractory to antibiotic treatment, and might reside latent for long periods of time within their host, resulting in the chronic manifestations of chlamydial infection. Such a persistent state can be induced during *in vitro* infection with agents such as penicillin, IFN-γ or by nutrient deprivation. It is unknown how far these conditions mimic *in vivo* situations that the bacteria may encounter, and the identification of the parameters sensed to enter the persistent state would constitute a crucial achievement to counteract chronic infections by *Chlamydia*.

The vacuole is not a safe hiding place: additional mechanisms to limit apoptosis and cellular immunity

Residing within the vacuole limits the exposure to several host defence processes, such as recognition by cytosolic proteins engaged in innate immune surveillance. However, experimental data show that microbial components can be recognized by cytosolic and membrane receptors, even for pathogens that reside within a vacuole (Buchholz and Stephens, 2008). Therefore, in addition to 'hiding' behind the vacuolar membrane, some bacteria have developed mechanisms to turn down or delay immediate host-cell defence mechanisms against intruders such as apoptosis, autophagy and inflammation (Radtke and O'Riordan, 2006). Some pathogens have also learned how to limit the efficiency of the adaptive immunity, enabling, in the long term, chronic infection or reinfection. Remaining silent might in some cases require a pause in the bacterial proliferation activity. All these strategies would require a chapter of its own and will only be briefly introduced here, with references to review articles for a more comprehensive view.

Inhibition of apoptosis

Apoptosis represents an ultimate defence strategy against intracellular pathogens: by engaging in programmed cell death, the host prevents the intracellular proliferation of the intruder. In the longer term, phagocytosis of apoptotic cells provides an opportunity to present microbial antigens and mount an adaptive immune response. Several activation pathways exist to engage the cell to commit apoptosis, each composed of several, mostly sequential steps. Not surprisingly, some intracellular pathogens have developed the ability to interrupt these molecular cascades at different steps, often via secreted effector proteins (Faherty and Maurelli, 2008). This is particularly important for bacteria with a slow division rate, or for *Chlamydia*, which needs to complete the full developmental cycle (2–3 days) to produce infectious progeny. *Chlamydia*-infected cells are refractory to pro-apoptotic signals, and several mechanisms are at work to protect the bacteria from premature death of its host (Zhong, 2009).

Intravacuolar bacteria do not always inhibit host cell death. In fact, many pathogens have the ability to shift the balance between death and survival in one direction or the other, sometimes depending on the cell type or the timing in infection. Bacteria such as *Salmonella* or *Legionella* can induce pyroptosis, a cell death distinct from apoptosis, characterized by the activation and release of pro-inflammatory cytokines (Fink and Cookson, 2007; Isberg *et al.*, 2009). Death of infected cells is also observed after *Chlamydia* has completed its developmental cycle and is one of the mechanisms of bacterial release into the extracellular space (Ying *et al.*, 2006; Hybiske and Stephens, 2008). An important consequence of pyroptosis is the release of free bacteria from the cell, which facilitates bacterial dissemination. However, host cell death also benefits the host by promoting bacterial clearance. Survival or death of the host cell is one of the most spectacular outcomes of the permanent struggle between the pathogen and its host, and it is not always easy to determine which side benefits most from this outcome.

Alteration of NF-κB signalling

NF-κB is a heterodimeric transcription factor that is central to innate immunity, acting as a key regulator of the inflammatory response. The signalling cascade of NF-κB is structurally and functionally conserved in vertebrates, insects and even plants. In mammals, it rapidly upregulates the synthesis of cytokines, adhesion molecules, inducible enzymes, and it is implicated in the prevention of apoptosis. Throughout evolution, the NF-κB pathway became an important target of different bacterial pathogens to modulate the inflammatory response of their cellular host. Like cell death, inflammation can be favourable or detrimental to infection, depending on the pathogen considered, the site and time of infection. Thus, bacteria have successfully developed multiple strategies to interfere with this central defence mechanism of the host, and one can observe both activation and inhibition of NF-κB signalling cascades. This is the case of *Salmonella* that can indirectly activate this pathway during cell invasion, through stimulation of Cdc42 by SPI-1 effectors SopE/SopE2/SopB (Bruno *et al.*, 2009); on the other hand, during later phases of infection, *Salmonella* can suppress that same immune response by blocking IκBα ubiquitylation and degradation, an activity described for at least two SPI-2 *Salmonella* effectors, SseL (Le Negrate *et al.*, 2008a) and AvrA (Ye *et al.*, 2007). Recent studies have highlighted the central importance of ubiquitylation in immune pathways, and pathogenic proteins directly connected to the ubiquitin pathway, usually E3 ligases and deubiquitinases, were identified. These bacterial proteins interfere with host proteins from immune signalling cascades, frequently stimulating their degradation by the proteasome. This seems to be a common feature used by a variety of pathogens to inhibit NF-κB activation, which includes *Chlamydia* deubiquitinase ChlaDub1 (Le Negrate *et al.*, 2008b) or *Shigella* OspG, also a deubiquitinase (Kim *et al.*, 2005) or IpaH, an E3 ligase (Rohde *et al.*, 2007a), to mention just a few.

A very different example of subversion within the NK-κB pathway is given by *Brucella* protein TcpB, which contains a Toll/IL-1 receptor (TIR) domain. TIR domain-containing proteins serve as important components of the host cell-signalling pathways that link the Toll-like receptors (TLR) to NF-κB (O'Neill and Bowie, 2007). As such, bacteria also evolved ways to interfere specifically with the TLR signalling in order to modulate immune response. TcpB is able to block TLR2- and TLR4-mediated induction of NK-κB expression by stimulating the degradation of the TLR signalling adapter MAL (Cirl *et al.*, 2008; Salcedo *et al.*, 2008; Radhakrishnan *et al.*, 2009).

Latency

The outcome of host–microbe interactions varies deeply depending on the particular host and pathogen involved, from mild asymptomatic infections to highly virulent diseases. During acute infection, pathogens usually replicate and disseminate rapidly. This strategy often elicits powerful immune responses from the host, with pathological consequences. In contrast, some bacterial pathogens maintain a relative equilibrium in order to minimize the degree of pathology and allow maintenance of long-term or even life-long infections. This is the case for pathogens like *Mycobacterium*, *Salmonella*, *Brucella* or *Chlamydia* that are a common cause of persistent infections.

What is the replicative and metabolic state of the bacteria during the persistent asymptomatic infection, and how do these organisms manage to escape clearance for so long in the presence of the host immune response? To answer these questions, both partners in the interaction need to be taken into account. On one hand, studies of *Salmonella* and *M. tuberculosis* infection of macrophages indicate that an intact adaptive immune system is important in the persistence of these pathogens. The immune system molecules INF-γ (Monack *et al.*, 2004a) and TNF-α (Netea *et al.*, 2003) seem to have a role in maintaining and controlling the level of bacterial replication in persistently infected hosts. IFN-γ- and IL-12 receptor-related genetic mutations predispose affected individuals to acute infections caused by both *Salmonella* and *Mycobacterium* (Casanova and Abel, 2002), supporting the idea that the adaptive immune system takes part in the establishment of persistence. On the other hand, it is not likely that pathogens are mere passive bystanders in terms of maintaining the balance between clearance and persistence. To establish a chronic infection or merely respond to changes in the intracellular host environment, pathogens can reprogramme gene expression to initiate dormancy. The two-component transcriptional regulatory system DosR/DosS of *Mycobacterium* was shown to be induced in response to hypoxia or nitric oxide stress (Kumar *et al.*, 2007; Sousa *et al.*, 2007). *Salmonella* is also able to produce a long-lasting infection inside the *Salmonella*-containing vacuole in fibroblasts, with very limited proliferation (Cano *et al.*, 2003), probably by a mechanism that restricts growth, but not survival, inside the vacuole. A recent report shows that even in macrophages, a large proportion of intracellular bacteria rapidly enter a non-replicative yet viable state (Helaine *et al.*, 2010). The same study suggests that these non-replicating bacteria are able to recognize host intracellular environmental changes that signal the initiation of an adaptive response, leading to bacterial dormancy. A genome-wide screen for *Salmonella* genes required for long-term systemic infection of the mouse revealed that a very diverse repertoire of virulence factors is involved, presumably reflecting the fact that bacteria sequentially encounter a variety of host environments and that *Salmonella* has evolved to respond to these selective forces in a way that permits both the bacteria and the host to survive (Lawley *et al.*, 2006).

In conclusion, dormancy appears as an interesting strategy for survival in response to host genetic factors and changes in the intracellular environment. Existence of persistence is not restricted to intracellular bacteria (Monack *et al.*, 2004b). However, for bacteria with an intracellular lifestyle, residing within a vacuole is clearly an advantage, with minimal exposure to the host and the possibility to create a niche adapted to persistence.

Exit

To infect neighbouring cells, or a new host, bacteria need to exit from their vacuole. Three main processes can be envisaged: fusion of the vacuolar membrane with the plasma membrane, rupture of the vacuolar membrane followed by that of the plasma membrane or extrusion of the vacuole out of the host cell. The last two models seem to be at work for the exit of *Chlamydia trachomatis* out of epithelial cells (Hybiske and Stephens, 2007). Because exit is an asynchronous event, difficult to observe, especially because of the high cell mortality rate, little is known about the mechanisms involved, and important discoveries probably still lie ahead. For instance, the observation that *M. tuberculosis* and *Mycobacterium leprae* are able to translocate from the phagolysosome to the cytosol in myeloid cells was highly unexpected (van der Wel *et al.*, 2007). In the 'surprise' category, although not related to 'exit' but to 'entry', comes also the recent discovery that enteropathogenic *Escherichia coli* (EPEC) is able to be internalized in non-phagocytic cells and reside within a vacuole, breaking the dogma of EPEC as an exclusive extracellular pathogen (Bulgin *et al.*, 2009).

Conclusion

Life in a vacuole takes as many different routes and shapes as the number of pathogens that have evolved the ability to develop in such an environment. Common to all is the necessity to develop redundant strategies to interact with the host cell, to ensure import of resources and adaptation to changes in the intracellular environment. In terms of therapeutic intervention, intracellular pathogens pose several difficulties. Since they are fully dependent on the host for growth, limiting pathogen access to an essential nutrient seems like a good strategy. However, pathogen starvation is likely to also affect the host to some extent, so that the window of intervention might be narrow. In addition, their high degree of adaptation to environmental changes might also undermine such approaches. The existence of redundant mechanisms to hijack host resources also limits the efficiency of drugs that would target only one mechanism. Furthermore, the intravacuolar position of the bacteria restricts access to some drugs that cannot cross membranes. Overall, present efforts to understand the mechanisms by which pathogens survive and even successfully replicate within a vacuole need to be pursued, to identify their Achilles' heel(s) and transform them into targets for therapeutic intervention.

References

Abdallah, A.M., Gey van Pittius, N.C., Champion, P.A., Cox, J., Luirink, J., Vandenbroucke-Grauls, C.M., Appelmelk, B.J., and Bitter, W. (2007). Type VII secretion-mycobacteria show the way. Nat. Rev. Microbiol. 5, 883–891.

Agranoff, D., Monahan, I.M., Mangan, J.A., Butcher, P.D., and Krishna, S. (1999). *Mycobacterium tuberculosis* expresses a novel pH-dependent divalent cation transporter belonging to the Nramp family. J. Exp. Med. *190*, 717–724.

Amer, A.O., and Swanson, M.S. (2005). Autophagy is an immediate macrophage response to *Legionella pneumophila*. Cell. Microbiol. 7, 765–778.

Arai, K., Shimaya, A., Hiratani, N., and Ohkuma, S. (1993). Purification and characterization of lysosomal H^+-ATPase. An anion-sensitive v-type H^+-ATPase from rat liver lysosomes. J. Biol. Chem. *268*, 5649–5660.

Armstrong, J.A., and Hart, P.D. (1971). Response of cultured macrophages to *Mycobacterium tuberculosis*, with observations on fusion of lysosomes with phagosomes. J. Exp. Med. *134*, 713–740.

Baca, O.G., Li, Y.P., and Kumar, H. (1994). Survival of the Q fever agent *Coxiella burnetii* in the phagolysosome. Trends Microbiol. *2*, 476–480.

Bànfi, B., Clark, R.A., Steger, K., and Krause, K.H. (2003). Two novel proteins activate superoxide generation by the NADPH oxidase NOX1. J. Biol. Chem. *278*, 3510–3513.

Bard, F., and Malhotra, V. (2006). The formation of TGN-to-plasma-membrane transport carriers. Annu. Rev. Cell Dev. Biol. 22, 439–455.

Baust, T., Anitei, M., Czupalla, C., Parshyna, I., Bourel, L., Thiele, C., Krause, E., and Hoflack, B. (2008). Protein networks supporting AP-3 function in targeting lysosomal membrane proteins. Mol. Biol. Cell *19*, 1942–1951.

Bedard, K., Lardy, B., and Krause, K.H. (2007). NOX family NADPH oxidases: not just in mammals. Biochimie *89*, 1107–1112.

Beuzón, C.R., Méresse, S., Unsworth, K.E., Ruiz-Albert, J., Garvis, S., Waterman, S.R., Ryder, T.A., Boucrot, E., and Holden, D.W. (2000). *Salmonella* maintains the integrity of its intracellular vacuole through the action of SifA. EMBO J. *19*, 3235–3249.

Birmingham, C.L., Smith, A.C., Bakowski, M.A., Yoshimori, T., and Brumell, J.H. (2006). Autophagy controls *Salmonella* infection in response to damage to the *Salmonella*-containing vacuole. J. Biol. Chem. *281*, 11374–11383.

Boncompain, G., Schneider, B., Delevoye, C., Kellermann, O., Dautry-Varsat, A., and Subtil, A. (2010). Production of reactive oxygen species is turned on and rapidly shut down in epithelial cells infected with *Chlamydia trachomatis*. Infect. Immun. *78*, 80–87.

Boschiroli, M.L., Ouahrani-Bettache, S., Foulongne, V., Michaux-Charachon, S., Bourg, G., Allardet-Servent, A., Cazevieille, C., Liautard, J.P., Ramuz, M., and O'Callaghan, D. (2002). The *Brucella suis virB* operon is induced intracellularly in macrophages. Proc. Natl. Acad. Sci. U.S.A. *99*, 1544–1549.

Bruno, V.M., Hannemann, S., Lara-Tejero, M., Flavell, R.A., Kleinstein, S.H., and Galán, J.E. (2009). *Salmonella* Typhimurium type III secretion effectors stimulate innate immune responses in cultured epithelial cells. PLoS Pathog. *5*, e1000538.

Brumell, J.H., and Scidmore, M.A. (2007). Manipulation of Rab GTPase function by intracellular bacterial pathogens. Microbiol. Mol. Biol. Rev. *71*, 636–652.

Buchholz, K.R., and Stephens, R.S. (2008). The cytosolic pattern recognition receptor NOD1 induces inflammatory interleukin-8 during *Chlamydia trachomatis* infection. Infect. Immun. *76*, 3150–3155.

Bulgin, R., Arbeloa, A., Goulding, D., Dougan, G., Crepin, V.F., Raymond, B., and Frankel, G. (2009). The T3SS effector EspT defines a new category of invasive enteropathogenic *E. coli* (EPEC) which form intracellular actin pedestals. PLoS Pathog. *5*, e1000683.

Cano, D.A., Pucciarelli, M.G., Martinez-Moya, M., Casadesús, J., and Garcià-del Portillo, F. (2003). Selection of small-colony variants of *Salmonella enterica* serovar Typhimurium in nonphagocytic eucaryotic cells. Infect. Immun. *71*, 3690–3698.

Casanova, J.L., and Abel, L. (2002). Genetic dissection of immunity to mycobacteria: the human model. Annu. Rev. Immunol. *20*, 581–620.

Cellier, M.F., Courville, P., and Campion, C. (2007). Nramp1 phagocyte intracellular metal withdrawal defense. Microbes Infect. *9*, 1662–1670.

Charpentier, X., Gabay, J. l. E., Reyes, M., Zhu, J.W., Weiss, A., and Shuman, H.A. (2009). Chemical genetics reveals bacterial and host cell functions critical for type

IV effector translocation by *Legionella pneumophila*. PLoS Pathog. *5*, e1000501.

Cheng, G., Cao, Z., Xu, X., van Meir, E.G., and Lambeth, J.D. (2001). Homologs of gp91phox: cloning and tissue expression of Nox3, Nox4, and Nox5. Gene *269*, 131–140.

Cirl, C., Wieser, A., Yadav, M., Duerr, S., Schubert, S., Fischer, H., Stappert, D., Wantia, N., Rodriguez, N., Wagner, H., *et al.* (2008). Subversion of Toll-like receptor signaling by a unique family of bacterial Toll/interleukin-1 receptor domain-containing proteins. Nat. Med. *14*, 399–406.

Clemens, D.L., and Horwitz, M.A. (1996). The *Mycobacterium tuberculosis* phagosome interacts with early endosomes and is accessible to exogenously administered transferrin. J. Exp. Med. *184*, 1349–1355.

Clifton, D.R., Fields, K.A., Grieshaber, N.A., Dooley, C.A., Fischer, E.R., Mead, D.J., Carabeo, R.A., and Hackstadt, T. (2004). A chlamydial type III translocated protein is tyrosine-phosphorylated at the site of entry and associated with recruitment of actin. Proc. Natl. Acad. Sci. USA *101*, 10166–10171.

Cocchiaro, J.L., Kumar, Y., Fischer, E.R., Hackstadt, T., and Valdivia, R.H. (2008). Cytoplasmic lipid droplets are translocated into the lumen of the *Chlamydia trachomatis* parasitophorous vacuole. Proc. Natl. Acad. Sci. USA *105*, 9379–9384.

Delevoye, C., Nilges, M., Dehoux, P., Paumet, F., Perrinet, S., Dautry-Varsat, A., and Subtil, A. (2008). SNARE protein mimicry by an intracellular bacterium. PLoS Pathog. *4*, e1000022.

Delrue, R.M., Deschamps, C., Léonard, S., Nijskens, C., Danese, I., Schaus, J.M., Bonnot, S., Ferooz, J., Tibor, A., De Bolle, X., and Letesson, J.J. (2005). A quorum-sensing regulator controls expression of both the type IV secretion system and the flagellar apparatus of *Brucella melitensis*. Cell. Microbiol. *7*, 1151–1161.

Dong, F., Pirbhai, M., Zhong, Y.M., and Zhong, G.M. (2004). Cleavage-dependent activation of a *Chlamydia*-secreted protease. Mol. Microbiol. *52*, 1487–1494.

Faherty, C.S., and Maurelli, A.T. (2008). Staying alive: bacterial inhibition of apoptosis during infection. Trends Microbiol. *16*, 173–180.

Fink, S.L., and Cookson, B.T. (2007). Pyroptosis and host cell death responses during *Salmonella* infection. Cell. Microbiol. *9*, 2562–2570.

Franco, I.S., Shuman, H.A., and Charpentier, X. (2009). The perplexing functions and surprising origins of *Legionella pneumophila* type IV secretion effectors. Cell. Microbiol. *11*, 1435–1443.

Fromme, J.C., Orci, L., and Schekman, R. (2008). Coordination of COPII vesicle trafficking by Sec23. Trends Cell Biol *18*, 330–336.

Galán, J.E., and Wolf-Watz, H. (2006). Protein delivery into eukaryotic cells by type III secretion machines. Nature *444*, 567–573.

Garciá-del Portillo, F. G.D., Nunez-Hernandez, C., Eisman, B., and Ramos-Vivas, J. (2008). Growth control in the *Salmonella*-containing vacuole. Curr. Opin. Microbiol. *11*, 46–52.

Grieshaber, S.S., Grieshaber, N.A., and Hackstadt, T. (2003). *Chlamydia trachomatis* uses host cell dynein to traffic to the microtubule-organizing center in a p50 dynamitin-independent process. J. Cell Sci. *116*, 3793–3802.

Grosshans, B.L., Ortiz, D., and Novick, P. (2006). Rabs and their effectors: achieving specificity in membrane traffic. Proc. Natl. Acad. Sci. U.S.A. *103*, 11821–11827.

Gu, F., and Gruenberg, J. (1999). Biogenesis of transport intermediates in the endocytic pathway. FEBS Lett. *452*, 61–66.

Guermonprez, P., Saveanu, L., Kleijmeer, M., Davoust, J., van Endert, P., and Amigorena, S. (2003). ER–phagosome fusion defines an MHC class I cross-presentation compartment in dendritic cells. Nature *425*, 397–402.

Gutierrez, M.G., Master, S.S., Singh, S.B., Taylor, G.A., Colombo, M.I., and Deretic, V. (2004). Autophagy is a defense mechanism inhibiting BCG and *Mycobacterium tuberculosis* survival in infected macrophages. Cell *119*, 753–766.

Hackstadt, T., Scidmore, M.A., and Rockey, D.D. (1995). Lipid metabolism in *Chlamydia trachomatis*-infected cells: directed trafficking of Golgi-derived sphingolipids to the chlamydial inclusion. Proc. Natl. Acad. Sci. U.S.A. 92, 4877–4881.

He, C., and Klionsky, D.J. (2009). Regulation mechanisms and signaling pathways of autophagy. Annu. Rev. Genet. *43*, 67–93.

Heinzen, R.A., and Hackstadt, T. (1997). The *Chlamydia trachomatis* parasitophorous vacuolar membrane is not passively permeable to low-molecular-weight compounds. Infect. Immun. *65*, 1088–1094.

Helaine, S., Thompson, J.A., Watson, K.G., Liu, M., Boyle, C., and Holden, D.W. (2010). Dynamics of intracellular bacterial replication at the single cell level. Proc. Natl. Acad. Sci. U S A *107*, 3746–3751.

Heuer, D., Brinkmann, V., Meyer, T.F., and Szczepek, A.J. (2003). Expression and translocation of chlamydial protease during acute and persistent infection of the epithelial HEp-2 cells with *Chlamydophila* (*Chlamydia*) *pneumoniae*. Cell. Microbiol. *5*, 315–322.

Heuer, D., Rejman Lipinski, A., Machuy, N., Karlas, A., Wehrens, A., Siedler, F., Brinkmann, V., and Meyer, T.F. (2009). *Chlamydia* causes fragmentation of the Golgi compartment to ensure reproduction. Nature *457*, 731–735.

Hoare, A., Timms, P., Bavoil, P., and Wilson, D. (2008). Spatial constraints within the chlamydial host cell inclusion predict interrupted development and persistence. BMC Microbiol. *8*, 5.

Howe, D., and Heinzen, R.A. (2006). *Coxiella burnetii* inhabits a cholesterol-rich vacuole and influences cellular cholesterol metabolism. Cell. Microbiol. *8*, 496–507.

Hybiske, K., and Stephens, R.S. (2007). Mechanisms of host cell exit by the intracellular bacterium *Chlamydia*. Proc. Natl. Acad. Sci. USA *104*, 11430–11435.

Hybiske, K., and Stephens, R.S. (2008). Exit strategies of intracellular pathogens. Nat. Rev. Microbiol. *6*, 99–110.

Isberg, R.R., O'Connor, T.J., and Heidtman, M. (2009). The *Legionella pneumophila* replication vacuole: making a cosy niche inside host cells. Nat. Rev. Microbiol. *7*, 13–24.

Jabado, N., Jankowski, A., Dougaparsad, S., Picard, V., Grinstein, S., and Gros, P. (2000). Natural resistance to intracellular infections: natural resistance-associated macrophage protein 1 (Nramp1) functions as a pH-dependent manganese transporter at the phagosomal membrane. J. Exp. Med. *192*, 1237–1248.

Jahn, R., and Scheller, R.H. (2006). SNAREs – engines for membrane fusion. Nat. Rev. Mol. Cell Biol. *7*, 631–643.

Kagan, J.C., Stein, M.P., Pypaert, M., and Roy, C.R. (2004). *Legionella* subvert the functions of Rab1 and Sec22b to create a replicative organelle. J. Exp. Med. *199*, 1201–1211.

Kim, D.W., Lenzen, G., Page, A.L., Legrain, P., Sansonetti, P.J., and Parsot, C. (2005). The *Shigella flexneri* effector OspG interferes with innate immune responses by targeting ubiquitin-conjugating enzymes. Proc. Natl. Acad. Sci. U S A *102*, 14046–14051.

Kirkin, V., McEwan, D.G., Novak, I., and Dikic, I. (2009). A role for ubiquitin in selective autophagy. Mol. Cell *34*, 259–269.

Klionsky, D.J. (2007). Autophagy: from phenomenology to molecular understanding in less than a decade. Nat. Rev. Mol. Cell Biol. *8*, 931–937.

Kobayashi, T., Stang, E., Fang, K.S., de Moerloose, P., Parton, R.G., and Gruenberg, J. (1998). A lipid associated with the antiphospholipid syndrome regulates endosome structure and function. Nature *392*, 193–197.

Kumar, Y., and Valdivia, R.H. (2008). Actin and intermediate filaments stabilize the *Chlamydia trachomatis* vacuole by forming dynamic structural scaffolds. Cell Host Microbe *4*, 159–169.

Kumar, Y., and Valdivia, R.H. (2009). Leading a sheltered life: intracellular pathogens and maintenance of vacuolar compartments. Cell Host Microbe *5*, 593–601.

Kumar, A., Toledo, J.C., Patel, R.P., Lancaster, J.R., Jr., and Steyn, A.J. (2007). *Mycobacterium tuberculosis* DosS is a redox sensor and DosT is a hypoxia sensor. Proc. Natl. Acad. Sci. U S A *104*, 11568–11573.

Lawley, T.D., Chan, K., Thompson, L.J., Kim, C.C., Govoni, G.R., and Monack, D.M. (2006). Genome-wide screen for *Salmonella* genes required for long-term systemic infection of the mouse. PLoS Pathog. *2*, e11.

Le Negrate, G., Faustin, B., Welsh, K., Loeffler, M., Krajewska, M., Hasegawa, P., Mukherjee, S., Orth, K., Krajewski, S., Godzik, A., *et al.* (2008a). *Salmonella* secreted factor L deubiquitinase of *Salmonella typhimurium* inhibits NF-κB, suppresses IκBα ubiquitination and modulates innate immune responses. J. Immunol. *180*, 5045–5056.

Le Negrate, G., Krieg, A., Faustin, B., Loeffler, M., Godzik, A., Krajewski, S., and Reed, J.C. (2008b). ChlaDub1 of *Chlamydia trachomatis* suppresses NF-κB activation and inhibits IκBα ubiquitination and degradation. Cell. Microbiol. *10*, 1879–1892.

Lerena, M.C., Vázquez, C.L., and Colombo, M.I. (2010). Bacterial pathogens and the autophagic response. Cell. Microbiol. *12*, 10–18.

Lingwood, D., and Simons, K. (2010). Lipid Rafts as a membrane-organizing principle. Science *327*, 46–50.

McMahon, H.T., and Mills, I.G. (2004). COP and clathrin-coated vesicle budding: different pathways, common approaches. Curr. Opin. Cell Biol. *16*, 379–391.

Marra, A., Blander, S.J., Horwitz, M.A., and Shuman, H.A. (1992). Identification of a *Legionella pneumophila* locus required for intracellular multiplication in human macrophages. Proc. Natl. Acad. Sci. U.S.A. *89*, 9607–9611.

van Meer, G., Voelker, D.R., and Feigenson, G.W. (2008). Membrane lipids: where they are and how they behave. Nat. Rev. Mol. Cell Biol. *9*, 112–124.

Méresse, S., Unsworth, K.E., Habermann, A., Griffiths, G., Fang, F., MartÌnez-Lorenzo, M.J., Waterman, S.R., Gorvel, J. P.E., and Holden, D.W. (2001). Remodelling of the actin cytoskeleton is essential for replication of intravacuolar *Salmonella*. Cell. Microbiol. *3*, 567–577.

Monack, D.M., Bouley, D.M., and Falkow, S. (2004a). *Salmonella typhimurium* persists within macrophages in the mesenteric lymph nodes of chronically infected *Nramp1* $^{+/+}$ mice and can be reactivated by IFNγ neutralization. J. Exp. Med. *199*, 231–241.

Monack, D.M., Mueller, A., and Falkow, S. (2004b). Persistent bacterial infections: the interface of the pathogen and the host immune system. Nat. Rev. Microbiol. *2*, 747–765.

Nagai, H., Kagan, J.C., Zhu, X., Kahn, R.A., and Roy, C.R. (2002). A bacterial guanine nucleotide exchange factor activates ARF on *Legionella* phagosomes. Science *295*, 679–682.

Netea, M.G., Radstake, T., Joosten, L.A., van der Meer, J.W., Barrera, P., and Kullberg, B.J. (2003). *Salmonella* septicemia in rheumatoid arthritis patients receiving anti-tumor necrosis factor therapy: association with decreased interferon-γ production and Toll-like receptor 4 expression. Arthritis Rheum. *48*, 1853–1857.

Norris, F.A., Wilson, M.P., Wallis, T.S., Galyov, E.E., and Majerus, P.W. (1998). SopB, a protein required for virulence of *Salmonella dublin*, is an inositol phosphate phosphatase. Proc. Natl. Acad. Sci. U.S.A. *95*, 14057–14059.

O'Neill, L.A., and Bowie, A.G. (2007). The family of five: TIR-domain-containing adaptors in Toll-like receptor signalling. Nat. Rev. Immunol. *7*, 353–364.

Pan, X., Lührmann, A., Satoh, A., Laskowski-Arce, M.A., and Roy, C.R. (2008). Ankyrin repeat proteins comprise a diverse family of bacterial type IV effectors. Science *320*, 1651–1654.

Parsons, D.A., and Heffron, F. (2005). *sciS*, an *icmF* homolog in *Salmonella enterica* serovar Typhimurium, limits intracellular replication and decreases virulence. Infect. Immun. *73*, 4338–4345.

Pethe, K., Swenson, D.L., Alonso, S., Anderson, J., Wang, C., and Russell, D.G. (2004). Isolation of *Mycobacterium tuberculosis* mutants defective in the arrest of phagosome maturation. Proc. Natl. Acad. Sci. U.S.A. *101*, 13642–13647.

Pukatzki, S., McAuley, S.B., and Miyata, S.T. (2009). The type VI secretion system: translocation of effectors and effector-domains. Curr. Opin. Microbiol. *12*, 11–17.

Purdy, G.E., and Russell, D.G. (2007). Lysosomal ubiquitin and the demise of *Mycobacterium tuberculosis*. Cell. Microbiol. *9*, 2768–2774.

Radhakrishnan, G.K., Yu, Q., Harms, J.S., and Splitter, G.A. (2009). *Brucella* TIR domain-containing protein mimics properties of the Toll-like receptor adaptor protein TIRAP.J. Biol. Chem. *284*, 9892–9898.

Radtke, A.L., and O'Riordan, M.X. D. (2006). Intracellular innate resistance to bacterial pathogens. Cell. Microbiol. *8*, 1720–1729.

Ramachandra, L., Simmons, D., and Harding, C.V. (2009). MHC molecules and microbial antigen processing in phagosomes. Curr. Opin. Immunol. *21*, 98–104.

Rambow-Larsen, A.A., Rajashekara, G., Petersen, E., and Splitter, G. (2008). Putative quorum-sensing regulator BlxR of *Brucella melitensis* regulates virulence factors including the type IV secretion system and flagella. J. Bacteriol. *190*, 3274–3282.

Rambow-Larsen, A.A., Petersen, E.M., Gourley, C.R., and Splitter, G.A. (2009). *Brucella* regulators: self-control in a hostile environment. Trends Microbiol. *17*, 371–377.

Rikihisa, Y. (2006). *Ehrlichia* subversion of host innate responses. Curr. Opin. Microbiol. *9*, 95–101.

Robertson, D.K., Gu, L., Rowe, R.K., and Beatty, W.L. (2009). Inclusion biogenesis and reactivation of persistent *Chlamydia trachomatis* requires host cell sphingolipid biosynthesis. PLoS Pathog. *5*, e1000664.

Robinson, M.S. (2004). Adaptable adaptors for coated vesicles. Trends Cell Biol. *14*, 167–174.

Rohde, J.R., Breitkreutz, A., Chenal, A., Sansonetti, P.J., and Parsot, C. (2007a). Type III secretion effectors of the IpaH family are E3 ubiquitin ligases. Cell Host Microbe *1*, 77–83.

Rohde, K., Yates, R.M., Purdy, G.E., and Russell, D.G. (2007b). *Mycobacterium tuberculosis* and the environment within the phagosome. Immunol. Rev. *219*, 37–54.

Rohde, K.H., Abramovitch, R.B., and Russell, D.G. (2007c). *Mycobacterium tuberculosis* invasion of macrophages: linking bacterial gene expression to environmental cues. Cell Host Microbe *2*, 352–364.

Roman, M.J., Coriz, P.D., and Baca, O.G. (1986). A proposed model to explain persistent infection of host cells with *Coxiella burnetii*. J. Gen. Microbiol. *132*, 1415–1422.

Romano, P.S., Gutierrez, M.G., Beron, W., Rabinovitch, M., and Colombo, M.I. (2007). The autophagic pathway is actively modulated by phase II *Coxiella burnetii* to efficiently replicate in the host cell. Cell. Microbiol. *9*, 891–909.

Roop, R.M., 2nd, Gaines, J.M., Anderson, E.S., Caswell, C.C., and Martin, D.W. (2009). Survival of the fittest: how *Brucella* strains adapt to their intracellular niche in the host. Med. Microbiol. Immunol. *198*, 221–238.

Salcedo, S.P., Marchesini, M.I., Lelouard, H., Fugier, E., Jolly, G., Balor, S., Muller, A., Lapaque, N., Demaria, O., Alexopoulou, L., *et al.* (2008). *Brucella* control of dendritic cell maturation is dependent on the TIR-containing protein Btp1. PLoS Pathog. *4*, e21.

Simeone, R., Bottai, D., and Brosch, R. (2009). ESX/type VII secretion systems and their role in host–pathogen interaction. Curr. Opin. Microbiol. *12*, 4–10.

Simons, K., and Ikonen, E. (1997). Functional rafts in cell membranes. Nature *387*, 569–572.

Söllner, T.H. (2003). Regulated exocytosis and SNARE function (Review). Mol. Membr. Biol. *20*, 209–220.

Sorkin, A. (2004). Cargo recognition during clathrin-mediated endocytosis: a team effort. Curr. Opin. Cell Biol. *16*, 392–399.

Sousa, E.H., Tuckerman, J.R., Gonzalez, G., and Gilles-Gonzalez, M.A. (2007). DosT and DevS are oxygen-switched kinases in *Mycobacterium tuberculosis*. Protein Sci. *16*, 1708–1719.

Steele-Mortimer, O. (2008). The *Salmonella*-containing vacuole: moving with the times. Curr. Opin. Microbiol. *11*, 38–45.

Stenmark, H. (2009). Rab GTPases as coordinators of vesicle traffic. Nat. Rev. Mol. Cell Biol. *10*, 513–525.

Sturgill-Koszycki, S., Schlesinger, P.H., Chakraborty, P., Haddix, P.L., Collins, H.L., Fok, A.K., Allen, R.D., Gluck, S.L., Heuser, J., and Russell, D.G. (1994). Lack of acidification in *Mycobacterium* phagosomes produced by exclusion of the vesicular proton-ATPase. Science *263*, 678–681.

Suchland, R.J., Rockey, D.D., Weeks, S.K., Alzhanov, D.T., and Stamm, W.E. (2005). Development of secondary inclusions in cells infected by *Chlamydia trachomatis*. Infect. Immun. *73*, 3954–3962.

Südhof, T.C., and Rothman, J.E. (2009). Membrane fusion: grappling with SNARE and SM proteins. Science *323*, 474–477.

Varlamov, O., Volchuk, A., Rahimian, V., Doege, C.A., Paumet, F., Eng, W.S., Arango, N., Parlati, F., Ravazzola, P., Orci, L., *et al.* (2004). i-SNAREs: inhibitory SNAREs that fine-tune the specificity of membrane fusion. J. Cell Biol. *164*, 79–88.

Voth, D.E., and Heinzen, R.A. (2007). Lounging in a lysosome: the intracellular lifestyle of *Coxiella burnetii*. Cell. Microbiol. *9*, 829–840.

Voth, D.E., and Heinzen, R.A. (2009). *Coxiella* type IV secretion and cellular microbiology. Curr. Opin. Microbiol. *12*, 74–80.

Wakeel, A., Zhu, B., Yu, X.J., and McBride, J.W. (2010). New insights into molecular *Ehrlichia chaffeensis*–host interactions. Microbes Infect. *12*, 337–345.

Weber, S.S., Ragaz, C., Reus, K., Nyfeler, Y., and Hilbi, H. (2006). *Legionella pneumophila* exploits PI(4)P to anchor secreted effector proteins to the replicative vacuole. PLoS Pathog. *2*, e46.

Weber, S.S., Ragaz, C., and Hilbi, H. (2009). Pathogen trafficking pathways and host phosphoinositide metabolism. Mol. Microbiol. *71*, 1341–1352.

van der Wel, N., Hava, D., Houben, D., Fluitsma, D., van Zon, M., Pierson, J., Brenner, M., and Peters, P.J. (2007). *M. tuberculosis* and *M. leprae* translocate from the phagolysosome to the cytosol in myeloid cells. Cell *129*, 1287–1298.

Wickner, W., and Schekman, R. (2008). Membrane fusion. Nat. Struct. Biol. *15*, 658–664.

Wylie, J.L., Hatch, G.M., and McClarty, G. (1997). Host cell phospholipids are trafficked to and then modified by *Chlamydia trachomatis*. J. Bacteriol. *179*, 7233–7242.

Ye, Z., Petrof, E.O., Boone, D., Claud, E.C., and Sun, J. (2007). *Salmonella* effector AvrA regulation of colonic epithelial cell inflammation by deubiquitination. Am. J. Pathol. *171*, 882–892.

Ying, S., Fischer, S.F., Pettengill, M., Conte, D., Paschen, S.A., Ojcius, D.M., and Häcker, G. (2006). Characterization of host cell death induced by *Chlamydia trachomatis*. Infect. Immun. *74*, 6057–6066.

Zhong, G. (2009). Killing me softly: chlamydial use of proteolysis for evading host defenses. Trends Microbiol. *17*, 467–474.

Zhong, G., Fan, P., Ji, H., Dong, F., and Huang, Y. (2001). Identification of a chlamydial protease-like activity factor responsible for the degradation of host transcription factors. J. Exp. Med. *193*, 935–942.

The Bacterial Life in the Cytosol

11

Serge Mostowy and Pascale Cossart

Abstract

To live inside host cells and replicate, pathogenic bacteria can either establish a niche in a vacuolar compartment or escape to the cytosol. In this chapter, we review our understanding of pathogenic mechanisms to escape and live in the cytosol, and highlight recent studies of bacterial survival strategies to counteract intracytosolic host defences.

Introduction

Upon infection, intracellular pathogens can be uptaken passively by macrophages (e.g. mycobacteria) or can actively invade epithelial cells (e.g. *Listeria* or *Shigella*). After internalization, bacteria are either transiently or definitively localized within a vacuolar compartment called the phagosome. The newly formed phagosome undergoes a series of events, including fusion with endosomes, in a process referred to as phagosome maturation. If the phagosome fuses with lysosomes giving rise to a phagolysosome, compartmentalized bacteria are destroyed. However, some pathogens, such as *Salmonella* or mycobacteria, interfere with the normal biogenesis of phagolysosomes, thus leading to the formation of replicative vacuoles. Other pathogens, such as *Listeria*, *Shigella*, or *Rickettsia*, have developed strategies to escape from phagosomes to the cytosol, and avoid destruction in phagolysosomes. Once in the cytosol, bacteria must get adequate nutrients from the intracellular milieu, evade intracytosolic immune responses, and spread to new host cells to sustain infection. Here, we focus on pathogenic bacteria that escape to the cytosol, and describe their mechanisms of vacuolar escape and intracytosolic survival. We also summarize the present knowledge on actin-based motility employed by bacteria for intercellular spread. Finally, we give an overview of innate immune process responding to intracellular pathogens, and the mechanisms evolved by bacteria to counteract these host defences.

Escape from the vacuole

Escape from the vacuole is a crucial step in the life cycle of cytosolic pathogens. This is a bacterially driven process, and the entry of pathogens into the cytosol relies on the production of secreted proteins triggered by specific environmental cues in the vacuole.

Listeria monocytogenes

L. monocytogenes is a Gram-positive bacterium responsible for foodborne infections, during which bacteria spread from the intestinal lumen to the liver and spleen, and ultimately, via a haematogenous route, to the brain and the placenta (Khelef *et al.*, 2006). The intracellular lifestyle of this pathogen has been well studied (Fig. 11.1). Listeria survives intracellularly by modifying and subsequently escaping from phagosomes using a combination of effectors (Hamon *et al.*, 2006). The cholesterol-dependent cytolysin listeriolysin O (LLO) inhibits the maturation of phagosomes by creating pores in the phagosomal membrane which inhibit lysosome fusion by altering pH and calcium (Shaughnessy *et al.*, 2006). LLO needs to be activated by acidification and/or by the

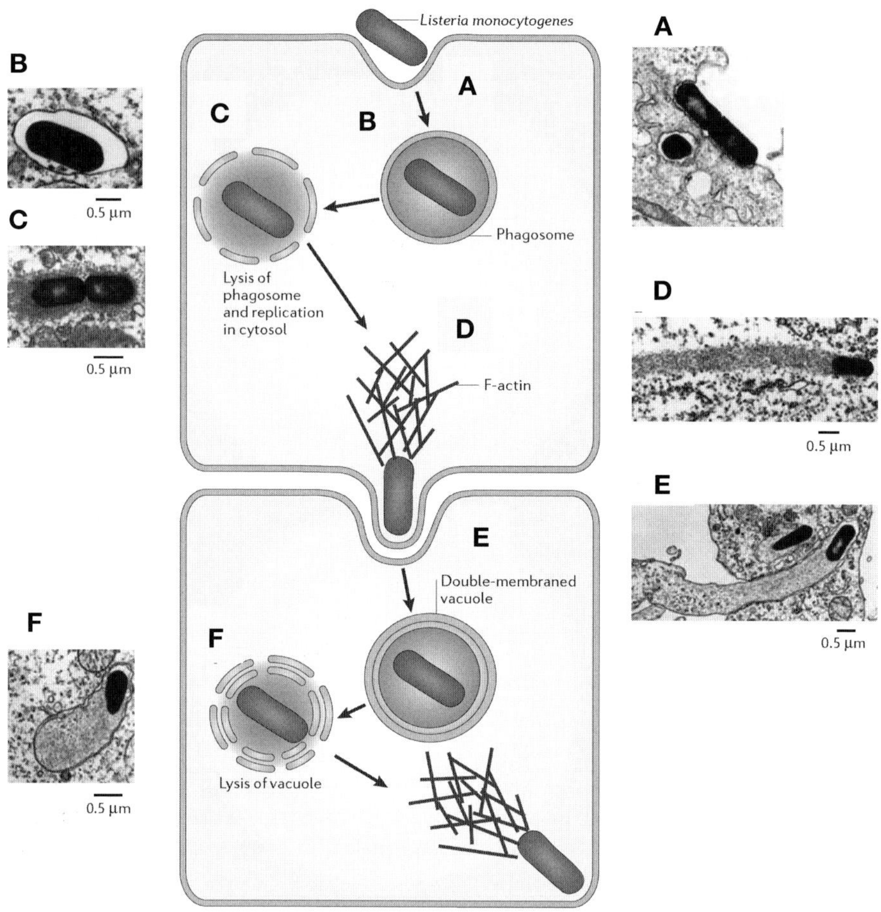

Figure 11.1 The intracellular lifestyle of *Listeria monocytogenes*. Schematic representation and electron micrographs of the *L. monocytogenes* life cycle. (A) *L. monocytogenes* induces its entry into a non-professional phagocyte. (B) Bacteria are internalized in a vacuole (i.e. a phagosome). (C) and (D) The membrane of the vacuole is disrupted by the secretion of phospholipases and LLO. Bacteria are released into the cytosol where they replicate and start to polymerize actin, as observed by the presence of actin tails. (E) Actin polymerization allows bacteria to propel into a neighbouring cell by forming protrusions in the plasma membrane. (F) Upon entry into the neighbouring cell, bacteria are present in a double-membrane vacuole, from which they can escape to continue their life cycle. This figure is reproduced with permission from Hamon *et al.* (2006). A colour version of this figure is located in the plate section at the back of the book.

host enzyme GILT (IFNγ-inducible lysosomal thiol reductase) that is found inside the phagosome (Singh *et al.*, 2008). *Listeria* expresses, in addition to LLO, two membrane-active phospholipase C enzymes (i.e. PI-PLC and PC-PLC) that contribute to rupture the membrane of the *L. monocytogenes*-containing phagosome and enable the bacteria to escape to the cytosol.

Although thought to reside primarily in the cytosol, *L. monocytogenes* can replicate, under some circumstances, inside macrophage vacuoles called 'spacious *Listeria*-containing phagosomes' or SLAPs (Birmingham *et al.*, 2008). The formation of SLAPs only occurs upon impaired LLO production. The rate of bacterial replication inside SLAPs is reduced compared with that in cytosol, yet it could contribute to the development of chronic *Listeria* infection.

Shigella flexneri

S. flexneri is a Gram-negative, foodborne pathogen that invades the colonic mucosa, causing inflammation and diarrhoea (Sansonetti, 2001). The infectious cycle is similar to that of *L. monocytogenes* (Fig. 11.1), but the mechanisms that *Shigella* uses to escape from the vacuole are poorly understood. The type III secretion system (TTSS) effector IpaB is required for vacuolar escape in macrophages and, like LLO, displays some haemolytic activity (High *et al.*, 1992). IpaB assembles with IpaC, another TTSS effector, into a pore complex that inserts into cell membranes (Picking *et al.*, 2005). Considering that the efficient insertion of IpaB/IpaC pores is involved in *Shigella* entry, it may also be important for disruption of the vacuolar membrane and subsequent bacterial escape (Osiecki *et al.*, 2001; Picking *et al.*, 2005). $IpaH_{7.8}$ is another bacterial factor shown to be involved in vacuolar escape, though its role remains unclear (Fernandez-Prada *et al.*, 2000).

Rickettsia species

Rickettsia species are obligate intracellular, Gram-negative bacteria that fall into two groups, the spotted fever group including *R. conorii* (the agent of the Mediterranean spotted fever) and *R. rickettsii* (the agent of Rocky Mountain spotted fever), and the typhus group including *R. prowazekii* (the agent of epidemic typhus) (Raoult *et al.*, 2005; Walker, 2007). There is little information about how bacteria escape from the vacuole but studies have suggested that haemolysin C and phospholipases produced by *Rickettsia* species play an important role (Winkler and Miller, 1982; Silverman *et al.*, 1992; Renesto *et al.*, 2003; Whitworth *et al.*, 2005).

Burkholderia pseudomallei

B. pseudomallei is a Gram-negative bacterium that is principally an environmental saprophyte, but can cause melioidosis in humans and other animals (Peacock, 2006). Following cellular uptake, *B. pseudomallei* can escape from the vacuole to the cytosol, where the bacteria can replicate (Kespichayawattana *et al.*, 2000). Vacuolar escape is attributed to the action of the *Burkholderia* secretion apparatus (Bsa) TTSS (Stevens *et al.*, 2002, 2004; Sun *et al.*, 2005; Suparak *et al.*, 2005), and BPSS1539, a protein of unknown function within the Bsa TTSS gene cluster, seems to be required (Pilatz *et al.*, 2006).

Francisella tularensis

F. tularensis is a Gram-negative facultative intracellular bacterium that causes tularaemia (i.e. rabbit fever) (Meibom and Charbit, 2010). The pathogenesis of *F. tularensis* depends on ability of the bacterium to survive and replicate within host cells, primarily macrophages (Oyston *et al.*, 2004; Santic *et al.*, 2006). Following entry, *F. tularensis* is localized within a *Francisella*-containing phagosome (FCP), but escapes from an acidified late-endosome-like vacuole before it matures into a lysosome (Santic *et al.*, 2006; Chong *et al.*, 2008). Lysis of the vacuole requires the acquisition of a vacuolar ATPase pump to transiently acidify the phagosome (Santic *et al.*, 2008). Francisella IglC and its regulator MglA are essential for lysis, albeit through unknown mechanisms (Santic *et al.*, 2005). An *F. tularensis* virulence factor FTT1103 has also been identified as necessary for vacuolar escape (Qin *et al.*, 2009).

After escaping from the vacuole, *Francisella* does not exhibit actin-based motility in the cytosol and instead proliferates rapidly. Eventually, the infected cell undergoes apoptosis, and the bacteria are released to continue infection. Alternatively, *Francisella* can also re-enter a membrane bound compartment called the *Francisella*-containing vacuole (FCV) after cytosolic replication, via a process mediated by autophagy (Checroun *et al.*, 2006).

Mycobacterium marinum

M. marinum is a close genetic relative of *M. tuberculosis*, the causative agent of human tuberculosis. *M. marinum* causes a systemic tuberculosis-like disease in fishes and frogs, and a localized disease in humans (Stinear *et al.*, 2008). *M. marinum* shares with *M. tuberculosis* the ability to infect host macrophages, as well as to induce tuberculous granulomas (Tobin and Ramakrishnan, 2008). Like most pathogenic mycobacteria, *M. marinum* survives within host macrophages at least in part by arresting phagosome maturation prior to phagolysosome fusion (Barker *et al.*, 1997; Rohde *et al.*, 2007). Yet *M. marinum* can also escape from the phagosome to the cytosol and initiate actin-based motility (Stamm *et al.*, 2003). The ESX (also called type VII) secretion system represents a specialized secretion pathway in mycobacteria (Bitter *et al.*, 2009). Mutations in the ESX-1 locus dramatically attenuates the virulence of *M. marinum* by abrogating phagosomal escape, cytosolic actin polymerization, and cell-to-cell spread (Stamm and Brown, 2004). While loss of ESX-1 activity also leads to failure of *M. tuberculosis* dissemination, the *in vivo* relevance of *M. tuberculosis* phagosomal escape remains controversial (van der Wel *et al.*, 2007).

Actin-based intra- and intercellular motility

After escaping from the phagosome to the cytosol, several intracellular bacterial pathogens share the ability to initiate actin-based motility (Gouin *et al.*, 2005). Actin polymerization propels the bacteria through the cytosol and into neighbouring cells. Direct spread from cell-to-cell allows these pathogens to circumvent host immune responses. All pathogens studied so far promote actin polymerization by exploiting the Arp2/3 complex (Fig. 11.2A), which is a conserved molecular machine used by host cells to nucleate actin filaments and organize them into branched arrays (Cossart, 2000).

L. monocytogenes

Actin assembly at the *L. monocytogenes* surface is mediated by ActA (Kocks *et al.*, 1992). ActA is necessary for bacterial motility in tissue culture cells, and is critical for pathogenesis in the murine model of infection (Brundage *et al.*, 1993). Moreover, ActA is sufficient for motility in the absence of other bacterial factors, as demonstrated by the observation that ActA-coated latex beads polymerize actin and undergo actin-based motility in cell extracts (Cameron *et al.*, 1999). The protein is enriched at the bacterial surface, where it displays a polarized distribution, with maximal expression at the pole where the comet tail forms (Kocks *et al.*, 1993). ActA acts as a nucleation promoting factor (NPF) for the Arp2/3 complex (Welch *et al.*, 1997). Interestingly, the N-terminal region of ActA contains consensus elements present in eukaryotic WASP family NPFs. Modulation and control of actin-based movements involve several other proteins such as capping protein which caps the barbed ends of actin filaments, profilin which helps to provide actin monomers to growing barbed ends, α-actinin which cross-links actin filaments, and VASP which binds to ActA and F-actin and modulates branch density and movement (Gouin *et al.*, 2005). Moving bacteria form comet tails consisting of filaments oriented with their fast-growing barbed ends facing the bacterial surface (Tilney *et al.*, 1992; Tilney and Portnoy, 1989), and the velocity of *Listeria* motility averages 10–15 μm/min (Dabiri *et al.*, 1990; Gouin *et al.*, 1999). The mechanistic role of motility in cell-to-cell spread has been confirmed by real time observation in tissue culture cells (Robbins *et al.*, 1999).

S. flexneri

Shigella escapes from the internalization vacuole and perform actin-based motility at rates of 15–20 μm/min (Goldberg and Theriot, 1995; Gouin *et al.*, 1999), slightly faster than *L. monocytogenes*. Actin polymerization is mediated by a bacterial protein, called IcsA (or VirG), which is both necessary and sufficient for this process (Goldberg and Theriot, 1995; Kocks *et al.*, 1995). IcsA is localized in the outer membrane in a polarized distribution, with the highest concentration at the pole where actin polymerizes (Goldberg *et al.*, 1993). In contrast to *Listeria* ActA, which evolved to mimic WASP family NPFs, IcsA has evolved the ability to bind the host NPF called N-WASP and recruit it to the bacterial surface (Egile *et al.*, 1999; Suzuki *et al.*, 1998). Host N-WASP co-localizes with IcsA at the bacterial pole from which actin polymerizes, and it is essential for S. flexneri motility in cell extracts and in infected cells (Lommel *et al.*, 2001; Suzuki *et al.*, 1998, 2002). Thus, *S. flexneri* has evolved a different mechanism for activating the Arp2/3 complex than *Listeria,* and it utilizes a distinct but overlapping set of actin binding proteins during motility (Gouin *et al.*, 2005). Recent work has shown that actin tail initiation by *S. flexneri* requires Toca-1 (transducer of Cdc42-dependent actin assembly) for the conversion of N-WASP from a closed inactive conformation to an open active one (Leung *et al.*, 2008). While N-WASP recruitment is dependent on IcsA, Toca-1 recruitment is instead mediated by TTSS effectors. Thus,

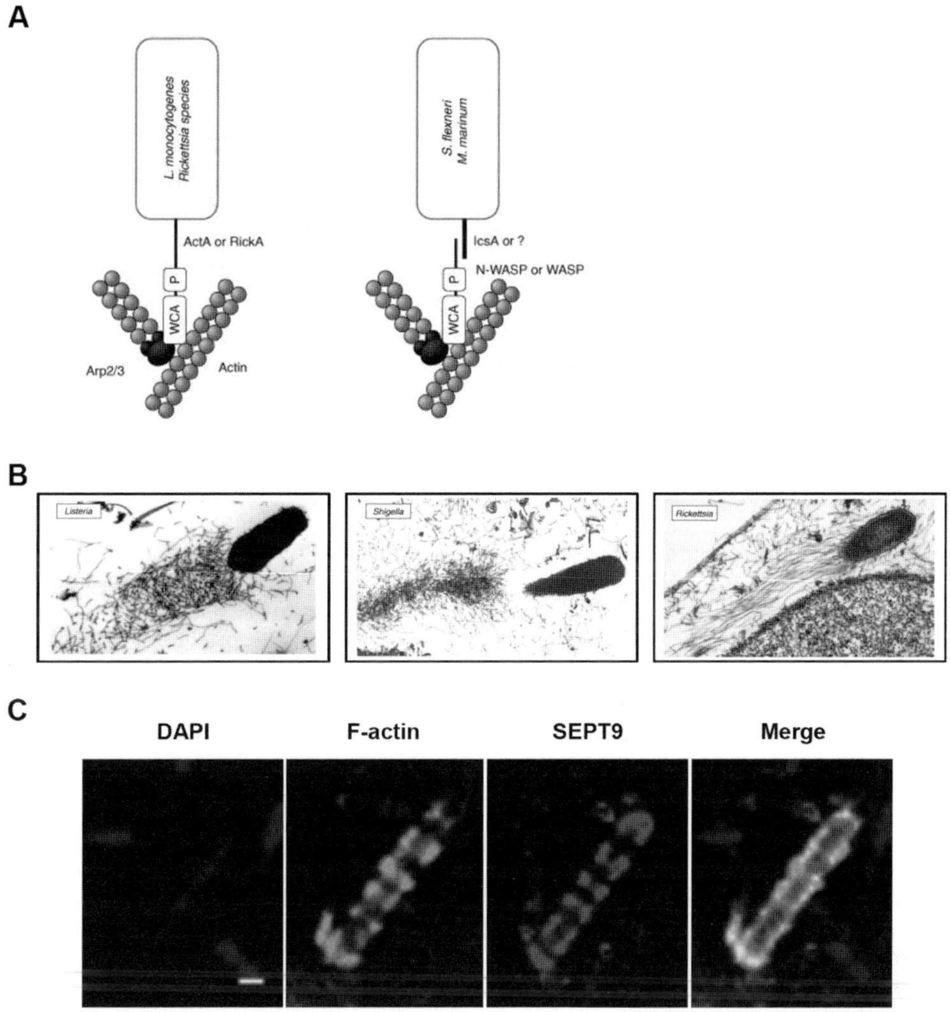

Figure 11.2 Actin tails and septin cages. (A) Schematic representation of the different types of Arp2/3 activation by *L. monocytogenes* and *Rickettsia* species (left panel) and *S. flexneri* and *M. marinum* (right panel). Most NPFs contain several conserved domains including a proline-rich region (P), a C-terminal WASP homology 2 (W) region that binds actin monomers, and C-terminal central and acidic regions (CA) that bind to the Arp2/3 complex. ActA of *L. monocytogenes* and RickA of *Rickettsia* species directly activate the Arp2/3 complex. IcsA of *Shigella* recruits N-WASP, which then activates the Arp2/3 complex. *M. marinum* recruits WASP, which then likely activates the Arp2/3 complex; however the mycobacterial protein(s) involved are unknown. These images are from Gouin *et al.* (2005). (B) Electron microscopy of actin tails. Unlike *Listeria* and *Shigella* tails, *Rickettsia* tails are made of long and unbranched filaments. These images are from Gouin *et al.* (2005). (C) Septin (SEPT9) recruitment to *S. flexneri* devoid of actin tails, i.e. the *Shigella*-septin cage. Of note, septins do not strictly colocalize with F-actin. Scale bar = 1 µm. These images are from Mostowy *et al.* (2010). A colour version of this figure is located in the plate section at the back of the book.

S. flexneri independently uses two nodes of the N-WASP actin assembly pathway to initiate actin tails. This mechanism is different from mechanisms evolved by other bacteria that polymerize actin tails.

Rickettsia species

After their escape into the cytosol, several *Rickettsia* species form actin tails made of long, unbranched actin filaments, which differ from those generated by ActA or IcsA/N-WASP (Fig.

11.2B) (Gouin *et al.*, 1999; Van Kirk *et al.*, 2000). The velocity of motility averages 5–8 μm/min, which is half the rate of *L. monocytogenes* (Gouin *et al.*, 1999). Genome sequence information revealed a gene in *R. conorii* that encodes a protein, called RickA, with sequence similarity to the human WASP family of NPFs (Gouin *et al.*, 2004). RickA is present in several spotted fever group *Rickettsia* species, but is absent from typhus group species such as *R. prowazekii* and *R. typhi* (Jeng *et al.*, 2004; McLeod *et al.*, 2004). RickA is expressed on the bacterial surface (Gouin *et al.*, 2004), but it lacks a signal sequence and the mode of its secretion is unknown. RickA behaves like *L. monocytogenes* ActA and other NPFs in its ability to activate the Arp2/3 complex *in vitro* (Gouin *et al.*, 2004; Jeng *et al.*, 2004). *Rickettsia* have also evolved a mechanism of polymerizing actin through functional mimicry of eukaryotic formin proteins by the bacterial surface protein Sca2 (Haglund *et al.*, 2010). The ability of Sca2 to nucleate long and unbranched filaments and protect them from capping *in vitro* corresponds with the long, parallel filament organization observed for *Rickettsia* tails. Thus, adaptation of a formin-dependent mechanism for actin assembly serves to distinguish *Rickettsia* from other intracellular pathogens which require the Arp2/3 complex for actin-based motility.

B. pseudomallei

Upon vacuolar escape to the cytosol, *B. pseudomallei* can induce actin polymerization and form actin tails (Rich *et al.*, 2003; Kespichayawattana *et al.*, 2000), and cell-to-cell spread has been demonstrated in tissue culture models of infection (Kespichayawattana *et al.*, 2000; Breitbach *et al.*, 2003). However, it is not clear whether *B. pseudomallei* expresses its own NPF for the Arp2/3 complex (like *L. monocytogenes*), recruits a host NPF to its surface (like *S. flexneri*), or uses an alternative mechanism for actin polymerization. Analysis of the *B. pseudomallei* genome sequence has led to the identification of BimA, a polarly distributed surface protein with sequence similarities to host and bacterial NPFs (Stevens *et al.*, 2005). In agreement with the possibility that BimA may act as an NPF mimic, bacteria do not utilize the host NPF N-WASP, as actin tails form normally in N-WASP-deficient cells (Breitbach *et al.*, 2003; Stevens *et al.*, 2005). However, work is still required to demonstrate whether the mechanism of actin polymerization by *B. pseudomallei* is similar to other pathogens, or whether it uses a so far undescribed mechanism of actin-based motility.

M. marinum

M. marinum has been shown to initiate actin polymerization by recruiting WASP, a host NPF that is specifically expressed in macrophages and other blood cells (Stamm *et al.*, 2003, 2005). WASP is concentrated at the bacterial pole where actin polymerizes, suggesting a role in *M. marinum* motility. Similarly, the Arp2/3 complex is localized throughout actin tails, as is the actin-binding protein VASP. Although these observations strongly suggest that *M. marinum* uses WASP and the Arp2/3 complex to nucleate actin assembly during motility, this hypothesis has not yet been directly tested. Nevertheless, *M. marinum* appears to have evolved the same mechanism as *S. flexneri* for promoting actin nucleation by recruiting a host NPF to the bacterial surface. Once in the cytosol, the velocity of *M. marinum* motility averages 11 μm/min (Stamm *et al.*, 2003).

Interestingly, another actin-based mode of mycobacterial spread has been recently described (Hagedorn *et al.*, 2009). Using the social amoeba *Dictyostelium* as a host, it has been discovered that *M. marinum* and *M. tuberculosis* are ejected from the cell through an actin-based structure called the mycobacteria ‘ejectosome’. Similar to actin tail polymerization, pathogenic mycobacteria require an intact ESX-1 secretion system to employ this conserved non-lytic spreading mechanism.

Cytoskeleton elements that interfere with actin-based motility: microtubules and septins

Despite the *in vitro* reconstitution of the actin tail with a restricted number of cytoskeleton proteins (Loisel *et al.*, 1999), movements of pathogenic bacteria may be influenced by a variety of cytoskeletal elements (Van Troys *et al.*, 2008). How these other cytoskeleton elements regulate actin-based motility in cellular environments is considerably less defined, yet this understanding will be imperative for the complete picture of infectious processes *in vivo*.

Pathogenic bacteria can modulate the dynamics of host cell microtubules during infection (Yoshida and Sasakawa, 2003). In the case of *Shigella*, bacteria have evolved a mechanism to destroy surrounding microtubules, and thus facilitate cytosolic movement, by secreting VirA, a TTSS effector protein (Yoshida *et al.*, 2006). VirA was originally proposed to be a protease that specifically cleaves α-tubulin, a major component of microtubules, and creates a tunnel inside the cytosol by breaking down the microtubule infrastructure. However, more recent structural and functional studies have indicated that VirA is not a protease, and does not directly destabilize microtubules (Germane *et al.*, 2008). The precise role of VirA in actin-based motility and microtubule destabilization therefore requires further investigation.

In the case of *Listeria*, ActA has been shown to recruit LaXp180 (Pfeuffer *et al.*, 2000). Interestingly, LaXp180 (or CC1) has been reported to be a binding partner of stathmin, a microtubule-sequestering protein of the host cell (Maucuer *et al.*, 1995). Whether an ActA–LaXp180–stathmin interaction destabilizes microtubules and helps bacterial movement in the cytosol has not been formally tested.

Septins are GTP-binding proteins essential for cell division that assemble as non-polar filaments (Sirajuddin *et al.*, 2007); yet their role as an unconventional component of the cytoskeleton remains poorly understood. As shown recently, septins prevent actin-tail formation and restrict the cell-to-cell spread of *Shigella* by entrapping actin-polymerizing bacteria in cage-like structures (Fig. 11.2C) (Mostowy *et al.*, 2010). Strikingly, *Shigella* either form actin tails or become compartmentalized in septin cages. Inactivation of septin caging increases *Shigella* actin tail formation and cell-to-cell spread. In contrast, stimulation of septin caging by a host factor produced during *Shigella* infection, TNF-α, prevents actin tail formation and cell-to-cell spread. These results have revealed septin caging as an unsuspected strategy of host defence to immobilize actin polymerizing bacteria, and ultimately restrict the dissemination of invasive pathogens.

Bacterial replication in the cytosol

Is the cytosol permissive for bacterial growth? The answer to this question is not yet clear, in part because the content of the cytosol and its variation in different cell types is poorly understood.

The identification of bacterial genes and growth requirements that are important for intracytosolic replication has provided some insight on nutrient availability in the cytosol (Ray *et al.*, 2009). To determine which nutrients are limiting in the cytosol, studies have used auxotrophic mutants that are defective for replication. Pathogens that lack an uptake system for a particular nutrient have also been used. In the case of *Listeria*, studies with auxotrophic mutants have suggested that aromatic amino acids, threonine, and adenine are limiting for cytosolic replication (Marquis *et al.*, 1993; Stritzker *et al.*, 2004). *L. monocytogenes* use a sugar uptake system, mediated by Hpt, which is required for intracellular growth (Chico-Calero *et al.*, 2002). In the case of *Shigella*, mutants that are unable to synthesize guanine (Noriega *et al.*, 1996), thymine (Cersini *et al.*, 2003), and *p*-aminobenzoic acid (Cersini *et al.*, 1998) fail to replicate intracellularly. *Shigella* that are auxotrophic for diaminopimelate are also unable to survive within the cytosol (Cersini *et al.*, 1998). In depth studies are still required to reveal a more complete picture of cytosolic constituents and the nutritional requirements of cytosolic pathogens.

Bacterial growth in the cytosol can also be dependent on the cell type (Goebel and Kuhn, 2000; O'Riordan and Portnoy, 2002). For example, studies have shown that *Salmonella* Typhimurium SifA [a TTSS effector required for maintenance of the *Salmonella*-containing vacuole (SCV)] mutants can replicate in the cytosol of epithelial cells (Beuzón *et al.*, 2000; Brumell *et al.*, 2002) but fail to multiply within macrophages (Brumell *et al.*, 2001; Beuzón *et al.*, 2002). This may in part explain why cytosolic bacteria occupy different cellular niches within the human host (Ray *et al.*, 2009). In this regard, it is interesting to consider that *F. tularensis* resides preferentially within macrophages (Santic *et al.*, 2006; Oyston, 2008), *Rickettsia* species replicate within endothelial cells (Walker, 2007), and *S. flexneri* grows inside human intestinal epithelial cells (Sansonetti, 2001). In contrast, both *L. monocytogenes* and

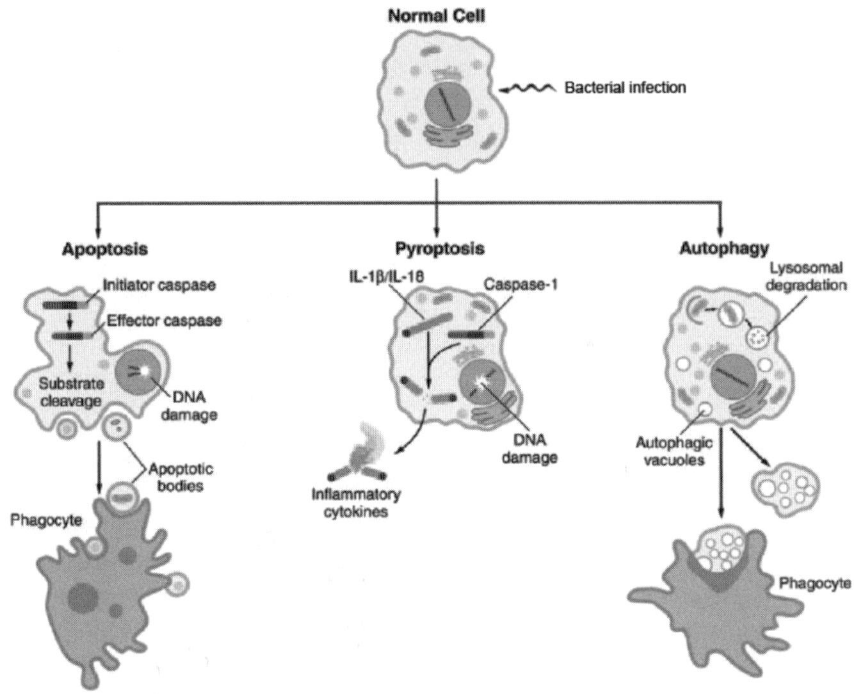

Figure 11.3 Cell responses to bacterial infection. Normal, healthy cells respond to bacterial infection by initiating a variety of molecular pathways. Apoptosis, a pathway leading to cell death, starts by the activation of initiator caspases that activate effector caspases to cleave cellular substrates. Apoptotic cells demonstrate cytoplasmic and nuclear condensation, DNA damage, formation of apoptotic bodies, maintenance of an intact plasma membrane, and exposure of surface molecules targeting cells for phagocytosis. Pyroptosis, a pro-inflammatory cell death pathway, is mediated by the activation of the inflammasome and consequently caspase-1, a protease that also activates the inflammatory cytokines, IL-1β, and IL-18. Pyroptosis features cell lysis and release of inflammatory cellular contents. Autophagy is a process in which the cell sequesters cytosolic constituents in a double membrane vesicle and delivers them to the lysosome for degradation. *In vivo*, cells undergoing autophagy can be phagocytized by neighbouring cells. These images are adapted from Fink and Cookson (2005). A colour version of this figure is located in the plate section at the back of the book.

B. pseudomallei can infect a range of phagocytic and non-phagocytic human cells during their life cycles (Wiersinga *et al.*, 2006; Cossart and Toledo-Arana, 2008).

Bacterial interaction with host defences

Once bacteria escape to the cytosol, they can interact with a variety of host cell defence mechanisms. Recognition by innate immune sensors can trigger apoptosis, pyroptosis, or autophagy (Fig. 11.3). Here, we will define these cell responses to intracellular pathogens and illustrate their regulation in response to bacterial infection.

Intracellular receptors

The innate immune system comprises several classes of pattern-recognition receptors, including Toll-like receptors (TLRs) and nucleotide binding oligomerization domain (NOD)-like receptors (NLRs) (Inohara *et al.*, 2005; Kawai and Akira, 2006, 2010). TLRs recognize microbes on the cell surface and in endosomes, whereas NLRs sense microbial molecules in the cytosol. In humans, the NLR family is made up of 23 cytosolic proteins characterized by the presence of a conserved NOD domain and leucine-rich repeats (LRRs) (Inohara *et al.*, 2005). The most studied NLRs are Nod1 and Nod2. Nod1 is prevalent in intestinal epithelial cells and shows specific recognition for muropeptides originating from the peptidoglycan of Gram-negative microorganisms (Chamaillard *et al.*, 2003; Girardin *et al.*, 2003a). Nod2 is a cytosolic receptor that recognizes peptidoglycans from any bacterial species (Girardin *et al.*, 2003b; Inohara *et al.*, 2003), particularly mycobacterial *N*-glycolyl muramyl dipeptide (Coulombe *et al.*, 2009).

Upon ligand recognition, Nod1 and Nod2 drive the activation of mitogen-activated protein kinase (MAPK) and NF-κB innate response. In contrast, other NLRs, namely Ipaf, Nalp1b, and Cryopyrin/Nalp3, are involved in the assembly of inflammasomes that are required for the activation of caspase-1. Mutations in NLR members, including Nod2 and Cryopyrin, are associated with the development of inflammatory disorders (Franchi *et al.*, 2008; Shaw *et al.*, 2008).

Apoptosis and pyroptosis

Cell death is a crucial process during development, homeostasis, and immune regulation, and its dysregulation is associated with several pathologies (Fink and Cookson, 2005; Galluzzi *et al.*, 2007). In particular, apoptosis is a cell death pathway that results in the controlled breakdown of the cell into apoptotic bodies, which are subsequently recognized and engulfed by surrounding cells and phagocytes (Samali *et al.*, 1999; Albert, 2004). Two main evolutionarily conserved protein families are involved in apoptosis, namely the Bcl-2 (B-cell lymphoma 2) family of proteins which control mitochondrial integrity (Youle and Strasser, 2008), and the caspases (i.e. cysteinyl aspartate-specific proteases) which mediate the execution phase of apoptosis (Fuentes-Prior and Salvesen, 2004). During bacterial infection, apoptosis of host cells functions as a defence mechanism by destroying the site of pathogen replication. However, pathogens have evolved several ways to prevent apoptosis, such as by protecting the mitochondria and preventing cytochrome c release, by activating cell survival pathways, or by preventing caspase activation (Faherty and Maurelli, 2008). Alternatively, pathogens can also benefit from apoptosis, either by inducing apoptosis of infected cells and thereby favouring systemic infection, or by killing uninfected immune cells and thereby evading the host defence (Philpott *et al.*, 2001).

Pyroptosis is a recently recognized form of regulated cell death, with morphological and biochemical properties distinct from apoptosis (Bergsbaken *et al.*, 2009). In contrast to the non-inflammatory nature and phagocytic uptake of membrane-bound apoptotic bodies that characterizes apoptosis (Albert, 2004), pyroptosis features rapid plasma-membrane rupture and release of pro-inflammatory intracellular contents. Pyroptosis is associated with antimicrobial responses during inflammation, where caspase-1 is activated by an inflammasome here called the pyroptosome (Fink and Cookson, 2007; Bergsbaken *et al.*, 2009). As caspase-1-deficient mice are more susceptible to a number of bacterial infections (Bergsbaken *et al.*, 2009), pyroptosis is considered a generalized protective host response to infection.

Autophagy

Autophagy is a general term for processes by which cytoplasmic materials reach lysosomes for degradation (Rubinsztein, 2006; Levine and Kroemer, 2008; Mizushima *et al.*, 2008). Among the three types of autophagy (macroautophagy, microautophagy, and chaperone-mediated autophagy), the most extensively studied is macroautophagy (Mizushima *et al.*, 2010). Chaperone-mediated autophagy involves the direct translocation of cytosolic proteins across the lysosomal membrane, which requires protein unfolding by chaperone proteins. Microautophagy involves inward invagination of lysosomal membrane, which delivers a small portion of cytoplasm into the lysosomal lumen. Macroautophagy (herein referred to as autophagy) is an evolutionarily conserved catabolic pathway that allows eukaryotes to degrade and recycle cellular components by sequestering proteins and organelles in specialized double-membrane vesicles designated autophagosomes (Klionsky and Emr, 2000; Levine and Deretic, 2007; Münz, 2009). So far, over 30 autophagy-related (Atg) genes have been identified in yeast, and several mammalian homologues have been isolated and functionally characterized (Levine and Klionsky, 2004; He and Klionsky, 2009). Basal levels of autophagy ensure the maintenance of intracellular homeostasis, and in addition, many studies have revealed its diverse functions in important cellular processes such as cellular stress, differentiation, development, and longevity (Levine and Klionsky, 2004; Vellai *et al.*, 2009). Not surprisingly, autophagy is linked to a wide range of disease states (Shintani and Klionsky, 2004; Levine and Kroemer, 2008; Mizushima *et al.*, 2008). Recent evidence has implicated the autophagic pathway in pathogenesis (Levine and Deretic, 2007), and ubiquitylation has emerged as central for autophagic recognition of intracytosolic pathogens (Thurston *et al.*, 2009; Yoshikawa *et al.*, 2009), i.e. ubiquitin-associated bacteria recognized by the autophagy machinery are trapped by autophagosomal membrane for ultimate delivery into the lytic compartment where they undergo degradation by autolysosomes. The autophagy receptors p62 (SQSTM1) (Pankiv *et al.*, 2007) and NDP52 (nuclear dot protein 52 kDa) (Thurston *et al.*, 2009) are proteins directly involved in detecting ubiquitylated bacteria and targeting them to autophagy.

Cell responses during bacterial infection

In the following section, we have selected certain pathogens to illustrate host immune responses during cytosolic invasion. In addition, we highlight the mechanisms evolved by cytosolic bacteria to evade host recognition and elimination.

S. Typhimurium

Salmonella induces intestinal epithelial cell death via apoptosis (Guiney, 2005). Infected, apoptotic cells are subsequently engulfed by incoming macrophages, thereby enabling *Salmonella* to spread from cell to cell. In contrast, *Salmonella* infection of macrophages triggers pyroptosis (Bergsbaken *et al.*, 2009). Recent studies have revealed that, in response to cellular infection by *Salmonella*, caspase-1 activation is mediated by the NLR Ipaf and the adaptor ASC, through the detection of bacterial flagellin (Mariathasan *et al.*, 2004; Franchi *et al.*, 2006; Miao *et al.*, 2006). Caspase-1 deficiency is associated with increased susceptibility to *Salmonella* (Lara-Tejero *et al.*, 2006), suggesting that Ipaf activation induces a host response that confers protection.

Following *Salmonella* invasion of host cells, most bacteria reside and replicate within the SCV. However, it is now appreciated that a significant fraction (~25%) of S. Typhimurium become cytosolic and can be targeted to autophagy (Perrin *et al.*, 2004; Thurston *et al.*, 2009). Recent work has demonstrated that NDP52 acts as an autophagy receptor that, together with a complex containing Tank-binding kinase (TBK1), can restrict the proliferation of ubiquitin-coated *Salmonella* in the cytosol (Thurston *et al.*, 2009). The unique features of NDP52 that are important for detecting cytosolic, ubiquitylated bacteria and activating autophagy are unknown.

Mycobacteria species

The pathogenicity of mycobacteria is largely attributed to its ability to survive within macrophages by arresting phagosomal maturation (Armstrong and Hart, 1971), although the strict confinement of mycobacteria within the phagosome has recently been challenged (van der Wel *et al.*, 2007). Studies have shown that TLRs and Nod2 mediate non-redundant recognition of pathogenic mycobacteria (Ferwerda *et al.*, 2005, 2007). Recent work has shown that Nod2 is particularly sensitive to mycobacterial peptidoglycan, suggesting that Nod2 may have evolved to favour this enhanced sensitivity (Coulombe *et al.*, 2009). To our knowledge, a Nod2-mediated immune response in response to mycobacteria having escaped from the vacuole has not yet been investigated.

In contrast to the well-documented resistance of *M. tuberculosis* to phagosome maturation and microbicidal effectors, several studies have now shown that autophagy can efficiently eliminate intracellular *M. tuberculosis* (Gutierrez *et al.*, 2004; Alonso *et al.*, 2007; Floto *et al.*, 2007; Xu *et al.*, 2007; Biswas *et al.*, 2008). It is thought that delivery of cytosolic components by the autophagy receptor p62 endows autophagosomes with unique antimicrobial properties to kill *M. tuberculosis* (Ponpuak *et al.*, 2010). On the other hand, it has recently been shown that autophagy modulation in macrophage cells is a general strategy employed by multiple types of *M. tuberculosis* isolates to evade the host immune system (Kumar *et al.*, 2010).

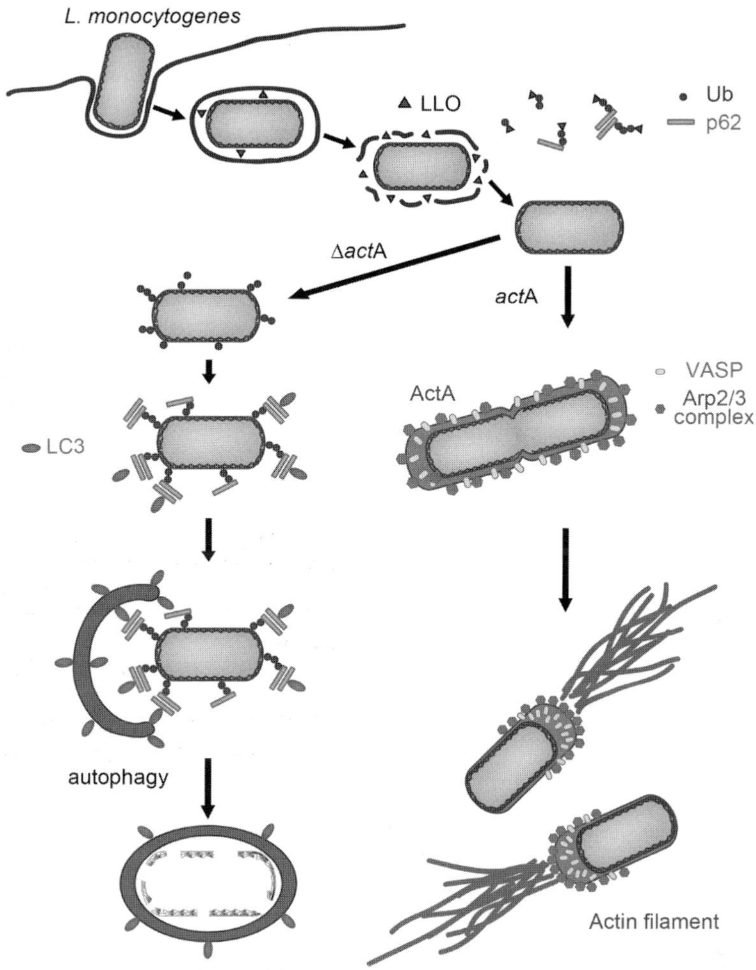

Figure 11.4 A model for autophagic recognition of *L. monocytogenes* lacking ActA. During intracellular growth, cytoplasmic *Listeria* expressing ActA can move by nucleating actin polymerization, while *Listeria* lacking ActA expression are ubiquitinated. This is followed by p62 binding to ubiquitin (Ub) via its ubiquitin-associated (UBA) domain in the vicinity of the bacterial surface. p62 then binds to LC3, an autophagosomal marker, leading to autophagosome formation around the bacterium. This figure is reproduced with permission from Yoshikawa *et al.* (2009). A colour version of this figure is located in the plate section at the back of the book.

S. flexneri

The intestinal epithelial barrier represents the first line of defence against *Shigella*. Experiments in human intestinal cell lines have shown that sensing of *Shigella* is primarily mediated by Nod1, which is required for the activation of NF-κB, c-Jun N-terminal kinase (JNK), and the secretion of the pro-inflammatory cytokine IL-8 (Girardin *et al.*, 2001). Macrophage is another component in the host defence against *Shigella*. Upon infection of macrophages, cytosolic *Shigella* triggers the activation of caspase-1 and the induction of pyroptosis (Hilbi *et al.*, 1998). Ipaf is the critical NLR responsible for caspase-1 activation in *Shigella*-infected macrophages (Suzuki *et al.*, 2007).

Autophagy is also induced by the presence of cytoplasmic *Shigella* (Ogawa *et al.*, 2005; Dupont *et al.*, 2009; Travassos *et al.*, 2009). In epithelial cells infected with *Shigella*, autophagy is triggered by the recognition of IcsA by Atg5, a protein critical for autophagome maturation (Ogawa *et al.*, 2005). The bacterium avoids the autophagic response via the bacterial effector protein IcsB, which prevents the recruitment of Atg5 to IcsA. In macrophages, however, the induction of autophagy occurs independently of IcsA, and is inhibited by Ipaf and caspase-1 (Suzuki *et al.*, 2007). Thus, NLR proteins not only play a role in the induction of the inflammatory response, but also appear to regulate the autophagic response, implicating an important intersection between these two host response pathways (Suzuki *et al.*, 2007; Cooney *et al.*, 2009; Travassos *et al.*, 2009).

L. monocytogenes

In vitro studies performed in several cell types have demonstrated the role of Nod1 in host immune responses to *Listeria* infection. In endothelial cells, invasive *Listeria* can induce activation of MAPK and IL-8 secretion via Nod1 (Opitz *et al.*, 2006). In macrophages, there is evidence for a redundant role for TLRs and Nod1/Nod2 in cytokine production induced by *Listeria* (Park *et al.*, 2007). Within macrophage vacuoles, an *N*-deacetylation mutant is rapidly destroyed and induces a massive IFN-β response in a TLR2 and Nod1-dependent manner, suggesting that peptidoglycan *N*-deacetylation is a highly efficient mechanism used by *Listeria* to evade innate host defences (Boneca *et al.*, 2007). LLO-dependent pore formation at the cell membrane (Hamon and Cossart, 2011) and cytosolic invasion by *Listeria* can induce caspase-1 activation (Mariathasan *et al.*, 2006; Ozören *et al.*, 2006; Franchi *et al.*, 2007), however the specific NLR involved in the regulation of caspase-1 activation remains controversial (Mariathasan *et al.*, 2004; Franchi *et al.*, 2006).

A variety of studies noticed that unless bacteria were forming actin tails, autophagic markers can accumulate around intracytosolic *L. monocytogenes* (Birmingham *et al.*, 2007; Rich *et al.*, 2003; Perrin *et al.*, 2004). Consequently, it was predicted that *L. monocytogenes* actin-tail formation helps the bacterium to evade ubiquitylation and subsequent autophagic recognition. The mechanism to explain how *Listeria* evades ubiquitin recognition and autophagy by expressing ActA has recently been described (Yoshikawa *et al.*, 2009): ActA disguises *Listeria* from ubiquitylation and autophagic recognition independent of its ability to mediate bacterial motility (Fig. 11.4). An unsuspected theme from studies of *Listeria* and *Shigella* is that proteins involved in actin-based motility are also connected with autophagy recognition. It will thus be of great interest to further study the link between actin polymerization, autophagy, and pathogen dissemination.

Conclusion and perspectives

An understanding of pathogenic mechanisms to escape and live in the cytosol is starting to emerge. A major issue is to validate the molecular and cellular events analysed *in vitro* during bacterial infection *in vivo* using relevant animal models. New microscopic techniques for improved visualization are required to improve the sensitivity and spatio-temporal resolution of events that take place during the infectious process. This information should provide vital clues towards understanding bacterial disease and for illuminating new therapeutic strategies. As a consequence, fascinating insights into the interaction between pathogens and host cells shall certainly continue.

References

Albert, M.L. (2004). Death-defying immunity: do apoptotic cells influence antigen processing and presentation? Nat. Rev. Immunol. *4*, 223–231.

Alonso, S., Pethe, K., Russell, D.G., and Purdy, G.E. (2007). Lysosomal killing of *Mycobacterium* mediated by ubiquitin-derived peptides is enhanced by autophagy. Proc Natl. Acad. Sci. U.S.A. *104*, 6031–6036.

Armstrong, J.A., and Hart, P.D. (1971). Response of cultured macrophages to *Mycobacterium tuberculosis*, with observations on fusion of lysosomes with phagosomes. J. Exp. Med. *134*, 713–740.

Barker, L.P., George, K.M., Falkow, S., and Small, P.L. (1997). Differential trafficking of live and dead *Mycobacterium marinum* organisms in macrophages. Infect. Immun. *65*, 1497–1504.

Bergsbaken, T., Fink, S.L., and Cookson, B.T. (2009). Pyroptosis: host cell death and inflammation. Nat. Rev. Microbiol. 7, 99–109.

Beuzón, C.R., Méresse, S., Unsworth, K.E., Ruíz-Albert, J., Garvis, S., Waterman, S.R., Ryder, T.A., Boucrot, E., and Holden, D.W. (2000). *Salmonella* maintains the integrity of its intracellular vacuole through the action of SifA. EMBO J. *19*, 3235–3249.

Beuzón, C.R., Salcedo, S.P., and Holden, D.W. (2002). Growth and killing of a *Salmonella enterica* serovar Typhimurium *sifA* mutant strain in the cytosol of different host cell lines. Microbiology *148*, 2705–2715.

Birmingham, C.L., Canadien, V., Gouin, E., Troy, E.B., Yoshimori, T., Cossart, P., Higgins, D.E., and Brumell, J.H. (2007). *Listeria monocytogenes* evades killing by autophagy during colonization of host cells. Autophagy 3, 442–451.

Birmingham, C.L., Canadien, V., Kaniuk, N.A., Steinberg, B.E., Higgins, D.E., and Brumell, J.H. (2008). Listeriolysin O allows *Listeria monocytogenes* replication in macrophage vacuoles. Nature *451*, 350–354.

Biswas, D., Qureshi, O.S., Lee, W.Y., Croudace, J.E., Mura, M., and Lammas, D.A. (2008). ATP-induced autophagy is associated with rapid killing of intracellular mycobacteria within human monocytes/macrophages. BMC Immunol. *9*, 35.

Bitter, W., Houben, E.N., Bottai, D., Brodin, P., Brown, E.J., Cox, J.S., Derbyshire, K., Fortune, S.M., Gao, L.Y., Liu, J., *et al.* (2009). Systematic genetic nomenclature for type VII secretion systems. PLoS Pathog. *5*, e1000507.

Boneca, I.G., Dussurget, O., Cabanes, D., Nahori, M.A., Sousa, S., Lecuit, M., Psylinakis, E., Bouriotis, V., Hugot, J.P., Giovannini, M., *et al.* (2007). A critical role for peptidoglycan N-deacetylation in *Listeria* evasion from the host innate immune system. Proc. Natl. Acad. Sci. U.S.A. *104*, 997–1002.

Breitbach, K., Rottner, K., Klocke, S., Rohde, M., Jenzora, A., Wehland, J., and Steinmetz, I. (2003). Actin-based motility of *Burkholderia pseudomallei* involves the Arp 2/3 complex, but not N-WASP and Ena/VASP proteins. Cell. Microbiol. *5*, 385–393.

Brumell, J.H., Rosenberger, C.M., Gotto, G.T., Marcus, S.L., and Finlay, B.B. (2001). SifA permits survival and replication of *Salmonella typhimurium* in murine macrophages. Cell. Microbiol. *3*, 75–84.

Brumell, J.H., Tang, P., Zaharik, M.L., and Finlay, B.B. (2002). Disruption of the *Salmonella*-containing vacuole leads to increased replication of *Salmonella enterica* serovar Typhimurium in the cytosol of epithelial cells. Infect. Immun. *70*, 3264–3270.

Brundage, R.A., Smith, G.A., Camilli, A., Theriot, J.A., and Portnoy, D.A. (1993). Expression and phosphorylation of the *Listeria monocytogenes* ActA protein in mammalian cells. Proc. Natl. Acad. Sci. U.S.A. *90*, 11890–11894.

Cameron, L.A., Footer, M.J., van Oudenaarden, A., and Theriot, J.A. (1999). Motility of ActA protein-coated microspheres driven by actin polymerization. Proc. Natl. Acad. Sci. U.S.A. *96*, 4908–4913.

Cersini, A., Salvia, A.M., and Bernardini, M.L. (1998). Intracellular multiplication and virulence of *Shigella flexneri* auxotrophic mutants. Infect. Immun. *66*, 549–557.

Cersini, A., Martino, M.C., Martini, I., Rossi, G., and Bernardini, M.L. (2003). Analysis of virulence and inflammatory potential of *Shigella flexneri* purine biosynthesis mutants. Infect. Immun. *71*, 7002–7013.

Chamaillard, M., Hashimoto, M., Horie, Y., Masumoto, J., Qiu, S., Saab, L., Ogura, Y., Kawasaki, A., Fukase, K., Kusumoto, S., *et al.* (2003). An essential role for NOD1 in host recognition of bacterial peptidoglycan containing diaminopimelic acid. Nat. Immunol. *4*, 702–707.

Checroun, C., Wehrly, T.D., Fischer, E.R., Hayes, S.F., and Celli, J. (2006). Autophagy-mediated reentry of *Francisella tularensis* into the endocytic compartment after cytoplasmic replication. Proc. Natl. Acad. Sci. U.S.A. *103*, 14578–14583.

Chico-Calero, I., Suárez, M., González-Zorn, B., Scortti, M., Slaghuis, J., Goebel, W., and Vázquez-Boland, J. (2002). Hpt, a bacterial homolog of the microsomal glucose- 6-phosphate translocase, mediates rapid intracellular proliferation in *Listeria*. Proc. Natl. Acad. Sci. U.S.A. *99*, 431–436.

Chong, A., Wehrly, T.D., Nair, V., Fischer, E.R., Barker, J.R., Klose, K.E., and Celli, J. (2008). The early phagosomal stage of *Francisella tularensis* determines optimal phagosomal escape and *Francisella* pathogenicity island protein expression. Infect. Immun. *76*, 5488–5499.

Cooney, R., Baker, J., Brain, O., Danis, B., Pichulik, T., Allan, P., Ferguson, D.J., Campbell, B.J., Jewell, D., and Simmons, A. (2009). NOD2 stimulation induces autophagy in dendritic cells influencing bacterial handling and antigen presentation. Nat. Med. *16*, 90–97.

Cossart, P. (2000). Actin-based motility of pathogens: the Arp2/3 complex is a central player. Cell. Microbiol 2, 195–205.

Cossart, P., and Toledo-Arana, A. (2008). *Listeria monocytogenes*, a unique model in infection biology: an overview. Microbes Infect. *10*, 1041–1050.

Coulombe, F., Divangahi, M., Veyrier, F., de Léséleuc, L., Gleason, J.L., Yang, Y., Kelliher, M.A., Pandey, A.K., Sassetti, C.M., Reed, M.B., *et al.* (2009). Increased NOD2-mediated recognition of N-glycolyl muramyl dipeptide. J. Exp. Med. *206*, 1709–1716.

Dabiri, G.A., Sanger, J.M., Portnoy, D.A., and Southwick, F.S. (1990). *Listeria monocytogenes* moves rapidly through the host-cell cytoplasm by inducing directional actin assembly. Proc. Natl. Acad. Sci. U.S.A. *87*, 6068–6072.

Dupont, N., Lacas-Gervais, S., Bertout, J., Paz, I., Freche, B., Tran Van Nhieu, G., van der Goot, F.G., Sansonetti, P.J., and Lafont, F. (2009). *Shigella* phagocytic vacuolar membrane remnants participate in the cellular response to pathogen invasion and are regulated by autophagy. Cell Host Microbes *6*, 137–149.

Egile, C., Loisel, T.P., Laurent, V., Li, R., Pantaloni, D., Sansonetti, P.J., and Carlier, M.F. (1999). Activation of the CDC42 effector N-WASP by the *Shigella flexneri* IcsA protein promotes actin nucleation by Arp2/3 complex and bacterial actin-based motility. J. Cell Biol. *146*, 1319–1332.

Faherty, C.S., and Maurelli, A.T. (2008). Staying alive: bacterial inhibition of apoptosis during infection. Trends Microbiol. *16*, 173–180.

Fernandez-Prada, C.M., Hoover, D.L., Tall, B.D., Hartman, A.B., Kopelowitz, J., and Venkatesan, M.M. (2000).

Shigella flexneri IpaH$_{7.8}$ facilitates escape of virulent bacteria from the endocytic vacuoles of mouse and human macrophages. Infect. Immun. *68*, 3608–3619.

Ferwerda, G., Girardin, S.E., Kullberg, B.J., Le Bourhis, L., de Jong, D.J., Langenberg, D.M., van Crevel, R., Adema, G.J., Ottenhoff, T.H., Van der Meer, J.W., *et al.* (2005). NOD2 and Toll-like receptors are nonredundant recognition systems of *Mycobacterium tuberculosis*. PLoS Pathog. *1*, e34.

Ferwerda, G., Kullberg, B.J., de Jong, D.J., Girardin, S.E., Langenberg, D.M., van Crevel, R., Ottenhoff, T.H., Van der Meer, J.W., and Netea, M.G. (2007). *Mycobacterium paratuberculosis* is recognized by Toll-like receptors and NOD2. J. Leukoc. Biol. *82*, 1011–1018.

Fink, S.L., and Cookson, B.T. (2005). Apoptosis, pyroptosis, and necrosis: mechanistic description of dead and dying eukaryotic cells. Infect. Immun. *73*, 1907–1916.

Fink, S.L., and Cookson, B.T. (2007). Pyroptosis and host cell death responses during *Salmonella* infection. Cell. Microbiol. *9*, 2562–2570.

Floto, R.A., Sarkar, S., Perlstein, E.O., Kampmann, B., Schreiber, S.L., and Rubinsztein, D.C. (2007). Small molecule enhancers of rapamycin-induced TOR inhibition promote autophagy, reduce toxicity in Huntington's disease models and enhance killing of mycobacteria by macrophages. Autophagy 3, 620–622.

Franchi, L., Amer, A., Body-Malapel, M., Kanneganti, T.D., Ozóren, N., Jagirdar, R., Inohara, N., Vandenabeele, P., Bertin, J., Coyle, A., *et al.* (2006). Cytosolic flagellin requires Ipaf for activation of caspase-1 and interleukin 1β in *Salmonella*-infected macrophages. Nat. Immunol. *7*, 576–582.

Franchi, L., Kanneganti, T.D., Dubyak, G.R., and Núñez, G. (2007). Differential requirement of P2X7 receptor and intracellular K^+ for caspase-1 activation induced by intracellular and extracellular bacteria. J. Biol. Chem. *282*, 18810–18818.

Franchi, L., Park, J.H., Shaw, M.H., Marina-Garcia, N., Chen, G., Kim, Y.G., and Núñez, G. (2008). Intracellular NOD-like receptors in innate immunity, infection and disease. Cell. Microbiol. *10*, 1–8.

Fuentes-Prior, P., and Salvesen, G.S. (2004). The protein structures that shape caspase activity, specificity, activation and inhibition. Biochem. J. *384*, 201–232.

Galluzzi, L., Maiuri, M.C., Vitale, I., Zischka, H., Castedo, M., Zitvogel, L., and Kroemer, G. (2007). Cell death modalities: classification and pathophysiological implications. Cell Death Differ. *14*, 1237–1243.

Germane, K.L., Ohi, R., Goldberg, M.B., and Spiller, B.W. (2008). Structural and functional studies indicate that *Shigella* VirA is not a protease and does not directly destabilize microtubules. Biochemistry *47*, 10241–10243.

Girardin, S.E., Tournebize, R., Mavris, M., Page, A.L., Li, X., Stark, G.R., Bertin, J., DiStefano, P.S., Yaniv, M., Sansonetti, P.J., *et al.* (2001). CARD4/Nod1 mediates NF-κB and JNK activation by invasive *Shigella flexneri*. EMBO Rep. 2, 736–742.

Girardin, S.E., Boneca, I.G., Carneiro, L.A., Antignac, A., Jéhanno, M., Viala, J., Tedin, K., Taha, M.K., Labigne, A., Zäthringer, U., *et al.* (2003a). Nod1 detects a unique muropeptide from Gram-negative bacterial peptidoglycan. Science *300*, 1584–1587.

Girardin, S.E., Boneca, I.G., Viala, J.R., Chamaillard, M., Labigne, A.S., Thomas, G., Philpott, D.J., and Sansonetti, P.J. (2003b). Nod2 is a general sensor of peptidoglycan through muramyl dipeptide (MDP) detection. J. Biol. Chem. *278*, 8869–8872.

Goebel, W., and Kuhn, M. (2000). Bacterial replication in the host cell cytosol. Curr. Opin. Microbiol. 3, 49–53.

Goldberg, M.B., Bârzu, O., Parsot, C., and Sansonetti, P.J. (1993). Unipolar localization and ATPase activity of IcsA, a *Shigella flexneri* protein involved in intracellular movement. J. Bacteriol. *175*, 2189–2196.

Goldberg, M.B., and Theriot, J.A. (1995). *Shigella flexneri* surface protein IcsA is sufficient to direct actin-based motility. Proc. Natl. Acad. Sci. U.S.A. *92*, 6572–6576.

Gouin, E., Gantelet, H., Egile, C., Lasa, I., Ohayon, H., Villiers, V., Gounon, P., Sansonetti, P.J., and Cossart, P. (1999). A comparative study of the actin-based motilities of the pathogenic bacteria *Listeria monocytogenes, Shigella flexneri* and *Rickettsia conorii*. J. Cell Sci. *112*, 1697–1708.

Gouin, E., Egile, C., Dehoux, P., Villiers, V., Adams, J., Gertler, F., Li, R., and Cossart, P. (2004). The RickA protein of *Rickettsia conorii* activates the Arp2/3 complex. Nature *427*, 457–461.

Gouin, E., Welch, M.D., and Cossart, P. (2005). Actin-based motility of intracellular pathogens. Curr. Opin. Microbiol. *8*, 35–45.

Guiney, D.G. (2005). The role of host cell death in *Salmonella* infections. Curr. Top Microbiol. Immunol. *289*, 131–150.

Gutierrez, M.G., Master, S.S., Singh, S.B., Taylor, G.A., Colombo, M.I., and Deretic, V. (2004). Autophagy is a defense mechanism inhibiting BCG and *Mycobacterium tuberculosis* survival in infected macrophages. Cell *119*, 753–766.

Hagedorn, M., Rohde, K.H., Russell, D.G., and Soldati, T. (2009). Infection by tubercular mycobacteria is spread by nonlytic ejection from their amoeba hosts. Science *323*, 1729–1733.

Haglund, C.M., Choe, J.E., Skau, C.T., Kovar, D.R., and Welch, M.D. (2010). *Rickettsia* Sca2 is a bacterial formin-like mediator of actin-based motility. Nat. Cell Biol. *12*, 1057–1063.

Hamon, M.A., and Cossart, P. (2011). K+ efflux, is required for histone-H3 dephosphorylation by *Listeria* LLO and other pore forming toxins. Infect. Immun. IAI.01243–01210.

Hamon, M., Bierne, H., and Cossart, P. (2006). *Listeria monocytogenes*: a multifaceted model. Nat. Rev. Microbiol. *4*, 423–434.

He, C., and Klionsky, D.J. (2009). Regulation mechanisms and signaling pathways of autophagy. Ann. Rev. Genet. *43*, 67–93.

High, N., Mounier, J., Prévost, M.C., and Sansonetti, P.J. (1992). IpaB of *Shigella flexneri* causes entry into epithelial cells and escape from the phagocytic vacuole. EMBO J. *11*, 1991–1999.

Hilbi, H., Moss, J.E., Hersh, D., Chen, Y., Arondel, J., Banerjee, S., Flavell, R.A., Yuan, J., Sansonetti, P.J., and Zychlinsky, A. (1998). *Shigella*-induced apoptosis is dependent on caspase-1 which binds to IpaB.J. Biol. Chem. *273*, 32895–32900.

Inohara, N., Ogura, Y., Fontalba, A., Gutierrez, O., Pons, F., Crespo, J., Fukase, K., Inamura, S., Kusumoto, S., Hashimoto, M., *et al.* (2003). Host recognition of bacterial muramyl dipeptide mediated through NOD2. Implication for Crohn's disease. J. Biol. Chem. *278*, 5509–5512.

Inohara, N., Chamaillard, M., McDonald, C., and Núñez, G. (2005). NOD-LRR proteins: role in host–microbial interactions and inflammatory disease. Ann. Rev. Biochem. *74*, 355–383.

Jeng, R.L., Goley, E.D., D'Alessio, J.A., Chaga, O.Y., Svitkina, T.M., Borisy, G.G., Heinzen, R.A., and Welch, M.D. (2004). A *Rickettsia* WASP-like protein activates the Arp2/3 complex and mediates actin-based motility. Cell. Microbiol. *6*, 761–769.

Kawai, T., and Akira, S. (2006). TLR signaling. Cell Death Differ. *13*, 816–825.

Kawai, T., and Akira, S. (2010). The role of pattern-recognition receptors in innate immunity: update on Toll-like receptors. Nat. Immunol. *11*, 373–384.

Kespichayawattana, W., Rattanachetkul, S., Wanun, T., Utaisincharoen, P., and Sirisinha, S. (2000). *Burkholderia pseudomallei* induces cell fusion and actin-associated membrane protrusion: a possible mechanism for cell-to-cell spreading. Infect. Immun. *68*, 5377–5384.

Khelef, N., Lecuit, M., Buchrieser, C., Cabanes, D., Dussurget, O., and Cossart, P. (2006). *Listeria monocytogenes* and the genus *Listeria*. In The Prokaryotes, Dwarkin, M., Falkow, S., Rosenberg, E., Schleifer, K.-H., and Stackebrandt, E., eds. (Springer-Verlag, New York), pp. 404–476.

Klionsky, D.J., and Emr, S.D. (2000). Autophagy as a regulated pathway of cellular degradation. Science *290*, 1717–1721.

Kocks, C., Gouin, E., Tabouret, M., Berche, P., Ohayon, H., and Cossart, P. (1992). *L. monocytogenes* induced actin assembly requires the *actA* gene product, a surface protein. Cell *68*, 521–531.

Kocks, C., Hellio, R., Gounon, P., Ohayon, H., and Cossart, P. (1993). Polarized distribution of *Listeria monocytogenes* surface protein ActA at the site of directional actin assembly. J. Cell Sci. *105*, 699–710.

Kocks, C., Marchand, J.B., Gouin, E., d'Hauteville, H., Sansonetti, P.J., Carlier, M.F., and Cossart, P. (1995). The unrelated surface proteins ActA of *Listeria monocytogenes* and IcsA of *Shigella flexneri* are sufficient to confer actin-based motility on *Listeria innocua* and *Escherichia coli* respectively. Mol. Microbiol. *18*, 413–423.

Kumar, D., Nath, L., Kamal, M.A., Varshney, A., Jain, A., Singh, S., and Rao, K.V. (2010). Genome-wide analysis of the host intracellular network that regulates survival of *Mycobacterium tuberculosis*. Cell *140*, 731–743.

Lara-Tejero, M., Sutterwala, F.S., Ogura, Y., Grant, E.P., Bertin, J., Coyle, A.J., Flavell, R.A., and Gálan, J.E. (2006). Role of the caspase-1 inflammasome in *Salmonella typhimurium* pathogenesis. J. Exp. Med. *203*, 1407–1412.

Leung, Y., Ally, S., and Goldberg, M.B. (2008). Bacterial actin assembly requires Toca-1 to relieve N-WASP autoinhibition. Cell Host Microb. *3*, 39–47.

Levine, B., and Deretic, V. (2007). Unveiling the roles of autophagy in innate and adaptive immunity. Nat. Rev. Immunol. *7*, 767–777.

Levine, B., and Klionsky, D.J. (2004). Development by self-digestion: molecular mechanisms and biological functions of autophagy. Dev. Cell *6*, 463–477.

Levine, B., and Kroemer, G. (2008). Autophagy in the pathogenesis of disease. Cell *132*, 27–42.

Loisel, T.P., Boujemaa, R., Pantaloni, D., and Carlier, M.F. (1999). Reconstitution of actin-based motility of *Listeria* and *Shigella* using pure proteins. Nature *401*, 613–616.

Lommel, S., Benesch, S., Rottner, K., Franz, T., Wehland, J., and Kühn, R. (2001). Actin pedestal formation by enteropathogenic *Escherichia coli* and intracellular motility of *Shigella flexneri* are abolished in N-WASP-defective cells. EMBO Rep. *2*, 850–857.

McLeod, M.P., Qin, X., Karpathy, S.E., Gioia, J., Highlander, S.K., Fox, G.E., McNeill, T.Z., Jiang, H., Muzny, D., Jacob, L.S., *et al.* (2004). Complete genome sequence of *Rickettsia typhi* and comparison with sequences of other *Rickettsiae*. J. Bacteriol *186*, 5842–5855.

Mariathasan, S., Newton, K., Monack, D.M., Vucic, D., French, D.M., Lee, W.P., Roose-Girma, M., Erickson, S., and Dixit, V.M. (2004). Differential activation of the inflammasome by caspase-1 adaptors ASC and Ipaf. Nature *430*, 213–218.

Mariathasan, S., Weiss, D.S., Newton, K., McBride, J., O'Rourke, K., Roose-Girma, M., Lee, W.P., Weinrauch, Y., Monack, D.M., and Dixit, V.M. (2006). Cryopyrin activates the inflammasome in response to toxins and ATP. Nature *440*, 228–232.

Marquis, H., Bouwer, H.G., Hinrichs, D.J., and Portnoy, D.A. (1993). Intracytoplasmic growth and virulence of *Listeria monocytogenes* auxotrophic mutants. Infect. Immun. *61*, 3756–3760.

Maucuer, A., Camonis, J.H., and Sobel, A. (1995). Stathmin interaction with a putative kinase and coiled-coil-forming protein domains. Proc. Natl. Acad. Sci. USA *92*, 3100–3104.

Meibom, K.L., and Charbit, A. (2010). The unraveling panoply of *Francisella tularensis* virulence attributes. Curr. Opin. Microbiol. *13*, 11–17.

Miao, E.A., Alpuche-Aranda, C.M., Dors, M., Clark, A.E., Bader, M.W., Miller, S.I., and Aderem, A. (2006). Cytoplasmic flagellin activates caspase-1 and secretion of interleukin 1β via Ipaf. Nat. Immunol. *7*, 569–575.

Mizushima, N., Levine, B., Cuervo, A.M., and Klionsky, D.J. (2008). Autophagy fights disease through cellular self-digestion. Nature *451*, 1069–1075.

Mizushima, N., Yoshimori, T., and Levine, B. (2010). Methods in mammalian autophagy research. Cell *140*, 313–326.

Mostowy, S., Bonazzi, M., Hamon, M., Mallet, A., Tham, T.N., Lelek, M., Gouin, E., Demangel, C., Brosch, R., Zimmer, C., *et al.* (2010). Entrapment of intracytosolic bacteria by septin cage-like structures. Cell Host Microbe *8*, 433–444.

Münz, C. (2009). Enhancing immunity through autophagy. Ann. Rev. Immunol. *27*, 423–449.

Noriega, F.R., Losonsky, G., Lauderbaugh, C., Liao, F.M., Wang, J.Y., and Levine, M.M. (1996). Engineered *ΔguaB-A ΔvirG Shigella flexneri* 2a strain CVD 1205: construction, safety, immunogenicity, and potential efficacy as a mucosal vaccine. Infect. Immun. *64*, 3055–3061.

O'Riordan, M., and Portnoy, D.A. (2002). The host cytosol: front-line or home front? Trends Microbiol. *10*, 361–364.

Ogawa, M., Yoshimori, T., Suzuki, T., Sagara, H., Mizushima, N., and Sasakawa, C. (2005). Escape of intracellular *Shigella* from autophagy. Science *307*, 727–731.

Opitz, B., Püschel, A., Beermann, W., Hocke, A.C., Förster, S., Schmeck, B., van Laak, V., Chakraborty, T., Suttorp, N., and Hippenstiel, S. (2006). *Listeria monocytogenes* activated p38 MAPK and induced IL-8 secretion in a nucleotide-binding oligomerization domain 1-dependent manner in endothelial cells. J. Immunol. *176*, 484–490.

Osiecki, J.C., Barker, J., Picking, N.L., Serfis, A.B., Berring, E., Shah, S., Harrington, A., and Picking, W.D. (2001). IpaC from *Shigella* and SipC from *Salmonella* possess similar biochemical properties but are functionally distinct. Mol. Microbiol. *42*, 469–481.

Oyston, P.C. (2008). *Francisella tularensis*: unravelling the secrets of an intracellular pathogen. J. Med. Microbiol. *57*, 921–930.

Oyston, P.C., Sjöstedt, A., and Titball, R.W. (2004). Tularaemia: bioterrorism defence renews interest in *Francisella tularensis*. Nat. Rev. Microbiol. *2*, 967–978.

Ozören, N., Masumoto, J., Franchi, L., Kanneganti, T.D., Body-Malapel, M., Ertürk, I., Jagirdar, R., Zhu, L., Inohara, N., Bertin, J., *et al.* (2006). Distinct roles of TLR2 and the adaptor ASC in IL-1β/IL-18 secretion in response to *Listeria monocytogenes*. J. Immunol. *176*, 4337–4342.

Pankiv, S., Clausen, T.H., Lamark, T., Brech, A., Bruun, J.A., Outzen, H., Øvervatn, A., Bjørkøy, G., and Johansen, T. (2007). p62/SQSTM1 binds directly to Atg8/LC3 to facilitate degradation of ubiquitinated protein aggregates by autophagy. J. Biol. Chem. *282*, 24131–24145.

Park, J.H., Kim, Y.G., McDonald, C., Kanneganti, T.D., Hasegawa, M., Body-Malapel, M., Inohara, N., and Núñez, G. (2007). RICK/RIP2 mediates innate immune responses induced through Nod1 and Nod2 but not TLRs. J. Immunol. *178*, 2380–2386.

Peacock, S.J. (2006). Melioidosis. Curr. Opin. Infect. Dis. *19*, 421–428.

Perrin, A.J., Jiang, X., Birmingham, C.L., So, N.S., and Brumell, J.H. (2004). Recognition of bacteria in the cytosol of mammalian cells by the ubiquitin system. Curr. Biol. *14*, 806–811.

Pfeuffer, T., Goebel, W., Laubinger, J., Bachmann, M., and Kuhn, M. (2000). LaXp180, a mammalian ActA-binding protein, identified with the yeast two-hybrid system, co-localizes with intracellular *Listeria monocytogenes*. Cell. Microbiol. *2*, 101–114.

Philpott, D.J., Girardin, S.E., and Sansonetti, P.J. (2001). Innate immune responses of epithelial cells following infection with bacterial pathogens. Curr. Opin. Immunol. *13*, 410–416.

Picking, W.L., Nishioka, H., Hearn, P.D., Baxter, M.A., Harrington, A.T., Blocker, A., and Picking, W.D. (2005). IpaD of *Shigella flexneri* is independently required for regulation of Ipa protein secretion and efficient insertion of IpaB and IpaC into host membranes. Infect. Immun. *73*, 1432–1440.

Pilatz, S., Breitbach, K., Hein, N., Fehlhaber, B., Schulze, J., Brenneke, B., Eberl, L., and Steinmetz, I. (2006). Identification of *Burkholderia pseudomallei* genes required for the intracellular life cycle and *in vivo* virulence. Infect. Immun. *74*, 3576–3586.

Ponpuak, M., Davis, A.S., Roberts, E.A., Delgado, M.A., Dinkins, C., Zhao, Z., Virgin, H.W., 4th, Kyei, G.B., Johansen, T., Vergne, I., *et al.* (2010). Delivery of cytosolic components by autophagic adaptor protein p62 endows autophagosomes with unique antimicrobial properties. Immunity *32*, 329–341.

Qin, A., Scott, D.W., Thompson, J.A., and Mann, B.J. (2009). Identification of an essential *Francisella tularensis* subsp. *tularensis* virulence factor. Infect. Immun. *77*, 152–161.

Raoult, D., Fournier, P.E., Eremeeva, M., Graves, S., Kelly, P.J., Oteo, J.A., Sekeyova, Z., Tamura, A., Tarasevich, I., and Zhang, L. (2005). Naming of Rickettsiae and rickettsial diseases. Ann. N. Y. Acad. Sci. *1063*, 1–12.

Ray, K., Marteyn, B., Sansonetti, P.J., and Tang, C.M. (2009). Life on the inside: the intracellular lifestyle of cytosolic bacteria. Nat. Rev. Microbiol. *7*, 333–340.

Renesto, P., Dehoux, P., Gouin, E., Touqui, L., Cossart, P., and Raoult, D. (2003). Identification and characterization of a phospholipase D-superfamily gene in Rickettsiae. J. Infect. Dis. *188*, 1276–1283.

Rich, K.A., Burkett, C., and Webster, P. (2003). Cytoplasmic bacteria can be targets for autophagy. Cell. Microbiol. *5*, 455–468.

Robbins, J.R., Barth, A.I., Marquis, H., de Hostos, E.L., Nelson, W.J., and Theriot, J.A. (1999). *Listeria monocytogenes* exploits normal host cell processes to spread from cell to cell. J. Cell Biol. *146*, 1333–1350.

Rohde, K., Yates, R.M., Purdy, G.E., and Russell, D.G. (2007). *Mycobacterium tuberculosis* and the environment within the phagosome. Immunol. Rev. *219*, 37–54.

Rubinsztein, D.C. (2006). The roles of intracellular protein-degradation pathways in neurodegeneration. Nature *443*, 780–786.

Samali, A., Zhivotovsky, B., Jones, D., Nagata, S., and Orrenius, S. (1999). Apoptosis: cell death defined by caspase activation. Cell Death Differ. *6*, 495–496.

Sansonetti, P.J. (2001). Rupture, invasion and inflammatory destruction of the intestinal barrier by

Shigella, making sense of prokaryote–eukaryote cross-talks. FEMS Microbiol. Rev. *25*, 3–14.

Santic, M., Molmeret, M., Klose, K.E., Jones, S., and Kwaik, Y.A. (2005). The *Francisella tularensis* pathogenicity island protein IglC and its regulator MglA are essential for modulating phagosome biogenesis and subsequent bacterial escape into the cytoplasm. Cell. Microbiol *7*, 969–979.

Santic, M., Molmeret, M., Klose, K.E., and Kwaik, Y.A. (2006). *Francisella tularensis* travels a novel, twisted road within macrophages. Trends Microbiol. *14*, 37–44.

Santic, M., Asare, R., Skrobonja, I., Jones, S., and Kwaik, Y.A. (2008). Acquisition of the vacuolar ATPase proton pump and phagosome acidification are essential for escape of *Francisella tularensis* into the macrophage cytosol. Infect. Immun. *76*, 2671–2677.

Shaughnessy, L.M., Hoppe, A.D., Christensen, K.A., and Swanson, J.A. (2006). Membrane perforations inhibit lysosome fusion by altering pH and calcium in *Listeria monocytogenes* vacuoles. Cell. Microbiol. *8*, 781–792.

Shaw, M.H., Reimer, T., Kim, Y.G., and Núñez, G. (2008). NOD-like receptors (NLRs): bona fide intracellular microbial sensors. Curr. Opin. Immunol. *20*, 377–382.

Shintani, T., and Klionsky, D.J. (2004). Autophagy in health and disease: a double-edged sword. Science *306*, 990–995.

Silverman, D.J., Santucci, L.A., Meyers, N., and Sekeyova, Z. (1992). Penetration of host cells by *Rickettsia rickettsii* appears to be mediated by a phospholipase of rickettsial origin. Infect. Immun. *60*, 2733–2740.

Singh, R., Jamieson, A., and Cresswell, P. (2008). GILT is a critical host factor for *Listeria monocytogenes* infection. Nature *455*, 1244–1247.

Sirajuddin, M., Farkasovsky, M., Hauer, F., Kühlmann, D., Macara, I.G., Weyand, M., Stark, H., and Wittinghofer, A. (2007). Structural insight into filament formation by mammalian septins. Nature *449*, 311–315.

Stamm, L.M., and Brown, E.J. (2004). *Mycobacterium marinum*: the generalization and specialization of a pathogenic *Mycobacterium*. Microbes Infect. *6*, 1418–1428.

Stamm, L.M., Morisaki, J.H., Gao, L.Y., Jeng, R.L., McDonald, K.L., Roth, R., Takeshita, S., Heuser, J., Welch, M.D., and Brown, E.J. (2003). *Mycobacterium marinum* escapes from phagosomes and is propelled by actin-based motility. J. Exp. Med. *198*, 1361–1368.

Stamm, L.M., Pak, M.A., Morisaki, J.H., Snapper, S.B., Rottner, K., Lommel, S., and Brown, E.J. (2005). Role of the WASP family proteins for *Mycobacterium marinum* actin tail formation. Proc. Natl. Acad. Sci. U.S.A. *102*, 14837–14842.

Stevens, M.P., Wood, M.W., Taylor, L.A., Monaghan, P., Hawes, P., Jones, P.W., Wallis, T.S., and Galyov, E.E. (2002). An Inv/Mxi-Spa-like type III protein secretion system in *Burkholderia pseudomallei* modulates intracellular behaviour of the pathogen. Mol. Microbiol. *46*, 649–659.

Stevens, M.P., Haque, A., Atkins, T., Hill, J., Wood, M.W., Easton, A., Nelson, M., Underwood-Fowler, C., Titball, R.W., Bancroft, G.J., *et al.* (2004). Attenuated virulence and protective efficacy of a *Burkholderia pseudomallei* bsa type III secretion mutant in murine models of melioidosis. Microbiology *150*, 2669–2676.

Stevens, M.P., Stevens, J.M., Jeng, R.L., Taylor, L.A., Wood, M.W., Hawes, P., Monaghan, P., Welch, M.D., and Galyov, E.E. (2005). Identification of a bacterial factor required for actin-based motility of *Burkholderia pseudomallei*. Mol. Microbiol. *56*, 40–53.

Stinear, T.P., Seemann, T., Harrison, P.F., Jenkin, G.A., Davies, J.K., Johnson, P.D., Abdellah, Z., Arrowsmith, C., Chillingworth, T., Churcher, C., *et al.* (2008). Insights from the complete genome sequence of *Mycobacterium marinum* on the evolution of *Mycobacterium tuberculosis*. Genome Res. *18*, 729–741.

Stritzker, J., Janda, J., Schoen, C., Taupp, M., Pilgrim, S., Gentschev, I., Schreier, P., Geginat, G., and Goebel, W. (2004). Growth, virulence, and immunogenicity of *Listeria monocytogenes aro* mutants. Infect. Immun. *72*, 5622–5629.

Sun, G.W., Lu, J., Pervaiz, S., Cao, W.P., and Gan, Y.H. (2005). Caspase-1 dependent macrophage death induced by *Burkholderia pseudomallei*. Cell. Microbiol. *7*, 1447–1458.

Suparak, S., Kespichayawattana, W., Haque, A., Easton, A., Damnin, S., Lertmemongkolchai, G., Bancroft, G.J., and Korbsrisate, S. (2005). Multinucleated giant cell formation and apoptosis in infected host cells is mediated by *Burkholderia pseudomallei* type III secretion protein BipB.J. Bacteriol. *187*, 6556–6560.

Suzuki, T., Miki, H., Takenawa, T., and Sasakawa, C. (1998). Neural Wiskott–Aldrich syndrome protein is implicated in the actin-based motility of *Shigella flexneri*. EMBO J. *17*, 2767–2776.

Suzuki, T., Mimuro, H., Suetsugu, S., Miki, H., Takenawa, T., and Sasakawa, C. (2002). Neural Wiskott-Aldrich syndrome protein (N-WASP) is the specific ligand for *Shigella* VirG among the WASP family and determines the host cell type allowing actin-based spreading. Cell. Microbiol. *4*, 223–233.

Suzuki, T., Franchi, L., Toma, C., Ashida, H., Ogawa, M., Yoshikawa, Y., Mimuro, H., Inohara, N., Sasakawa, C., and Núñez, G. (2007). Differential regulation of caspase-1 activation, pyroptosis, and autophagy via Ipaf and ASC in *Shigella*-infected macrophages. PLoS Pathog. *3*, e111.

Thurston, T.L., Ryzhakov, G., Bloor, S., von Muhlinen, N., and Randow, F. (2009). The TBK1 adaptor and autophagy receptor NDP52 restricts the proliferation of ubiquitin-coated bacteria. Nat. Immunol. *10*, 1215–1221.

Tilney, L.G., and Portnoy, D.A. (1989). Actin filaments and the growth, movement, and spread of the intracellular bacterial parasite, *Listeria monocytogenes*. J. Cell Biol. *109*, 1597–1608.

Tilney, L.G., DeRosier, D.J., Weber, A., and Tilney, M.S. (1992). How *Listeria* exploits host cell actin to form its own cytoskeleton. II. Nucleation, actin filament polarity, filament assembly, and evidence for a pointed end capper. J. Cell Biol. *118*, 83–93.

Tobin, D.M., and Ramakrishnan, L. (2008). Comparative pathogenesis of *Mycobacterium marinum* and

Mycobacterium tuberculosis. Cell. Microbiol. *10*, 1027–1039.

Travassos, L.H., Carneiro, L.A., Ramjeet, M., Hussey, S., Kim, Y.G., Magalhães, J.G., Yuan, L., Soares, F., Chea, E., Le Bourhis, L., *et al.* (2009). Nod1 and Nod2 direct autophagy by recruiting ATG16L1 to the plasma membrane at the site of bacterial entry. Nat. Immunol. *11*, 55–62.

Van Kirk, L.S., Hayes, S.F., and Heinzen, R.A. (2000). Ultrastructure of *Rickettsia rickettsii* actin tails and localization of cytoskeletal proteins. Infect. Immun. *68*, 4706–4713.

Van Troys, M., Lambrechts, A., David, V., Demol, H., Puype, M., Pizarro-Cerdá, J., Gevaert, K., Cossart, P., and Vandekerckhove, J. (2008). The actin propulsive machinery: the proteome of *Listeria monocytogenes* tails. Biochem. Biophys. Res. Commun *375*, 194–199.

Vellai, T., Takács-Vellai, K., Sass, M., and Klionsky, D.J. (2009). The regulation of aging: does autophagy underlie longevity? Trends Cell Biol. *19*, 487–494.

Walker, D.H. (2007). *Rickettsiae* and rickettsial infections: the current state of knowledge. Clin. Infect. Dis. *45*, S39–44.

van der Wel, N., Hava, D., Houben, D., Fluitsma, D., van Zon, M., Pierson, J., Brenner, M., and Peters, P.J. (2007). *M. tuberculosis* and *M. leprae* translocate from the phagolysosome to the cytosol in myeloid cells. Cell *129*, 1287–1298.

Welch, M.D., Iwamatsu, A., and Mitchison, T.J. (1997). Actin polymerization is induced by Arp 2/3 protein complex at the surface of *Listeria monocytogenes*. Nature *385*, 265–269.

Whitworth, T., Popov, V.L., Yu, X.J., Walker, D.H., and Bouyer, D.H. (2005). Expression of the *Rickettsia prowazekii pld* or *tlyC* gene in *Salmonella enterica* serovar Typhimurium mediates phagosomal escape. Infect. Immun. *73*, 6668–6673.

Wiersinga, W.J., van der Poll, T., White, N.J., Day, N.P., and Peacock, S.J. (2006). Melioidosis: insights into the pathogenicity of *Burkholderia pseudomallei*. Nat. Rev. Microbiol. *4*, 272–282.

Winkler, H.H., and Miller, E.T. (1982). Phospholipase A and the interaction of *Rickettsia prowazekii* and mouse fibroblasts (L-929 cells). Infect. Immun. *38*, 109–113.

Xu, Y., Jagannath, C., Liu, X.D., Sharafkhaneh, A., Kolodziejska, K.E., and Eissa, N.T. (2007). Toll-like receptor 4 is a sensor for autophagy associated with innate immunity. Immunity *27*, 135–144.

Yoshida, S., and Sasakawa, C. (2003). Exploiting host microtubule dynamics: a new aspect of bacterial invasion. Trends Microbiol. *11*, 139–143.

Yoshida, S., Handa, Y., Suzuki, T., Ogawa, M., Suzuki, M., Tamai, A., Abe, A., Katayama, E., and Sasakawa, C. (2006). Microtubule-severing activity of *Shigella* is pivotal for intercellular spreading. Science *314*, 985–989.

Yoshikawa, Y., Ogawa, M., Hain, T., Yoshida, M., Fukumatsu, M., Kim, M., Mimuro, H., Nakagawa, I., Yanagawa, T., Ishii, T., *et al.* (2009). *Listeria monocytogenes* ActA-mediated escape from autophagic recognition. Nat. Cell Biol. *11*, 1233–1240.

Youle, R.J., and Strasser, A. (2008). The BCL-2 protein family: opposing activities that mediate cell death. Nat. Rev. Mol. Cell Biol. *9*, 47–59.

Part V

Bacterial Evasion of Host Defences

Bacterial Handling of Host Nutrients: the Iron Paradigm

12

Klaus Hantke

Abstract

Pathogenic bacteria must compete with their host for iron. The host lowers serum iron levels during inflammation and binds iron with high affinity to transferrin and lactoferrin, thereby limiting iron availability for the pathogens. Under such iron-limiting conditions, many pathogens, like their non-pathogenic relatives, secrete iron chelators, called siderophores, which mobilize even traces of iron for the bacteria. The host combats catecholate siderophores of various pathogens by avidly binding the chelator to the host protein siderocalin. Pathogenic bacteria can evade this host mechanism in one or more ways: secreting other siderophores that are not bound by siderocalin, utilizing haem as an iron source, and transporting ferrous iron. All these uptake systems are regulated by the general iron needs of bacteria and fine tuned according to the surrounding supply of iron-containing compounds. Iron limitation is also a signal for many pathogens to produce toxins and other virulence factors. When one iron supply system is disabled, virulence is often only mildly attenuated due to the redundancy of the systems. Although it has long been known that iron is required for the infection process, no therapeutics directed at this have been marketed yet.

Introduction

Bacteria, like all organisms, are highly adapted to their environment. However, they have only limited possibilities to determine their habitat. Many bacteria live closely associated with eukaryotic hosts in a commensal or mutualistic partnership. In some cases, this relationship has developed to parasitism, in which the host is damaged by its resident.

Bacteria living in close association with eukaryotes sometimes have to compete with their host for nutrients. Iron (Fe), with its rather special chemistry and its inherent toxicity, is a nutrient that is handled with great care by both the host and the parasite. Bacteria living in the digestive tract can acquire iron from the food of their host. Bacteria, located elsewhere in the body, are forced to take the iron from their host. Free soluble iron, i.e. iron not chelated to proteins or small molecule ligands, such as glutathione, is scarcely available; therefore, bacteria obtain the metal from iron-bearing molecules, e.g. haem (all oxidation states), haemoglobin, transferrin, lactoferrin and ferritin. The type of binding is important for the reactivity of the iron. Redox reactions of Fe(II) with oxygen generate harmful radicals and oxidative stress. Therefore, every organism living in an oxic environment has mechanisms to store the iron securely and to repair oxidative damage. These Janus-faced properties of iron have led to a rather complicated regulatory system in which iron uptake, storage, and mobilization, and the biosynthesis and degradation of iron-containing proteins are balanced to achieve a well-running metabolism under different conditions.

In each section of this review, iron transport systems found in both pathogenic and apathogenic bacteria will be discussed, followed by a discussion of the more specialized iron transport systems found only in pathogenic bacteria. A short introduction of the regulation of the bacterial iron metabolism will be given. At the end of the chapter, the host responses to infecting bacteria with regard to iron metabolism are introduced. These topics on iron and pathogenic bacteria have been nicely reviewed in the book *Iron Transport in Bacteria* (Crosa, 2004), but remarkable progress has been made since then. The short overview presented here will provide a balanced treatment of the fundamentals and summarizes the new developments.

Siderophores–small secreted iron carriers

Many microorganisms growing under oxic conditions in iron-poor medium produce low molecular weight chelators, called siderophores, which avidly bind Fe(III), the stable form of iron in the presence of oxygen. At neutral pH, Fe(III) is insoluble; the minor amounts in solution do not suffice for bacterial growth. The siderophores form water-soluble iron complexes and may dissolve precipitated iron hydroxide following the law of mass action. These Fe(III)–siderophore complexes can be taken up by bacteria via special transport systems.

Several such transport systems were first studied in depth in *Escherichia coli*. Similar transport systems are found in Gram-positive and other Gram-negative bacteria. Most bacteria have, in addition to their cognate siderophore utilization system, the capacity to use xenosiderophores, i.e. siderophores produced by other organisms.

In contrast to the usual unity in biochemistry, siderophores produced by different groups of bacteria differ greatly. About 280 different siderophores have been described (Hider and Kong, 2010). Siderophores chelate ferric iron with hydroxamates [e.g. ferrichrome Fig. 12.1(1)], catecholates [e.g. enterochelin Fig. 12.1(4)], carboxylates [e.g. citrate, Fig. 12.1(3)] or mixed residues [Fig. 12.1(7), Fig. 12.1(8), Fig. 12.1(9)].

The iron-laden siderophores have molecular weights of 450–2000 Da. Molecules of this size cannot easily pass through the porins in the outer membrane of Gram-negative bacteria such as *E. coli*. This may be one reason why specific transport proteins are embedded in the outer membrane. These transport proteins bind their specific siderophores with high affinity, which is necessary because of the scarcity of the siderophores. In an energy-dependent process, the siderophore is transported into the periplasm, where usually a specific, binding-protein-dependent transport system takes over (Fig. 12.2). The binding protein delivers the siderophore via an integral dimeric cytoplasmic membrane protein into the cytoplasm. The transport process is energized by two membrane-associated ATPases. This general scheme is found in nearly all Gram-negative bacteria living in an oxic environment. Gram-positive bacteria have homologous systems; only the outer membrane transporter with its energy-transducing device is missing (Fig. 12.2).

Inside the cell, reductases reduce the Fe(III) to Fe(II) and take part in removing the iron from the siderophore. In certain bacteria (e.g. in *Pseudomonas aeruginosa*), the Fe(III)-siderophore is transported into the periplasm, where iron is reduced and removed; only the iron is then transported into the cell, while the empty siderophore is excreted (Imperi *et al.*, 2009). In those cases where the siderophore is unloaded in the cytoplasm, the siderophore is either secreted and used again, or degraded or modified and then secreted.

The haem uptake systems have nearly the same blueprint (see below). The systems in Gram-positive bacteria differ in that a cascade of proteins embedded in the peptidoglycan (murein) layer transports haem to a membrane-bound binding protein (Fig. 12.2). Peptidoglycan-anchored proteins that bind siderophores have not been detected. A possible reason for this difference may lie in the hydrophobicity of haem, which may hinder it to pass through the peptidoglycan layer. Iron siderophore complexes, in contrast, are hydrophilic and can pass through the peptidoglycan layer by diffusion.

Iron transport systems of *E. coli*

The ferrichrome transport system: architecture of a typical siderophore transport system

The *E. coli* ferrichrome siderophore transport system was one of the first to be studied in detail. One reason for this was the awareness that a group of phages (bacterial viruses) use the outer

(1) Ferrichrome

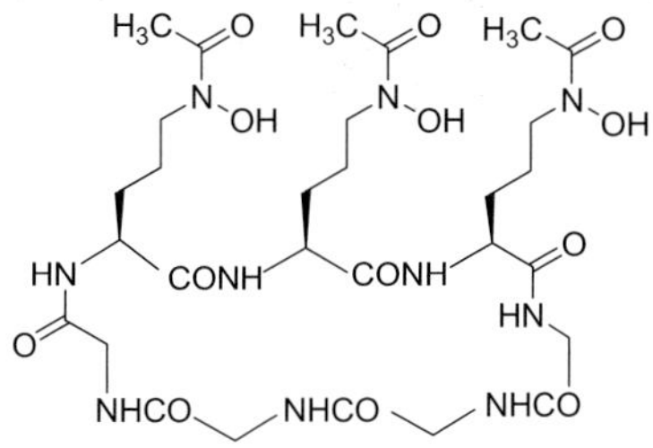

(2) Aerobactin

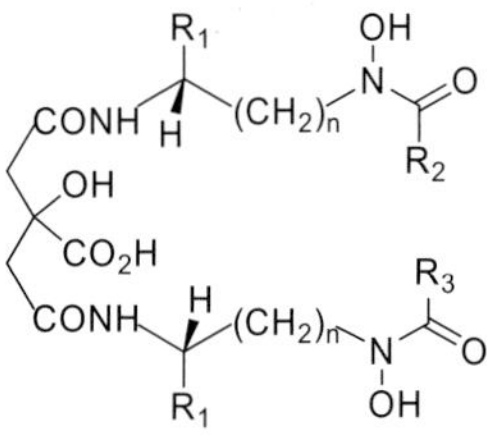

(3) Citrate

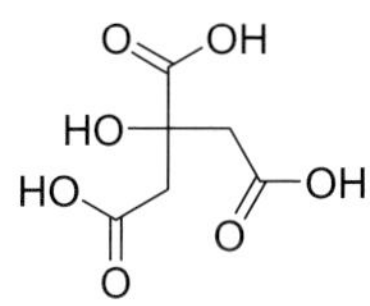

(4) Enterochelin

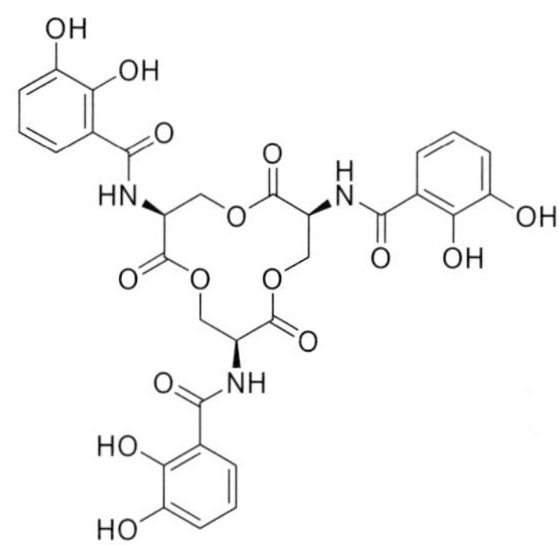

(5) Salmochelin

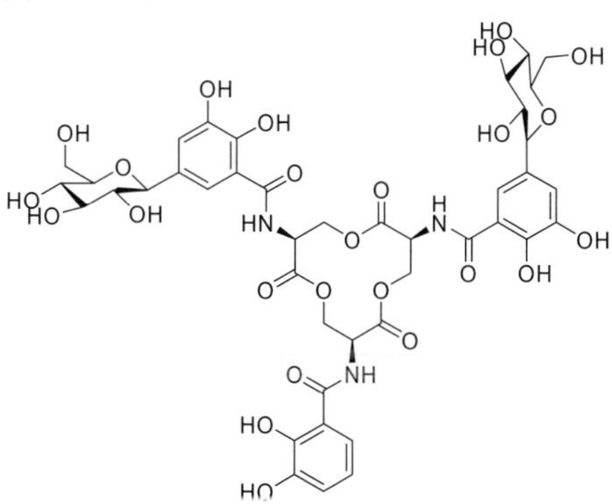

(6) Petrobactin (R1 = R2 = H)

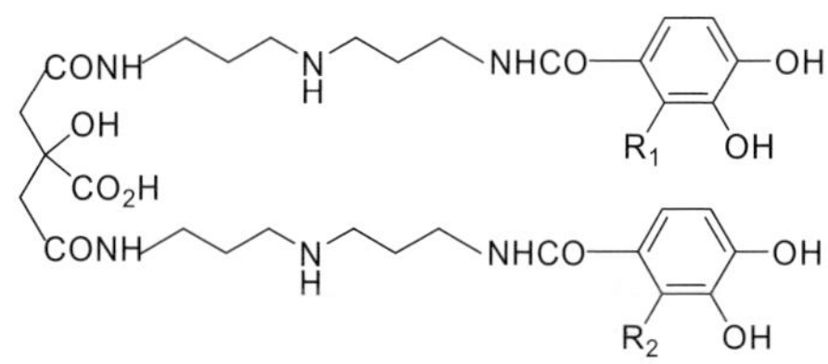

(7) Yersiniabactin (R1 = R2 = H, R3 = OH))

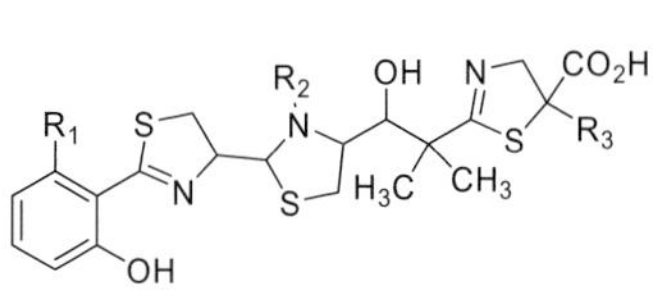

(8) Anguibactin (R=H; X=S)

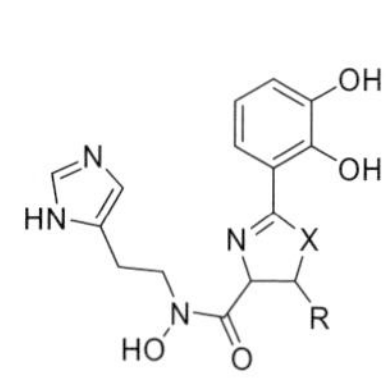

Figure 12.1 Structures of some siderophores: (1) ferrichrome, from fungi; (2) aerobactin, from pathogenic enterobacteria; (3) citrate; (4) enterochelin, from enterobacteria; (5) salmochelin, from pathogenic enterobacteria; (6) petrobactin, from pathogenic *Bacillus* spp.; (7) yersiniabactin, from pathogenic enterobacteria; (8) anguibactin, from *V. anguillarum* O1 (continued overleaf).

(9) Pyoverdin R

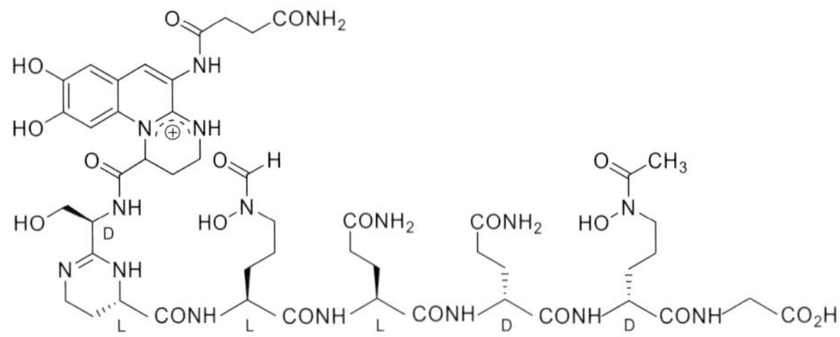

(10) Pyochelin (R1=R2=H)

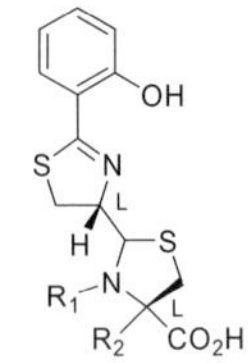

(11) Carboxymycobactin (R= $(CH_2)_nCO_2H$ (n=2-9) ; R3 = H; R4 = CH3; R5 = H)

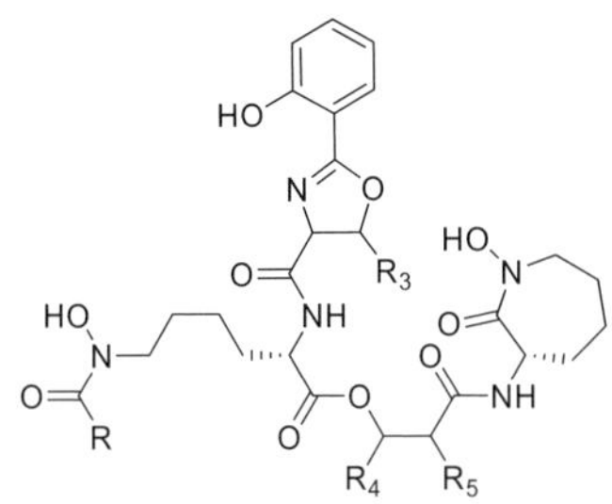

(12) Ornibactin (R=CH_3)

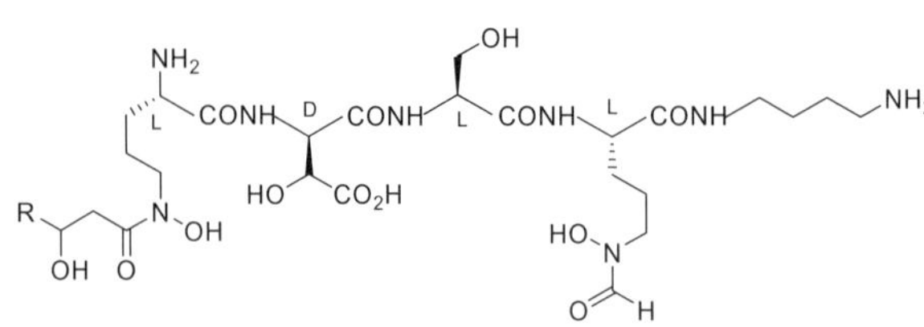

(13) Cepabactin

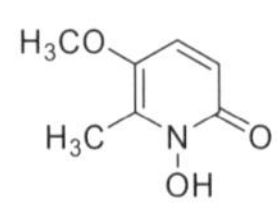

(14) Albomycin δ_2 X =

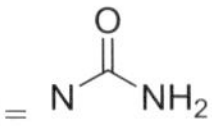

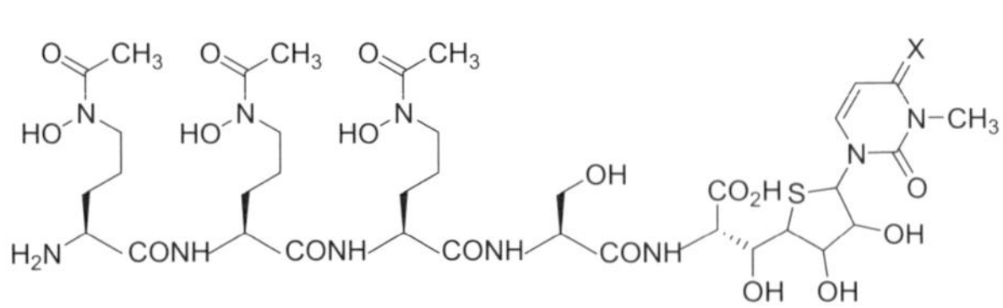

Figure 12.1 (continued). Structures of some siderophores: (9) pyoverdin, from *P. aeruginosa*; (10) pyochelin, from *Pseudomonas* spp. and *Burkholderia* spp.; (11) carboxymycobactin, from *Mycobacterium* spp.; (12) ornibactin, from *Burkholderia* spp.; (13) cepabactin *Burkholderia* spp.; (14) albomycin δ_2, from *Streptomyces* spp. All structures are from Hider and Kong (2010).

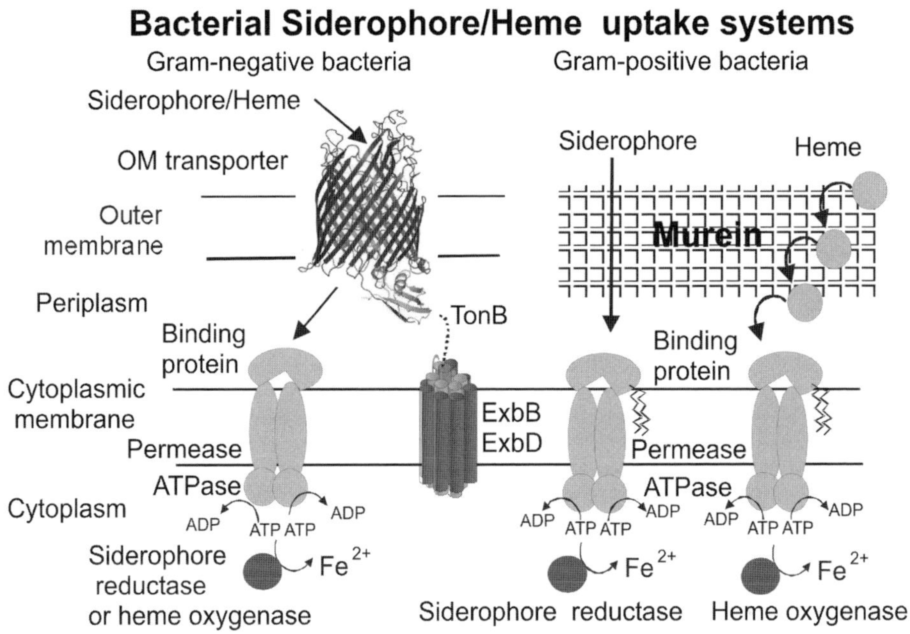

Figure 12.2 Siderophore/haem uptake systems of Gram-negative and Gram-positive bacteria. In Gram-negative bacteria, the siderophore is bound to the outer membrane receptor and transported with the help of TonB into the periplasm. A binding protein delivers the siderophore through a permease into the cytoplasm. The process is driven by ATP. The same scheme applies for haem uptake. Gram-positive bacteria lack an outer membrane and have a thick peptidoglycan layer. The transport systems through the cytoplasmic membrane are homologous to those of Gram-negative bacteria. The binding protein is anchored via a diacylglycerol lipid to the cytoplasmic membrane. Haem-uptake systems in Gram-positive bacteria often contain a cascade of proteins covalently bound to the peptidoglycan; these proteins transport haem to the binding protein (Maresso and Schneewind, 2006). Such binding proteins have not been described for siderophores; however, in mycobacteria, such intermediary proteins may be found in the cell wall.

membrane protein FhuA (old designation: TonA) as receptor during infection. The main function of this protein was later found to be the transport of ferrichrome into the cell (Braun, 2009). Like many other bacteria, *E. coli* is able to utilize the fungal siderophore ferrichrome (Fig. 12.1(1)), a cyclic hydroxamate, as an iron source. The FhuA protein binds ferrichrome with high affinity and delivers it in an energy-dependent process into the periplasm. The energy of the membrane potential required for the translocation is provided by the TonB protein, which is anchored in a complex with proteins ExbB and ExbD in the cytoplasmic membrane (Braun, 2009). In the periplasm, the protein FhuD, as part of an ABC-transporter, binds the siderophore. The integral cytoplasmic membrane protein FhuB, with its two sets of ten membrane-spanning helices, forms a channel for the siderophore, and two FhuC ATPases supply the energy for the transport into the cytoplasm. FhuF, an unusual iron–sulfur protein, is involved in reduction and mobilization of the iron from the iron–ferrichrome complex (Matzanke *et al.*, 2004). The schematic overview given in Fig. 12.2 is valid for many siderophore uptake systems. In the case of ferrichrome, an additional, uncharacterized enzyme acylates desferri-ferrichrome before it is secreted (Hartmann and Braun, 1980). The acyl-ferrichrome does not have a high affinity for iron. The acyl transferase of *P. aeruginosa* has been recently identified (Hannauer *et al.*, 2010).

The crystal structure of FhuA reveals a porin-like structure with 22 antiparallel β-sheets that form a relatively large barrel compared with outer membrane porins, whose barrel is formed of 16 β-sheets (Fig. 12.3). In contrast to the porins, an N-terminal globular domain is found inside the

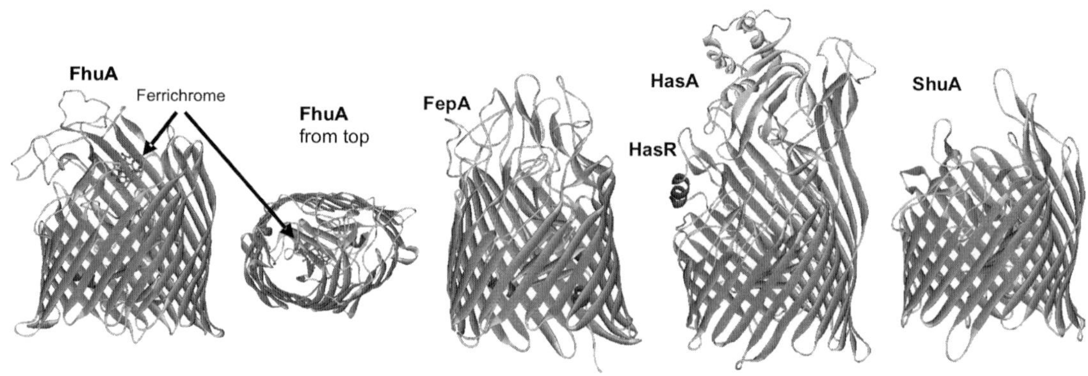

Figure 12.3 Crystal structures of TonB-dependent outer membrane siderophore/haem transporters all have the same basic structure. FhuA with the ligand ferrichrome, side view, and view from top (Ferguson *et al.*, 1998); the enterochelin transporter FepA (Buchanan *et al.*, 1999); the haem transporter HasR with the haem-binding protein HasA on top (Krieg *et al.*, 2009), and the haem transporter ShuA (Cobessi *et al.*, 2010).

barrel of these TonB-dependent transporters. This plug or cork seems to block the passage of ferrichrome (Fig. 12.3). The binding sites on the plug and inside the barrel have been identified in crystal structures with the ligand ferrichrome. Other substrates of FhuA, e.g. the antibiotics albomycin and a rifamycin derivative, bind to the same site (Ferguson *et al.*, 2000). The few available structures of these outer membrane transporters are similar, regardless of whether they transport siderophore, haem or vitamin B12 (Fig. 12.3; the vitamin B12 BtuB receptor is not shown).

As mentioned above, energy is necessary to move the bound siderophore into the periplasm. Phage T1 adsorption studies revealed that FhuA is a competent phage receptor only in an energized cell; when the membrane potential is destroyed, the phage reversibly binds to FhuA (Hancock and Braun, 1976). A stretch of about seven amino acid residues near the FhuA N-terminus is responsible for the interaction with the TonB protein (Cadieux and Kadner, 1999). These amino acids are conserved and found not only in TonB-dependent outer membrane transporters, but also in certain TonB-dependent colicins, which require energy for invasion and killing of the bacterium (Braun *et al.*, 2002). The crystal structure of a complex between the C-terminal domain of TonB and an outer membrane transporter has been solved (Shultis *et al.*, 2006).

Other iron transport systems of *E. coli*

Under iron-limiting conditions, faecal *E. coli* strains produce the siderophore enterochelin (also called enterobactin). Enterochelin is a cyclic triester of serine with three 2,3-dihydroxy-benzoyl residues hooked to the amino groups of serine (Fig. 12.1(4)). Enterochelin complexes Fe(III) with very high affinity. The uptake system for Fe(III)-enterochelin has the same blueprint as the ferrichrome uptake system. The outer membrane receptor FepA binds iron–enterochelin with high affinity. The complex is then transferred into the periplasm in a TonB-dependent process, where the binding protein FepB takes over and delivers the iron complex via the two membrane components FepD and FepG, driven by ATP, via FepC into the cytoplasm. In the cytoplasm, an ester bond of enterochelin is hydrolysed by Fes, and the iron is reduced to Fe(II) and used in the cellular metabolism (Earhart, 2004). The iron-free linear form of enterochelin is secreted and used for another uptake cycle. In this process, dimers and monomers of 2,3-dihydroxybenzoylserine are generated. These compounds also contribute to iron solubilization and uptake, but they have a much lower affinity for ferric iron than enterochelin.

Some strains of *E. coli* also have an iron uptake system for ferric dicitrate. This system is induced under iron-limiting conditions in the presence of 1 mM citrate [Fig. 12.1(3)], i.e. in the presence of ferric dicitrate. The sigma factor FecI and the signal transducer FecR have become a paradigm for the regulation of envelope cell functions (extracytoplasmic sigma factor, ECF) (Braun *et al.*, 2006). The genes of the Fec transport system are efficiently transcribed with the help of the FecI sigma factor. This sigma factor is activated when ferric dicitrate is bound by the FecA transporter and internalized with the assistance of TonB. An N-terminal extension of FecA interacts with FecR in the periplasm. The FecR protein is anchored in the cytoplasmic membrane and transmits the signal from FecA to the sigma factor FecI in the cytoplasm (see also below).

Iron transport systems in pathogenic bacteria

Additional siderophore transport systems in pathogenic enterobacteria

As part of the mammalian innate immune system, siderocalin (NGAL-lipocalin) is secreted during inflammation by macrophages and epithelial cells. A mouse carrying a siderocalin knock-out mutation is highly sensitive to *E. coli* infections (Flo *et al.*, 2004). When recombinant siderocalin is expressed in *E. coli*, the purified protein is pink; the red colour is caused by binding of enterochelin to siderocalin (Goetz *et al.*, 2002). These results indicate that in mammals, a catecholate-specific siderophore-binding protein, i.e. siderocalin, is used for defence against invading bacteria (see also below). In response, several pathogens produce other siderophores that seem to help them, avoiding the bacteriostatic activity of siderocalin. The siderophores discussed below are often found in known pathogens and are considered as virulence factors; they are not bound by siderocalin.

Aerobactin is a hydroxamate siderophore [Fig. 12.1(2)]. Its biosynthesis and uptake are encoded on pathogenicity islands on the chromosome or on plasmids. Aerobactin is found in clinical isolates of *E. coli*, *Shigella flexneri*, and *Shigella boydii*, as well as in some clinical isolates of *Salmonella enterica* and *Klebsiella*. The genes *iucABCD* encode the enzymes necessary to synthesize aerobactin from citrate and lysine. IutA is the TonB-dependent transporter that binds ferric-aerobactin and guides it through the outer membrane. In the periplasm, the hydroxamate-binding protein FhuD takes over and allows the transport via FhuB through the cytoplasmic membrane energized by the FhuC ATPase. IutA is encoded by a gene located next to the aerobactin operon; FhuCDB are encoded with FhuA in a separate operon in all *E. coli* strains (Payne and Mey, 2004).

Yersiniabactin (Fig. 12.1(7)) is the principal siderophore of *Yersinia pestis* and is an important virulence factor for this organism. The biosynthesis and uptake systems are encoded on a high pathogenicity island. This island is found not only in *Y. pestis* and some *Yersinia pseudotuberculosis* and *Yersinia enterocolitica* strains, but also in *S. enterica* subsp. III and VI, and in extraintestinal pathogenic *E. coli* (ExPEC) strains, e.g. uropathogenic *E. coli* strains (Perry, 2004). The TonB-dependent outer membrane transporter FyuA (Psn) delivers yersiniabactin into the periplasm. Unfortunately, identical proteins in *Y. enterocolitica* and *Y. pestis* were given different names. No periplasmic binding protein has been identified in this system. An ABC transporter composed of two proteins, YbtP (Irp6) and YbtQ (Irp7), each with a membrane permease domain and an ATPase domain, transport the iron complex into the cell.

Salmochelin is a characteristic siderophore of *S. enterica* strains. In fact, it is a modified form of enterochelin. The siderophore is C-glucosylated at two positions (Fig. 12.1(5)) (Bister *et al.*, 2004), which drastically reduces the hydrophobic character of enterochelin and allows evasion from siderocalin (Fischbach *et al.*, 2006; Valdebenito *et al.*, 2007; see above). Since the synthesis of salmochelin is dependent on functional enterochelin biosynthesis, only one other gene product, IroB, is necessary for the glucosylation of enterochelin. The esterases IroD and IroE are necessary for the cleavage of the ester bonds in the molecule; Fes does not cleave the bonds. The IroC protein mainly exports salmochelin through the cytoplasmic membrane. The fifth gene of the *iro* gene cluster, *iroN*, encodes an outer membrane transporter for the uptake of iron–salmochelin (Hantke *et al.*, 2003).

Most faecal *E. coli* strains produce only one siderophore – enterochelin. Uropathogenic *E. coli* strains (UTI), however, also produce two or three of the siderophores described above (aerobactin, yersiniabactin, and salmochelin). UTI strains also encode about 18 mostly unassigned TonB-dependent outer membrane transporters that are assumed to recognize siderophores or haem-compounds, in contrast to *E. coli* K-12, which encodes six dedicated transporters. The higher number of transporters synthesized by UTI strains is in accord with the very low iron content of urine, i.e. the bacteria must make a greater effort to obtain the necessary iron (Valdebenito *et al.*, 2006). Enteropathogenic *E. coli* (EPEC) and enterotoxigenic *E. coli* (ETEC) strains may also have one of the transport systems mentioned above, but they generally have fewer transporters than the UTI strains. The highly specialized parasite *Y. pestis* seems to produce only one siderophore –y ersiniabactin – whereas many *Salmonella* strains produce only enterochelin and salmochelin. However, certain *Salmonella* strains also produce aerobactin or yersiniabactin. A similar spectrum of siderophores is produced by *Citrobacter* and *Klebsiella* strains.

Siderophore transport systems of other pathogenic bacteria

Anguibactin

Anguibactin [Fig. 12.1(8)], a siderophore produced by the fish pathogen *Vibrio anguillarum,* is an important virulence factor (Crosa, 2004). Other serotypes of often less-virulent *V. anguillarum* strains produce only vanchrobactin, the synthesis of which is chromosomally encoded. In the O1 serotype strains, vanchrobactin is not synthesized because of a mutation in the biosynthesis genes; only plasmid-encoded anguibactin is secreted.

Pyoverdin(e)s

Pyoverdin(e)s are typical siderophores from *Pseudomonas* strains. They are composed of a dihydroxyquinoline chromophore with an attached peptide of up to 12 amino acids in length [Fig. 12.1(9)]. Pyoverdin contains L- and D-amino acids and *N*-hydroxy-*N*-acyl amino acid derivatives, which supply binding sites for ferric iron. The fluorescence of the dihydroxyquinoline chromophore is strongly influenced by the bound metal cation: iron quenches the fluorescence, and gallium and aluminium enhance the fluorescence (Greenwald *et al.*, 2008). The importance of siderophore production for virulence has been demonstrated in various animal model systems: a mouse model of ascending urinary tract infection (Mittal *et al.*, 2008), a *Caenorhabditis elegans* model (Zaborin *et al.*, 2009), and an endocarditis model (Xiong *et al.*, 2000). In a mouse model, also the second siderophore produced by *P. aeruginosa* and some *Burkholderia* strains – pyochelin (Fig. 12.1(10))– was shown to contribute to the virulence of the *P. aeruginosa* strain tested (Takase *et al.*, 2000). One can generally assume that different iron transport systems contribute to the fitness of a virulent strain.

In addition to pyochelin, several other siderophores, e.g. cepabactin [Fig. 12.1(13)] and ornibactin [Fig. 12.1(12)], are produced by various *Burkholderia cepacia* strains (Thomas, 2007). These opportunistic pathogens are very difficult to treat because of their high antibiotic resistance.

Xenosiderophores

The enterochelin receptor CfrA of *Campylobacter jejuni* is necessary for successful infection of chickens. However, enterochelin is not produced by the organism, but is supplied by gut bacteria, most likely enterobacteria (Zeng *et al.*, 2009). CfrA also allows norepinephrin-bound iron to be utilized, although noradrenaline has a lower affinity for ferric iron than enterochelin. Three molecules of noradrenaline are needed for a tri-catecholate complex. This complex is clearly less stable than the enterochelin complex with three catecholates connected by covalent linkages. Future studies will reveal which property of CfrA is more important for virulence, the iron supply by enterochelin or the recognition of noradrenaline.

Carboxymycobactin

Mycobacterium tuberculosis produces two siderophores: the very hydrophobic mycobactin and the more hydrophilic carboxymycobactin [Fig. 12.1(11)] (Ratledge, 2004). The more soluble carboxymycobactin is secreted; mycobactin allows storage of iron in the waxy envelope of the bacterium. Iron bound to one of the mycobactins might be taken up by the ABC transporter IrtAB, in which the ATPase domain is fused to the membrane component; no binding protein of this system has been found. Likewise, no binding protein has been found for the yersiniabactin ABC transporter encoded by *ybtPQ*. However, IrtA has an N-terminal extension that binds flavin-adenine-dinucleotide, which may be involved in reducing the siderophore-bound iron and which may help to release Fe(II) from the siderophore (Ryndak *et al.*, 2010). How the iron is transported through the lipid-rich outer layer is not clear. The ligand may be donated from carboxymycobactin to mycobactin, or carboxymycobactin may reach the cytoplasm via IrtAB. Mycobactin synthesis is necessary for growth within macrophages (De Voss *et al.*, 2000), although a mycobactin mutant grows as well as the parent strain in an iron-replete medium. There is evidence that *M. tuberculosis* experiences low iron in macrophages and that iron overload predisposes the host for an *M. tuberculosis* infection (Timm *et al.*, 2003).

Petrobactin

Petrobactin [Fig. 12.1(6)] was first isolated from *Marinobacter hydrocarbonoclasticus* (Barbeau *et al.*, 2002), a bacterium which had gained interest because it participates in the degradation of oil spills in the sea. Later, petrobactin was found as a siderophore produced in the early growth phase of *Bacillus anthracis* (Wilson *et al.*, 2009). *B. anthracis* also produces bacillibactin, a catechol siderophore typically produced by various *Bacillus* strains. It is assumed that petrobactin, like additional siderophores produced by other pathogens, is additionally produced by *B. anthracis* and *Bacillus cereus* to help the pathogen evade the host defence; bacillibactin is bound by the host siderocalin, whereas petrobactin is not.

Staphyloferrin

Staphylococcus aureus can produce two siderophores: staphyloferrin A and B. The biosynthesis of the siderophores has recently been elucidated (Cheung *et al.*, 2009; Cotton *et al.*, 2009), but their role in virulence remains to be determined.

Transferrin and lactoferrin

Utilization of human iron compounds

Transferrin, the iron transporter in blood plasma, supplies cells with the necessary iron. Transferrin tightly binds iron, which is therefore not easily accessible for bacteria. Since transferrin is only about 40% loaded with ferric iron, it has the capacity to keep the free iron level low. The related protein lactoferrin is found mainly in secretory fluids. Its function is to scavenge free iron and to inhibit bacterial growth. In contrast to transferrin, lactoferrin is able to avidly bind ferric iron also under acidic condition.

Some highly specialized pathogens have evolved ways to tap this iron source in the body. The best-studied examples are *Neisseria* and *Haemophilus,* which have specific receptors in their outer membrane to bind transferrin, release the iron, and transport the ferric iron through the cytoplasmic membrane.

Interestingly, the TonB-dependent protein TbpA, together with the outer membrane lipoprotein TbpB, binds transferrin. These proteins from *Neisseria gonorrhoeae* and *Neisseria meningitidis* recognize only human or primate transferrin, which explains why there is no good animal model to study these pathogens. However, transgenic mice expressing human transferrin were recently created in which a meningococcal infection could be observed; in contrast, in wild-type mice, *N. meningitidis* were rapidly cleared (Zarantonelli *et al.*, 2007). The TonB-dependent receptor TbpA together with the highly surface-exposed TbpB protein have to structurally change the bound transferrin for iron to be released through the barrel of TbpA into the periplasm. There, the ferric-binding protein FbpA of an ABC transporter takes over and delivers the iron into the cytoplasm via the transmembrane permease FbpB and the energizing ATPase FbpC (Perkins-Balding *et al.*, 2004).

Another mechanistically very similar transport system is found in *Neisseria* spp.; LbtA and LbtB transport lactoferrin. In *Haemophilus* spp., only transferrin-specific transporters have been demonstrated. Also *N. meningitidis* recognizes the environment through the availability of the various iron sources. Haemoglobin, lactoferrin, or transferrin as an iron source leads to very different patterns of gene expression (Jordan and Saunders, 2009).

Haem and haem-protein utilization systems

Haem transport

In humans, 2 g of haem iron in haemoglobin are a potentially dangerous iron reservoir and, indeed, many pathogenic bacteria possess haemolysins (which often also lyse other cell types) and are able to use haem. The first haem uptake system studied in molecular detail was that of *Y. enterocolitica* (Stojiljkovic and Hantke, 1994). Surprisingly, it showed the same characteristics as the siderophore transport systems. HemR is the TonB-dependent outer membrane receptor, HemT is the periplasmic binding protein, HemU is the cytoplasmic permease, and HemV is the ATPase energizing the uptake through the cytoplasmic membrane. A nearly identical transport system was found in *Y. pestis* (Perry, 2004), and closely related systems were found mostly in pathogenic *E. coli* and *Shigella* strains, but not in *Salmonella* strains. A different nomenclature is used for the *Shigella* system: *Shigella* haem uptake *shuTWXYUV* and *shuAS*. *shuA* encodes the outer membrane receptor, which has recently been crystallized (Fig. 12.3). The receptor has the same general 3-D structure as the siderophore receptors, with a 22-stranded β-barrel and a plug. It is assumed that haem is bound by a histidine at the apex of the plug and a histidine in loop 7 of the barrel (Cobessi *et al.*, 2010). The genes *shuTUV* encode proteins that transport hemin through the cytoplasmic membrane, very similar to the transport proteins in *Y. enterocolotica* (Stojiljkovic and Hantke, 1994). However, the haem dependency of an *E. coli* K-12 mutant strain can be complemented solely by an outer membrane haem receptor, which indicates the presence of a haem transport system in the cytoplasmic membrane. *In silico* searches for *hemTUV* homologues in the K-12 genome were not successful, and another haem transport system had to be postulated. Indeed, it was later shown that the known dipeptide transport system encoded by *dppBCDF* was additionally able to transport haem. The two periplasmic binding proteins of this transport system, the dipeptide-binding protein DppA and the L-alanyl-γ-D-glutamyl-meso-diaminopimelate-binding protein MppA, were shown to bind haem in competition with their respective peptide substrate (Létoffé *et al.*, 2006).

In *Sinorhizobium meliloti*, the *hmuTUV* genes encode a haem-specific ABC transporter in the cytoplasmic membrane. HemT is the periplasmic binding protein, HemU is the permease, and HemV is the ATPase. Interestingly, HemU and HemV are also components of the ferrichrome and ferrioxamine transport system, with FhuP as the periplasmic binding protein (Cuiv *et al.*, 2008).

Haem utilization by Gram-negative bacteria

Enterobacteria

It has long been assumed that the iron can only be removed from haem by destroying haem, e.g. by a haem oxygenase. As mentioned above, an outer membrane TonB-dependent haem transporter is sufficient to supply *E. coli* with the necessary haem-derived iron for growth. However, how the iron is removed from haem remained a mystery, since an intensive search for any haem-degrading enzyme in *E. coli* failed. The dogma of haem destruction has recently been challenged by the detection of two proteins, YfeX and EfeB, in *E. coli* that are able to remove iron from the tetrapyrrol backbone (Létoffé *et al.*, 2009). Protein YfeX is found in the cytoplasm. The periplasmic EfeB protein is part of the iron transport system EfeUOB (Grosse *et al.*, 2006; Cao *et al.*, 2007), which is mainly active under acidic conditions and has been characterized as an Fe(II) transporter (see below). However, the main function of EfeUOB may be the evacuation of iron from haem. Homologues of YfeX and EfeB are found in many other bacteria and may be important for ordinary haem turnover.

The extra haem-utilization systems in enterobacteria contain four additional proteins that have only partly been characterized. These proteins most likely help to remove the iron from haem. ShuS is a cytoplasmic haem-binding protein. The apo-protein binds to DNA (Kaur and Wilks, 2007). The ShuS homologue PhuS from *Pseudomonas fluorescence* probably has a dual function: (i) presenting haem to the haem oxygenase HemO (also called PigA) and (ii) influencing the iron and haem homeostasis by sensing haem (Kaur *et al.*, 2009). PhuS with bound haem no longer binds to DNA.

The function of the *shuWXY* gene products and homologues in other bacteria is unknown. In *E. coli* K-12, the *shuWXY* genes are not necessary for haem uptake and utilization, as mentioned above. ChuX is structurally similar to HemS and is able to bind two haem per ChuX dimer (Suits *et al.*, 2009). However, ShuW seems to be distantly related to HemN of *E. coli*, which is an oxygen-independent coproporphyrinogen III oxidase (Layer *et al.*, 2006). Therefore, ShuW is a candidate for an alternative haem-degrading enzyme.

Bordetella pertussis

Bordetella spp. cause respiratory tract infections in mammals. *Bordetella pertussis* is the human-adapted agent of whooping cough. Its haem transport system, encoded by *bhuRSTV*, is similar to the haem transport systems of enterobacteria. A peculiarity is the regulation by an extracytoplasmic function (ECF) sigma factor (see below).

Serratia marcescens, *Y. pestis* and *P. aeruginosa*

Some bacteria have an interesting variation on the haem theme. They secrete proteins–the hemophores–which avidly bind haem and function like siderophores in that they collect haem from the surroundings. Owing to its hydrophobicity, haem is insoluble and sticks to hydrophobic surfaces and membranes. *Serratia marcescens, Y. pestis,* and *P. aeruginosa* secrete the HasA protein, which takes haem out of haemoglobin and delivers it to the outer membrane receptor HasR. The molecular details of this transport process have been better elucidated than many the other systems. Haem is bound with high affinity to the hemophore HasA and is transferred to the low-affinity binding site in the receptor by the HasA–HasR complex formation. During the complex formation, the axial coordination of haem to a HasA His residue and a HasA Tyr residue is broken, and the haem is bound by a His residue from the HasR loop 7 and a His from the apex of the HasR plug domain (Krieg *et al.*, 2009). The HasR outer membrane receptor has the same general 3-D structure as siderophore receptors (Fig. 12.3), and its transport activity depends on a TonB protein that slightly differs from a second TonB used for siderophore uptake in *Serratia*.

Haemophilus influenzae

Haemophilus influenzae is a highly host-adapted pathogen. For growth, it requires in addition to NAD also protoporphyrin IX or haem. *H. influenzae* has developed highly specialized systems to utilize all haem sources in the human body. This redundancy may allow the organism to live without synthesizing haem itself.

In mammals, haem is sequestered by hemopoxin with high affinity from lysed cells and lysed erythrocytes in the plasma. Hemopoxin thereby protects the body from oxidative damage by free haem. *H. influenzae* secretes the hemophore HxuA, which binds the haem–hemopoxin complex. This complex is then most likely bound by HxuC, a TonB-dependent outer membrane transporter. Here the haem is extracted and delivered in the periplasm to the haem-binding protein HbpA, which is a homologue of DppA, the haem- and dipeptide-binding protein of *E. coli.* HbpA cooperates with the ABC dipeptide transporter DppBCDF in the cytoplasmic membrane to transport haem into the cytoplasm (Morton *et al.*, 2009). Whether HbpA also functions as a dipeptide-binding protein is not known.

The human body also produces the haemoglobin-binding protein haptoglobin, which binds the haemoglobin set free from lysed erythrocytes. *H. influenzae* strains have up to four different outer membrane binding proteins for this complex. The expression of these Hgp proteins is phase variable owing to a CCAA repeat near the N-terminus. The variability of the Hgp may be explained by the polymorphism of the haptoglobins (Morton *et al.*, 2006). Again, these proteins show the signature of TonB-dependent proteins.

Porphyromonas gingivalis

The Gram-negative, anaerobic, periodontal pathogen *Porphyromonas gingivalis* is able to grow in the presence of 6% oxygen (Lewis *et al.*, 2009). It lacks the protoporphyrin IX (PPIX) biosynthetic pathway and thus requires exogenous haem for growth. The bacterium secretes a set of unique cysteine proteases called gingipains, which degrade haem-containing proteins such as haemoglobin, haptoglobin, myoglobin and hemopoxin. The lysine- and arginine-specific gingipains have domains with specific affinity for haem and seem to allow dimerization of haem (Smalley *et al.*, 2007). *P. gingivalis* produces a characteristic black pigment on blood agar plates. The pigment is composed of the μ-oxo bis-haem complex of Fe(III) protoporphyrin IX (μ-oxo oligomer, dimeric haem). Production of the pigment is a means of detoxifying surplus haem. When iron is scarce, the gingipains help to utilize haem.

The proteins encoded in the haem uptake operon *hmuYRSTUV* are quite different from the haem transport systems in enterobacteria. Only two of the proteins have been studied in some detail: the outer membrane lipoprotein HmuY and the TonB-dependent outer membrane receptor HmuR. HmuY is cleaved off the membrane by gingipain Kgp (K specific) at the N-terminal lipid anchor lip-CGKKK↓DE and may serve as a hemophore (Wójtowicz *et al.*, 2009) that delivers the bound haem to HmuR. The binding of haem to dimeric HmuY leads to HmuY tetramerization, which may protect the bound haem from host haem-scavenging binding proteins. How the haem is transferred from its cage in the HmuY tetramer to HmuR – which has a tenfold lower affinity for haem than HmuY – remains to be determined. It is not known whether the proteins HmuSTUV are involved in haem transport across the cytoplasmic membrane since they are very different from the respective proteins in enterobacteria. The naming of the proteins is therefore misleading since HmuT and HmuV are predicted membrane proteins, and HmuS is a member of the CobN/magnesium chelatase protein family.

In addition, the *htrABCD* genes encoding an ABC transporter with similarity to the haem-specific enterobacterial ABC transporters has been identified; this system is assumed to transport haem through the cytoplasmic membrane (Slakeski *et al.*, 2000). Downstream from the *htrABCD* operon, the Tlr TonB-dependent receptor is encoded. This receptor has a postulated specificity for haem and seems to back up the HmuYR system. Unexpectedly, a *tlr* mutant as well as an *hmuR* mutant grows less well in a haem-poor medium than the parent strain (Aduse-Opoku *et al.*, 1997; Simpson *et al.*, 2000).

Neisseria

N. gonorrhoeae and *N. meningitidis* are highly human-adapted pathogens. Although their main iron source is transferrin, they have, in addition, outer membrane haem transport systems, which mainly use haemoglobin as a haem source. Both pathogens sometimes carry a two-component haemoglobin–haptoglobin outer membrane uptake system consisting of HpuA and HpuB. HpuA is a 36-kDa outer membrane lipoprotein, and HpuB is a TonB-dependent receptor. Both proteins are necessary for haem utilization from the haemoglobin–haptoglobin complex, and again humans provide the preferred sources (Rohde and Dyer, 2004). HmbR is a TonB-dependent outer membrane receptor found only in *N. meningitidis;* HmbR specifically recognizes human haemoglobin (Stojiljkovic *et al.*, 1996). Nothing is known about haem transport through the cytoplasmic membrane. In contrast to the apathogenic *Neisseria sicca* and *Neisseria mucosa,* no homologue of the ABC transporter HemTUV or of the peptide/haem-binding proteins DppA and MppA is encoded in the genomes of *N. gonorrhoeae* and *N. meningitidis.*

In human male volunteers, the haem-uptake system seems to be less important than the transferrin iron-uptake system (Cornelissen *et al.*, 1998). The expression of the *hpu* genes is phase variable, and in most freshly isolated *N. gonorrhoeae* strains, the Hpu proteins are not expressed. During the early phase of the menstrual cycle in women, when haemoglobin is readily available from menstrual blood, the expression of the *hpu* genes is favoured (Anderson *et al.*, 2001).

Bartonella

Bartonella spp. are intracellular pathogens that are able to parasitize erythrocytes and trigger pathological angiogenesis. These bacteria are typically transmitted by haematophagous arthropod vectors, e.g. lice and fleas. *Bartonella* spp. seem to have a special relationship with blood, and it is not surprising that very high amounts of haem are required for their cultivation. Five haem-binding proteins have been identified in the outer membrane of *Bartonella henselae* (Zimmermann *et al.*, 2003). Homologues are found in other *Bartonella* spp. and in *Brucella*. In *Bartonella quintana,* the expression of these proteins is regulated by temperature, oxygen, and haem, but their function is unknown (Battisti *et al.*, 2006). One of these proteins, Pap31, binds fibronectin (Dabo *et al.*, 2006).

Further analysis of the genome of *B. quintana* revealed an operon predicted to encode HutA (TonB-dependent outer membrane receptor), HemS (haem-binding factor), HutB (periplasmic haem-binding protein), HutC (ABC transporter permease), and HmuV (ABC transporter ATPase) (Parrow *et al.*, 2009). As to be expected in an α-proteobacterium, the operon is regulated by IrrA, a member of the Fur protein family, which binds haem and has a special role in haem homeostasis regulation.

Haem utilization by Gram-positive bacteria

Corynebacterium diphtheriae

The first haem-specific ABC transporter with similarity to the haem transporters in enterobacteria was identified in *Corynebacterium diphtheriae* (Drazek *et al.*, 2000). The binding protein HmuT has the typical lipid anchor of Gram-positive 'periplasmic' binding proteins, HmuU is the permease, and HmuV is the energy-providing ATPase of the complex. Two additional haem-binding proteins, HtaA and HtaB, have been identified; however, in contrast to those from staphylococci which are bound to peptidoglycan (see below), these proteins are anchored by a C-terminal transmembrane region to the cytoplasmic membrane (Allen and Schmitt, 2009). Deletion of *htaB* does not impair growth on haem-iron-limiting medium. The *htaA* and *hmuABC* genes are necessary for good growth on haem as sole iron source, but since the mutants nevertheless grow slightly, the presence of additional, less-efficient haem transport systems have been postulated. Interestingly, an HtaA homologue has been found in *Propionibacterium acnes*; the protein may be covalently anchored to the cell-wall peptidoglycan, since it contains a C-terminal signal recognized by sortases (see below) (Allen and Schmitt, 2009). In *C. diphtheriae,* haem is degraded by the haem oxygenase HmuO (Schmitt, 1997). A complex sensory system regulates *hmuO*. Each of the two-component systems ChrA/ChrS and HrrA/HrrS activate *hmuO* transcription. Both regulators are also responsible for the haemoglobin-dependent repression of *hemA*; HemA is involved in haem biosynthesis (Bibb *et al.*, 2007). Recently, the haem-dependent phosphorylation of the sensor kinase ChrS has been demonstrated (Ito *et al.*, 2009).

S. aureus

The highly human-adapted pathogen *S. aureus* produces two siderophore synthesis and uptake systems, and it preferentially utilizes haem from different sources (Skaar *et al.*, 2004). However, how this Gram-positive bacterium obtains haem differs from the means of Gram-negative bacteria. TonB-dependent receptor proteins at the cell surface are lacking. Instead, Isd (iron-regulated surface determinant) proteins are anchored to the peptidoglycan by sortases. The sortase SrtA recognizes the C-terminal sequence LPXTG, which is found in many proteins anchored in the cell wall. Sortase SrtB encoded in the *isd* operon specifically recognizes the C-terminal sequence NPQTN of certain haem-transport proteins.

In the first step of haem-iron acquisition, haemoglobin is bound to the peptidoglycan-anchored receptor IsdB, and haemoglobin–haptoglobin is bound to IsdH. IsdB and IsdH then remove haem and pass it to IsdA, which then passes it to IsdC (Maresso and Schneewind, 2006). IsdC, in turn, passes haem to the haem-binding protein IsdE. The peptidoglycan-anchored proteins IsdBHAC all contain a so-called NEAT (near iron transporter) domain, which specifically binds haem. These proteins deliver haem in an ordered cascade to IsdE (Muryoi *et al.*, 2008). IsdE is a lipid-anchored binding protein and part of a typical ABC transporter, which also contains the membrane permease IsdF and the ATPase IsdD. In the cytoplasm, haem is degraded by the haem oxygenases IsdG and IsdI (Maresso *et al.*, 2008). However, haem can be toxic if it is not degraded or appropriately bound. *S. aureus* senses high levels of haem via the two-component regulatory system HssRS and up-regulates the expression of the haem-regulated transporter HrtAB. Strains lacking *hrtAB* are sensitive to high levels of haem (Stauff *et al.*, 2008; Stauff and Skaar, 2009) since they are unable to export haem.

Systems homologous to the Isd haem-utilization system are found in other pathogenic Gram-positive bacteria, e.g. *B. anthracis, B. cereus, Listeria* spp., and *Streptococcus* spp. Interestingly, *B. anthracis* secretes the hemophore IsdX1, a haem-binding protein with a NEAT domain that scavenges haem from haemoglobin and delivers it to the peptidoglycan-anchored IsdC (Fabian *et al.*, 2009). The sequences of the hemophores HasA from *S. marcescens,* HmuY from *P. gingivalis,* and IsdX1 from *B. anthracis* are quite different, which suggests that hemophore systems have evolved independently, at least three times.

Streptococcus pyogenes

Streptococcus pyogenes has a haem-specific ABC transporter encoded by *htsABC* (= *siaABC*) (Bates *et al.*, 2003), consisting of the lipoprotein HtsA, the membrane permease HtsB, and the ATPase HtsC. The two surface proteins Shr and Shp are also involved in haem transport. Haem bound to Shr (1275 amino acids) is efficiently transferred to Shp (291 amino acids). Both proteins have a putative C-terminal membrane anchor (Zhu *et al.*, 2008).

Lactic acid bacteria depend mainly on manganese and need very little or no iron. However, the mostly harmless, only sometimes opportunistic pathogenic lactic acid bacterium *Enterococcus faecalis* seems to utilize haem when grown in blood, as suggested by recent transcriptome studies (Vebø *et al.*, 2009). A possible contribution of haem utilization to virulence is assumed.

Other iron transporters

Ferrous iron is stable under anoxic conditions and is more soluble than ferric iron in the physiological pH range. Therefore, it is not surprising that many bacteria have transporters for ferrous iron.

Feo

One of the first bacterial ferrous iron transporters identified was FeoB, an inner membrane protein with a permease domain and a GTPase domain (Kammler *et al.*, 1993; Stojiljkovic *et al.*, 1993). It is widely distributed in the bacterial kingdom. Often, one or two small genes, *feoA* and *feoC,* are found upstream and downstream of *feoB*. FeoA is predicted to have a regulatory function, and FeoC may function as a FeS-dependent regulator of the system (Cartron *et al.*, 2006). In *P. gingivalis,* a FeoB homologue mediates manganese uptake (Dashper *et al.*, 2005).

FeoB is the only known bacterial transporter containing a small GTPase domain with similarity to the eukaryotic G proteins (Hantke, 2003). Recently, several crystal structures of the soluble FeoB GTPase domain have been published; the structures nicely show the canonical RAS-like GTPase fold (Guilfoyle *et al.*, 2009; Hattori *et al.*, 2009; Köster *et al.*, 2009). In contrast to ABC transporters in which the hydrolysis of the nucleotide triphosphate leads to a structural change and coupling to transport, in the G proteins, the binding of the nucleotide triphosphate leads to the activating structural change (Marlovits *et al.*, 2002). The GTPase seems to control the influx of iron via the membrane part of FeoB. The linker between the GTPase domain and membrane permease functions as a guanine nucleotide dissociation inhibitor (GDI-like domain), as indicated by the substantial stabilization of GDP binding in the presence of the membrane part (Eng *et al.*, 2008).

Sit

SitABCD is a widely distributed ABC transporter for divalent cations, such as manganese (Mn) and Fe(II). The proteins have been studied in *S. enterica*. The periplasmic binding protein SitA is related to the manganese-binding proteins Psa from *Streptococcus pneumoniae* and TroA from *Treponema pallidum* (Desrosiers *et al.*, 2007). SitB is the ATPase component, and SitC and SitD are the two transmembrane components of the ABC transporter. The metal specificity of these binding proteins can differ in different species, and their function *in vivo* has frequently been debated; mainly Mn(II) and Fe(II) are recognized and transported. TroA may also bind zinc, but this may not be relevant for the organism, since a zinc (Zn)-specific Zur-type transport system is also synthesized (Desrosiers *et al.*, 2007).

Interestingly, *sit*-like genes have been found mainly in pathogenic enterobacteria, e.g. *Salmonella, Shigella,* pathogenic *E. coli, Klebsiella, Citrobacter,* and *Yersinia* strains. In many cases, the importance of this system for full virulence has been demonstrated. For *S. enterica,* not only the iron supply, but also the manganese supply is necessary for full virulence (Boyer *et al.*, 2002). The α-proteobacteria *Rhizobium, Agrobacterium,* and *Bartonella* also encode SitA binding proteins that have a high specificity for Mn(II) and which are very similar to SitA of *Salmonella* (up to 72% identity).

Other divalent metal ions certainly strongly influence the iron homeostasis in the cell. For example, in *E. coli,* the MntH manganese transporter supports growth in iron-limited medium (Anjem *et al.*, 2009). One reason for this is that some iron-containing enzymes may be substituted by manganese-containing enzymes, e.g. under iron-limiting conditions, iron superoxide dismutase SodB is replaced by manganese superoxide dismutase SodA. Manganese can also substitute iron in the activation of certain iron enzymes. MntH is also strongly induced under oxidative stress, and mutants devoid of MntH are more sensitive to oxidative stress (Anjem *et al.*, 2009).

Efe

The first oxidase-dependent iron transporter, Ftr1p, was studied in yeast. Later, homologues in *E. coli* strains were identified. The *E. coli* K-12 homologue EfeU, however, is not functional because of a frameshift mutation; in some other *E. coli* strains, the transporter is functional (Grosse *et al.*, 2006). The two periplasmic proteins EfeO and EfeB are necessary for a functional Fe(II) transport system. EfeO has a cupredoxin domain, and EfeB is a haem-oxidase-like protein (Cao *et al.*, 2007). Recently EfeB was shown to have a deferrochelatase activity (see above, Létoffé *et al.*, 2009). It channels the iron from haem to the transport system. The transport system is mainly active at the acidic pH 5, where ferrous iron has an increased stability in the presence of oxygen. EfeO contains copper, which makes this iron-uptake pathway copper dependent, as also observed in the Ftr1-Fet3p iron-transport system of yeast.

Iron export

Humans are unable to secrete iron; surplus iron is usually stored in ferritins. The only natural ways to get rid of iron is by shedding of epithelia and blood loss from wounds. The iron content of the body is only regulated by the uptake. In bacteria, very little is known about iron secretion, except for those bacteria that utilize iron as an energy source in redox processes. In *E. coli,* surplus iron is also stored in ferritins and in the Dps protein. However, in *E. coli* the cation diffusion facilitator FieF – a proton antiporter – catalyses the export of Fe(II) and Zn(II), but a $FieF^-$ mutant phenotype is only observed in a Fur mutant, in which oxidative stress caused by iron is enhanced (Grass *et al.*, 2005). This indicates that, at least in *E. coli,* the export of the potentially toxic iron is only of minor importance.

Regulation of iron uptake: global regulators

Fur and DtxR

Both Fur and DtxR are used by pathogens to regulate not only the nutritional supply but also certain virulence determinants. Since the mammalian host keeps free iron levels very low and has developed strategies of iron withholding from uninvited guests, low iron supply is recognized by many pathogenic bacteria as an important environmental signal which is answered by the induction of virulence factors. These virulence factors are not always directly related to the iron supply. The repression of diphtheria toxin production by iron has been known since 1936 (Pappenheimer and Johnson, 1936) and that of Shiga toxin, since 1946 (Dubos and Geiger, 1946). Other virulence factors regulated by iron are, e.g. various haemolysins, exotoxin A from *P.*

aeruginosa, elastase from *P. aeruginosa*, various fimbria, and biofilm formation.

Fur

In the exemplary *E. coli* K-12, the global regulator for iron is Fur (ferric iron uptake regulator), a protein with an unusual N-terminal helix–turn–helix motif for binding to DNA and a C-terminal dimerization and metal-binding domain. With Fe(II) as a corepressor, this protein represses iron uptake systems. Fur also influences the oxidative stress response and iron homeostasis. Panina *et al.* (2001) predicted that Fur in *E. coli* regulates about 70 genes, which has been confirmed by microarray studies (McHugh *et al.*, 2003). Fur-like proteins are found in many Gram-negative and in some Gram-positive bacteria, and many of these proteins with Fe(II) as a cofactor mainly repress genes involved in iron homeostasis and the oxidative stress response. Some puzzling observations that indicated that Fur acts also as a positive regulator were resolved when RyhB was detected, a small antisense RNA negatively regulated by Fur (Massé *et al.*, 2005). Under low iron conditions, Fur repression of *ryhB* is relieved, and the small RNA (90 nucleotides) is produced. This RNA binds with the help of Hfq (an RNA-binding protein) to the mRNA of certain genes, e.g. *sdhCDAB*, which encodes the iron-containing succinate dehydrogenase complex. The mRNA is degraded and the protein, e.g. succinate dehydrogenase, is not produced. This is reasonable since under iron-poor conditions, the apo-proteins cannot be loaded with iron–sulfur clusters and cytochrome. Some proteins of the Fur family regulate zinc uptake with zinc as corepressor (Zur), manganese uptake (Mur), nickel uptake (Nur), or the oxidative stress response (Per) or haem homeostasis (Irr) (Lee and Helmann, 2007).

DtxR

A second general regulator for 'iron genes' was first found in *C. diphtheriae*. DtxR with iron as a corepressor represses diphtheria toxin production. This protein is also composed of an N-terminal helix–turn–helix motif and a C-terminal metal-binding domain, but there is no sequence similarity with Fur. Interestingly, this protein has a homologue with different metal specificity: MntR. With manganese as corepressor, MntR regulates manganese uptake in several Gram-negative bacteria, e.g. *E. coli,* and in Gram-positive bacteria, e.g. *S. aureus* and *B. anthracis*.

Specific regulators

Besides these global regulators, certain systems have additional specific regulatory devices to fine tune the response of the organism. Especially for the utilization of xenosiderophores, it is advantageous for the organism to express the uptake system only when the siderophore is present in the surroundings.

ECF sigma factors

A paradigm of this type of regulation is the well-studied citrate–iron transport system of *E. coli* (Braun *et al.*, 2006), which belongs to the ECF regulatory systems. The FecA siderophore receptor has an extension of about 80 amino acid residues on the N-terminal side of the TonB box. This domain interacts with the C-terminal domain of FecR, a membrane-anchored periplasmic protein. The TonB-dependent ferric citrate transport through FecA is recognized by this interaction, and this signal makes FecR a substrate for the RseP protease. RseP is part of a periplasmic stress signal-transduction cascade in the cytoplasmic membrane, which leads to the activation of sigma factor E. An N-terminal fragment of FecR is cleaved off by RseP and serves as an activator of the sigma factor FecI. The activated FecI recruits the RNA polymerase, directs it to the promoter of the *fecABCDE* operon and stimulates transcription. The *fecIR* genes are regulated by Fur, which ensures that the Fec transport system is only expressed when iron is limiting. Similar systems have been observed in pseudomonads, where many xenosiderophore iron uptake systems are ECF regulated.

As mentioned above, the *bhuRSTUV* genes for haem uptake in *B. pertussis* are regulated by the ECF sigma factor HurIR (Vanderpool and Armstrong, 2004). The BhuR protein is the haem receptor which has, like FecR, an N-terminal extension of about 70 amino acids. This extension is assumed to signal via HurR the TonB-dependent haem transport through BhuR. Activated

HurR somehow stimulates the sigma factor HurI, which activates transcription of *bhuRSTUV*. The transcription of the *hurP* gene, which encodes a homologue of the site 2 protease RseP of *E. coli*, is also activated by HurI. Interestingly, a *hurP* mutant is unable to induce the HurR haem receptor in the presence of haem.

A particular feature of the siderophore ECF sigma factors seems to be that the proteolytically generated fragment of the 'anti-sigma factor' activates the sigma factor, in contrast to other anti-sigma factors, where the degradation sets an active sigma factor free from the anti-sigma factor/sigma factor complex (Mettrick and Lamont, 2009).

AraC-like regulators

AraC-like regulators specifically regulate certain siderophore biosynthesis and transport systems. Well-known examples are the pyochelin regulator PchR, the alcaligin regulator AlcR (Brickman *et al.*, 2001), and the yersiniabactin regulator YbtA (Anisimov *et al.*, 2005). An AraC-type regulator binds as a repressor to the promoter region of the gene to be regulated; however, in the presence of an effector molecule, another DNA binding site is recognized and transcription of the corresponding gene is stimulated. For the above-mentioned proteins, the cognate siderophores are the effectors, which stimulate transcription of biosynthesis and uptake genes.

Two-component histidine kinase response regulators

Two-component histidine sensor kinases and their cognate response regulators are widely distributed signal transducers that allow the cells to recognize and respond to stimuli in the medium. In only a few cases in Gram-positive bacteria is this signal transduction pathway used to induce the degradation or secretion of haem.

C. diphtheriae has two haem-dependent two-component systems: ChrSR and HrrAS. Both systems are necessary for full induction of the haem oxygenase HmuO (Bibb *et al.*, 2007). The genes necessary for uptake of haem seem to be regulated mainly by the global regulator DtxR and iron.

In *S. aureus*, the two-component system HssRS induces expression of the haem export system HrtAB, which is an ABC-type efflux pump that protects the cells from a toxic haem surplus. Orthologous Hss and Hrt systems have been found *in silico* in *B. anthracis*, *Listeria monocytogenes* and *E. faecalis*, which suggests a conserved regulatory system by which these Gram-positive pathogens sense haem and protect themselves from excess haem (Torres *et al.*, 2007).

Iron-withholding strategies of the host

It has long been known that after infection the iron levels in serum are lowered; this has been interpreted as a measure of the host against the invading parasite (Ganz, 2009). The high susceptibility of patients with iron overload to infections and animal models in which iron supply enhances the severity of the infection supports the assumption that lowering the level of serum iron is a defence strategy of the organism against invaders.

Iron must generally be cautiously handled by the body because iron not properly bound will form radicals with oxygen and lead to oxidative stress. Free haem similarly catalyses the generation of oxygen radicals. Of the iron in serum, about 90% is bound to transferrin and 10% is bound to ferritin; haem is bound to hemopoxin; and hemoglobin is bound to haptoglobin. Therefore, the dangerous molecules iron and haem are bound in a form that is not easily accessible to most free-living bacteria. Commensals and parasites have tricks to tap these resources. The components of the innate immune system that participate in the iron-deprivation strategy of the body are described in the following sections.

Transferrin

Transferrin in serum is only saturated to about 30–40% and is therefore always able to complex a small surplus of iron. During inflammation, macrophages, which recycle erythrocytes, decrease the release of iron-loaded transferrin and thereby decrease the levels of circulating iron, which leads to hypoferraemia. The level of the transferrin receptor Tfr1 in macrophages is lowered, with the consequence that the growth of intracellular pathogens, e.g. *Legionella pneumophila*, is inhibited (Byrd and Horwitz, 1989).

Lactoferrin

Lactoferrin is secreted by neutrophils and mucosal epithelia. Its secretion is stimulated by inflammation. The protein is able to complex ferric iron under more acidic conditions than transferrin. These measures lower the iron level in the bloodstream. Degradation products of lactoferrin have additional defensin activity (Jenssen and Hancock, 2009).

Siderocalin

Since siderophores are able to compete with transferrin for iron, it is interesting that during inflammation, a protein that binds catecholate siderophores specifically and with high affinity is secreted by neutrophils, epithelia, and hepatocytes. This protein, siderocalin, has also been called NGAL lipocalin or lipocalin 2 in humans, 24p3/uterocalin in mice, and neu-related lipocalin in rats. Experiments with knockout mice deficient in siderocalin have demonstrated that the production of siderocalin indeed protects the mouse from infection by a faecal *E. coli* strain that produces only enterochelin (Flo *et al.*, 2004). Enterochelin perfectly fits into the cup-like cavity formed by the β-sheets of siderocalin (Goetz *et al.*, 2002). Consequently, enterochelin production is not an important virulence trait. Pathogens such as *S. enterica* and virulent *E. coli* produce the di-glucosylated enterochelin called salmochelin or other siderophores, e.g. aerobactin and yersiniabactin, which are not bound by siderocalin (Fischbach *et al.*, 2006; Valdebenito *et al.*, 2007). Although carboxymycobactins from mycobacteria have a quite different structure [Fig. 12.1(11)] they are also avidly bound by siderocalin (Holmes *et al.*, 2005), which may inhibit the establishment of a *M. tuberculosis* infection (Saiga *et al.*, 2008).

The stressed enteric nervous system stimulates release of noradrenaline, which has been shown *in vitro* and in animal model systems to stimulate growth of many pathogens. Besides a quorum sensing effect (Clarke *et al.*, 2006), noradrenalin and related catecholamines support the growth of pathogens by improving the iron supply under iron limiting conditions (Methner *et al.*, 2008; Pullinger *et al.*, 2010). In this context siderocalin may have an additional function in scavenging iron–noradrenaline complexes from the circulations (Miethke and Skerra, 2010).

Nramp1

In macrophages, the Nramp1 transporter for divalent cations, such as Fe(II), Co(II) and Mn(II), functions in defence. Mouse Nramp1 mutants are more susceptible to *Salmonella* and *M. tuberculosis* infections than the wild-type animal. It is assumed that the Nramp1 transporter in the phagosomal membrane depletes the phagosome of iron and manganese, thereby inhibiting the growth of pathogens. However, a defective Nramp1 has additional effects on iron homeostasis in the macrophage (Fritsche *et al.*, 2007).

Hemopexin

Hemopexin has the highest haem-binding affinity in the plasma and mainly functions in scavenging haem from the bloodstream. The hydrophobic haem may intercalate into membranes and produce cell-damaging hydroxyl radicals. Levels of hemopoxin, as an acute-phase protein, increase during inflammation. Many pathogenic bacteria possess haem-uptake systems and are able to recruit haem from haemoglobin; however, only a few specialists are able to utilize haem from hemopoxin (see above). Therefore, hemopoxin has, in addition to its protecting function against oxidative stress, also a function in the iron-withholding strategy of the host. Similarly, haptoglobin binds haemoglobin in the bloodstream to remove this, in free form, toxic protein. In this case also, only bacterial specialists are able to remove the haem from this complex. Therefore, also haptoglobin aids in withholding iron from pathogens (Ascenzi *et al.*, 2005).

Concluding remarks

Iron uptake and virulence

The human body has recognized iron as a much-sought nutrient of many bacterial invaders. It has developed strategies to use iron deprivation as a weapon against intruders to protect against a plethora of bacteria. However, pathogens have evolved mechanisms to bypass this host defence mechanism. As illustrated by the many examples given above, pathogens have developed various tools to gain iron. Therefore, it is often difficult to assess whether a specific iron-transport system is decisive for the virulence of a specific bacterial strain. In this context, it is interesting to note that most pathogenic bacteria–in contrast to apathogenic bacteria–have developed haem-uptake and -utilization systems.

At different stages of the infection route, different iron sources may be available. It is plausible that in these various stages, different iron-transport systems become important for the outcome of the infection. For example, in *Y. pestis,* the yersiniabactin system is absolutely necessary during the lymphatic stage of bubonic plague, whereas the Yfe transport system, which is specific for Fe(II) and Mn(II), is essential in later stages of the infection (Perry, 2004). Interestingly, several other iron-transport systems are encoded in the genome of *Y. pestis* strains, but a closer inspection reveals that owing to mutations these systems are no longer functional, as has been shown for the aerobactin system (Forman *et al.*, 2007).

Owing to the redundancy of iron-transport systems in many bacteria, the absence of a certain transport system leads in some cases only to a relatively mild attenuation of the pathogen.

Therapeutic applications?

Although we have been aware for decades of the battle for iron between host and invader, no efficient therapeutics have been developed along these lines. Some natural antibiotics have the structure of a siderophore with an antibiotically active compound attached. An example is albomycin (Fig 12.1), a hydroxamate siderophore produced by certain *Streptomyces* strains. Albomycin is accumulated as a siderophore by *E. coli* via FhuA. In the cytoplasm, the peptidase PepN cleaves off the active compound, which is an inhibitor of the seryl-tRNA synthetase (Braun *et al.*, 2009). This 'Trojan horse' principle found in nature has also been applied to combine catechols with β-lactams (Curtis *et al.*, 1988). Other siderophores have also been employed as carriers of antimicrobial compounds (Miller *et al.*, 2009). These antibiotics have the advantage that they are actively accumulated by efficient transport systems that are strongly induced in the host. Consequently, often very low minimal inhibitory concentrations are observed. In addition, resistance is less likely to occur, since mutations in these uptake systems often lower the fitness of the bacterium in the host. However, to date, no antibiotic of this type has appeared on the market. Another trial in the pipeline is the new siderophore β-lactam, BAL30072 (Mushtaq *et al.*, 2010).

One siderophore, ferrioxamine B, sold under the trade name Desferal, has been marketed and produced since 1970 in large amounts. However, it is not sold as a therapeutic against infections, but is rather used against iron poisoning and hemochromatosis. Iron in the body is chelated by Desferal, and the iron complex is excreted via faeces and urine. Since the chelator has to be administered by injection, alternative siderophores or chelators that can be applied orally have been sought. In the last years, two synthetic chelators, deferiprone and deferasirox (Paradkar *et al.*, 2008), have appeared on the market. Both have been developed as oral drugs, and their iron complexes are secreted via the faeces. In comparison to ferrioxamine B, they seem to induce a different means of cellular ferritin degradation (De Domenico *et al.*, 2009).

Acknowledgements

I thank Volkmar Braun and Karen A. Brune for critical reading of the manuscript. I am very grateful to Robert C. Hider (London) for the generously allowing the siderophore structures shown in Fig. 12.1 to be used. The work of the author has been continually supported by the DFG (HA1186).

References

Aduse-Opoku, J., Slaney, J.M., Rangarajan, M., Muir, J., Young, K.A., and Curtis, M.A. (1997). The Tla protein of *Porphyromonas gingivalis* W50: a homolog of the RI protease precursor (PrpRI) is an outer membrane receptor required for growth on low levels of hemin. J. Bacteriol. *179*, 4778–4788.

Allen, C.E., and Schmitt, M.P. (2009). HtaA is an iron-regulated hemin binding protein involved in the utilization of heme iron in *Corynebacterium diphtheriae.* J. Bacteriol. *191*, 2638–2648.

Anderson, J.E., Leone, P.A., Miller, W.C., Chen, C., Hobbs, M.M., and Sparling, P.F. (2001). Selection for expression of the gonococcal hemoglobin receptor during menses. J. Infect. Dis. *184*, 1621–1623.

Anisimov, R., Brem, D., Heesemann, J., and Rakin, A. (2005). Transcriptional regulation of high pathogenicity island iron uptake genes by YbtA. Int. J. Med. Microbiol. *295*, 19–28.

Anjem, A., Varghese, S., and Imlay, J.A. (2009). Manganese import is a key element of the OxyR response to hydrogen peroxide in *Escherichia coli*. Mol. Microbiol. *72*, 844–858.

Ascenzi, P., Bocedi, A., Visca, P., Altruda, F., Tolosano, E., Beringhelli, T., and Fasano, M. (2005). Hemoglobin and heme scavenging. IUBMB Life *57*, 749–759.

Barbeau, K., Zhang, G., Live, D.H., and Butler, A. (2002). Petrobactin, a photoreactive siderophore produced by the oil-degrading marine bacterium *Marinobacter hydrocarbonoclasticus*. J. Am. Chem. Soc. *124*, 378–379.

Bates, C.S., Montañez, G.E., Woods, C.R., Vincent, R.M., and Eichenbaum, Z. (2003). Identification and characterization of a *Streptococcus pyogenes* operon involved in binding of hemoproteins and acquisition of iron. Infect. Immun. *71*, 1042–1055.

Battisti, J.M., Sappington, K.N., Smitherman, L.S., Parrow, N.L., and Minnick, M.F. (2006). Environmental signals generate a differential and coordinated expression of the heme receptor gene family of *Bartonella quintana*. Infect. Immun. *74*, 3251–3261.

Bibb, L.A., Kunkle, C.A., and Schmitt, M.P. (2007). The ChrA-ChrS and HrrA-HrrS signal transduction systems are required for activation of the *hmuO* promoter and repression of the *hemA* promoter in *Corynebacterium diphtheriae*. Infect. Immun. *75*, 2421–2431.

Bister, B., Bischoff, D., Nicholson, G.J., Valdebenito, M., Schneider, K., Winkelmann, G., Hantke, K., and Süssmuth, R.D. (2004). The structure of salmochelins: C-glucosylated enterobactins of *Salmonella enterica*. Biometals *17*, 471–481.

Boyer, E., Bergevin, I., Malo, D., Gros, P., and Cellier, M.F. (2002). Acquisition of Mn(II) in addition to Fe(II) is required for full virulence of *Salmonella enterica* serovar Typhimurium. Infect. Immun. *70*, 6032–6042.

Braun, V. (2009). FhuA (TonA), the career of a protein. J. Bacteriol. *191*, 3431–3436.

Braun, V., Patzer, S.I., and Hantke, K. (2002). Ton-dependent colicins and microcins: modular design and evolution. Biochimie *84*, 365–380.

Braun, V., Mahren, S., and Sauter, A. (2006). Gene regulation by transmembrane signaling. Biometals *19*, 103–113.

Braun, V., Pramanik, A., Gwinner, T., Köberle, M., and Bohn, E. (2009). Sideromycins: tools and antibiotics. Biometals *22*, 3–13.

Brickman, T.J., Kang, H.Y., and Armstrong, S.K. (2001). Transcriptional activation of *Bordetella* alcaligin siderophore genes requires the AlcR regulator with alcaligin as inducer. J. Bacteriol. *183*, 483–489.

Buchanan, S.K., Smith, B.S., Venkatramani, L., Xia, D., Esser, L., Palnitkar, M., Chakraborty, R., van der Helm, D., and Deisenhofer, J. (1999). Crystal structure of the outer membrane active transporter FepA from *Escherichia coli*. Nat. Struct. Biol. *6*, 56–63.

Byrd, T.F., and Horwitz, M.A. (1989). Interferon gamma-activated human monocytes downregulate transferrin receptors and inhibit the intracellular multiplication of *Legionella pneumophila* by limiting the availability of iron. J. Clin. Invest. *83*, 1457–1465.

Cadieux, N., and Kadner, R.J. (1999). Site-directed disulfide bonding reveals an interaction site between energy-coupling protein TonB and BtuB, the outer membrane cobalamin transporter. Proc. Natl. Acad. Sci. U.S.A. *96*, 10673–10678.

Cao, J., Woodhall, M.R., Alvarez, J., Cartron, M.L., and Andrews, S.C. (2007). EfeUOB (YcdNOB) is a tripartite, acid-induced and CpxAR-regulated, low-pH $Fe2^+$ transporter that is cryptic in *Escherichia coli* K-12 but functional in *E. coli* O157:H7. Mol. Microbiol. *65*, 857–875.

Cartron, M.L., Maddocks, S., Gillingham, P., Craven, C.J., and Andrews, S.C. (2006). Feo-transport of ferrous iron into bacteria. Biometals *19*, 143–157.

Cheung, J., Beasley, F.C., Liu, S., Lajoie, G.A., and Heinrichs, D.E. (2009). Molecular characterization of staphyloferrin B biosynthesis in *Staphylococcus aureus*. Mol. Microbiol. *74*, 594–608.

Clarke, M.B., Hughes, D.T., Zhu, C., Boedeker, E.C., and Sperandio, V. (2006). The QseC sensor kinase: a bacterial adrenergic receptor. Proc. Natl. Acad. Sci. U.S.A. *103*, 10420–10425.

Cobessi, D., Meksem, A., and Brillet, K. (2010). Structure of the heme/hemoglobin outer membrane receptor ShuA from *Shigella dysenteriae*: heme binding by an induced fit mechanism. Proteins *78*, 286–294.

Cornelissen, C.N., Kelley, M., Hobbs, M.M., Anderson, J.E., Cannon, J.G., Cohen, M.S., and Sparling, P.F. (1998). The transferrin receptor expressed by gonococcal strain FA1090 is required for the experimental infection of human male volunteers. Mol. Microbiol. *27*, 611–616.

Cotton, J.L., Tao, J., and Balibar, C.J. (2009). Identification and characterization of the *Staphylococcus aureus* gene cluster coding for staphyloferrin A. Biochemistry *48*, 1025–1035.

Crosa, J.H., Mey, A.R., and Payne, S.M., eds. (2004). Iron Transport in Bacteria (ASM Press, Washington DC).

Cuiv, P.O., Keogh, D., Clarke, P., and O'Connell, M. (2008). The *hmu*UV genes of *Sinorhizobium meliloti* 2011 encode the permease and ATPase components of an ABC transport system for the utilization of both heme and the hydroxamate siderophores, ferrichrome and ferrioxamine B. Mol. Microbiol. *70*, 1261–1273.

Curtis, N.A., Eisenstadt, R.L., East, S.J., Cornford, R.J., Walker, L.A., and White, A.J. (1988). Iron-regulated outer membrane proteins of *Escherichia coli* K-12 and mechanism of action of catechol-substituted cephalosporins. Antimicrob. Agents Chemother. 32, 1879–1886.

Dabo, S.M., Confer, A.W., Anderson, B.E., and Gupta, S. (2006). *Bartonella henselae* Pap31, an extracellular matrix adhesin, binds the fibronectin repeat III13 module. Infect. Immun. *74*, 2513–2521.

Dashper, S.G., Butler, C.A., Lissel, J.P., Paolini, R.A., Hoffmann, B., Veith, P.D., O'Brien-Simpson, N.M., Snelgrove, S.L., Tsiros, J.T., and Reynolds, E.C. (2005). A novel *Porphyromonas gingivalis* FeoB plays a role in manganese accumulation. J. Biol. Chem. *280*, 28095–28102.

De Domenico, I., Ward, D.M., and Kaplan, J. (2009). Specific iron chelators determine the route of ferritin degradation. Blood *114*, 4546–4551.

De Voss, J.J., Rutter, K., Schroeder, B.G., Su, H., Zhu, Y., and Barry, C.E., 3rd (2000). The salicylate-derived mycobactin siderophores of *Mycobacterium tuberculosis* are essential for growth in macrophages. Proc. Natl. Acad. Sci. U.S.A. *97*, 1252–1257.

Desrosiers, D.C., Sun, Y.C., Zaidi, A.A., Eggers, C.H., Cox, D.L., and Radolf, J.D. (2007). The general transition metal (Tro) and Zn^{2+} (Znu) transporters in *Treponema pallidum*: analysis of metal specificities and expression profiles. Mol. Microbiol. *65*, 137–152.

Drazek, E.S., Hammack, C.A., and Schmitt, M.P. (2000). *Corynebacterium diphtheriae* genes required for acquisition of iron from haemin and haemoglobin are homologous to ABC haemin transporters. Mol Microbiol *36*, 68–84.

Dubos, R.J., and Geiger, J.W. (1946). Preparation and properties of Shiga Toxin and Toxoid. J. Exp. Med. *84*, 143–156.

Earhart, C.F. (2004). Iron uptake via the enterobactin system. In Iron Transport in Bacteria, Crosa, J.H., Mey, A.R., and Payne, S.M., eds. (ASM Press, Washington DC), pp. 133–147.

Eng, E.T., Jalilian, A.R., Spasov, K.A., and Unger, V.M. (2008). Characterization of a novel prokaryotic GDP dissociation inhibitor domain from the G protein coupled membrane protein FeoB.J. Mol. Biol. *375*, 1086–1097.

Fabian, M., Solomaha, E., Olson, J.S., and Maresso, A.W. (2009). Heme transfer to the bacterial cell envelope occurs via a secreted hemophore in the Gram-positive pathogen *Bacillus anthracis*. J. Biol. Chem. *284*, 32138–32146.

Ferguson, A.D., Hofmann, E., Coulton, J.W., Diederichs, K., and Welte, W. (1998). Siderophore-mediated iron transport: crystal structure of FhuA with bound lipopolysaccharide. Science *282*, 2215–2220.

Ferguson, A.D., Braun, V., Fiedler, H.P., Coulton, J.W., Diederichs, K., and Welte, W. (2000). Crystal structure of the antibiotic albomycin in complex with the outer membrane transporter FhuA. Protein Sci. *9*, 956–963.

Fischbach, M.A., Lin, H., Zhou, L., Yu, Y., Abergel, R.J., Liu, D.R., Raymond, K.N., Wanner, B.L., Strong, R.K., Walsh, C.T., *et al.* (2006). The pathogen-associated *iroA* gene cluster mediates bacterial evasion of lipocalin 2. Proc. Natl. Acad. Sci. U.S.A. *103*, 16502–16507.

Flo, T.H., Smith, K.D., Sato, S., Rodriguez, D.J., Holmes, M.A., Strong, R.K., Akira, S., and Aderem, A. (2004). Lipocalin 2 mediates an innate immune response to bacterial infection by sequestrating iron. Nature *432*, 917–921.

Forman, S., Nagiec, M.J., Abney, J., Perry, R.D., and Fetherston, J.D. (2007). Analysis of the aerobactin and ferric hydroxamate uptake systems of *Yersinia pestis*. Microbiology *153*, 2332–2341.

Fritsche, G., Nairz, M., Theurl, I., Mair, S., Bellmann-Weiler, R., Barton, H.C., and Weiss, G. (2007). Modulation of macrophage iron transport by Nramp1 (Slc11a1). Immunobiology *212*, 751–757.

Ganz, T. (2009). Iron in innate immunity: starve the invaders. Curr. Opin. Immunol. *21*, 63–67.

Goetz, D.H., Holmes, M.A., Borregaard, N., Bluhm, M.E., Raymond, K.N., and Strong, R.K. (2002). The neutrophil lipocalin NGAL is a bacteriostatic agent that interferes with siderophore-mediated iron acquisition. Mol. Cell *10*, 1033–1043.

Grass, G., Otto, M., Fricke, B., Haney, C.J., Rensing, C., Nies, D.H., and Munkelt, D. (2005). FieF (YiiP) from *Escherichia coli* mediates decreased cellular accumulation of iron and relieves iron stress. Arch. Microbiol. *183*, 9–18.

Greenwald, J., Zeder-Lutz, G., Hagege, A., Celia, H., and Pattus, F. (2008). The metal dependence of pyoverdine interactions with its outer membrane receptor FpvA.J. Bacteriol. *190*, 6548–6558.

Grosse, C., Scherer, J., Koch, D., Otto, M., Taudte, N., and Grass, G. (2006). A new ferrous iron-uptake transporter, EfeU (YcdN), from *Escherichia coli*. Mol. Microbiol. *62*, 120–131.

Guilfoyle, A., Maher, M.J., Rapp, M., Clarke, R., Harrop, S., and Jormakka, M. (2009). Structural basis of GDP release and gating in G protein coupled Fe^{2+} transport. EMBO J. *28*, 2677–2685.

Hancock, R.W., and Braun, V. (1976). Nature of the energy requirement for the irreversible adsorption of bacteriophages T1 and phi80 to *Escherichia coli*. J. Bacteriol. *125*, 409–415.

Hannauer, M., Barda, Y., Mislin, G.L., Shanzer, A., and Schalk, I.J. (2010). The ferrichrome uptake pathway in *Pseudomonas aeruginosa* involves an iron release mechanism with acylation of the siderophore and a recycling of the modified desferrichrome. J. Bacteriol. *192*, 1212–1220.

Hantke, K. (2003). Is the bacterial ferrous iron transporter FeoB a living fossil? Trends Microbiol. *11*, 192–195.

Hantke, K., Nicholson, G., Rabsch, W., and Winkelmann, G. (2003). Salmochelins, siderophores of *Salmonella enterica* and uropathogenic *Escherichia coli* strains, are recognized by the outer membrane receptor IroN. Proc. Natl. Acad. Sci. U.S.A. *100*, 3677–3682.

Hartmann, A., and Braun, V. (1980). Iron transport in *Escherichia coli*: uptake and modification of ferrichrome. J. Bacteriol. *143*, 246–255.

Hattori, M., Jin, Y., Nishimasu, H., Tanaka, Y., Mochizuki, M., Uchiumi, T., Ishitani, R., Ito, K., and Nureki, O. (2009). Structural basis of novel interactions between the small-GTPase and GDI-like domains in prokaryotic FeoB iron transporter. Structure *17*, 1345–1355.

Hider, R.C., and Kong, X. (2010). Chemistry and biology of siderophores. Nat. Prod. Rep. *27*, 637–657.

Holmes, M.A., Paulsene, W., Jide, X., Ratledge, C., and Strong, R.K. (2005). Siderocalin (Lcn 2) also binds carboxymycobactins, potentially defending against mycobacterial infections through iron sequestration. Structure *13*, 29–41.

Imperi, F., Tiburzi, F., and Visca, P. (2009). Molecular basis of pyoverdine siderophore recycling in

Pseudomonas aeruginosa. Proc. Natl. Acad. Sci. U S A *106*, 20440–20445.

Ito, Y., Nakagawa, S., Komagata, A., Ikeda-Saito, M., Shiro, Y., and Nakamura, H. (2009). Heme-dependent autophosphorylation of a heme sensor kinase, ChrS, from *Corynebacterium diphtheriae* reconstituted in proteoliposomes. FEBS Lett. *583*, 2244–2248.

Jenssen, H., and Hancock, R.E. (2009). Antimicrobial properties of lactoferrin. Biochimie *91*, 19–29.

Jordan, P.W., and Saunders, N.J. (2009). Host iron binding proteins acting as niche indicators for *Neisseria meningitidis*. PLoS One *4*, e5198.

Kammler, M., Schön, C., and Hantke, K. (1993). Characterization of the ferrous iron uptake system of *Escherichia coli*. J. Bacteriol. *175*, 6212–6219.

Kaur, A.P., and Wilks, A. (2007). Heme inhibits the DNA binding properties of the cytoplasmic heme binding protein of *Shigella dysenteriae* (ShuS). Biochemistry *46*, 2994–3000.

Kaur, A.P., Lansky, I.B., and Wilks, A. (2009). The role of the cytoplasmic heme-binding protein (PhuS) of *Pseudomonas aeruginosa* in intracellular heme trafficking and iron homeostasis. J. Biol. Chem. *284*, 56–66.

Köster, S., Wehner, M., Herrmann, C., Kühlbrandt, W., and Yildiz, O. (2009). Structure and function of the FeoB G-domain from *Methanococcus jannaschii*. J. Mol. Biol. *392*, 405–419.

Krieg, S., Huché, F., Diederichs, K., Izadi-Pruneyre, N., Lecroisey, A., Wandersman, C., Delepelaire, P., and Welte, W. (2009). Heme uptake across the outer membrane as revealed by crystal structures of the receptor–hemophore complex. Proc. Natl. Acad. Sci. U.S.A. *106*, 1045–1050.

Layer, G., Pierik, A.J., Trost, M., Rigby, S.E., Leech, H.K., Grage, K., Breckau, D., Astner, I., Jänsch, L., Heathcote, P., *et al.* (2006). The substrate radical of *Escherichia coli* oxygen-independent coproporphyrinogen III oxidase HemN.J. Biol. Chem. *281*, 15727–15734.

Lee, J.W., and Helmann, J.D. (2007). Functional specialization within the Fur family of metalloregulators. Biometals *20*, 485–499.

Létoffé, S., Delepelaire, P., and Wandersman, C. (2006). The housekeeping dipeptide permease is the *Escherichia coli* heme transporter and functions with two optional peptide binding proteins. Proc. Natl. Acad. Sci. U.S.A. *103*, 12891–12896.

Létoffé, S., Heuck, G., Delepelaire, P., Lange, N., and Wandersman, C. (2009). Bacteria capture iron from heme by keeping tetrapyrrol skeleton intact. Proc. Natl. Acad. Sci. U.S.A. *106*, 11719–11724.

Lewis, J.P., Iyer, D., and Anaya-Bergman, C. (2009). Adaptation of *Porphyromonas gingivalis* to microaerophilic conditions involves increased consumption of formate and reduced utilization of lactate. Microbiology *155*, 3758–3774.

McHugh, J.P., Rodriguez-Quinoñes, F., Abdul-Tehrani, H., Svistunenko, D.A., Poole, R.K., Cooper, C.E., and Andrews, S.C. (2003). Global iron-dependent gene regulation in *Escherichia coli*. A new mechanism for iron homeostasis. J. Biol. Chem. *278*, 29478–29486.

Maresso, A.W., and Schneewind, O. (2006). Iron acquisition and transport in *Staphylococcus aureus*. Biometals *19*, 193–203.

Maresso, A.W., Garufi, G., and Schneewind, O. (2008). *Bacillus anthracis* secretes proteins that mediate heme acquisition from hemoglobin. PLoS Pathog. *4*, e1000132.

Marlovits, T.C., Haase, W., Herrmann, C., Aller, S.G., and Unger, V.M. (2002). The membrane protein FeoB contains an intramolecular G protein essential for Fe(II) uptake in bacteria. Proc. Natl. Acad. Sci. U.S.A. *99*, 16243–16248.

Massé, E., Vanderpool, C.K., and Gottesman, S. (2005). Effect of RyhB small RNA on global iron use in *Escherichia coli*. J. Bacteriol. *187*, 6962–6971.

Matzanke, B.F., Anemüller, S., Schünemann, V., Trautwein, A.X., and Hantke, K. (2004). FhuF, part of a siderophore-reductase system. Biochemistry *43*, 1386–1392.

Methner, U., Rabsch, W., Reissbrodt, R., and Williams, P.H. (2008). Effect of norepinephrine on colonisation and systemic spread of *Salmonella enterica* in infected animals: role of catecholate siderophore precursors and degradation products. Int. J. Med. Microbiol. *298*, 429–439.

Mettrick, K.A., and Lamont, I.L. (2009). Different roles for anti-sigma factors in siderophore signalling pathways of *Pseudomonas aeruginosa*. Mol. Microbiol. *74*, 1257–1271.

Miethke, M., and Skerra, A. (2010). Neutrophil gelatinase-associated lipocalin possesses antimicrobial activity by interfering with L-norepinephrine-mediated bacterial iron acquisition. Antimicrob. Agents Chemother. *54*, 1580–1589.

Miller, M.J., Zhu, H., Xu, Y., Wu, C., Walz, A.J., Vergne, A., Roosenberg, J.M., Moraski, G., Minnick, A.A., McKee-Dolence, J., *et al.* (2009). Utilization of microbial iron assimilation processes for the development of new antibiotics and inspiration for the design of new anticancer agents. Biometals *22*, 61–75.

Mittal, R., Sharma, S., Chhibber, S., and Harjai, K. (2008). Iron dictates the virulence of *Pseudomonas aeruginosa* in urinary tract infections. J. Biomed. Sci. *15*, 731–741.

Morton, D.J., VanWagoner, T.M., Seale, T.W., Whitby, P.W., and Stull, T.L. (2006). Differential utilization by *Haemophilus influenzae* of haemoglobin complexed to the three human haptoglobin phenotypes. FEMS Immunol. Med. Microbiol. *46*, 426–432.

Morton, D.J., Seale, T.W., Vanwagoner, T.M., Whitby, P.W., and Stull, T.L. (2009). The dppBCDF gene cluster of *Haemophilus influenzae*: role in heme utilization. BMC Res. Notes *2*, 166.

Muryoi, N., Tiedemann, M.T., Pluym, M., Cheung, J., Heinrichs, D.E., and Stillman, M.J. (2008). Demonstration of the iron-regulated surface determinant (Isd) heme transfer pathway in *Staphylococcus aureus*. J. Biol. Chem. *283*, 28125–28136.

Mushtaq, S., Warner, M., and Livermore, D. (2010). Activity of the siderophore monobactam BAL30072 against multiresistant non-fermenters. J. Antimicrob. Chemother. *65*, 266–270.

Panina, E.M., Mironov, A.A., and Gelfand, M.S. (2001). Comparative analysis of FUR regulons in gamma-proteobacteria. Nucleic Acids Res. *29*, 5195–5206.

Pappenheimer, A.M., and Johnson, S.J. (1936). Studies in diphtheria toxin production. 1.The effect of iron and copper. Brit. J. Exp. Pathol. *17*, 335–341.

Paradkar, P.N., De Domenico, I., Durchfort, N., Zohn, I., Kaplan, J., and Ward, D.M. (2008). Iron depletion limits intracellular bacterial growth in macrophages. Blood *112*, 866–874.

Parrow, N.L., Abbott, J., Lockwood, A.R., Battisti, J.M., and Minnick, M.F. (2009). Function, regulation, and transcriptional organization of the hemin utilization locus of *Bartonella quintana*. Infect. Immun. *77*, 307–316.

Payne, S.M., and Mey, A.R. (2004). Pathogenic *Escherichia coli, Shigella* and *Salmonella* In Iron Transport in Bacteria, Crosa, J.H., Mey, A.R., and Payne, S.M., eds. (ASM Press, Washington DC), pp. 199–218.

Perkins-Balding, D., Ratliff-Griffin, M., and Stojiljkovic, I. (2004). Iron transport systems in *Neisseria meningitidis*. Microbiol. Mol. Biol. Rev. *68*, 154–171.

Perry, R.D. (2004). *Yersinia*. In Iron Transport in Bacteria, Crosa, J.H., Mey, A.R., and Payne, S.M., eds. (ASM Press, Washington DC), pp. 219–240.

Pullinger, G.D., Carnell, S.C., Sharaff, F.F., van Diemen, P.M., Dziva, F., Morgan, E., Lyte, M., Freestone, P.P., and Stevens, M.P. (2010). Norepinephrine augments *Salmonella enterica*-induced enteritis in a manner associated with increased net replication but independent of the putative adrenergic sensor kinases QseC and QseE. Infect. Immun. *78*, 372–380.

Ratledge, C. (2004). Iron, mycobacteria and tuberculosis. Tuberculosis (Edinb) *84*, 110–130.

Rohde, K.H., and Dyer, D.W. (2004). Analysis of haptoglobin and hemoglobin–haptoglobin interactions with the *Neisseria meningitidis* TonB-dependent receptor HpuAB by flow cytometry. Infect. Immun. *72*, 2494–2506.

Ryndak, M.B., Wang, S., Smith, I., and Rodriguez, G.M. (2010). The *Mycobacterium tuberculosis* high-affinity iron importer, IrtA, contains an FAD-binding domain. J. Bacteriol. *192*, 861–869.

Saiga, H., Nishimura, J., Kuwata, H., Okuyama, M., Matsumoto, S., Sato, S., Matsumoto, M., Akira, S., Yoshikai, Y., Honda, K., *et al.* (2008). Lipocalin 2-dependent inhibition of mycobacterial growth in alveolar epithelium. J. Immunol. *181*, 8521–8527.

Schmitt, M.P. (1997). Utilization of host iron sources by *Corynebacterium diphtheriae*: identification of a gene whose product is homologous to eukaryotic heme oxygenases and is required for acquisition of iron from heme and hemoglobin. J. Bacteriol. *179*, 838–845.

Shultis, D.D., Purdy, M.D., Banchs, C.N., and Wiener, M.C. (2006). Outer membrane active transport: structure of the BtuB:TonB complex. Science *312*, 1396–1399.

Simpson, W., Olczak, T., and Genco, C.A. (2000). Characterization and expression of HmuR, a TonB-dependent hemoglobin receptor of *Porphyromonas gingivalis*. J. Bacteriol. *182*, 5737–5748.

Skaar, E.P., Humayun, M., Bae, T., DeBord, K.L., and Schneewind, O. (2004). Iron-source preference of *Staphylococcus aureus* infections. Science *305*, 1626–1628.

Slakeski, N., Dashper, S.G., Cook, P., Poon, C., Moore, C., and Reynolds, E.C. (2000). A *Porphyromonas gingivalis* genetic locus encoding a heme transport system. Oral Microbiol. Immunol. *15*, 388–392.

Smalley, J.W., Birss, A.J., Szmigielski, B., and Potempa, J. (2007). Sequential action of R- and K-specific gingipains of *Porphyromonas gingivalis* in the generation of the heme-containing pigment from oxyhaemoglobin. Arch. Biochem. Biophys. *465*, 44–49.

Stauff, D.L., and Skaar, E.P. (2009). The heme sensor system of *Staphylococcus aureus*. Contrib. Microbiol. *16*, 120–135.

Stauff, D.L., Bagaley, D., Torres, V.J., Joyce, R., Anderson, K.L., Kuechenmeister, L., Dunman, P.M., and Skaar, E.P. (2008). *Staphylococcus aureus* HrtA is an ATPase required for protection against heme toxicity and prevention of a transcriptional heme stress response. J. Bacteriol. *190*, 3588–3596.

Stojiljkovic, I., Cobeljic, M., and Hantke, K. (1993). *Escherichia coli* K-12 ferrous iron uptake mutants are impaired in their ability to colonize the mouse intestine. FEMS Microbiol. Lett. *108*, 111–115.

Stojiljkovic, I., and Hantke, K. (1994). Transport of haemin across the cytoplasmic membrane through a haemin-specific periplasmic binding-protein-dependent transport system in *Yersinia enterocolitica*. Mol. Microbiol. *13*, 719–732.

Stojiljkovic, I., Larson, J., Hwa, V., Anic, S., and So, M. (1996). HmbR outer membrane receptors of pathogenic *Neisseria* spp.: iron-regulated, hemoglobin-binding proteins with a high level of primary structure conservation. J. Bacteriol. *178*, 4670–4678.

Suits, M.D., Lang, J., Pal, G.P., Couture, M., and Jia, Z. (2009). Structure and heme binding properties of *Escherichia coli* O157:H7 ChuX. Protein Sci. *18*, 825–838.

Takase, H., Nitanai, H., Hoshino, K., and Otani, T. (2000). Impact of siderophore production on *Pseudomonas aeruginosa* infections in immunosuppressed mice. Infect. Immun. *68*, 1834–1839.

Thomas, M.S. (2007). Iron acquisition mechanisms of the *Burkholderia cepacia* complex. Biometals *20*, 431–452.

Timm, J., Post, F.A., Bekker, L.G., Walther, G.B., Wainwright, H.C., Manganelli, R., Chan, W.T., Tsenova, L., Gold, B., Smith, I., *et al.* (2003). Differential expression of iron-, carbon-, and oxygen-responsive mycobacterial genes in the lungs of chronically infected mice and tuberculosis patients. Proc. Natl. Acad. Sci. U.S.A. *100*, 14321–14326.

Torres, V.J., Stauff, D.L., Pishchany, G., Bezbradica, J.S., Gordy, L.E., Iturregui, J., Anderson, K.L., Dunman, P.M., Joyce, S., and Skaar, E.P. (2007). A *Staphylococcus aureus* regulatory system that responds to host heme and modulates virulence. Cell Host Microbe *1*, 109–119.

Valdebenito, M., Crumbliss, A.L., Winkelmann, G., and Hantke, K. (2006). Environmental factors influence the production of enterobactin, salmochelin, aerobactin, and yersiniabactin in *Escherichia coli* strain Nissle 1917. Int. J. Med. Microbiol. *296*, 513–520.

Valdebenito, M., Müller, S.I., and Hantke, K. (2007). Special conditions allow binding of the siderophore salmochelin to siderocalin (NGAL-lipocalin). FEMS Microbiol. Lett. *277*, 182–187.

Vanderpool, C.K., and Armstrong, S.K. (2004). Integration of environmental signals controls expression of *Bordetella* heme utilization genes. J. Bacteriol. *186*, 938–948.

Vebø, H.C., Snipen, L., Nes, I.F., and Brede, D.A. (2009). The transcriptome of the nosocomial pathogen *Enterococcus faecalis* V583 reveals adaptive responses to growth in blood. PLoS One *4*, e7660.

Wilson, M.K., Abergel, R.J., Arceneaux, J.E., Raymond, K.N., and Byers, B.R. (2009). Temporal production of the two *Bacillus anthracis* siderophores, petrobactin and bacillibactin. Biometals *23*, 129–134.

Wójtowicz, H., Guevara, T., Tallant, C., Olczak, M., Sroka, A., Potempa, J., Solà, M., Olczak, T., and Gomis-Rüth, F.X. (2009). Unique structure and stability of HmuY, a novel heme-binding protein of *Porphyromonas gingivalis*. PLoS Pathog. *5*, e1000419.

Xiong, Y.Q., Vasil, M.L., Johnson, Z., Ochsner, U.A., and Bayer, A.S. (2000). The oxygen- and iron-dependent sigma factor *pvdS* of *Pseudomonas aeruginosa* is an important virulence factor in experimental infective endocarditis. J. Infect. Dis. *181*, 1020–1026.

Zaborin, A., Romanowski, K., Gerdes, S., Holbrook, C., Lepine, F., Long, J., Poroyko, V., Diggle, S.P., Wilke, A., Righetti, K., *et al.* (2009). Red death in *Caenorhabditis elegans* caused by *Pseudomonas aeruginosa* PAO1. Proc. Natl. Acad. Sci. U.S.A. *106*, 6327–6332.

Zarantonelli, M.L., Szatanik, M., Giorgini, D., Hong, E., Huerre, M., Guillou, F., Alonso, J.M., and Taha, M.K. (2007). Transgenic mice expressing human transferrin as a model for meningococcal infection. Infect. Immun. *75*, 5609–5614.

Zeng, X., Xu, F., and Lin, J. (2009). Molecular, antigenic, and functional characteristics of ferric enterobactin receptor CfrA in *Campylobacter jejuni*. Infect. Immun. *77*, 5437–5448.

Zhu, H., Liu, M., and Lei, B. (2008). The surface protein Shr of *Streptococcus pyogenes* binds heme and transfers it to the streptococcal heme-binding protein Shp. BMC Microbiol. *8*, 15.

Zimmermann, R., Kempf, V.A., Schiltz, E., Oberle, K., and Sander, A. (2003). Hemin binding, functional expression, and complementation analysis of Pap 31 from *Bartonella henselae*. J. Bacteriol. *185*, 1739–1744.

Bacterial Escape from the Complement System

13

Marta Biedzka-Sarek and Mikael Skurnik

Abstract

Bacterial infections represent a global health problem. To establish infection bacteria need to defeat the action of the non-specific immune system machinery. Its essential component, activated immediately upon pathogen entry, is the complement system. Complement activation (through the classical, the lectin, and the alternative pathways) tags microbes for destruction by phagocytic cells, causes microbial lysis, and leads to generation of downstream pro-inflammatory responses. Moreover, vaccination or pre-exposure to a given microbe is not required for the complement system to eliminate the intruder. Many bacteria have, however, developed strategies to evade the complement system. Investigating these processes should further our understanding of host–bacterium interactions and contribute to prevention and treatment of bacterial infections.

The serum bactericidal activity – overview of the complement system

The complement system is known to comprise approximately 35 interacting plasma and membrane-bound proteins (Table 13.1). Activation of complement is achieved through the classical (CP), lectin (LP), and alternative pathways (AP) that differ considerably in their initiation steps. The classical and lectin pathways are initiated by specific recognition of microbes, whereas the alternative pathway is triggered spontaneously. Once the complement is activated, the pathways converge in the critical step of C3 cleavage to C3a and C3b. The cleavage is followed by the lytic pathway and the formation of the multiprotein complex called membrane-attack complex (MAC). MAC is a pore-like structure that inserts into a target membrane and disrupts it causing the cell lysis.

Classical pathway

The CP is initiated by the recognition of the antigen–antibody complexes on a microbial surface by a pattern-recognition molecule C1q. C1q is a 460-kDa hexameric protein, and each hexamer is composed of 29-, 27- and 22-kDa subunits (Table 13.1). The C1q hexamer resembles in structure a bunch of tulips with globular 'heads' and collagen-like triple helical stalks or stems (Calcott and Müller-Eberhard, 1972; Reid and Porter, 1976; Schumaker *et al.*, 1982; Strang *et al.*, 1982). Calcium-dependent interactions between C1q and two C1r and two C1s serine protease molecules lead to formation of the C1 complex (Loos *et al.*, 1980; Busby and Ingham, 1990; Thielens *et al.*, 1990). Six globular heads of C1q bind to Fc-regions of IgM or multiple IgG molecules deposited on the microbial surface (Ishizaka *et al.*, 1966; Schumaker *et al.*, 1986). In addition to immunoglobulins, C-reactive protein (CRP), nucleic acids, lipopolysaccharide and immune complexes, can serve as C1q ligands (Loos, 1982; Jiang *et al.*, 1992a,b; Nauta *et al.*, 2002). Initially, C1q binding to a target causes autoactivation of the associated C1r molecules (Gaboriaud *et al.*, 2004), which in turn activate C1s. The two activated C1s subunits cleave C4 into a small anaphylatoxic peptide C4a and a larger C4b

Table 13.1 Components of the complement system

Component	Structure	Serum conc. (μg/ml)
C1	3 subunits: C1q (460 kDa), C1r (80 kDa), and C1s (80 kDa) in a complex (C1qr2s2)	180
MBL	Polymer of 32-kDa chains	0–5
MASP-1	Single chain, 100 kDa	1.5–12
MASP-2	Single chain, 76 kDa	Not determined
C2	Single chain, 102 kDa	20
C3	2 chains: α, 110 kDa; β, 75 kDa	1300
C4	3 chains: α, 97 kDa; β, 75 kDa; γ, 33 kDa	600
Factor B	Single chain, 93 kDa	210
Factor D	Single chain, 24 kDa	2
Properdin	Oligomers of 53-kDa chains	5
C5	2 chains: 115 kDa, 75 kDa	70
C6	Single chain, 120 kDa	65
C7	Single chain, 110 kDa	55
C8	3 chains: α, 65 kDa; β, 65 kDa; γ, 22 kDa	55
C9	Single chain, 69 kDa	60

(Fig. 13.1) (Patrick *et al.*, 1970; Gorski *et al.*, 1979). Upon cleavage, an internal thioester group of C4 becomes exposed in the C4b fragment and mediates an immediate and covalent attachment of C4b to the microbial surface (Schreiber and Müller-Eberhard, 1974; Janatova and Tack, 1981; Dodds *et al.*, 1996). In the presence of Mg^{2+}, C4b acts as a C2 receptor, and activated C1s on the nearby C1 complex cleaves C2 into C2a and C2b (Nagasawa and Stroud, 1977). C2a remains complexed with C4b, whereas C2b is released into the microenvironment (Fig. 13.1). The C4b2a complex is known as CP C3 convertase (Kerr, 1980), in which the C2a catalytic subunit cleaves the C3 component. The C3 convertase is unstable and undergoes a time- and temperature-dependent decay upon which the C2a fragment dissociates, whereas C4b is ready to take up another C2 molecule (Kerr, 1980).

C3 is the most abundant complement protein in serum (Table 13.1). Native C3, like C4, contains an internal thioester bond that is hidden within a hydrophobic pocket (Janssen *et al.*, 2006; Law and Dodds, 1990, 1997). C3 cleavage releases an anaphylatoxin C3a into the environment (Hugli, 1975) and the membrane-bound C3b fragment (Fig. 13.1). C3b plays a role in opsonization by binding to nearby surfaces via the exposed thioester (Law and Levine, 1977; Sim *et al.*, 1981; Law and Dodds, 1997). The CP C3 convertase (C4b2a) may bind a C3b molecule and form the C4b2a3b complex known as the CP C5 convertase (Fig. 13.1) (Takata *et al.*, 1987).

Alternative pathway

The AP is initiated by the binding of C3b to a target membrane in the absence of specific antibodies. In addition to C3b generated during CP or LP activation, C3b is also generated as a result of spontaneous low-rate hydrolysis of the internal thioester bond in C3 (tick-over mechanism) (Lachmann and Hughes-Jones, 1984) whereby the native C3 is hydrolysed to an active intermediate known as $C3(H_2O)$ (Pangburn *et al.*, 1981). Analogous to membrane-bound C3b, $C3(H_2O)$ binds factor B (Fig. 13.1). This makes the latter susceptible for the factor D cleavage (Lesavre and Müller-Eberhard, 1978) into fragments Ba and Bb. The $C3(H_2O)Bb$ complex, known as the AP C3 convertase, generates the same C3 fragments as the C3 convertase of the CP (Fig. 13.1). Ultimately, C3b deposits on bacterial surfaces via the exposed thioester group and the AP activation proceeds (Law and Dodds, 1990 and 1997).

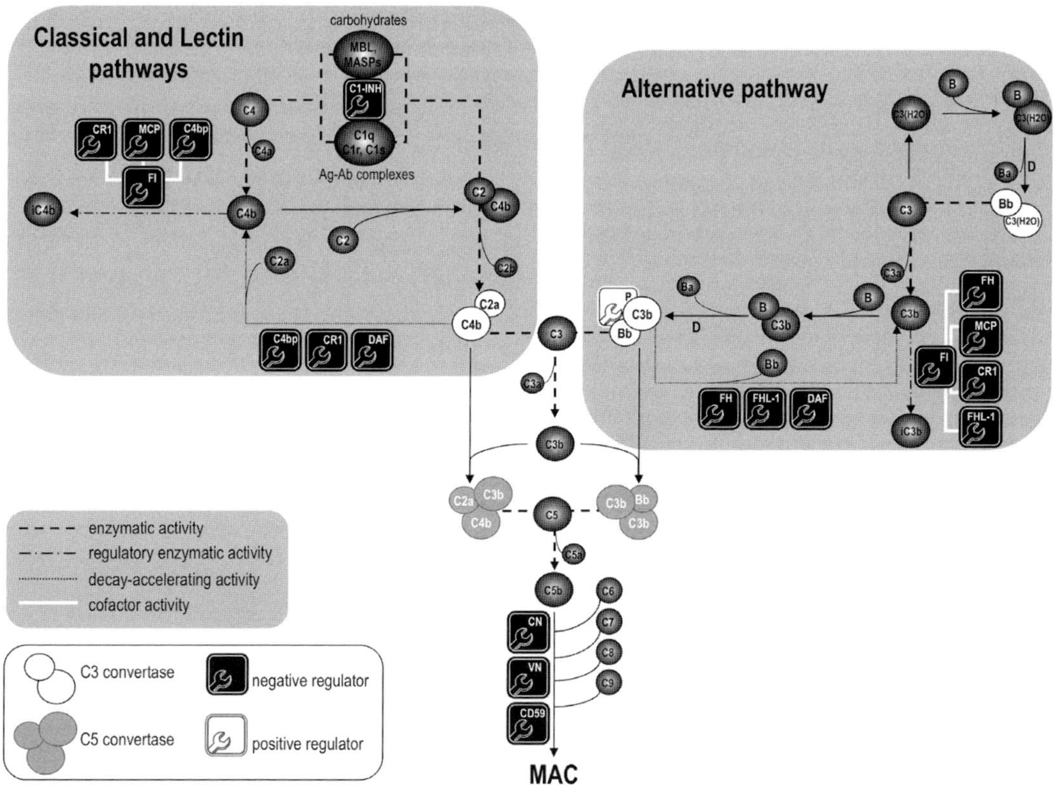

Figure 13.1 Overview of the complement activation cascade.

Magnesium-dependent binding of factor B to C3b leads to the formation of the C3bB complex. The complex is transformed by factor D into AP C3 convertase, C3bBb, and stabilized from decay by properdin (Fig. 13.1). The Bb molecule contains the enzymatic site of the AP C3 convertase.

The AP is an amplification loop for the CP and LP. The serine protease Bb of the AP C3 convertase generates enormous amounts of C3b molecules. These can rapidly opsonize the target cell or contribute to the formation of the C5 convertases of all three pathways via binding to generated C3 convertases (Lachmann and Hughes-Jones, 1984).

Lectin pathway

Mannan-binding lectin (MBL) and the structurally related ficolins are the trigger molecules of the LP. Upon recognition of certain mannose- or *N*-acetyl-glucosamine-rich polysaccharides on numerous bacteria, these molecules interact with MBL-associated serine proteases: MASP-1 and MASP-2 (Sato *et al.*, 1994; Thiel *et al.*, 1997). The MBL–MASP1–MASP2 complex is analogous to the C1 complex of the CP. Upon activation, the MASPs cleave and activate the downstream complement components C4 and C2 (Matsushita and Fujita, 1992; Thiel *et al.*, 1997). From this point onward, the LP overlaps with the CP (Fig. 13.1).

Lytic pathway

All three pathways lead to a common terminal pathway known as the lytic pathway that culminates in MAC that forms a transmembrane pore (Fig. 13.1). The C5 convertases (C3b4b2a, C3bBbC3b) perform the last proteolytic reaction of the cascade (Fig. 13.1). They cleave C5 into C5a, a potent anaphylatoxic and chemotactic fragment, and C5b, the initiating constituent of MAC (Cochrane and Müller-Eberhard, 1968; Ward and Newman, 1969; Cooper and Müller-Eberhard, 1970; DiScipio *et al.*, 1983). C5b instantaneously binds C6, and the resulting C5b6 complex acquires C7 to yield a hydrophobic C5b67 complex (Fig. 13.1) (DiScipio, 1988, 1992a). The C5b67 complex attaches to nearby membranes without harming the cells. Addition of C8 to the C5b67 complex results in the formation of the C5b678 complex that inserts into the membrane (Tamura *et al.*, 1972; Kolb and Müller-Eberhard, 1973). Binding of a C9 molecule to C5b678 initiates a process of C9 polymerization and leads to the formation of MAC (Fig. 13.1) (Tschopp *et al.*, 1982). Transmembrane MAC creates channels responsible for lytic events of complement activation. According to a 'doughnut theory', the 110-Å-wide pore has a hydrophilic lumen through which cell contents leak out and water is drawn inside (Lachmann, 2006). The osmotic pressure thereby generated ruptures the cell membrane, and in consequence the cell dies. The leakage through the pore, however, could also be assisted by the passage of the solutes and water through the 'leaky patches' formed in the membrane upon MAC insertion (Lachmann, 2006). It can be assumed that serum contains optimal concentrations of complement components to ascertain that complement activation leads to the most efficient and rapid deposition of MAC on intruding pathogens. Rawal and Pangburn (2003) reported that the C5 convertase is formed very rapidly, in less than 1 minute under *in vitro* conditions, and has a half-life of 2–3 min (Rawal and Pangburn, 2003). On the other hand, it was reported that C3b deposition on complement-susceptible human melanoma cells was saturated within 5 min, and MAC deposition reached a maximum in 10 min (Ollert *et al.*, 1994). Thus, any pathogen able to resist complement must express complement resistance factors prior to encountering complement, as there is hardly time for *de novo* synthesis of the factors.

Complement regulation

Activation of the complement system, pivotal for anti-microbial resistance, may well have deleterious consequences for the host if not tightly controlled. The labile nature of activated complement molecules, such as C3b, C4b, or C5b, ensures local activation of complement and features in protection of distant host cells against complement-mediated damage. Supporting this, a battery of membrane-bound and circulating complement regulators protects the host cells from an accidental complement attack.

The regulators of complement activation (RCA) are built of a variable number of short consensus repeat (SCR) domains of approximately 60 amino acids each (Norman *et al.*, 1991; Casasnovas *et al.*, 1999). Additionally, a few structurally unrelated proteins add to the complement-regulatory potential of the host.

Although complement regulation occurs at various stages of the cascade, most regulators act at the C3 level (Fig. 13.1). To inactivate complement at this converging point for all three pathways, the RCA proteins inhibit the formation of C3 convertases, accelerate their decay, or act as cofactors for factor I (FI)-mediated inactivation of C3b and C4b. FI is a plasma protease (Ruddy and Austen, 1969) that cleaves C3b and C4b when they are complexed to complement regulators (Fig. 13.2). Accordingly, membrane cofactor protein (MCP, CD46), complement receptor type 1 (CR1, CD35) and C4b-binding protein (C4bp) contribute to the cleavage of both C3b and C4b, whereas Factor H (FH) participates exclusively in C3b cleavage (Pangburn *et al.*, 1977; Fujita and Nussenzweig, 1979; Seya *et al.*, 1986; Ahearn and Fearon, 1989; Blom *et al.*, 2003). FI-mediated cleavage of the C3b α'-chain results in the generation of iC3b, C3c or C3dg, whereas the cleavage of the C4b α-chain generates fragments C4c and C4d (Fig. 13.2) (Nagasawa *et al.*, 1980; Davis and Harrison, 1982; Ross *et al.*, 1982; Medicus *et al.*, 1983; Lambris *et al.*, 1996).

Membrane-bound regulators

The membrane-bound regulators of complement are widely expressed by human cells and include decay-accelerating factor (DAF, CD55), MCP, CR1, and CD59 (protectin) (Morgan and Harris, 1999). Although DAF, MCP and CR1 belong to the RCA family, they differ with respect of attachment to the host cell membranes and modes of action. MCP and CR1 are transmembrane glycoproteins that act as cofactors for FI-mediated cleavage of C3b and C4b (Fig. 13.1) (Iida and

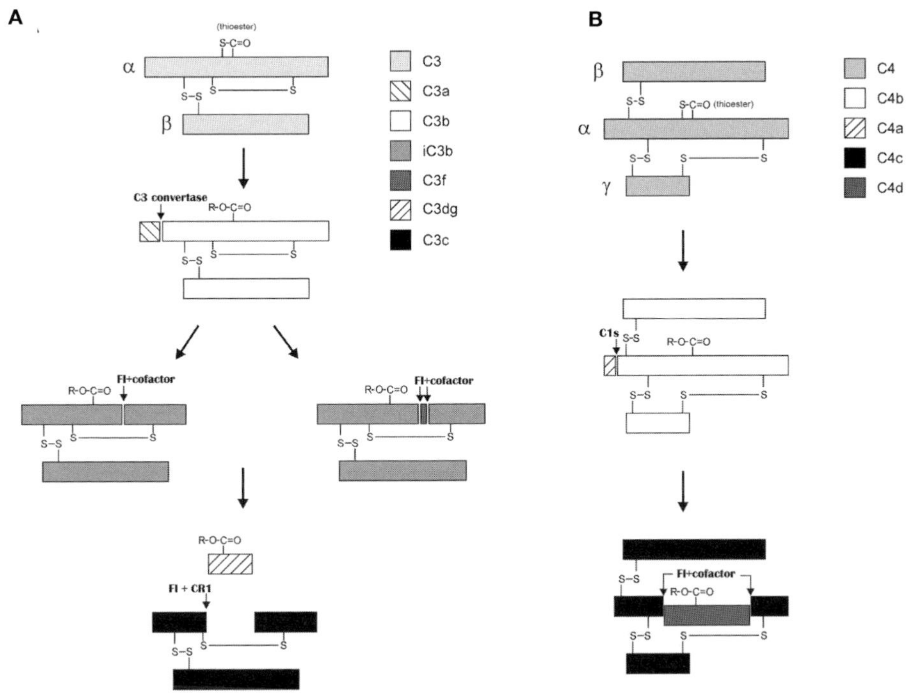

Figure 13.2 Schematic representation of the C3 (A) and C4 (B) cleavage.

Nussenzweig, 1981; Seya *et al.*, 1986). CR1, in addition, acts as cofactor in iC3b cleavage to C3c and C3d (Fig. 13.2), and accelerates decay of C3 convertases (Fearon, 1979). The latter activity is also displayed by DAF, a GPI-anchored glycoprotein (Fujita *et al.*, 1987).

CD59 has no sequence or structural similarity to the proteins of RCA family. This GPI-anchored protein is the only regulator of the lytic pathway of complement activation on host cell membranes. CD59 interferes with MAC formation by association with the C5b-8 and C5b-9 complexes (Fig. 13.1) (Meri *et al.*, 1990; Rollins and Sims, 1990).

Soluble regulators

C1-inhibitor

The C1 inhibitor (C1-INH) is a glycoprotein that belongs to a family of serine protease inhibitors (serpins). It is an acute-phase protein, thus its concentration rises dramatically during inflammatory states (Harrison, 1983). C1-INH inactivates C1 complexes in a two-fold manner. It blocks the active sites on serine proteases C1r and C1s (Harpel and Cooper, 1975) or removes the proteases from the complex (Ziccardi and Cooper, 1979). In addition, C1-INH regulates C1 activation. It attaches non-covalently to its proenzymes and prevents C1 autoactivation (Ziccardi, 1982, 1985). Moreover, C1-INH inhibits proteins of the MBL pathway of complement activation and prevents formation of the AP C3 convertase by competing with factor B for C3b binding (Matsushita *et al.*, 2000; Jiang *et al.*, 2001).

Factor H

FH is a soluble single-chain glycoprotein that down-regulates complement activity, both in the fluid phase and on cell membranes. FH accelerates the decay of the AP C3 convertase, competes with FB for C3b binding, and acts as a cofactor for FI-mediated C3b inactivation (Fig. 13.1) (Weiler *et al.*, 1976; Whaley and Ruddy, 1976; Pangburn *et al.*, 1977). FH is composed of 20 SCRs that are arranged in a continuous fashion (Fig. 13.3) (Ripoche *et al.*, 1988; Discipio, 1992b).

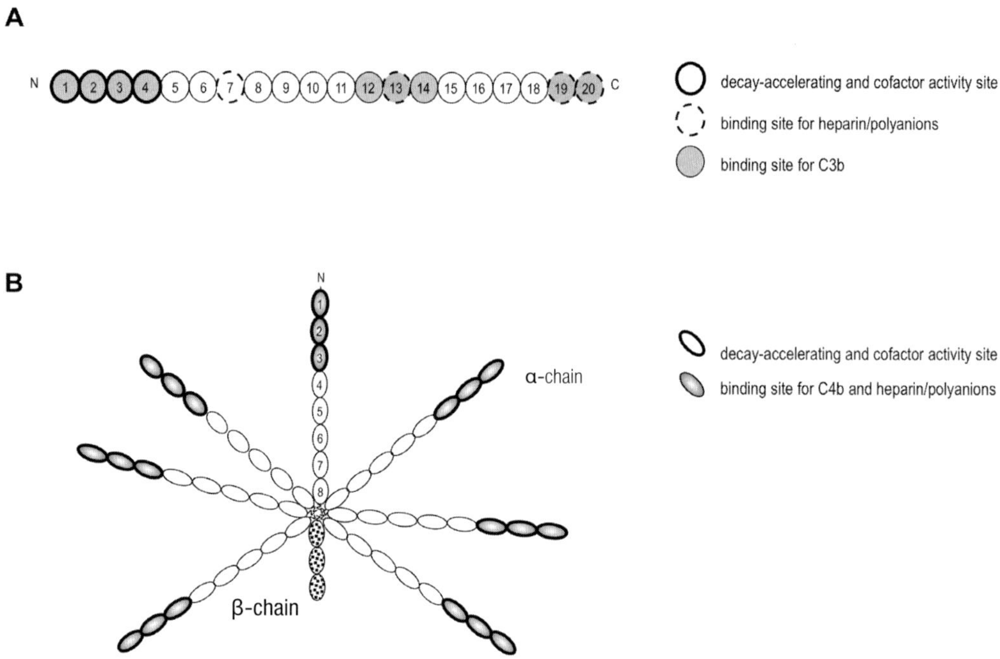

Figure 13.3 Schematic structure of FH (A) and C4bp (B). Sites for co-factor and decay-accelerating activity, as well as the binding sites for C3b, C4b, and heparin/polyanions are indicated.

Decay-accelerating and cofactor functions, as well as ability to bind C3b, heparin and polyanions have been ascribed to specific SCRs of this glycoprotein (Fig. 13.3) (Alsenz *et al.*, 1985; Pangburn *et al.*, 1991; Gordon *et al.*, 1995; Kühn *et al.*, 1995; Blackmore *et al.*, 1996, 1998b; Jokiranta *et al.*, 1996, 2000; Kühn and Zipfel, 1996; Sharma and Pangburn, 1996; Prodinger *et al.*, 1998).

Discrimination between self and non-self is crucial for FH action and is based on recognition of polyanionic substances, such as sialic acid and glycosaminoglycans (Fearon, 1978; Meri and Pangburn, 1990). They cover the host cells but, in general, are missing from foreign cell membranes. Therefore, inactivation of the surface-bound C3b by FH depends on the presence of the polyanions in the near vicinity of the opsonin. Accordingly, in the absence of the polyanions on the C3b-coated surface, AP activation proceeds as FH affinity for C3b dramatically decreases (Kazatchkine *et al.*, 1979; Pangburn *et al.*, 1983).

Factor H-like protein

Factor H-like protein (FHL-1, reconectin) is an alternatively spliced variant of the FH-encoding gene (Estaller *et al.*, 1991). FHL-1 comprises seven N-terminal SCRs of FH and a unique C-terminal sequence of four amino acids. Not surprisingly, FHL-1, like FH, displays cofactor and decay-accelerating activities and presumably plays a role in down-regulating complement *in vivo* (Kühn and Zipfel, 1996; Zipfel and Skerka, 1999). However, the plasma concentration of FHL-1 is approximately one-tenth of that of FH.

C4b-binding protein

The C4b-binding protein (C4bp) is a spider-like glycoprotein that circulates complexed with the vitamin K-dependent protein S (Dahlbäck, 1983; Chung *et al.*, 1985). Protein S helps in directing C4bp to negatively charged surfaces and apoptotic cells (Schwalbe *et al.*, 1990; Webb *et al.*, 2002). C4bp consists of seven identical α-chains and one β-chain linked together by disulfide bonds at their most C-terminal parts (Fig. 13.3). The chains are composed of eight and three SCRs, respectively (Hillarp and Dahlbäck, 1990; Kask *et al.*, 2002).

C4bp inhibits the CP and LP activation steps that involve C4b (Fig. 13.1) (Barnum, 1991). Additionally, C4bp in high concentrations contributes to the decay of the AP C3 convertase (Blom *et al.*, 2003). Binding of C4bp to C4b renders the latter susceptible for FI-mediated cleavage and inactivation (Kask *et al.*, 2002). Furthermore, C4bp prevents the assembly of the CP/LP C3-convertase (C4b2a) and accelerates a natural decay of the preformed convertases by dissociating C2a from the complex (Gigli *et al.*, 1979).

Clusterin

Clusterin (SP-40,40, apolipoprotein J) is a multifunctional, heterodimeric, and highly glycosylated protein that regulates the terminal pathway of complement activation (Tsuruta *et al.*, 1990; Tobe *et al.*, 1991). It prevents proper MAC insertion into cell membranes by binding to the terminal complement complexes C5b-7, C5b-8 and C5b-9 (Fig. 13.1) (Choi *et al.*, 1989; Murphy *et al.*, 1989; Tschopp *et al.*, 1993). It is also a constituent of a non-lytic, soluble MAC along with vitronectin, C5b and C6-C9 (Müller-Eberhard, 1988).

Vitronectin

Vitronectin (S-protein) is a glycoprotein with a carbohydrate moiety constituting up to 30% of its molecular mass (Podack and Müller-Eberhard, 1979, Barnes and Silnutzer, 1983; Shaffer *et al.*, 1984; Preissner *et al.*, 1986). Similar to clusterin, vitronectin interferes with the terminal pathway of complement activation (Fig. 13.1). It binds to the C5b-7 complex and blocks its insertion into the cell membrane (Podack and Müller-Eberhard, 1979).

Properdin

Properdin, a large oligomeric plasma protein (2–4 identical subunits), was the first AP-specific complement component identified (Pillemer *et al.*, 1954). Properdin uniquely functions as a positive regulator of complement. It binds to the C3b portion of the AP C3 convertase to strengthen its activity and to stabilize it via inhibition of spontaneous and accelerated decay (Fig. 13.1) (Fearon and Austen, 1975; Medicus *et al.*, 1976; Smith *et al.*, 1984).

Functions of the complement system

The importance of the complement system in host defence against invading pathogens is reflected by the predisposition of individuals deficient in complement components to infectious diseases. For example, deficiencies in late complement components or factor D are associated with recurrent invasive infections caused by *Neisseria meningitidis* or *Neisseria gonorrhoeae*, whereas deficiencies in C2 or MASP-2 predispose to *Streptococcus pneumoniae* infection (Sjöholm *et al.*, 2006).

MAC formation and membrane lysis of the bacterium are just one important outcome of the complement activation. The anti-bacterial potential of the complement system stems also from the ability to enhance microbial uptake by phagocytic cells and to release inflammatory mediators. Furthermore, the complement activation products, as well as complement receptors form a complex immunological network that mediates a cross-talk between the innate and acquired immunity.

Opsonin-mediated clearance of microbes and host debris

The complement is known to contribute to phagocytosis of pathogens and to the cleaning up of host debris. Phagocytic cells bear complement receptors, such as CR1 (CD35), CR3 (CD11b/CD18) and CR4 (CD11c/CD18) that recognize certain complement components or complement cleavage products. CR1 binds to targets coated with C1q, C3b, C4b and iC3b, whereas CR3 and CR4 are iC3b-specific (Newman *et al.*, 1984; Klickstein *et al.*, 1988, 1997; Krych *et al.*, 1992; Krych-Goldberg and Atkinson, 2001). Ligation of complement receptors on phagocytic cells with their ligands deposited on a microbial surface ensures efficient uptake and destruction of pathogens.

Induction of inflammation

Complement cleavage products, such as C3a, C4a, and C5a, promote clearance of pathogens by inducing a local inflammatory response (Hugli, 1981). Of the three, C5a is known to have the highest biological activity. The anaphylatoxins act as stimulators and chemoattractants for myeloid cells bearing anaphylatoxin-specific receptors. Furthermore, they increase vascular permeability, and induce the expression of adhesin molecules on endothelial cells to recruit phagocytic cells to the site of inflammation (Gerard and Gerard, 1994; Foreman *et al.*, 1996; DiScipio *et al.*, 1999).

Shaping the specific immune response

Complement-derived fragments play an important role in humoral immunity by influencing the antibody response. On B cells, CD21 is complexed with CD19 to form the B cell co-receptor complex. Binding of C3d (Fig. 13.2) to CD21 triggers CD19-mediated signalling and in consequence B cell activation (Dempsey *et al.*, 1996; Dempsey and Fearon, 1996).

Bacterial mechanisms of complement evasion

The race between the bacteria to escape the immune surveillance, and the immune system to find new ways of eliminating the harmful microbes, never ends. As the complement system is an ancient, first-line defence mechanism that links the innate immunity with adaptive immunity, many pathogens have developed strategies that focus on complement evasion. These strategies, often shared by different bacteria, include interference with the initial complement activation, C3 convertase functions, or MAC insertion. The reader is also advised to read an excellent review on this subject by Lambris *et al.* (2008).

Interference with initial complement activation

Non-immune binding of immunoglobulins

Immunoglobulins are built of variable Fab fragments recognizing antigens and constant Fc effector regions. The Fc region triggers activation of the CP and mediates an interaction with the Fc receptors on phagocytic cells. Not surprisingly, a non-immune binding of immunoglobulins via the Fc portion is a virulence feature of many bacteria, including staphylococci, streptococci, *Fusobacterium nucleatum*, *Aggregatibacter actinomycetemcomitans*, or *Prevotella intermedia* (Forsgren and Sjöquist, 1966; Björck and Kronvall, 1984; Mintz and Fives-Taylor, 1994; Labbe and Grenier, 1995; Guo *et al.*, 2000). Bacterial Fc receptors block complement activation by competing with C1q for the binding to the Fc region of immunoglobulins. Furthermore, non-immune binding of immunoglobulins results in inverted tagging that, in consequence, prevents bacterial phagocytosis. The best characterized bacterial Fc receptors are protein G of streptococci and protein A of *Staphylococcus aureus* (Forsgren and Sjöquist, 1966; Björck and Kronvall, 1984).

Inactivation of immunoglobulins and C3/C3b

The degradation of complement components into smaller, non-functional fragments is a strategy of some bacteria to evade the complement attack. IdeS (IgG-degrading enzyme of *Streptococcus pyogenes*) and SpeB (streptococcal pyrogenic exotoxin B), cysteine proteinases of group A streptococci, cleave the heavy chain of IgG in the flexible hinge region to generate two Fab fragments and one Fc fragment (von Pawel-Rammingen and Björck, 2003). Although SpeB expression is repressed in stationary phase bacteria, constitutive expression of IdeS provides means to cleave IgGs as soon as they appear on the streptococcal surface (Chaussee *et al.*, 1997; Unnikrishnan *et al.*, 1999; von Pawel-Rammingen *et al.*, 2002). The *Pseudomonas aeruginosa* elastase (PaE) and alkaline protease (PaAP) degrade C3, whereas the *Porphyromonas gingivalis prtH*-encoded protease degrades both IgG and C3 (Hong and Ghebrehiwet, 1992; Schenkein *et al.*, 1995). Interestingly, *S. aureus* degrades these complement-activation initiators indirectly by utilizing the proteolytic activity of human plasmin. Initially, the bacterium acquires the host plasminogen and converts it into plasmin with the help of staphylokinase. In consequence, plasmin attached to the bacterial surface degrades both, IgG and C3b (Cabellos *et al.*, 1992).

Furthermore, *S. aureus* can also exclude C3 molecules from participating in downstream activation events. This is achieved by the fibrinogen-binding protein (Efb) and Efb-homologous protein (Ehp) that bind to the thioester-containing C3d domain of the C3 molecule, thereby preventing its cleavage (Lee *et al.*, 2004; Hammel *et al.*, 2007a,b).

Acquisition of C1-INH and C1q degradation

Anti-complement activities utilized by bacteria include also acquisition of C1-INH. For example, *Bordetella pertussis,* the aetiological agent for the whooping cough, recruits C1-INH from the human serum to block the complement activation at the initial point of the cascade (Marr *et al.*, 2007). An even more sophisticated strategy is utilized by enterohaemorrhagic *Escherichia coli* O157:H7. This bacterium secretes a protease of the C1 esterase inhibitor (StcE) that not only recruits C1-INH to the bacterial surface but also cleaves it to potentiate its complement-inhibitory function (Lathem *et al.*, 2004). Another mechanism to interfere with the CP initiation step is represented by *P. aeruginosa* that utilizes the proteases PaE and PaAP to degrade the CP recognition molecule C1q (Hong and Ghebrehiwet, 1992).

Inhibition of C3 convertases

Direct interference with C3 convertases

S. aureus, in addition to interference with the initial complement activation steps, interferes with the C3 convertase function to suppress downstream actions of complement. Staphylococcal complement inhibitor (SCIN) is a secreted protein that binds to C3 convertases of all three pathways and impairs their enzymatic activity. As the binding stabilizes C3 convertases, the formation of new convertases is also prevented (Rooijakkers *et al.*, 2005). Additionally, Efb and Ehp bind to the C3b constituent of the AP C3 convertases to prevent the cleavage of C3 (Jongerius *et al.*, 2007). Another mechanism of interference with the C3 convertase function is exploited by group A streptococci that utilize SpeB to degrade the host properdin. This inhibits the formation of the functional AP C3-convertase and prevents opsonophagocytosis (Tsao *et al.*, 2006).

Microbial factors binding host complement regulators

The acquisition of complement regulators, present in serum in high concentrations, is the most widely disseminated strategy bacteria use to prevent the complement attack. Several bacterial species have been described to acquire FH or FHL-1 to gain the ability to inactivate AP C3 convertases and C3b deposits on their surfaces (Fig. 13.4 and Table 13.2).

As the interaction of FH with sialic acid inhibits the AP activation on the host cells (Fearon, 1978), group B streptococci and *N. gonorrhoeae* cover their surfaces with sialic acid (Marques *et al.*, 1992; Ram *et al.*, 1998b) to mimic the host and benefit from complement-protective potential of FH. Instead of mimicking the host, some bacteria express specialized FH and/or FHL-1 receptors. These include (i) FH-binding proteins, such as BbCRASP-3 to -5 of *Borrelia burgdorferi,* β protein of group B *Streptococcus,* Por1A of *N. gonorrhoeae,* Hic and PspC of *S. pneumoniae,* YadA and Ail of *Yersinia enterocolitica;* (ii) FHL-1-specific BaCRASP-3 of *Borrelia afzelii;* and (iii) FH/FHL-1-binding M protein and fibronectin-binding protein (Fba) of *S. pyogenes* or BbCRASP-1, and BbCRASP-2 of *B. burgdorferi* (Horstmann *et al.*, 1988; Ram *et al.*, 1998a; Janulczyk *et al.*, 2000; Dave *et al.*, 2001; Kotarsky *et al.*, 2001; Kraiczy *et al.*, 2001; Areschoug *et al.*, 2002; Pandiripally *et al.*, 2002; Biedzka-Sarek *et al.*, 2008). Different regions on FH have been described to interact with these bacterial proteins (Fig. 13.4). For example, BbCRASP-1, -3 and -5 target C-terminus, β protein, PspC, and Hic bind to the middle part, whereas M protein, Fba, and Ail interact with the N-terminal portion of FH (Blackmore *et al.*, 1998a; Kotarsky *et al.*, 1998; Kraiczy *et al.*, 2001; Duthy *et al.*, 2002; Jarva *et al.*, 2002, 2004; Pandiripally *et al.*, 2003; Dave *et al.*, 2004; Biedzka-Sarek *et al.*, 2008). Uniquely, the YadA–FH interaction has been shown to engage the SCRs of the entire polypeptide chain of FH (Fig. 13.4) (Biedzka-Sarek *et al.*, 2008).

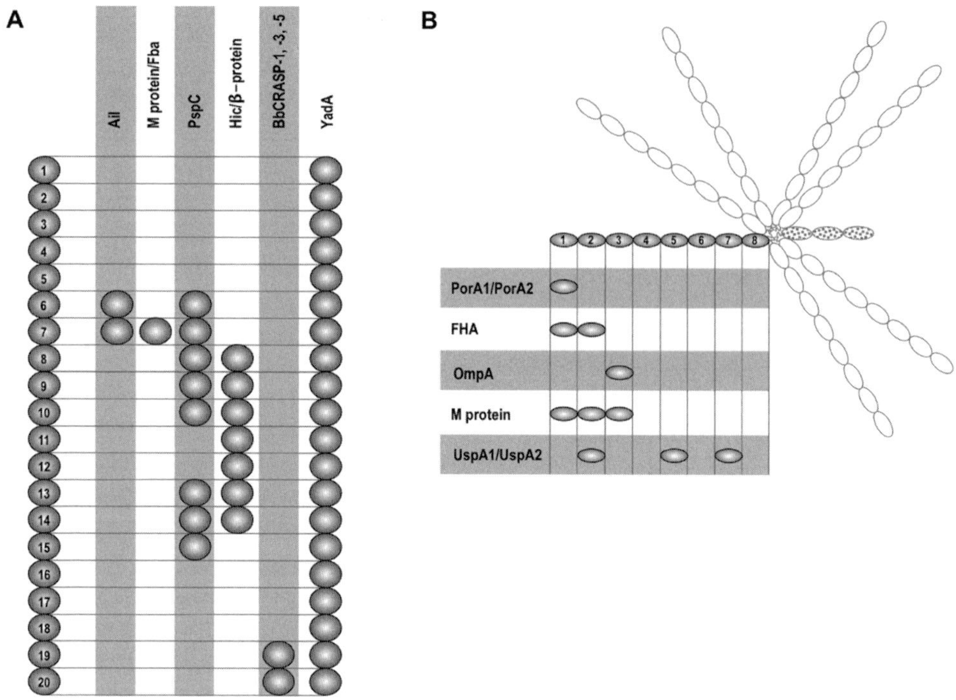

Figure 13.4 Schematic representation of FH (A) and C4bp (B) domains interacting with bacterial proteins.

Table 13.2 Complement regulators targeted by pathogenic bacteria

Regulator	Size (kDa)	Serum conc. (μg/ml)	Function	Bacterium
Soluble complement regulators				
C1-INH	105	200	Blocks active sites on C1s and C1r or removes them from the C1 complex, prevents C1 autoactivation, inhibits formation of AP C3 convertase	Acquired by: *B. pertussis*, *E. coli* O157:H7
FH	150	500	Accelerates decay of AP C3 convertase, inhibits formation of AP C3 convertase, co-factor for FI	Acquired by: *N. gonorrhoeae*, *B. burgdorferi*, group *B Streptococcus*, *S. pyogenes*, *S. pneumoniae*, *Y. enterocolitica*
C4bp	570	200	Accelerates decay of CP/LP C3 convertase, inhibits formation of CP/LP C3 convertase, co-factor for FI	Acquired by: *S. pyogenes*, *N. gonorrhoeae*, *M. catarrhalis*, *Y. enterocolitica*, *B. pertussis*, *Y. pestis*
Properdin	210	26	Stabilizes AP C3 convertase	Degraded by: group A streptococci
Vitronectin (S-protein)	75	500	Prevents MAC insertion by binding to terminal complement complexes C5b-7	Acquired by: *M. catarrhalis*
Membrane-bound regulators				
CD59	20		Prevents interaction between C9 and C5b-8	Acquired by: *E. coli*, *H. pylori*

In addition to FH and FHL-1, C4bp is acquired by bacteria to inactivate surface-deposited CP/LP C3 convertases and C4b (Fig. 13.4 and Table 13.2). Specialized bacterial proteins functioning as C4bp receptors include the M protein of *S. pyogenes,* Por1A and Por1B of *N. gonorrhoeae,* OmpA of *E. coli* K1, UspA1 and UspA2 of *Moraxella catarrhalis,* YadA and Ail of *Y. enterocolitica,* and FHA of *B. pertussis* (Johnsson *et al.,* 1996; Berggård *et al.,* 1997; Ram *et al.,* 2001; Prasadarao *et al.,* 2002; Nordström *et al.,* 2004; Kirjavainen *et al.,* 2008). These bacterial proteins predominantly interact with N-terminal SCRs of the C4bp α-chains, exceptions being UspA-1 and UspA-2 that, in addition to N-terminal SCR2, target C-terminal SCR5 and SCR7 of the C4bp α-chains (Fig. 13.4) (Accardo *et al.,* 1996; Berggård *et al.,* 2001; Ram *et al.,* 2001; Prasadarao *et al.,* 2002; Nordström *et al.,* 2004).

Interference with the terminal pathway

Physical barriers

Many bacteria are able to prevent MAC-triggered cytolysis via steric hindrance. All the Gram-positive bacteria use the thick and impenetrable peptidoglycan layer as a shield that blocks the access of C5b-9 to the cytoplasmic membrane (Joiner *et al.,* 1983). Some Gram-negative bacteria, however, also evolved a mechanism to keep the MAC complex at a safe distance from their outer membrane. For example, *Salmonella* Montevideo expresses long lipopolysaccharide (LPS) chains that hinder the access of MAC to the outer membrane by binding MAC to the remote tips of the polysaccharide (Joiner *et al.,* 1986).

Microbial proteins interfering with MAC insertion

Resistance to the terminal pathway of the complement activation is a prerequisite for Gram-negative bacteria to establish infection. The *E. coli* outer membrane protein TraT is a good example of the terminal pathway inhibitor. TraT inhibitory action most likely either involves interference with the formation of the C5b6 complex or a structural modification of the complex leading to its inactivation (Pramoonjago *et al.,* 1992). Another example of a bacterial protein that interferes with the terminal pathway of complement activation is the *B. burgdorferi* CD59-like protein that, like human CD59, prevents C9 polymerization and the generation of MAC (Pausa *et al.,* 2003).

Acquisition of host terminal pathway regulators

In addition to the regulators of the initial complement activation steps, bacteria are also able to bind complement regulators of the terminal pathway of complement activation (Table 13.2). *M. catarrhalis* UspA2 binds vitronectin to block MAC insertion into the cell membrane, whereas *E. coli* and *Helicobacter pylori* bind CD59 from the fluid phase to resist the MAC-induced lysis (Rautemaa *et al.,* 1998, 2001; Attia *et al.,* 2006). *E. coli* and *H. pylori* are the only bacteria described thus far that recruit normally membrane-bound terminal pathway regulators from serum.

Interference with generation and function of the complement-derived chemoattractants

Although the complement system does not directly kill the Gram-positive bacteria, it still controls the course of Gram-positive infections by its ability to tag the bacteria for phagocytic destruction (Moffitt and Frank, 1994). Not surprisingly, to avoid phagocytosis Gram-positive bacteria need to interfere not only with the complement opsonization but also with the generation of complement-derived chemoattractants. To achieve this goal, *S. aureus* produces the superantigen-related protein, SSL7, that binds to C5 and, in consequence, blocks its cleavage into C5a and C5b (Langley *et al.,* 2005). Additionally, the staphylococcal Efb and Ehp proteins bind to C5 convertases of all three pathways and block their activity, thereby limiting C5a generation and neutrophil activation (Jongerius *et al.,* 2007). Finally, *S. aureus* has developed ways to directly interfere with neutrophil and monocyte responses to C5a. Namely, the chemotaxis inhibitory protein of *S. aureus* (CHIPS) competes with C5a for the binding to C5a/formylated peptide receptors on the phagocytic cells. This antagonizes the C5a receptors and, in consequence, any response of the phagocytic cells to C5a is impaired (de Haas *et al.*, 2004). Another mechanism of interference with downstream C5a inflammatory responses involves a direct cleavage of C5a. Group A and group B streptococci use cell wall-anchored C5a peptidase, whereas *Serratia marcescens* produces a C5a protease to inactivate already formed C5a (Wexler and Cleary, 1985; Wexler *et al.*, 1985; Oda *et al.*, 1990; Cleary *et al.*, 1992; Chmouryguina *et al.*, 1996).

Concluding remarks

Taken together, the complement system has a powerful anti-bacterial potential. Not surprisingly, every pathogen needs to develop more than one strategy to block the complement activation at different levels and prevent downstream events of complement activation. It appears that only the multiplicity of the anti-complement strategies will ensure efficient protection against this early immune defence and enable successful colonization of the host.

Detailed studies on the interactions between the bacteria and the complement system should provide insights that will allow us to understand bacterial pathogenesis. Moreover, they should create ways of treatment of bacterial infections by fine-tuning of the complement-based reactions.

References

Accardo, P., Sánchez-Corral, P., Criado, O., Garcia, E., and Rodriguez de Córdoba, S. (1996). Binding of human complement component C4b-binding protein (C4BP) to *Streptococcus pyogenes* involves the C4b-binding site. J. Immunol. *157*, 4935–4939.

Ahearn, J.M., and Fearon, D.T. (1989). Structure and function of the complement receptors, CR1 (CD35) and CR2 (CD21). Adv. Immunol. *46*, 183–219.

Alsenz, J., Schulz, T.F., Lambris, J.D., Sim, R.B., and Dierich, M.P. (1985). Structural and functional analysis of the complement component factor H with the use of different enzymes and monoclonal antibodies to factor H. Biochem. J. *232*, 841–850.

Areschoug, T., Stålhammar-Carlemalm, M., Karlsson, I., and Lindahl, G. (2002). Streptococcal β protein has separate binding sites for human factor H and IgA-Fc. J. Biol. Chem. *277*, 12642–12648.

Attia, A.S., Ram, S., Rice, P.A., and Hansen, E.J. (2006). Binding of vitronectin by the *Moraxella catarrhalis* UspA2 protein interferes with late stages of the complement cascade. Infect. Immun. *74*, 1597–1611.

Barnes, D.W., and Silnutzer, J. (1983). Isolation of human serum spreading factor. J. Biol. Chem. *258*, 12548–12552.

Barnum, S.R. (1991). C4b-binding protein, a regulatory protein of complement. Immunol. Res. *10*, 28–42.

Berggård, K., Johnsson, E., Mooi, F.R., and Lindahl, G. (1997). *Bordetella pertussis* binds the human complement regulator C4BP: role of filamentous hemagglutinin. Infect. Immun. *65*, 3638–3643.

Berggård, K., Lindahl, G., Dahlbäck, B., and Blom, A.M. (2001). *Bordetella pertussis* binds to human C4b-binding protein (C4BP) at a site similar to that used by the natural ligand C4b. Eur. J. Immunol. *31*, 2771–2780.

Biedzka-Sarek, M., Jarva, H., Hyytiäinen, H., Meri, S., and Skurnik, M. (2008). Characterization of complement factor H binding to *Yersinia enterocolitica* serotype O:3. Infect. Immun. *76*, 4100–4109.

Björck, L., and Kronvall, G. (1984). Purification and some properties of streptococcal protein G, a novel IgG-binding reagent. J. Immunol. *133*, 969–974.

Blackmore, T.K., Sadlon, T.A., Ward, H.M., Lublin, D.M., and Gordon, D.L. (1996). Identification of a heparin binding domain in the seventh short consensus repeat of complement factor H. J. Immunol. *157*, 5422–5427.

Blackmore, T.K., Fischetti, V.A., Sadlon, T.A., Ward, H.M., and Gordon, D.L. (1998a). M protein of the group A *Streptococcus* binds to the seventh short consensus repeat of human complement factor H. Infect. Immun. *66*, 1427–1431.

Blackmore, T.K., Hellwage, J., Sadlon, T.A., Higgs, N., Zipfel, P.F., Ward, H.M., and Gordon, D.L. (1998b). Identification of the second heparin-binding domain in human complement factor H. J. Immunol. *160*, 3342–3348.

Blom, A.M., Kask, L., and Dahlbäck, B. (2003). CCP1–4 of the C4b-binding protein α-chain are required for factor I mediated cleavage of complement factor C3b. Mol. Immunol. *39*, 547–556.

Busby, T.F., and Ingham, K.C. (1990). NH2-terminal calcium-binding domain of human complement C1s mediates the interaction of C1r with C1q. Biochemistry *29*, 4613–4618.

Cabellos, C., Macintyre, D.E., Forrest, M., Burroughs, M., Prasad, S., and Tuomanen, E. (1992). Differing roles for platelet-activating factor during inflammation of the lung and subarachnoid space. The special case of *Streptococcus pneumoniae*. J. Clin. Invest. *90*, 612–618.

Calcott, M.A., and Müller-Eberhard, H.J. (1972). C1q protein of human complement. Biochemistry *11*, 3443–3450.

Casasnovas, J.M., Larvie, M., and Stehle, T. (1999). Crystal structure of two CD46 domains reveals an extended measles virus-binding surface. EMBO J. *18*, 2911–2922.

Chaussee, M.S., Phillips, E.R., and Ferretti, J.J. (1997). Temporal production of streptococcal erythrogenic toxin B (streptococcal cysteine proteinase) in response to nutrient depletion. Infect. Immun. *65*, 1956–1959.

Chmouryguina, I., Suvorov, A., Ferrieri, P., and Cleary, P.P. (1996). Conservation of the C5a peptidase genes in group A and B streptococci. Infect. Immun. *64*, 2387–2390.

Choi, N.H., Mazda, T., and Tomita, M. (1989). A serum protein SP40,40 modulates the formation of membrane attack complex of complement on erythrocytes. Mol. Immunol. *26*, 835–840.

Chung, L.P., Bentley, D.R., and Reid, K.B. (1985). Molecular cloning and characterization of the cDNA coding for C4b-binding protein, a regulatory protein of the classical pathway of the human complement system. Biochem. J. *230*, 133–141.

Cleary, P.P., Prahbu, U., Dale, J.B., Wexler, D.E., and Handley, J. (1992). Streptococcal C5a peptidase is a highly specific endopeptidase. Infect. Immun. *60*, 5219–5223.

Cochrane, C.G., and Müller-Eberhard, H.J. (1968). The derivation of two distinct anaphylatoxin activities from the third and fifth components of human complement. J. Exp. Med. *127*, 371–386.

Cooper, N.R., and Müller-Eberhard, H.J. (1970). The reaction mechanism of human C5 in immune hemolysis. J. Exp. Med. *132*, 775–793.

Dahlbäck, B. (1983). Purification of human C4b-binding protein and formation of its complex with vitamin K-dependent protein S. Biochem. J. *209*, 847–856.

Dave, S., Brooks-Walter, A., Pangburn, M.K., and McDaniel, L.S. (2001). PspC, a pneumococcal surface protein, binds human factor H. Infect. Immun. *69*, 3435–3437.

Dave, S., Carmicle, S., Hammerschmidt, S., Pangburn, M.K., and McDaniel, L.S. (2004). Dual roles of PspC, a surface protein of *Streptococcus pneumoniae*, in binding human secretory IgA and factor H.J. Immunol. *173*, 471–477.

Davis, A.E., 3rd, and Harrison, R.A. (1982). Structural characterization of factor I mediated cleavage of the third component of complement. Biochemistry *21*, 5745–5749.

Dempsey, P.W., and Fearon, D.T. (1996). Complement: instructing the acquired immune system through the CD21/CD19 complex. Res. Immunol. *147*, 71–75; 119–120.

Dempsey, P.W., Allison, M.E., Akkaraju, S., Goodnow, C.C., and Fearon, D.T. (1996). C3d of complement as a molecular adjuvant: bridging innate and acquired immunity. Science *271*, 348–350.

DiScipio, R.G. (1992a). Formation and structure of the C5b–7 complex of the lytic pathway of complement. J. Biol. Chem. *267*, 17087–17094.

Discipio, R.G. (1992b). Ultrastructures and interactions of complement Factor H and I.J. Immunol. *149*, 2592–2599.

DiScipio, R.G., Smith, C.A., Müller-Eberhard, H.J., and Hugli, T.E. (1983). The activation of human complement component C5 by a fluid phase C5 convertase. J. Biol. Chem. *258*, 10629–10636.

DiScipio, R.G., Chakravarti, D.N., Müller-Eberhard, H.J., and Fey, G.H. (1988). The structure of human complement component C7 and the C5b–7 complex. J. Biol. Chem. *263*, 549–560.

DiScipio, R.G., Daffern, P.J., Jagels, M.A., Broide, D.H., and Sriramarao, P. (1999). A comparison of C3a and C5a-mediated stable adhesion of rolling eosinophils in postcapillary venules and transendothelial migration *in vitro* and *in vivo*. J. Immunol. *162*, 1127–1136.

Dodds, A.W., Ren, X.D., Willis, A.C., and Law, S.K. (1996). The reaction mechanism of the internal thioester in the human complement component C4. Nature *379*, 177–179.

Duthy, T.G., Ormsby, R.J., Giannakis, E., Ogunniyi, A.D., Stroeher, U.H., Paton, J.C., and Gordon, D.L. (2002). The human complement regulator factor H binds pneumococcal surface protein PspC via short consensus repeats 13 to 15. Infect. Immun. *70*, 5604–5611.

Estaller, C., Koistinen, V., Schwaeble, W., Dierich, M.P., and Weiss, E.H. (1991). Cloning of the 1.4-kb mRNA species of human complement factor H reveals a novel member of the short consensus repeat family related to the carboxy terminal of the classical 150-kDa molecule. J. Immunol. *146*, 3190–3196.

Fearon, D.T. (1978). Regulation by membrane sialic acid of β1H-dependent decay-dissociation of amplification C3 convertase of the alternative complement pathway. Proc. Natl. Acad. Sci. U.S.A. *75*, 1971–1975.

Fearon, D.T. (1979). Regulation of the amplification C3 convertase of human complement by an inhibitory protein isolated from human erythrocyte membrane. Proc. Natl. Acad. Sci. U.S.A. *76*, 5867–5871.

Fearon, D.T., and Austen, K.F. (1975). Properdin: initiation of alternative complement pathway. Proc. Natl. Acad. Sci. U.S.A. *72*, 3220–3224.

Foreman, K.E., Glovsky, M.M., Warner, R.L., Horvath, S.J., and Ward, P.A. (1996). Comparative effect of C3a and C5a on adhesion molecule expression on neutrophils and endothelial cells. Inflammation *20*, 1–9.

Forsgren, A., and Sjöquist, J. (1966). 'Protein A' from *S. aureus*. I. Pseudo-immune reaction with human γ-globulin. J. Immunol. *97*, 822–827.

Fujita, T., and Nussenzweig, V. (1979). The role of C4-binding protein and β1H in proteolysis of C4b and C3b. J. Exp. Med. *150*, 267–276.

Fujita, T., Inoue, T., Ogawa, K., Iida, K., and Tamura, N. (1987). The mechanism of action of decay-accelerating factor (DAF). DAF inhibits the assembly of C3 convertases by dissociating C2a and Bb. J. Exp. Med. *166*, 1221–1228.

Gaboriaud, C., Thielens, N.M., Gregory, L.A., Rossi, V., Fontecilla-Camps, J.C., and Arlaud, G.J. (2004). Structure and activation of the C1 complex of complement: unraveling the puzzle. Trends Immunol. *25*, 368–373.

Gerard, C., and Gerard, N.P. (1994). C5a anaphylatoxin and its seven transmembrane-segment receptor. Annu. Rev. Immunol. *12*, 775–808.

Gigli, I., Fujita, T., and Nussenzweig, V. (1979). Modulation of the classical pathway C3 convertase by plasma proteins C4 binding protein and C3b inactivator. Proc. Natl. Acad. Sci. U.S.A. *76*, 6596–6600.

Gordon, D.L., Kaufman, R.M., Blackmore, T.K., Kwong, J., and Lublin, D.M. (1995). Identification of complement regulatory domains in human factor H.J. Immunol. *155*, 348–356.

Gorski, J.P., Hugli, T.E., and Müller-Eberhard, H.J. (1979). C4a: the third anaphylatoxin of the human complement system. Proc. Natl. Acad. Sci. U.S.A. *76*, 5299–5302.

Guo, M., Han, Y.W., Sharma, A., and De Nardin, E. (2000). Identification and characterization of human immunoglobulin G Fc receptors of *Fusobacterium nucleatum*. Oral Microbiol. Immunol. *15*, 119–123.

de Haas, C.J., Veldkamp, K.E., Peschel, A., Weerkamp, F., Van Wamel, W.J., Heezius, E.C., Poppelier, M.J., Van Kessel, K.P., and van Strijp, J.A. (2004). Chemotaxis inhibitory protein of *Staphylococcus aureus*, a bacterial antiinflammatory agent. J. Exp. Med. *199*, 687–695.

Hammel, M., Sfyroera, G., Pyrpassopoulos, S., Ricklin, D., Ramyar, K.X., Pop, M., Jin, Z., Lambris, J.D., and Geisbrecht, B.V. (2007a). Characterization of Ehp, a secreted complement inhibitory protein from *Staphylococcus aureus*. J. Biol. Chem. *282*, 30051–30061.

Hammel, M., Sfyroera, G., Ricklin, D., Magotti, P., Lambris, J.D., and Geisbrecht, B.V. (2007b). A structural basis for complement inhibition by *Staphylococcus aureus*. Nat. Immunol. *8*, 430–437.

Harpel, P.C., and Cooper, N.R. (1975). Studies on human plasma C1 inactivator–enzyme interactions. I. Mechanisms of interaction with C1s, plasmin, and trypsin. J. Clin. Invest. *55*, 593–604.

Harrison, R.A. (1983). Human C1 inhibitor: improved isolation and preliminary structural characterization. Biochemistry *22*, 5001–5007.

Hillarp, A., and Dahlbäck, B. (1990). Cloning of cDNA coding for the β chain of human complement component C4b-binding protein: sequence homology with the α chain. Proc. Natl. Acad. Sci. U.S.A. *87*, 1183–1187.

Hong, Y.Q., and Ghebrehiwet, B. (1992). Effect of *Pseudomonas aeruginosa* elastase and alkaline protease on serum complement and isolated components C1q and C3. Clin. Immunol. Immunopathol. *62*, 133–138.

Horstmann, R.D., Sievertsen, H.J., Knobloch, J., and Fischetti, V.A. (1988). Antiphagocytic activity of streptococcal M protein: selective binding of complement control protein factor H. Proc. Natl. Acad. Sci. U.S.A. *85*, 1657–1661.

Hugli, T.E. (1975). Human anaphylatoxin (C3a) from the third component of complement. Primary structure. J. Biol. Chem. *250*, 8293–8301.

Hugli, T.E. (1981). The structural basis for anaphylatoxin and chemotactic functions of C3a, C4a, and C5a. Crit. Rev. Immunol. *1*, 321–366.

Iida, K., and Nussenzweig, V. (1981). Complement receptor is an inhibitor of the complement cascade. J. Exp. Med. *153*, 1138–1150.

Ishizaka, T., Ishizaka, K., Borsos, T., and Rapp, H. (1966). C'1 fixation by human isoagglutinins: fixation of C'1 by γG and γM but not by γA antibody. J. Immunol. *97*, 716–726.

Janatova, J., and Tack, B.F. (1981). Fourth component of human complement: studies of an amine-sensitive site comprised of a thiol component. Biochemistry *20*, 2394–2402.

Janssen, B.J., Christodoulidou, A., McCarthy, A., Lambris, J.D., and Gros, P. (2006). Structure of C3b reveals conformational changes that underlie complement activity. Nature *444*, 213–216.

Janulczyk, R., Iannelli, F., Sjöholm, A.G., Pozzi, G., and Björck, L. (2000). Hic, a novel surface protein of *Streptococcus pneumoniae* that interferes with complement function. J. Biol. Chem. *275*, 37257–37263.

Jarva, H., Janulczyk, R., Hellwage, J., Zipfel, P.F., Björck, L., and Meri, S. (2002). *Streptococcus pneumoniae* evades complement attack and opsonophagocytosis by expressing the *pspC* locus-encoded Hic protein that binds to short consensus repeats 8–11 of factor H.J. Immunol. *168*, 1886–1894.

Jarva, H., Hellwage, J., Jokiranta, T.S., Lehtinen, M.J., Zipfel, P.F., and Meri, S. (2004). The group B streptococcal β and pneumococcal Hic proteins are structurally related immune evasion molecules that bind the complement inhibitor factor H in an analogous fashion. J. Immunol. *172*, 3111–3118.

Jiang, H., Cooper, B., Robey, F.A., and Gewurz, H. (1992a). DNA binds and activates complement via residues 14–26 of the human C1q A chain. J. Biol. Chem. *267*, 25597–25601.

Jiang, H., Robey, F.A., and Gewurz, H. (1992b). Localization of sites through which C-reactive protein binds and activates complement to residues 14–26 and 76–92 of the human C1q A chain. J. Exp. Med. *175*, 1373–1379.

Jiang, H., Wagner, E., Zhang, H., and Frank, M.M. (2001). Complement 1 inhibitor is a regulator of the alternative complement pathway. J. Exp. Med. *194*, 1609–1616.

Johnsson, E., Thern, A., Dahlbäck, B., Hedén, L.O., Wikström, M., and Lindahl, G. (1996). A highly variable region in members of the streptococcal M protein family binds the human complement regulator C4BP.J. Immunol. *157*, 3021–3129.

Joiner, K., Brown, E., Hammer, C., Warren, K., and Frank, M. (1983). Studies on the mechanism of bacterial resistance to complement-mediated killing. III. C5b-9 deposits stably on rough and type 7 *S. pneumoniae* without causing bacterial killing. J. Immunol. *130*, 845–849.

Joiner, K.A., Grossman, N., Schmetz, M., and Leive, L. (1986). C3 binds preferentially to long-chain lipopolysaccharide during alternative pathway activation by *Salmonella montevideo*. J. Immunol. *136*, 710–715.

Jokiranta, T.S., Zipfel, P.F., Hakulinen, J., Kühn, S., Pangburn, M.K., Tamerius, J.D., and Meri, S. (1996). Analysis of the recognition mechanism of the alternative pathway of complement by monoclonal anti-factor H antibodies: evidence for multiple interactions between H and surface bound C3b. FEBS Lett. *393*, 297–302.

Jokiranta, T.S., Hellwage, J., Koistinen, V., Zipfel, P.F., and Meri, S. (2000). Each of the three binding sites on complement factor H interacts with a distinct site on C3b. J. Biol. Chem. *275*, 27657–27662.

Jongerius, I., Köhl, J., Pandey, M.K., Ruyken, M., van Kessel, K.P., van Strijp, J.A., and Rooijakkers, S.H. (2007). Staphylococcal complement evasion by various convertase-blocking molecules. J. Exp. Med. *204*, 2461–2471.

Kask, L., Hillarp, A., Ramesh, B., Dahlbäck, B., and Blom, A.M. (2002). Structural requirements for the intracellular subunit polymerization of the complement inhibitor C4b-binding protein. Biochemistry *41*, 9349–9357.

Kazatchkine, M.D., Fearon, D.T., and Austen, K.F. (1979). Human alternative complement pathway: membrane-associated sialic acid regulates the competition between B and betaβ1H for cell-bound C3b. J. Immunol. *122*, 75–81.

Kerr, M.A. (1980). The human complement system: assembly of the classical pathway C3 convertase. Biochem. J. *189*, 173–181.

Kirjavainen, V., Jarva, H., Biedzka-Sarek, M., Blom, A.M., Skurnik, M., and Meri, S. (2008). *Yersinia enterocolitica* serum resistance proteins YadA and Ail bind the complement regulator C4b-binding protein. PLoS Pathog. *4*, e1000140.

Klickstein, L.B., Bartow, T.J., Miletic, V., Rabson, L.D., Smith, J.A., and Fearon, D.T. (1988). Identification of distinct C3b and C4b recognition sites in the human C3b/C4b receptor (CR1, CD35) by deletion mutagenesis. J. Exp. Med. *168*, 1699–1717.

Klickstein, L.B., Barbashov, S.F., Liu, T., Jack, R.M., and Nicholson-Weller, A. (1997). Complement receptor type 1 (CR1, CD35) is a receptor for C1q. Immunity *7*, 345–355.

Kolb, W.P., and Müller-Eberhard, H.J. (1973). The membrane attack mechanism of complement. Verification of a stable C5–9 complex in free solution. J. Exp. Med. *138*, 438–451.

Kotarsky, H., Hellwage, J., Johnsson, E., Skerka, C., Svensson, H.G., Lindahl, G., Sjöbring, U., and Zipfel, P.F. (1998). Identification of a domain in human factor H and factor H-like protein-1 required for the interaction with streptococcal M proteins. J. Immunol. *160*, 3349–3354.

Kotarsky, H., Gustafsson, M., Svensson, H.G., Zipfel, P.F., Truedsson, L., and Sjöbring, U. (2001). Group A streptococcal phagocytosis resistance is independent of complement factor H and factor H-like protein 1 binding. Mol. Microbiol. *41*, 817–826.

Kraiczy, P., Skerka, C., Brade, V., and Zipfel, P.F. (2001). Further characterization of complement regulator-acquiring surface proteins of *Borrelia burgdorferi*. Infect. Immun. *69*, 7800–7809.

Krych-Goldberg, M., and Atkinson, J.P. (2001). Structure–function relationships of complement receptor type 1. Immunol. Rev. *180*, 112–122.

Krych, M., Atkinson, J.P., and Holers, V.M. (1992). Complement receptors. Curr. Opin. Immunol. *4*, 8–13.

Kühn, S., and Zipfel, P.F. (1996). Mapping of the domains required for decay acceleration activity of the human factor H-like protein 1 and factor H. Eur. J. Immunol. *26*, 2383–2387.

Kühn, S., Skerka, C., and Zipfel, P.F. (1995). Mapping of the complement regulatory domains in the human factor H-like protein 1 and in factor H1. J. Immunol. *155*, 5663–5670.

Labbe, S., and Grenier, D. (1995). Characterization of the human immunoglobulin G Fc-binding activity in *Prevotella intermedia*. Infect. Immun. *63*, 2785–2789.

Lachmann, P. (2006). Complement before molecular biology. Mol. Immunol. *43*, 496–508.

Lachmann, P.J., and Hughes-Jones, N.C. (1984). Initiation of complement activation. Springer Semin. Immunopathol. *7*, 143–162.

Lambris, J.D., Lao, Z., Oglesby, T.J., Atkinson, J.P., Hack, C.E., and Becherer, J.D. (1996). Dissection of CR1, factor H, membrane cofactor protein, and factor B binding and functional sites in the third complement component. J. Immunol. *156*, 4821–4832.

Lambris, J.D., Ricklin, D., and Geisbrecht, B.V. (2008). Complement evasion by human pathogens. Nat. Rev. Microbiol. *6*, 132–142.

Langley, R., Wines, B., Willoughby, N., Basu, I., Proft, T., and Fraser, J.D. (2005). The staphylococcal superantigen-like protein 7 binds IgA and complement C5 and inhibits IgA-Fc alpha RI binding and serum killing of bacteria. J. Immunol. *174*, 2926–2933.

Lathem, W.W., Bergsbaken, T., and Welch, R.A. (2004). Potentiation of C1 esterase inhibitor by StcE, a metalloprotease secreted by *Escherichia coli* O157:H7. J. Exp. Med. *199*, 1077–1087.

Law, S.K., and Dodds, A.W. (1990). C3, C4 and C5: the thioester site. Biochem. Soc. Trans. *18*, 1155–1159.

Law, S.K., and Dodds, A.W. (1997). The internal thioester and the covalent binding properties of the complement proteins C3 and C4. Protein Sci. *6*, 263–274.

Law, S.K., and Levine, R.P. (1977). Interaction between the third complement protein and cell surface macromolecules. Proc. Natl. Acad. Sci. U.S.A. *74*, 2701–2705.

Lee, L.Y., Liang, X., Hook, M., and Brown, E.L. (2004). Identification and characterization of the C3 binding domain of the *Staphylococcus aureus* extracellular fibrinogen-binding protein (Efb). J. Biol. Chem. *279*, 50710–50716.

Lesavre, P.H., and Müller-Eberhard, H.J. (1978). Mechanism of action of factor D of the alternative complement pathway. J. Exp. Med. *148*, 1498–1509.

Loos, M. (1982). Antibody-independent activation of C1, the first component of complement. Ann. Immunol. (Paris) *133C*, 165–179.

Loos, M., Laurell, A.B., Sjöholm, A.G., Mårtensson, U., and Berkel, A.I. (1980). Immunochemical and functional analysis of a complete C1q deficiency in man: evidence that C1r and C1s are in the native form, and that they reassociate with purified C1q to form macromolecular C1. J. Immunol. *124*, 59–63.

Marques, M.B., Kasper, D.L., Pangburn, M.K., and Wessels, M.R. (1992). Prevention of C3 deposition by capsular polysaccharide is a virulence mechanism of type III group B streptococci. Infect. Immun. *60*, 3986–3993.

Marr, N., Luu, R.A., and Fernandez, R.C. (2007). *Bordetella pertussis* binds human C1 esterase inhibitor during the virulent phase, to evade complement-mediated killing. J. Infect. Dis. *195*, 585–588.

Matsushita, M., and Fujita, T. (1992). activation of the classical complement pathway by mannose- binding protein in association with a novel C1s-like serine protease. J. Exp. Med. *176*, 1497–1502.

Matsushita, M., Thiel, S., Jensenius, J.C., Terai, I., and Fujita, T. (2000). Proteolytic activities of two types of mannose-binding lectin-associated serine protease. J. Immunol. *165*, 2637–2642.

Medicus, R.G., Götze, O., and Müller-Eberhard, H.J. (1976). Alternative pathway of complement: recruitment of precursor properdin by the labile C3/C5 convertase and the potentiation of the pathway. J. Exp. Med. *144*, 1076–1093.

Medicus, R.G., Melamed, J., and Arnaout, M.A. (1983). Role of human factor I and C3b receptor in the cleavage of surface-bound C3bi molecules. Eur. J. Immunol. *13*, 465–470.

Meri, S., and Pangburn, M.K. (1990). Discrimination between activators and nonactivators of the alternative pathway of complement: regulation via a sialic acid/polyanion binding site on factor H. Proc. Natl. Acad. Sci. U.S.A. *87*, 3982–3986.

Meri, S., Morgan, B.P., Davies, A., Daniels, R.H., Olavesen, M.G., Waldmann, H., and Lachmann, P.J. (1990). Human protectin (CD59), an 18,000–20,000 MW complement lysis restricting factor, inhibits C5b-8 catalysed insertion of C9 into lipid bilayers. Immunology *71*, 1–9.

Mintz, K.P., and Fives-Taylor, P.M. (1994). Identification of an immunoglobulin Fc receptor of *Actinobacillus actinomycetemcomitans*. Infect. Immun. *62*, 4500–4505.

Moffitt, M.C., and Frank, M.M. (1994). Complement Resistance in Microbes. Springer Semin. Immunopathol. *15* 327–344.

Morgan, B.P., and Harris, C.L. (1999). Complement Regulatory Proteins (Academic Press, San Diego).

Murphy, B.F., Saunders, J.R., O'Bryan, M.K., Kirszbaum, L., Walker, I.D., and d'Apice, A.J. (1989). SP-40,40 is an inhibitor of C5b-6-initiated haemolysis. Int. Immunol. *1*, 551–554.

Müller-Eberhard, H.J. (1988). Molecular organization and function of the complement system. Annu. Rev. Biochem. *57*, 321–347.

Nagasawa, S., and Stroud, R.M. (1977). Cleavage of C2 by C1s into the antigenically distinct fragments C2a and C2b: demonstration of binding of C2b to C4b. Proc. Natl. Acad. Sci. U.S.A. *74*, 2998–3001.

Nagasawa, S., Ichihara, C., and Stroud, R.M. (1980). Cleavage of C4b by C3b inactivator: production of a nicked form of C4b, C4b', as an intermediate cleavage product of C4b by C3b inactivator. J. Immunol. *125*, 578–582.

Nauta, A.J., Trouw, L.A., Daha, M.R., Tijsma, O., Nieuwland, R., Schwaeble, W.J., Gingras, A.R., Mantovani, A., Hack, E.C., and Roos, A. (2002). Direct binding of C1q to apoptotic cells and cell blebs induces complement activation. Eur. J. Immunol. *32*, 1726–1736.

Newman, S.L., Devery-Pocius, J.E., Ross, G.D., and Henson, P.M. (1984). Phagocytosis by human monocyte-derived macrophages. Independent function of receptors for C3b (CR1) and iC3b (CR3). Complement *1*, 213–227.

Nordström, T., Blom, A.M., Forsgren, A., and Riesbeck, K. (2004). The emerging pathogen *Moraxella catarrhalis* interacts with complement inhibitor C4b binding protein through ubiquitous surface proteins A1 and A2. J. Immunol. *173*, 4598–4606.

Norman, D.G., Barlow, P.N., Baron, M., Day, A.J., Sim, R.B., and Campbell, I.D. (1991). Three-dimensional structure of a complement control protein module in solution. J. Mol. Biol. *219*, 717–725.

Oda, T., Kojima, Y., Akaike, T., Ijiri, S., Molla, A., and Maeda, H. (1990). Inactivation of chemotactic activity of C5a by the serratial 56-kilodalton protease. Infect. Immun. *58*, 1269–1272.

Ollert, M.W., Kadlec, J.V., David, K., Petrella, E.C., Bredehorst, R., and Vogel, C.W. (1994). Antibody-mediated complement activation on nucleated cells. A quantitative analysis of the individual reaction steps. J. Immunol. *153*, 2213–2221.

Pandiripally, V., Gregory, E., and Cue, D. (2002). Acquisition of regulators of complement activation by *Streptococcus pyogenes* serotype M1. Infect. Immun. *70*, 6206–6214.

Pandiripally, V., Wei, L., Skerka, C., Zipfel, P.F., and Cue, D. (2003). Recruitment of complement factor H-like protein 1 promotes intracellular invasion by group A streptococci. Infect. Immun. *71*, 7119–7128.

Pangburn, M.K., Schreiber, R.D., and Müller-Eberhard, H.J. (1977). Human complement C3b inactivator: isolation, characterization, and demonstration of an absolute requirement for the serum protein β1H for cleavage of C3b and C4b in solution. J. Exp. Med. *146*, 257–270.

Pangburn, M.K., Schreiber, R.D., and Müller-Eberhard, H.J. (1981). Formation of the initial C3 convertase of the alternative complement pathway. Acquisition of C3b-like activities by spontaneous hydrolysis of the putative thioester in native C3. J. Exp. Med. *154*, 856–867.

Pangburn, M.K., Schreiber, R.D., and Müller-Eberhard, H.J. (1983). C3b deposition during activation of the alternative complement pathway and the effect of deposition on the activating surface. J. Immunol. *131*, 1930–1935.

Pangburn, M.K., Atkinson, M.A., and Meri, S. (1991). Localization of the heparin-binding site on complement factor H. J. Biol. Chem. *266*, 16847–16853.

Patrick, R.A., Taubman, S.B., and Lepow, I.H. (1970). Cleavage of the fourth component of human complement (C4) by activated Cls. Immunochemistry *7*, 217–225.

Pausa, M., Pellis, V., Cinco, M., Giulianini, P.G., Presani, G., Perticarari, S., Murgia, R., and Tedesco, F. (2003).

Serum resistant strains of *Borrelia burgdorferi* evade complement-mediated killing by expressing a CD59-like complement inhibitory molecule. J. Immunol. *170*, 3214–3222.

von Pawel-Rammingen, U., and Björck, L. (2003). IdeS and SpeB: immunoglobulin-degrading cysteine proteinases of *Streptococcus pyogenes*. Curr. Opin. Microbiol. *6*, 50–55.

von Pawel-Rammingen, U., Johansson, B.P., and Bjorck, L. (2002). IdeS, a novel streptococcal cysteine proteinase with unique specificity for immunoglobulin G. EMBO J. *21*, 1607–1615.

Pillemer, L., Blum, L., Lepow, I.H., Ross, O.A., Todd, E.W., and Wardlaw, A.C. (1954). The properdin system and immunity. I. Demonstration and isolation of a new serum protein, properdin, and its role in immune phenomena. Science *120*, 279–285.

Podack, E.R., and Müller-Eberhard, H.J. (1979). Isolation of human S-protein, an inhibitor of the membrane attack complex of complement. J. Biol. Chem. *254*, 9908–9914.

Pramoonjago, P., Kaneko, M., Kinoshita, T., Ohtsubo, E., Takeda, J., Hong, K., Inagi, R., and Inoue, K. (1992). Role of TraT protein, an anticomplementary protein produced in *Escherichia coli* by R100 factor, in serum resistance. J. Immunol. *148*, 827–836.

Prasadarao, N.V., Blom, A.M., Villoutreix, B.O., and Linsangan, L.C. (2002). A novel interaction of outer membrane protein A with C4b binding protein mediates serum resistance of *Escherichia coli* K1. J. Immunol. *169*, 6352–6360.

Preissner, K.T., Heimburger, N., Anders, E., and Müller-Berghaus, G. (1986). Physicochemical, immunochemical and functional comparison of human S-protein and vitronectin. Evidence for the identity of both plasma proteins. Biochem. Biophys. Res. Commun. *134*, 951–956.

Prodinger, W.M., Hellwage, J., Spruth, M., Dierich, M.P., and Zipfel, P.F. (1998). The C-terminus of factor H: monoclonal antibodies inhibit heparin binding and identify epitopes common to factor H and factor H-related proteins. Biochem. J. *331*, 41–47.

Ram, S., McQuillen, D.P., Gulati, S., Elkins, C., Pangburn, M.K., and Rice, P.A. (1998a). Binding of complement factor H to loop 5 of porin protein 1A: a molecular mechanism of serum resistance of nonsialylated *Neisseria gonorrhoeae*. J. Exp. Med. *188*, 671–680.

Ram, S., Sharma, A.K., Simpson, S.D., Gulati, S., Mcquillen, D.P., Pangburn, M.K., and Rice, P.A. (1998b). A novel sialic acid binding site on factor H mediates serum resistance of sialylated *Neisseria gonorrhoeae*. J. Exp. Med. *187*, 743–752.

Ram, S., Cullinane, M., Blom, A.M., Gulati, S., McQuillen, D.P., Boden, R., Monks, B.G., O'Connell, C., Elkins, C., Pangburn, M.K., *et al.* (2001). C4bp binding to porin mediates stable serum resistance of *Neisseria gonorrhoeae*. Int. Immunopharmacol. *1*, 423–432.

Rautemaa, R., Jarvis, G.A., Marnila, P., and Meri, S. (1998). Acquired resistance of *Escherichia coli* to complement lysis by binding of glycophosphoinositol-anchored protectin (CD59). Infect. Immun. *66*, 1928–1933.

Rautemaa, R., Rautelin, H., Puolakkainen, P., Kokkola, A., Kärkkäinen, P., and Meri, S. (2001). Survival of *Helicobacter pylori* from complement lysis by binding of GPI-anchored protectin (CD59). Gastroenterology *120*, 470–479.

Rawal, N., and Pangburn, M.K. (2003). Formation of high affinity C5 convertase of the classical pathway of complement. J. Biol. Chem. *278*, 38476–38483.

Reid, K.B., and Porter, R.R. (1976). Subunit composition and structure of subcomponent C1q of the first component of human complement. Biochem. J. *155*, 19–23.

Ripoche, J., Day, A.J., Harris, T.J.R., and Sim, R.B. (1988). The complete amino acid sequence of human complement factor H. Biochem. J. *249*, 593–602.

Rollins, S.A., and Sims, P.J. (1990). The complement-inhibitory activity of CD59 resides in its capacity to block incorporation of C9 into membrane C5b-9. J. Immunol. *144*, 3478–3483.

Rooijakkers, S.H., Ruyken, M., Roos, A., Daha, M.R., Presanis, J.S., Sim, R.B., van Wamel, W.J., van Kessel, K.P., and van Strijp, J.A. (2005). Immune evasion by a staphylococcal complement inhibitor that acts on C3 convertases. Nat. Immunol. *6*, 920–927.

Ross, G.D., Lambris, J.D., Cain, J.A., and Newman, S.L. (1982). Generation of three different fragments of bound C3 with purified factor I or serum. I. Requirements for factor H vs CR1 cofactor activity. J. Immunol. *129*, 2051–2060.

Ruddy, S., and Austen, K.F. (1969). C3 inactivator of man. I. Hemolytic measurement by the inactivation of cell-bound C3. J. Immunol. *102*, 533–543.

Sato, T., Endo, Y., Matsushita, M., and Fujita, T. (1994). Molecular characterization of a novel serine protease involved in activation of the complement system by mannose-binding protein. Int. Immunol. *6*, 665–669.

Schenkein, H.A., Fletcher, H.M., Bodnar, M., and Macrina, F.L. (1995). Increased opsonization of a *prtH*-defective mutant of *Porphyromonas gingivalis* W83 is caused by reduced degradation of complement-derived opsonins. J. Immunol. *154*, 5331–5337.

Schreiber, R.D., and Müller-Eberhard, H.J. (1974). Fourth component of human complement: description of a three polypeptide chain structure. J. Exp. Med. *140*, 1324–1335.

Schumaker, V.N., Strang, C.J., Siegel, R.L., Phillips, M.L., and Poon, P.H. (1982). Electron microscopy of the first component of human complement. Surv. Immunol. Res. *1*, 305–315.

Schumaker, V.N., Hanson, D.C., Kilchherr, E., Phillips, M.L., and Poon, P.H. (1986). A molecular mechanism for the activation of the first component of complement by immune complexes. Mol. Immunol. *23*, 557–565.

Schwalbe, R., Dahlbäck, B., Hillarp, A., and Nelsestuen, G. (1990). Assembly of protein S and C4b-binding protein on membranes. J. Biol. Chem. *265*, 16074–16081.

Seya, T., Turner, J.R., and Atkinson, J.P. (1986). Purification and characterization of a membrane protein (gp45–70) that is a cofactor for cleavage of C3b and C4b. J. Exp. Med. *163*, 837–855.

Shaffer, M.C., Foley, T.P., and Barnes, D.W. (1984). Quantitation of spreading factor in human biologic fluids. J. Lab. Clin. Med; *103*, 783–791.

Sharma, A.K., and Pangburn, M.K. (1996). Identification of three physically and functionally distinct binding sites for C3b in human complement factor H by deletion mutagenesis. Proc. Natl. Acad. Sci. USA *93*, 10996–11001.

Sim, R.B., Twose, T.M., Paterson, D.S., and Sim, E. (1981). The covalent-binding reaction of complement component C3. Biochem. J. *193*, 115–127.

Sjöholm, A.G., Jönsson, G., Braconier, J.H., Sturfelt, G., and Truedsson, L. (2006). Complement deficiency and disease: an update. Mol. Immunol. *43*, 78–85.

Smith, C.A., Pangburn, M.K., Vogel, C.W., and Müller-Eberhard, H.J. (1984). Molecular architecture of human properdin, a positive regulator of the alternative pathway of complement. J. Biol. Chem. *259*, 4582–4588.

Strang, C.J., Siegel, R.C., Phillips, M.L., Poon, P.H., and Schumaker, V.N. (1982). Ultrastructure of the first component of human complement: electron microscopy of the crosslinked complex. Proc. Natl. Acad. Sci. USA *79*, 586–590.

Takata, Y., Kinoshita, T., Kozono, H., Takeda, J., Tanaka, E., Hong, K., and Inoue, K. (1987). Covalent association of C3b with C4b within C5 convertase of the classical complement pathway. J. Exp. Med. *165*, 1494–1507.

Tamura, N., Shimada, A., and Chang, S. (1972). Further evidence for immune cytolysis by antibody and the first eight components of complement in the absence of C9. Immunology *22*, 131–140.

Thiel, S., Vorup-Jensen, T., Stover, C.M., Schwaeble, W., Laursen, S.B., Poulsen, K., Willis, A.C., Eggleton, P., Hansen, S., Holmskov, U., *et al.* (1997). A second serine protease associated with mannan-binding lectin that activates complement. Nature *386*, 506–510.

Thielens, N.M., Aude, C.A., Lacroix, M.B., Gagnon, J., and Arlaud, G.J. (1990). Ca^{2+} binding properties and Ca^{2+}-dependent interactions of the isolated NH_2-terminal α fragments of human complement proteases C1r and C1s. J. Biol. Chem. *265*, 14469–14475.

Tobe, T., Minoshima, S., Yamase, S., Choi, N.H., Tomita, M., and Shimizu, N. (1991). Assignment of a human serum glycoprotein SP-40,40 gene (CLI) to chromosome 8. Cytogenet. Cell Genet. 57, 193–195.

Tsao, N., Tsai, W.H., Lin, Y.S., Chuang, W.J., Wang, C.H., and Kuo, C.F. (2006). Streptococcal pyrogenic exotoxin B cleaves properdin and inhibits complement-mediated opsonophagocytosis. Biochem. Biophys. Res. Commun. *339*, 779–784.

Tschopp, J., Müller-Eberhard, H.J., and Podack, E.R. (1982). Formation of transmembrane tubules by spontaneous polymerization of the hydrophilic complement protein C9. Nature *298*, 534–538.

Tschopp, J., Chonn, A., Hertig, S., and French, L.E. (1993). Clusterin, the human apolipoprotein and complement inhibitor, binds to complement C7, C8β, and the b domain of C9. J. Immunol. *151*, 2159–2165.

Tsuruta, J.K., Wong, K., Fritz, I.B., and Griswold, M.D. (1990). Structural analysis of sulphated glycoprotein 2 from amino acid sequence. Relationship to clusterin and serum protein 40,40. Biochem. J. *268*, 571–578.

Unnikrishnan, M., Cohen, J., and Sriskandan, S. (1999). Growth-phase-dependent expression of virulence factors in an M1T1 clinical isolate of *Streptococcus pyogenes*. Infect. Immun. *67*, 5495–5499.

Ward, P.A., and Newman, L.J. (1969). A neutrophil chemotactic factor from human C'5. J. Immunol. *102*, 93–99.

Webb, J.H., Blom, A.M., and Dahlbäck, B. (2002). Vitamin K-dependent protein S localizing complement regulator C4b-binding protein to the surface of apoptotic cells. J. Immunol. *169*, 2580–2586.

Weiler, J.M., Daha, M.R., Austen, K.F., and Fearon, D.T. (1976). Control of the amplification convertase of complement by the plasma protein β1H. Proc. Natl. Acad. Sci. USA *73*, 3268–3272.

Wexler, D.E., and Cleary, P.P. (1985). Purification and characteristics of the streptococcal chemotactic factor inactivator. Infect. Immun. *50*, 757–764.

Wexler, D.E., Chenoweth, D.E., and Cleary, P.P. (1985). Mechanism of action of the group A streptococcal C5a inactivator. Proc. Natl. Acad. Sci. U.S.A. *82*, 8144–8148.

Whaley, K., and Ruddy, S. (1976). Modulation of the alternative complement pathways by β1H globulin. J. Exp. Med. *144*, 1147–1163.

Ziccardi, R.J. (1982). A new role for C-1-inhibitor in homeostasis: control of activation of the first component of human complement. J. Immunol. *128*, 2505–2508.

Ziccardi, R.J. (1985). Demonstration of the interaction of native C1 with monomeric immunoglobulins and C1 inhibitor. J. Immunol. *134*, 2559–2563.

Ziccardi, R.J., and Cooper, N.R. (1979). Active disassembly of the first complement component, C-1, by C-1 inactivator. J. Immunol. *123*, 788–792.

Zipfel, P.F., and Skerka, C. (1999). FHL-1/reconectin: a human complement and immune regulator with cell-adhesive function. Immunol. Today *20*, 135–140.

Bacterial Resistance to Antimicrobial Peptides

14

John D. F. Hale

Abstract

Host defence peptides (HDP) are small cationic, amphipathic molecules produced by all organisms as a first line of defence against microbial invasion and are commonly found at host–microbe interfaces, such as epithelial layers. Although initially identified to control microbial levels through direct antimicrobial activity, their role has been expanded to include the multitude of other functions such as modulation of innate defence factors. While HDPs are active against a range of microbes, the best understood resistance mechanisms are in bacteria. Many pathogenic bacteria have evolved resistance mechanisms to ensure their own survival and an increased virulence following continual exposure to HDPs. Although individual species can affect resistance differently, they still fall under certain themes, such as repulsion, degradation, sequestration and expulsion, all of which are based the exploitation of the common features of HDPs, such as charge or their peptide nature. Despite this diverse range of resistance mechanisms, bacteria have yet to achieve full resistance to the large repertoire of HDPs.

Introduction

All living organisms must coexist with microorganisms. This cohabitation for the most part is mutualistic and, in many cases, even beneficial. However, certain pathogenic microbes can capitalize on this relationship with their host and cause infection. In response to microbial threat, all organisms (including bacteria themselves) have developed various mechanisms to control microbial presence. Among the most interesting is a class of evolutionary ancient peptides referred to as cationic antimicrobial peptides (AMPs) (Nizet, 2006) or CAMPs (Kraus and Peschel, 2006). While this nomenclature referred to their initial identification as antimicrobial agents, further *in vitro* studies with these peptides and the use of deficient transgenic mice have helped to identify immunomodulatory functions including (but not exclusively) macrophage activation, chemokine release and blocking of pathogen receptors on target cells (Hamill *et al.*, 2008). Consequently, their name has been revised to <u>h</u>ost <u>d</u>efence <u>p</u>eptides (HDPs) to reflect the broad range of critical and complex roles these peptides play within an organism. However, the name 'AMP' is still commonly used when referring to the specific antimicrobial function of these peptides.

HDPs have been identified in nearly all living organisms, ranging from bacteria (where HDPs are more commonly known as bacteriocins), protozoa, insects, amphibians, fish, plants, to the higher order organisms such as mammals (Jenssen *et al.*, 2006). This ubiquity suggests an important evolutionary role in protecting the host from microbial invasion and, despite the diversity of origin, HDPs share many features [they are 12–50 amino acids long, (most) have a positive charge, between +2 to +9], while their expression can be either constitutive, e.g. human β-defensin 1 (HβD1) or inducible, e.g. human β-defensin 2 (HβD2) or LL-37 (Bowdish *et al.*, 2006). Antimicrobial HDPs demonstrate a unique broad inhibitory spectrum that can include both Gram-positive and Gram-negative bacteria, fungi and protozoa, which makes them attractive molecules for further development as potential therapeutic agents (Zasloff, 2002). Currently >1000 HDPs have been identified and lodged in databases, such as http://aps.unmc.edu/AP/main.php (Wang *et al.*, 2009), http://www.bbcm.univ.trieste.it/~tossi/pag2.htm, AMPer (http://marray.cmdr.ubc.ca/cgi-bin/amp.pl) (Fjell *et al.*, 2007) and Collection of Antimicrobial Peptides (CAMP) http://www.bicnirrh.res.in/antimicrobial (Thomas *et al.*, 2010). From this large number of peptides, it is apparent that they have evolved with significant diversity in their primary sequences although similarity is found in HDPs produced by closely related species (Hancock and Sahl, 2006). Despite this, HDPs still demonstrate a conserved nature at the secondary structure level and conform into four basic structural groups: α-helical; β-sheet; loop and extended (Jenssen *et al.*, 2006) (Fig. 14.1). Of these, (especially) α-helical and β-sheet are the most prevalent (Peschel and Sahl, 2006).

Antimicrobial HDPs have been identified to kill bacterial cells through interactions with multiple targets. Traditionally, the most common target identified has been the cytoplasmic membrane. However, there may be some bias in this, as studies have shown that certain HDPs (e.g. pleurocidin) exhibit a concentration-dependent effect, where at high concentrations [multiple times the minimum inhibitory concentration (MIC)] pleurocidin was found to cause membrane lysis in *Escherichia coli*, while at lower concentrations (1 × MIC) no effect on the membrane was seen but, instead, cell death occurred through interruption of macromolecular synthesis (Patrzykat *et al.*, 2002). Subsequently, a number of studies using HDPs at their MIC have identified a multitude of alternative (non-membrane) targets, including chaperones and nucleic acids (Park *et al.*, 1998; Otvos *et al.*, 2000; Hale and Hancock, 2007). Because of this unique promiscuous antimicrobial effect, HDPs have been referred to as 'dirty drugs' (Peschel and Sahl, 2006). Regardless of their mechanism(s) of action, all cationic HDPs initially interact with a bacterial cell through electrostatic attraction.

Resistance to AMPs

Resistance to HDPs confers a massive advantage to a pathogen, increasing its survival chances and/or pathogenesis (Kraus and Peschel, 2006).

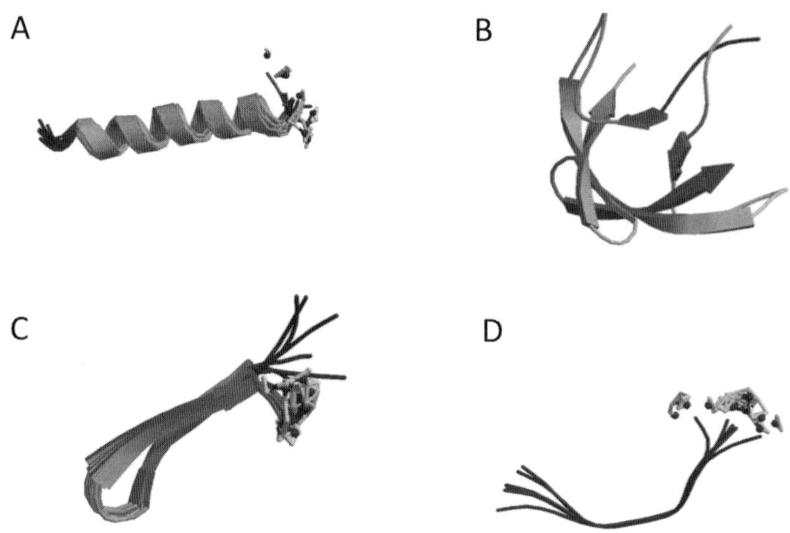

Figure 14.1 Different structural types of host-defence peptides (HDPs). (A) α-helical, e.g. magainin 2 [PDB ID: 2MAG (Gesell *et al.*, 1997)]. (B) β- sheet, e.g. human neutrophil peptide-1 (HNP-1) [PDB ID: 3HJ2 (Wei *et al.*, unpublished data, *What Dictates the Multifaceted Functions of the Human alpha-Defensin HNP1?*)]. (C) β-turn, e.g. protegrin-1 [PDB ID: 1PG1 (Fahrner *et al.*, 1996)]. (D) Extended structure, e.g. indolicidin [PDB ID 1G89 (Rozek *et al.*, 2000)].

Given the prevalence of HDPs and the continual exposure to them, it could be assumed that there is a high selective pressure on bacteria to develop resistance mechanisms against them. Surprisingly, it seems that this is not the case with most bacteria (with the exceptions of *Burkholderia* spp. and some *Proteus* and *Serratia* spp.; Hancock and Patrzykat, 2002), demonstrating an intrinsic susceptibility to their effects. Furthermore, serial passaging of bacteria at sub-MIC concentrations of protegrin-1 failed to select any resistant isolates (Steinberg *et al.*, 1997; Devine and Hancock, 2002). However, a number of examples of adaptive resistance have evolved to counteract the effects of HDPs, but unlike resistance mechanisms to many classical antibiotics these are generally not located on transferable elements such as transposons or plasmids (Kraus and Peschel, 2006). Indeed, it appears that the co-evolution of increasingly complex HDPs versus more elaborate resistance mechanisms emulate a 'cat and mouse' scenario. Despite this, simpler HDPs, such as α-helical peptides, still appear in abundance. The purpose of this review is to present an overview of the different HDP resistance mechanisms that have been identified in bacteria. Each is discussed in some detail below and is also summarized in Fig. 14.2.

How bacteria resist HDPs

Repulsion by altering the cell's net charge

Charge modification at the outer surface

The first step in antimicrobial activity of HDPs is their electrostatic attraction to the bacterial cell. Bacteria contain a net negative charge on their cell surface owing to the presence of anionic molecules. In contrast, mammalian cells have evolved with an outer leaflet comprising of uncharged or zwitterionic lipids (Devaux and Morris, 2004), such that no electrostatic attraction occurs. Therefore, a major mechanism of resistance for bacteria is to reduce this attraction effectively, leading to

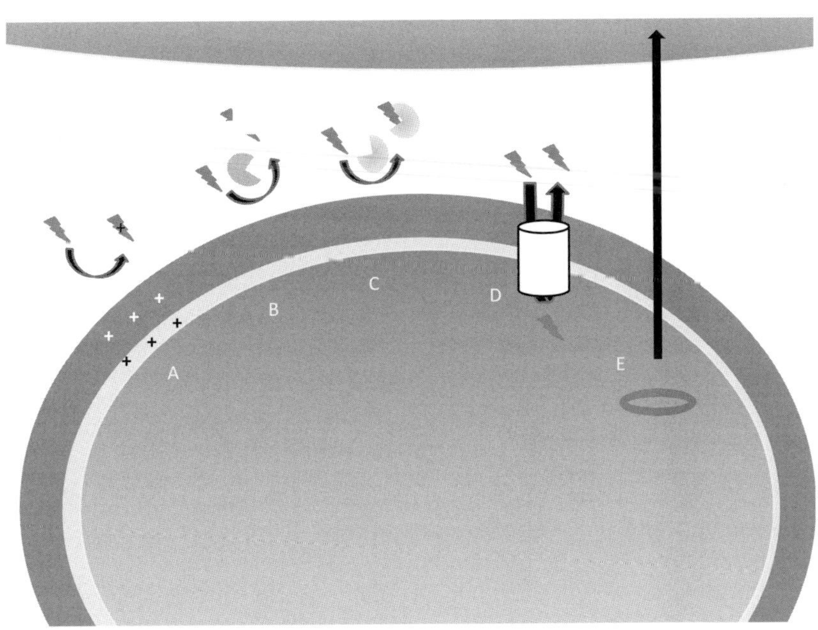

Figure 14.2 Mechanisms of host defence peptide (HDP) resistance in bacteria (grey circle). (A) repulsion through alteration of net cell charge at the outer membrane (Gram-negative bacteria) or cell wall (white '+') or at the cytoplasmic membrane (black '+'); (B) degradation by proteases and peptidases; (C) sequestration of HDPs by secreted or cell-surface bound molecules, or reduced HDP uptake by capsule; (D) expulsion from the cell; (E) suppression of HDP expression in mammalian cells (large sphere) by plasmid DNA. Not shown, biofilm.

the repulsion of cationic HDPs. This is achieved by altering their net surface charge through the addition of cationic charged molecules (Ernst *et al.*, 2001; Peschel, 2002).

In Gram-positive bacteria, the anionic cell-surface charge is due to the presence of a complex matrix of peptidoglycan and teichoic acids (consisting of both cell-wall attached teichoic acids and cell-membrane anchored lipoteichoic acids) (Neuhaus and Baddiley, 2003). In teichoic acids, the anionic charge is due to the large number of phosphate groups that are present in glycerophosphate and ribitolphosphate groups. Modification of this charge occurs through the incorporation of cationic D-alanine residues onto the phosphate groups on the teichoic acid backbone (Peschel *et al.*, 1999; Collins *et al.*, 2002), which was first identified by mutagenesis studies to be encoded by the *dltABCD* operon in *Staphylococcus aureus* (Peschel *et al.*, 1999). Other studies confirmed that this mechanism is more widely spread in Gram-positive species including *Bacillus subtilis* (Wecke *et al.*, 1997), *Streptococcus agalactiae* (Poyart *et al.*, 2003), *Streptococcus pyogenes* (Kristian *et al.*, 2005) and *Streptococcus pneumoniae* (Wecke *et al.*, 1997; Kovács *et al.*, 2006), indicating its importance as a mechanism of resistance to HDPs (Weidenmaier *et al.*, 2003). More recent studies with staphylococci have identified that regulation of this pathway occurs through a three-component regulatory system termed antimicrobial peptide sensor (Aps), which consists of a sensor kinase (ApsS), response regulator (ApsR) and AspX, a third protein of still-yet undefined role (Li *et al.*, 2007a,b).

Although structurally different, Gram-negative bacteria also present an anionic cell-surface charge due to the presence of lipopolysaccharide (LPS) (also known as endotoxin) on the outer leaflet of the outer membrane (Raetz and Whitfield, 2002). LPS comprises three domains: O antigen, variable core oligosaccharide, and lipid A, which anchors the molecule to the outer membrane. The anionic charge occurs because of the presence of phosphate groups on lipid A. Similarly to Gram-positive bacteria, Gram-negative species can modify these phosphate groups with the addition of cationic molecules, such as aminoarabinose and/or phosphoethanolamine, effectively repelling cationic HDPs (Gunn *et al.*, 1998; Nummila *et al.*, 1995; Lee *et al.*, 2004). This resistance is encoded by the *pmr* (*pmrHFIJKLM* also known as *arnBCADTEF*) operon and appears to be a likely common mechanism of resistance, as it is now identified in a growing number of Gram-negative bacteria including *E. coli*, *Pseudomonas aeruginosa* and *Salmonella enterica* serovar Typhi (Devine and Hancock, 2002; Miller *et al.*, 2005). Lipid A modification is under the control of two two-component regulatory systems: PmrA-PmrB and PhoP-PhoQ. The presence of most HDPs (or under Mg^{2+} limiting conditions) causes the autophosphorylation of two membrane histidine kinases PhoQ (in *Salmonella* but not *Pseudomonas*) and PmrB, which can then activate their cognate response regulators PhoP and PmrA, resulting in the transcriptional upregulation of their own operons and the lipid A-modifying operon (Gooderham and Hancock, 2009).

Charge modification of the cytoplasmic membrane

The cytoplasmic membrane also exhibits a net negative charge due to the presence of anionic phospholipid head groups. Modification of this charge can also confer resistance to HDPs. A large unique integral membrane protein named multiple resistance factor (MprF), identified as a lysyl-phosphatidylglycerol synthase, acts to reduce the negative charge through the addition of cationic L-lysine to phospholipid phosphatidylglycerol (Peschel *et al.*, 2001; Oku *et al.*, 2004). While originally identified through mutagenesis studies in staphylococci (Peschel *et al.*, 2001), homologues of MprF have been identified in many other Gram-positive species including *Listeria monocytogenes* (Thedieck *et al.*, 2006) and *Bacillus anthracis* (Samant *et al.*, 2009), as well as a predicted homologue identified in the Gram-negative species, *P. aeruginosa* (Klein *et al.*, 2009). Interestingly, two paralogues of MprF (MprF1 and MprF2) were found in the Gram-positive pathogen *Clostridium perfringens* SM101, and both aminoacylate phosphatidylglycerol, but with different amino acids. MprF1 catalyses the addition of alanine, while MprF2 is specific for lysine (Roy and Ibba, 2008). This apparent growing ubiquity of MprF signifies its importance as a resistance determinant.

Degradation by proteases

Perhaps the most direct method to resist AMPs is through their inactivation by proteases and peptidases. Examples have been identified from members of both Gram-positive and Gram-negative bacteria, and their effect is thought to confer increased pathogenicity. The simpler structured HDPs (α-helical and extended), which lack any disulfide bonds, are more susceptible to proteases (Peschel and Sahl, 2006) than complex β-sheet and looped HDPs (Rozek *et al.*, 2003). Interestingly, α-helical HDPs are still the most common peptides and are conserved throughout many species (Peschel and Sahl, 2006). The incorporation of prolines into AMPs (as seen in many cathelecidins) can also increase resistance to serine proteases (Shinnar *et al.*, 2003).

In Gram-negative bacteria, the omptins (the name derived from the best characterized example, OmpT; Kukkonen and Korhonen, 2004) are a family of structurally similar outer membrane proteases, some of which have been identified to degrade HDPs. Examples include *E. coli* OmpT (Stumpe *et al.*, 1998), *Salmonella enterica* PgtE (Guina *et al.*, 2000), *Yersinia pestis* Pla (Sodeinde and Goguen, 1989) and *Citrobacter rodentium* CroP (Le Sage *et al.*, 2009). The omptins contain conserved catalytic residues, and their mechanism of proteolysis appears novel, with proteolytic cleavage occurring between two successive basic amino acid residues or between one basic amino acid and a subsequent nonpolar amino acid (Sugimura and Nishihara, 1988; Kukkonen and Korhonen, 2004). Proteolytic activity against HDPs has been demonstrated for the omptins including OmpT degrading colicin (Cavard and Lazdunski, 1990) and protamine (Stumpe *et al.*, 1998), OmpP cleaving protamine (Hwang *et al.*, 2007) and PgtE inactivating human cathelicidin LL37 and the mouse equivalent CRAMP amongst others (Guina *et al.*, 2000). Despite exhibiting high sequence similarity between each other, different members of this family may in fact be involved in many different roles in pathogenesis. For example, Pla from *Y. pestis* acts as an adhesin but can also proteolytically activate plasminogen into plasmin, which is an activator of cell migration and fibrinolysis (Kukkonen and Korhonen, 2004). Interestingly, different omptins may act synergistically. For example, while both OmpT and OmpP can degrade the HDP protamine, they achieve this through the action at different cleavage sites (Hwang *et al.*, 2007). Therefore, the expression of two proteases could act to enhance proteolytic resistance through an expedited response or perhaps act to broaden their action against a wider range of HDPs with a certain level of crossing over.

Other proteases have been identified in Gram-negative bacteria, confirming their importance as a resistance mechanism. *P. aeruginosa, Proteus mirabilis, Porphyromonas gingavalis* (Schmidtchen *et al.*, 2002) and *Burkholderia cenocepacia* [a member of the *Burkholderia cepacia* complex (Bcc)(Kooi and Sokol, 2009)] all produce a protease capable of degrading the human HDP, LL-37. Additionally, members of Bcc have been shown to produce proteases that also degrade magainin II (Thwaite *et al.*, 2009), HβD-1 and protamine (Kooi and Sokol, 2009).

Gram-positive bacteria are also capable of degrading HDPs. *S aureus* has been identified to produce at least two different proteases: the serine protease (glutamylendopeptidase) V8 and the metalloprotease aureolysin, both of which are capable of cleaving LL-37 (Sieprawska-Lupa *et al.*, 2004). While aureolysin proteolysis completely abrogates LL-37 activity, protease V8 proteolysis generates a C-terminal compound (LL-17–37) that is as active as the LL-37. Thus in this case, it appears that aureolysin is the more effective protease in generating resistance to HDPs. *B. anthracis* also produces a metalloprotease capable of degrading LL-37 (Thwaite *et al.*, 2006). While most examples of protease activity are against cationic HDPs, anionic-degrading proteases exist. *Staphylococcus epidermidis* produces a metalloprotease SepA, which can degrade the anionic HDP, dermicidin (Lai *et al.*, 2007).

Blocking of HDP uptake

Effect of capsule

Many pathogenic bacteria produce an extracellular capsule commonly comprises anionic capsular polysaccharide (CPS) to resist host immune responses, such as phagocytosis and complement. Mutagenesis studies in *Neisseria meningitidis* (Spinosa *et al.*, 2007; Jones *et al.*, 2009) and *Klebsiella pneumoniae* (Campos *et al.*, 2004) identified that capsule-deficient mutants are more sensitive to HDPs, including LL-37, than their parent strains, suggesting that the capsule provides a protective shield that prevents the passage of HDPs to the cytoplasmic membrane. However, this protection does not appear to be universal, with other studies demonstrating that the presence of capsule does not necessarily ensure resistance. For example, mutagenesis studies in *Campylobacter jejuni* found that the loss of CPS resulted in no difference in susceptibility to the HDPs HβD2 and HβD3 (Zilbauer *et al.*, 2005), while further studies using the Gram-positive pathogen *S. pneumoniae* found that encapsulated strains were actually more susceptible to the α-defensins HNP 1–3 than their respective capsule-deficient mutants (Beiter *et al.*, 2008). In this latter case, it is not fully understood how encapsulation enhances the effects of α-defensins killing, but it was suggested that the capsule could actually mask other resistance mechanisms, such as electrostatic repulsion. Other Gram-positive bacteria, *S. epidermidis* and *B. subtilis*, produce a capsule-like extracellular polymer, called poly γ-DL-glutamic acid (PGA). Although the function of this polymer is still not fully understood, it has been shown to somehow play a role in resistance to the human HDPs LL-37 and HβD3 (Kocianova *et al.*, 2005). At this stage, it appears that, for certain bacterial species, capsule could play some role in resistance to HDPs. However, this may not be the sole mechanism, as many bacteria harbour resistance genes for additional mechanisms (Tzeng *et al.*, 2005; Spinosa *et al.*, 2007).

Sequestration by secreted proteins

The passage of HDPs across the outer cell layers to the cytoplasmic membrane can be inhibited through designated molecules that bind to them. For example, in addition to its role within the capsule, the effect of free CPS on HDPs was also investigated. Using purified CPS from *K. pneumoniae*, *S. pneumoniae* and *P. aeruginosa*, it was found that free CPS effectively binds HNP-1 and another antimicrobial peptide, polymyxin B, resulting in an increased MIC, suggesting its possible role as a decoy molecule that can neutralize HDPs (Llobet *et al.*, 2008).

In staphylococci, certain strains harbour the prophage-encoded exoprotein staphylokinase (SAK) (Bokarewa *et al.*, 2006). SAK was originally identified to facilitate the binding and activation of host plasminogen into plasmin. However, an additional function was assigned when it was shown that SAK can directly bind to α-defensins, effectively abrogating any bactericidal effect (Jin *et al.*, 2004). Furthermore, SAK's efficiency is increased by the presence of multiple binding sites for α-defensins (Bokarewa *et al.*, 2006). *S. epidermidis* produces a cationic extracellular polysaccharide, called polysaccharide intracellular adhesin (PIA) (Vuong *et al.*, 2004). PIA-deficient strains were shown to be more susceptible to various human HDPs, such as LL-37, HβD3 and anionic dermicidin, as well as to phagocytosis, demonstrating its importance in protecting bacteria against both cationic and anionic peptides.

Sequestration by surface attached molecules

S. pyogenes produces at least two different HDP-binding proteins: *Streptococcus* inhibitor of complement (SIC) and M protein. SIC is a major virulence factor produced in certain (especially the prevalent and virulent M1) serotypes. Although originally identified to prevent complement insertion into the bacterial membrane (Fernie-King *et al.*, 2001), SIC was later found to inhibit host innate defence mechanisms through sequestration of (albeit at a low affinity) lysozyme, secretory leucocyte proteinase inhibitor and HDPs, such as LL-37 and HNP-1 (Frick *et al.*, 2003), HβD2 and HβD3 (Fernie-King *et al.*, 2004). *S. pyogenes* also produces the highly variable cellular attached virulence factor, M protein. Although over 100 M protein serotypes have been identified, a screening study found that the M1 serotype was most resistant to LL-37 (Nizet, 2006). Resistance owing to the M1 protein was further confirmed through mutagenesis of its structural gene, *emm1*, and also by heterologous expression of the M1 protein in either a second (non-M1) serotype *S. pyogenes* strain (M49) or in *Lactococcus lactis*. Binding was found to occur to a novel region in a hypervariable region of M1 that does not exist in the M49 protein.

Expulsion from cell

Energy-dependent efflux pumps play an important role in the expulsion of antibiotics from bacterial cells (Zgurskaya, 2002). Like repulsion, this mechanism offers broad resistance to a range of antibiotics, including, in some cases, HDPs. The sexually transmitted pathogen *Neisseria gonorrhoeae* expresses an energy-dependent efflux pump MtrCDE that was shown to confer resistance to different hydrophobic antimicrobial agents (Veal *et al.*, 2002). This mechanism was responsible for the efflux of structurally different HDPs, including porcine protegrin-1 and human LL-37 (Shafer *et al.*, 1998), while further studies showed that its presence contributes to increased virulence in a mouse infection model, presumably in part by counteracting HDPs (Jerse *et al.*, 2003). *S. enterica* serovar Typhimurium is a causative agent of human gastroenteritis. Mutagenesis studies in this species identified the sensitive to antimicrobial peptides (*sap*) operon that is involved in resistance to HDPs, such as protamine (Groisman *et al.*, 1992). In addition to transporting K^{+}, it is thought that the *sap* operon confers resistance through expulsion of HDPs from the cell (Parra-Lopez *et al.*, 1994). Resistance to the rabbit HDP thrombin-induced platelet microbicidal protein 1 (tPMP-1) was found to be mediated by the plasmid-encoded QacA efflux pump in *S. aureus* (Kupferwasser *et al.*, 1999). However, resistance appears specific to tPMP-1, as other structurally diverse HDPs tested were not affected.

Suppression of host HDP genes

In contrast to the above mechanisms, which focus on changes in bacterial cell appearance or the production of molecules that directly interact with HDPs, a novel approach of aggressive resistance was identified in *Shigella* spp. (Islam *et al.*, 2001). Using semi-quantitative analyses of gut tissues from patients infected with *Shigella*, it was found that the expression of two human HDPs, LL-37 and HβD3 was much lower in infected patients compared with healthy controls. Reductive analysis found that this suppression was somehow due to the presence of *Shigella* plasmid DNA (Islam *et al.*, 2001). Unfortunately, it is not fully understood how plasmid DNA can confer this resistance.

Effect of biofilm

A biofilm is a complex, dynamic, mixed microbial community embedded in a macromolecular matrix (Sutherland, 2001). Biofilm formation has been shown to cause a significant increase in resistance to antibiotics and innate immune defence, such as HDPs (Costerton *et al.*, 1999). How this resistance is achieved is still not totally understood but is probably due to a combination of reduced HDP penetration, sequestration of peptide and an altered metabolic state for some cells. While certain classical antibiotics, such as vancomycin and rifampicin, can penetrate biofilms (Dunne *et al.*, 1993), it appears that HDPs do not have the ability to do so. A recent study shed some light on this: sub-inhibitory concentrations of LL-37 were shown to not only inhibit existing *P. aeruginosa* biofilms but to also inhibit their formation (Overhage *et al.*, 2008). LL-37 was identified to affect biofilm formation in three ways: (i) a lowered attachment of cells and (ii) an increased cell motility, which both reduce the chances of biofilm initiation, while it also (iii) causes the down-regulation of two quorum-sensing mechanisms (Las and Rhl), both of which regulate key genes involved in biofilm formation.

Conclusions

HDPs are successful first-line defence mechanisms employed by living organisms to combat the threat of microbial invasion. With >1000 HDPs identified representing vast sequence variation, they still typically fall into only four structural classes. In response, many pathogens have evolved a variety of mechanisms to ensure their own survival and increased virulence. Yet, no bacteria has developed full resistance to all HDPs (Kraus and Peschel, 2006). This may be a result of the high metabolic burden required by a bacterial cell to become resistant. Of the currently known mechanisms, it is apparent that certain resistance themes, such as repulsion or degradation, are employed by bacteria across both major bacterial classes, suggesting their effectiveness in resistance. Furthermore, the molecular mechanisms behind these themes, such as DltABCD, may themselves become a target for novel antimicrobial therapies, as researchers look for new ways to control pathogenic bacteria. Alternatively, novel therapies must be investigated to circumvent these defences. Currently a better understanding of how HDPs induce killing and of the responses by bacteria can only assist in the design of more effective therapeutic agents.

References

Beiter, K., Wartha, F., Hurwitz, R., Normark, S., Zychlinsky, A., and Henriques-Normark, B. (2008). The capsule sensitizes *Streptococcus pneumoniae* to α-defensins human neutrophil proteins 1 to 3. Infect. Immun. *76*, 3710–3716.

Bokarewa, M.I., Jin, T., and Tarkowski, A. (2006). *Staphylococcus aureus*: Staphylokinase. Int. J. Biochem. Cell Biol. *38*, 504–509.

Bowdish, D.M., Davidson, D.J., and Hancock, R.E. (2006). Immunomodulatory properties of defensins and cathelicidins. Curr. Top. Microbiol. Immunol. *306*, 27–66.

Campos, M.A., Vargas, M.A., Regueiro, V., Llompart, C.M., Albertí, S., and Bengoechea, J.A. (2004). Capsule polysaccharide mediates bacterial resistance to antimicrobial peptides. Infect. Immun. *72*, 7107–7114.

Cavard, D., and Lazdunski, C. (1990). Colicin cleavage by OmpT during entry into and release from *Escherichia coli*. J. Bacteriol. *172*, 648–652.

Collins, L.V., Kristian, S.A., Weidenmaier, C., Faigle, M., Van Kessel, K.P., Van Strijp, J.A., Götz, F., Neumeister, B., and Peschel, A. (2002). *Staphylococcus aureus* strains lacking D-alanine modifications of teichoic acids are highly susceptible to human neutrophil killing and are virulence attenuated in mice. J. Infect. Dis. *186*, 214–219.

Costerton, J.W., Stewart, P.S., and Greenberg, E.P. (1999). Bacterial biofilms: a common cause of persistent infections. Science *284*, 1318–1322.

Devaux, P.F., and Morris, R. (2004). Transmembrane asymmetry and lateral domains in biological membranes. Traffic *5*, 241–246.

Devine, D.A., and Hancock, R.E. (2002). Cationic peptides: distribution and mechanisms of resistance. Curr. Pharm. Des. *8*, 703–714.

Dunne, W.M., Jr., Mason, E.O., Jr., and Kaplan, S.L. (1993). Diffusion of rifampin and vancomycin through a *Staphylococcus epidermidis* biofilm. Antimicrob. Agents Chemother. *37*, 2522–2526.

Ernst, R.K., Guina, T., and Miller, S.I. (2001). *Salmonella typhimurium* outer membrane remodeling: role in resistance to host innate immunity. Microbes Infect. *3*, 1327–1334.

Fahrner, R.L., Dieckmann, T., Harwig, S.S., Lehrer, R.I., Eisenberg, D., and Feigon, J. (1996). Solution structure of protegrin-1, a broad-spectrum antimicrobial peptide from porcine leukocytes. Chem. Biol. *3*, 543–550.

Fernie-King, B.A., Seilly, D.J., Willers, C., Würzner, R., Davies, A., and Lachmann, P.J. (2001). Streptococcal inhibitor of complement (SIC) inhibits the membrane attack complex by preventing uptake of C567 onto cell membranes. Immunology *103*, 390–398.

Fernie-King, B.A., Seilly, D.J., and Lachmann, P.J. (2004). The interaction of streptococcal inhibitor of complement (SIC) and its proteolytic fragments with the human β-defensins. Immunology *111*, 444–452.

Fjell, C.D., Hancock, R.E., and Cherkasov, A. (2007). AMPer: a database and an automated discovery tool for antimicrobial peptides. Bioinformatics *23*, 1148–1155.

Frick, I.M., Akesson, P., Rasmussen, M., Schmidtchen, A., and Björck, L. (2003). SIC, a secreted protein of *Streptococcus pyogenes* that inactivates antibacterial peptides. J. Biol. Chem. *278*, 16561–16566.

Gesell, J., Zasloff, M., and Opella, S.J. (1997). Two-dimensional 1H NMR experiments show that the 23-residue magainin antibiotic peptide is an α-helix in dodecylphosphocholine micelles, sodium dodecylsulfate micelles, and trifluoroethanol/water solution. J. Biomol. NMR *9*, 127–135.

Gooderham, W.J., and Hancock, R.E. (2009). Regulation of virulence and antibiotic resistance by two-component regulatory systems in *Pseudomonas aeruginosa*. FEMS Microbiol. Rev. *33*, 279–294.

Groisman, E.A., Parra-Lopez, C., Salcedo, M., Lipps, C.J., and Heffron, F. (1992). Resistance to host antimicrobial peptides is necessary for *Salmonella* virulence. Proc. Natl. Acad. Sci. U.S.A. *89*, 11939–11943.

Guina, T., Yi, E.C., Wang, H., Hackett, M., and Miller, S.I. (2000). A PhoP-regulated outer membrane protease of *Salmonella enterica* serovar Typhimurium promotes resistance to alpha-helical antimicrobial peptides. J. Bacteriol. *182*, 4077–4086.

Gunn, J.S., Lim, K.B., Krueger, J., Kim, K., Guo, L., Hackett, M., and Miller, S.I. (1998). PmrA-PmrB-regulated genes necessary for 4-aminoarabinose lipid A modification and polymyxin resistance. Mol. Microbiol. *27*, 1171–1182.

Hale, J.D., and Hancock, R.E. (2007). Alternative mechanisms of action of cationic antimicrobial peptides on bacteria. Expert Rev. Anti Infect. Ther. *5*, 951–959.

Hamill, P., Brown, K., Jenssen, H., and Hancock, R.E. (2008). Novel anti-infectives: is host defence the answer? Curr. Opin. Biotechnol. *19*, 628–636.

Hancock, R.E., and Patrzykat, A. (2002). Clinical development of cationic antimicrobial peptides: from natural to novel antibiotics. Curr. Drug Targets Infect. Disord. *2*, 79–83.

Hancock, R.E., and Sahl, H.G. (2006). Antimicrobial and host-defense peptides as new anti-infective therapeutic strategies. Nat. Biotechnol. *24*, 1551–1557.

Hwang, B.Y., Varadarajan, N., Li, H., Rodriguez, S., Iverson, B.L., and Georgiou, G. (2007). Substrate specificity of the *Escherichia coli* outer membrane protease OmpP.J. Bacteriol. *189*, 522–530.

Islam, D., Bandholtz, L., Nilsson, J., Wigzell, H., Christensson, B., Agerberth, B., and Gudmundsson, G. (2001). Downregulation of bactericidal peptides in enteric infections: a novel immune escape mechanism with bacterial DNA as a potential regulator. Nat. Med. *7*, 180–185.

Jenssen, H., Hamill, P., and Hancock, R.E. (2006). Peptide antimicrobial agents. Clin. Microbiol. Rev. *19*, 491–511.

Jerse, A.E., Sharma, N.D., Simms, A.N., Crow, E.T., Snyder, L.A., and Shafer, W.M. (2003). A gonococcal efflux pump system enhances bacterial survival in a female mouse model of genital tract infection. Infect. Immun. *71*, 5576–5582.

Jin, T., Bokarewa, M., Foster, T., Mitchell, J., Higgins, J., and Tarkowski, A. (2004). *Staphylococcus aureus* resists human defensins by production of staphylokinase, a novel bacterial evasion mechanism. J. Immunol. *172*, 1169–1176.

Jones, A., Geörg, M., Maudsdotter, L., and Jonsson, A.B. (2009). Endotoxin, capsule, and bacterial attachment contribute to *Neisseria meningitidis* resistance to the human antimicrobial peptide LL-37. J. Bacteriol. *191*, 3861–3868.

Klein, S., Lorenzo, C., Hoffmann, S., Walther, J.M., Storbeck, S., Piekarski, T., Tindall, B.J., Wray, V., Nimtz, M., and Moser, J. (2009). Adaptation of *Pseudomonas aeruginosa* to various conditions includes tRNA-dependent formation of alanyl-phosphatidylglycerol. Mol. Microbiol. *71*, 551–565.

Kocianova, S., Vuong, C., Yao, Y., Voyich, J.M., Fischer, E.R., DeLeo, F.R., and Otto, M. (2005). Key role of poly γ-dl-glutamic acid in immune evasion and virulence of *Staphylococcus epidermidis*. J. Clin. Invest. *115*, 688–694.

Kooi, C., and Sokol, P.A. (2009). *Burkholderia cenocepacia* zinc metalloproteases influence resistance to antimicrobial peptides. Microbiology *155*, 2818–2825.

Kovács, M., Halfmann, A., Fedtke, I., Heintz, M., Peschel, A., Vollmer, W., Hakenbeck, R., and Brückner, R. (2006). A functional *dlt* operon, encoding proteins required for incorporation of D-alanine in teichoic acids in Gram-positive bacteria, confers resistance to cationic antimicrobial peptides in *Streptococcus pneumoniae*. J. Bacteriol. *188*, 5797–5805.

Kraus, D., and Peschel, A. (2006). Molecular mechanisms of bacterial resistance to antimicrobial peptides. Curr. Top. Microbiol. Immunol. *306*, 231–250.

Kristian, S.A., Datta, V., Weidenmaier, C., Kansal, R., Fedtke, I., Peschel, A., Gallo, R.L., and Nizet, V. (2005). D-alanylation of teichoic acids promotes group A *Streptococcus* antimicrobial peptide resistance, neutrophil survival, and epithelial cell invasion. J. Bacteriol. *187*, 6719–6725.

Kukkonen, M., and Korhonen, T.K. (2004). The omptin family of enterobacterial surface proteases/adhesins: from housekeeping in *Escherichia coli* to systemic spread of *Yersinia pestis*. Int. J. Med. Microbiol. *294*, 7–14.

Kupferwasser, L.I., Skurray, R.A., Brown, M.H., Firth, N., Yeaman, M.R., and Bayer, A.S. (1999). Plasmid-mediated resistance to thrombin-induced platelet microbicidal protein in staphylococci: role of the *qacA* locus. Antimicrob. Agents Chemother. *43*, 2395–2399.

Lai, Y., Villaruz, A.E., Li, M., Cha, D.J., Sturdevant, D.E., and Otto, M. (2007). The human anionic antimicrobial peptide dermcidin induces proteolytic defence mechanisms in staphylococci. Mol. Microbiol. *63*, 497–506.

Le Sage, V., Zhu, L., Lepage, C., Portt, A., Viau, C., Daigle, F., Gruenheid, S., and Le Moual, H. (2009). An outer membrane protease of the omptin family prevents activation of the *Citrobacter rodentium* PhoPQ two-component system by antimicrobial peptides. Mol. Microbiol. *74*, 98–111.

Lee, H., Hsu, F.F., Turk, J., and Groisman, E.A. (2004). The PmrA-regulated *pmrC* gene mediates phosphoethanolamine modification of lipid A and polymyxin resistance in *Salmonella enterica*. J. Bacteriol. *186*, 4124–4133.

Li, M., Cha, D.J., Lai, Y., Villaruz, A.E., Sturdevant, D.E., and Otto, M. (2007a). The antimicrobial peptide-sensing system *aps* of *Staphylococcus aureus*. Mol. Microbiol. *66*, 1136–1147.

Li, M., Lai, Y., Villaruz, A.E., Cha, D.J., Sturdevant, D.E., and Otto, M. (2007b). Gram-positive three-component antimicrobial peptide-sensing system. Proc. Natl. Acad. Sci. U.S.A. *104*, 9469–9474.

Llobet, E., Tomás, J.M., and Bengoechea, J.A. (2008). Capsule polysaccharide is a bacterial decoy for antimicrobial peptides. Microbiology *154*, 3877–3886.

Miller, S.I., Ernst, R.K., and Bader, M.W. (2005). LPS, TLR4 and infectious disease diversity. Nat. Rev. Microbiol. *3*, 36–46.

Neuhaus, F.C., and Baddiley, J. (2003). A continuum of anionic charge: structures and functions of D-alanyl-teichoic acids in Gram-positive bacteria. Microbiol. Mol. Biol. Rev. *67*, 686–723.

Nizet, V. (2006). Antimicrobial peptide resistance mechanisms of human bacterial pathogens. Curr. Issues Mol. Biol. *8*, 11–26.

Nummila, K., Kilpeläinen, I., Zähringer, U., Vaara, M., and Helander, I.M. (1995). Lipopolysaccharides of polymyxin B-resistant mutants of *Escherichia coli* are extensively substituted by 2-aminoethyl pyrophosphate and contain aminoarabinose in lipid A. Mol. Microbiol. *16*, 271–278.

Oku, Y., Kurokawa, K., Ichihashi, N., and Sekimizu, K. (2004). Characterization of the *Staphylococcus aureus mprF* gene, involved in lysinylation of phosphatidylglycerol. Microbiology *150*, 45–51.

Otvos, L., Jr., O, I., Rogers, M.E., Consolvo, P.J., Condie, B.A., Lovas, S., Bulet, P., and Blaszczyk-Thurin, M. (2000). Interaction between heat shock proteins and antimicrobial peptides. Biochemistry *39*, 14150–14159.

Overhage, J., Campisano, A., Bains, M., Torfs, E.C., Rehm, B.H., and Hancock, R.E. (2008). Human host defense peptide LL-37 prevents bacterial biofilm formation. Infect. Immun. *76*, 4176–4182.

Park, C.B., Kim, H.S., and Kim, S.C. (1998). Mechanism of action of the antimicrobial peptide buforin II: buforin II kills microorganisms by penetrating the cell membrane and inhibiting cellular functions. Biochem. Biophys. Res. Commun. *244*, 253–257.

Parra-Lopez, C., Lin, R., Aspedon, A., and Groisman, E.A. (1994). A *Salmonella* protein that is required for resistance to antimicrobial peptides and transport of potassium. EMBO J. *13*, 3964–3972.

Patrzykat, A., Friedrich, C.L., Zhang, L., Mendoza, V., and Hancock, R.E. (2002). Sublethal concentrations of pleurocidin-derived antimicrobial peptides inhibit macromolecular synthesis in *Escherichia coli*. Antimicrob. Agents Chemother. *46*, 605–614.

Peschel, A. (2002). How do bacteria resist human antimicrobial peptides? Trends Microbiol. *10*, 179–186.

Peschel, A., and Sahl, H.G. (2006). The co-evolution of host cationic antimicrobial peptides and microbial resistance. Nat. Rev. Microbiol. *4*, 529–536.

Peschel, A., Otto, M., Jack, R.W., Kalbacher, H., Jung, G., and Götz, F. (1999). Inactivation of the *dlt* operon in *Staphylococcus aureus* confers sensitivity to defensins, protegrins, and other antimicrobial peptides. J. Biol. Chem. *274*, 8405–8410.

Peschel, A., Jack, R.W., Otto, M., Collins, L.V., Staubitz, P., Nicholson, G., Kalbacher, H., Nieuwenhuizen, W.F., Jung, G., Tarkowski, A., *et al.* (2001). *Staphylococcus aureus* resistance to human defensins and evasion of neutrophil killing via the novel virulence factor MprF is based on modification of membrane lipids with l-lysine. J. Exp. Med. *193*, 1067–1076.

Poyart, C., Pellegrini, E., Marceau, M., Baptista, M., Jaubert, F., Lamy, M.C., and Trieu-Cuot, P. (2003). Attenuated virulence of *Streptococcus agalactiae* deficient in D-alanyl-lipoteichoic acid is due to an increased susceptibility to defensins and phagocytic cells. Mol. Microbiol. *49*, 1615–1625.

Raetz, C.R., and Whitfield, C. (2002). Lipopolysaccharide endotoxins. Annu. Rev. Biochem. *71*, 635–700.

Roy, H., and Ibba, M. (2008). RNA-dependent lipid remodeling by bacterial multiple peptide resistance factors. Proc. Natl. Acad. Sci. U.S.A. *105*, 4667–4672.

Rozek, A., Friedrich, C.L., and Hancock, R.E. (2000). Structure of the bovine antimicrobial peptide indolicidin bound to dodecylphosphocholine and sodium dodecyl sulfate micelles. Biochemistry *39*, 15765–15774.

Rozek, A., Powers, J.P., Friedrich, C.L., and Hancock, R.E. (2003). Structure-based design of an indolicidin peptide analogue with increased protease stability. Biochemistry *42*, 14130–14138.

Samant, S., Hsu, F.F., Neyfakh, A.A., and Lee, H. (2009). The *Bacillus anthracis* protein MprF is required for synthesis of lysylphosphatidylglycerols and for resistance to cationic antimicrobial peptides. J. Bacteriol. *191*, 1311–1319.

Schmidtchen, A., Frick, I.M., Andersson, E., Tapper, H., and Björck, L. (2002). Proteinases of common pathogenic bacteria degrade and inactivate the antibacterial peptide LL-37. Mol. Microbiol. *46*, 157–168.

Shafer, W.M., Qu, X., Waring, A.J., and Lehrer, R.I. (1998). Modulation of *Neisseria gonorrhoeae* susceptibility to vertebrate antibacterial peptides due to a member of the resistance/nodulation/division efflux pump family. Proc. Natl. Acad. Sci. U.S.A. *95*, 1829–1833.

Shinnar, A.E., Butler, K.L., and Park, H.J. (2003). Cathelicidin family of antimicrobial peptides: proteolytic processing and protease resistance. Bioorg. Chem. *31*, 425–436.

Sieprawska-Lupa, M., Mydel, P., Krawczyk, K., Wójcik, K., Puklo, M., Lupa, B., Suder, P., Silberring, J., Reed, M., Pohl, J., *et al.* (2004). Degradation of human antimicrobial peptide LL-37 by *Staphylococcus aureus*-derived proteinases. Antimicrob. Agents Chemother. *48*, 4673–4679.

Sodeinde, O.A., and Goguen, J.D. (1989). Nucleotide sequence of the plasminogen activator gene of *Yersinia pestis*: relationship to *ompT* of *Escherichia coli* and gene *E* of *Salmonella typhimurium*. Infect. Immun. *57*, 1517–1523.

Spinosa, M.R., Progida, C., Talà, A., Cogli, L., Alifano, P., and Bucci, C. (2007). The *Neisseria meningitidis* capsule is important for intracellular survival in human cells. Infect. Immun. *75*, 3594–3603.

Steinberg, D.A., Hurst, M.A., Fujii, C.A., Kung, A.H., Ho, J.F., Cheng, F.C., Loury, D.J., and Fiddes, J.C. (1997). Protegrin-1: a broad-spectrum, rapidly microbicidal peptide with *in vivo* activity. Antimicrob. Agents Chemother. *41*, 1738–1742.

Stumpe, S., Schmid, R., Stephens, D.L., Georgiou, G., and Bakker, E.P. (1998). Identification of OmpT as the protease that hydrolyzes the antimicrobial peptide protamine before it enters growing cells of *Escherichia coli*. J. Bacteriol. *180*, 4002–4006.

Sugimura, K., and Nishihara, T. (1988). Purification, characterization, and primary structure of *Escherichia coli* protease VII with specificity for paired basic residues: identity of protease VII and OmpT.J. Bacteriol. *170*, 5625–5632.

Sutherland, I.W. (2001). The biofilm matrix-an immobilized but dynamic microbial environment. Trends Microbiol. *9*, 222–227.

Thedieck, K., Hain, T., Mohamed, W., Tindall, B.J., Nimtz, M., Chakraborty, T., Wehland, J., and Jänsch, L. (2006). The MprF protein is required for lysinylation of phospholipids in listerial membranes and confers resistance to cationic antimicrobial peptides (CAMPs) on *Listeria monocytogenes*. Mol. Microbiol. *62*, 1325–1339.

Thomas, S., Karnik, S., Barai, R.S., Jayaraman, V.K., and Idicula-Thomas, S. (2010). CAMP: a useful resource for research on antimicrobial peptides. Nucleic Acids Res. *38*, D774–780.

Thwaite, J.E., Hibbs, S., Titball, R.W., and Atkins, T.P. (2006). Proteolytic degradation of human antimicrobial peptide LL-37 by *Bacillus anthracis* may contribute to virulence. Antimicrob. Agents Chemother. *50*, 2316–2322.

Thwaite, J.E., Humphrey, S., Fox, M.A., Savage, V.L., Laws, T.R., Ulaeto, D.O., Titball, R.W., and Atkins, H.S. (2009). The cationic peptide magainin II is antimicrobial for *Burkholderia cepacia*-complex strains. J. Med. Microbiol. *58*, 923–929.

Tzeng, Y.L., Ambrose, K.D., Zughaier, S., Zhou, X., Miller, Y.K., Shafer, W.M., and Stephens, D.S. (2005). Cationic antimicrobial peptide resistance in *Neisseria meningitidis*. J. Bacteriol. *187*, 5387–5396.

Veal, W.L., Nicholas, R.A., and Shafer, W.M. (2002). Overexpression of the MtrC-MtrD-MtrE efflux pump due to an *mtrR* mutation is required for chromosomally mediated penicillin resistance in *Neisseria gonorrhoeae*. J. Bacteriol. *184*, 5619–5624.

Vuong, C., Voyich, J.M., Fischer, E.R., Braughton, K.R., Whitney, A.R., DeLeo, F.R., and Otto, M. (2004). Polysaccharide intercellular adhesin (PIA) protects *Staphylococcus epidermidis* against major components of the human innate immune system. Cell. Microbiol. *6*, 269–275.

Wang, G., Li, X., and Wang, Z. (2009). APD2: the updated antimicrobial peptide database and its application in peptide design. Nucleic Acids Res. *37*, D933–937.

Wecke, J., Madela, K., and Fischer, W. (1997). The absence of D-alanine from lipoteichoic acid and wall teichoic acid alters surface charge, enhances autolysis and increases susceptibility to methicillin in *Bacillus subtilis*. Microbiology *143*, 2953–2960.

Weidenmaier, C., Kristian, S.A., and Peschel, A. (2003). Bacterial resistance to antimicrobial host defenses-an emerging target for novel antiinfective strategies? Curr. Drug Targets *4*, 643–649.

Zasloff, M. (2002). Antimicrobial peptides of multicellular organisms. Nature *415*, 389–395.

Zgurskaya, H.I. (2002). Molecular analysis of efflux pump-based antibiotic resistance. Int. J. Med. Microbiol. *292*, 95–105.

Zilbauer, M., Dorrell, N., Boughan, P.K., Harris, A., Wren, B.W., Klein, N.J., and Bajaj-Elliott, M. (2005). Intestinal innate immunity to *Campylobacter jejuni* results in induction of bactericidal human beta-defensins 2 and 3. Infect. Immun. *73*, 7281–7289.

Bacteria-induced Host Cell Death

15

Scott D. Kobayashi, Kevin M. Rigby and Frank R. DeLeo

Abstract

The interaction of bacteria with host cells largely dictates the outcome of infection. Whether following a breach in the epithelium or by dissemination from colonized tissues, bacteria interact directly with immune and non-immune host cells. Therefore, it is perhaps not surprising that bacteria implement multiple strategies to circumvent destruction or avoid detection by the host. One such strategy is the ability of bacteria to alter host cell apoptosis or cause other forms of cell death. Although some bacterial pathogens induce or delay cell death and thereby promote disease, bacteria-induced cell death can also be beneficial to the host. This topic is reviewed in the present chapter.

Introduction

Bacterial infections are either cleared by the innate immune system or the invading microorganisms circumvent host defence and thereby cause disease. Bacterial pathogens employ numerous strategies to avoid destruction by the innate immune system, including utilization of host nutrients, inhibition of chemotaxis and phagocytosis, and escape from the complement system, resistance to reactive oxygen species and antimicrobial peptides, and induction or inhibition of host cell apoptosis or other forms of cell death.

Most eukaryotic cell turnover is accomplished by a cell death process known as apoptosis. Notably, apoptosis is essential for maintaining immune system homeostasis (Erwig and Henson, 2008). Characteristics of apoptosis include rounding of the cell, plasma membrane blebbing, decrease in cell volume, fragmentation and condensation of the nucleus, and uptake and removal by neighbouring cells or phagocytes (Kroemer *et al.*, 2009) (Fig. 15.1). During infection or at sites of inflammation, apoptotic cells are removed primarily by phagocytic leucocytes, such as dendritic cells and macrophages (Savill *et al.*, 1990; Erwig and Henson, 2008). Importantly, removal of apoptotic cells is usually considered as a non-inflammatory process and prevents host tissue damage that could be caused by secondary lysis of effete cells such as neutrophils (Savill *et al.*, 1989b, 2002). Therefore, apoptosis and removal of apoptotic cells is a means to clear spent cells from sites of infection and promote the resolution of inflammation (Savill *et al.*, 1989b, 2002; Schwab *et al.*, 2007; Serhan *et al.*, 2008).

In general, apoptosis occurs by death-receptor-mediated (extrinsic) or mitochondrial (intrinsic) pathways that have been well studied over the past 20 years (Hotchkiss *et al.*, 2009). Ligation of death receptors (FAS, TNF, TRAIL) on the host cell surface is the impetus for assembly of the death-inducing signalling complex (DISC), which includes an adaptor molecule named FAS-associated via death domain (FADD or MORT1) and caspase 8 (Danial and Korsmeyer, 2004; Hotchkiss *et al.*, 2009). Caspase 8 in turn activates caspase 3 and other effector caspases, which cleave dozens of cellular substrates, leading to apoptosis (Fischer *et al.*, 2003; Danial and Korsmeyer, 2004; Hotchkiss *et al.*, 2009). Apoptosis that occurs by the intrinsic pathway is initiated by cell stress, such as that caused by reactive oxygen species

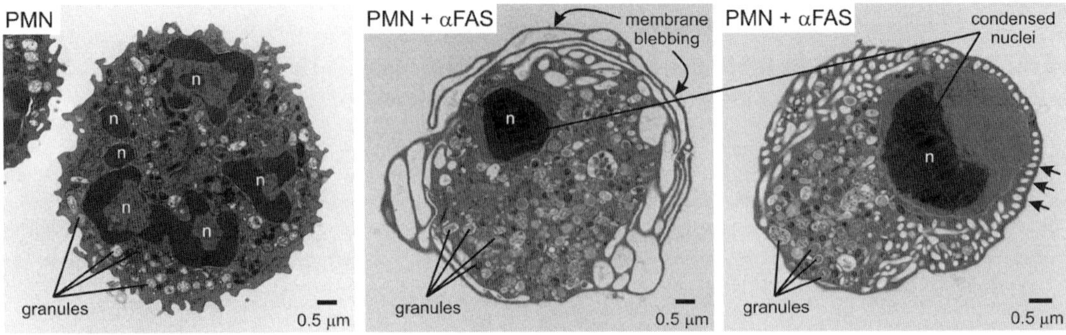

Figure 15.1 Human PMN (neutrophil) apoptosis. Left panel, quiescent human PMN. Middle and right panels, human PMN induced to undergo apoptosis using 500 ng/ml anti-FAS antibody (αFAS). Note extensive membrane blebbing in the cell in the middle panel, cytosolic vacuoles in the cell in the right panel (arrows), and condensed nuclei in both apoptotic cells. Neutrophil granules are apparent in all three panels as indicated. n, nucleus.

(ROS) or DNA damage (Hotchkiss *et al.*, 2009). This type of apoptosis involves B-cell CLL/lymphoma 2 (BCL2) protein family members, which promote or inhibit apoptosis, and the mitochondrion. BCL2, BCL2-like 1 (BCL2L1/BCL2-XL), BCL2-like 2 (BCL2L2/BCL2-W), myeloid cell leukaemia sequence 1 (MCL1), and BCL2-related protein A1 (BCL2A1/BFL1) prevent or regulate apoptosis by binding and sequestering pro-apoptotic molecules, such as BCL2-associated X protein (BAX) or BCL2 antagonist/killer 1 (BAK), which form pores in mitochondria (Brunelle and Letai, 2009; Hotchkiss *et al.*, 2009; Levine *et al.*, 2008). Perforated mitochondria release second mitochondria-derived activator of caspase (SMAC/DIABLO) and cytochrome c, which subsequently activate caspases to promote apoptosis (Bergman *et al.*, 2005; Brunelle and Letai, 2009). Thus, the intrinsic pathway of apoptosis is regulated by a delicate balance between anti- and pro-apoptosis molecules. Inasmuch as control of apoptosis, especially that of neutrophils and other phagocytes, is a key component to host defence, it is perhaps not surprising that bacteria can induce or delay host cell apoptosis or cause alternative forms of host cell death.

As an alternative to apoptosis, bacteria can induce necrosis, autophagy-associated cell death, or a hybrid of apoptosis and necrosis called pyroptosis (Fig. 15.2). Cellular necrosis is death of cells through injury. Necrosis is characterized by swelling of the cell and cytoplasmic organelles, rupture of the plasma membrane and release of intracellular contents into the surrounding environment, which for many cell types is pro-inflammatory (Kroemer *et al.*, 2009) (Fig. 15.2). Necrosis is typically considered as accidental or passive cell death and can have many causes. For example, bacterial pore-forming toxins cause necrosis by perforating the plasma membrane of host cells, thereby promoting osmotic imbalance and subsequent cytolysis. Interestingly, a recent report by Hitomi *et al.* (2008) indicates that necrosis can be regulated or programmed. Programmed necrosis – called necroptosis – involves death receptors and receptor-interacting serine-threonine kinase 1 (RIPK1 or RIP1) and 3 (RIPK3 or RIP3) (Christofferson and Yuan, 2010). Although some bacterial pathogens are known to cause host cell necrosis, whether this cell death process is necroptosis remains to be determined. Notably, necrosis is pro-inflammatory (Shi *et al.*, 2003) and lysis of some cell types such as neutrophils causes non-specific host tissue damage.

Autophagy is a process whereby cells recycle their own cytosolic constituents for use as metabolites during periods of stress (Hotchkiss *et al.*, 2009). During autophagy, an autophagic vesicle or autophagosome sequesters cytosol and damaged organelles (Fink and Cookson, 2005; Hotchkiss *et al.*, 2009; Kroemer *et al.*, 2009). This vesicle then fuses with lysosomes and the contents are digested for use as an energy source or as building blocks for new macromolecules. The most current

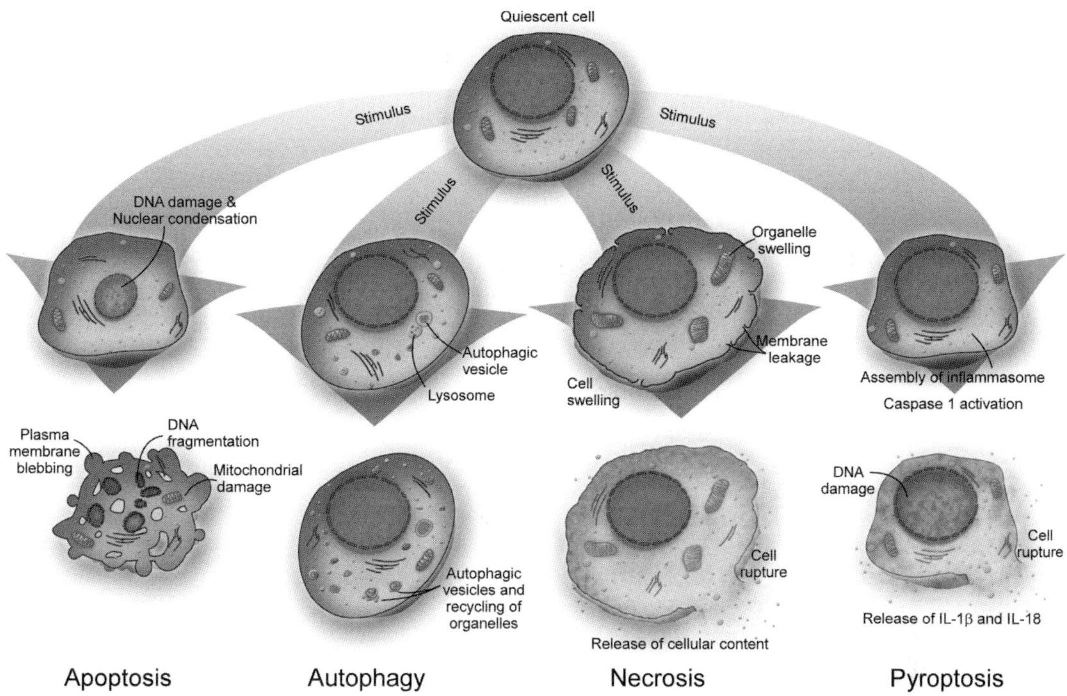

Figure 15.2 Cell fate pathways. A colour version of this figure is located in the plate section at the back of the book.

notion is that autophagy does not promote or execute cell death; rather it simply accompanies certain types of cell demise (Kroemer and Levine, 2008). Because of this recent line of thinking, cell death accompanied by autophagy has been termed autophagy-associated cell death or cell death with autophagy (Hotchkiss *et al.*, 2009; Kroemer *et al.*, 2009; Kroemer and Levine, 2008). Although some bacteria induce autophagy following uptake by host cells, this phenomenon presumably accompanies another form of cell death (Kroemer and Levine, 2008; Labbé and Saleh, 2008). Therefore, autophagy and autophagy-associated cell death are not discussed further in this chapter and we refer the reader to recent reviews on the subject (Levine and Yuan, 2005; Kroemer and Levine, 2008; Labbé and Saleh, 2008; Hotchkiss *et al.*, 2009; Kroemer *et al.*, 2009).

Pyroptosis is a caspase 1-dependent programmed cell death that ultimately leads to lysis of the host cell and release of pro-inflammatory molecules (Bergsbaken *et al.*, 2009). This form of cell death occurs in mononuclear phagocytes and is induced following phagocytosis of selected bacterial pathogens (Bergsbaken *et al.*, 2009). Cells undergoing pyroptosis have morphological characteristics of both apoptosis and necrosis. As such, some of the early reports of bacteria-induced macrophage apoptosis and associated pro-inflammatory molecule release likely describe pyroptosis (Zychlinsky *et al.*, 1992, 1994a; Sansonetti *et al.*, 2000; Raupach *et al.*, 2006). Pyroptosis is characterized by osmotic lysis of host cells, much like necrosis, except that formation of plasma membrane pores is dependent on activation of caspase 1 (Fink and Cookson, 2006). During pyroptosis, a multi-protein complex, called inflammasome, forms in mononuclear phagocytes after interaction with- or phagocytosis of certain bacteria (Franchi *et al.*, 2009). The inflammasome is formed/activated in response to bacterial products, such as flagellin or peptidoglycan fragments, that are detected in the cytosol by Nod-like receptors (NLRs) (Franchi *et al.*, 2009). Importantly, caspase 1 is activated by the inflammasome and in turn processes pro-interleukin 1β (pro-IL-1β) and pro-IL-18 into mature forms, which are released through membrane pores and during cell rupture (Bergsbaken *et al.*, 2009; Franchi *et al.*, 2009). Whether pyroptosis and associated release of pro-inflammatory IL-1β and IL-18 are beneficial to the host or pathogen is addressed below.

There is an extensive body of published literature that indicates defects in cell fate, primarily those of apoptosis, result in human disease. Notably, leukaemia and other types of cancer are caused by defects in apoptosis, and apoptosis of multiple cell types has been shown to accompany sepsis (Pinheiro da Silva and Nizet, 2009). Therefore, modulation of host cell apoptosis or induction of other types of cell death by bacteria likely has a profound effect on the outcome of infection. Here, we review the ability of bacteria and molecules secreted by bacteria to alter host cell apoptosis or cause host cell death.

Bacteria and host cell death

Bacterial toxins and other secreted molecules

Bacterial pathogenesis is a highly complex process that involves pathogen entry into the host, evasion of the immune response, which frequently results in dissemination of microbes. Invading microorganisms invariably encounter a diversity of host cell types throughout the course of infection. Bacteria produce a litany of secreted molecules and toxins, in addition to cell surface determinants, that facilitate the establishment of disease. Although the mechanisms of action and delivery systems for bacterial toxins are heterogeneous, most bacterial toxins – with few exceptions – are similarly deleterious to host cells. To this end, there are numerous host signal-transduction pathways that modulate host cell fate and many of these are 'hijacked' by bacterial toxins (Gao and Kwaik, 2000; Bhavsar *et al.*, 2007) (Table 15.1). Enhanced purification methods have facilitated increased understanding of toxin mechanism of action and structure. Expanding molecular genetic tools have facilitated studies on toxin activity through creation of genetic deletion strains that can be used to study host–pathogen interaction in both *in vitro* and *in vivo* models of infection. However, many bacteria are capable of producing multiple toxins, and in some cases the toxins may have opposite effects on host cell fate, thus confounding interpretation of the contribution of apoptosis induction/delay to bacterial pathogenesis. It is clear that many toxins are capable of inducing catastrophic host cell death (necrosis). Despite numerous studies describing the ability of toxins to induce a programmed cell death (Table 15.1), the mechanism of induction is often unclear, and the overall relevance of these phenomena in bacterial pathogenesis remains uncertain. The effects of bacterial toxins on host cell fate will be discussed below.

Cytolytic toxins

Bacterial toxins can be broadly grouped by function, such as the ability to lyse host cells. Bacterial cytolytic toxins comprise a diversity of molecules, including several members of the RTX (repeats in toxin) family of proteins, cholesterol-binding cytolysins, and other bacterial pore-forming toxins. Soluble pore-forming proteins generally fall into two primary classes, consisting of either membrane-interacting β sheet configurations or those forming *a-helical* pores (Bayley, 2009). Pore formation is most often initiated by target recognition by toxin subunits, which often determines toxin host cell tropism. The abundance of target/receptor can also determine the degree of host cell intoxication. In contrast to bacterial toxins with intracellular targets, the primary function of pore-forming toxins is arguably to irreversibly damage the integrity of host cell membranes. High doses of pore-forming toxins are generally accompanied by rapid target cell lysis and death. At sublytic concentrations, however, most pore-forming toxins induce programmed cell death in association with the reduced membrane perturbation (Table 15.1). In addition, several pore-forming toxins are involved in activating pro-inflammatory signal-transduction pathways in immune cells. For example, cholesterol-dependent cytolysins such as anthrolycin O, listeriolysin O, pneumolysin, and perfringolysin O are TLR4-agonists and have been shown to activate macrophages (Park *et al.*, 2004). Taken together, these findings indicate that pore-forming toxins initiate apoptosis through the induction of a general cellular stress response or directly activate signal-transduction cascades linked to programmed cell death.

Toxins with intracellular enzymatic activity

The primary mechanism of action of several bacterial toxins is inhibition of host intracellular signal-transduction pathways. One of the most common enzymatic properties of bacterial toxins is the ability to mono-ADP-ribosylate host cellular proteins (Koch-Nolte *et al.*, 2006). ADP-ribosyltransferase activity is a characteristic shared by diphtheria, cholera, pertussis, and several clostridial toxins and is essential for pathogenesis. Bacterial ADP-ribosylating toxins target specific host proteins such as the Rho GTPase, monomeric actin, elongation factor 2, and the α-subunit of heterotrimeric GTP-binding proteins (Corda and Di Girolamo, 2003). Host cell apoptosis is associated with the activities of some ADP-ribosylating toxins and most notably *Pseudomonas aeruginosa* exotoxin A and the heat-labile enterotoxin of *Escherichia coli* (Fiorentini *et al.*, 2003). In addition, toxins with targets similar to those of the ADP-ribosylating enzymes are involved in the induction of host programmed cell death. The *Clostridium difficile* toxins A (TcdA) and B (TcdB) are Rho GTP-ase glycosylating toxins that induce disorganization of the actin cytoskeleton and promote host cell apoptosis (Popoff and Bouvet, 2009). The induction of host cell apoptosis is also associated with the enzymatic activity of Shiga toxins. Shiga toxins are members of the AB_5 toxin family and are produced by *Shigella dysenteriae* serotype 1 and enterohaemorrhagic *E. coli* strains (Table 15.1). Shiga toxins function as *N*-glycosidases and catalytically cleave adenine from the 28S ribosomal RNA of eukaryotic ribosomes, thus inhibiting peptide elongation. The ability of Shiga toxins to promote apoptosis in a wide variety of cell types is well documented (Cherla *et al.*, 2003). The mechanism for Shiga toxin-mediated apoptosis is cell type-dependent and can involve both intrinsic and extrinsic pathways of programmed cell death (Johannes and Römer, 2010).

Cytolethal distending toxins

Cytolethal distending toxins (CDT) are a family of heat-labile proteins that possess a DNase I-like activity that induces DNA double strand breaks and eukaryotic cell cycle arrest at the G2/M stage (Ge *et al.*, 2008). CDTs are widely distributed among Gram-negative bacteria, such as *E. coli*, *Campylobacter jejuni*, *Haemophilus ducreyi*, *Salmonella enterica* serovar Typhi, *S. dysenteriae*, and *Aggregatibacter actinomycetemcomitans* (Table 15.1). CDT is a tripartite holotoxin consisting of subunits CdtA and CdtC that are complexed with the nuclease CdtB (Nesic *et al.*, 2004). CDT induces cytopathic effects on host cells in part through cytoskeletal changes promoted by a generalized stress response to DNA damage (Ceelen *et al.*, 2006). Similar to other toxins studied, various reports suggest that the ability of CDT to induce apoptosis is both cell type- and pathogen-specific. The Gram-negative opportunistic pathogen *A. actinomycetemcomitans* CDT induces apoptosis in diversity of cells, including peripheral blood mononuclear cells, murine B cells (HS-72), Jurkat and MOULT-4 T-cell lines, and epithelial carcinoma cells (Ca9–22) (Smith and Bayles, 2006). *H. ducreyi* CDT is reported to have greater effects on apoptosis of immune cells, such as T cells and monocytes, compared with non-immune cells. CDT treatment elicits activation of caspase 3, 8, and 9 (Shenker *et al.*, 2001), and apoptosis is inhibited by over-expression of Bcl-2 (Shenker *et al.*, 2006).

Superantigens

Superantigens that target T cells bind directly to MHC class II molecules and the TCR β-chain and thus, bypass classical antigen-presentation mechanisms governing peptide-MHC recognition (Fraser and Proft, 2008). Superantigens are potent T-cell mitogens and induce a supraclonal response. In addition, superantigens induce the production of exhaustive quantities of interleukin-2, interferon-γ, and tumour necrosis factor (TNF)-α. Bacterial superantigens comprise a family of structurally similar molecules produced by *Staphylococcus aureus* (e.g. staphylococcal enterotoxins and toxic shock syndrome toxin-1) and *Streptococcus pyogenes* (e.g. streptococcal

Table 15.1 Bacterial effector molecules that modulate host cell death

Bacterium	Effector	Cell death	References
Type III secretion systems			
Salmonella spp.	SipB	Apoptosis, pyroptosis	Hersh *et al.* (1999), Brennan and Cookson (2000), Jesenberger *et al.* (2000, 2001), Monack *et al.* (2000)
Shigella flexneri	IpaB	Apoptosis, pyroptosis	Zychlinsky *et al.* (1994b), Chen *et al.* (1996b), Hilbi *et al.* (1998)
Yersinia spp.	YopJ/P	Apoptosis, pyroptosis	Mills *et al.* (1997), Monack *et al.* (1997, 1998), Schesser *et al.* (1998), Orth *et al.* (1999), Gröbner *et al.* (2006), Mittal *et al.* (2006), Zauberman *et al.* (2006), Bergsbaken and Cookson (2007), Brodsky and Medzhitov (2008), Zhang *et al.* (2008)
Pseudomonas spp.	ExoS	Apoptosis	Bruno *et al.* (2000), Epelman *et al.* (2002), Jia *et al.* (2003, 2006), Alaoui-El-Azher *et al.* (2006), Jansson *et al.* (2006)
	ExoT	Apoptosis	Shafikhani *et al.* (2008)
Cytolytic toxins			
Actinobacillus (Aggregatibacter) actinomycetemcomitans	Leukotoxin (LtxA)	Apoptosis, necrosis	Korostoff *et al.* (1998, 2000), Yamaguchi *et al.* (2001, 2004), Kelk *et al.* (2003)
Aeromonas hydrophila	Cytotoxic enterotoxin (Act)	Apoptosis, necrosis	Falcon *et al.* (2001), Galindo *et al.* (2004, 2005, 2006), di Pietro *et al.* (2005), Krzyminska *et al.* (2009)
	Aerolysin	Apoptosis	Nelson *et al.* (1999)
Bacillus anthracis	Anthrolysin O	Apoptosis	Park *et al.* (2004)
	Lethal toxin (Letx)	Apoptosis, necrosis, pyroptosis	Peek, Jr. *et al.* (1997), Koo *et al.* (2002), Popov *et al.* (2002a,b), Kirby (2004), Pandey and Warburton (2004), Alileche *et al.* (2005), Kassam *et al.* (2005), Drysdale *et al.* (2007), Muehlbauer *et al.* (2007), Squires *et al.* (2007), Nour *et al.* (2009), Bradburne *et al.* (2008), Fink *et al.* (2008), Hammamieh *et al.* (2008), Wickliffe *et al.* (2008a,b), Wu *et al.* (2009), Ngai *et al.* (2010), Terra *et al.* (2010)
Bordetella pertussis	Adenylate cyclase-haemolysin	Apoptosis	Khelef *et al.* (1993, 2001), Boyd *et al.* (2005), Cheung *et al.* (2009)
Clostridium perfringens	Enterotoxin (CPE)	Apoptosis, necrosis	Chakrabarti *et al.* (2003), Chakrabarti and McClane (2005)
Clostridium septicum	α-Toxin	Necrosis	Kennedy *et al.* (2009)
Escherichia coli	Haemolysin (Hly)	Apoptosis, necrosis	Russo *et al.* (2005)
Helicobacter pylori	Vacuolating cytotoxin A (VaCA)	Apoptosis	Peek, Jr. *et al.* (1997), Kuck *et al.* (2001), Cover *et al.* (2003), Willhite and Blanke (2004)
	CagA	Apoptosis	Oldani *et al.* (2009)

Listeria monocytogenes	Listeriolysin O	Apoptosis	Carrero *et al.* (2004, 2008), Parra *et al.* (2008)
Mannheimia haemolytica	Leukotoxin (LktA)	Apoptosis, necrosis	Sun *et al.* (2000), Cudd *et al.* (2001), Jeyaseelan *et al.* (2001), Marcatili *et al.* (2002), Thumbikat *et al.* (2005)
Plesiomonas shigelloides	ComP	Apoptosis	Tsugawa *et al.* (2008)
Staphylococcus aureus	α-haemolysin (Hla)	Apoptosis, necrosis	Menzies and Kourteva (2000), Bantel *et al.* (2001), Essmann *et al.* (2003), Haslinger *et al.* (2003), Haslinger-Löffler *et al.* (2006), Liang and Ji (2007)
	Panton–Valentine leukocidin (PVL)	Apoptosis, necrosis	Panton and Valentine (1932), Szmigielski *et al.* (1999), Genestier *et al.* (2005), Woodin (1961, 1970)
Streptococcus agalactiae	β-haemolysin	Apoptosis, necrosis	Hensler *et al.* (2008)
Streptococcus pyogenes	Streptolysin O/S (Slo, Sls)	Apoptosis, necrosis, pyroptosis	Cywes *et al.* (2005), Miyoshi-Akiyama *et al.* (2005), Cortes and Wessels (2009), Goldmann *et al.* (2009), Harder *et al.* (2009), Timmer *et al.* (2009)
Streptococcus pneumoniae	Pneumolysin	Apoptosis	Braun *et al.* (2002, 2007), Bermpohl *et al.* (2005), Srivastava *et al.* (2005)
Vibrio cholerae	Cytolysin (HlyA)	Apoptosis	Saka *et al.* (2008)
Vibrio parahaemolyticus	Thermostable direct haemolysin (TDH)	Apoptosis	Naim *et al.* (2001)
Vibrio vulnificus	Cytolysin (VVC)	Apoptosis	Zhao *et al.* (2009)
	Haemolysin (VVH)	Apoptosis	Kashimoto *et al.* (2009)
	RtxA toxin	Apoptosis	Lee *et al.* (2008a)
Superantigens			
Peptostreptococcus magnus	Protein I (superantigen)	Apoptosis	Goodyear *et al.* (2004)
Staphylococcus aureus	Staphylococcal enterotoxins (SEs)	Apoptosis	Moulding *et al.* (1999), Renno *et al.* (1999), Lavoie *et al.* (2001), Takahashi *et al.* (2001), Gorak-Stolinska *et al.* (2002), Lin *et al.* (2003), Goodyear and Silverman (2004), McKallip *et al.* (2005), Batisson *et al.* (2006), Park *et al.* (2006), Goodyear *et al.* (2007), Ionin *et al.* (2008), Dauwalder *et al.* (2009)
Toxins with enzymatic activity			
Actinobacillus (Aggregatibacter) actinomycetemcomitans	Cytolethal distending toxin	Apoptosis	Ohguchi *et al.* (1998), Shenker *et al.* (2001, 2006), Ohara *et al.* (2004, 2008), Rabin *et al.* (2009)
Bordetella pertussis	Pertussis toxin	Apoptosis	Ramirez *et al.* (1994)
Campylobacter jejuni	Cytolethal distending toxin	Apoptosis	Hickey *et al.* (2005)

Table 15.1 continued

Bacterium	Effector	Cell death	References
Clostridium botulinum	C2 toxin	Apoptosis	Heine *et al.* (2008)
Clostridium difficile	Toxin A/toxin B	Apoptosis	Mahida *et al.* (1996, 1998), Fiorentini *et al.* (1998a), Qa'Dan *et al.* (2002), Kim *et al.* (2005, 2007), Solomon *et al.* (2005), Carneiro *et al.* (2006), Matarrese *et al.* (2007), Na *et al.* (2008), Matte *et al.* (2009)
Clostridium perfringens	Iota-toxin	Apoptosis	Hilger *et al.* (2009)
Clostridium sordellii	Lethal toxin (TcsL)	Apoptosis	Petit *et al.* (2003), Voth and Ballard (2007), Schulz *et al.* (2009)
Corynebacterium diphtheriae	Diphtheria toxin	Apoptosis, necrosis	Chang *et al.* (1989a,b), Morimoto and Bonavida (1992)
Escherichia coli	Verocytotoxin 1	Apoptosis	Kodama *et al.* (1999)
	Shiga toxin 1/Shiga toxin 2	Apoptosis	Uchida *et al.* (1999), Yoshida *et al.* (1999), Ching *et al.* (2002), Fujii *et al.* (2003, 2008), Harrison *et al.* (2005), Lee *et al.* (2005, 2007, 2008b, 2009), Wilson *et al.* (2005), Creydt *et al.* (2006), Brigotti *et al.* (2008), Takahashi *et al.* (2008), Psotka *et al.* (2009)
	Cytotoxic necrotizing factor (CNF1)	Apoptosis	Fiorentini *et al.* (1997, 1998b), Mills *et al.* (2000), Russo *et al.* (2005), Miraglia *et al.* (2007)
	Cytolethal distending toxin	Apoptosis	Bielaszewska *et al.* (2005)
	Heat-labile enterotoxin (EtxB or LT)	Apoptosis	Soriani *et al.* (2001), Salmond *et al.* (2002, 2003, 2004), Fraser *et al.* (2003), Arce *et al.* (2005), Tamayo *et al.* (2005)
	Subtilase cytotoxin (SubAB)	Apoptosis	Wang *et al.* (2007)
Haemophilus ducreyi	Cytolethal distending toxin	Apoptosis	Gelfanova *et al.* (1999), Cortes-Bratti *et al.* (2001), Nesic *et al.* (2004), Xu *et al.* (2004), Wising *et al.* (2005)
Pseudomonas aeruginosa	Exotoxin A	Apoptosis, necrosis	Huang *et al.* (2003), Jenkins *et al.* (2004), Plotkowski *et al.* (2002), Schümann *et al.* (1998)
Streptococcus pyogenes	Streptococcal pyrogenic exotoxin B (SpeB)	Apoptosis	Kuo *et al.* (1999), Viera *et al.* (2001), Tsai *et al.* (2004, 2008), Chang *et al.* (2007, 2009)
Vibrio cholerae	Cholera toxin	Apoptosis	Arce *et al.* (2005), Pessina *et al.* (2006)
Other toxins			
Mycobacterium ulcerans	Mycolactone	Apoptosis	Coutanceau *et al.* (2005)
Pseudomonas aeruginosa	Pyocyanin	Apoptosis	Usher *et al.* (2002), Muller (2006), Prince *et al.* (2008)

pyrogenic exotoxins) with relatively few exceptions. In addition, bacterial pathogens can produce toxins that have B cell mitogenic properties, such as *S. aureus* protein A and *Peptostreptococcus magnus* protein L (Goodyear *et al.*, 2004). Superantigens for B cells are expressed as multiple copies of homologous Fab-binding domains and function through crosslinking of B-cell receptors (Silverman and Goodyear, 2006).

The ability of superantigens to promote activation-induced death of T cells is well described (Kabelitz and Janssen, 1997). T cell death is a normal process for peripheral deletion of expanded clonal cell populations and is governed by the FAS-mediated apoptosis pathway. There is significant evidence that superantigen-mediated T-cell death is dependent on the extrinsic mediator CD95/FASL (Ulett and Adderson, 2006). However, experiments performed in cells lacking the proapoptotic BCL-2-interacting mediator of cell death (Bim) suggest a role for the intrinsic pathway in superantigen-induced death in both T cells (Hildeman *et al.*, 2002) and B cells (Goodyear *et al.*, 2007). Thus, while it is clear that superantigens are capable of eliminating large populations of lymphocytes, the specific signal-transduction pathways that govern superantigen-induced cell death are simply not yet defined.

Type III effector molecules

The type III secretion system (T3SS) is a highly specialized secretion system commonly employed by Gram-negative bacterial pathogens to introduce toxins into the host cell cytosol. T3SS are functionally analogous systems comprising a flagellum-like basal body, a delivery apparatus, translocators, regulatory elements and chaperone proteins. Interaction of the T3SS with the target eukaryotic cell results in formation of a translocation pore for delivery of the effector molecules (or toxins). Although the formation of membrane pores is often sufficient to initiate apoptosis (similar to pore-forming toxins), several T3SS effectors target specific signal-transduction pathways to induce programmed cell death of host cells. The T3SS of *Yersinia* spp., *S. enterica* serovar Typhimurium, *Shigella flexneri*, *E. coli* and *P. aeruginosa* have clearly established roles in the development of human disease, and several of the effector molecules are implicated in promoting programmed death of host cells (Table 15.1).

Pathogenic *Yersinia* spp. are among the first pathogens described to promote apoptosis in immune cells. There is significant evidence that the T3SS effector YopJ (YopP in *Yersinia enterocolitica*) plays a primary role in the induction of *Yersinia*-mediated apoptosis (Mills *et al.*, 1997; Monack *et al.*, 1997). YopJ is a deubiquitinating cysteine protease (Orth *et al.*, 2000; Zhou *et al.*, 2005) that also functions as an acetyltransferase (Mittal *et al.*, 2006; Mukherjee *et al.*, 2006). In addition, YopJ is required for cleavage of the caspase-8 substrate BID (Denecker *et al.*, 2001; Gröbner *et al.*, 2007a) and is a potent inhibitor of the MAPK and NF-κB signalling pathways (Viboud and Bliska, 2005). YopJ-dependent apoptosis requires signalling through Toll-like receptor 4 (Zhang and Bliska, 2003). Although virtually identical, there are differences in cytotoxicity between *Y. enterocolitica* YopP and either *Yersinia pestis* or *Yersinia pseudotuberculosis* YopJ. The increased cytotoxicity of YopP is attributed to its increased translocation by the T3SS of *Y. enterocolitica* (Zauberman *et al.*, 2006; Brodsky and Medzhitov, 2008).

The T3SS of *P. aeruginosa* secretes four characterized effector proteins: ExoS, ExoT, ExoU, and ExoY (Cornelis, 2006). The ExoS and ExoT effector toxins exert GTPase-activating protein and ADP-ribosyltransferase activity, and the presence of either protein is sufficient to induce apoptosis (Hauser, 2009). Although the exact mechanism for *P. aeruginosa* T3SS-mediated induction of apoptosis remains to be determined, several factors have been implicated in the process. The ADP-ribosyltransferase activity of ExoS is required for *P. aeruginosa*-induced apoptosis (Kaufman *et al.*, 2000). In addition, host cell survival signal pathways, such as ERK1/2 and p38 activation, are inhibited, whereas proapoptotic signals are activated through JNK1/2 (Jia *et al.*, 2003). Host cell intoxication with ExoS inhibits phosphorylation of FOXO3a, and induces cytochrome c release and activation of caspases 3, 8 and 9 (Alaoui-El-Azher *et al.*, 2006; Jansson *et al.*, 2006). The primary target(s) of the ExoS ADP-ribosyltransferase responsible for apoptosis induction has not been identified yet.

S. flexneri and *S. enterica* contain a similar T3SS encoded by the invasion antigen plasmid in *Shigella* and the *Salmonella* pathogen island 1 (SPI1) in the latter (Cornelis, 2006). Both of these enteric pathogens share the ability to induce programmed cell death in part through activities of analogous translocases: IpaB in *S. flexneri* and SipB in *S. enterica* (Zychlinsky *et al.*, 1992; Hersh *et al.*, 1999). IpaB and SipB bind and activate caspase1, and the role of macrophage programmed cell death in pathogenesis has been reviewed recently (Haimovich and Venkatesan, 2006).

Apoptosis of non-immune cells

Some of the first studies that demonstrate the ability of bacterial pathogens to induce or delay apoptosis or cause necrosis of non-immune cells were published in the late 1990s (Clifton *et al.*, 1998; Fan *et al.*, 1998; Kim *et al.*, 1998; Crane *et al.*, 1999). It is now known that bacteria can alter apoptosis or cause necrosis of many different host cell types, including endothelial and epithelial cells, fibroblasts, hepatocytes, and splenocytes (Tables 15.2 and 15.3). Here, we focus discussion on a few selected example pathogens that can invade host non-immune cells and thereby alter host cell death.

Host cells infected with viruses or bacteria may undergo apoptosis as a means to prevent the spread of infection (Fratazzi *et al.*, 1997; Galluzzi *et al.*, 2008). This phenomenon poses a problem for bacteria that are obligate intracellular organisms, as the ability to maintain host cell viability is critical for pathogen replication and ultimately, dissemination. Indeed, many of these organisms inhibit host cell apoptosis (Table 15.2). For example, *Rickettsia rickettsii*, the causative agent of Rocky Mountain spotted fever, targets vascular endothelial cells during infection (Sahni and Rydkina, 2009), and endothelial cells support growth and replication of the pathogen *in vitro* (Silverman and Bond, 1984). Infection of endothelial cells by *R. rickettsii* activates the global transcription factor NF-κB (Clifton *et al.*, 1998), which in turn prevents activation of caspases 3, 8 and 9, thereby inhibiting apoptosis (Joshi *et al.*, 2003). By comparison, epithelial cells infected with *Chlamydia* spp. are protected against apoptosis-inducing agents by the degradation of pro-apoptotic proteins Bim, Puma and Bad (Fischer *et al.*, 2004). Although the mechanisms for preventing apoptosis are distinct, the outcomes are similar and in favour of the pathogen.

Epithelial cells infected with enteroinvasive *E. coli* or *Salmonella* spp. undergo apoptosis, but cell death is delayed (12–28 h post-infection) compared with that of infected myeloid cells (Monack *et al.*, 1996; Kim *et al.*, 1998; Paesold *et al.*, 2002). Apoptosis of these host cells requires bacterial invasion and replication, is regulated by TNF-α and nitric oxide, and involves caspase 3 (Kim *et al.*, 1998; Paesold *et al.*, 2002). Genes on the *Salmonella* pathogenicity island 2 (SPI2), which encode a T3SS, and the plasmid-encoded *spv* virulence locus are important for triggering apoptosis in the infected epithelial cells (Paesold *et al.*, 2002). It is curious that apoptosis-inducing events occur early after *Salmonella* infection, such as production of host TNF-α and nitric oxide, and induction of bacterial SPI2 and *spv*, but activation of caspase-3 and apoptosis are delayed (Kim *et al.*, 1998; Paesold *et al.*, 2002). In subsequent work, Knodler *et al.* (2005) found that the *Salmonella* effector SpoB (also SigD) induces sustained activation of Akt, a pro-survival serine-threonine kinase, and thereby prolongs survival of *Salmonella*-infected epithelial cells. Here, one can make a case for late onset apoptosis as being beneficial to the host, since infected cells have sufficient time to elicit pro-inflammatory signals and would be cleared eventually, thus limiting the infection. This idea extends to *S. flexneri*-infected epithelial cells as well, albeit these host cells undergo late programmed necrosis rather than apoptosis (Carneiro *et al.*, 2009). On the other hand, a delay in apoptosis might allow sufficient time for these enteric pathogens to replicate and disseminate. It is interesting that *Salmonella* and *Shigella* have adapted to prolong host cell survival or induce death depending on the type of host cell – e.g. rapid death in macrophages versus late onset apoptosis in epithelial cells.

Compared with studies of bacteria-induced apoptosis in myeloid cells, our understanding of cell death mechanisms in non-immune cells caused by bacterial invasion is relatively limited and merits further investigation.

Table 15.2 Bacteria or bacterial molecules that delay apoptosis

Bacterium	Host cell	Origin	References
Anaplasma phagocytophilum	PMNs, promyelocytic cells (NB4)	Human, ovine	Yoshiie *et al.* (2000), Scaife *et al.* (2003), Borjesson *et al.* (2005), Choi *et al.* (2005), Ge *et al.* (2005), Pedra *et al.* (2005), Lee and Goodman (2006)
Bartonella spp.	Endothelial cells (HUVECs), dermal microvascular cells, monocyte cell line (Mono Mac 6)	Human	Kirby and Nekorchuk (2002), Liberto *et al.* (2003, 2004), Kempf *et al.* (2005), Schmid *et al.* (2006)
Brucella abortus	J774.A1, lymphocytes, monocytes, RAW264.7	Bovine, murine	Galdiero *et al.* (2000), Wang *et al.* (2001)
Brucella suis	Monocytes, PBMCs, THP-1	Human	Gross *et al.* (2000)
Chlamydiae spp.	Endothelial cells (HUVECs), HeLa, Hep-2, HL60, Jurkat T-cells, L929, MCF-7, U937	Human	Fan *et al.* (1998), Fischer *et al.* (2004), Greene *et al.* (2004), Xiao *et al.* (2004), Dong *et al.* (2005), Pirbhai *et al.* (2006), Rajalingam *et al.* (2006)
Chlamydophila pneumoniae	PMNs, THP-1, U937	Human	Airenne *et al.* (2002), Carratelli *et al.* (2002), van Zandbergen *et al.* (2004)
Clostridium perfringens	Kidney epithelial cells (Vero)	Monkey	Hilger *et al.* (2009)
Coxiella burnetii	Primary AMs, CHO, THP-1	Hamster, human, monkey	Lührmann and Roy (2007), Voth *et al.* (2007), Voth and Heinzen (2009)
Ehrlichia chaffensis	THP-1	Human	Zhang *et al.* (2004)
Escherichia coli verotoxin 2	PMNs	Human	Liu *et al.* (1999, 2002)
Helicobacter pylori	B-cells, B-cell line (IM-9), gastric adenocarcinoma cells (MKN45)	Human	Yanai *et al.* (2003), Ohmae *et al.* (2005), Zhu *et al.* (2007)
Helicobacter pylori lipopolysaccharide	PMNs	Human, murine	Feterowski *et al.* (2001), Hofman *et al.* (2001)
Helicobacter pylori water-soluble surface proteins	PMNs	Human	Kim *et al.* (2001)
Legionella pneumophila	MDMs	Human	Abu-Zant *et al.* (2007)
Mycobacterium bovis BCG	Monocytes, PMNs	Human	Kremer *et al.* (1997), Suttmann *et al.* (2003)
Neisseria gonorrhoeae	HeLa, endocervical epithelial cells, fallopian tube primary epithelial cells, HPV E6- or E7-transformed urethral epithelial cells (THUECs), PMNs, primary urethral epithelial cells	Human	Binnicker *et al.* (2003, 2004), Follows *et al.* (2009), Morales *et al.* (2006), Simons *et al.* (2006)
Neisseria meningitidis	Epithelial cells (Hec-1-B, HeLa), Jurkat T-cells	Human	Massari *et al.* (2000, 2003), Deghmane *et al.* (2009)

Table 15.2 continued

Bacterium	Host cell	Origin	References
Orientia (Rickettsia) tsutsugamushi	Lymphocytes, THP-1	Human, murine	Kasuya *et al.* (1996), Kim *et al.* (2002)
Porphyromonas gingivalis	Gingival epithelial cells	Human	Nakhjiri *et al.* (2001), Yilmaz *et al.* (2004)
Porphyromonas gingivalis fimbriae	THP-1	Human	Ozaki and Hanazawa (2001)
Porphyromonas gingivalis lipopolysaccharide	PMNs	Human	Hiroi *et al.* (1998), Preshaw *et al.* (1999)
Pseudomonas aeruginosa	Kupffer cells	Human	Ashare *et al.* (2007)
Rickettsia rickettsii	Endothelial cells (HUVECs)	Human	Clifton *et al.* (1998), Joshi *et al.* (2003, 2004)
Salmonella enterica serovar Typhimurium	Epithelial cells	Murine	Vijay-Kumar *et al.* (2006)
Staphylococcus aureus enterotoxins A, B, C and TSST1	B cells, eosinophils, monocytes, PMNs	Human	Hofer *et al.* (1996), Bratton *et al.* (1999), Moulding *et al.* (1999), Wedi *et al.* (2002)
Staphylococcus aureus lipotechoic acid	PMNs	Human	Lotz *et al.* (2004)
Staphylococcus aureus phenol soluble modulins	PMNs, monocytes	Human	Liles *et al.* (2001)

AM, alveolar macrophage; CHO, Chinese hamster ovary; HUVEC, human umbilical vein endothelial cell; MDM, monocyte-derived macrophage; PBMC, peripheral blood mononuclear cell; PMN, polymorphonuclear leukocyte; J774.A1 and RAW264.7, murine macrophage-like cells; THP-1, human promonocytic cells.

Table 15.3 Bacteria that induce host cell death

Bacterium	Host cell	Origin	Cell death	References
Bacillus anthracis	BMDMs, endothelial cells (DMVECs, HUVECs), J774A.1, melanocytes, PBMCs, peritoneal macrophages, RAW264.7	Human, murine	Apoptosis, necrosis, pyroptosis	Lin *et al.* (1996), Koo *et al.* (2002), Park *et al.* (2002, 2005), Popov *et al.* (2002a), Hsu *et al.* (2004, 2008), Kirby (2004), Bergman *et al.* (2005), Pétrilli *et al.* (2007), Ngai *et al.* (2010)
Bacillus anthracis lethal toxin	J774A.1, RAW264.7	Murine	Apoptosis	Lin *et al.* (1996), Popov *et al.* (2002a,b)
Borrelia hermsii	PMNs	Human	Apoptosis, PICD	Kobayashi *et al.* (2003a)
Bordetella bronchiseptica	L2 (lung epithelial cells), J774A.1	Murine, rat	Apoptosis, necrosis	Yuk *et al.* (1998, 2000)
Brucella abortus	J774A.1, RAW264.7	Murine	Necrosis	Pei and Ficht (2004), Pei *et al.* (2006, 2008)
Brucella melitensis	Monocytes	Human	Apoptosis	Fernandez-Prada *et al.* (2003)
Burkholderia cepacia/ cenopacia	Airway epithelia, A549, PMNs, J774.2	Human, murine	Apoptosis, necrosis	Hutchison *et al.* (1998), Kobayashi *et al.* (2003a), Punj *et al.* (2003), Bylund *et al.* (2005), Cheung, Jr. *et al.* (2007), Moura *et al.* (2008)
Burkholderia pseudomallei	BMDMs, PMNs, RAW264.7, J774A.1, THP-1, MDMs, monocytes	Human, murine	Apoptosis, pyroptosis	Sun *et al.* (2005), Suparak *et al.* (2005), Lengwehasatit *et al.* (2008), Breitbach *et al.* (2009), Chanchamroen *et al.* (2009)
Campylobacter jejuni	Kidney fibroblasts (COS-1), THP-1	Human, monkey	Apoptosis	Lara-Tejero and Galán (2000), Siegesmund *et al.* (2004)
Chlamydophila psittaci	J774	Murine	Apoptosis	Ojcius *et al.* (1998)
Chlamydia trachomatis	Monocytes, PMNs	Human	Apoptosis	Goth and Stephens (2001)
Coxiella burnetii	THP-1	Human	Apoptosis	Dellacasagrande *et al.* (1999, 2002)
Escherichia coli	Epithelial cells (HeLa, HT-29, Caco-2, Vero-C1 and T84), J774, PMNs, PBMCs, peritoneal macrophages, RAW264.7, T lymphocytes, urothelium (TEU-2), U937	Human, murine	Apoptosis, necrosis	Jonas *et al.* (1993), Baran *et al.* (1996), Watson *et al.* (1996a,b), Fernandez-Prada *et al.* (1998), Kim *et al.* (1998), Colamussi *et al.* (1999), Crane *et al.* (1999), Lai *et al.* (1999, 2000), Matsuda *et al.* (1999), Rodrigues *et al.* (1999), Stravodimos *et al.* (1999), Häcker *et al.* (2002), Blomgran *et al.* (2004), Klumpp *et al.* (2006), Flynn and Buret (2008), Matsuura *et al.* (2009)
Escherichia coli lipopolysaccharide	PMNs	Bovine	Apoptosis	Yagi *et al.* (2002)
Francisella novicida	Macrophages, αβ T cells	Murine	Apoptosis, pyroptosis	Henry *et al.* (2007), Sharma *et al.* (2009)
Francisella tularensis	B-cells, BMDMs, DCs, J774, J774A.1, J774.2,	Human, murine	Apoptosis	Lai *et al.* (2001), Lai and Sjöstedt (2003), Hrstka *et al.* (2005), Mariathasan *et al.* (2005), Krocová *et al.* (2008), Wickstrum *et al.* (2009)

Table 15.3 continued

Bacterium	Host cell	Origin	Cell death	References
Fusobacterium necrophorium leukotoxin	PMNs	Bovine	Apoptosis, necrosis	Narayanan *et al.* (2002)
Fusobacterium nucleatum	PBMCs, PMNs	Human	Apoptosis	Jewett *et al.* (2000)
Haemophilus somnus	PMNs	Bovine	Apoptosis	Yang *et al.* (1998)
Helicobacter pylori	B-cells (primary, EBV-transformed cells), gastric epithelial cells (Kato III, NCI-N87, AGS), peritoneal macrophages, RAW264.7	Human, murine	Apoptosis	Moss *et al.* (1996, 2001), Chen *et al.* (1997), Jones *et al.* (1997, 1999), Wagner *et al.* (1997), Kohda *et al.* (1999), Konturek *et al.* (1999), Gobert *et al.* (2002), Menaker *et al.* (2004), Cheng *et al.* (2005), Ding *et al.* (2007), Calvino-Fernandez *et al.* (2008), Tsao *et al.* (2009)
Helicobacter pylori cecropin-like peptide	PBMCs	Human	Apoptosis	Betten *et al.* (2001)
Legionella pneumophila	BMDMs, DCs, HL60, J774A.1, Jurkat T cells, macrophages, monocytes, MonoMac6, THP-1, U937	Human, murine	Apoptosis, necrosis, pyroptosis	Müller *et al.* (1996), Hägele *et al.* (1998), Gao and Kwaik (1999a,b), Alli *et al.* (2000), Walz *et al.* (2000), Neumeister *et al.* (2002), Zink *et al.* (2002), Abu-Zant *et al.* (2005), Amer *et al.* (2006), Fischer *et al.* (2006), Laguna *et al.* (2006), Molofsky *et al.* (2006), Zamboni *et al.* (2006), Lightfield *et al.* (2008), Akhter *et al.* (2009), Case *et al.* (2009), Nogueira *et al.* (2009)
Legionella longbeachae	HL60	Human	Apoptosis	Arakaki *et al.* (2002)
Leptospira interrogans	Hepatocytes, J774A.1	Murine	Apoptosis	Merien *et al.* (1997, 1998)
Listeria monocytogenes	BMDMs, DCs, IC-21, lymphocytes, PMNs, RAW264	Human, murine	Apoptosis, necrosis, pyroptosis, PICD	Guzmán *et al.* (1996), Barsig and Kaufmann (1997), Merrick *et al.* (1997), Kobayashi *et al.* (2003a), Cervantes *et al.* (2008)
Mannheimia haemolytica	PMNs	Bovine	Apoptosis	Cudd *et al.* (2001)
Mycobacterium avium	DCs, J774, macrophages, MDMs, monocytes	Bovine, human, murine	Apoptosis	Fratazzi *et al.* (1997), Balcewicz-Sablinska *et al.* (1999), Bermudez *et al.* (1999), Mohagheghpour *et al.* (2000), Allen *et al.* (2001), Feng *et al.* (2002), Bhattacharyya *et al.* (2003)
Mycobacterium avium sonicate	MDMs, THP-1	Human	Apoptosis	Hayashi *et al.* (1997)
Mycobacterium bovis-BCG	AMs, DCs, J774, MDMs, monocytes, peritoneal macrophages, PMNs, RAW264.7, THP-1	Bovine, human, murine	Apoptosis	Molloy *et al.* (1994), Ahmad *et al.* (1999), Keane *et al.* (2000), Fairbairn *et al.* (2001), Kausalya *et al.* (2001), Gutiérrez-Pabello *et al.* (2002), Riendeau and Kornfeld (2003), Schaible *et al.* (2003)
Mycobacterium kansasii	AMs, monocytes, THP-1	Human	Apoptosis	Keane *et al.* (2000), Guérardel *et al.* (2003)

Mycobacterium leprae	MDMs	Human	Apoptosis	Hernandez *et al.* (2003)
Mycobacterium ulcerans toxin, mycolactone	J774	Murine	Apoptosis	George *et al.* (2000)
Mycobacterium tuberculosis	AMs, BMDMs, B10R cells, macrophages, peritoneal macrophages, MDMs, monocyte, PBMCs, PMNs, RAW 264.7, THP-1, U937	Human, murine	Apoptosis, necrosis, cell death with autophagy	Durrbaum-Landmann *et al.* (1996), Keane *et al.* (1997, 2000, 2002), Klingler *et al.* (1997), Placido *et al.* (1997), Rojas *et al.* (1997, 1998, 1999, 2000, 2002), Kasahara *et al.* (1998), Ragno *et al.* (1998), Oddo *et al.* (1998), Mustafa *et al.* (1999), Ciaramella *et al.* (2000, 2002), Santucci *et al.* (2000), Duan *et al.* (2001, 2002), Alemán *et al.* (2002), Kisich *et al.* (2002), Perskvist *et al.* (2002), Danelishvili *et al.* (2003), López *et al.* (2003), Riendeau and Kornfeld (2003), Schaible *et al.* (2003), Sly *et al.* (2003), Spira *et al.* (2003), Gutierrez *et al.* (2004), Lee *et al.* (2006), Gan *et al.* (2008)
Mycoplasma fermentans	HL60, U937	Human	Apoptosis	Hall *et al.* (2000)
Mycoplasma fermentans-lipoproteins	HL60	Human	Apoptosis	Into *et al.* (2002)
Neisseria gonorrhoeae	HeLa, JOSK-M, Jurkat T cells, U937	Human	Apoptosis	Kepp *et al.* (2009), Kozjak-Pavlovic *et al.* (2009), Müller *et al.* (1999, 2000)
Parachlamydia acanthamoeba	MDMs	Human	Apoptosis	Greub *et al.* (2003)
Pasteurella haemolytica leukotoxin	PMNs	Bovine	Apoptosis	Stevens and Czuprynski (1996)
Photobacterium damselae subsp. *piscicida*	Sea bass PMNs and macrophages	Sea bass	Apoptosis	do Vale *et al.* (2003)
Pseudomonas aeruginosa	AMs, BMDMs, DCs, epithelial cells, NK cells (NK92), J774, J774A.1, Kupffer cells, mast cells, monocytes, PMNs, U937	Human, murine	Apoptosis, necrosis, pyroptosis, PICD	Baran *et al.* (1996), Dacheux *et al.* (1999, 2000), Hauser and Engel (1999), Grassmé *et al.* (2000), Zaborina *et al.* (2000), Hotchkiss *et al.* (2001), Worgall *et al.* (2002), Zhang *et al.* (2003b), Jenkins *et al.* (2006), Franchi *et al.* (2007), Sutterwala *et al.* (2007), Chai *et al.* (2008), Miao *et al.* (2008), Prince *et al.* (2008), Chung *et al.* (2009), Rodriguez *et al.* (2009)
Pseudomonas aeruginosa azurin	J774	Murine	Apoptosis	Yamada *et al.* (2002), Goto *et al.* (2003)
Pseudomonas aeruginosa autoinducer N-3-oxododecanoyl homoserine lactone	PMNs, macrophages	Murine	Apoptosis	Tateda *et al.* (2003)

Table 15.3 continued

Bacterium	Host cell	Origin	Cell death	References
Pseudomonas aeruginosa exotoxin A	Peritoneal macrophages	Rat	Apoptosis, necrosis	Huang *et al.* (2003)
Pseudomonas aeruginosa pyocyanin	PMNs	Human	Apoptosis	Usher *et al.* (2002)
Rhodococcus equi	J774E, footpad cells	Murine	Apoptosis	Kanaly *et al.* (1999), Lührmann *et al.* (2004)
Salmonella Choleraesuis	Peritoneal macrophages	Murine	Apoptosis	Arai *et al.* (1995)
Salmonella enterica serovar Dublin	Epithelial cells (HT-29 and Caco-2), J774.2, AMs, THP-1	Bovine, human, murine	Apoptosis, necrosis	Kim *et al.* (1998), Watson *et al.* (2000), Valle and Guiney (2005)
Salmonella enterica serovar Enteritidis	Epithelial cells (HeLa and T84), monocytes, Peyer's patches	Human	Apoptosis	Baran *et al.* (1996), Crane *et al.* (1999), Cerquetti *et al.* (2002)
Salmonella enterica serovar Typhi	J774A.1, MDMs, U937	Human, murine	Apoptosis	Schwan *et al.* (2000), Basso *et al.* (2002)
Salmonella enterica serovar Typhimurium	AMs, BMDMs, DCs, epithelial cells (HeLa and IEC-6), J774.2, macrophages and MDMs, RAW264.7, U937	Bovine, human, murine, rat	Apoptosis, necrosis, pyroptosis	Chen *et al.* (1996a), Monack *et al.* (1996, 2000, 2001), Richter-Dahlfors *et al.* (1997), Hersh *et al.* (1999), Lundberg *et al.* (1999), Brennan and Cookson (2000), Jesenberger *et al.* (2000, 2001), Libby *et al.* (2000), van der Velden *et al.* (2000, 2003), Watson *et al.* (2000), Yrlid and Wick (2000), Zhou *et al.* (2000), Santos *et al.* (2001), Collier-Hyams *et al.* (2002), Forsberg *et al.* (2003), Mariathasan *et al.* (2004), Knodler *et al.* (2005), Franchi *et al.* (2006), Miao *et al.* (2006), Lara-Tejero *et al.* (2006), Grant *et al.* (2008)
Salmonella enterica serovar Typhimurium, lipopolyaccharide	T cells	Murine	Apoptosis	Castro *et al.* (1998)
Shigella dysenteriae	Macrophages, PMNs, T cells	Human	Apoptosis	Raqib *et al.* (2002)
Shigella flexneri	BMDCs, epithelial cells (HeLa, HEK293T), macrophages, MDMs, monocytes (PBMCs), peritoneal macrophages, PMNs	Human, murine	Apoptosis, necrosis, pyroptosis, cell death with autophagy	Zychlinsky *et al.* (1992, 1993, 1994a,b), Chen *et al.* (1996b), Fernandez-Prada *et al.* (1997), Guichon and Zychlinsky (1997), Hilbi *et al.* (1997, 1998, 2000), Nonaka *et al.* (1999, 2003), François *et al.* (2000), Yu *et al.* (2001), Hathaway *et al.* (2002), Suzuki *et al.* (2005, 2007), Clark and Maurelli (2007), Tattoli *et al.* (2008), Carneiro *et al.* (2009), Faherty and Maurelli (2009), Lee *et al.* (2009)
Sphingobacterium spiritivorum-derived ceramides	HL60	Human	Apoptosis	Minamino *et al.* (2003)

Staphylococcus aureus	mammary epithelial cell line (MAC-T), monocytes, PBMCs, PMNs, umbilical vein endothelial	Bovine, human	Apoptosis, PICD	Jonas *et al.* (1994), Baran *et al.* (1996), Bayles *et al.* (1998), Menzies and Kourteva (1998, 2000), Wesson *et al.* (1998, 2000), Guzik *et al.* (1999), Lundqvist-Gustafsson *et al.* (2001), Yamamoto *et al.* (2002), Kobayashi *et al.* (2003a), Chau *et al.* (2009)
Staphylococcus aureus α-toxin	T cells (primary and Jurkat)	Human	Apoptosis, necrosis	Jonas *et al.* (1994), Bantel *et al.* (2001)
Streptococcus agalactiae (group B *Streptococcus*)	J774, J774A.1, MDMs, microglia, monocytes, peritoneal macrophages	Human, murine	Apoptosis	Fettucciari *et al.* (2000, 2006), Buratta *et al.* (2002), Ulett *et al.* (2003, 2005), Liu *et al.* (2004), Ulett and Adderson (2005), Lehnardt *et al.* (2007)
Streptococcus iniae	Pronephros phagocytes	Fish	Apoptosis	Zlotkin *et al.* (2003)
Streptococcus pneumoniae	AMs, BMDMs, BMDCs, brain microvascular endothelial cells, epithelial cells (BEAS-2B, A549), HEK293, MDMs, microglia, PBMCs, peritoneal macrophages, PMNs, RAW264.7, splenocytes, splenic leukocytes, thymocytes	Human, murine, rat	Apoptosis	Zysk *et al.* (2000), Braun *et al.* (2001, 2002), Dockrell *et al.* (2001, 2003), Engelich *et al.* (2001), Kemp *et al.* (2002), Ali *et al.* (2003), Colino and Snapper (2003), Marriott *et al.* (2004, 2005, 2006, 2007), Schmeck *et al.* (2004), Wellmer *et al.* (2004), Bermpohl *et al.* (2005), Srivastava *et al.* (2005), Dockrell and Whyte (2006), Marriott and Dockrell (2006), Schreiber *et al.* (2006)
Streptococcus pyogenes	BMDM, DCs, HeLa, embryonic stem cells, pharyngeal epithelial cells (Hep-2), J774, PBMCs, PMNs, PBMCs, OKP7 soft palate keratinocytes, THP-1, U937	Human, murine	Apoptosis, necrosis, cell death with autophagy, PICD	Kuo *et al.* (1999), Tsai *et al.* (1999), Nakagawa *et al.* (2001, 2004), Kobayashi *et al.* (2003a), Cywes *et al.* (2005), Miyoshi-Akiyama *et al.* (2005), Cortes and Wessels (2009), Timmer *et al.* (2009)
Vibrio vulnificus	J774A.1, peritoneal macrophages	Murine	Apoptosis	Kashimoto *et al.* (2003)
Yersinia enterocolitica	BMDCs, J774A.1, MDMs, peritoneal macrophages, thymocytes	Human, murine	Apoptosis, necrosis	Mills *et al.* (1997), Ruckdeschel *et al.* (1997, 1998, 2001a,b, 2002), Lin *et al.* (1998), Denecker *et al.* (2001), Ruckdeschel and Richter (2002), Haase *et al.* (2003), Erfurth *et al.* (2004), Grobner *et al.* (2006, 2007a,b)
Yersinia pestis	BMDMs, epithelial cells (HeLa), J774, J774A.1, Jurkat T-cells, macrophages, PMNs, RAW264.7	Human, murine, rat	Apoptosis, pyroptosis	Orth *et al.* (1999), Weeks *et al.* (2002), Ng *et al.* (2003), Lukaszewski *et al.* (2005), Marketon *et al.* (2005), Lemaitre *et al.* (2006), Zauberman *et al.* (2006), Bergsbaken and Cookson (2007), Park *et al.* (2007), Brodsky and Medzhitov (2008), Lilo *et al.* (2008)
Yersinia pseudotuberculosis	BMDMs, J774A.1, macrophages, RAW264.7, T cells	Human, murine	Apoptosis	Monack *et al.* (1997, 1998), Arencibia *et al.* (2002), Zhang and Bliska (2003), Zhang *et al.* (2005, 2008)

AM, alveolar macrophage; BMDC, bone marrow-derived dendritic cell; BMDM, bone marrow-derived macrophage; DC, dendritic cell; MDM, monocyte-derived macrophage; PBMC, peripheral blood mononuclear cell; PICO, phagocytosis-induced cell death; PMN, polymorphonuclear leukocyte. HL60 are human promyelocytic leukaemia cells; J774, J774.A1 and RAW264.7 are murine macrophage-like cells; THP-1 and U937 are human promonocytic leukaemia cells.

Bacteria-induced death of phagocytic leucocytes

Macrophages and intracellular pathogens

Macrophages are relatively long-lived phagocytes that perform a wide range of functions in the host, such as initiation of the inflammatory response, phagocytosis and killing of invading microorganisms, presentation of antigen (thereby functioning as a bridge between innate and acquired immunity), and non-inflammatory removal of apoptotic host cells and cell debris (Gordon, 2008; Medzhitov, 2008). The ability of macrophages to perform these diverse functions and maintain tissue and immune system homeostasis is offset by their relatively limited microbicidal capacity compared with neutrophils. As such, macrophages are amenable to modulation by bacterial pathogens.

A number of bacterial pathogens, including *Bordetella, Brucella, Burkholderia, Chlamydia, Coxiella, Francisella, Legionella, Listeria, Mycobacterium,* and *Salmonella,* have been shown to survive and replicate within macrophages Celli, 2006; Flannagan *et al.*, 2009). The ability of bacterial pathogens to survive and replicate within macrophages, and ultimately alter apoptosis or cell death, begins with the mode of uptake. For example, virulent strains of *Coxiella burnetti* selectively avoid/inhibit the engagement of the CR3 host cell surface receptor to redirect uptake through a specific mechanism (Flannagan *et al.*, 2009). Survival of *Brucella* spp. in macrophages is dependent on lipid-raft-mediated uptake, which may involve the lipopolysaccharide (LPS) O-chain of *Brucella* (Celli, 2006). Mutant strains lacking the LPS O-chain enter macrophages in a lipid-raft-independent manner and are unable to survive. Once inside the macrophage, bacterial pathogens proficiently alter biogenesis and maturation of the phagosome in an effort to manipulate the environment amenable to replication (Flannagan *et al.*, 2009). *Mycobacterium* spp. inhibit early phagosome maturation by preventing phagosome–lysosome fusion (Flannagan *et al.*, 2009). As a result, this vacuole resembles an early endosome and is conducive to pathogen replication. *Brucella* and *Legionella* spp. utilize secreted effector proteins to inhibit phagosome–lysosome fusion, avoiding endocytic trafficking (Celli, 2006; Flannagan *et al.*, 2009). In this case, the resulting replicative vesicles contain endoplasmic reticulum-derived markers. Infection of macrophages with *Helicobacter pylori* stalls early phagosome maturation, resulting in megasomes that contain viable bacteria (Allen and McCaffrey, 2007). This process is dependent on urease and VacA, both secreted effector proteins. *F. tularensis* are found in late-endosomal vesicles, while *Salmonella* spp. are found in atypical late-stage vesicles derived from *trans* Golgi membranes (Salcedo and Holden, 2005; Santic *et al.*, 2006). Interestingly, *C. burnetii* do not subvert endosomal trafficking, but instead adapt to the harsh conditions of the mature phagosome, a process also dependent on secreted effector proteins (Flannagan *et al.*, 2009). Bacterial pathogens also possess mechanisms to escape the phagosome and utilize the cytosol as a replicative niche and this phenomenon is reviewed elsewhere (Flannagan *et al.*, 2009).

Using conditions that represent late stages of infection or use a high multiplicity of infection (MOI) *in vitro*, many bacterial pathogens induce macrophage death (Table 15.3). It is not always clear whether this phenomenon is active modulation of cell death by the pathogen or if it is a specific host response to infection. Here, we discuss these two possibilities. Namely, that (i) active modulation of macrophage fate by bacterial pathogens is a component of pathogenesis, or (ii) cell death is a desired host response important for pathogen clearance and the resolution of infection.

Inhibition of macrophage apoptosis

Several bacterial pathogens are known to prevent macrophage apoptosis, and the process can occur by distinct mechanisms (Table 15.2). Mononuclear phagocytes (monocytes, macrophages or macrophage cell lines) infected with pathogens such as *Brucella* spp., *Chlamydia* spp., *C. burnetii*, or *E. coli* are resistant to apoptosis induced by exogenous stimuli, e.g. staurosporine or Fas ligand (Gross *et al.*, 2000; Airenne *et al.*, 2002; Sukumaran *et al.*, 2004; Voth *et al.*, 2007) (Table 15.2). In most cases, increased resistance to apoptosis is dependent on intracellular survival of the bacteria. For instance, *Brucella* promotes intracellular survival by modulating expression of *BCL2A1* (Gross *et al.*, 2000), whereas *Burkholderia* spp. actively evade autophagy, a process mediated by the secreted effector protein, BopA. *C. burnetii*, *Chlamydia* spp. and *E. coli* inhibit the release of cytochrome c from mitochondria and activation of caspase 3, thereby preventing apoptosis (Airenne *et al.*, 2002; Sukumaran *et al.*, 2004; Lührmann and Roy, 2007). *C. burnetti* and *E. coli* also alter the expression of anti-apoptosis and anti-inflammation genes, which contributes to macrophage survival (Voth *et al.*, 2007; Sukumaran *et al.*, 2004). *C. burnetti* affects Beclin-1/Bcl-2 interplay to inhibit induction of apoptosis and autophagy (Vázquez and Colombo, 2010). There is insufficient space here for a comprehensive discussion of the mechanisms used by each pathogen to prevent macrophage apoptosis, but it is clear that pathogens actively prevent macrophage apoptosis (Table 15.2).

Bacteria-induced macrophage death

Infection of macrophages with bacteria typically induces cell death, which can be apoptosis, pyroptosis, or lysis/necrosis (Fig. 15.2). For example, *Bordetella, Burkholderia, Chlamydia, E. coli, Francisella, Helicobacter, Legionella, Listeria, Mycobacterium, Salmonella, Shigella, Streptococcus pneumoniae,* and *Yersinia* each have the capacity to induce macrophage death (Table 15.3). It is intriguing that some of the bacterial pathogens that inhibit macrophage apoptosis (e.g. *Burkholderia pseudomallei, Chlamydia* spp., *E. coli,* and *S. pneumoniae*), also have capacity to induce host cell death (Tables 15.2 and 15.3). Induction of cell death in macrophages typically occurs under conditions resembling late stages of infection, e.g. with high MOIs and long-term pathogen exposure. In addition to being dependent on the specific pathogen or bacterial species and type of host cell (tissue-specific macrophages or macrophage-like cell lines), the mode of cell death appears sensitive to the conditions *in vitro,* as the ability to influence the extent or type of macrophage death is MOI-dependent for some pathogens. *E. coli, Francisella tularensis, Legionella* spp., *Mycobacterium* spp., and *Salmonella* spp. can induce apoptosis through the mitochondrial pathway, yet under specific conditions *in vitro* cause a caspase1-dependent, pro-inflammatory, macrophage death (Table 15.3). It is noteworthy that terminology used to describe cell death is often varied and subjective, and can be limited by the detection methods used by any given laboratory. To address this issue, the Nomenclature Committee on Cell Death 2009 suggests specific guidelines for interpretation of cell death assays and classifying modes of cell death (Galluzzi *et al.*, 2009; Kroemer *et al.*, 2009).

B. pseudomallei, E. coli, F. tularensis, Legionella pneumophila, Listeria monocytogenes, Mycobacterium tuberculosis, Salmonella spp., and *Shigella* spp., each cause macrophage pyroptosis and the process is typically mediated by specific pathogen effector molecules (Table 15.1 and 15.3). Alternatively, some human pathogens that induce pyroptosis also induce caspase 1-independent macrophage death at higher MOIs (Labbé and Saleh, 2008). The hierarchy among overlapping bacteria-induced cell death pathways remains incompletely determined and is an active area of investigation for many laboratories.

In light of the evidence presented above, it is tempting to postulate that the delay or induction of macrophage death is advantageous to the pathogen, since these processes can promote pathogen survival. However, the capacity of some pathogens to both inhibit and promote apoptosis in macrophages does not clearly imply a pathogenic advantage, as the nature and extent of induced cell death appears to be dose-dependent. Rather, induction of macrophage apoptosis by bacteria is likely a desired host response, as the process has been shown to facilitate efficient pathogen clearance, implying that the induction of apoptosis is advantageous to the host. Consistent with this notion, *Mycobacterium* spp. that induce macrophage apoptosis are typically less virulent than those that survive and replicate within these phagocytes. It has been proposed that a temporal trigger may exist, signalling the bacterial pathogen to switch from a state of survival and replication to a mode of escape. This model has also been proposed for *Shigella* spp. and *Salmonella* spp., which utilize effector proteins that cause activation of caspase 1, resulting in pyroptosis and dissemination of surviving bacteria (Table 15.1). Alternatively, the pro-inflammatory response following pyroptosis presumably involves recruitment of neutrophils, which would promote pathogen clearance. There is evidence that some bacterial pathogens have devised means to inhibit caspase 1-dependent death. For example, *L. pneumophila* down-regulates production of flagellin in an effort to avoid activation of caspase1 and subsequent pyroptosis (Bergsbaken *et al.*, 2009). *M. tuberculosis* contains a Zn-metalloprotease that prevents the activation of caspase 1 through an unknown mechanism (Bergsbaken *et al.*, 2009). Consistent with this idea, mice deficient in caspase 1 have increased susceptibility to infection by pathogens that cause pyroptosis *in vitro* (Labbé and Saleh, 2008). These and other lines of evidence suggest that macrophage pyroptosis can be beneficial to the host. The overarching observation that cell death can have positive or negative outcomes for the host is probably a paradigm for the ongoing adaptation of pathogens to the host.

PMNs and innate host defence

PMNs (also called granulocytes or neutrophils) are the most abundant white cell in humans, and ~60% of haematopoiesis is dedicated to their production (Bainton *et al.*, 1971). These terminally differentiated phagocytes are also typically the first white cells recruited to sites of infection. Their abundance, rapid turnover and adept ability to rapidly destroy invading microorganisms has earned them notoriety as the 'executioners' of the innate immune system. Neutrophils maintain high microbicidal capacity, surpassing that of macrophages. PMNs have the capacity to produce high levels of ROS (Quinn *et al.*, 2006), and contain numerous antimicrobial peptides and proteases in lysosome-like vesicles known as granules (Borregaard *et al.*, 2007). In essence, the primary function of PMNs is the elimination of invading microorganisms. The average life-span of a human neutrophil is 9–10 days – 1–2 days are spent in circulation – and the turnover rate in humans is on the order of 10^{11} cells per day (Kennedy and DeLeo, 2009). Inasmuch as PMNs contain or produce an abundance of cytotoxic molecules or proteases that can cause damage to host tissues should the cell rupture, it is critical that there is timely removal of effete cells. To that end, neutrophil homeostasis and turnover is regulated by spontaneous (constitutive) apoptosis and subsequent removal of apoptotic neutrophils by macrophages (Savill *et al.*, 1989a; Haslett *et al.*, 1990). Induction of neutrophil apoptosis is accompanied by a general decrease in cell function and pro-inflammatory capacity (Whyte *et al.*, 1993; Kobayashi *et al.*, 2005). Taken together, high microbicidal capacity, abundance at the site of infection, and short life-span make neutrophils an extremely proficient primary defence against invading microorganisms.

Bacteria-induced PMN apoptosis

During the inflammatory response, PMNs are recruited to sites of infection by host-derived chemokines and cytokines, such as IL-8 and granulocyte-macrophage colony-stimulating factor (GM-CSF). In addition to priming neutrophils for enhanced function, these molecules delay constitutive neutrophil apoptosis (Colotta *et al.*, 1992). Additionally, bacterial-derived products such as LPS and lipoteichoic acid are potent priming agents and inhibitors of neutrophil apoptosis (Guthrie *et al.*, 1984; Colotta *et al.*, 1992; Lotz *et al.*, 2004). Acute hypoxia also enhances the inflammatory response and reduces neutrophil apoptosis, reflecting the ability of the extracellular environment to influence neutrophil survival. The delay in spontaneous PMN apoptosis ensures an efficient and robust initial response to invading microorganisms.

In contrast to macrophages, phagocytosis ultimately induces apoptosis in human neutrophils (Coxon *et al.*, 1996; Gamberale *et al.*, 1998; Kobayashi *et al.*, 2002, 2003b; Zhang *et al.*, 2003a). Phagocytosis of *Borrelia hermsii, Burkholderia cepacia, E. coli, L. monocytogenes, M. tuberculosis, Neisseria gonorrhoeae, S. aureus, S. pneumoniae* and *S. pyogenes* accelerates neutrophil apoptosis (DeLeo, 2004) (Table 15.3). Importantly, the production of ROS following phagocytosis has been linked to the subsequent induction of cell death and potentially the resolution of the inflammatory response (Coxon *et al.*, 1996; Kobayashi *et al.*, 2002, 2003b, 2004). This form of cell death, which has characteristics of apoptosis, is also known as phagocytosis-induced cell death (PICD) (Zhang *et al.*, 2003a). The signals for PICD override any host- or bacterial-derived signals that may serve to otherwise delay apoptosis (Watson *et al.*, 1996a).

It is of note that ROS production, to an extent, is dictated by the bacteria-to-neutrophil ratio, which in turn may influence neutrophil PICD. For example, interaction of *E. coli* or *N. gonorrhoeae* with neutrophils at a low MOI (1) delays apoptosis, presumably owing to the effects of LPS and insufficient ROS (Watson *et al.*, 1996b; Simons *et al.*, 2006). By comparison, MOIs of 10 or more–which elicit significant ROS production–triggers apoptosis/PICD (Watson *et al.*, 1996b; Simons *et al.*, 2006).

Phagocytosis of bacterial pathogens, including *B. hermsii, B. cepacia, L. monocytogenes, S. aureus* and *S. pyogenes*, is followed by global changes in PMN gene expression, which we have named an apoptosis differentiation programme (Kobayashi *et al.*, 2003a). The apoptosis differentiation programme contributes to down-regulation of pro-inflammatory capacity and induction of apoptosis. In general, this process promotes resolution of the inflammatory response.

Delayed PMN apoptosis

Although many bacteria-derived molecules can delay PMN apoptosis (Table 15.2), short-lived neutrophils are not amenable to invasion and replication strategies employed by most intracellular pathogens. As such, only a few bacterial pathogens are known to prolong neutrophil survival, and thereby promote pathogenesis (Table 15.2). *Anaplasma phagocytophilum*, the causative agent of human granulocytic anaplasmosis (formerly human granulocytic ehrlichiosis), was the first bacterial pathogen reported to delay PMN apoptosis (Yoshiie *et al.*, 2000). *A. phagocytophilum* survives and replicates within an endosomal compartment in human neutrophils (Webster *et al.*, 1998). The pathogen inhibits PMN ROS production, alters normal granule trafficking, and fails to induce the apoptosis differentiation programme, even at relatively high MOIs (up to 100) (Carlyon *et al.*, 2004; Borjesson *et al.*, 2005; Carlyon and Fikrig, 2006). *Chlamydophila pneumoniae* are readily phagocytosed by neutrophils in an opsonin-independent manner, prolong host cell survival, and survive and replicate intracellularly (van Zandbergen *et al.*, 2004). Delayed apoptosis in these infected PMNs is associated with decreased processing of caspase 3, the presence of LPS in the cytosol, and autocrine production of IL-8. In the two examples provided here, the delay in PMN apoptosis is important for pathogen survival and transmission to new host cells.

PMN necrosis

Induction of PMN apoptosis following phagocytosis of bacteria can be considered as an important component of the resolution of infection. Therefore, it is not surprising that human pathogens produce molecules, such as cytolytic toxins, that cause lysis of neutrophils *in vitro* (see above and Table 15.1). However, *S. pyogenes* and *S. aureus*, most notably strains of community-associated methicillin-resistant *S. aureus* (CA-MRSA), cause overt lysis of human PMNs following phagocytosis or during phagocytic interaction (Rogers and Tompsett, 1952; Kobayashi *et al.*, 2003a; Voyich *et al.*, 2005). Interestingly, this process is accompanied by signs of apoptosis, but occurs rapidly after phagocytosis (within 4–6 h). PMN lysis *in vitro* ultimately promotes pathogen survival (Rogers and Tompsett, 1952; Kobayashi *et al.*, 2003a; Voyich *et al.*, 2005). Therefore, PMN lysis following phagocytosis can be considered a component of bacterial pathogenesis, and more work is needed to understand this interesting phenomenon.

Concluding remarks

Host cells can undergo apoptosis in response to bacteria or bacterial effector molecules as a mechanism to clear infected or effete cells and promote the resolution of the inflammatory response. On the other hand, bacterial pathogens can alter normal host cell turnover as a means to cause infection. Therefore, the fate of host cells during or after interaction with bacteria – and/or the molecules they produce – can be considered as a fulcrum that dictates at least in part whether invading microorganisms cause disease.

Acknowledgements

We thank Anita Mora and Austin Athman (Rocky Mountain Laboratories, NIAID, USA) for help with generating Fig. 15.2, and David Dorward (Rocky Mountain Laboratories, NIAID, USA) for performing transmission electron microscopy. This article was supported by the Intramural Research Program of the National Institute of Allergy and Infectious Diseases, National Institutes of Health.

References

Abu-Zant, A., Santic, M., Molmeret, M., Jones, S., Helbig, J., and Kwaik, Y.A. (2005). Incomplete activation of macrophage apoptosis during intracellular replication of *Legionella pneumophila*. Infect. Immun. *73*, 5339–5349.

Abu-Zant, A., Jones, S., Asare, R., Suttles, J., Price, C., Graham, J., and Kwaik, Y.A. (2007). Anti-apoptotic signalling by the Dot/Icm secretion system of *L. pneumophila*. Cell. Microbiol. *9*, 246–264.

Ahmad, A., Khan, M., Raykundalia, C., and Catty, D. (1999). Study of the mechanisms of killing of *Mycobacterium bovis* BCG by apoptosis in J774 murine macrophages. J. Pak. Med. Assoc. *49*, 273–278.

Airenne, S., Surcel, H.M., Tuukkanen, J., Leinonen, M., and Saikku, P. (2002). *Chlamydia pneumoniae* inhibits apoptosis in human epithelial and monocyte cell lines. Scand. J. Immunol. *55*, 390–398.

Akhter, A., Gavrilin, M.A., Frantz, L., Washington, S., Ditty, C., Limoli, D., Day, C., Sarkar, A., Newland, C., Butchar, J., *et al.* (2009). Caspase-7 activation by the Nlrc4/Ipaf inflammasome restricts *Legionella pneumophila* infection. PLoS Pathog. *5*, e1000361.

Alaoui-El-Azher, M., Jia, J., Lian, W., and Jin, S. (2006). ExoS of *Pseudomonas aeruginosa* induces apoptosis through a Fas receptor/caspase 8-independent pathway in HeLa cells. Cell. Microbiol. *8*, 326–338.

Alemán, M., García, A., Saab, M.A., De La Barrera, S.S., Finiasz, M., Abbate, E., and Sasiain, M.C. (2002). *Mycobacterium tuberculosis*-induced activation accelerates apoptosis in peripheral blood neutrophils from patients with active tuberculosis. Am. J. Respir. Cell Mol. Biol. *27*, 583–592.

Ali, F., Lee, M.E., Iannelli, F., Pozzi, G., Mitchell, T.J., Read, R.C., and Dockrell, D.H. (2003). *Streptococcus pneumoniae*-associated human macrophage apoptosis after bacterial internalization via complement and Fcgamma receptors correlates with intracellular bacterial load. J. Infect. Dis. *188*, 1119–1131.

Alileche, A., Serfass, E.R., Muehlbauer, S.M., Porcelli, S.A., and Brojatsch, J. (2005). Anthrax lethal toxin-mediated killing of human and murine dendritic cells impairs the adaptive immune response. PLoS Pathog. *1*, e19.

Allen, L.A.,and McCaffrey, R.L. (2007). To activate or not to activate: distinct strategies used by *Helicobacter pylori* and *Francisella tularensis* to modulate the NADPH oxidase and survive in human neutrophils. Immunol. Rev. *219*, 103–117.

Allen, S., Sotos, J., Sylte, M.J., and Czuprynski, C.J. (2001). Use of Hoechst 33,342 staining to detect apoptotic changes in bovine mononuclear phagocytes infected with *Mycobacterium avium* subsp. *paratuberculosis*. Clin. Diagn. Lab. Immunol. *8*, 460–464.

Alli, O.A., Gao, L.Y., Pedersen, L.L., Zink, S., Radulic, M., Doric, M., and Kwaik, Y.A. (2000). Temporal pore formation-mediated egress from macrophages and alveolar epithelial cells by *Legionella pneumophila*. Infect. Immun. *68*, 6431–6440.

Amer, A., Franchi, L., Kanneganti, T.D., Body-Malapel, M., Ozoren, N., Brady, G., Meshinchi, S., Jagirdar, R., Gewirtz, A., Akira, S., *et al.* (2006). Regulation of *Legionella* phagosome maturation and infection through flagellin and host Ipaf. J. Biol. Chem. *281*, 35217–35223.

Arai, T., Hiromatsu, K., Nishimura, H., Kimura, Y., Kobayashi, N., Ishida, H., Nimura, Y., and Yoshikai, Y. (1995). Endogenous interleukin 10 prevents apoptosis in macrophages during *Salmonella* infection. Biochem. Biophys. Res. Commun. *213*, 600–607.

Arakaki, N., Higa, F., Koide, M., Tateyama, M., and Saito, A. (2002). Induction of apoptosis of human macrophages *in vitro* by *Legionella longbeachae* through activation of the caspase pathway. J. Med. Microbiol. *51*, 159–168.

Arce, S., Nawar, H.F., Russell, M.W., and Connell, T.D. (2005). Differential binding of *Escherichia coli* enterotoxins LT-IIa and LT-IIb and of cholera toxin elicits differences in apoptosis, proliferation, and activation of lymphoid cells. Infect. Immun. *73*, 2718–2727.

Arencibia, I., Frankel, G., and Sundqvist, K.G. (2002). Induction of cell death in T lymphocytes by invasin via beta1-integrin. Eur. J. Immunol. *32*, 1129–1138.

Ashare, A., Monick, M.M., Nymon, A.B., Morrison, J.M., Noble, M., Powers, L.S., Yarovinsky, T.O., Yahr, T.L., and Hunninghake, G.W. (2007). *Pseudomonas aeruginosa* delays Kupffer cell death via stabilization of the X-chromosome-linked inhibitor of apoptosis protein. J. Immunol. *179*, 505–513.

Bainton, D.F., Ullyot, J.L., and Farquhar, M.G. (1971). The development of neutrophilic polymorphonuclear leukocytes in human bone marrow. J. Exp. Med. *134*, 907–934.

Balcewicz-Sablinska, M.K., Gan, H., and Remold, H.G. (1999). Interleukin 10 produced by macrophages inoculated with *Mycobacterium avium* attenuates mycobacteria-induced apoptosis by reduction of TNF-α activity. J. Infect. Dis. *180*, 1230–1237.

Bantel, H., Sinha, B., Domschke, W., Peters, G., Schulze-Osthoff, K., and Jänicke, R.U. (2001). α-Toxin is a mediator of *Staphylococcus aureus*-induced cell death and activates caspases via the intrinsic death pathway independently of death receptor signaling. J. Cell. Biol. *155*, 637–648.

Baran, J., Guzik, K., Hryniewicz, W., Ernst, M., Flad, H.D., and Pryjma, J. (1996). Apoptosis of monocytes and prolonged survival of granulocytes as a result of phagocytosis of bacteria. Infect. Immun. *64*, 4242–4248.

Barsig, J., and Kaufmann, S.H. (1997). The mechanism of cell death in *Listeria monocytogenes*-infected murine macrophages is distinct from apoptosis. Infect. Immun. *65*, 4075–4081.

Basso, H., Rharbaoui, F., Staendner, L.H., Medina, E., Garciá-del Portillo, F., and Guzmán, C.A. (2002).

Characterization of a novel intracellularly activated gene from *Salmonella enterica* serovar Typhi. Infect. Immun. *70*, 5404–5411.

Batisson, M., Strazielle, N., Hejmadi, M., Thomas, D., Ghersi-Egea, J.F., Etienne, J., Vandenesch, F., and Lina, G. (2006). Toxic shock syndrome toxin-1 challenges the neuroprotective functions of the choroidal epithelium and induces neurotoxicity. J. Infect. Dis. *194*, 341–349.

Bayles, K.W., Wesson, C.A., Liou, L.E., Fox, L.K., Bohach, G.A., and Trumble, W.R. (1998). Intracellular *Staphylococcus aureus* escapes the endosome and induces apoptosis in epithelial cells. Infect. Immun. *66*, 336–342.

Bayley, H. (2009). Membrane-protein structure: piercing insights. Nature *459*, 651–652.

Bergman, N.H., Passalacqua, K.D., Gaspard, R., Shetron-Rama, L.M., Quackenbush, J., and Hanna, P.C. (2005). Murine macrophage transcriptional responses to *Bacillus anthracis* infection and intoxication. Infect. Immun. *73*, 1069–1080.

Bergsbaken, T., and Cookson, B.T. (2007). Macrophage activation redirects *Yersinia*-infected host cell death from apoptosis to caspase-1-dependent pyroptosis. PLoS Pathog. *3*, e161.

Bergsbaken, T., Fink, S.L., and Cookson, B.T. (2009). Pyroptosis: host cell death and inflammation. Nat. Rev. Microbiol. *7*, 99–109.

Bermpohl, D., Halle, A., Freyer, D., Dagand, E., Braun, J.S., Bechmann, I., Schroder, N.W., and Weber, J.R. (2005). Bacterial programmed cell death of cerebral endothelial cells involves dual death pathways. J. Clin. Invest. *115*, 1607–1615.

Bermudez, L.E., Parker, A., and Petrofsky, M. (1999). Apoptosis of *Mycobacterium avium*-infected macrophages is mediated by both tumour necrosis factor (TNF) and Fas, and involves the activation of caspases. Clin. Exp. Immunol. *116*, 94–99.

Betten, A., Bylund, J., Christophe, T., Boulay, F., Romero, A., Hellstrand, K., and Dahlgren, C. (2001). A pro-inflammatory peptide from *Helicobacter pylori* activates monocytes to induce lymphocyte dysfunction and apoptosis. J. Clin. Invest. *108*, 1221–1228.

Bhattacharyya, A., Pathak, S., Basak, C., Law, S., Kundu, M., and Basu, J. (2003). Execution of macrophage apoptosis by *Mycobacterium avium* through apoptosis signal-regulating kinase 1/p38 mitogen-activated protein kinase signaling and caspase 8 activation. J. Biol. Chem. *278*, 26517–26525.

Bhavsar, A.P., Guttman, J.A., and Finlay, B.B. (2007). Manipulation of host-cell pathways by bacterial pathogens. Nature *449*, 827–834.

Bielaszewska, M., Sinha, B., Kuczius, T., and Karch, H. (2005). Cytolethal distending toxin from Shiga toxin-producing *Escherichia coli* O157 causes irreversible G2/M arrest, inhibition of proliferation, and death of human endothelial cells. Infect. Immun. *73*, 552–562.

Binnicker, M.J., Williams, R.D., and Apicella, M.A. (2003). Infection of human urethral epithelium with *Neisseria gonorrhoeae* elicits an upregulation of host anti-apoptotic factors and protects cells from staurosporine-induced apoptosis. Cell. Microbiol. *5*, 549–560.

Binnicker, M.J., Williams, R.D., and Apicella, M.A. (2004). Gonococcal porin IB activates NF-κB in human urethral epithelium and increases the expression of host antiapoptotic factors. Infect. Immun. *72*, 6408–6417.

Blomgran, R., Zheng, L., and Stendahl, O. (2004). Uropathogenic *Escherichia coli* triggers oxygen-dependent apoptosis in human neutrophils through the cooperative effect of type 1 fimbriae and lipopolysaccharide. Infect. Immun. *72*, 4570–4578.

Borjesson, D.L., Kobayashi, S.D., Whitney, A.R., Voyich, J.M., Argue, C.M., and DeLeo, F.R. (2005). Insights into pathogen immune evasion mechanisms: *Anaplasma phagocytophilum* fails to induce an apoptosis differentiation program in human neutrophils. J. Immunol. *174*, 6364–6372.

Borregaard, N., Sorensen, O.E., and Theilgaard-Monch, K. (2007). Neutrophil granules: a library of innate immunity proteins. Trends Immunol. *28*, 340–345.

Boyd, A.P., Ross, P.J., Conroy, H., Mahon, N., Lavelle, E.C., and Mills, K.H. (2005). *Bordetella pertussis* adenylate cyclase toxin modulates innate and adaptive immune responses: distinct roles for acylation and enzymatic activity in immunomodulation and cell death. J. Immunol. *175*, 730–738.

Bradburne, C., Chung, M.C., Zong, Q., Schlauch, K., Liu, D., Popova, T., Popova, A., Bailey, C., Soppet, D., and Popov, S. (2008). Transcriptional and apoptotic responses of THP-1 cells to challenge with toxigenic, and non-toxigenic *Bacillus anthracis*. BMC Immunol. *9*, 67.

Bratton, D.L., May, K.R., Kailey, J.M., Doherty, D.E., and Leung, D.Y. (1999). Staphylococcal toxic shock syndrome toxin-1 inhibits monocyte apoptosis. J. Allergy Clin. Immunol. *103*, 895–900.

Braun, J.S., Novak, R., Murray, P.J., Eischen, C.M., Susin, S.A., Kroemer, G., Halle, A., Weber, J.R., Tuomanen, E.I., and Cleveland, J.L. (2001). Apoptosis-inducing factor mediates microglial and neuronal apoptosis caused by pneumococcus. J. Infect. Dis. *184*, 1300–1309.

Braun, J.S., Sublett, J.E., Freyer, D., Mitchell, T.J., Cleveland, J.L., Tuomanen, E.I., and Weber, J.R. (2002). Pneumococcal pneumolysin and H_2O_2 mediate brain cell apoptosis during meningitis. J. Clin. Invest. *109*, 19–27.

Braun, J.S., Hoffmann, O., Schickhaus, M., Freyer, D., Dagand, E., Bermpohl, D., Mitchell, T.J., Bechmann, I., and Weber, J.R. (2007). Pneumolysin causes neuronal cell death through mitochondrial damage. Infect. Immun. *75*, 4245–4254.

Breitbach, K., Sun, G.W., Köhler, J., Eske, K., Wongprompitak, P., Tan, G., Liu, Y., Gan, Y.H., and Steinmetz, I. (2009). Caspase-1 mediates resistance in murine melioidosis. Infect. Immun. *77*, 1589–1595.

Brennan, M.A., and Cookson, B.T. (2000). *Salmonella* induces macrophage death by caspase-1-dependent necrosis. Mol. Microbiol. *38*, 31–40.

Brigotti, M., Carnicelli, D., Ravanelli, E., Barbieri, S., Ricci, F., Bontadini, A., Tozzi, A.E., Scavia, G., Caprioli, A., and Tazzari, P.L. (2008). Interactions between Shiga toxins and human polymorphonuclear leukocytes. J. Leukoc. Biol. *84*, 1019–1027.

Brodsky, I.E., and Medzhitov, R. (2008). Reduced secretion of YopJ by *Yersinia* limits *in vivo* cell death but enhances bacterial virulence. PLoS Pathog. *4*, e1000067.

Brunelle, J.K., and Letai, A. (2009). Control of mitochondrial apoptosis by the Bcl-2 family. J. Cell Sci. *122*, 437–441.

Bruno, T.F., Woods, D.E., and Mody, C.H. (2000). Exoenzyme S from *Pseudomonas aeruginosa* induces apoptosis in T lymphocytes. J. Leukoc. Biol. *67*, 808–816.

Buratta, S., Fettucciari, K., Mambrini, R., Fetriconi, I., Marconi, P., and Mozzi, R. (2002). Group B *Streptococcus* (GBS) modifies macrophage phosphatidylserine metabolism during induction of apoptosis. FEBS Lett. *520*, 68–72.

Bylund, J., Campsall, P.A., Ma, R.C., Conway, B.A., and Speert, D.P. (2005). *Burkholderia cenocepacia* induces neutrophil necrosis in chronic granulomatous disease. J. Immunol. *174*, 3562–3569.

Calvino-Fernandez, M., Benito-Martinez, S., and Parra-Cid, T. (2008). Oxidative stress by *Helicobacter pylori* causes apoptosis through mitochondrial pathway in gastric epithelial cells. Apoptosis *13*, 1267–1280.

Carlyon, J.A., and Fikrig, E. (2006). Mechanisms of evasion of neutrophil killing by *Anaplasma phagocytophilum*. Curr. Opin. Hematol. *13*, 28–33.

Carlyon, J.A., Abdel-Latif, D.A., Pypaert, M., Lacy, P., and Fikrig, E. (2004). *Anaplasma phagocytophilum* utilizes multiple host evasion mechanisms to thwart NADPH oxidase-mediated killing during neutrophil infection. Infect. Immun. *72*, 4772–4783.

Carneiro, B.A., Fujii, J., Brito, G.A., Alcantara, C., Oria, R.B., Lima, A.A., Obrig, T., and Guerrant, R.L. (2006). Caspase and Bid involvement in *Clostridium difficile* toxin A-induced apoptosis and modulation of toxin A effects by glutamine and alanyl-glutamine *in vivo* and *in vitro*. Infect. Immun. *74*, 81–87.

Carneiro, L.A., Travassos, L.H., Soares, F., Tattoli, I., Magalhaes, J.G., Bozza, M.T., Plotkowski, M.C., Sansonetti, P.J., Molkentin, J.D., Philpott, D.J., *et al.* (2009). *Shigella* induces mitochondrial dysfunction and cell death in nonmyleoid cells. Cell Host Microbe *5*, 123–136.

Carratelli, C.R., Rizzo, A., Catania, M.R., Gallè, F., Losi, E., Hasty, D.L., and Rossano, F. (2002). *Chlamydia pneumoniae* infections prevent the programmed cell death on THP-1 cell line. FEMS Microbiol. Lett. *215*, 69–74.

Carrero, J.A., Calderon, B., and Unanue, E.R. (2004). Listeriolysin O from *Listeria monocytogenes* is a lymphocyte apoptogenic molecule. J. Immunol. *172*, 4866–4874.

Carrero, J.A., Vivanco-Cid, H., and Unanue, E.R. (2008). Granzymes drive a rapid listeriolysin O-induced T cell apoptosis. J. Immunol. *181*, 1365–1374.

Case, C.L., Shin, S., and Roy, C.R. (2009). Asc and Ipaf Inflammasomes direct distinct pathways for caspase-1 activation in response to *Legionella pneumophila*. Infect. Immun. *77*, 1981–1991.

Castro, A., Bemer, V., Nóbrega, A., Coutinho, A., and Truffa-Bachi, P. (1998). Administration to mouse of endotoxin from Gram-negative bacteria leads to activation and apoptosis of T lymphocytes. Eur. J. Immunol. *28*, 488–495.

Ceelen, L.M., Decostere, A., Ducatelle, R., and Haesebrouck, F. (2006). Cytolethal distending toxin generates cell death by inducing a bottleneck in the cell cycle. Microbiol. Res. *161*, 109–120.

Celli, J. (2006). Surviving inside a macrophage: the many ways of *Brucella*. Res. Microbiol. *157*, 93–98.

Cerquetti, M.C., Goren, N.B., Ropolo, A.J., Grasso, D., Giacomodonato, M.N., and Vaccaro, M.I. (2002). Nitric oxide and apoptosis induced in Peyer's patches by attenuated strains of *Salmonella enterica* serovar Enteritidis. Infect. Immun. *70*, 964–969.

Cervantes, J., Nagata, T., Uchijima, M., Shibata, K., and Koide, Y. (2008). Intracytosolic *Listeria monocytogenes* induces cell death through caspase-1 activation in murine macrophages. Cell. Microbiol. *10*, 41–52.

Chai, W.S., Zhu, X.M., Li, S.H., Fan, J.X., and Chen, B.Y. (2008). Role of Bcl-2 family members in caspase-3/9-dependent apoptosis during *Pseudomonas aeruginosa* infection in U937 cells. Apoptosis *13*, 833–843.

Chakrabarti, G., and McClane, B.A. (2005). The importance of calcium influx, calpain and calmodulin for the activation of CaCo-2 cell death pathways by *Clostridium perfringens* enterotoxin. Cell. Microbiol. *7*, 129–146.

Chakrabarti, G., Zhou, X., and McClane, B.A. (2003). Death pathways activated in CaCo-2 cells by *Clostridium perfringens* enterotoxin. Infect. Immun. *71*, 4260–4270.

Chanchamroen, S., Kewcharoenwong, C., Susaengrat, W., Ato, M., and Lertmemongkolchai, G. (2009). Human polymorphonuclear neutrophil responses to *Burkholderia pseudomallei* in healthy and diabetic subjects. Infect. Immun. *77*, 456–463.

Chang, C.W., Tsai, W.H., Chuang, W.J., Lin, Y.S., Wu, J.J., Liu, C.C., Tsai, P.J., and Lin, M.T. (2007). The fate of SPE B after internalization and its implication in SPEB-induced apoptosis. J. Biomed. Sci. *14*, 419–427.

Chang, C.W., Tsai, W.H., Chuang, W.J., Lin, Y.S., Wu, J.J., Liu, C.C., Tsai, P.J., and Lin, M.T. (2009). Procaspase 8 and Bax are up-regulated by distinct pathways in streptococcal pyrogenic exotoxin B-induced apoptosis. J. Biol. Chem. *284*, 33195–33205.

Chang, M.P., Baldwin, R.L., Bruce, C., and Wisnieski, B.J. (1989a). Second cytotoxic pathway of diphtheria toxin suggested by nuclease activity. Science *246*, 1165–1168.

Chang, M.P., Bramhall, J., Graves, S., Bonavida, B., and Wisnieski, B.J. (1989b). Internucleosomal DNA cleavage precedes diphtheria toxin-induced cytolysis. Evidence that cell lysis is not a simple consequence of translation inhibition. J. Biol. Chem. *264*, 15261–15267.

Chau, T.A., McCully, M.L., Brintnell, W., An, G., Kasper, K.J., Vines, E.D., Kubes, P., Haeryfar, S.M., McCormick, J.K., Cairns, E., *et al.* (2009). Toll-like receptor 2 ligands on the staphylococcal cell wall downregulate superantigen-induced T cell activation and prevent toxic shock syndrome. Nat. Med. *15*, 641–648.

Chen, G., Sordillo, E.M., Ramey, W.G., Reidy, J., Holt, P.R., Krajewski, S., Reed, J.C., Blaser, M.J., and Moss, S.F. (1997). Apoptosis in gastric epithelial cells is induced by *Helicobacter pylori* and accompanied by increased expression of BAK. Biochem. Biophys. Res. Commun. *239*, 626–632.

Chen, L.M., Kaniga, K., and Galán, J.E. (1996a). *Salmonella* spp. are cytotoxic for cultured macrophages. Mol. Microbiol. *21*, 1101–1115.

Chen, Y., Smith, M.R., Thirumalai, K., and Zychlinsky, A. (1996b). A bacterial invasin induces macrophage apoptosis by binding directly to ICE. EMBO J. *15*, 3853–3860.

Cheng, Y., Chaturvedi, R., Asim, M., Bussière, F.I., Scholz, A., XU, H., Casero, R.A., Jr., and Wilson, K.T. (2005). *Helicobacter pylori*-induced macrophage apoptosis requires activation of ornithine decarboxylase by c-Myc. J. Biol. Chem. *280*, 22492–22496.

Cherla, R.P., Lee, S.Y., and Tesh, V.L. (2003). Shiga toxins and apoptosis. FEMS Microbiol. Lett. *228*, 159–166.

Cheung, G.Y., Kelly, S.M., Jess, T.J., Prior, S., Price, N.C., Parton, R., and Coote, J.G. (2009). Functional and structural studies on different forms of the adenylate cyclase toxin of *Bordetella pertussis*. Microb. Pathog. *46*, 36–42.

Cheung, K.J., Jr., Li, G., Urban, T.A., Goldberg, J.B., Griffith, A., Lu, F., and Burns, J.L. (2007). Pilus-mediated epithelial cell death in response to infection with *Burkholderia cenocepacia*. Microbes Infect. *9*, 829–837.

Ching, J.C., Jones, N.L., Ceponis, P.J., Karmali, M.A., and Sherman, P.M. (2002). *Escherichia coli* Shiga-like toxins induce apoptosis and cleavage of poly(ADP-ribose) polymerase via *in vitro* activation of caspases. Infect. Immun. *70*, 4669–4677.

Choi, K.S., Park, J.T., and Dumler, J.S. (2005). *Anaplasma phagocytophilum* delay of neutrophil apoptosis through the p38 mitogen-activated protein kinase signal pathway. Infect. Immun. *73*, 8209–8218.

Christofferson, D.E., and Yuan, J. (2010). Necroptosis as an alternative form of programmed cell death. Curr. Opin. Cell. Biol. *22*, 263–268.

Chung, J.W., Piao, Z.H., Yoon, S.R., Kim, M.S., Jeong, M., Lee, S.H., Min, J.K., Kim, J.W., Cho, Y.H., Kim, J.C., Ahn, *et al.* (2009). *Pseudomonas aeruginosa* eliminates natural killer cells via phagocytosis-induced apoptosis. PLoS Pathog. *5*, e1000561.

Ciaramella, A., Martino, A., Cicconi, R., Colizzi, V., and Fraziano, M. (2000). Mycobacterial 19-kDa lipoprotein mediates *Mycobacterium tuberculosis*-induced apoptosis in monocytes/macrophages at early stages of infection. Cell Death Differ. *7*, 1270–1272.

Ciaramella, A., Cavone, A., Santucci, M.B., Amicosante, M., Martino, A., Auricchio, G., Pucillo, L.P., Colizzi, V., and Fraziano, M. (2002). Pro-inflammatory cytokines in the course of *Mycobacterium tuberculosis*-induced apoptosis in monocytes/macrophages. J. Infect. Dis. *186*, 1277–1282.

Clark, C.S., and Maurelli, A.T. (2007). *Shigella flexneri* inhibits staurosporine-induced apoptosis in epithelial cells. Infect. Immun. *75*, 2531–2539.

Clifton, D.R., Goss, R.A., Sahni, S.K., van Antwerp, D., Baggs, R.B., Marder, V.J., Silverman, D.J., and Sporn, L.A. (1998). NF-κB-dependent inhibition of apoptosis is essential for host cell survival during *Rickettsia rickettsii* infection. Proc. Natl. Acad. Sci. U.S.A. *95*, 4646–4651.

Colamussi, M.L., White, M.R., Crouch, E., and Hartshorn, K.L. (1999). Influenza A virus accelerates neutrophil apoptosis and markedly potentiates apoptotic effects of bacteria. Blood *93*, 2395–2403.

Colino, J., and Snapper, C.M. (2003). Two distinct mechanisms for induction of dendritic cell apoptosis in response to intact *Streptococcus pneumoniae*. J. Immunol. *171*, 2354–2365.

Collier-Hyams, L.S., Zeng, H., Sun, J., Tomlinson, A.D., Bao, Z.Q., Chen, H., Madara, J.L., Orth, K., and Neish, A.S. (2002). Cutting edge: *Salmonella* AvrA effector inhibits the key pro-inflammatory, anti-apoptotic NF-κB pathway. J. Immunol. *169*, 2846–2850.

Colotta, F., Re, F., Polentarutti, N., Sozzani, S., and Mantovani, A. (1992). Modulation of granulocyte survival and programmed cell death by cytokines and bacterial products. Blood *80*, 2012–2020.

Corda, D., and Di Girolamo, M. (2003). Functional aspects of protein mono-ADP-ribosylation. EMBO J. *22*, 1953–1958.

Cornelis, G.R. (2006). The type III secretion injectisome. Nat. Rev. Microbiol. *4*, 811–825.

Cortes, G., and Wessels, M.R. (2009). Inhibition of dendritic cell maturation by group A *Streptococcus*. J. Infect. Dis. *200*, 1152–1161.

Cortes-Bratti, X., Karlsson, C., Lagergård, T., Thelestam, M., and Frisan, T. (2001). The *Haemophilus ducreyi* cytolethal distending toxin induces cell cycle arrest and apoptosis via the DNA damage checkpoint pathways. J. Biol. Chem. *276*, 5296–5302.

Coutanceau, E., Marsollier, L., Brosch, R., Perret, E., Goossens, P., Tanguy, M., Cole, S.T., Small, P.L., and Demangel, C. (2005). Modulation of the host immune response by a transient intracellular stage of *Mycobacterium ulcerans*: the contribution of endogenous mycolactone toxin. Cell. Microbiol. *7*, 1187–1196.

Cover, T.L., Krishna, U.S., Israel, D.A., and Peek, R.M., Jr. (2003). Induction of gastric epithelial cell apoptosis by *Helicobacter pylori* vacuolating cytotoxin. Cancer Res. *63*, 951–957.

Coxon, A., Rieu, P., Barkalow, F.J., Askari, S., Sharpe, A.H., von Adrian, U.H., Arnaout, M.A., and Mayadas, T.N. (1996). A novel role for the β2 integrin CD11b/CD18 in neutrophil apoptosis: ahomeostatic mechanism in inflammation. Immunity *5*, 653–666.

Crane, J.K., Majumdar, S., and Pickhardt, D.F., 3rd, (1999). Host cell death due to enteropathogenic *Escherichia coli* has features of apoptosis. Infect. Immun. *67*, 2575–2584.

Creydt, V.P., Silberstein, C., Zotta, E., and Ibarra, C. (2006). Cytotoxic effect of Shiga toxin-2 holotoxin and its B subunit on human renal tubular epithelial cells. Microbes Infect. *8*, 410–419.

Cudd, L.A., Ownby, C.L., Clarke, C.R., Sun, Y., and Clinkenbeard, K.D. (2001). Effects of *Mannheimia haemolytica* leukotoxin on apoptosis and oncosis of bovine neutrophils. Am. J. Vet. Res. *62*, 136–141.

Cywes Bentley C., Hakansson, A., Christianson, J., and Wessels, M.R. (2005). Extracellular group A *Streptococcus* induces keratinocyte apoptosis by dysregulating calcium signalling. Cell. Microbiol. *7*, 945–955.

Dacheux, D., Attree, I., Schneider, C., and Toussaint, B. (1999). Cell death of human polymorphonuclear neutrophils induced by a *Pseudomonas aeruginosa* cystic fibrosis isolate requires a functional type III secretion system. Infect. Immun. *67*, 6164–6167.

Dacheux, D., Toussaint, B., Richard, M., Brochier, G., Croizé, J., and Attree, I. (2000). *Pseudomonas aeruginosa* cystic fibrosis isolates induce rapid, type III secretion-dependent, but ExoU-independent, oncosis of macrophages and polymorphonuclear neutrophils. Infect. Immun. *68*, 2916–2924.

Danelishvili, L., McGarvey, J., Li, Y.J., and Bermudez, L.E. (2003). *Mycobacterium tuberculosis* infection causes different levels of apoptosis and necrosis in human macrophages and alveolar epithelial cells. Cell. Microbiol. *5*, 649–660.

Danial, N.N., and Korsmeyer, S.J. (2004). Cell death: critical control points. Cell *116*, 205–219.

Dauwalder, O., Pachot, A., Cazalis, M.A., Paye, M., Faudot, C., Badiou, C., Mougin, B., Vandenesch, F., Etienne, J., Lina, G., *et al.* (2009). Early kinetics of the transcriptional response of human leukocytes to staphylococcal superantigenic enterotoxins A and G. Microb. Pathog. *47*, 171–176.

Deghmane, A.E., Veckerlé, C., Giorgini, D., Hong, E., Ruckly, C., and Taha, M.K. (2009). Differential modulation of TNF-α-induced apoptosis by *Neisseria meningitidis*. PLoS Pathog. *5*, e1000405.

DeLeo, F.R. (2004). Modulation of phagocyte apoptosis by bacterial pathogens. Apoptosis *9*, 399–413.

Dellacasagrande, J., Capo, C., Raoult, D., and Mege, J.L. (1999). IFN-γ-mediated control of *Coxiella burnetii* survival in monocytes: the role of cell apoptosis and TNF.J. Immunol. *162*, 2259–2265.

Dellacasagrande, J., Ghigo, E., Raoult, D., Capo, C., and Mege, J.L. (2002). IFN-γ-induced apoptosis and microbicidal activity in monocytes harboring the intracellular bacterium *Coxiella burnetii* require membrane TNF and homotypic cell adherence. J. Immunol. *169*, 6309–6315.

Denecker, G., Declercq, W., Geuijen, C.A., Boland, A., Benabdillah, R., van Gurp, M., Sory, M.P., Vandenabeele, P., and Cornelis, G.R. (2001). *Yersinia enterocolitica* YopP-induced apoptosis of macrophages involves the apoptotic signaling cascade upstream of Bid. J. Biol. Chem. *276*, 19706–19714.

Ding, S.Z., Minohara, Y., Fan, X.J., Wang, J., Reyes, V.E., Patel, J., Dirden-Kramer, B., Boldogh, I., Ernst, P.B., and Crowe, S.E. (2007). *Helicobacter pylori* infection induces oxidative stress and programmed cell death in human gastric epithelial cells. Infect. Immun. *75*, 4030–4039.

Dockrell, D.H., and Whyte, M.K. (2006). Regulation of phagocyte lifespan in the lung during bacterial infection. J. Leukoc. Biol. *79*, 904–908.

Dockrell, D.H., Lee, M., Lynch, D.H., and Read, R.C. (2001). Immune-mediated phagocytosis and killing of *Streptococcus pneumoniae* are associated with direct and bystander macrophage apoptosis. J. Infect. Dis. *184*, 713–722.

Dockrell, D.H., Marriott, H.M., Prince, L.R., Ridger, V.C., Ince, P.G., Hellewell, P.G., and Whyte, M.K. (2003). Alveolar macrophage apoptosis contributes to pneumococcal clearance in a resolving model of pulmonary infection. J. Immunol. *171*, 5380–5388.

Dong, F., Pirbhai, M., Xiao, Y., Zhong, Y., Wu, Y., and Zhong, G. (2005). Degradation of the proapoptotic proteins Bik, Puma, and Bim with Bcl-2 domain 3 homology in *Chlamydia trachomatis*-infected cells. Infect. Immun. 73, 1861–1864.

Drysdale, M., Olson, G., Koehler, T.M., Lipscomb, M.F., and Lyons, C.R. (2007). Murine innate immune response to virulent toxigenic and nontoxigenic *Bacillus anthracis* strains. Infect. Immun. *75*, 1757–1764.

Duan, L., Gan, H., Arm, J., and Remold, H.G. (2001). Cytosolic phospholipase A2 participates with TNF-α in the induction of apoptosis of human macrophages infected with *Mycobacterium tuberculosis* H37Ra. J. Immunol. *166*, 7469–7476.

Duan, L., Gan, H., Golan, D.E., and Remold, H.G. (2002). Critical role of mitochondrial damage in determining outcome of macrophage infection with *Mycobacterium tuberculosis*. J. Immunol. *169*, 5181–5187.

Durrbaum-Landmann, I., Gercken, J., Flad, H.D., and Ernst, M. (1996). Effect of *in vitro* infection of human monocytes with low numbers of *Mycobacterium tuberculosis* bacteria on monocyte apoptosis. Infect. Immun. *64*, 5384–5389.

Engelich, G., White, M., and Hartshorn, K.L. (2001). Neutrophil survival is markedly reduced by incubation with influenza virus and *Streptococcus pneumoniae*: role of respiratory burst. J. Leukoc. Biol. *69*, 50–56.

Epelman, S., Neely, G.G., Ma, L.L., Gjomarkaj, M., Pace, E., Melis, M., Woods, D.E., and Mody, C.H. (2002). Distinct fates of monocytes and T cells directly activated by *Pseudomonas aeruginosa* exoenzyme S.J. Leukoc. Biol. *71*, 458–468.

Erfurth, S.E., Gröbner, S., Kramer, U., Gunst, D.S., Soldanova, I., Schaller, M., Autenrieth, I.B., and Borgmann, S. (2004). *Yersinia enterocolitica* induces apoptosis and inhibits surface molecule expression and cytokine production in murine dendritic cells. Infect. Immun. *72*, 7045–7054.

Erwig, L.P., and Henson, P.M. (2008). Clearance of apoptotic cells by phagocytes. Cell Death Differ. *15*, 243–250.

Essmann, F., Bantel, H., Totzke, G., Engels, I.H., Sinha, B., Schulze-Osthoff, K., and Jänicke, R.U. (2003). *Staphylococcus aureus* α-toxin-induced cell death: predominant necrosis despite apoptotic caspase activation. Cell Death Differ. *10*, 1260–1272.

Faherty, C.S., and Maurelli, A.T. (2009). Spa15 of *Shigella flexneri* is secreted through the type III secretion system and prevents staurosporine-induced apoptosis. Infect. Immun. *77*, 5281–5290.

Fairbairn, I.P., Stober, C.B., Kumararatne, D.S., and Lammas, D.A. (2001). ATP-mediated killing of intracellular mycobacteria by macrophages is a P2X(7)-dependent process inducing bacterial death by phagosome–lysosome fusion. J. Immunol. *167*, 3300–3307.

Falcon, R.M., Carvalho, H.F., Joazeiro, P.P., Gatti, M.S., and Yano, T. (2001). Induction of apoptosis in HT29 human intestinal epithelial cells by the cytotoxic enterotoxin of *Aeromonas hydrophila*. Biochem. Cell Biol. *79*, 525–531.

Fan, T., Lu, H., Hu, H., Shi, L., McClarty, G.A., Nance, D.M., Greenberg, A.H., and Zhong, G. (1998). Inhibition of apoptosis in *Chlamydia*-infected cells: blockade of mitochondrial cytochrome c release and caspase activation. J. Exp. Med. *187*, 487–496.

Feng, C.G., Kullberg, M.C., Jankovic, D., Cheever, A.W., Caspar, P., Coffman, R.L., and Sher, A. (2002). Transgenic mice expressing human interleukin-10 in the antigen-presenting cell compartment show increased susceptibility to infection with *Mycobacterium avium* associated with decreased macrophage effector function and apoptosis. Infect. Immun. *70*, 6672–6679.

Fernandez-Prada, C., Tall, B.D., Elliott, S.E., Hoover, D.L., Nataro, J.P., and Venkatesan, M.M. (1998). Hemolysin-positive enteroaggregative and cell-detaching *Escherichia coli* strains cause oncosis of human monocyte-derived macrophages and apoptosis of murine J774 cells. Infect. Immun. *66*, 3918–3924.

Fernandez-Prada, C.M., Hoover, D.L., Tall, B.D., and Venkatesan, M.M. (1997). Human monocyte-derived macrophages infected with virulent *Shigella flexneri in vitro* undergo a rapid cytolytic event similar to oncosis but not apoptosis. Infect. Immun. *65*, 1486–1496.

Fernandez-Prada, C.M., Zelazowska, E.B., Nikolich, M., Hadfield, T.L., Roop, R.M., 2rd., Robertson, G.L., and Hoover, D.L. (2003). Interactions between *Brucella melitensis* and human phagocytes: bacterial surface O-Polysaccharide inhibits phagocytosis, bacterial killing, and subsequent host cell apoptosis. Infect. Immun. *71*, 2110–2119.

Feterowski, C., Weighardt, H., Emmanuilidis, K., Hartung, T., and Holzmann, B. (2001). Immune protection against septic peritonitis in endotoxin-primed mice is related to reduced neutrophil apoptosis. Eur. J. Immunol. *31*, 1268–1277.

Fettucciari, K., Rosati, E., Scaringi, L., Cornacchione, P., Migliorati, G., Sabatini, R., Fetriconi, I., Rossi, R., and Marconi, P. (2000). Group B *Streptococcus* induces apoptosis in macrophages. J. Immunol. *165*, 3923–3933.

Fettucciari, K., Fetriconi, I., Mannucci, R., Nicoletti, I., Bartoli, A., Coaccioli, S., and Marconi, P. (2006). Group B *Streptococcus* induces macrophage apoptosis by calpain activation. J. Immunol. *176*, 7542–7556.

Fink, S.L., and Cookson, B.T. (2005). Apoptosis, pyroptosis, and necrosis: mechanistic description of dead and dying eukaryotic cells. Infect. Immun. *73*, 1907–1916.

Fink, S.L., and Cookson, B.T. (2006). Caspase-1-dependent pore formation during pyroptosis leads to osmotic lysis of infected host macrophages. Cell. Microbiol. *8*, 1812–1825.

Fink, S.L., Bergsbaken, T., and Cookson, B.T. (2008). Anthrax lethal toxin and *Salmonella* elicit the common cell death pathway of caspase-1-dependent pyroptosis via distinct mechanisms. Proc. Natl. Acad. Sci. U.S.A. *105*, 4312–4317.

Fiorentini, C., Fabbri, A., Matarrese, P., Falzano, L., Boquet, P., and Malorni, W. (1997). Hinderance of apoptosis and phagocytic behaviour induced by *Escherichia coli* cytotoxic necrotizing factor 1: two related activities in epithelial cells. Biochem. Biophys. Res. Commun. *241*, 341–346.

Fiorentini, C., Fabbri, A., Falzano, L., Fattorossi, A., Matarrese, P., Rivabene, R., and Donelli, G. (1998a). *Clostridium difficile* toxin B induces apoptosis in intestinal cultured cells. Infect. Immun. *66*, 2660–2665.

Fiorentini, C., Matarrese, P., Straface, E., Falzano, L., Donelli, G., Boquet, P., and Malorni, W. (1998b). Rho-dependent cell spreading activated by *E.coli* cytotoxic necrotizing factor 1 hinders apoptosis in epithelial cells. Cell Death Differ. *5*, 921–929.

Fiorentini, C., Falzano, L., Travaglione, S., and Fabbri, A. (2003). Hijacking Rho GTPases by protein toxins and apoptosis: molecular strategies of pathogenic bacteria. Cell Death Differ. *10*, 147–152.

Fischer, S.F., Vier, J., Kirschnek, S., Klos, A., Hess, S., Ying, S., and Häcker, G. (2004). *Chlamydia inhibit* host cell apoptosis by degradation of proapoptotic BH3-only proteins. J. Exp. Med. *200*, 905–916.

Fischer, S.F., Vier, J., Müller-Thomas, C., and Häcker, G. (2006). Induction of apoptosis by *Legionella pneumophila* in mammalian cells requires the mitochondrial pathway for caspase activation. Microbes Infect. *8*, 662–669.

Fischer, U., Jänicke, R.U., and Schulze-Osthoff, K. (2003). Many cuts to ruin: a comprehensive update of caspase substrates. Cell Death. Differ. *10*, 76–100.

Flannagan, R.S., Cosio, G., and Grinstein, S. (2009). Antimicrobial mechanisms of phagocytes and bacterial evasion strategies. Nat. Rev. Microbiol. *7*, 355–366.

Flynn, A.N., and Buret, A.G. (2008). Caspases-3, -8, and -9 are required for induction of epithelial cell apoptosis by enteropathogenic *E. coli* but are dispensable for increased paracellular permeability. Microb. Pathog. *44*, 311–319.

Follows, S.A., Murlidharan, J., Massari, P., Wetzler, L.M., and Genco, C.A. (2009). *Neisseria gonorrhoeae* infection protects human endocervical epithelial cells from apoptosis via expression of host antiapoptotic proteins. Infect. Immun. *77*, 3602–3610.

Forsberg, M., Blomgran, R., Lerm, M., Särndahl, E., Sebti, S.M., Hamilton, A., Stendahl, O., and Zheng, L. (2003). Differential effects of invasion by and phagocytosis of *Salmonella typhimurium* on apoptosis in human macrophages: potential role of Rho-GTPases and Akt. J. Leukoc. Biol. *74*, 620–629.

Franchi, L., Amer, A., Body-Malapel, M., Kanneganti, T.D., Ozören, N., Jagirdar, R., Inohara, N., Vandenabeele, P., Bertin, J., Coyle, A., Grant, E.P., and Núñez, G. (2006). Cytosolic flagellin requires Ipaf for activation of caspase-1 and interleukin 1β in *Salmonella*-infected macrophages. Nat. Immunol. *7*, 576–582.

Franchi, L., Stoolman, J., Kanneganti, T.D., Verma, A., Ramphal, R., and Núñez, G. (2007). Critical role for Ipaf in *Pseudomonas aeruginosa*-induced caspase-1 activation. Eur. J. Immunol. *37*, 3030–3039.

Franchi, L., Eigenbrod, T., Muñoz-Planillo, R., and Núñez, G. (2009). The inflammasome: a caspase-1-activation platform that regulates immune responses and disease pathogenesis. Nat. Immunol. *10*, 241–247.

François, M., Le,Cabec, V, Dupont, M.A., Sansonetti, P.J., and Maridonneau-Parini, I. (2000). Induction of necrosis in human neutrophils by *Shigella flexneri* requires type III secretion, IpaB and IpaC invasins, and actin polymerization. Infect. Immun. *68*, 1289–1296.

Fraser, J.D., and Proft, T. (2008). The bacterial superantigen and superantigen-like proteins. Immunol. Rev. *225*, 226–243.

Fraser, S.A., de, H.L., Hearn, A.R., Bone, H.K., Salmond, R.J., Rivett, A.J., Williams, N.A., and Hirst, T.R. (2003). Mutant *Escherichia coli* heat-labile toxin B subunit that separates toxoid-mediated signaling and immunomodulatory action from trafficking and delivery functions. Infect. Immun. *71*, 1527–1537.

Fratazzi, C., Arbeit, R.D., Carini, C., and Remold, H.G. (1997). Programmed cell death of *Mycobacterium avium* serovar 4-infected human macrophages prevents the mycobacteria from spreading and induces mycobacterial growth inhibition by freshly added, uninfected macrophages. J. Immunol. *158*, 4320–4327.

Fujii, J., Matsui, T., Heatherly, D.P., Schlegel, K.H., Lobo, P.I., Yutsudo, T., Ciraolo, G.M., Morris, R.E., and Obrig, T. (2003). Rapid apoptosis induced by Shiga toxin in HeLa cells. Infect. Immun. *71*, 2724–2735.

Fujii, J., Wood, K., Matsuda, F., Carneiro-Filho, B.A., Schlegel, K.H., Yutsudo, T., Binnington-Boyd, B., Lingwood, C.A., Obata, F., Kim, K.S., *et al.* (2008). Shiga toxin 2 causes apoptosis in human brain microvascular endothelial cells via C/EBP homologous protein. Infect. Immun. *76*, 3679–3689.

Galdiero, E., Romano, C.C., Vitiello, M., Nuzzo, I., Del Vecchio, E., Bentivoglio, C., Perillo, G., and Galdiero, F. (2000). HSP and apoptosis in leukocytes from infected or vaccinated animals by *Brucella abortus*. New Microbiol. *23*, 271.

Galindo, C.L., Fadl, A.A., Sha, J., Gutierrez, C., Jr., Popov, V.L., Boldogh, I., Aggarwal, B.B., and Chopra, A.K. (2004). *Aeromonas hydrophila* cytotoxic enterotoxin activates mitogen-activated protein kinases and induces apoptosis in murine macrophages and human intestinal epithelial cells. J. Biol. Chem. *279*, 37597–37612.

Galindo, C.L., Fadl, A.A., Sha, J., Pillai, L., Gutierrez, C., Jr., and Chopra, A.K. (2005). Microarray and proteomics analyses of human intestinal epithelial cells treated with the *Aeromonas hydrophila* cytotoxic enterotoxin. Infect. Immun. *73*, 2628–2643.

Galindo, C.L., Gutierrez, C., Jr., and Chopra, A.K. (2006). Potential involvement of galectin-3 and SNAP23 in *Aeromonas hydrophila* cytotoxic enterotoxin-induced host cell apoptosis. Microb. Pathog. *40*, 56–68.

Galluzzi, L., Brenner, C., Morselli, E., Touat, Z., and Kroemer, G. (2008). Viral control of mitochondrial apoptosis. PLoS Pathog. *4*, e1000018.

Galluzzi, L., Aaronson, S.A., Abrams, J., Alnemri, E.S., Andrews, D.W., Baehrecke, E.H., Bazan, N.G., Blagosklonny, M.V., Blomgren, K., Borner, C., *et al.* (2009). Guidelines for the use and interpretation of assays for monitoring cell death in higher eukaryotes. Cell Death. Differ. *16*, 1093–1107.

Gamberale, R., Giordano, M., Trevani, A.S., Andonegui, G., and Geffner, J.R. (1998). Modulation of human neutrophil apoptosis by immune complexes. J. Immunol. *161*, 3666–3674.

Gan, H., Lee, J., Ren, F., Chen, M., Kornfeld, H., and Remold, H.G. (2008). *Mycobacterium tuberculosis* blocks crosslinking of annexin-1 and apoptotic envelope formation on infected macrophages to maintain virulence. Nat. Immunol. *9*, 1189–1197.

Gao, L.Y., and Kwaik, Y.A. (1999a). Activation of caspase 3 during *Legionella pneumophila*-induced apoptosis. Infect. Immun. *67*, 4886–4894.

Gao, L.Y., and Kwaik, Y.A. (1999b). Apoptosis in macrophages and alveolar epithelial cells during early stages of infection by *Legionella pneumophila* and its role in cytopathogenicity. Infect. Immun. *67*, 862–870.

Gao, L., and Kwaik, Y.A. (2000). Hijacking of apoptotic pathwaysby bacterial pathogens. Microbes Infect. *2*, 1705–1719.

Ge, Y., Yoshiie, K., Kuribayashi, F., Lin, M., and Rikihisa, Y. (2005). *Anaplasma phagocytophilum* inhibits human neutrophil apoptosis via upregulation of *bfl-1*, maintenance of mitochondrial membrane potential and prevention of caspase 3 activation. Cell. Microbiol. *7*, 29–38.

Ge, Z., Schauer, D.B., and Fox, J.G. (2008). *In vivo* virulence properties of bacterial cytolethal-distending toxin. Cell. Microbiol. *10*, 1599–1607.

Gelfanova, V., Hansen, E.J., and Spinola, S.M. (1999). Cytolethal distending toxin of *Haemophilus ducreyi* induces apoptotic death of Jurkat T cells. Infect. Immun. *67*, 6394–6402.

Genestier, A.L., Michallet, M.C., Prévost, G., Bellot, G., Chalabreysse, L., Peyrol, S., Thivolet, F., Etienne, J., Lina, G., Vallette, F.M., *et al.* (2005). *Staphylococcus aureus* Panton-Valentine leukocidin directly targets mitochondria and induces Bax-independent apoptosis of human neutrophils. J. Clin. Invest. *115*, 3117–3127.

George, K.M., Pascopella, L., Welty, D.M., and Small, P.L. (2000). A *Mycobacterium ulcerans* toxin, mycolactone, causes apoptosis in guinea pig ulcers and tissue culture cells. Infect. Immun. *68*, 877–883.

Gobert, A.P., Cheng, Y., Wang, J.Y., Boucher, J.L., Iyer, R.K., Cederbaum, S.D., Casero, R.A., Jr., Newton, J.C., and Wilson, K.T. (2002). *Helicobacter pylori* induces macrophage apoptosis by activation of arginase II.J. Immunol. *168*, 4692–4700.

Goldmann, O., Sastalla, I., Wos-Oxley, M., Rohde, M., and Medina, E. (2009). *Streptococcus pyogenes* induces oncosis in macrophages through the activation of an inflammatory programmed cell death pathway. Cell. Microbiol. *11*, 138–155.

Goodyear, C.S., and Silverman, G.J. (2004). Staphylococcal toxin induced preferential and prolonged *in vivo* deletion of innate-like B lymphocytes. Proc. Natl. Acad. Sci. USA *101*, 11392–11397.

Goodyear, C.S., Narita, M., and Silverman, G.J. (2004). *In vivo* VL-targeted activation-induced apoptotic supraclonal deletion by a microbial B cell toxin. J. Immunol. *172*, 2870–2877.

Goodyear, C.S., Corr, M., Sugiyama, F., Boyle, D.L., and Silverman, G.J. (2007). Bim is required for superantigen-mediated B cell death. J. Immunol. *178*, 2636–2640.

Gorak-Stolinska, P., Kemeny, D.M., and Noble, A. (2002). Activation-induced cell death in human T cells is a suicidal process regulated by cell density but superantigen induces T cell fratricide. Cell. Immunol. *219*, 98–107.

Gordon, S. (2008). Macrophages and phagocytosis. In Fundamental Immunology, Paul, W.E.,ed. (Lippincott Williams and Wilkins, Philadelphia), pp. 548–569.

Goth, S.R., and Stephens, R.S. (2001). Rapid, transient phosphatidylserine externalization induced in host cells by infection with *Chlamydia* spp. Infect. Immun. *69*, 1109–1119.

Goto, M., Yamada, T., Kimbara, K., Horner, J., Newcomb, M., Gupta, T.K., and Chakrabarty, A.M. (2003). Induction of apoptosis in macrophages by *Pseudomonas aeruginosa* azurin: tumour-suppressor protein p53 and reactive oxygen species, but not redox activity, as critical elements in cytotoxicity. Mol. Microbiol. *47*, 549–559.

Grant, A.J., Sheppard, M., Deardon, R., Brown, S.P., Foster, G., Bryant, C.E., Maskell, D.J., and Mastroeni, P. (2008). Caspase-3-dependent phagocyte death during systemic *Salmonella enterica* serovar Typhimurium infection of mice. Immunology *125*, 28–37.

Grassmé, H., Kirschnek, S., Riethmueller, J., Riehle, A., vonKünthy, G., Lang, F., Weller, M., and Gulbins, E. (2000). CD95/CD95 ligand interactions on epithelial cells in host defense to *Pseudomonas aeruginosa*. Science *290*, 527–530.

Greene, W., Xiao, Y., Huang, Y., McClarty, G., and Zhong, G. (2004). *Chlamydia*-infected cells continue to undergo mitosis and resist induction of apoptosis. Infect. Immun. *72*, 451–460.

Greub, G., Mege, J.L., and Raoult, D. (2003). *Parachlamydia acanthamoebae* enters and multiplies within human macrophages and induces their apoptosis. Infect. Immun. *71*, 5979–5985.

Gröbner, S., Autenrieth, S.E., Soldanova, I., Gunst, D.S., Schaller, M., Bohn, E., Müller, S., Leverkus, M., Wesselborg, S., Autenrieth, I.B., *et al.* (2006). *Yersinia* YopP-induced apoptotic cell death in murine dendritic cells is partially independent from action of caspases and exhibits necrosis-like features. Apoptosis *11*, 1959–1968.

Gröbner, S., Adkins, I., Schulz, S., Richter, K., Borgmann, S., Wesselborg, S., Ruckdeschel, K., Micheau, O., and Autenrieth, I.B. (2007a). Catalytically active *Yersinia* outer protein P induces cleavage of RIP and caspase-8 at the level of the DISC independently of death receptors in dendritic cells. Apoptosis *12*, 1813–1825.

Gröbner, S., Schulz, S., Soldanova, I., Gunst, D.S., Waibel, M., Wesselborg, S., Borgmann, S., and Autenrieth, I.B. (2007b). Absence of Toll-like receptor 4 signaling results in delayed *Yersinia enterocolitica* YopP-induced cell death of dendritic cells. Infect. Immun. *75*, 512–517.

Gross, A., Terraza, A., Ouahrani-Bettache, S., Liautard, J.P., and Dornand, J. (2000). *In vitro Brucella suis* infection prevents the programmed cell death of human monocytic cells. Infect. Immun. *68*, 342–351.

Guérardel, Y., Maes, E., Briken, V., Chirat, F., Leroy, Y., Locht, C., Strecker, G., and Kremer, L. (2003). Lipomannan and lipoarabinomannan from a clinical isolate of *Mycobacterium kansasii*: novel structural features and apoptosis-inducing properties. J. Biol. Chem. *278*, 36637–36651.

Guichon, A., and Zychlinsky, A. (1997). Clinical isolates of *Shigella* species induce apoptosis in macrophages. J. Infect. Dis. *175*, 470–473.

Guthrie, L.A., McPhail, L.C., Henson, P.M., and Johnston, R.B., Jr. (1984). Priming of neutrophils for enhanced release of oxygen metabolites by bacterial lipopolysaccharide. Evidence for increased activity of the superoxide-producing enzyme. J. Exp. Med. *160*, 1656–1671.

Gutiérrez-Pabello, J.A., McMurray, D.N., and Adams, L.G. (2002). Upregulation of thymosin β–10 by *Mycobacterium bovis* infection of bovine macrophages is associated with apoptosis. Infect. Immun. *70*, 2121–2127.

Gutiérrez, M.G., Master, S.S., Singh, S.B., Taylor, G.A., Colombo, M.I., and Deretic, V. (2004). Autophagy is a defense mechanism inhibiting BCG and *Mycobacterium tuberculosis* survival in infected macrophages. Cell *119*, 753–766.

Guzik, K., Bzowska, M., Dobrucki, J., and Pryjma, J. (1999). Heat-shocked monocytes are resistant to *Staphylococcus aureus*-induced apoptotic DNA fragmentation due to expression of HSP72. Infect. Immun. *67*, 4216–4222.

Guzmán, C.A., Domann, E., Rohde, M., Bruder, D., Darji, A., Weiss, S., Wehland, J., Chakraborty, T., and Timmis, K.N. (1996). Apoptosis of mouse dendritic cells is triggered by listeriolysin, the major virulence determinant of *Listeria monocytogenes*. Mol. Microbiol. *20*, 119–126.

Haase, R., Kirschning, C.J., Sing, A., Schröttner, P., Fukase, K., Kusumoto, S., Wagner, H., Heesemann, J., and

Ruckdeschel, K. (2003). A dominant role of Toll-like receptor 4 in the signaling of apoptosis in bacteria-faced macrophages. J. Immunol. *171*, 4294–4303.

Häcker, H., Fürmann, C., Wagner, H., and Häcker, G. (2002). Caspase-9/-3 activation and apoptosis are induced in mouse macrophages upon ingestion and digestion of *Escherichia coli* bacteria. J. Immunol. *169*, 3172–3179.

Hägele, S., Hacker, J., and Brand, B.C. (1998). *Legionella pneumophila* kills human phagocytes but not protozoan host cells by inducing apoptotic cell death. FEMS Microbiol. Lett. *169*, 51–58.

Haimovich, B., and Venkatesan, M.M. (2006). *Shigella* and *Salmonella*: death as a means of survival. Microbes Infect. *8*, 568–577.

Hall, R.E., Agarwal, S., and Kestler, D.P. (2000). Induction of leukemia cell differentiation and apoptosis by recombinant P48, a modulin derived from *Mycoplasma fermentans*. Biochem. Biophys. Res. Commun. *269*, 284–289.

Hammamieh, R., Ribot, W.J., Abshire, T.G., Jett, M., and Ezzell, J. (2008). Activity of the *Bacillus anthracis* 20 kDa protective antigen component. BMC. Infect. Dis. *8*, 124.

Harder, J., Franchi, L., Muñoz-Planillo, R., Park, J.H., Reimer, T., and Núñez, G. (2009). Activation of the Nlrp3 inflammasome by *Streptococcus pyogenes* requires streptolysin O and NF-κB activation but proceeds independently of TLR signaling and P2X7 receptor. J. Immunol. *183*, 5823–5829.

Harrison, L.M., Cherla, R.P., van den Hoogen, C., van Haaften, W.C., Lee, S.Y., and Tesh, V.L. (2005). Comparative evaluation of apoptosis induced by Shiga toxin 1 and/or lipopolysaccharides in human monocytic and macrophage-like cells. Microb. Pathog. *38*, 63–76.

Haslett, C., Savill, J., and Meagher, L. (1990). Macrophage recognition of senescent granulocytes. Biochem. Soc. Trans. *18*, 225–227.

Haslinger, B., Strangfeld, K., Peters, G., Schulze-Osthoff, K., and Sinha, B. (2003). *Staphylococcus aureus* α-toxin induces apoptosis in peripheral blood mononuclear cells: role of endogenous tumour necrosis factor-α and the mitochondrial death pathway. Cell. Microbiol. *5*, 729–741.

Haslinger-Löffler, B., Wagner, B., Brück, M., Strangfeld, K., Grundmeier, M., Fischer, U., Volker, W., Peters, G., Schulze-Osthoff, K., and Sinha, B. (2006). *Staphylococcus aureus* induces caspase-independent cell death in human peritoneal mesothelial cells. Kidney Int. *70*, 1089–1098.

Hathaway, L.J., Griffin, G.E., Sansonetti, P.J., and Edgeworth, J.D. (2002). Human monocytes kill *Shigella flexneri* but then die by apoptosis associated with suppression of pro-inflammatory cytokine production. Infect. Immun. *70*, 3833–3842.

Hauser, A.R., (2009). The type III secretion system of *Pseudomonas aeruginosa*: infection by injection. Nat. Rev. Microbiol. 7, 654–665.

Hauser, A.R., and Engel, J.N. (1999). *Pseudomonas aeruginosa* induces type-III-secretion-mediated apoptosis of macrophages and epithelial cells. Infect. Immun. *67*, 5530–5537.

Hayashi, T., Catanzaro, A., and Rao, S.P. (1997). Apoptosis of human monocytes and macrophages by *Mycobacterium avium* sonicate. Infect. Immun. *65*, 5262–5271.

Heine, K., Pust, S., Enzenmüller, S., and Barth, H. (2008). ADP-ribosylation of actin by the *Clostridium botulinum* C2 toxin in mammalian cells results in delayed caspase-dependent apoptotic cell death. Infect. Immun. *76*, 4600–4608.

Henry, T., Brotcke, A., Weiss, D.S., Thompson, L.J., and Monack, D.M. (2007). Type I interferon signaling is required for activation of the inflammasome during *Francisella* infection. J. Exp. Med. *204*, 987–994.

Hensler, M.E., Miyamoto, S., and Nizet, V. (2008). Group B streptococcal β-hemolysin/cytolysin directly impairs cardiomyocyte viability and function. PLoS One *3*, e2446.

Hernandez, M.O., Neves, I., Sales, J.S., Carvalho, D.S., Sarno, E.N., and Sampaio, E.P. (2003). Induction of apoptosis in monocytes by *Mycobacterium leprae in vitro*: a possible role for tumour necrosis factor-α. Immunology *109*, 156–164.

Hersh, D., Monack, D.M., Smith, M.R., Ghori, N., Falkow, S., and Zychlinsky, A. (1999). The *Salmonella* invasin SipB induces macrophage apoptosis by binding to caspase-1. Proc. Natl. Acad. Sci. U.S.A. *96*, 2396–2401.

Hickey, T.E., Majam, G., and Guerry, P. (2005). Intracellular survival of *Campylobacter jejuni* in human monocytic cells and induction of apoptotic death by cytholethal distending toxin. Infect. Immun. *73*, 5194–5197.

Hilbi, H., Chen, Y., Thirumalai, K., and Zychlinsky, A. (1997). The interleukin 1β-converting enzyme, caspase 1, is activated during *Shigella flexneri*-induced apoptosis in human monocyte-derived macrophages. Infect. Immun. *65*, 5165–5170.

Hilbi, H., Moss, J.E., Hersh, D., Chen, Y., Arondel, J., Banerjee, S., Flavell, R.A., Yuan, J., Sansonetti, P.J., and Zychlinsky, A. (1998). *Shigella*-induced apoptosis is dependent on caspase-1 which binds to IpaB.J. Biol. Chem. *273*, 32895–32900.

Hilbi, H., Puro, R.J., and Zychlinsky, A. (2000). Tripeptidyl peptidase II promotes maturation of caspase-1 in *Shigella flexneri*-induced macrophage apoptosis. Infect. Immun. *68*, 5502–5508.

Hildeman, D.A., Zhu, Y., Mitchell, T.C., Bouillet, P., Strasser, A., Kappler, J., and Marrack, P. (2002). Activated T cell death *in vivo* mediated by proapoptotic Bcl-2 family member Bim. Immunity *16*, 759–767.

Hilger, H., Pust, S., von Figura, G., Kaiser, E., Stiles, B.G., Popoff, M.R., and Barth, H. (2009). The long-lived nature of *Clostridium perfringens* iota toxin in mammalian cells induces delayed apoptosis. Infect. Immun. *77*, 5593–5601.

Hiroi, M., Shimojima, T., Kashimata, M., Miyata, T., Takano, H., Takahama, M., and Sakagami, H. (1998). Inhibition by *Porphyromonas gingivalis* LPS of apoptosis induction in human peripheral blood polymorphonuclear leukocytes. Anticancer Res. *18*, 3475–3479.

Hitomi, J., Christofferson, D.E., Ng, A., Yao, J., Degterev, A., Xavier, R.J., and Yuan, J. (2008). Identification of a molecular signaling network that regulates a cellular necrotic cell death pathway. Cell *135*, 1311–1323.

Hofer, M.F., Newell, K., Duke, R.C., Schlievert, P.M., Freed, J.H., and Leung, D.Y. (1996). Differential effects of staphylococcal toxic shock syndrome toxin-1 on B cell apoptosis. Proc. Natl. Acad. Sci. USA *93*, 5425–5430.

Hofman, V., Ricci, V., Mograbi, B., Brest, P., Luciano, F., Boquet, P., Rossi, B., Auberger, P., and Hofman, P. (2001). *Helicobacter pylori* lipopolysaccharide hinders polymorphonuclear leucocyte apoptosis. Lab. Invest. *81*, 375–384.

Hotchkiss, R.S., Dunne, W.M., Swanson, P.E., Davis, C.G., Tinsley, K.W., Chang, K.C., Buchman, T.G., and Karl, I.E. (2001). Role of apoptosis in *Pseudomonas aeruginosa* pneumonia. Science *294*, 1783.

Hotchkiss, R.S., Strasser, A., McDunn, J.E., and Swanson, P.E. (2009). Cell death. N. Engl. J. Med. *361*, 1570–1583.

Hrstka, R., Stulík, J., and Vojtesek, B. (2005). The role of MAPK signal pathways during *Francisella tularensis* LVS infection-induced apoptosis in murine macrophages. Microbes Infect. *7*, 619–625.

Hsu, L.C., Ali, S.R., McGillivray, S., Tseng, P.H., Mariathasan, S., Humke, E.W., Eckmann, L., Powell, J.J., Nizet, V., Dixit, V.M., *et al.* (2008). A NOD2–NALP1 complex mediates caspase-1-dependent IL-1β secretion in response to *Bacillus anthracis* infection and muramyl dipeptide. Proc. Natl. Acad. Sci. U.S.A. *105*, 7803–7808.

Hsu, L.C. Park, J.M., Zhang, K., Luo, J.L., Maeda, S., Kaufman, R.J., Eckmann, L., Guiney, D.G., and Karin, M. (2004). The protein kinase PKR is required for macrophage apoptosis after activation of Toll-like receptor 4. Nature *428*, 341–345.

Huang, Y.T., Jeng, C.R., Cheng, C.H., Chueh, L.L., Liu, J.J., and Pang, V.F. (2003). Morphological and immunological evidence of a unique selective production and endoplasmic reticular accumulation of interleukin-1α in rat peritoneal macrophages induced by *Pseudomonas aeruginosa* exotoxin A. Cell. Immunol. *221*, 143–156.

Hutchison, M.L., Poxton, I.R., and Govan, J.R. (1998). *Burkholderia cepacia* produces a hemolysin that is capable of inducing apoptosis and degranulation of mammalian phagocytes. Infect. Immun. *66*, 2033–2039.

Into, T., Nodasaka, Y., Hasebe, A., Okuzawa, T., Nakamura, J., Ohata, N., and Shibata, K. (2002). Mycoplasmal lipoproteins induce Toll-like receptor 2- and caspases-mediated cell death in lymphocytes and monocytes. Microbiol. Immunol. *46*, 265–276.

Ionin, B., Hammamieh, R., Shupp, J.W., Das, R., Pontzer, C.H., and Jett, M. (2008). Staphylococcal enterotoxin B causes differential expression of Rnd3 and RhoA in renal proximal tubule epithelial cells while inducing actin stress fiber assembly and apoptosis. Microb. Pathog. *45*, 303–309.

Jansson, A.L., Yasmin, L., Warne, P., Downward, J., Palmer, R.H., and Hallberg, B. (2006). Exoenzyme S of *Pseudomonas aeruginosa* is not able to induce apoptosis when cells express activated proteins, such as Ras or protein kinase B/Akt. Cell. Microbiol. *8*, 815–822.

Jenkins, C.E., Swiatoniowski, A., Issekutz, A.C., and Lin, T.J. (2004). *Pseudomonas aeruginosa* exotoxin A induces human mast cell apoptosis by a caspase-8 and -3-dependent mechanism. J. Biol. Chem. *279*, 37201–37207.

Jenkins, C.E., Swiatoniowski, A., Power, M.R., and Lin, T.J. (2006). *Pseudomonas aeruginosa*-induced human mast cell apoptosis is associated with up-regulation of endogenous Bcl-xS and down-regulation of Bcl-xL.J. Immunol. *177*, 8000–8007.

Jesenberger, V., Procyk, K.J., Yuan, J., Reipert, S., and Baccarini, M. (2000). *Salmonella*-induced caspase-2 activation in macrophages: a novel mechanism in pathogen-mediated apoptosis. J. Exp. Med. *192*, 1035–1046.

Jesenberger, V., Procyk, K.J., Rüth, J., Schreiber, M., Theussl, H.C., Wagner, E.F., and Baccarini, M. (2001). Protective role of Raf-1 in *Salmonella*-induced macrophage apoptosis. J. Exp. Med. *193*, 353–364.

Jewett, A., Hume, W.R., Le, H., Huynh, T.N., Han, Y.W., Cheng, G., and Shi, W. (2000). Induction of apoptotic cell death in peripheral blood mononuclear and polymorphonuclear cells by an oral bacterium, *Fusobacterium nucleatum*. Infect. Immun. *68*, 1893–1898.

Jeyaseelan, S., Kannan, M.S., Hsuan, S.L., Singh, A.K., Walseth, T.F., and Maheswaran, S.K. (2001). *Pasteurella (Mannheimia) haemolytica* leukotoxin-induced cytolysis of bovine leukocytes: role of arachidonic acid and its regulation. Microb. Pathog. *30*, 59–69.

Jia, J., Alaoui-El-Azher, M., Chow, M., Chambers, T.C., Baker, H., and Jin, S. (2003). c-Jun NH2-terminal kinase-mediated signaling is essential for *Pseudomonas aeruginosa* ExoS-induced apoptosis. Infect. Immun. *71*, 3361–3370.

Jia, J., Wang, Y., Zhou, L., and Jin, S. (2006). Expression of *Pseudomonas aeruginosa* toxin ExoS effectively induces apoptosis in host cells. Infect. Immun. *74*, 6557–6570.

Johannes, L., and Römer, W. (2010). Shiga toxins-from cell biology to biomedical applications. Nat. Rev. Microbiol. *8*, 105–116.

Jonas, D., Schultheis, B., Klas, C., Krammer, P.H., and Bhakdi, S. (1993). Cytocidal effects of *Escherichia coli* hemolysin on human T lymphocytes. Infect. Immun. *61*, 1715–1721.

Jonas, D., Walev, I., Berger, T., Liebetrau, M., Palmer, M., and Bhakdi, S. (1994). Novel path to apoptosis: small transmembrane pores created by staphylococcal α-toxin in T lymphocytes evoke internucleosomal DNA degradation. Infect. Immun. *62*, 1304–1312.

Jones, N.L., Shannon, P.T., Cutz, E., Yeger, H., and Sherman, P.M. (1997). Increase in proliferation and apoptosis of gastric epithelial cells early in the natural history of *Helicobacter pylori* infection. Am. J. Pathol. *151*, 1695–1703.

Jones, N.L., Day, A.S., Jennings, H.A., and Sherman, P.M. (1999). *Helicobacter pylori* induces gastric epithelial cell apoptosis in association with increased Fas receptor expression. Infect. Immun. *67*, 4237–4242.

Joshi, S.G., Francis, C.W., Silverman, D.J., and Sahni, S.K. (2003). Nuclear factor κ B protects against host cell apoptosis during *Rickettsia rickettsii* infection by inhibiting activation of apical and effector caspases and maintaining mitochondrial integrity. Infect. Immun. *71*, 4127–4136.

Joshi, S.G., Francis, C.W., Silverman, D.J., and Sahni, S.K. (2004). NF-κB activation suppresses host cell apoptosis during *Rickettsia rickettsii* infection via regulatory effects on intracellular localization or levels of apoptogenic and anti-apoptotic proteins. FEMS Microbiol. Lett. *234*, 333–341.

Kabelitz, D. and Janssen, O. (1997). Antigen-induced death of T-lymphocytes. Front. Biosci. *2*, d61-d77.

Kanaly, S.T., Nashleanas, M., Hondowicz, B., and Scott, P. (1999). TNF receptor p55 is required for elimination of inflammatory cells following control of intracellular pathogens. J. Immunol. *163*, 3883–3889.

Kasahara, K., Sato, I., Ogura, K., Takeuchi, H., Kobayashi, K., and Adachi, M. (1998). Expression of chemokines and induction of rapid cell death in human blood neutrophils by *Mycobacterium tuberculosis*. J. Infect. Dis. *178*, 127–137.

Kashimoto, T., Ueno, S., Hanajima, M., Hayashi, H., Akeda, Y., Miyoshi, S., Hongo, T., Honda, T., and Susa, N. (2003). *Vibrio vulnificus* induces macrophage apoptosis *in vitro* and *in vivo*. Infect. Immun. *71*, 533–535.

Kashimoto, T., Ueno, S., Ehara, H., Fukudome, S., Komai, M., and Susa, N. (2009). Oligomerization is essential for apoptotic activity of *Vibrio vulnificus* hemolysin. J. Vet. Med. Sci. *71*, 1403–1406.

Kassam, A., Der, S.D., and Mogridge, J. (2005). Differentiation of human monocytic cell lines confers susceptibility to *Bacillus anthracis* lethal toxin. Cell. Microbiol. *7*, 281–292.

Kasuya, S., Nagano, I., Ikeda, T., Goto, C., Shimokawa, K., and Takahashi, Y. (1996). Apoptosis of lymphocytes in mice induced by infection with *Rickettsia tsutsugamushi*. Infect. Immun. *64*, 3937–3941.

Kaufman, M.R., Jia, J., Zeng, L., Ha, U., Chow, M., and Jin, S. (2000). *Pseudomonas aeruginosa* mediated apoptosis requires the ADP-ribosylating activity of exoS. Microbiology *146)*, 2531–2541.

Kausalya, S., Somogyi, R., Orlofsky, A., and Prystowsky, M.B. (2001). Requirement of A1-a for bacillus Calmette-Guerin-mediated protection of macrophages against nitric oxide-induced apoptosis. J. Immunol. *166*, 4721–4727.

Keane, J., Balcewicz-Sablinska, M.K., Remold, H.G., Chupp, G.L., Meek, B.B., Fenton, M.J., and Kornfeld, H. (1997). Infection by *Mycobacterium tuberculosis* promotes human alveolar macrophage apoptosis. Infect. Immun. *65*, 298–304.

Keane, J., Remold, H.G., and Kornfeld, H. (2000). Virulent *Mycobacterium tuberculosis* strains evade apoptosis of infected alveolar macrophages. J. Immunol. *164*, 2016–2020.

Keane, J., Shurtleff, B., and Kornfeld, H. (2002). TNF-dependent BALB/c murine macrophage apoptosis following *Mycobacterium tuberculosis* infection inhibits bacillary growth in an IFN-γ independent manner. Tuberculosis (Edinb.) *82*, 55–61.

Kelk, P., Johansson, A., Claesson, R., Hänstrom, L., and Kalfas, S. (2003). Caspase 1 involvement in human monocyte lysis induced by *Actinobacillus actinomycetemcomitans* leukotoxin. Infect. Immun. *71*, 4448–4455.

Kemp, K., Bruunsgaard, H., Skinhøj, P., and Klarlund Petersen, B. (2002). Pneumococcal infections in humans are associated with increased apoptosis and trafficking of type 1 cytokine-producing T cells. Infect. Immun. *70*, 5019–5025.

Kempf, V.A., Schairer, A., Neumann, D., Grassl, G.A., Lauber, K., Lebiedziejewski, M., Schaller, M., Kyme, P., Wesselborg, S., and Autenrieth, I.B. (2005). *Bartonella henselae* inhibits apoptosis in Mono Mac 6 cells. Cell. Microbiol. *7*, 91–104.

Kennedy, A.D., and DeLeo, F.R. (2009). Neutrophil apoptosis and the resolution of infection. Immunol. Res. *43*,25–61.

Kennedy, C.L., Smith, D.J., Lyras, D., Chakravorty, A., and Rood, J.I. (2009). Programmed cellular necrosis mediated by the pore-forming α-toxin from *Clostridium septicum*. PLoS Pathog. *5*, e1000516.

Kepp, O., Gottschalk, K., Churin, Y., Rajalingam, K., Brinkmann, V., Machuy, N., Kroemer, G., and Rudel, T. (2009). Bim and Bmf synergize to induce apoptosis in *Neisseria gonorrhoeae* infection. PLoS Pathog. *5*, e1000348.

Khelef, N., Gounon, P., and Guiso, N. (2001). Internalization of *Bordetella pertussis* adenylate cyclase-haemolysin into endocytic vesicles contributes to macrophage cytotoxicity. Cell. Microbiol. *3*, 721–730.

Khelef, N., Zychlinsky, A., and Guiso, N. (1993). *Bordetella pertussis* induces apoptosis in macrophages: role of adenylate cyclase-hemolysin. Infect. Immun. *61*, 4064–4071.

Kim, H., Kokkotou, E., Na, X., Rhee, S.H., Moyer, M.P., Pothoulakis, C., and Lamont, J.T. (2005). *Clostridium difficile* toxin A-induced colonocyte apoptosis involves p53-dependent p21(WAF1/CIP1) induction via p38 mitogen-activated protein kinase. Gastroenterology *129*, 1875–1888.

Kim, H., Rhee, S.H., Pothoulakis, C., and Lamont, J.T. (2007). Inflammation and apoptosis in *Clostridium difficile* enteritis is mediated by PGE2 up-regulation of Fas ligand. Gastroenterology *133*, 875–886.

Kim, J.M., Eckmann, L., Savidge, T.C., Lowe, D.C., Witthöft, T., and Kagnoff, M.F. (1998). Apoptosis of human intestinal epithelial cells after bacterial invasion. J. Clin. Invest. *102*, 1815–1823.

Kim, J.S., Kim, J.M., Jung, H.C., Song, I.S., and Kim, C.Y. (2001). Inhibition of apoptosis in human neutrophils by *Helicobacter pylori* water-soluble surface proteins. Scand. J. Gastroenterol. *36*, 589–600.

Kim, M.K., Seong, S.Y., Seoh, J.Y., Han, T.H., Song, H.J., Lee, J.E., Shin, J.H., Lim, B.U., and Kang, J.S. (2002). *Orientia tsutsugamushi* inhibits apoptosis of macrophages by retarding intracellular calcium release. Infect. Immun. *70*, 4692–4696.

Kirby, J.E. (2004). Anthrax lethal toxin induces human endothelial cell apoptosis. Infect. Immun. *72*, 430–439.

Kirby, J.E., and Nekorchuk, D.M. (2002). *Bartonella*-associated endothelial proliferation depends on inhibition of apoptosis. Proc. Natl. Acad. Sci. U.S.A. *99*, 4656–4661.

Kisich, K.O., Higgins, M., Diamond, G., and Heifets, L. (2002). Tumor necrosis factor -α stimulates killing of *Mycobacterium tuberculosis* by human neutrophils. Infect. Immun. *70*, 4591–4599.

Klingler, K., Tchou-Wong, K.M., Brändli, O., Aston, C., Kim, R., Chi, C., and Rom, W.N. (1997). Effects of mycobacteria on regulation of apoptosis in mononuclear phagocytes. Infect. Immun. *65*, 5272–5278.

Klumpp, D.J., Rycyk, M.T., Chen, M.C., Thumbikat, P., Sengupta, S., and Schaeffer, A.J. (2006). Uropathogenic *Escherichia coli* induces extrinsic and intrinsic cascades to initiate urothelial apoptosis. Infect. Immun. *74*, 5106–5113.

Knodler, L.A., Finlay, B.B., and Steele-Mortimer, O. (2005). The *Salmonella* effector protein SopB protects epithelial cells from apoptosis by sustained activation of Akt. J. Biol. Chem. *280*, 9058–9064.

Kobayashi, S.D., Voyich, J.M., Buhl, C.L., Stahl, R.M., and DeLeo, F.R. (2002). Global changes in gene expression by human polymorphonuclear leukocytes during receptor-mediated phagocytosis: cell fate is regulated at the level of gene expression. Proc. Natl. Acad. Sci. USA *99*, 6901–6906.

Kobayashi, S.D., Braughton, K.R., Whitney, A.R., Voyich, J.M., Schwan, T.G., Musser, J.M., and DeLeo, F.R. (2003a). Bacterial pathogens modulate an apoptosis differentiation program in human neutrophils. Proc. Natl. Acad. Sci. USA. *100*, 10948–10953.

Kobayashi, S.D., Voyich, J.M., Braughton, K.R., and DeLeo, F.R. (2003b). Down-regulation of pro-inflammatory capacity during apoptosis in human polymorphonuclear leukocytes. J. Immunol. *170*, 3357–3368.

Kobayashi, S.D., Voyich, J.M., Braughton, K.R., Whitney, A.R., Nauseef, W.M., Malech, H.L., and DeLeo, F.R. (2004). Gene expression profiling provides insight into the pathophysiology of chronic granulomatous disease. J. Immunol. *172*, 636–643.

Kobayashi, S.D., Voyich, J.M., Whitney, A.R., and DeLeo, F.R. (2005). Spontaneous neutrophil apoptosis and regulation of cell survival by granulocyte macrophage-colony stimulating factor. J. Leukoc. Biol. *78*, 1408–1418.

Koch-Nolte, F., Adriouch, S., Bannas, P., Krebs, C., Scheuplein, F., Seman, M., and Haag, F. (2006). ADP-ribosylation of membrane proteins: unveiling the secrets of a crucial regulatory mechanism in mammalian cells. Ann. Med. *38*, 188–199.

Kodama, T., Nagayama, K., Yamada, K., Ohba, Y., Akeda, Y., and Honda, T. (1999). Induction of apoptosis in human renal proximal tubular epithelial cells by *Escherichia coli* verocytotoxin 1 *in vitro*. Med. Microbiol. Immunol. *188*, 73–78.

Kohda, K., Tanaka, K., Aiba, Y., Yasuda, M., Miwa, T., and Koga, Y. (1999). Role of apoptosis induced by *Helicobacter pylori* infection in the development of duodenal ulcer. Gut *44*, 456–462.

Konturek, P.C., Pierzchalski, P., Konturek, S.J., Meixner, H., Faller, G., Kirchner, T., and Hahn, E.G. (1999). *Helicobacter pylori* induces apoptosis in gastric mucosa through an upregulation of Bax expression in humans. Scand. J. Gastroenterol. *34*, 375–383.

Koo, H.M., VanBrocklin, M., McWilliams, M.J., Leppla, S.H., Duesbery, N.S., and Woude, G.F. (2002). Apoptosis and melanogenesis in human melanoma cells induced by anthrax lethal factor inactivation of mitogen-activated protein kinase kinase. Proc. Natl. Acad. Sci. U.S.A. *99*, 3052–3057.

Korostoff, J., Wang, J.F., Kieba, I., Miller, M., Shenker, B.J., and Lally, E.T. (1998). *Actinobacillus actinomycetemcomitans* leukotoxin induces apoptosis in HL-60 cells. Infect. Immun. *66*, 4474–4483.

Korostoff, J., Yamaguchi, N., Miller, M., Kieba, I., and Lally, E.T. (2000). Perturbation of mitochondrial structure and function plays a central role in *Actinobacillus actinomycetemcomitans* leukotoxin-induced apoptosis. Microb. Pathog. *29*, 267–278.

Kozjak-Pavlovic, V., Dian-Lothrop, E.A., Meinecke, M., Kepp, O., Ross, K., Rajalingam, K., Harsman, A., Hauf, E., Brinkmann, V., Günther, D., *et al.* (2009). Bacterial porin disrupts mitochondrial membrane potential and sensitizes host cells to apoptosis. PLoS Pathog. *5*, e1000629.

Kremer, L., Estaquier, J., Brandt, E., Ameisen, J.C., and Locht, C. (1997). *Mycobacterium bovis* Bacillus Calmette Guérin infection prevents apoptosis of resting human monocytes. Eur. J. Immunol. *27*, 2450–2456.

Krocová, Z., Härtlova, A., Souckova, D., Zivna, L., Kroca, M., Rudolf, E., Macela, A., and Stulik, J. (2008). Interaction of B cells with intracellular pathogen *Francisella tularensis*. Microb. Pathog. *45*, 79–85.

Kroemer, G., and Levine, B. (2008). Autophagic cell death: the story of a misnomer. Nat. Rev. Mol. Cell Biol. *9*, 1004–1010.

Kroemer, G., Galluzzi, L., Vandenabeele, P., Abrams, J., Alnemri, E.S., Baehrecke, E.H., Blagosklonny, M.V., El-Deiry, W.S., Golstein, P., Green, D.R., *et al.* (2009). Classification of cell death: recommendations of the Nomenclature Committee on Cell Death 2009. Cell Death Differ. *16*, 3–11.

Krzyminska, S., Kaznowski, A., and Chodysz, M. (2009). *Aeromonas* spp. human isolates induce apoptosis of murine macrophages. Curr. Microbiol. *58*, 252–257.

Kuck, D., Kolmerer, B., Iking-Konert, C., Krammer, P.H., Stremmel, W., and Rudi, J. (2001). Vacuolating cytotoxin of *Helicobacter pylori* induces apoptosis in the human gastric epithelial cell line AGS. Infect. Immun. *69*, 5080–5087.

Kuo, C.F., Wu, J.J., Tsai, P.J., Kao, F.J., Lei, H.Y., Lin, M.T., and Lin, Y.S. (1999). Streptococcal pyrogenic exotoxin B induces apoptosis and reduces phagocytic activity in U937 cells. Infect. Immun. *67*, 126–130.

Labbé, K. and Saleh, M. (2008). Cell death in the host response to infection. Cell Death Differ. *15*, 1339–1349.

Laguna, R.K., Creasey, E.A., Li, Z., Valtz, N., and Isberg, R.R. (2006). A *Legionella pneumophila*-translocated substrate that is required for growth within macrophages and protection from host cell death. Proc. Natl. Acad. Sci. U.S.A. *103*, 18745–18750.

Lai, X.H., and Sjöstedt, A. (2003). Delineation of the molecular mechanisms of *Francisella tularensis*-induced apoptosis in murine macrophages. Infect. Immun. *71*, 4642–4646.

Lai, X.H., Xu, J.G., Melgar, S., and Uhlin, B.E. (1999). An apoptotic response by J774 macrophage cells is common upon infection with diarrheagenic *Escherichia coli*. FEMS Microbiol. Lett. *172*, 29–34.

Lai, X.H., Arencibia, I., Johansson, A., Wai, S.N., Oscarsson, J., Kalfas, S., Sundqvist, K.G., Mizunoe, Y., Sjöstedt, A., and Uhlin, B.E. (2000). Cytocidal and apoptotic effects of the ClyA protein from *Escherichia coli* on primary and cultured monocytes and macrophages. Infect. Immun. *68*, 4363–4367.

Lai, X.H., Golovliov, I., and Sjöstedt, A. (2001). *Francisella tularensis* induces cytopathogenicity and apoptosis in murine macrophages via a mechanism that requires intracellular bacterial multiplication. Infect. Immun. *69*, 4691–4694.

Lara-Tejero, M., and Galán, J.E. (2000). A bacterial toxin that controls cell cycle progression as a deoxyribonuclease I-like protein. Science *290*, 354–357.

Lara-Tejero, M., Sutterwala, F.S., Ogura, Y., Grant, E.P., Bertin, J., Coyle, A.J., Flavell, R.A., and Galán, J.E. (2006). Role of the caspase-1 inflammasome in *Salmonella typhimurium* pathogenesis. J. Exp. Med. *203*, 1407–1412.

Lavoie, P.M., McGrath, H., Shoukry, N.H., Cazenave, P.A., Sékaly, R.P., and Thibodeau, J. (2001). Quantitative relationship between MHC class II–superantigen complexes and the balance of T cell activation versus death. J. Immunol. *166*, 7229–7237.

Lee, B.C., Choi, S.H., and Kim, T.S. (2008a). *Vibrio vulnificus* RTX toxin plays an important role in the apoptotic death of human intestinal epithelial cells exposed to *Vibrio vulnificus*. Microbes Infect. *10*, 1504–1513.

Lee, H.C., and Goodman, J.L. (2006). *Anaplasma phagocytophilum* causes global induction of antiapoptosis in human neutrophils. Genomics *88*, 496–503.

Lee, J., Remold, H.G., Ieong, M.H., and Kornfeld, H. (2006). Macrophage apoptosis in response to high intracellular burden of *Mycobacterium tuberculosis* is mediated by a novel caspase-independent pathway. J. Immunol. *176*, 4267–4274.

Lee, M.S., Cherla, R.P., Leyva-Illades, D., and Tesh, V.L. (2009). Bcl-2 regulates the onset of Shiga toxin 1-induced apoptosis in THP-1 cells. Infect. Immun. *77*, 5233–5244.

Lee, S.Y., Cherla, R.P., Caliskan, I., and Tesh, V.L. (2005). Shiga toxin 1 induces apoptosis in the human myelogenous leukemia cell line THP-1 by a caspase-8-dependent, tumor necrosis factor receptor-independent mechanism. Infect. Immun. *73*, 5115–5126.

Lee, S.Y., Cherla, R.P., and Tesh, V.L. (2007). Simultaneous induction of apoptotic and survival signaling pathways in macrophage-like THP-1 cells by Shiga toxin 1. Infect. Immun. *75*, 1291–1302.

Lee, S.Y., Lee, M.S., Cherla, R.P., and Tesh, V.L. (2008b). Shiga toxin 1 induces apoptosis through the endoplasmic reticulum stress response in human monocytic cells. Cell. Microbiol. *10*, 770–780.

Lehnardt, S., Wennekamp, J., Freyer, D., Liedtke, C., Krueger, C., Nitsch, R., Bechmann, I., Weber, J.R., and Henneke, P. (2007). TLR2 and caspase-8 are essential for group B *Streptococcus*-induced apoptosis in microglia. J. Immunol. *179*, 6134–6143.

Lemaître, N., Sebbane, F., Long, D., and Hinnebusch, B.J. (2006). *Yersinia pestis* YopJ suppresses tumor necrosis factor α induction and contributes to apoptosis of immune cells in the lymph node but is not required for virulence in a rat model of bubonic plague. Infect. Immun. *74*, 5126–5131.

Lengwehasatit, I., Nuchtas, A., Tungpradabkul, S., Sirisinha, S., and Utaisincharoen, P. (2008). Involvement of *B. pseudomallei* RpoS in apoptotic cell death in mouse macrophages. Microb. Pathog. *44*, 238–245.

Levine, B., and Yuan, J. (2005). Autophagy in cell death: an innocent convict? J. Clin. Invest. *115*, 2679–2688.

Levine, B., Sinha, S., and Kroemer, G. (2008). Bcl-2 family members: dual regulators of apoptosis and autophagy. Autophagy *4*, 600–606.

Liang, X., and Ji, Y. (2007). Involvement of $\alpha_5\beta_1$-integrin and TNF-α in *Staphylococcus aureus* α-toxin-induced death of epithelial cells. Cell. Microbiol. *9*, 1809–1821.

Libby, S.J., Lesnick, M., Hasegawa, P., Weidenhammer, E., and Guiney, D.G. (2000). The *Salmonella* virulence plasmid *spv* genes are required for cytopathology in human monocyte-derived macrophages. Cell. Microbiol. *2*, 49–58.

Liberto, M.C., Matera, G., Lamberti, A.G., Barreca, G.S., Quirino, A., and Focá, A. (2003). *In vitro Bartonella quintana* infection modulates the programmed cell death and inflammatory reaction of endothelial cells. Diagn. Microbiol. Infect. Dis *45*, 107–115.

Liberto, M.C., Matera, G., Lamberti, A.G., Barreca, G.S., Foca, D., Quirino, A., Soria, M.R., and Foca, A. (2004). *Bartonella quintana*-induced apoptosis inhibition of human endothelial cells is associated with p38 and SAPK/JNK modulation and with stimulation of mitosis. Diagn. Microbiol. Infect. Dis. *50*, 159–166.

Lightfield, K.L. Persson, J., Brubaker, S.W., Witte, C.E., von Moltke, J., Dunipace, E.A., Henry, T., Sun, Y.H., Cado, D., Dietrich, W.F., *et al.* (2008). Critical function for

Naip5 in inflammasome activation by a conserved carboxy-terminal domain of flagellin. Nat. Immunol. *9*, 1171–1178.

Liles, W.C., Thomsen, A.R., O'Mahony, D.S., and Klebanoff, S.J. (2001). Stimulation of human neutrophils and monocytes by staphylococcal phenol-soluble modulin. J. Leukoc. Biol. *70*, 96–102.

Lilo, S., Zheng, Y., and Bliska, J.B. (2008). Caspase-1 activation in macrophages infected with *Yersinia pestis* KIM requires the type III secretion system effector YopJ. Infect. Immun. *76*, 3911–3923.

Lin, C.G., Kao, Y.T., Liu, W.T., Huang, H.H., Chen, K.C., Wang, T.M., and Lin, H.C. (1996). Cytotoxic effects of anthrax lethal toxin on macrophage-like cell line J774A.1. Curr. Microbiol. *33*, 224–227.

Lin, Y.S., Chen, K.H., Kuo, C.F., Huang, K.J., and Wu, J.J. (1998). Induction of thymocyte apoptosis in mice by *Yersinia enterocolitica* products. J. Med. Microbiol. *47*, 447–454.

Lin, Y.T., Wang, C.T., Hsu, C.T., Wang, L.F., Shau, W.Y., Yang, Y.H., and Chiang, B.L. (2003). Differential susceptibility to staphylococcal superantigen (SsAg)-induced apoptosis of CD4 $^{+}$ T cells from atopic dermatitis patients and healthy subjects: the inhibitory effect of IL-4 on SsAg-induced apoptosis. J. Immunol. *171*, 1102–1108.

Liu, G.Y., Doran, K.S., Lawrence, T., Turkson, N., Puliti, M., Tissi, L., and Nizet, V. (2004). Sword and shield: linked group B streptococcal β-hemolysin/cytolysin and carotenoid pigment function to subvert host phagocyte defense. Proc. Natl. Acad. Sci. USA *101*, 14491–14496.

Liu, J., Akahoshi, T., Sasahana, T., Kitasato, H., Namai, R., Sasaki, T., Inoue, M., and Kondo, H. (1999). Inhibition of neutrophil apoptosis by verotoxin 2 derived from *Escherichia coli* O157:H7. Infect. Immun. *67*, 6203–6205.

Liu, J., He, T., He, Y., Zhang, Z., Akahoshi, T., Kondo, H., and Zhong, S. (2002). Prolongation of functional life-span of neutrophils by recombinant verotoxin 2. Chin. Med. J. (Engl.) *115*, 900–903.

López, M., Sly, L.M., Luu, Y., Young, D., Cooper, H., and Reiner, N.E. (2003). The 19-kDa *Mycobacterium tuberculosis* protein induces macrophage apoptosis through Toll-like receptor-2. J. Immunol. *170*, 2409–2416.

Lotz, S., Aga, E., Wilde, I., van Zandbergen, G., Hartung, T., Solbach, W., and Laskay, T. (2004). Highly purified lipoteichoic acid activates neutrophil granulocytes and delays their spontaneous apoptosis via CD14 and TLR2. J. Leukoc. Biol. *75*, 467–477.

Lührmann, A., and Roy, C.R. (2007). *Coxiella burnetii* inhibits activation of host cell apoptosis through a mechanism that involves preventing cytochrome c release from mitochondria. Infect. Immun. *75*, 5282–5289.

Lührmann, A., Mauder, N., Sydor, T., Fernandez-Mora, E., Schulze-Luehrmann, J., Takai, S., and Haas, A. (2004). Necrotic death of *Rhodococcus equi*-infected macrophages is regulated by virulence-associated plasmids. Infect. Immun. *72*, 853–862.

Lukaszewski, R.A., Kenny, D.J., Taylor, R., Rees, D.G., Hartley, M.G., and Oyston, P.C. (2005). Pathogenesis of *Yersinia pestis* infection in BALB/c mice: effects on host macrophages and neutrophils. Infect. Immun. *73*, 7142–7150.

Lundberg, U., Vinatzer, U., Berdnik, D., von Gabain, A., and Baccarini, M. (1999). Growth phase-regulated induction of *Salmonella*-induced macrophage apoptosis correlates with transient expression of SPI-1 genes. J. Bacteriol. *181*, 3433–3437.

Lundqvist-Gustafsson, H., Norrman, S., Nilsson, J., and Wilsson, A. (2001). Involvement of p38-mitogen-activated protein kinase in *Staphylococcus* aureus-induced neutrophil apoptosis. J. Leukoc. Biol. *70*, 642–648.

McKallip, R.J., Fisher, M., Gunthert, U., Szakal, A.K., Nagarkatti, P.S., and Nagarkatti, M. (2005). Role of CD44 and its v7 isoform in staphylococcal enterotoxin B-induced toxic shock: CD44 deficiency on hepatic mononuclear cells leads to reduced activation-induced apoptosis that results in increased liver damage. Infect. Immun. *73*, 50–61.

Mahida, Y.R., Makh, S., Hyde, S., Gray, T., and Borriello, S.P. (1996). Effect of *Clostridium difficile* toxin A on human intestinal epithelial cells: induction of interleukin 8 production and apoptosis after cell detachment. Gut *38*, 337–347.

Mahida, Y.R., Galvin, A., Makh, S., Hyde, S., Sanfilippo, L., Borriello, S.P., and Sewell, H.F. (1998). Effect of *Clostridium difficile* toxin A on human colonic lamina propria cells: early loss of macrophages followed by T-cell apoptosis. Infect. Immun. *66*, 5462–5469.

Marcatili, A., D'Isanto, M., Vitiello, M., Galdiero, R., and Galdiero, M. (2002). p53 and c-Myc activation by *Pasteurella haemolytica* leukotoxin is correlated with bovine mononuclear cells apoptosis. New Microbiol. *25*, 195–204.

Mariathasan, S., Newton, K., Monack, D.M., Vucic, D., French, D.M., Lee, W.P., Roose-Girma, M., Erickson, S., and Dixit, V.M. (2004). Differential activation of the inflammasome by caspase-1 adaptors ASC and Ipaf. Nature *430*, 213–218.

Mariathasan, S., Weiss, D.S., Dixit, V.M., and Monack, D.M. (2005). Innate immunity against *Francisella tularensis* is dependent on the ASC/caspase–1 axis. J. Exp. Med. *202*, 1043–1049.

Marketon, M.M., DePaolo, R.W., DeBord, K.L., Jabri, B., and Schneewind, O. (2005). Plague bacteria target immune cells during infection. Science *309*, 1739–1741.

Marriott, H.M., and Dockrell, D.H. (2006). *Streptococcus pneumoniae*: the role of apoptosis in host defense and pathogenesis. Int. J. Biochem. Cell Biol. *38*, 1848–1854.

Marriott, H.M., Ali, F., Read, R.C., Mitchell, T.J., Whyte, M.K., and Dockrell, D.H. (2004). Nitric oxide levels regulate macrophage commitment to apoptosis or necrosis during pneumococcal infection. FASEB J. *18*, 1126–1128.

Marriott, H.M., Bingle, C.D., Read, R.C., Braley, K.E., Kroemer, G., Hellewell, P.G., Craig, R.W., Whyte,

M.K., and Dockrell, D.H. (2005). Dynamic changes in Mcl-1 expression regulate macrophage viability or commitment to apoptosis during bacterial clearance. J. Clin. Invest. *115*, 359–368.

Marriott, H.M., Hellewell, P.G., Cross, S.S., Ince, P.G., Whyte, M.K., and Dockrell, D.H. (2006). Decreased alveolar macrophage apoptosis is associated with increased pulmonary inflammation in a murine model of pneumococcal pneumonia. J. Immunol. *177*, 6480–6488.

Marriott, H.M., Hellewell, P.G., Whyte, M.K., and Dockrell, D.H. (2007). Contrasting roles for reactive oxygen species and nitric oxide in the innate response to pulmonary infection with *Streptococcus pneumoniae*. Vaccine *25*, 2485–2490.

Massari, P., Ho, Y., and Wetzler, L.M. (2000). *Neisseria meningitidis* porin PorB interacts with mitochondria and protects cells from apoptosis. Proc. Natl. Acad. Sci. U.S.A. *97*, 9070–9075.

Massari, P., King, C.A., Ho, A.Y., and Wetzler, L.M. (2003). Neisserial PorB is translocated to the mitochondria of HeLa cells infected with *Neisseria meningitidis* and protects cells from apoptosis. Cell. Microbiol. *5*, 99–109.

Matarrese, P., Falzano, L., Fabbri, A., Gambardella, L., Frank, C., Geny, B., Popoff, M.R., Malorni, W., and Fiorentini, C. (2007). *Clostridium difficile* toxin B causes apoptosis in epithelial cells by thrilling mitochondria. Involvement of ATP-sensitive mitochondrial potassium channels. J. Biol. Chem. *282*, 9029–9041.

Matsuda, T., Saito, H., Inoue, T., Fukatsu, K., Lin, M.T., Han, I., Furukawa, S., Ikeda, S., and Muto, T. (1999). Ratio of bacteria to polymorphonuclear neutrophils (PMNs) determines PMN fate. Shock *12*, 365–372.

Matsuura, G., Morinaga, N., Yahiro, K., Komine, R., Moss, J., Yoshida, H., and Noda, M. (2009). Novel subtilase cytotoxin produced by Shiga-toxigenic *Escherichia coli* induces apoptosis in vero cells via mitochondrial membrane damage. Infect. Immun. *77*, 2919–2924.

Matte, I., Lane, D., Côté, E., Asselin, A.E., Fortier, L.C., Asselin, C., and Piché, A. (2009). Antiapoptotic proteins Bcl-2 and Bcl-XL inhibit *Clostridium difficile* toxin A-induced cell death in human epithelial cells. Infect. Immun. *77*, 5400–5410.

Medzhitov, R. (2008). The innate immune system. In Fundamental Immunology, Paul, W.E., ed. (Lippincott Williams and Wilkins, Philadelphia), pp. 428–450.

Menaker, R.J., Ceponis, P.J., and Jones, N.L. (2004). *Helicobacter pylori* induces apoptosis of macrophages in association with alterations in the mitochondrial pathway. Infect. Immun. *72*, 2889–2898.

Menzies, B.E., and Kourteva, I. (1998). Internalization of *Staphylococcus aureus* by endothelial cells induces apoptosis. Infect. Immun. *66*, 5994–5998.

Menzies, B.E., and Kourteva, I. (2000). *Staphylococcus aureus* α-toxin induces apoptosis in endothelial cells. FEMS Immunol. Med. Microbiol. *29*, 39–45.

Merien, F., Baranton, G., and Perolat, P. (1997). Invasion of Vero cells and induction of apoptosis in macrophages by pathogenic *Leptospira interrogans* are correlated with virulence. Infect. Immun. *65*, 729–738.

Merien, F., Truccolo, J., Rougier, Y., Baranton, G., and Perolat, P. (1998). *In vivo* apoptosis of hepatocytes in guinea pigs infected with *Leptospira interrogans* serovar icterohaemorrhagiae. FEMS Microbiol. Lett. *169*, 95–102.

Merrick, J.C., Edelson, B.T., Bhardwaj, V., Swanson, P.E., and Unanue, E.R. (1997). Lymphocyte apoptosis during early phase of *Listeria* infection in mice. Am. J. Pathol. *151*, 785–792.

Miao, E.A., Alpuche-Aranda, C.M., Dors, M., Clark, A.E., Bader, M.W., Miller, S.I., and Aderem, A. (2006). Cytoplasmic flagellin activates caspase-1 and secretion of interleukin 1β via Ipaf. Nat. Immunol. *7*, 569–575.

Miao, E.A., Ernst, R.K., Dors, M., Mao, D.P., and Aderem, A. (2008). *Pseudomonas aeruginosa* activates caspase 1 through Ipaf. Proc. Natl. Acad. Sci. U.S.A. *105*, 2562–2567.

Mills, M., Meysick, K.C., and O'Brien, A.D. (2000). Cytotoxic necrotizing factor type 1 of uropathogenic *Escherichia coli* kills cultured human uroepithelial 5637 cells by an apoptotic mechanism. Infect. Immun. *68*, 5869–5880.

Mills, S.D., Boland, A., Sory, M.P., van der Smissen, P., Kerbourch, C., Finlay, B.B., and Cornelis, G.R. (1997). *Yersinia enterocolitica* induces apoptosis in macrophages by a process requiring functional type III secretion and translocation mechanisms and involving YopP, presumably acting as an effector protein. Proc. Natl. Acad. Sci. U.S.A. *94*, 12638–12643.

Minamino, M., Sakaguchi, I., Naka, T., Ikeda, N., Kato, Y., Tomiyasu, I., Yano, I., and Kobayashi, K. (2003). Bacterial ceramides and sphingophospholipids induce apoptosis of human leukaemic cells. Microbiology *149*, 2071–2081.

Miraglia, A.G., Travaglione, S., Meschini, S., Falzano, L., Matarrese, P., Quaranta, M.G., Viora, M., Fiorentini, C., and Fabbri, A. (2007). Cytotoxic necrotizing factor 1 prevents apoptosis via the Akt/IκB kinase pathway: role of nuclear factor-κB and Bcl-2. Mol. Biol. Cell *18*, 2735–2744.

Mittal, R., Peak-Chew, S.Y., and McMahon, H.T. (2006). Acetylation of MEK2 and IκB kinase (IKK) activation loop residues by YopJ inhibits signaling. Proc. Natl. Acad. Sci. U.S.A. *103*, 18574–18579.

Miyoshi-Akiyama, T., Takamatsu, D., Koyanagi, M., Zhao, J., Imanishi, K., and Uchiyama, T. (2005). Cytocidal effect of *Streptococcus pyogenes* on mouse neutrophils *in vivo* and the critical role of streptolysin S.J. Infect. Dis. *192*, 107–116.

Mohagheghpour, N., van Vollenhoven, A., Goodman, J., and Bermudez, L.E. (2000). Interaction of *Mycobacterium avium* with human monocyte-derived dendritic cells. Infect. Immun. *68*, 5824–5829.

Molloy, A., Laochumroonvorapong, P., and Kaplan, G. (1994). Apoptosis, but not necrosis, of infected monocytes is coupled with killing of intracellular bacillus Calmette-Guerin. J. Exp. Med. *180*, 1499–1509.

Molofsky, A.B., Byrne, B.G., Whitfield, N.N., Madigan, C.A., Fuse, E.T., Tateda, K., and Swanson, M.S. (2006). Cytosolic recognition of flagellin by mouse macrophages restricts *Legionella pneumophila* infection. J. Exp. Med. *203*, 1093–1104.

Monack, D.M., Raupach, B., Hromockyj, A.E., and Falkow, S. (1996). *Salmonella typhimurium* invasion induces apoptosis in infected macrophages. Proc. Natl. Acad. Sci. USA *93*, 9833–9838.

Monack, D.M., Mecsas, J., Ghori, N., and Falkow, S. (1997). *Yersinia* signals macrophages to undergo apoptosis and YopJ is necessary for this cell death. Proc. Natl. Acad. Sci. U.S.A. *94*, 10385–10390.

Monack, D.M., Mecsas, J., Bouley, D., and Falkow, S. (1998). *Yersinia*-induced apoptosis *in vivo* aids in the establishment of a systemic infection of mice. J. Exp. Med. *188*, 2127–2137.

Monack, D.M., Hersh, D., Ghori, N., Bouley, D., Zychlinsky, A., and Falkow, S. (2000). *Salmonella* exploits caspase-1 to colonize Peyer's patches in a murine typhoid model. J. Exp. Med. *192*, 249–258.

Monack, D.M., Detweiler, C.S., and Falkow, S. (2001). *Salmonella* pathogenicity island 2-dependent macrophage death is mediated in part by the host cysteine protease caspase-1. Cell. Microbiol. *3*, 825–837.

Morales, P., Reyes, P., Vargas, M., Rios, M., Imarai, M., Cardenas, H., Croxatto, H., Orihuela, P., Vargas, R., Fuhrer, J., *et al.* (2006). Infection of human fallopian tube epithelial cells with *Neisseria gonorrhoeae* protects cells from tumor necrosis factor α-induced apoptosis. Infect. Immun. *74*, 3643–3650.

Morimoto, H. and Bonavida, B. (1992). Diphtheria toxin- and *Pseudomonas* A toxin-mediated apoptosis. ADP ribosylation of elongation factor-2 is required for DNA fragmentation and cell lysis and synergy with tumor necrosis factor-α. J. Immunol. *149*, 2089–2094.

Moss, S.F., Calam, J., Agarwal, B., Wang, S., and Holt, P.R. (1996). Induction of gastric epithelial apoptosis by *Helicobacter pylori*. Gut *38*, 498–501.

Moss, S.F., Sordillo, E.M., Abdalla, A.M., Makarov, V., Hanzely, Z., Perez-Perez, G.I., Blaser, M.J., and Holt, P.R. (2001). Increased gastric epithelial cell apoptosis associated with colonization with *cagA* + *Helicobacter pylori* strains. Cancer Res. *61*, 1406–1411.

Moulding, D.A., Walter, C., Hart, C.A., and Edwards, S.W. (1999). Effects of staphylococcal enterotoxins on human neutrophil functions and apoptosis. Infect. Immun. *67*, 2312–2318.

Moura, J.A., Cristina de Assis, A.M., Ventura, G.C., Saliba, A.M., Gonzaga, L., Jr., Si-Tahar, M., Marques Ede, A., and Plotkowski, M.C. (2008). Differential interaction of bacterial species from the *Burkholderia cepacia* complex with human airway epithelial cells. Microbes Infect. *10*, 52–59.

Muehlbauer, S.M., Evering, T.H., Bonuccelli, G., Squires, R.C., Ashton, A.W., Porcelli, S.A., Lisanti, M.P., and Brojatsch, J. (2007). Anthrax lethal toxin kills macrophages in a strain-specific manner by apoptosis or caspase-1-mediated necrosis. Cell Cycle *6*, 758–766.

Mukherjee, S., Keitany, G., Li, Y., Wang, Y., Ball, H.L., Goldsmith, E.J., and Orth, K. (2006). *Yersinia* YopJ acetylates and inhibits kinase activation by blocking phosphorylation. Science *312*, 1211–1214.

Müller, A., Hacker, J., and Brand, B.C. (1996). Evidence for apoptosis of human macrophage-like HL-60 cells by *Legionella pneumophila* infection. Infect. Immun. *64*, 4900–4906.

Müller, A., Günther, D., Düx, F., Naumann, M., Meyer, T.F., and Rudel, T. (1999). Neisserial porin (PorB) causes rapid calcium influx in target cells and induces apoptosis by the activation of cysteine proteases. EMBO J. *18*, 339–352.

Müller, A., Günther, D., Brinkmann, V., Hurwitz, R., Meyer, T.F., and Rudel, T. (2000). Targeting of the pro-apoptotic VDAC-like porin (PorB) of *Neisseria gonorrhoeae* to mitochondria of infected cells. EMBO J. *19*, 5332–5343.

Muller, M. (2006). Premature cellular senescence induced by pyocyanin, a redox-active *Pseudomonas aeruginosa* toxin. Free Radic. Biol. Med. *41*, 1670–1677.

Mustafa, T., Phyu, S., Nilsen, R., Bjune, G., and Jonsson, R. (1999). Increased expression of Fas ligand on *Mycobacterium tuberculosis* infected macrophages: a potential novel mechanism of immune evasion by *Mycobacterium tuberculosis*? Inflammation *23*, 507–521.

Na, X., Kim, H., Moyer, M.P., Pothoulakis, C., and LaMont, J.T. (2008). gp96 is a human colonocyte plasma membrane binding protein for *Clostridium difficile* toxin A. Infect. Immun. *76*, 2862–2871.

Naim, R., Yanagihara, I., Iida, T., and Honda, T. (2001). *Vibrio parahaemolyticus* thermostable direct hemolysin can induce an apoptotic cell death in Rat-1 cells from inside and outside of the cells. FEMS Microbiol. Lett. *195*, 237–244.

Nakagawa, I., Nakata, M., Kawabata, S., and Hamada, S. (2001). Cytochrome c-mediated caspase-9 activation triggers apoptosis in *Streptococcus pyogenes*-infected epithelial cells. Cell. Microbiol. *3*, 395–405.

Nakagawa, I., Amano, A., Mizushima, N., Yamamoto, A., Yamaguchi, H., Kamimoto, T., Nara, A., Funao, J., Nakata, M., Tsuda, K., *et al.* (2004). Autophagy defends cells against invading group A *Streptococcus*. Science *306*, 1037–1040.

Nakhjiri, S.F., Park, Y., Yilmaz, O., Chung, W.O., Watanabe, K., El-Sabaeny, A., Park, K., and Lamont, R.J. (2001). Inhibition of epithelial cell apoptosis by *Porphyromonas gingivalis*. FEMS Microbiol. Lett. *200*, 145–149.

Narayanan, S., Stewart, G.C., Chengappa, M.M., Willard, L., Shuman, W., Wilkerson, M., and Nagaraja, T.G. (2002). *Fusobacterium necrophorum* leukotoxin induces activation and apoptosis of bovine leukocytes. Infect. Immun. *70*, 4609–4620.

Nelson, K.L., Brodsky, R.A., and Buckley, J.T. (1999). Channels formed by subnanomolar concentrations of the toxin aerolysin trigger apoptosis of T lymphomas. Cell. Microbiol. *1*, 69–74.

Nesic, D., Hsu, Y., and Stebbins, C.E. (2004). Assembly and function of a bacterial genotoxin. Nature *429*, 429–433.

Neumeister, B., Faigle, M., Lauber, K., Northoff, H., and Wesselborg, S. (2002). *Legionella pneumophila* induces apoptosis via the mitochondrial death pathway. Microbiology *148*, 3639–3650.

Ng, L.C., Forslund, O., Koh, S., Kuoppa, K., and Sjöstedt, A. (2003). The response of murine macrophages to infection with *Yersinia pestis* as revealed by DNA microarray analysis. Adv. Exp. Med. Biol. *529*, 155–160.

Ngai, S., Batty, S., Liao, K.C., and Mogridge, J. (2010). An anthrax lethal factor mutant that is defective at causing pyroptosis retains proapoptotic activity. FEBS J. *277*, 119–127.

Nogueira, C.V., Lindsten, T., Jamieson, A.M., Case, C.L., Shin, S., Thompson, C.B., and Roy, C.R. (2009). Rapid pathogen-induced apoptosis: a mechanism used by dendritic cells to limit intracellular replication of *Legionella pneumophila*. PLoS Pathog. *5*, e1000478.

Nonaka, T., Kuwae, A., Sasakawa, C., and Imajoh-Ohmi, S. (1999). *Shigella flexneri* YSH6000 induces two types of cell death, apoptosis and oncosis, in the differentiated human monoblastic cell line U937. FEMS Microbiol. Lett. *174*, 89–95.

Nonaka, T., Kuwabara, T., Mimuro, H., Kuwae, A., and Imajoh-Ohmi, S. (2003). *Shigella*-induced necrosis and apoptosis of U937 cells and J774 macrophages. Microbiology *149*, 2513–2527.

Nour, A.M., Yeung, Y.G., Santambrogio, L., Boyden, E.D., Stanley, E.R., and Brojatsch, J. (2009). Anthrax lethal toxin triggers the formation of a membrane-associated inflammasome complex in murine macrophages. Infect. Immun. *77*, 1262–1271.

Oddo, M., Renno, T., Attinger, A., Bakker, T., MacDonald, H.R., and Meylan, P.R. (1998). Fas ligand-induced apoptosis of infected human macrophages reduces the viability of intracellular *Mycobacterium tuberculosis*. J. Immunol. *160*, 5448–5454.

Ohara, M., Hayashi, T., Kusunoki, Y., Miyauchi, M., Takata, T., and Sugai, M. (2004). Caspase-2 and caspase-7 are involved in cytolethal distending toxin-induced apoptosis in Jurkat and MOLT-4 T-cell lines. Infect. Immun. *72*, 871–879.

Ohara, M., Hayashi, T., Kusunoki, Y., Nakachi, K., Fujiwara, T., Komatsuzawa, H., and Sugai, M. (2008). Cytolethal distending toxin induces caspase-dependent and -independent cell death in MOLT-4 cells. Infect. Immun. *76*, 4783–4791.

Ohguchi, M., Ishisaki, A., Okahashi, N., Koide, M., Koseki, T., Yamato, K., Noguchi, T., and Nishihara, T. (1998). *Actinobacillus actinomycetemcomitans* toxin induces both cell cycle arrest in the G2/M phase and apoptosis. Infect. Immun. *66*, 5980–5987.

Ohmae, T., Hirata, Y., Maeda, S., Shibata, W., Yanai, A., Ogura, K., Yoshida, H., Kawabe, T., and Omata, M. (2005). *Helicobacter pylori* activates NF-κB via the alternative pathway in B lymphocytes. J. Immunol. *175*, 7162–7169.

Ojcius, D.M., Souque, P., Perfettini, J.L., and Dautry-Varsat, A. (1998). Apoptosis of epithelial cells and macrophages due to infection with the obligate intracellular pathogen *Chlamydia psittaci*. J. Immunol. *161*, 4220–4226.

Oldani, A., Cormont, M., Hofman, V., Chiozzi, V., Oregioni, O., Canonici, A., Sciullo, A., Sommi, P., Fabbri, A., Ricci, V., *et al.* (2009). *Helicobacter pylori* counteracts the apoptotic action of its VacA toxin by injecting the CagA protein into gastric epithelial cells. PLoS Pathog. *5*, e1000603.

Orth, K., Palmer, L.E., Bao, Z.Q., Stewart, S., Rudolph, A.E., Bliska, J.B., and Dixon, J.E. (1999). Inhibition of the mitogen-activated protein kinase kinase superfamily by a *Yersinia* effector. Science *285*, 1920–1923.

Orth, K., Xu, Z., Mudgett, M.B., Bao, Z.Q., Palmer, L.E., Bliska, J.B., Mangel, W.F., Staskawicz, B., and Dixon, J.E. (2000). Disruption of signaling by *Yersinia* effector YopJ, a ubiquitin-like protein protease. Science *290*, 1594–1597.

Ozaki, K. and Hanazawa, S. (2001). *Porphyromonas gingivalis* fimbriae inhibit caspase-3-mediated apoptosis of monocytic THP-1 cells under growth factor deprivation via extracellular signal-regulated kinase-dependent expression of p21 Cip/WAF1. Infect. Immun. *69*, 4944–4950.

Paesold, G., Guiney, D.G., Eckmann, L., and Kagnoff, M.F. (2002). Genes in the *Salmonella* pathogenicity island 2 and the *Salmonella* virulence plasmid are essential for *Salmonella*-induced apoptosis in intestinal epithelial cells. Cell. Microbiol. *4*, 771–781.

Pandey, J. and Warburton, D. (2004). Knock-on effect of anthrax lethal toxin on macrophages potentiates cytotoxicity to endothelial cells. Microbes Infect. *6*, 835–843.

Panton, P.N., and Valentine, F.C.O. (1932). Staphylococcal toxin. Lancet *219*, 506–508.

Park, H., Teja, K., O'Shea, J.J., and Siegel, R.M. (2007). The *Yersinia* effector protein YpkA induces apoptosis independently of actin depolymerization. J. Immunol. *178*, 6426–6434.

Park, J.M., Greten, F.R., Li, Z.W., and Karin, M. (2002). Macrophage apoptosis by anthrax lethal factor through p38 MAP kinase inhibition. Science *297*, 2048–2051.

Park, J.M., Ng, V.H., Maeda, S., Rest, R.F., and Karin, M. (2004). Anthrolysin O and other Gram-positive cytolysins are Toll-like receptor 4 agonists. J. Exp. Med. *200*, 1647–1655.

Park, J.M., Greten, F.R., Wong, A., Westrick, R.J., Arthur, J.S., Otsu, K., Hoffmann, A., Montminy, M., and Karin, M. (2005). Signaling pathways and genes that inhibit pathogen-induced macrophage apoptosis-CREB and NF-κB as key regulators. Immunity *23*, 319–329.

Park, Y.H., Lee, S.U., Ferens, W.A., Samuels, S., Davis, W.C., Fox, L.K., Ahn, J.S., Seo, K.S., Chang, B.S., Hwang, S.Y., *et al.* (2006). Unique features of bovine lymphocytes exposed to a staphylococcal enterotoxin. J. Vet. Sci. *7*, 233–239.

Parra, M.C., Baquero, F., and Perez-Diaz, J.C. (2008). The role of apoptosis in *Listeria monocytogenes* neural infection: listeriolysin O interaction with neuroblastoma Neuro-2a cells. Infect. Genet. Evol. *8*, 59–67.

Pedra, J.M., Sukumaran, B., Carlyon, J.A., Berliner, N., and Fikrig, E. (2005). Modulation of NB4 promyelocytic leukemic cell machinery by *Anaplasma phagocytophilum*. Genomics *86*, 365–377.

Peek, R.M., Jr., Moss, S.F., Tham, K.T., Pérez-Pérez, G.I., Wang, S., Miller, G.G., Atherton, J.C., Holt, P.R., and Blaser, M.J. (1997). *Helicobacter pylori cagA* + strains and dissociation of gastric epithelial cell proliferation from apoptosis. J. Natl. Cancer Inst. *89*, 863–868.

Pei, J., and Ficht, T.A. (2004). *Brucella abortus* rough mutants are cytopathic for macrophages in culture. Infect. Immun. *72*, 440–450.

Pei, J., Turse, J.E., Wu, Q., and Ficht, T.A. (2006). *Brucella abortus* rough mutants induce macrophage oncosis that requires bacterial protein synthesis and direct interaction with the macrophage. Infect. Immun. *74*, 2667–2675.

Pei, J., Wu, Q., Kahl-McDonagh, M., and Ficht, T.A. (2008). Cytotoxicity in macrophages infected with rough *Brucella* mutants is type IV secretion system dependent. Infect. Immun. *76*, 30–37.

Perskvist, N., Long, M., Stendahl, O., and Zheng, L. (2002). *Mycobacterium tuberculosis* promotes apoptosis in human neutrophils by activating caspase-3 and altering expression of Bax/Bcl-xL via an oxygen-dependent pathway. J. Immunol. *168*, 6358–6365.

Pessina, A., Croera, C., Savalli, N., Bonomi, A., Cavicchini, L., Turlizzi, E., Guizzardi, F., Guido, L., Daprai, L., and Neri, M.G. (2006). Bcl-2 down modulation in WEHI-3B/CTRES cells resistant to cholera toxin (CT)-induced apoptosis. Cell Res. *16*, 306–312.

Petit, P., Breard, J., Montalescot, V., El Hadj, N.B., Levade, T., Popoff, M., and Geny, B. (2003). Lethal toxin from *Clostridium sordellii* induces apoptotic cell death by disruption of mitochondrial homeostasis in HL-60 cells. Cell. Microbiol. *5*, 761–771.

Pétrilli, V., Papin, S., Dostert, C., Mayor, A., Martinon, F., and Tschopp, J. (2007). Activation of the NALP3 inflammasome is triggered by low intracellular potassium concentration. Cell Death Differ. *14*, 1583–1589.

di Pietro, A., Picerno, I., Visalli, G., Chirico, C., Spataro, P., Cannavò, G., and Scoglio, M.E. (2005). *Aeromonas hydrophila* exotoxin induces cytoplasmic vacuolation and cell death in VERO cells. New Microbiol. *28*, 251–259.

Pinheiro da Silva F., and Nizet, V. (2009). Cell death during sepsis: integration of disintegration in the inflammatory response to overwhelming infection. Apoptosis *14*, 509–521.

Pirbhai, M., Dong, F., Zhong, Y., Pan, K.Z., and Zhong, G. (2006). The secreted protease factor CPAF is responsible for degrading pro-apoptotic BH3-only proteins in *Chlamydia trachomatis*-infected cells. J. Biol. Chem. *281*, 31495–31501.

Placido, R., Mancino, G., Amendola, A., Mariani, F., Vendetti, S., Piacentini, M., Sanduzzi, A., Bocchino, M.L., Zembala, M., and Colizzi, V. (1997). Apoptosis of human monocytes/macrophages in *Mycobacterium tuberculosis* infection. J. Pathol. *181*, 31–38.

Plotkowski, M.C., Póvoa, H.C., Zahm, J.M., Lizard, G., Pereira, G.M., Tournier, J.M., and Puchelle, E. (2002). Early mitochondrial dysfunction, superoxide anion production, and DNA degradation are associated with non-apoptotic death of human airway epithelial cells induced by *Pseudomonas aeruginosa* exotoxin A. Am. J. Respir. Cell Mol. Biol. *26*, 617–626.

Popoff, M.R., and Bouvet, P. (2009). Clostridial toxins. Future Microbiol. *4*, 1021–1064.

Popov, S.G., Villasmil, R., Bernardi, J., Grene, E., Cardwell, J., Popova, T., Wu, A., Alibek, D., Bailey, C., and Alibek, K. (2002a). Effect of *Bacillus anthracis* lethal toxin on human peripheral blood mononuclear cells. FEBS Lett. *527*, 211–215.

Popov, S.G., Villasmil, R., Bernardi, J., Grene, E., Cardwell, J., Wu, A., Alibek, D., Bailey, C., and Alibek, K. (2002b). Lethal toxin of *Bacillus anthracis* causes apoptosis of macrophages. Biochem. Biophys. Res. Commun. *293*, 349–355.

Preshaw, P.M., Schifferle, R.E., and Walters, J.D. (1999). *Porphyromonas gingivalis* lipopolysaccharide delays human polymorphonuclear leukocyte apoptosis *in vitro*. J. Periodontal Res. *34*, 197–202.

Prince, L.R., Bianchi, S.M., Vaughan, K.M., Bewley, M.A., Marriott, H.M., Walmsley, S.R., Taylor, G.W., Buttle, D.J., Sabroe, I., Dockrell, D.H., *et al.* (2008). Subversion of a lysosomal pathway regulating neutrophil apoptosis by a major bacterial toxin, pyocyanin. J. Immunol. *180*, 3502–3511.

Psotka, M.A., Obata, F., Kolling, G.L., Gross, L.K., Saleem, M.A., Satchell, S.C., Mathieson, P.W., and Obrig, T.G. (2009). Shiga toxin 2 targets the murine renal collecting duct epithelium. Infect. Immun. *77*, 959–969.

Punj, V., Sharma, R., Zaborina, O., and Chakrabarty, A.M. (2003). Energy-generating enzymes of *Burkholderia cepacia* and their interactions with macrophages. J. Bacteriol. *185*, 3167–3178.

Qa'Dan, M., Ramsey, M., Daniel, J., Spyres, L.M., Safiejko-Mroczka, B., Ortiz-Leduc, W., and Ballard, J.D. (2002). *Clostridium difficile* toxin B activates dual caspase-dependent and caspase-independent apoptosis in intoxicated cells. Cell. Microbiol. *4*, 425–434.

Quinn, M.T., Ammons, M.C., and DeLeo, F.R. (2006). The expanding role of NADPH oxidases in health and disease: no longer just agents of death and destruction. Clin. Sci. (Lond.) *111*, 1–20.

Rabin, S.D., Flitton, J.G., and Demuth, D.R. (2009). *Aggregatibacter actinomycetemcomitans* cytolethal distending toxin induces apoptosis in nonproliferating macrophages by a phosphatase-independent mechanism. Infect. Immun. *77*, 3161–3169.

Ragno, S., Estrada-Garcia, I., Butler, R., and Colston, M.J. (1998). Regulation of macrophage gene expression by *Mycobacterium tuberculosis*: down-regulation of mitochondrial cytochrome c oxidase. Infect. Immun. *66*, 3952–3958.

Rajalingam, K., Sharma, M., Paland, N., Hurwitz, R., Thieck, O., Oswald, M., Machuy, N., and Rudel, T. (2006). IAP–IAP complexes required for apoptosis resistance of C. *trachomatis*-infected cells. PLoS Pathog. *2*, e114.

Ramirez, R., Carracedo, J., Zamzami, N., Castedo, M., and Kroemer, G. (1994). Pertussis toxin inhibits activation-induced cell death of human thymocytes, pre-B leukemia cells and monocytes. J. Exp. Med. *180*, 1147–1152.

Raqib, R., Ekberg, C., Sharkar, P., Bardhan, P.K., Zychlinsky, A., Sansonetti, P.J., and Andersson, J. (2002). Apoptosis in acute shigellosis is associated with increased production of Fas/Fas ligand, perforin, caspase-1, and caspase-3 but reduced production of Bcl-2 and interleukin-2. Infect. Immun. *70*, 3199–3207.

Raupach, B., Peuschel, S.K., Monack, D.M., and Zychlinsky, A. (2006). Caspase-1-mediated activation of interleukin-1β (IL-1 β) and IL-18 contributes to innate immune defenses against *Salmonella enterica* serovar Typhimurium infection. Infect. Immun. *74*, 4922–4926.

Renno, T., Attinger, A., Locatelli, S., Bakker, T., Vacheron, S., and MacDonald, H.R. (1999). Apoptosis of superantigen-activated T cells occurs preferentially after a discrete number of cell divisions *in vivo*. J. Immunol. *162*, 6312–6315.

Richter-Dahlfors, A., Buchan, A.M., and Finlay, B.B. (1997). Murine salmonellosis studied by confocal microscopy: *Salmonella typhimurium* resides intracellularly inside macrophages and exerts a cytotoxic effect on phagocytes *in vivo*. J. Exp. Med. *186*, 569–580.

Riendeau, C.J., and Kornfeld, H. (2003). THP-1 cell apoptosis in response to mycobacterial infection. Infect. Immun. *71*, 254–259.

Rodrigues, V.S., Vidotto, M.C., Felipe, I., Santos, D.S., and Gaziri, L.C. (1999). Apoptosis of murine peritoneal macrophages induced by an avian pathogenic strain of *Escherichia coli*. FEMS Microbiol. Lett. *179*, 73–78.

Rodriguez, S., Chora, A., Goumnerov, B., Mumaw, C., Goebel, W.S., Fernandez, L., Baydoun, H., HogenEsch, H., Dombkowski, D.M., Karlewicz, C.A., *et al.* (2009). Dysfunctional expansion of hematopoietic stem cells and block of myeloid differentiation in lethal sepsis. Blood *114*, 4064–4076.

Rogers, D.E., and Tompsett, R. (1952). The survival of staphylococci within human leukocytes. J. Exp. Med. *95*, 209–230.

Rojas, M., Barrera, L.F., Puzo, G., and Garcia, L.F. (1997). Differential induction of apoptosis by virulent *Mycobacterium tuberculosis* in resistant and susceptible murine macrophages: role of nitric oxide and mycobacterial products. J. Immunol. *159*, 1352–1361.

Rojas, M., Barrera, L.F., and Garcia, L.F. (1998). Induction of apoptosis in murine macrophages by *Mycobacterium tuberculosis* is reactive oxygen intermediates-independent. Biochem. Biophys. Res. Commun. *247*, 436–442.

Rojas, M., Olivier, M., Gros, P., Barrera, L.F., and Garcia, L.F. (1999). TNF-α and IL-10 modulate the induction of apoptosis by virulent *Mycobacterium tuberculosis* in murine macrophages. J. Immunol. *162*, 6122–6131.

Rojas, M., Garcia, L.F., Nigou, J., Puzo, G., and Olivier, M. (2000). Mannosylated lipoarabinomannan antagonizes *Mycobacterium tuberculosis*-induced macrophage apoptosis by altering Ca^{2+}-dependent cell signaling. J. Infect. Dis. *182*, 240–251.

Rojas, M., Olivier, M., and Garcia, L.F. (2002). Activation of JAK2/STAT1-α-dependent signaling events during *Mycobacterium tuberculosis*-induced macrophage apoptosis. Cell. Immunol. *217*, 58–66.

Ruckdeschel, K., and Richter, K. (2002). Lipopolysaccharide desensitization of macrophages provides protection against *Yersinia enterocolitica*-induced apoptosis. Infect. Immun. *70*, 5259–5264.

Ruckdeschel, K., Roggenkamp, A., Lafont, V., Mangeat, P., Heesemann, J., and Rouot, B. (1997). Interaction of *Yersinia enterocolitica* with macrophages leads to macrophage cell death through apoptosis. Infect. Immun. *65*, 4813–4821.

Ruckdeschel, K., Harb, S., Roggenkamp, A., Hornef, M., Zumbihl, R., Köhler, S., Heesemann, J., and Rouot, B. (1998). *Yersinia enterocolitica* impairs activation of transcription factor NF-κB: involvement in the induction of programmed cell death and in the suppression of the macrophage tumor necrosis factor alpha production. J. Exp. Med. *187*, 1069–1079.

Ruckdeschel, K., Mannel, O., Richter, K., Jacobi, C.A., Trülzsch, K., Rouot, B., and Heesemann, J. (2001a). *Yersinia* outer protein P of *Yersinia enterocolitica* simultaneously blocks the nuclear factor-κB pathway and exploits lipopolysaccharide signaling to trigger apoptosis in macrophages. J. Immunol. *166*, 1823–1831.

Ruckdeschel, K., Richter, K., Mannel, O., and Heesemann, J. (2001b). Arginine-143 of *Yersinia enterocolitica* YopP crucially determines isotype-related NF-κB suppression and apoptosis induction in macrophages. Infect. Immun. *69*, 7652–7662.

Ruckdeschel, K., Mannel, O., and Schröttner, P. (2002). Divergence of apoptosis-inducing and preventing signals in bacteria-faced macrophages through myeloid differentiation factor 88 and IL-1 receptor-associated kinase members. J. Immunol. *168*, 4601–4611.

Russo, T.A., Davidson, B.A., Genagon, S.A., Warholic, N.M., Macdonald, U., Pawlicki, P.D., Beanan, J.M., Olson, R., Holm, B.A., and Knight, P.R.,3rd. (2005). *E. coli* virulence factor hemolysin induces neutrophil apoptosis and necrosis/lysis *in vitro* and necrosis/lysis and lung injury in a rat pneumonia model. Am. J. Physiol. Lung Cell Mol. Physiol. *289*, L207-L216.

Sahni, S.K., and Rydkina, E. (2009). Host–cell interactions with pathogenic *Rickettsia* species. Future Microbiol. *4*, 323–339.

Saka, H.A., Bidinost, C., Sola, C., Carranza, P., Collino, C., Ortiz, S., Echenique, J.R., and Bocco, J.L. (2008). *Vibrio cholerae* cytolysin is essential for high enterotoxicity and apoptosis induction produced by a cholera toxin gene-negative *V. cholerae* non-O1, non-O139 strain. Microb. Pathog. *44*, 118–128.

Salcedo, S.P., and Holden, D.W. (2005). Bacterial interactions with the eukaryotic secretory pathway. Curr. Opin. Microbiol. *8*, 92–98.

Salmond, R.J., Pitman, R.S., Jimi, E., Soriani, M., Hirst, T.R., Ghosh, S., Rincon, M., and Williams, N.A. (2002). $CD8^+$ T cell apoptosis induced by *Escherichia coli* heat-labile enterotoxin B subunit occurs via a novel pathway involving NF-κB-dependent caspase activation. Eur. J. Immunol. *32*, 1737–1747.

Salmond, R.J., Williams, R., Hirst, T.R., and Williams, N.A. (2003). Selective induction of $CD8^+ CD4^-$ thymocyte apoptosis mediated by the B-subunit of *Escherichia coli* heat-labile enterotoxin. Immunol. Lett. *88*, 43–46.

Salmond, R.J., Williams, R., Hirst, T.R., and Williams, N.A. (2004). The B subunit of *Escherichia coli* heat-labile enterotoxin induces both caspase-dependent and -independent cell death pathways in $CD8^+$ T cells. Infect. Immun. *72*, 5850–5857.

Sansonetti, P.J., Phalipon, A., Arondel, J., Thirumalai, K., Banerjee, S., Akira, S., Takeda, K., and Zychlinsky, A. (2000). Caspase-1 activation of IL-1β and IL-18 are essential for *Shigella flexneri*-induced inflammation. Immunity *12*, 581–590.

Santic, M., Molmeret, M., Klose, K.E., and Kwait,Y.A. (2006). *Francisella tularensis* travels a novel, twisted road within macrophages. Trends Microbiol. *14*, 37–44.

Santos, R.L., Tsolis, R.M., Bäumler, A.J., Smith, R., 3rd, and Adams, L.G. (2001). *Salmonella enterica* serovar Typhimurium induces cell death in bovine monocyte-derived macrophages by early *sipB*-dependent and delayed *sipB*-independent mechanisms. Infect. Immun. *69*, 2293–2301.

Santucci, M.B., Amicosante, M., Cicconi, R., Montesano, C., Casarini, M., Giosuè, S., Bisetti, A., Colizzi, V., and Fraziano, M. (2000). *Mycobacterium tuberculosis*-induced apoptosis in monocytes/macrophages: early membrane modifications and intracellular mycobacterial viability. J. Infect. Dis. *181*, 1506–1509.

Savill, J.S., Henson, P.M., and Haslett, C. (1989a). Phagocytosis of aged human neutrophils by macrophages is mediated by a novel 'charge-sensitive' recognition mechanism. J. Clin. Invest. *84*, 1518–1527.

Savill, J.S., Wyllie, A.H., Henson, J.E., Walport, M.J., Henson, P.M., and Haslett, C. (1989b). Macrophage phagocytosis of aging neutrophils in inflammation. Programmed cell death in the neutrophil leads to its recognition by macrophages. J. Clin. Invest. *83*, 865–875.

Savill, J., Dransfield, I., Hogg, N., and Haslett, C. (1990). Vitronectin receptor-mediated phagocytosis of cells undergoing apoptosis. Nature *343*, 170–173.

Savill, J., Dransfield, I., Gregory, C., and Haslett, C. (2002). A blast from the past: clearance of apoptotic cells regulates immune responses. Nat. Rev. Immunol. *2*, 965–975.

Scaife, H., Woldehiwet, Z., Hart, C.A., and Edwards, S.W. (2003). *Anaplasma phagocytophilum* reduces neutrophil apoptosis *in vivo*. Infect. Immun. *71*, 1995–2001.

Schaible, U.E., Winau, F., Sieling, P.A., Fischer, K., Collins, H.L., Hagens, K., Modlin, R.L., Brinkmann, V., and Kaufmann, S.H. (2003). Apoptosis facilitates antigen presentation to T lymphocytes through MHC-I and CD1 in tuberculosis. Nat. Med. *9*, 1039–1046.

Schesser, K., Spiik, A.K., Dukuzumuremyi, J.M., Neurath, M.F., Pettersson, S., and Wolf-Watz, H. (1998). The *yopJ* locus is required for *Yersinia*-mediated inhibition of NF-κB activation and cytokine expression: YopJ contains a eukaryotic SH2-like domain that is essential for its repressive activity. Mol. Microbiol. *28*, 1067–1079.

Schmeck, B., Gross, R., N'guessan, P.D., Hocke, A.C., Hammerschmidt, S., Mitchell, T.J., Rosseau, S., Suttorp, N., and Hippenstiel, S. (2004). *Streptococcus pneumoniae*-induced caspase 6-dependent apoptosis in lung epithelium. Infect. Immun. *72*, 4940–4947.

Schmid, M.C., Scheidegger, F., Dehio, M., Balmelle-Devaux, N., Schulein, R., Guye, P., Chennakesava, C.S., Biedermann, B., and Dehio, C. (2006). A translocated bacterial protein protects vascular endothelial cells from apoptosis. PLoS Pathog. *2*, e115.

Schreiber, T., Swanson, P.E., Chang, K.C., Davis, C.C., Dunne, W.M., Karl, I.E., Reinhart, K., and Hotchkiss, R.S. (2006). Both Gram-negative and Gram-positive experimental pneumonia induce profound lymphocyte but not respiratory epithelial cell apoptosis. Shock *26*, 271–276.

Schulz, F., Just, I., and Genth, H. (2009). Prevention of *Clostridium sordellii* lethal toxin-induced apoptotic cell death by tauroursodeoxycholic acid. Biochemistry *48*, 9002–9010.

Schümann, J., Angermüller, S., Bang, R., Lohoff, M., and Tiegs, G. (1998). Acute hepatotoxicity of *Pseudomonas aeruginosa* exotoxin A in mice depends on T cells and TNF. J. Immunol. *161*, 5745–5754.

Schwab, J.M., Chiang, N., Arita, M., and Serhan, C.N. (2007). Resolvin E1 and protectin D1 activate inflammation-resolution programmes. Nature *447*, 869–874.

Schwan, W.R., Huang, X.Z., Hu, L., and Kopecko, D.J. (2000). Differential bacterial survival, replication, and apoptosis-inducing ability of *Salmonella* serovars within human and murine macrophages. Infect. Immun. *68*, 1005–1013.

Serhan, C.N., Chiang, N., and Van Dyke, T.E. (2008). Resolving inflammation: dual anti-inflammatory and pro-resolution lipid mediators. Nat. Rev. Immunol. *8*, 349–361.

Shafikhani, S.H., Morales, C., and Engel, J. (2008). The *Pseudomonas aeruginosa* type III secreted toxin ExoT is necessary and sufficient to induce apoptosis in epithelial cells. Cell. Microbiol. *10*, 994–1007.

Sharma, J., Li, Q., Mishra, B.B., and Teale, J.M. (2009). Lethal pulmonary infection with *Francisella novicida* causes depletion of αβ T cells from lungs. Cell. Immunol. *257*, 1–4.

Shenker, B.J., Demuth, D.R., and Zekavat, A. (2006). Exposure of lymphocytes to high doses of *Actinobacillus actinomycetemcomitans* cytolethal distending toxin induces rapid onset of apoptosis-mediated DNA fragmentation. Infect. Immun. *74*, 2080–2092.

Shenker, B.J., Hoffmaster, R.H., Zekavat, A., Yamaguchi, N., Lally, E.T., and Demuth, D.R. (2001). Induction of apoptosis in human T cells by *Actinobacillus actinomycetemcomitans* cytolethal distending toxin is a consequence of G2 arrest of the cell cycle. J. Immunol. *167*, 435–441.

Shi, Y., Evans, J.E., and Rock, K.L. (2003). Molecular identification of a danger signal that alerts the immune system to dying cells. Nature *425*, 516–521.

Siegesmund, A.M., Konkel, M.E., Klena, J.D., and Mixter, P.F. (2004). *Campylobacter jejuni* infection of differentiated THP-1 macrophages results in interleukin 1β release and caspase-1-independent apoptosis. Microbiology *150*, 561–569.

Silverman, D.J., and Bond, S.B. (1984). Infection of human vascular endothelial cells by *Rickettsia rickettsii*. J. Infect. Dis. *149*, 201–206.

Silverman, G.J., and Goodyear, C.S. (2006). Confounding B-cell defences: lessons from a staphylococcal superantigen. Nat. Rev. Immunol. *6*, 465–475.

Simons, M.P., Nauseef, W.M., Griffith, T.S., and Apicella, M.A. (2006). *Neisseria gonorrhoeae* delays the onset of apoptosis in polymorphonuclear leukocytes. Cell. Microbiol. *8*, 1780–1790.

Sly, L.M., Hingley-Wilson, S.M., Reiner, N.E., and McMaster, W.R. (2003). Survival of *Mycobacterium tuberculosis* in host macrophages involves resistance to apoptosis dependent upon induction of antiapoptotic Bcl-2 family member Mcl-1. J. Immunol. *170*, 430–437.

Smith, J.L., and Bayles, D.O. (2006). The contribution of cytolethal distending toxin to bacterial pathogenesis. Crit. Rev. Microbiol. *32*, 227–248.

Solomon, K., Webb, J., Ali, N., Robins, R.A., and Mahida, Y.R. (2005). Monocytes are highly sensitive to *Clostridium difficile* toxin A-induced apoptotic and nonapoptotic cell death. Infect. Immun. *73*, 1625–1634.

Soriani, M., Williams, N.A., and Hirst, T.R. (2001). *Escherichia coli* enterotoxin B subunit triggers apoptosis of CD8 + T cells by activating transcription factor c-Myc. Infect. Immun. *69*, 4923–4930.

Spira, A., Carroll, J.D., Liu, G., Aziz, Z., Shah, V., Kornfeld, H., and Keane, J. (2003). Apoptosis genes in human alveolar macrophages infected with virulent or attenuated *Mycobacterium tuberculosis*: a pivotal role for tumor necrosis factor. Am. J. Respir. Cell Mol. Biol. *29*, 545–551.

Squires, R.C., Muehlbauer, S.M., and Brojatsch, J. (2007). Proteasomes control caspase-1 activation in anthrax lethal toxin-mediated cell killing. J. Biol. Chem. *282*, 34260–34267.

Srivastava, A., Henneke, P., Visintin, A., Morse, S.C., Martin, V., Watkins, C., Paton, J.C., Wessels, M.R., Golenbock, D.T., and Malley, R. (2005). The apoptotic response to pneumolysin is Toll-like receptor 4 dependent and protects against pneumococcal disease. Infect. Immun. *73*, 6479–6487.

Stevens, P.K., and Czuprynski, C.J. (1996). *Pasteurella haemolytica* leukotoxin induces bovine leukocytes to undergo morphologic changes consistent with apoptosis *in vitro*. Infect. Immun. *64*, 2687–2694.

Stravodimos, K.G., Singhal, P.C., Sharma, S., Reddy, K., and Smith, A.D. (1999). *Escherichia coli* promotes macrophage apoptosis. J. Endourol. *13*, 273–277.

Sukumaran, S.K., Selvaraj, S.K., and Prasadarao, N.V. (2004). Inhibition of apoptosis by *Escherichia coli* K1 is accompanied by increased expression of BclXL and blockade of mitochondrial cytochrome c release in macrophages. Infect. Immun. *72*, 6012–6022.

Sun, G.W., Lu, J., Pervaiz, S., Cao, W.P., and Gan, Y.H. (2005). Caspase-1 dependent macrophage death induced by *Burkholderia pseudomallei*. Cell. Microbiol. *7*, 1447–1458.

Sun, Y., Clinkenbeard, K.D., Ownby, C.L., Cudd, L., Clarke, C.R., and Highlander, S.K. (2000). Ultrastructural characterization of apoptosis in bovine lymphocytes exposed to *Pasteurella haemolytica* leukotoxin. Am. J. Vet. Res. *61*, 51–56.

Suparak, S., Kespichayawattana, W., Haque, A., Easton, A., Damnin, S., Lertmemongkolchai, G., Bancroft, G.J., and Korbsrisate, S. (2005). Multinucleated giant cell formation and apoptosis in infected host cells is mediated by *Burkholderia pseudomallei* type III secretion protein BipB.J. Bacteriol. *187*, 6556–6560.

Sutterwala, F.S., Mijares, L.A., Li, L., Ogura, Y., Kazmierczak, B.I., and Flavell, R.A. (2007). Immune recognition of *Pseudomonas aeruginosa* mediated by the IPAF/NLRC4 inflammasome. J. Exp. Med. *204*, 3235–3245.

Suttmann, H., Lehan, N., Böhle, A., and Brandau, S. (2003). Stimulation of neutrophil granulocytes with *Mycobacterium bovis* bacillus Calmette-Guérin induces changes in phenotype and gene expression and inhibits spontaneous apoptosis. Infect. Immun. *71*, 4647–4656.

Suzuki, T., Nakanishi, K., Tsutsui, H., Iwai, H., Akira, S., Inohara, N., Chamaillard, M., Nuñez, G., and Sasakawa, C. (2005). A novel caspase-1/Toll-like receptor 4-independent pathway of cell death induced by cytosolic *Shigella* in infected macrophages. J. Biol. Chem. *280*, 14042–14050.

Suzuki, T., Franchi, L., Toma, C., Ashida, H., Ogawa, M., Yoshikawa, Y., Mimuro, H., Inohara, N., Sasakawa, C., and Nuñez, G. (2007). Differential regulation of caspase-1 activation, pyroptosis, and autophagy via Ipaf and ASC in *Shigella*-infected macrophages. PLoS Pathog. *3*, e111.

Szmigielski, S., Prevost, G., Monteil, H., Colin, D.A., and Jeljaszewicz, J. (1999). Leukocidal toxins of staphylococci. Zentralbl. Bakteriol. *289*, 185–201.

Takahashi, K., Funata, N., Ikuta, F., and Sato, S. (2008). Neuronal apoptosis and inflammatory responses in the central nervous system of a rabbit treated with Shiga toxin-2. J. Neuroinflammation *5*, 11.

Takahashi, M., Takahashi, M., Shinohara, F., Takada, H., and Rikiishi, H. (2001). Effects of superantigen and lipopolysaccharide on induction of CD80 through apoptosis of human monocytes. Infect. Immun. *69*, 3652–3657.

Tamayo, E., Merino, R., González-Rojas, J., Marquina, R., Santiuste, I., Amado, J.A., Rappuoli, R., Del Giudice, G., and Merino, J. (2005). The *Escherichia coli* heat-labile enterotoxin induces apoptosis of immature lymphocytes *in vivo* via a glucocorticoid-dependent pathway. Eur. J. Immunol. *35*, 3505–3515.

Tateda, K., Ishii, Y., Horikawa, M., Matsumoto, T., Miyairi, S., Pechere, J.C., Standiford, T.J., Ishiguro, M., and Yamaguchi, K. (2003). The *Pseudomonas aeruginosa* autoinducer N-3-oxododecanoyl homoserine lactone accelerates apoptosis in macrophages and neutrophils. Infect. Immun. *71*, 5785–5793.

Tattoli, I., Lembo-Fazio, L., Nigro, G., Carneiro, L.A., Ferraro, E., Rossi, G., Martino, M.C., de Stefano, M.E., Cecconi, F., Girardin, S.E., *et al.* (2008). Intracellular bacteriolysis triggers a massive apoptotic cell death in *Shigella*-infected epithelial cells. Microbes. Infect. *10*, 1114–1123.

Terra, J.K., Cote, C.K., France, B., Jenkins, A.L., Bozue, J.A., Welkos, S.L., LeVine, S.M., and Bradley, K.A. (2010). Resistance to *Bacillus anthracis* infection mediated by a lethal toxin sensitive allele of *Nalp1b/Nlrp1b*. J. Immunol. *184*, 17–20.

Thumbikat, P., Dileepan, T., Kannan, M.S., and Maheswaran, S.K. (2005). Mechanisms underlying *Mannheimia haemolytica* leukotoxin-induced oncosis and apoptosis of bovine alveolar macrophages. Microb. Pathog. *38*, 161–172.

Timmer, A.M., Timmer, J.C., Pence, M.A., Hsu, L.C., Ghochani, M., Frey, T.G., Karin, M., Salvesen, G.S., and Nizet, V. (2009). Streptolysin O promotes group A *Streptococcus* immune evasion by accelerated macrophage apoptosis. J. Biol. Chem. *284*, 862–871.

Tsai, P.J., Lin, Y.S., Kuo, C.F., Lei, H.Y., and Wu, J.J. (1999). Group A *Streptococcus* induces apoptosis in human epithelial cells. Infect. Immun. *67*, 4334–4339.

Tsai, W.H., Chang, C.W., Chuang, W.J., Lin, Y.S., Wu, J.J., Liu, C.C., Chang, W.T., and Lin, M.T. (2004). Streptococcal pyrogenic exotoxin B-induced apoptosis in A549 cells is mediated by a receptor- and mitochondrion-dependent pathway. Infect. Immun. *72*, 7055–7062.

Tsai, W.H., Chang, C.W., Lin, Y.S., Chuang, W.J., Wu, J.J., Liu, C.C., Tsai, P.J., and Lin, M.T. (2008). Streptococcal pyrogenic exotoxin B-induced apoptosis in A549 cells is mediated through $\alpha_v\beta_3$ integrin and Fas. Infect. Immun. *76*, 1349–1357.

Tsao, M.Y., Lin, T.L., Hsieh, P.F., and Wang, J.T. (2009). The 3′-to-5′ exoribonuclease (encoded by HP1248) of *Helicobacter pylori* regulates motility and apoptosis-inducing genes. J. Bacteriol. *191*, 2691–2702.

Tsugawa, H., Ogawa, A., Takehara, S., Kimura, M., and Okawa, Y. (2008). Primary structure and function of a cytotoxic outer-membrane protein (ComP) of *Plesiomonas shigelloides*. FEMS Microbiol. Lett. *281*, 10–16.

Uchida, H., Kiyokawa, N., Taguchi, T., Horie, H., Fujimoto, J., and Takeda, T. (1999). Shiga toxins induce apoptosis in pulmonary epithelium-derived cells. J. Infect. Dis. *180*, 1902–1911.

Ulett, G.C., and Adderson, E.E. (2005). Nitric oxide is a key determinant of group B *Streptococcus*-induced murine macrophage apoptosis. J. Infect. Dis. *191*, 1761–1770.

Ulett, G.C., and Adderson, E.E. (2006). Regulation of apoptosis by Gram-positive bacteria: mechanistic diversity and consequences for immunity. Curr. Immunol. Rev. *2*, 119–141.

Ulett, G.C., Bohnsack, J.F., Armstrong, J., and Adderson, E.E. (2003). β-hemolysin-independent induction of apoptosis of macrophages infected with serotype III group B *Streptococcus*. J. Infect. Dis. *188*, 1049–1053.

Ulett, G.C., Maclean, K.H., Nekkalapu, S., Cleveland, J.L., and Adderson, E.E. (2005). Mechanisms of group B streptococcal-induced apoptosis of murine macrophages. J. Immunol. *175*, 2555–2562.

Usher, L.R., Lawson, R.A., Geary, I., Taylor, C.J., Bingle, C.D., Taylor, G.W., and Whyte, M.K. (2002). Induction of neutrophil apoptosis by the *Pseudomonas aeruginosa* exotoxin pyocyanin: a potential mechanism of persistent infection. J. Immunol. *168*, 1861–1868.

do Vale, A., Marques, F., and Silva, M.T. (2003). Apoptosis of sea bass (*Dicentrarchus labrax L.*) neutrophils and macrophages induced by experimental infection with *Photobacterium damselae* subsp. *piscicida*. FishShellfish Immunol. *15*, 129–144.

Valle, E., and Guiney, D.G. (2005). Characterization of *Salmonella*-induced cell death in human macrophage-like THP-1 cells. Infect. Immun. *73*, 2835–2840.

van der Velden, A.W., Lindgren, S.W., Worley, M.J., and Heffron, F. (2000). *Salmonella* pathogenicity island 1-independent induction of apoptosis in infected macrophages by *Salmonella enterica* serotype Typhimurium. Infect. Immun. *68*, 5702–5709.

van der Velden, A.W., Velasquez, M., and Starnbach, M.N. (2003). *Salmonella* rapidly kill dendritic cells via a caspase-1-dependent mechanism. J. Immunol. *171*, 6742–6749.

Vázquez, C.L., and Colombo, M.I. (2010). *Coxiella burnetii* modulates Beclin 1 and Bcl-2, preventing host cell apoptosis to generate a persistent bacterial infection. Cell Death Differ. *17*, 421–438.

Viboud, G.I., and Bliska, J.B. (2005). *Yersinia* outer proteins: role in modulation of host cell signaling responses and pathogenesis. Annu. Rev. Microbiol. *59*, 69–89.

Viera, N.T., Romero, M.J., Montero, M.K., Rincon, J., and Mosquera, J.A. (2001). Streptococcal erythrogenic toxin B induces apoptosis and proliferation in human leukocytes. Kidney Int. *59*, 950–958.

Vijay-Kumar, M., Wu, H., Jones, R., Grant, G., Babbin, B., King, T.P., Kelly, D., Gewirtz, A.T., and Neish, A.S. (2006). Flagellin suppresses epithelial apoptosis and limits disease during enteric infection. Am. J. Pathol. *169*, 1686–1700.

Voth, D.E., and Ballard, J.D. (2007). Critical intermediate steps in *Clostridium sordellii* lethal toxin-induced apoptosis. Biochem. Biophys. Res. Commun. *363*, 959–964.

Voth, D.E., and Heinzen, R.A. (2009). Sustained activation of Akt and Erk1/2 is required for *Coxiella burnetii* antiapoptotic activity. Infect. Immun. *77*, 205–213.

Voth, D.E., Howe, D., and Heinzen, R.A. (2007). *Coxiella burnetii* inhibits apoptosis in human THP-1 cells and monkey primary alveolar macrophages. Infect. Immun. *75*, 4263–4271.

Voyich, J.M., Braughton, K.R., Sturdevant, D.E., Whitney, A.R., Saïd-Salim, B., Porcella, S.F., Long, R.D., Dorward, D.W., Gardner, D.J., Kreiswirth, B.N., *et al.* (2005). Insights into mechanisms used by *Staphylococcus aureus* to avoid destruction by human neutrophils. J. Immunol. *175*, 3907–3919.

Wagner, S., Beil, W., Westermann, J., Logan, R.P., Bock, C.T., Trautwein, C., Bleck, J.S., and Manns, M.P. (1997). Regulation of gastric epithelial cell growth by *Helicobacter pylori*: offdence for a major role of apoptosis. Gastroenterology *113*, 1836–1847.

Walz, J.M., Gerhardt, H., Faigle, M., Wolburg, H., and Neumeister, B. (2000). *Legionella* species of different human prevalence induce different rates of apoptosis in human monocytic cells. APMIS *108*, 398–408.

Wang, H., Paton, J.C., and Paton, A.W. (2007). Pathologic changes in mice induced by subtilase cytotoxin, a potent new *Escherichia coli* AB5 toxin that targets the endoplasmic reticulum. J. Infect. Dis. *196*, 1093–1101.

Wang, M., Qureshi, N., Soeurt, N., and Splitter, G. (2001). High levels of nitric oxide production decrease early but increase late survival of *Brucella abortus* in macrophages. Microb. Pathog. *31*, 221–230.

Watson, P.R., Gautier, A.V., Paulin, S.M., Bland, A.P., Jones, P.W., and Wallis, T.S. (2000). *Salmonella enterica* serovars Typhimurium and Dublin can lyse macrophages by a mechanism distinct from apoptosis. Infect. Immun. *68*, 3744–3747.

Watson, R.W., Redmond, H.P., Wang, J.H., and Bouchier-Hayes, D. (1996a). Bacterial ingestion, tumor necrosis factor-α, and heat induce programmed cell death in activated neutrophils. Shock *5*, 47–51.

Watson, R.W., Redmond, H.P., Wang, J.H., Condron, C., and Bouchier-Hayes, D. (1996b). Neutrophils undergo apoptosis following ingestion of *Escherichia coli*. J. Immunol. *156*, 3986–3992.

Webster, P., Ijdo, J.W., Chicoine, L.M., and Fikrig, E. (1998). The agent of human granulocytic ehrlichiosis resides in an endosomal compartment. J. Clin. Invest. *101*, 1932–1941.

Wedi, B., Wieczorek, D., Stünkel, T., Breuer, K., and Kapp, A. (2002). Staphylococcal exotoxins exert pro-inflammatory effects through inhibition of eosinophil apoptosis, increased surface antigen expression (CD11b, CD45, CD54, and CD69), and enhanced cytokine-activated oxidative burst, thereby triggering allergic inflammatory reactions. J. Allergy Clin. Immunol. *109*, 477–484.

Weeks, S., Hill, J., Friedlander, A., and Welkos, S. (2002). Anti-V antigen antibody protects macrophages from *Yersinia pestis*-induced cell death and promotes phagocytosis. Microb. Pathog. *32*, 227–237.

Wellmer, A., von Mering, M., Spreer, A., Diem, R., Eiffert, H., Noeske, C., Bunkowski, S., Gold, R., and Nau, R. (2004). Experimental pneumococcal meningitis: impaired clearance of bacteria from the blood due to increased apoptosis in the spleen in Bcl-2-deficient mice. Infect. Immun. *72*, 3113–3119.

Wesson, C.A., Liou, L.E., Todd, K.M., Bohach, G.A., Trumble, W.R., and Bayles, K.W. (1998). *Staphylococcus aureus* Agr and Sar global regulators influence internalization and induction of apoptosis. Infect. Immun. *66*, 5238–5243.

Wesson, C.A., Deringer, J., Liou, L.E., Bayles, K.W., Bohach, G.A., and Trumble, W.R. (2000). Apoptosis induced by *Staphylococcus aureus* in epithelial cells utilizes a mechanism involving caspases 8 and 3. Infect. Immun. *68*, 2998–3001.

Whyte, M.K., Meagher, L.C., MacDermot, J., and Haslett, C. (1993). Impairment of function in aging neutrophils is associated with apoptosis. J. Immunol. *150*, 5124–5134.

Wickliffe, K.E., Leppla, S.H., and Moayeri, M. (2008a). Anthrax lethal toxin-induced inflammasome formation and caspase-1 activation are late events dependent on ion fluxes and the proteasome. Cell. Microbiol. *10*, 332–343.

Wickliffe, K.E., Leppla, S.H., and Moayeri, M. (2008b). Killing of macrophages by anthrax lethal toxin: involvement of the N-end rule pathway. Cell. Microbiol. *10*, 1352–1362.

Wickstrum, J.R., Bokhari, S.M., Fischer, J.L., Pinson, D.M., Yeh, H.W., Horvat, R.T., and Parmely, M.J. (2009). *Francisella tularensis* induces extensive caspase-3 activation and apoptotic cell death in the tissues of infected mice. Infect. Immun. *77*, 4827–4836.

Willhite, D.C., and Blanke, S.R. (2004). *Helicobacter pylori* vacuolating cytotoxin enters cells, localizes to the mitochondria, and induces mitochondrial membrane permeability changes correlated to toxin channel activity. Cell. Microbiol. *6*, 143–154.

Wilson, C., Foster, G.H., and Bitzan, M. (2005). Silencing of Bak ameliorates apoptosis of human proximal tubular epithelial cells by *Escherichia coli*-derived Shiga toxin 2. Infection 33, 362–367.

Wising, C., Azem, J., Zetterberg, M., Svensson, L.A., Ahlman, K., and Lagergård, T. (2005). Induction of apoptosis/necrosis in various human cell lineages by *Haemophilus ducreyi* cytolethal distending toxin. Toxicon. *45*, 767–776.

Woodin, A.M., (1961). The effect of staphylococcal leucocidin on the leucocyte. Biochem. J. *80*, 562–572.

Woodin, A.M. (1970). Staphylococcal leukocidin. In Microbial Toxins, Montje, T., Kadis, S., and Ajl, S., eds. (Academic Press, Inc., New York), pp. 327–355.

Worgall, S., Martushova, K., Busch, A., Lande, L., and Crystal, R.G. (2002). Apoptosis induced by *Pseudomonas aeruginosa* in antigen presenting cells is diminished by genetic modification with CD40 ligand. Pediatr. Res. *52*, 636–644.

Wu, W., Mehta, H., Chakrabarty, K., Booth, J.L., Duggan, E.S., Patel, K.B., Ballard, J.D., Coggeshall, K.M., and Metcalf, J.P. (2009). Resistance of human alveolar macrophages to *Bacillus anthracis* lethal toxin. J. Immunol. *183*, 5799–5806.

Xiao, Y., Zhong, Y., Greene, W., Dong, F., and Zhong, G. (2004). *Chlamydia trachomatis* infection inhibits both Bax and Bak activation induced by staurosporine. Infect. Immun. *72*, 5470–5474.

Xu, T., Lundqvist, A., Ahmed, H.J., Eriksson, K., Yang, Y., and Lagergard, T. (2004). Interactions of *Haemophilus ducreyi* and purified cytolethal distending toxin with human monocyte-derived dendritic cells, macrophages and CD4 + T cells. Microbes Infect. *6*, 1171–1181.

Yagi, Y., Shiono, H., Shibahara, T., Chikayama, Y., Nakamura, I., and Ohnuma, A. (2002). Increase in apoptotic polymorphonuclear neutrophils in peripheral blood after intramammary infusion of *Escherichia coli* lipopolysaccharide. Vet. Immunol. Immunopathol. *89*, 115–125.

Yamada, T., Goto, M., Punj, V., Zaborina, O., Kimbara, K., Das Gupta, T.K., and Chakrabarty, A.M. (2002). The bacterial redox protein azurin induces apoptosis in J774 macrophages through complex formation and stabilization of the tumor suppressor protein p53. Infect. Immun. *70*, 7054–7062.

Yamaguchi, N., Kieba, I.R., Korostoff, J., Howard, P.S., Shenker, B.J., and Lally, E.T. (2001). Maintenance of oxidative phosphorylation protects cells from *Actinobacillus actinomycetemcomitans* leukotoxin-induced apoptosis. Cell. Microbiol. *3*, 811–823.

Yamaguchi, N., Kubo, C., Masuhiro, Y., Lally, E.T., Koga, T., and Hanazawa, S. (2004). Tumor necrosis factor alpha enhances *Actinobacillus actinomycetemcomitans* leukotoxin-induced HL-60 cell apoptosis by stimulating lymphocyte function-associated antigen 1 expression. Infect. Immun. *72*, 269–276.

Yamamoto, A., Taniuchi, S., Tsuji, S., Hasui, M., and Kobayashi, Y. (2002). Role of reactive oxygen species in neutrophil apoptosis following ingestion of heat-killed *Staphylococcus aureus*. Clin. Exp. Immunol. *129*, 479–484.

Yanai, A., Hirata, Y., Mitsuno, Y., Maeda, S., Shibata, W., Akanuma, M., Yoshida, H., Kawabe, T., and Omata, M. (2003). *Helicobacter pylori* induces antiapoptosis through nuclear factor-κB activation. J. Infect. Dis. *188*, 1741–1751.

Yang, Y.F., Sylte, M.J., and Czuprynski, C.J. (1998). Apoptosis: a possible tactic of *Haemophilus somnus* for evasion of killing by bovine neutrophils? Microb. Pathog. *24*, 351–359.

Yilmaz, O., Jungas, T., Verbeke, P., and Ojcius, D.M. (2004). Activation of the phosphatidylinositol 3-kinase/Akt pathway contributes to survival of primary epithelial cells infected with the periodontal pathogen *Porphyromonas gingivalis*. Infect. Immun. *72*, 3743–3751.

Yoshida, T., Fukada, M., Koide, N., Ikeda, H., Sugiyama, T., Kato, Y., Ishikawa, N., and Yokochi, T. (1999). Primary cultures of human endothelial cells are susceptible to low doses of Shiga toxins and undergo apoptosis. J. Infect. Dis. *180*, 2048–2052.

Yoshiie, K., Kim, H.Y., Mott, J., and Rikihisa, Y. (2000). Intracellular infection by the human granulocytic ehrlichiosis agent inhibits human neutrophil apoptosis. Infect. Immun. *68*, 1125–1133.

Yrlid, U., and Wick, M.J. (2000). *Salmonella*-induced apoptosis of infected macrophages results in presentation of a bacteria-encoded antigen after uptake by bystander dendritic cells. J. Exp. Med. *191*, 613–624.

Yu, J., Oragui, E.E., Stephens, A., Kroll, J.S., and Venkatesan, M.M. (2001). Inactivation of DsbA alters the behaviour of *Shigella flexneri* towards murine and human-derived macrophage-like cells. FEMS Microbiol. Lett. *204*, 81–88.

Yuk, M.H., Harvill, E.T., and Miller, J.F. (1998). The BvgAS virulence control system regulates type III secretion in *Bordetella bronchiseptica*. Mol. Microbiol. *28*, 945–959.

Yuk, M.H., Harvill, E.T., Cotter, P.A., and Miller, J.F. (2000). Modulation of host immune responses, induction of apoptosis and inhibition of NF-κB activation by the *Bordetella* type III secretion system. Mol. Microbiol. *35*, 991–1004.

Zaborina, O., Dhiman, N., Ling Chen, M., Kostal, J., Holder, I.A., and Chakrabarty, A.M. (2000). Secreted products of a nonmucoid *Pseudomonas aeruginosa* strain induce two modes of macrophage killing: external-ATP-dependent, P2Z-receptor-mediated necrosis and ATP-independent, caspase-mediated apoptosis. Microbiology *146)*, 2521–2530.

Zamboni, D.S., Kobayashi, K.S., Kohlsdorf, T., Ogura, Y., Long, E.M., Vance, R.E., Kuida, K., Mariathasan, S., Dixit, V.M., Flavell, R.A., *et al.* (2006). The Birc1e cytosolic pattern-recognition receptor contributes to the detection and control of *Legionella pneumophila* infection. Nat. Immunol. *7*, 318–325.

van Zandbergen, G., Gieffers, J., Kothe, H., Rupp, J., Bollinger, A., Aga, E., Klinger, M., Brade, H., Dalhoff, K., Maass, M., *et al.* (2004). *Chlamydia pneumoniae* multiply in neutrophil granulocytes and delay their spontaneous apoptosis. J. Immunol. *172*, 1768–1776.

Zauberman, A., Cohen, S., Mamroud, E., Flashner, Y., Tidhar, A., Ber, R., Elhanany, E., Shafferman, A., and Velan, B. (2006). Interaction of *Yersinia* pestis with macrophages: limitations in YopJ-dependent apoptosis. Infect. Immun. *74*, 3239–3250.

Zhang, B., Hirahashi, J., Cullere, X., and Mayadas, T.N. (2003a). Elucidation of molecular events leading to neutrophil apoptosis following phagocytosis: cross-talk between caspase 8, reactive oxygen species, and MAPK/ERK activation. J. Biol. Chem. *278*, 28443–28454.

Zhang, J., Takayama, H., Matsuba, T., Jiang, R., and Tanaka, Y. (2003b). Induction of apoptosis in macrophage cell line, J774, by the cell-free supernatant from *Pseudomonas aeruginosa*. Microbiol. Immunol. *47*, 199–206.

Zhang, J.Z., Sinha, M., Luxon, B.A., and Yu, X.J. (2004). Survival strategy of obligately intracellular *Ehrlichia chaffeensis*: novel modulation of immune response and host cell cycles. Infect. Immun. *72*, 498–507.

Zhang, Y., and Bliska, J.B. (2003). Role of Toll-like receptor signaling in the apoptotic response of macrophages to *Yersinia* infection. Infect. Immun. *71*, 1513–1519.

Zhang, Y., Ting, A.T., Marcu, K.B., and Bliska, J.B. (2005). Inhibition of MAPK and NF-κB pathways is necessary for rapid apoptosis in macrophages infected with *Yersinia*. J. Immunol. *174*, 7939–7949.

Zhang, Y., Murtha, J., Roberts, M.A., Siegel, R.M., and Bliska, J.B. (2008). Type III secretion decreases bacterial and host survival following phagocytosis of *Yersinia pseudotuberculosis* by macrophages. Infect. Immun. *76,* 4299–4310.

Zhao, J.F., Sun, A.H., Ruan, P., Zhao, X.H., Lu, M.Q., and Yan, J. (2009). *Vibrio vulnificus* cytolysin induces apoptosis in HUVEC, SGC-7901 and SMMC-7721 cells via caspase-9/3-dependent pathway. Microb. Pathog. *46,* 194–200.

Zhou, H., Monack, D.M., Kayagaki, N., Wertz, I., Yin, J., Wolf, B., and Dixit, V.M. (2005). *Yersinia* virulence factor YopJ acts as a deubiquitinase to inhibit NF-κB activation. J. Exp. Med. *202,* 1327–1332.

Zhou, X., Mantis, N., Zhang, X.R., Potoka, D.A., Watkins, S.C., and Ford, H.R. (2000). *Salmonella typhimurium* induces apoptosis in human monocyte-derived macrophages. Microbiol. Immunol. *44,* 987–995.

Zhu, Y., Wang, C., Huang, J., Ge, Z., Dong, Q., Zhong, X., Su, Y., and Zheng, S. (2007). The *Helicobacter pylori* virulence factor CagA promotes Erk1/2-mediated Bad phosphorylation in lymphocytes: a mechanism of CagA-inhibited lymphocyte apoptosis. Cell. Microbiol. *9,* 952–961.

Zink, S.D., Pedersen, L., Cianciotto, N.P., and Abu-Kwaik, Y. (2002). The Dot/Icm type IV secretion system of *Legionella pneumophila* is essential for the induction of apoptosis in human macrophages. Infect. Immun. *70,* 1657–1663.

Zlotkin, A., Chilmonczyk, S., Eyngor, M., Hurvitz, A., Ghittino, C., and Eldar, A. (2003). Trojan horse effect: phagocyte-mediated *Streptococcus iniae* infection of fish. Infect. Immun. *71,* 2318–2325.

Zychlinsky, A., Prevost, M.C., and Sansonetti, P.J. (1992). *Shigella flexneri* induces apoptosis in infected macrophages. Nature *358,* 167–169.

Zychlinsky, A., Kenny, B., Prévost, M.C., Holland, I.B., and Sansonetti, P.J. (1993). The *ipaB* gene of *Shigella flexneri* and macrophage-programmed cell death. Infect. Agents Dis. *2,* 212–214.

Zychlinsky, A., Fitting, C., Cavaillon, J.M., and Sansonetti, P.J. (1994a). Interleukin 1 is released by murine macrophages during apoptosis induced by *Shigella flexneri*. J. Clin. Invest. *94,* 1328–1332.

Zychlinsky, A., Kenny, B., Ménard, R., Prévost, M.C., Holland, I.B., and Sansonetti, P.J. (1994b). IpaB mediates macrophage apoptosis induced by *Shigella flexneri*. Mol. Microbiol. *11,* 619–627.

Zysk, G., Bejo, L., Schneider-Wald, B.K., Nau, R., and Heinz, H. (2000). Induction of necrosis and apoptosis of neutrophil granulocytes by *Streptococcus pneumoniae*. Clin. Exp. Immunol. *122,* 61–66.

Index

A

B

C

M

N

O

P

R

S

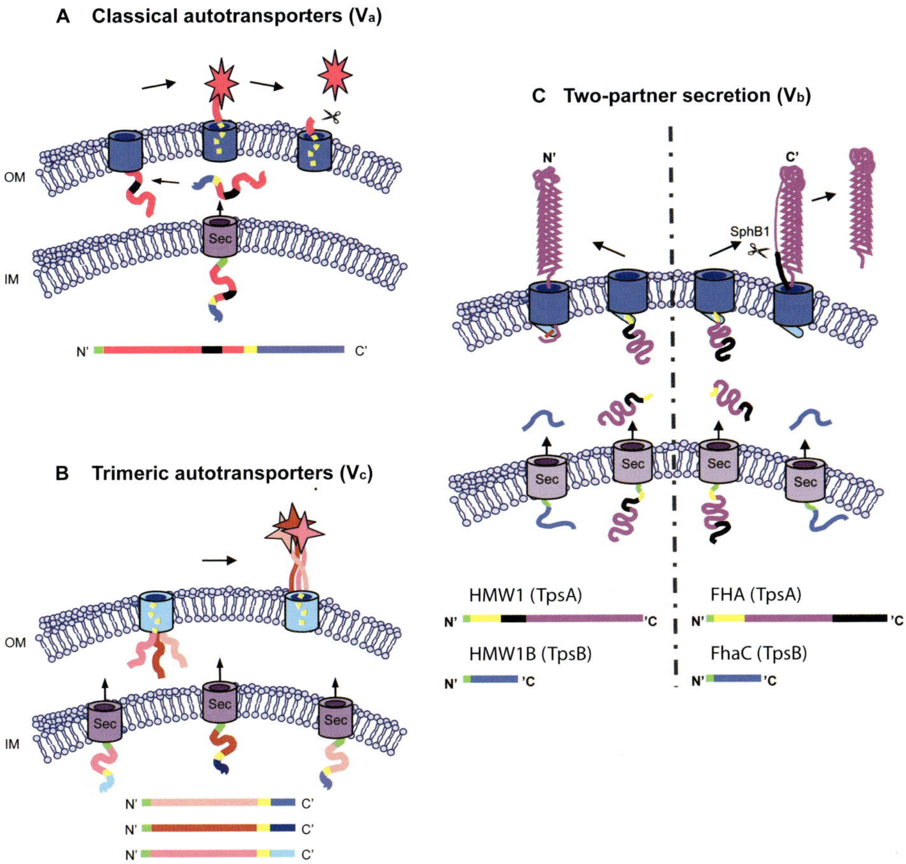

Plate 5.1 Gram-negative non-pilus adhesins. (A) Classical autotransporters. An N-terminal signal peptide (green) directs translocation of the protein across the cytoplasmic membrane and is cleaved. A C-terminal β-domain (blue) targets the autotransporter to the outer membrane, forms a β-barrel pore, and is occluded by a transmembrane α-helical linker region (yellow) that connects with the extracellular passenger domain (pink). A number of classical autotransporter adhesins have an intramolecular chaperone domain (black) that is important for folding of the passenger domain, which mediates binding to host receptors. In virtually all conventional autotransporters, the passenger domain is cleaved from the translocator domain and is either tethered non-covalently to the bacterial surface or released extracellularly. (B) Trimeric autotransporters. The C-terminal domain from each monomer contributes four β -strands which trimerize to form a 12-stranded β-barrel pore. All known trimeric autotransporters harbour adhesive activity in the passenger domain, lack an apparent intramolecular chaperone domain, form heat-resistant oligomers, and remain uncleaved at the cell surface. Oftentimes these proteins form short, thin surface fibres that are visible by transmission electron microscopy. (C) Two-partner secretion. The extracellular effector TpsA proteins contain four functional domains: the signal peptide (green), the TPS domain (yellow), the pro-domain (HMW1: yellow and black, FHA: black), and the extracellular effector domain (purple). The outer membrane translocator TpsB proteins contain three functional domains: the signal peptide (green), the periplasmic domain, and the β-barrel domain (blue). N-terminal signal peptides direct translocation of TpsA and TpsB across the cytoplasmic membrane by the Sec secretion system. Each TpsB protein inserts into the outer membrane and forms a β-barrel. The TPS domain targets the TpsA pro-protein to the cognate TpsB protein, interacting with the periplasmic domain and initiating translocation across the outer membrane to the extracellular environment where it folds into a β-helix or β-solenoid. The prototype adhesins of this family, *H. influenzae* HMW1 and *B. pertussis* FHA, exhibit slight differences with respect to TPS sequence and orientation of the mature protein and may represent two distinct TPS subfamilies. Secretion of HMW1: the N-terminal TPS-containing pro-domain is cleaved and released extracellularly, generating mature HMW1. The C-terminus undergoes disulfide bond formation (red), which serves to anchor mature HMW1 to HMW1B. Surface anchoring of HMW1 also requires glycosylation, which is mediated by the HMW1C glycosyltransferase. Secretion of FHA: A C-terminal pro-domain is cleaved at the cell surface by the serine protease SphB1 and other unidentified factors, releasing a significant amount of mature FHA into the extracellular environment.

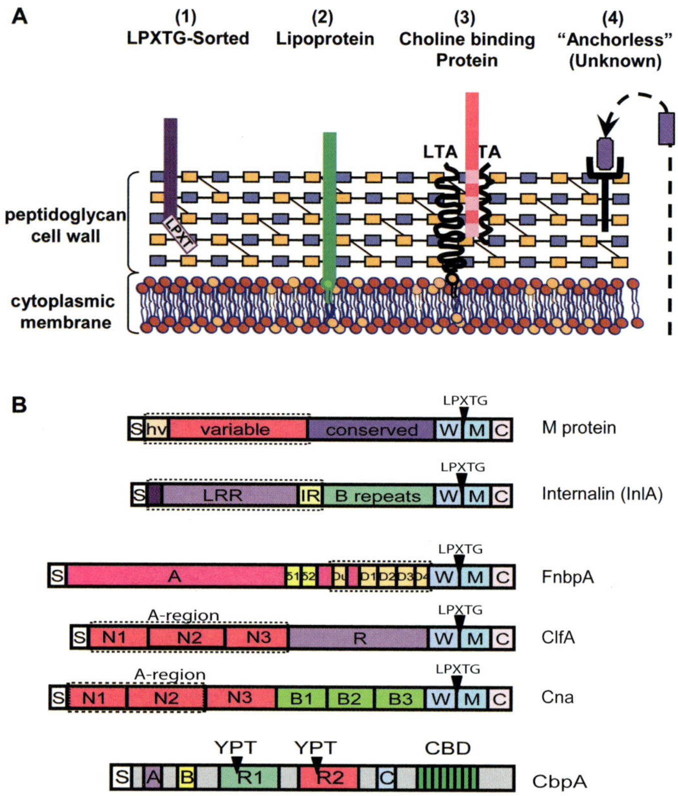

Plate 5.2 Gram-positive non-pilus adhesins. (A) Surface protein anchoring mechanisms. (i) The majority of Gram-positive bacterial adhesins are anchored to the cell wall by a sortase. Each of these proteins contains an N-terminal signal peptide (S), which directs secretion across the cytoplasmic membrane. The C-terminus of these adhesins contains features typical of surface-anchored proteins, a proline-rich wall-spanning region (W), a hydrophobic membrane-spanning (M), and a positively charged tail (C). An LPXTG motif near the C-terminus is recognized by sortase, which covalently links the protein to the peptidoglycan layer of the cell wall. (2) Lipoproteins are anchored in the cytoplasmic membrane. Pre-prolipoproteins contain a lipobox sequence motif located in the C-terminal part of the leader peptide, with a conserved cysteine serving as the target amino acid for lipidation. (3) Members of the choline-binding protein (Cbp) family of proteins are non-covalently bound to a phosphorylcholine moiety on cell wall teichoic acids and membrane-bound lipoteichoic acids via a choline-binding repeat domain. (4) Anchorless proteins are secreted and re-associate with the bacterial cell surface by an as-yet-unknown mechanism. (B) Domain organization of prototype adhesins: M protein of group A streptococci (*S. pyogenes*) contains an N-proximal variable domain with a hypervariable tip (hv) and a C-terminal conserved domain. The variable domain is associated with adhesive activity and mediates binding to a variety of substrates. Internalin (InlA) of *L. monocytogenes* contains two repeat regions, the leucine-rich repeats (LRR) and the B repeats, which are connected by the inter-repeat region. Structure studies indicate that the E-cadherin-binding LRR domain forms a bowed tube with a calcium-binding cap and that the IR domain forms an Ig-like fold. *S. aureus* ECM-binding adhesins (MSCRAMMs): FnbpA of *S. aureus* is a prototype fibronectin-binding adhesin and is characterized by N-proximal non-repetitive sequence containing two almost identical 30 amino acid repeats of unknown function (δ1, δ2) and five fibronectin-binding units (Du-D4). ClfA of *S. aureus* is a plasma clumping factor that binds fibrinogen. The ligand-binding A-region contains three subdomains (N1, N2 and N3) and is followed by unique dipeptide repeat region R consisting of mainly aspartic acid and serine residues. Cna of *S. aureus* contains an N-terminal collagen-binding A-region and a 187-residue motif repeated 1–3 times (B1, B2, B3), which varies from strain to strain. For each protein, adhesive regions are highlighted by dashed boxes. *S. pneumoniae* choline-binding protein CbpA: Leucine zipper motifs cluster within five domains: A, B, and C form coiled-coil dimers, and each R-domain adopts a unique 3-α-helix, antiparallel raft-like structure. The R-domains bind the secretory IgA receptor (pIgR), enabling pneumocci to adhere to and invade respiratory epithelial cells. The locations of critical YPT motifs are noted. The choline-binding repeats (CBD) associate non-covalently with phosphocholine on cell-wall teichoic acids and membrane-bound lipoteichoic acids.

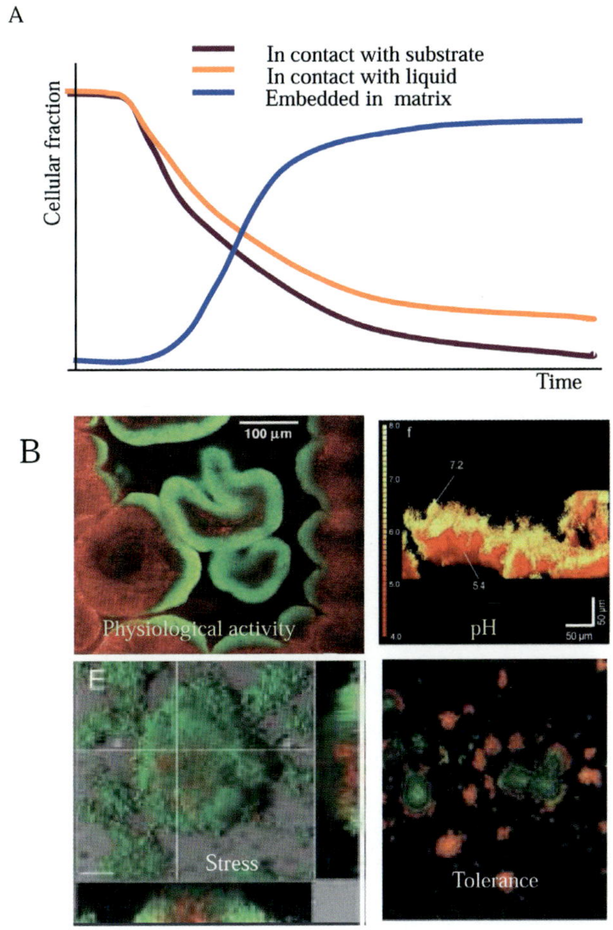

Plate 6.4 Heterogeneity generated within bacterial biofilms. Scheme inspired from Spormann, 2008 describing the increasing proportion of the microbial community embedded in matrix while cells in contact with substrate or liquid are decreasing with time during this bacterial life cycle. Examples of metabolic (Werner *et al.*, 2004), pH (Hidalgo *et al.*, 2009), stress (Ito *et al.*, 2009) and tolerance to antimicrobial substances (from Folkesson *et al.*, 2008) heterogeneity within the microcommunity observed with two-colour markers.

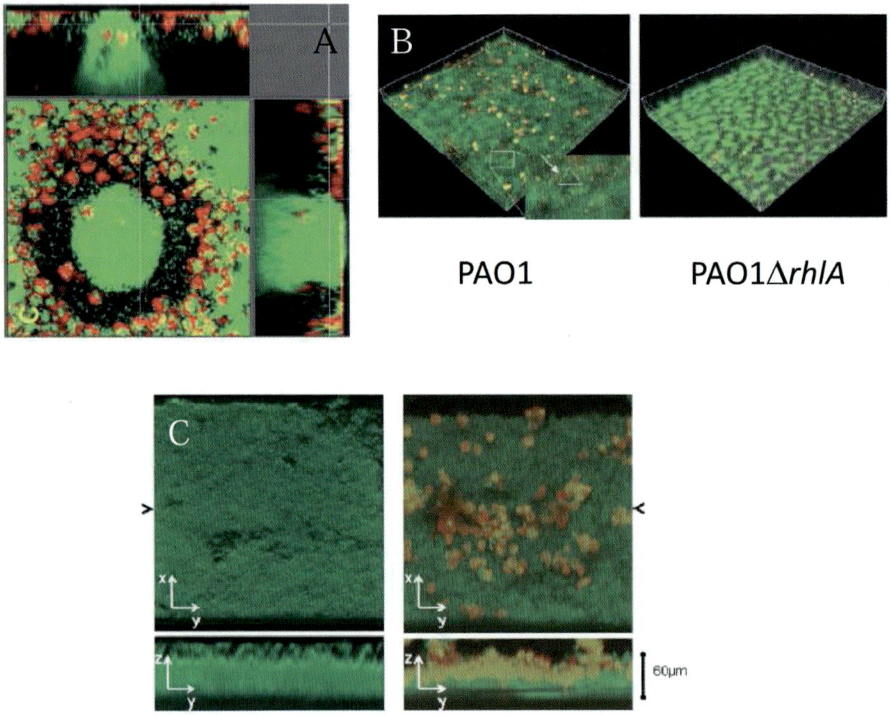

Plate 6.5 Biofilm against host immune defence. (A) Confocal microscopic observation of *Pseudomonas aeruginosa* GFP tagged and PMN stained with cyto62. Note that intact PMN are never identified with the top layer of the biofilm (Alhede *et al.*, 2009). (B) Confocal microscopic observation of *Pseudomonas aeruginosa* GFP-tagged inducing necrosis of PMN stained with ethidium bromide (red) whereas biofilm of the *rhlA* mutant does not (Van Gennip *et al.*, 2009). (C) *Staphylococcus aureus* biofilms SytoBC-labelled are rapidly infiltrated by PMN which move as a front, and turn yellow, the latter indicative again of biofilm uptake biofilms in red. PMN are labelled with antiCD66b and Cy3 conjugated antibody (Günther *et al.*, 2009).

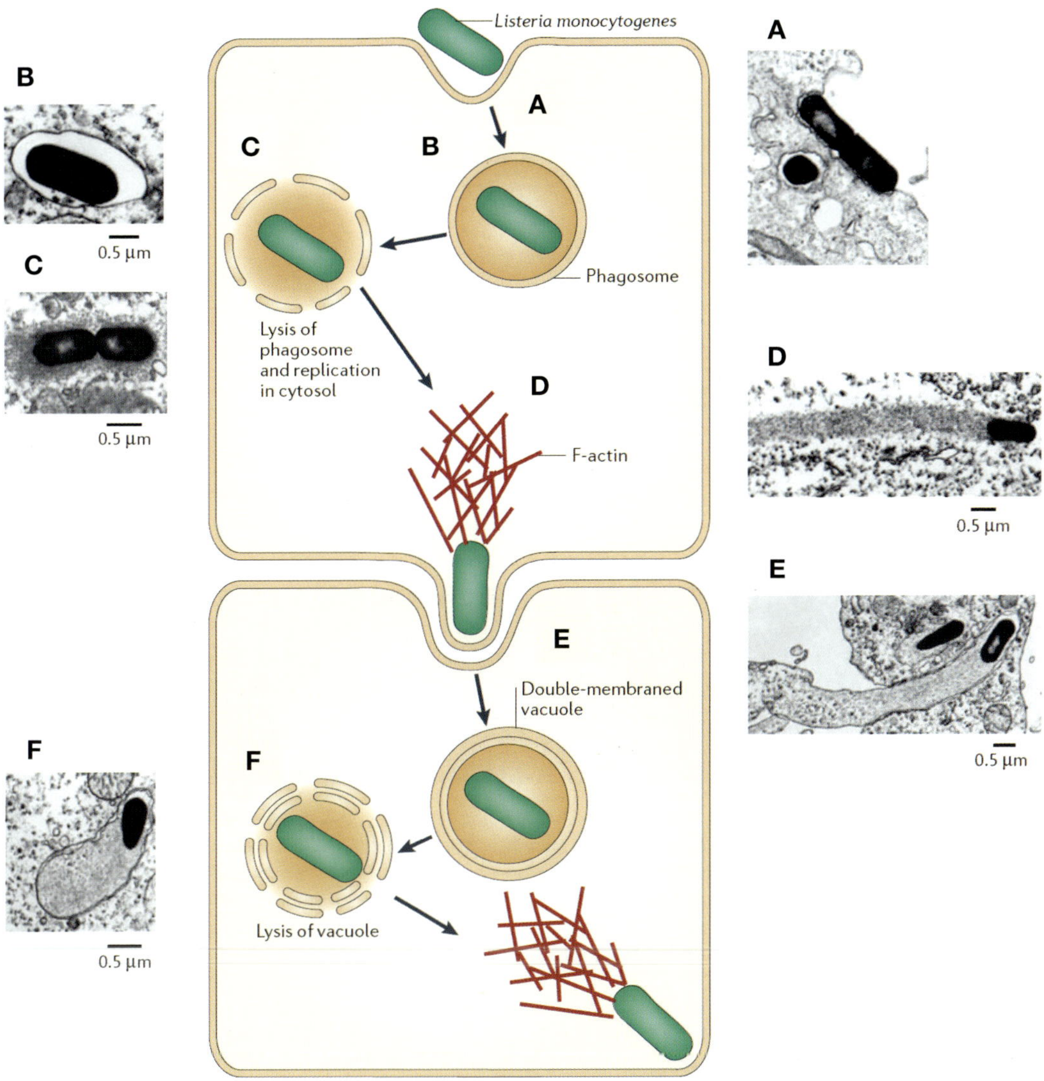

Plate 11.1 The intracellular lifestyle of *Listeria monocytogenes*. Schematic representation and electron micrographs of the *L. monocytogenes* life cycle. (A) *L. monocytogenes* induces its entry into a non-professional phagocyte. (B) Bacteria are internalized in a vacuole (i.e. a phagosome). (C) and (D) The membrane of the vacuole is disrupted by the secretion of phospholipases and LLO. Bacteria are released into the cytosol where they replicate and start to polymerize actin, as observed by the presence of actin tails. (E) Actin polymerization allows bacteria to propel into a neighbouring cell by forming protrusions in the plasma membrane. (F) Upon entry into the neighbouring cell, bacteria are present in a double-membraned vacuole, from which they can escape to continue their life cycle. This figure is reproduced with permission from Hamon *et al.* (2006).

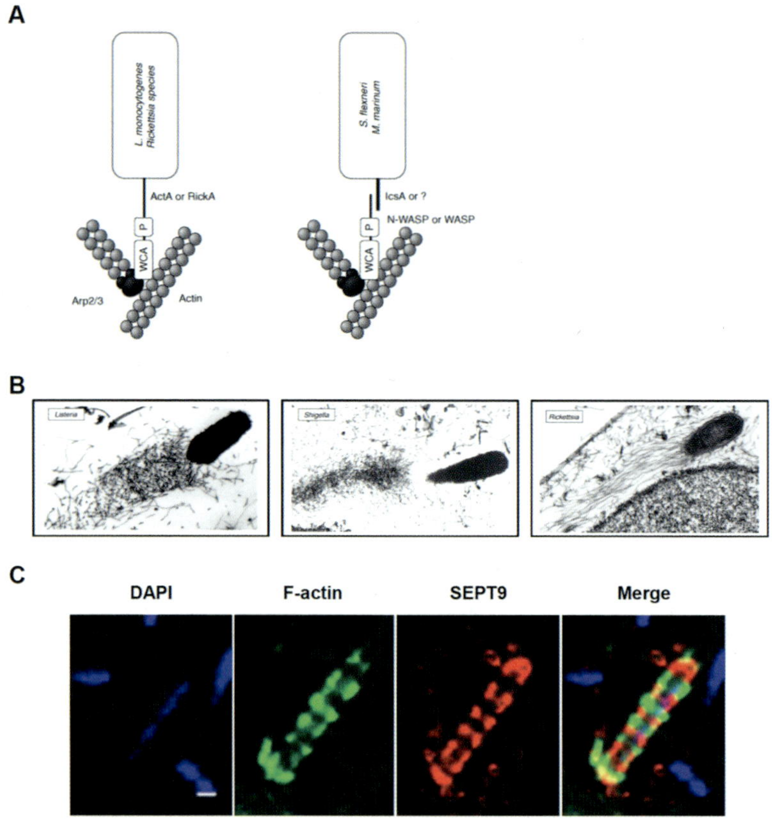

Plate 11.2 Actin tails and septin cages. (A) Schematic representation of the different types of Arp2/3 activation by *L. monocytogenes* and *Rickettsia* species (left panel) and *S. flexneri* and *M. marinum* (right panel). Most NPFs contain several conserved domains including a proline-rich region (P), a C-terminal WASP homology 2 (W) region that binds actin monomers, and C-terminal central and acidic regions (CA) that bind to the Arp2/3 complex. ActA of *L. monocytogenes* and RickA of *Rickettsia* species directly activate the Arp2/3 complex. IcsA of *Shigella* recruits N-WASP, which then activates the Arp2/3 complex. *M. marinum* recruits WASP, which then likely activates the Arp2/3 complex; however the mycobacterial protein(s) involved are unknown. These images are adapted from Gouin *et al.* (2005). (B) Electron microscopy of actin tails. Unlike *Listeria* and *Shigella* tails, *Rickettsia* tails are made of long and unbranched filaments. These images are adapted from Gouin *et al.* (2005). (C) Septin (SEPT9) recruitment to *S. flexneri* devoid of actin tails, i.e. the *Shigella*-septin cage. Of note, septins do not strictly colocalize with F-actin. Scale bar = 1 μm. These images are adapted from Mostowy *et al.* (2010).

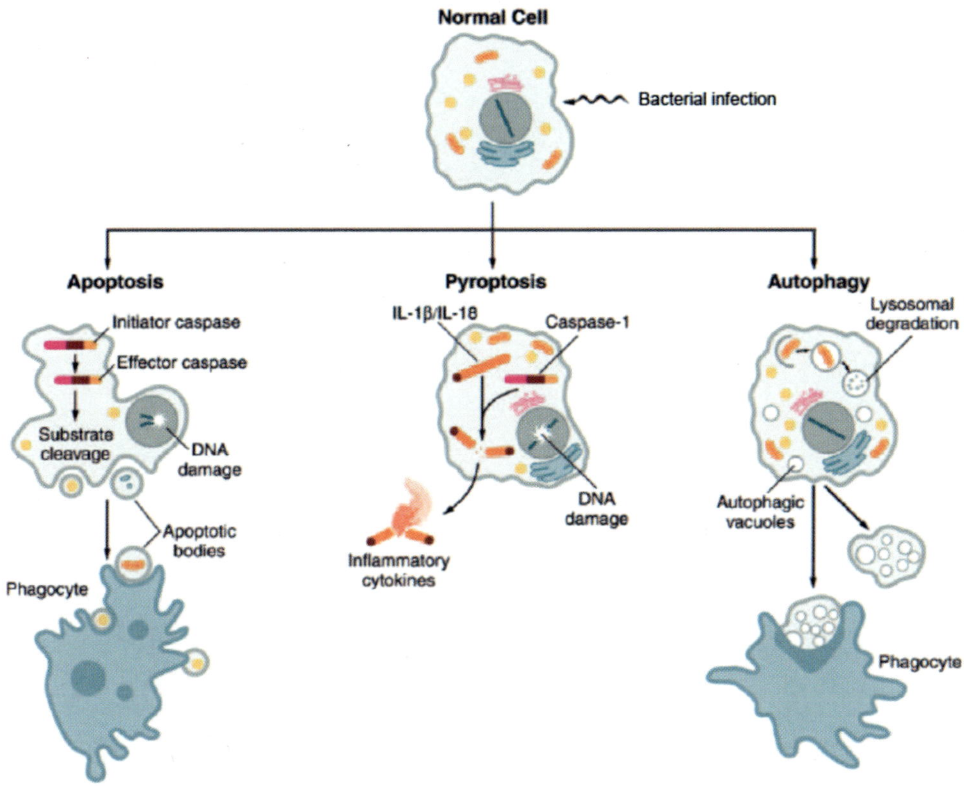

Plate 11.3 Cell responses to bacterial infection. Normal, healthy cells respond to bacterial infection by initiating a variety of molecular pathways. Apoptosis, a pathway leading to cell death, features the activation of initiator caspases that activate effector caspases to cleave cellular substrates. Apoptotic cells demonstrate cytoplasmic and nuclear condensation, DNA damage, formation of apoptotic bodies, maintenance of an intact plasma membrane, and exposure of surface molecules targeting cells for phagocytosis. Pyroptosis, a pro-inflammatory cell death pathway, is mediated by the activation of caspase-1, a protease that also activates the inflammatory cytokines, IL-1β, and IL-18. Pyroptosis also features cell lysis and release of inflammatory cellular contents. Autophagy is a process in which the cell sequesters cytosolic constituents in a double membrane vesicle and delivers them to the lysosome for degradation. *In vivo*, cells undergoing autophagy can be phagocytized by neighbouring cells. These images are adapted from Fink and Cookson (2005).

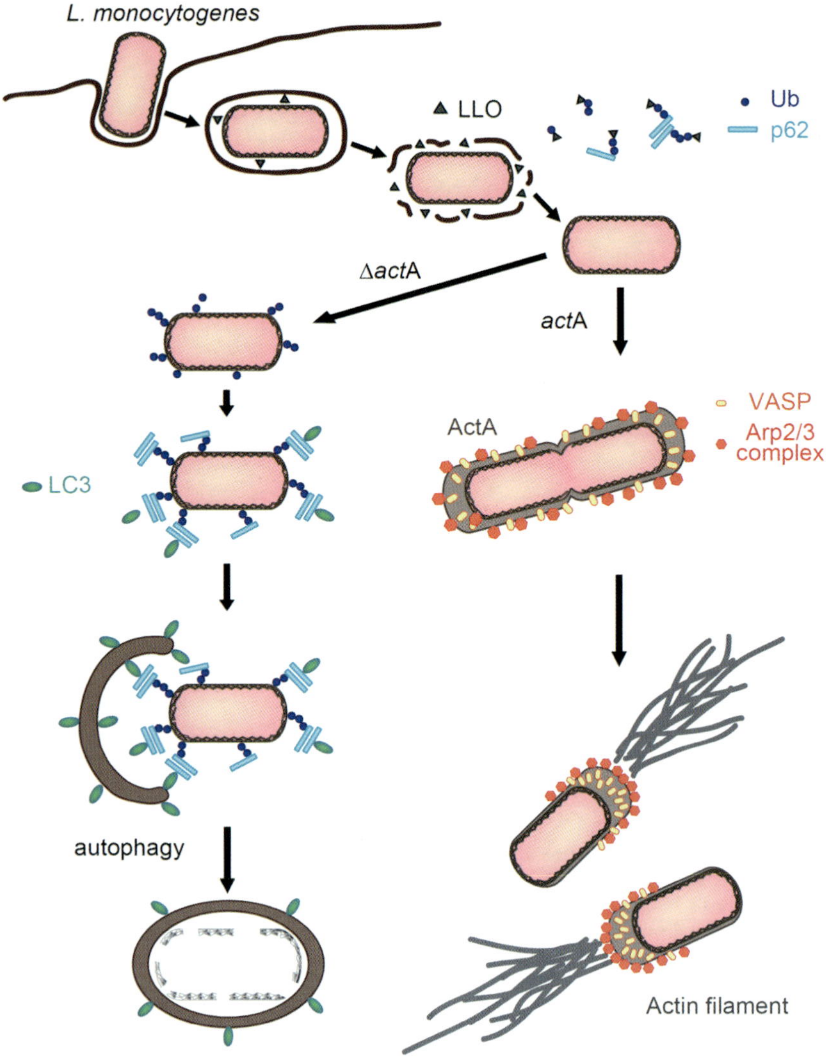

Plate 11.4 A model for autophagic recognition of *L. monocytogenes* lacking ActA. During intracellular growth, cytoplasmic *Listeria* expressing ActA can move by nucleating actin polymerization, while *Listeria* lacking ActA expression are ubiquitinated. This is followed by p62 binding to ubiquitin (Ub) via its ubiquitin-associated (UBA) domain in the vicinity of the bacterial surface. p62 then binds to LC3, an autophagosomal marker, leading to autophagosome formation around the bacterium. This figure is reproduced with permission from Yoshikawa *et al.* (2009).

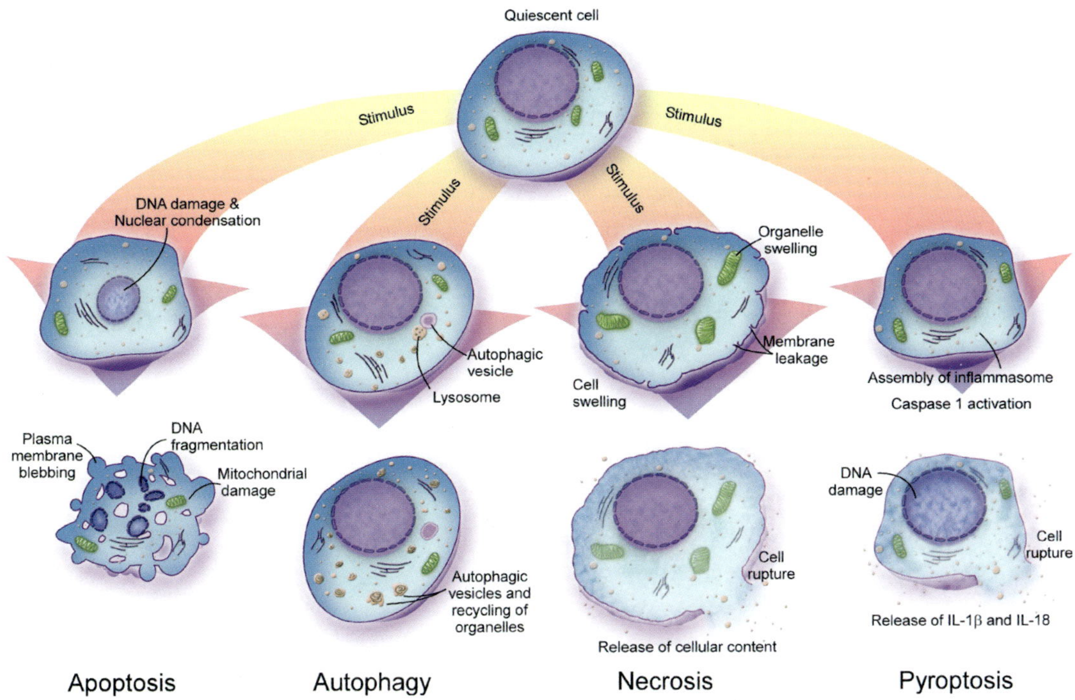

Plate 15.2 Cell fate pathways.

Current Books of Interest

Small DNA Tumour Viruses	2012
Extremophiles: Microbiology and Biotechnology	2012
Bacillus: Cellular and Molecular Biology	2012
Microbial Biofilms: Current Research and Applications	2012
Bacterial Glycomics: Current Research, Technology and Applications	2012
Non-coding RNAs and Epigenetic Regulation of Gene Expression	2012
Brucella: Molecular Microbiology and Genomics	2012
Molecular Virology and Control of Flaviviruses	2012
Bacterial Pathogenesis: Molecular and Cellular Mechanisms	2012
Bunyaviridae: Molecular and Cellular Biology	2011
Emerging Trends in Antibacterial Discovery: Answering the Call to Arms	2011
Epigenetics: A Reference Manual	2011
Metagenomics: Current Innovations and Future Trends	2011
Nitrogen Cycling in Bacteria: Molecular Analysis	2011
The Biology of Paramyxoviruses	2011
Helicobacter pylori	2011
Microbial Bioremediation of Non-metals: Current Research	2011
Lactic Acid Bacteria and Bifidobacteria: Current Progress in Advanced Research	2011
Viruses and Interferon: Current Research	2011
Essentials of Veterinary Parasitology	2011
Hepatitis C: Antiviral Drug Discovery and Development	2011
Streptomyces: Molecular Biology and Biotechnology	2011
Alphaherpesviruses: Molecular Virology	2011
Recent Advances in Plant Virology	2011
Vaccine Design: Innovative Approaches and Novel Strategies	2011
Salmonella: From Genome to Function	2011
PCR Troubleshooting and Optimization: The Essential Guide	2011
Insect Virology	2010
Environmental Microbiology: Current Technology and Water Applications	2011
Sensory Mechanisms in Bacteria: Molecular Aspects of Signal Recognition	2010
Bifidobacteria: Genomics and Molecular Aspects	2010
Molecular Phylogeny of Microorganisms	2010
Nanotechnology in Water Treatment Applications	2010
Iron Uptake and Homeostasis in Microorganisms	2010
Caliciviruses: Molecular and Cellular Virology	2010
Epstein-Barr Virus: Latency and Transformation	2010
Anaerobic Parasitic Protozoa: Genomics and Molecular Biology	2010
Lentiviruses and Macrophages: Molecular and Cellular Interactions	2010
Microbial Population Genetics	2010
Borrelia: Molecular and Cellular Biology	2010
Influenza: Molecular Virology	2010